physica status solidi (b)

VOLUME 65 · NUMBER 1 · 1974

AKADEMIE-VERLAG · BERLIN

THEORIA CUM PRAXI

Classification Scheme

1. Structure of Crystalline Solids
    1.1 Perfectly Periodic Structure
    1.2 Solid-State Phase Transformations
    1.3 Alloys. Metallurgy
    1.4 Microstructure (Magnetic Domains See 18; Ferroelectric Domains See 14.4.1)
    1.5 Films
    1.6 Surfaces
2. Non-Crystalline State
3. Crystal Growth
4. Bonding Properties
5. Mössbauer Spectroscopy
6. Lattice Dynamics. Phonons
7. Acoustic Properties
8. Thermal Properties
9. Diffusion
10. Defect Properties (Irradiation Defects See 11)
    10.1 Metals
    10.2 Non-Metals
11. Irradiation Effects (X-Ray Diffraction Investigations See 1 and 10)
12. Mechanical Properties (Plastic Deformations See 10)
    12.1 Metals
    12.2 Non-Metals
13. Electron States
    13.1 Band Structure
    13.2 Fermi Surfaces
    13.3 Surface and Interface States
    13.4 Impurity and Defect States
    13.5 Elementary Excitations (Phonons See 6)
        13.5.1 Excitons
        13.5.2 Plasmons
        13.5.3 Polarons
        13.5.4 Magnons
14. Electrical Properties. Transport Phenomena
    14.1 Metals. Semi-Metals
    14.2 Superconductivity. Superconducting Materials and Devices
    14.3 Semiconductors
        14.3.1 Films
        14.3.2 Surfaces and Interfaces
        14.3.3 Devices. Junctions (Contact Problems See 14.3.4)
        14.3.4 High-Field Phenomena, Space-Charge Effects, Inhomogeneities, Injected Carriers (Electroluminescence See 20.3; Junctions See 14.3.3)
    14.4 Dielectrics
        14.4.1 Ferroelectrics
15. Thermoelectric and Thermomagnetic Properties
16. Photoconductivity. Photovoltaic Effects
17. Emission of Electrons and Ions
    17.1 Field Emission Microscope Investigations
18. Magnetic Properties
    18.1 Paramagnetic Properties
    18.2 Ferromagnetic Properties
        18.2.1 Ferromagnetic Films
    18.3 Ferrimagnetic Properties
    18.4 Antiferromagnetic Properties

(Continued on cover three)

Author Index

physica status solidi (b)

basic research

Volume 65 · Number 1 · Pages 1 to 428, K1 to K78, and A1 to A8

September 1, 1974

PSSB 65(1) 1—428, K1—K78, A1—A8 (1974)

AKADEMIE-VERLAG · BERLIN

Subscriptions and orders for single copies should be addressed to
AKADEMIE-VERLAG, 108 Berlin, Leipziger Straße 3 – 4 or to Buchhandlung KUNST UND WISSEN, Erich Bieber, 7 Stuttgart 1, Wilhelmstr. 4 – 6 or to Deutsche Buch-Export und -Import GmbH, 701 Leipzig, Postschließfach 160

Editorial Note:

"physica status solidi (b)" undertakes that an original paper *accepted for publication* before the 8th of any month will be published within 50 days of this date unless the author requests a postponement. In special cases there may be some delay between *receipt* and *acceptance* of a paper due to the review and, if necessary, revision of the paper.

Schriftleiter und verantwortlich für den Inhalt: Professor Dr. Dr. h. c. P. Görlich, 102 Berlin, Neue Schönhauser Str. 20 bzw. 69 Jena, Humboldtstr. 26. Chefredakteur: Dr. H.-J. Hänsch. Redaktionskollegium: Prof. Dr. E. Gutsche, Dr. H.-J. Hänsch, Dr. H. Lange, Dr. S. Oberländer. Anschrift der Schriftleitung: 102 Berlin, Neue Schönhauser Str. 20, Fernruf: 4223380. Verlag: Akademie-Verlag, 108 Berlin, Leipziger Str. 3—4, Fernruf: 220441, Telex-Nr. 112020, Postscheckkonto: Berlin 35021. — Die Zeitschrift „physica status solidi (b)" erscheint jeweils am 1. des Monats. Bezugspreis eines Bandes 140,— M; Sonderpreis für die DDR 120,— M. Bestellnummer dieses Bandes: 1068/65. Jeder Band enthält zwei Hefte. Gesamtherstellung: VEB Druckerei „Thomas Müntzer" Bad Langensalza.
Veröffentlicht unter der Lizenznummer 1310 des Presseamtes beim Vorsitzenden des Ministerrates der Deutschen Demokratischen Republik.
Printed in the German Democratic Republic.

Contents

Page

Short Notes

Systematic List

Contents of Volume 65 Continued on Page 431

Review Article

phys. stat. sol. (b) **65**, 11 (1974)

Subject classification: 13.5 and 18; 13.1; 13.5.3; 13.5.4; 18.2; 18.4; 22

Ministry of Electrotechnical Industry of the USSR, Moscow

Spin Polaron Theory for Magnetic Semiconductors with Narrow Bands

By

E. L. Nagaev

Contents

1. Introduction

Subject of this review is a theory of magnetic conductors with narrow bands, their magnetic properties being determined by the localized magnetic moments of atoms with partially filled d- or f-shells. On the whole, semiconductors and semimetals are considered, but some results are also valid for metals. Magnetic conductors of the type under consideration are very interesting from the point of view of physical processes occurring in them. Many features typical for solids in general become manifest especially clearly in them — for example, the motion of charge carriers through the crystal together with a deformation of the periodic structure of the crystal being produced by the carriers themselves (in this case a deformation of its magnetic structure). Cooperative phenomena such as phase transitions that lead to changes in the magnetic ordering or in the type of conductivity, are also of great interest.

Although such materials can in principle be described with the aid of Vonsovskii's s–d model [1], the theory of narrow-band magnetic conductors had been staying for a long time at such a level that only a few qualitative results could be obtained by means of a semi-intuitive treatment. The reason for this was the lack of mathematical methods appropriate to describe the interaction between conduction electrons and localized magnetic moments in the case under consideration. In this article, a method to treat narrow-band magnetic conductors is developed. This method has made it possible to find out many interesting properties of magnetic conductors, which are stated further.

First of all it was proved that in magnetic semiconductors special types of charge carrier states may occur which have no analogues in other solids. The reason for this is that the interaction between charge charriers and magnons in magnetic semiconductors is either nonlocal or local but nonlinear. As an example for charged quasi-particles which may exist in magnetic semiconductors, the so-called ferrons should be mentioned. These are characterized by the fact that a charge carrier changes the type of magnetic ordering in its vicinity producing a microregion with a giant magnetic moment. This is an essentially new state of a conduction electron in the crystal: the electron is localized in the microregion of another phase produced in the crystal by the electron itself.

In magnetic semiconductors, charged quasi-particles of another type, the so-called quasi-oscillators, may exist. Their main specific feature is that a conduction electron oscillates relative to an equilibrium position which is moving itself through the crystal. These oscillations are accompanied by oscillations of the number of excited magnons, which increases as the electron moves off from equilibrium position and decreases as it returns back.

When the conduction electron density is large enough, the charge carriers may strongly influence the magnetic properties of the crystal as a whole, carrying out the indirect exchange between the localized magnetic moments of the atoms. Certainly such exchange in narrow-band conductors is non-Heisenbergian, i.e., the Ruderman-Kittel theory is surely not applicable to the indirect exchange in those materials. This leads to a number of qualitatively new properties of such materials as compared with Heisenberg magnets, for instance to new types of magnetic structures (noncollinear antiferromagnetism, nonsaturated ferromagnetism).

With increasing temperature, some heavily doped ferromagnetic semiconductors go over from the conducting into the insulating state ("the cooperative ferron state").

In a number of cases we succeed not only to find out the spectra of elementary excitations in magnetic conductors, but also to calculate their transport coefficients. In particular it should be noted that the latter have been found in the high-temperature limit when the concept of charge carriers as quasi-particles with well defined quasi-momenta is meaningless. Many results obtained here are confirmed by experimental data.

2. Use of the s–d Model for Narrow-Band Magnetic Conductors

2.1 s–d model

As is well known, the strong correlation between electrons of partially filled d- or f-shells is typical for magnetic conductors (i.e. transition or rare-earth metals and their compounds). We shall treat here only the case of extremely strong correlation. Then each d- (f-) electron can be regarded as localized at an atom. For this reason, they cannot give any contribution to the electric current flowing through the crystal. The conductivity of such materials is thus nonzero only if apart from the localized d- or f-electrons, there are collectivized electrons in the crystal. These ideas are the basis of Vonsovskii's s–d model of magnetic conductors [1] which turns out to be very useful for the modern physics of magnetism. This model assumes the orbital states of localized d- or f-electrons as well as the total spin magnitude of d- or f-electrons at each atom, S, to be fixed, the only quantity for them to change being the atomic total spin projection. This makes it possible to use the operators of atomic total spins, S_g, instead of d-(f-) electron operators, where g is a magnetic atom label.

The collectivized electrons are described within the framework of the s–d model just like in the band model. Though in the s–d model it is the custom to call them s-electrons, actually the electrons in d- (f-) states can play their role under some circumstances. As an example let us assume an equal number of d-electrons to be at each atom at first. If we increase the electron number of some atom by one, the surplus electron will be able to go over to another magnetic atom in the state with the same energy since it interacts with the same number of electrons at each atom. Thus the surplus electron should get collectivized, i.e., it should participate in the charge transport through the crystal. It is also possible that in some cases the d-multiplet may be split by the crystal field so strongly that the electrons of the lower component are localized unlike the electrons of the higher component which are collectivized. Such a crystal is then a conductor even when no surplus electrons are present.

In the nearest-neighbour approximation, the s–d model Hamiltonian takes the form

$$\mathcal{H}_{\text{sd}} = \mathcal{H}_{\text{B}} + \mathcal{H}_{\text{A}} + \mathcal{H}_{\text{M}} \; ; \tag{2.1}$$

$$\mathcal{H}_{\text{B}} = B \sum a^*_{g\sigma} a_{g+\Delta\sigma} \, , \tag{2.2}$$

$$\mathcal{H}_{\text{A}} = - A \sum (\boldsymbol{S}_g \cdot \mathbf{s})_{\sigma\sigma'} \, a^*_{g\sigma} a_{g\sigma'} \, , \tag{2.3}$$

$$\mathcal{H}_{\text{M}} = - \frac{I}{2} \sum (\boldsymbol{S}_g \cdot \boldsymbol{S}_{g+\Delta}) \, , \tag{2.4}$$

where $a^*_{g\sigma}$, $a_{g\sigma}$ are the creation and annihilation operators of conduction electron with spin projection σ at the magnetic atom with the label $\boldsymbol{g} = (g_x, g_y, g_z)$, $\mathbf{s}$ and $\boldsymbol{S}_g$ being the spin operators of the conduction electron and the g-th magnetic atom, respectively. Further, B is the Bloch integral, A the s–d exchange integral, and I the d–d exchange integral. The index Δ numerates the z nearest neighbours of an atom.

Evidently the Hamiltonians $\mathcal{H}_{\text{B}}$, $\mathcal{H}_{\text{A}}$, and $\mathcal{H}_{\text{M}}$ respectively describe free s-electrons, free d-electrons, and the exchange interaction between the former and the latter. The sign of the Bloch integral B, unlike those of the exchange integrals, does not affect the final results. It is more convenient to take the negative sign for B.

The assumed form of the Hamiltonian (2.1 to 2.4) does not take into account the possibility of orbital degeneracy of conduction electrons. This is not essential, and the corresponding generalization can be easily carried out. Further the Hamiltonian (2.1 to 2.4) is valid even if the charge carriers are d-electrons, so that their wave functions are constructed of the same atomic orbitals as those of the localized electrons. In other words, a far more simple s–d Hamiltonian can be used for semiconductors instead of the Hubbard Hamiltonian. The proof of this statement is given in [2]. The Hamiltonian (2.1 to 2.4) can be also easily generalized to the case of localized electrons orbital moments being not frozen out.

The principal feature of the s–d model which limits its applicability consists in neglecting electron transitions inside the partically filled d- or f-shells. Such transitions can be regarded as the creation or the annihilation of Frenkel excitons at magnetic atoms. Neglecting exciton transitions is justified for exciton energies high enough. In cases where exciton transitions are essential, qualitatively new effects can arise which are impossible within the framework of the s–d model. These effects were investigated in papers [3] to [9].[1]) The consideration presented in what follows is completely based on the s–d model, and effects produced by the electron–exciton interaction are not considered here.

We are interested only in conductors with narrow bands, for instance, when the charge carriers are electrons in d-type states. Let us present the estimates of the order of magnitude of parameters in the Hamiltonian (2.1 to 2.4). The s–d exchange energy AS being of zeroth order in d-orbital overlapping amounts to 1 to 10 eV according to [10]. The conduction band width $W = 2z|B|$ is of first order in d-orbital overlapping and, consequently, should be of the order of 0.1 to 1 eV. The energy of d–d exchange zIS is of the second order, so that it is of the order of 10^{-3} to 10^{-1} eV, and in some cases even smaller. Henceforth the inequality

$$|A|S \gg z|B| \gg z|I|S^2 \tag{2.5}$$

is assumed to be fulfilled. This inequality is the quantitative criterion of the band narrowness of a magnetic conductor.

One should emphasize that so far the s–d model practically was used only in the opposite limit of wide bands $z|B| \gg |A|S$, for which the s–d exchange Hamiltonian $\mathcal{H}_A$ can be regarded as a perturbation. Under the condition (2.5) the treatment of the problem should, of course, be quite different, i.e., on the contrary, $\mathcal{H}_A$ should be regarded as the zeroth-approximation Hamiltonian and $\mathcal{H}_B$ as a perturbation.

2.2 *Two-atomic problem*

For a better understanding of the narrow-band specifics, it is advisable first of all to treat a system which consists of an electron and two magnetic atoms ($g = 1$ or -1). For the sake of simplicity the direct exchange between the magnetic atoms is disregarded ($I = 0$). Qualitatively a system of such kind was treated by Zener [11], who pointed out that electron transitions between atoms tend to align spins of atoms parallel to each other (the double exchange). Anderson and Hasegawa [12] calculated the energy spectrum of the two atoms–one electron system, but their method cannot be employed for systems consisting of a larger number of magnetic atoms or electrons. In what follows,

[1]) See also a paper by Nagaev and Sokolova to be published in phys. stat. sol. (b).

a method is presented which makes it possible to treat also systems consisting of arbitrarily large numbers of atoms and electrons.

The calculation is carried out in the first approximation with respect to B/A. First of all it is necessary to find the eigenstates of the system in the zeroth approximation in this small parameter. In this approximation the electron may be regarded as localized at one of the atoms, and both atoms do not interact with each other. For this reason, the wave function of the system can be represented as the product of the wave functions of atom g, at which the electron is localized, and of the second atom $(-g)$.

The latter is given by the expression

$$\overset{0}{\psi}_{(-g)m}(S^z_{(-g)}) = \delta(S^z_{(-g)}, m_{(-g)}) ,$$

where $\delta(x, y)$ is the delta function of a discrete argument which is equal to 1 if $x = y$ and 0 if $x \neq y$. The projection $S^z_{(-g)}$ of the spin to an arbitrarily chosen axis z plays the role of the variable, while the index of the state $m_{(-g)}$ is a fixed number, belonging to the set of allowed values of the spin projection. Obviously such a wave function corresponds to a state in which the spin projection $m_{(-g)}$ is conserved (its analogue in the coordinate representation being Dirac's delta function $\delta(x - x_0)$).

Due to the isotropy of the exchange interaction (2.3), a state of the subsystem "electron + atom" is determined by the magnitude of their total spin S_t, which is equal to $S + \frac{1}{2}$ or $S - \frac{1}{2}$, and its projection M on the axis z. The wave function of the subsystem as a whole is expressed in terms of those of the atom and the electron in the usual manner with the aid of Clebsch-Gordan coefficients:

$$\varphi^+_{gM} = \frac{1}{\sqrt{2S+1}} \left\{ \sqrt{S + \frac{1}{2} + M}\, \delta\left(S^z_g, M - \frac{1}{2}\right) a^*_{g\uparrow} + \right.$$

$$\left. + \sqrt{S + \frac{1}{2} - M}\, \delta\left(S^z_g, M + \frac{1}{2}\right) a^*_{g\downarrow} \right\} |0\rangle \tag{2.6}$$

for $S_t = S + \frac{1}{2}$,

$$\varphi^-_{gM} = \frac{1}{\sqrt{2S+1}} \left\{ \sqrt{S + \frac{1}{2} + M}\, \delta\left(S^z_g, M + \frac{1}{2}\right) a^*_{g\downarrow} - \right.$$

$$\left. - \sqrt{S + \frac{1}{2} - M}\, \delta\left(S^z_g, M - \frac{1}{2}\right) a^*_{g\uparrow} \right\} |0\rangle \tag{2.7}$$

for $S_t = S - \frac{1}{2}$, where $|0\rangle$ is the vacuum electron wave function. Substituting the expressions (2.6) and (2.7) into the Schrödinger equation with the Hamiltonian $\mathcal{H}_A$ (2.3), one easily finds that the s–d exchange energy E_A depends only on the magnitude of the total spin S_t, but not on its projection:

$$E^+_A = -\frac{AS}{2} \quad \text{for} \quad S_t = S + \frac{1}{2}, \tag{2.8}$$

$$E^-_A = \frac{A(S+1)}{2} \quad \text{for} \quad S_t = S - \frac{1}{2}. \tag{2.9}$$

In this section, we shall confine ourselves like the authors of [12] to the case of ferromagnetic s–d coupling ($A > 0$). Then it is sufficient to take into account only the states (2.6) with total spin $S_t = S + \frac{1}{2}$ in order to obtain the energy in the first approximation with respect to B/A, that is with allowance for electron transitions between atoms, which depend on the spin directions.

If we take now into account electron transitions between atoms, the total spin of the system "two atoms + one electron", j, and its projection m will be integrals of motion. Obviously the energy of the system does not depend on m. Further, as the Hamiltonian (2.1) is invariant with respect to the permutation of the atoms, the wave function of the system should be either symmetrical or antisymmetrical with respect to this transformation. Thus the correct zeroth approximation of the wave function of the system, ψ_{jm}, with the aid of which the first approximation energy ΔE should be calculated, is given by

$$\psi_{jm}^{\pm} = (-1)^{m+\frac{1}{2}} \sqrt{j+\tfrac{1}{2}} \sum_{m'} \begin{pmatrix} S+\frac{1}{2} & S & j \\ m' & m-m' & -m \end{pmatrix} \times \\ \times \left[\varphi_{1m'}^{+}\varphi_{(-1)m-m'}^{0} \pm \varphi_{(-1)m'}^{+}\varphi_{1m-m'}^{0}\right], \qquad (2.10)$$

the coefficients of which are $3j$ Wigner symbols. According to (2.10) the energy ΔE is

$$\Delta E_{jm}^{\pm} = (\psi_{jm}^{\pm}, \mathcal{H}_B \psi_{jm}^{\pm}) = \pm B(2j+1) \sum_{m'm''} \begin{pmatrix} S+\frac{1}{2} & S & j \\ m' & m-m' & -m \end{pmatrix}^{*} \times \\ \times \begin{pmatrix} S+\frac{1}{2} & S & j \\ m'' & m-m'' & -m \end{pmatrix} (m', m-m'\,|\mathcal{H}_B|\, m-m'', m''), \qquad (2.11)$$

where with the aid of (2.3) and (2.6)

$$(m', m-m'\,|\mathcal{H}_B|\, m-m'', m'') \equiv (\varphi_{1m'}^{+}\varphi_{(-1)m-m'}^{0}\,|\mathcal{H}_B|\,\varphi_{1m-m''}^{0}\varphi_{(-1)m''}^{+}) = \\ = \frac{1}{2S+1}\left\{\sqrt{\left(S+\frac{1}{2}+m'\right)\left(S+\frac{1}{2}+m''\right)}\,\delta\left(m-m'', m'-\frac{1}{2}\right)+ \\ + \sqrt{\left(S+\frac{1}{2}-m'\right)\left(S+\frac{1}{2}-m''\right)}\,\delta\left(m-m'', m'+\frac{1}{2}\right)\right\}. \qquad (2.12)$$

If the atomic spins are parallel to each other ($j = 2S + \frac{1}{2}$), assuming $m = 2S + 1$ and making use of the expression for $3j$ symbols presented in [13] we obtain from (2.12) that, as should be the case,

$$\Delta E_{2S+\frac{1}{2}}^{\pm} = \pm B\,. \qquad (2.13)$$

If the atomic spins are antiparallel ($j = \frac{1}{2}$), assuming $m = -\frac{1}{2}$ we obtain

$$\Delta E_{1/2}^{\pm} = \pm \frac{B}{2(S+1)}. \qquad (2.14)$$

Thus, according to (2.13) and (2.14), the minimum energy of the system increases with the angle θ between the spins. In the limit $2S \gg 1$, Anderson and Hasegawa proposed to describe the energy dependence on the angle θ in terms of the effective Bloch integral:

$$\Delta E(\theta) = \pm B(\theta) = \pm B \cos\frac{\theta}{2}. \qquad (2.15)$$

De Gennes [14] suggested a generalization of the result (2.15) to a system consisting of a arbitrarily large number of magnetic atoms. Namely, it was assumed that the Hamiltonian, describing the electron motion through the crystal, can be represented as a matrix with elements $B(g - g') \cos \frac{\theta_{gg'}}{2}$, where $\theta_{gg'}$ is the angle between the spins of atoms g and g'. It should be emphasized that such generalization is strictly speaking unjustified[2]), and in practice it leads not only to quantitative but in some cases to qualitative errors, as will be shown further.

In fact, the result (2.15) actually gives the total energy of the system, but not the matrix elements of the Hamiltonian $\mathcal{H}_B$ for electron transition between atoms. The total energy (2.15) has been found under the condition that only the total spin projection of two atoms and the electron is conserved, whereas the spin of each separate atom can in general rotate in space. But in the crystal, the total spin of a pair of atoms and an electron certainly cannot be an integral of motion because of their interaction with other atoms. In other words, in the crystal the atomic spins cannot separately freely rotate in space due to such interaction.

As to the true matrix element for the electron transition between two atoms, it cannot be described by a function similar to (2.15). Really, for the case of antiparallel spins, assuming $m = \frac{1}{2}$, $m' = S + \frac{1}{2}$, $m'' = \frac{1}{2} - S$ in (2.12), we obtain the matrix element to be equal to $1/\sqrt{2S+1}$. Thus, though it is smaller than that for parallel spins but unlike $\Delta E_{1/2}$ (2.15), it cannot be neglected when $2S \gg 1$ (for real spin values it cannot be smaller than $\frac{1}{3}$).

3. Ferromagnetic Conductors

3.1 Hamiltonian of a ferromagnetic conductor in the spin-wave approximation

In this section, the method developed in Section 2.2 for two magnetic atoms will be generalized to the case of ferromagnetic conductors at low temperatures. In accordance with the foregoing, in the zeroth approximation with respect to W/AS, when the conduction electron can be regarded as fixed at one of the magnetic atoms, its spin is rigidly bound to the spin of that atom. The total spin of the atom at which the conduction electron is located, is equal to $S + \frac{1}{2}$ for $A > 0$ and $S - \frac{1}{2}$ for $A < 0$. In the first approximation in W/AS, the motion of the electron through the crystal is possible. It occurs by means of electron transitions from one atom to another, and at any time the electron spin is rigidly bound to the spin of the atom at which the electron is momentarily located. Thus the motion of a conduction electron through the crystal is equivalent to the motion of an irregular spin which differs from the regular spins by $\frac{1}{2}$. In some respects such quasi-particles resemble polarons in ionic crystals, and for this reason they will be called spin polarons henceforth. The similarity and difference between spin polarons and polarons will be discussed further in more detail.

Due to such a strong coupling of the electron spin to the spins of magnetic atoms, the type of magnetic ordering in the crystal will drastically influence the motion of charge carriers. Certainly the influence of thermal fluctuations of the crystal magnetic moment will also be very strong. In the limit of weak s-d coupling, one can say that charge carriers are scattered by spin waves which are thought to be waves of deviations of localized d-moments. In the limit under

[2]) Conditions for its validity are presented in Section 6.

consideration, however, it makes only sense to consider waves of oscillations of the total moment of the system which consists not only of the d-spins but also of the spins of the conduction electrons. Thus, the problem to describe their interaction with the conduction electrons arises.

In the next section, a general method will be presented which makes it possible to describe a spin polaron in the crystal for arbitrary spin configurations and signs of the s–d exchange integral. But to make the method more clear it is advisable to present it at first for the most simple case when $A > 0$, $I = 0$, and spin deviations from the complete ferromagnetic ordering are small.

To use the perturbation theory with respect to W/AS, a respresentation should be chosen in which the Hamiltonian $\mathcal{H}_A$ (2.3) is diagonal. However, the matrix form of the operators is extremely inconvenient for the application of modern methods of the quantum theory of solids. For this reason, we are left with the task to construct the representation indicated above such that the operators be of the usual (nonmatrix) form. To a certain degree this problem resembles that of the small polaron in ionic crystals, for which it is typical that the electron–phonon interaction is taken into account in the zeroth approximation in the Bloch integral [15 to 17]. As is shown in [17], there is a canonical transformation which replaces the operators of the electron by those of the polaron, and the operators of phonons with fixed equilibrium positions by those with equilibrium positions following the motion of the electron through the crystal.

In the problem considered, the conduction electron does not change the magnon equilibrium positions, but changes the magnitude of atomic spins. The total spin operator of the atom at which the conduction electron is located, cannot be constructed from the operators of its proper spin and the electron spin, i.e., a canonical transformation of the type mentioned above is impossible. The reason for it is that the exchange interaction being essentially nonlinear does not allow such a simple treatment as the electron–phonon interaction which is linear in the phonon operators.

In what follows the spin polaron–magnon Hamiltonian will be constructred by means of its expansion in powers of operators of spin deviations. The coefficients of the expansion will be found by comparing with the s–d Hamiltonian in the matrix representation. It should be noted at once that the procedure presented further is valid to the first approximation in W/AS.

First of all, introducing the operators of spin deviations itself is nontrivial. Usually they are determined in terms of spin operators by means of the well known relations

$$S_g^z = S - b_g^* b_g, \quad S_g^+ \equiv S_g^x + i S_g^y = \sqrt{2S}\, b_g\,, \quad S_g^- \equiv S_g^x - i S_g^y = \sqrt{2S}\, b_g^* \tag{3.1}$$

which assume the magnitudes of spins to be fixed. But in the case considered, the spin of an atom may have the magnitudes $S + \frac{1}{2}$ or S depending on whether or not a conduction electron is located at it. For this reason the relations (3.1) are useless here.

The creation operator of a spin deviation will be defined as one changing the state with maximum spin projection to the state with a projection less by a unit. Such a definition is invariant with respect to the spin magnitude and is appropriate even if the latter is not fixed. The spin deviation introduced in that way is obviously analogous to the phonon the equilibrium position of which is not fixed, but follows the electron.

Since the total spin projection of the system as a whole is an integral of motion, the number of spin deviations introduced with allowance for conduction electrons should remain constant. Thus the expansion of the Hamiltonian in powers of spin deviations should be of the form

$$\left.\begin{aligned}
\mathcal{H} &= \mathcal{H}_0 + \mathcal{H}_2 + O(b^*b^*bb);\\
\mathcal{H}_0 &= B \sum \alpha_g^* \alpha_{g+\Delta} - \frac{AS}{2} \sum \alpha_g^* \alpha_g ,\\
\mathcal{H}_2 &= B\{X \sum \alpha_{g+\Delta}^* \alpha_g (b_g^* b_g + b_{g+\Delta}^* b_{g+\Delta}) + Y \sum \alpha_{g+\Delta}^* \alpha_g b_{g+\Delta}^* b_g +\\
&\quad + Z \sum \alpha_{g+\Delta}^* \alpha_g b_g^* b_{g+\Delta}\} ,
\end{aligned}\right\} \tag{3.2}$$

where α_g^*, α_g are Fermion creation and annihilation operators of a spin polaron at the atom with label g.

The Hamiltonian $\mathcal{H}_0$ corresponds to complete ferromagnetic ordering. In this case the eigenfunction of the total Hamiltonian $\mathcal{H}_{\text{sd}}$ (2.1) is simultaneously the eigenfunction of the s–d exchange Hamiltonian $\mathcal{H}_{\text{A}}$ (2.3) and of the band Hamiltonian $\mathcal{H}_{\text{B}}$ (2.2), separately. Thus, the ferromagnetic ordering being complete and A positive, the spin polaron moves as a usual band electron, its energy being shifted by $(AS/2)$.

When writing the Hamiltonian $\mathcal{H}_2$ quadratic in the deviation operators, we have taken into account that for $I = 0$ it arises only due to the Hamiltonian $\mathcal{H}_{\text{B}}$. For this reason the Hamiltonian $\mathcal{H}_2$, similar to $\mathcal{H}_{\text{B}}$, should have a form corrresponding to the nearest-neighbour approximation, terms of the type $\alpha_g^* \alpha_g b_g^* b_{g+\Delta}$ being excluded.

The Hamiltonian (3.2), being quadratic in the spin deviation operators, differs essentially from the usual s–d Hamiltonian, if the spin deviations are thought to be deviations only of d-spins. In such a case, there are terms linear in the spin deviations in the s–d Hamiltonian, as is seen from (2.1) and (3.1).

To find the unknown coefficients x, y, z let us consider an eigenstate of one electron system, such that the spin projection of the system is less by a unit than its maximum value. In the spin polaron representation, the wave function of the state under consideration is given by

$$\psi = \sum c_{gh} \alpha_g^* b_h^* |v\rangle , \tag{3.3}$$

where $|v\rangle$ is the vacuum state for spin polarons and spin deviations. Then the wave equation, making use of (3.2), can be represented as follows:

$$\left.\begin{aligned}
&B \sum_{\Delta} c_{m+\Delta, n} + BX \sum_{\Delta} c_{nn} \, \delta(m + \Delta, n) + BZ \sum_{\Delta} c_{n, n+\Delta} \, \delta(n + \Delta, m) =\\
&\qquad = \left(E + \frac{AS}{2}\right) c_{mn} ; \qquad m \neq n ,\\
&B(1 + X) \sum_{\Delta} c_{m+\Delta, m} + BY \sum_{\Delta} c_{m+\Delta, m+\Delta} = \left(E + \frac{AS}{2}\right) c_{mm} .
\end{aligned}\right\} \tag{3.4}$$

On the other hand, if we proceed directly from the Hamiltonian (2.1), the wave function of the system should be constructed as follows:

$$\psi = \sum_{g \neq h} c_{gh} \overset{+}{\varphi}_{g, S+\frac{1}{2}} \delta(S_h^z, S - 1) \prod_{f \neq g, h} \delta(S_f^z, S) + \sum c_{gg} \overset{+}{\varphi}_{g, S-\frac{1}{2}} \prod_{f \neq g} \delta(S_f^z, S) , \tag{3.5}$$

where the function φ_{gM}^{+} is given by (2.6). Evidently, formula (3.5) is a generalization of the expression (2.10) for a system consisting of an arbitrarily large number of magnetic atoms. Taking it and formulae (2.1) to (2.4) into account, we notice that the Schrödinger equation takes the form

$$\left.\begin{aligned} &B \sum_{\Delta,\, m+\Delta \neq n} c_{m+\Delta,\, n} + \sqrt{\frac{2S}{2S+1}}\, B \sum_{\Delta} c_{nn} \delta(n+\Delta, m) = \\ &= \left(E + \frac{AS}{2}\right) c_{nm}\,; \qquad n \neq m\,, \\ &B \sqrt{\frac{2S}{2S+1}} \sum_{\Delta} c_{m+\Delta,\, m} + \frac{B}{2S+1} \sum c_{m+\Delta,\, m+\Delta} = \\ &= \left(E + \frac{AS}{2}\right) c_{mm}\,. \end{aligned}\right\} \tag{3.6}$$

Comparing (3.4) with (3.6) the following expressions for the sought cofficients are obtained:

$$X = \sqrt{\frac{2S}{2S+1}} - 1\,, \qquad Y = \frac{1}{2S+1}\,, \qquad Z = 0\,. \tag{3.7}$$

This completes constructing the spin polaron–magnon Hamiltonian in the spin-wave approximation for $A > 0$.

3.2 Spin polaron Hamiltonian in the general case

The ideas presented in the previous section can be generalized, so that the results may be valid to any order in W/AS and any degree of magnetic disorder. As before, the starting point is the fact that the charge carriers are spin polarons. As mentioned above, the main difficulty in constructing the spin polaron Hamiltonian is the fact of the atomic spin magnitude to be not fixed. To avoid it, a special representation is used in which the same spin magnitude is prescribed to all atoms independently of whether or not conduction electrons are located at them. The real difference in the spin magnitudes for the former and the latter is taken into account by the structure of operators in this representation [18, 19].

To go over to it, we shall change describing the state of an atom bearing an electron. Instead of its total spin projection M, we shall use one of two values m of its "proper" spin projection allowed for a given M. Namely, we put $M = m - \frac{1}{2}$ for the state φ_{gM}^{-} (2.7) and $M = m + \frac{1}{2}$ for the state φ_{gM}^{+} (2.6), where m can take $(2S + 1)$ different values.

Evidently, in the former case the number of allowed values of m exceeds that of M by one. In reality the extra degree of freedom is fictitious, since the wave function φ^{-} (2.7) will vanish for $m = -S$ (i.e., $M = -S - \frac{1}{2}$).

In the latter case, the number of allowed values m is less than that of M by one. Our failure to take into account that state with $M = -S - \frac{1}{2}$ can appreciably influence the final result only when the following conditions are fulfilled simultaneously: 1. The s–d couping should be ferromagnetic ($A > 0$); 2. the temperature should be comparable with or be above the Curie point; 3. the spin S should not be large (if these conditions are fulfilled, the relative weight of the state not taken into account is $1/2(S + 1)$).[3])

[3]) In [71] (see Section 6) a method is developed to take into account all the states.

Let us rewrite expressions (2.6) taking into account the changes in the indices mentioned above, introducing the raising and lowering operators of the spin projections, $S^{\pm} = S^x \pm iS^y$, and changing the indices of states from m to S^z_{g0}[4]):

$$\varphi^{\pm}_{gM} \equiv \varphi^{\pm}_g (S^z_{g0}) = A^*_{g\pm}(S^z_{g0})\, \delta(S^z_g, S^z_{g0})\, |0\rangle \,, \tag{3.8}$$

where

$$A^*_{g+}(S^z_{g0}) = \frac{1}{\sqrt{2S+1}} \left\{ \sqrt{S + S^z_{g0} + 1}\, a^*_{g\uparrow} + \frac{a^*_{g\downarrow} S^+_g}{\sqrt{S + S^z_{g0} + 1}} \right\},$$

$$A^*_{g-}(S^z_{g0}) = \frac{1}{\sqrt{2S+1}} \left\{ \sqrt{S + S^z_{g0}}\, a^*_{g\downarrow} - \frac{a^*_{g\uparrow} S^-_g}{\sqrt{S + S^z_{g0}}} \right\}.$$

When writing (3.8), we make use of the relations [13]

$$\left. \begin{aligned} S^+\, \delta(S^z, m) &= \sqrt{(S + m + 1)\,(S - m)}\, \delta(S^z, m + 1)\,, \\ S^-\, \delta(S^z, m) &= \sqrt{(S + m)\,(S - m + 1)}\, \delta(S^z, m - 1)\,. \end{aligned} \right\} \tag{3.9}$$

Since we are not interested in the interactions between charge carriers, it suffices to consider the one-electron problem. To go over the to spin polaron representation, we expand an eigenfunction ψ of the Hamiltonian $\mathscr{H}_{sd}$ (2.1) in terms of those of the Hamiltonian $\mathscr{H}_A$. The latter can be represented in the form

$$\varphi^{\pm}_{gM_g} \prod_{f \neq g} \delta(S^z_f, S^z_{f0}) \tag{3.10}$$

since the electron interacts only with the atom at which it is located.

Obviously the set of functions (3.10) is complete in the space of functions depending on the electron coordinates g and the spin projections for the electron and all magnetic atoms. As is indicated above, no substantial errors arise if we omit the functions of the type $\varphi^+_{g,-S-\frac{1}{2}} \prod_{f \neq g} \delta(S^z_f, S^z_{f0})$. Thus such an expansion with the aid of (3.8) can be represented in the form

$$\psi = \sum c^{\mu}_g(S^z_0)\, \chi^{\mu}_g(S^z_0)\,, \tag{3.11}$$

$$\chi^{\mu}_g(S^z_0) = A^*_{g\mu}(S^z_{g0})\, |0\rangle\, |S^z_0\rangle\,, \tag{3.12}$$

where μ is + or −, S^z_0 is the whole set of quantities S^z_{f0}, and the notation

$$|S^z_0\rangle = \prod_f \delta(S^z_f, S^z_{f0}) \tag{3.13}$$

is used. In (3.11) the summation is to be carried out over all g, μ, and S^z_{f0}. Since the base functions $\chi^{\mu}_g(S^z_0)$ are of the type of a delta function, the coefficients $c^{\mu}_g(S^z_0)$ represent the envelope of the wave function of the system. Substituting (3.11) into (2.1) gives the Schrödinger equation in matrix form:

$$c^{(-)}_g(S^z_1) \left[\frac{A(S+1)}{2} - E \right] = \sum_{f\mu S^z_0} c^{\mu}_f(S^z_0)\, [\chi^{(-)}_g(S^z_1), (\mathscr{H}_B + \mathscr{H}_M)\, \chi^{\mu}_f(S^z_0)]\,, \tag{3.14}$$

$$c^{(+)}_g(S^z_1) \left[-\frac{AS}{2} - E \right] = \sum_{f\mu S^z_0} c^{\mu}_f(S^z_0)\, [\chi^{(+)}_g(S^z_1), (\mathscr{H}_B + \mathscr{H}_M)\, \chi^{\mu}_f(S^z_0)]\,. \tag{3.15}$$

---

[4]) Thus S^z_{g0} is a value taken by the spin projection S^z_g of atom g.

We have used the expressions (2.8) and (2.9) for the eigenvalues of $\mathcal{H}_A$ when writing (3.14) and (3.15).

To go over from (3.14) and (3.15) to the wave equation in the operator form, we introduce the operators S_{g0} acting in the space of functions of the parameters S_{g0}^z in the same manner as the corresponding operators S_g in the space of functions of the true variables S_g^z. To show in what a manner such a transition is done, let us consider the first term in the right-hand side of equation (3.14):

$$\sum_{S_0^z f} c_f^{(-)}(S_0^z)\,[\chi_g^{(-)}(S_1^z),\, \mathcal{H}_B \chi_f^{(-)}(S_0^z)] \equiv X_1 + X_2\,. \tag{3.16}$$

Using (2.2), (3.11) to (3.13), and the orthonormality of the set of functions $|S_0^z\rangle$, we obtain

$$X_1 \equiv \frac{B}{2S+1} \sum_{S_0^z \Delta} \sqrt{(S + S_{g1}^z)\,(S + S_{g+\Delta 0}^z)}\,\langle S_1^z | S_0^z\rangle\, c_{g+\Delta}^{(-)}(S_0^z) =$$

$$= \frac{B}{2S+1} \sum_{\Delta} \sqrt{(S + S_{g1}^z)\,(S + S_{g+\Delta 1}^z)}\, c_{g+\Delta}^{(-)}(S_1^z)\,, \tag{3.17}$$

since

$$\langle S_1^z | S_0^z\rangle = \prod_f \sum_{S_f^z} \delta(S_f^z, S_{f1}^z)\,\delta(S_f^z, S_{f0}^z) = \prod_f \delta(S_{f1}^z, S_{f0}^z)\,. \tag{3.18}$$

When calculating X_2 we twice make use of the relation

$$S^-\,\delta(S^z, S_0^z) = S_0^+\,\delta(S^z, S_0^z) \tag{3.19}$$

and of that conjugate to it, so that, taking into account (3.18), we obtain

$$X_2 \equiv \frac{B}{2S+1} \sum_{S_0^z \Delta} \frac{\langle S_1^z|\, S_g^+ S_{g+\Delta}^- \,|S_0^z\rangle\, c_{g+\Delta}^{(-)}(S_0^z)}{\sqrt{(S + S_{g1}^z)\,(S + S_{g+\Delta 0}^z)}} =$$

$$= \frac{B}{2S+1} \frac{1}{\sqrt{S + S_{g1}^z}} \sum_{S_0^z \Delta} \frac{[S_{g0}^- S_{g+\Delta 0}^+ \langle S_1^z|S_0^z\rangle]\, c_{g+\Delta}^{(-)}(S_0^z)}{\sqrt{S + S_{g+\Delta 0}^z}} =$$

$$= \frac{B}{2S+1} \frac{1}{\sqrt{S + S_{g1}^z}} \sum_{\Delta} S_{g1}^+ S_{g+\Delta 1}^- \sum_{S_0^z} \frac{\langle S_1^z|S_0^z\rangle\, c_{g+\Delta}^{(-)}(S_0^z)}{\sqrt{S + S_{g+\Delta 0}^z}} =$$

$$= \frac{B}{2S+1} \sum_{\Delta} \frac{1}{\sqrt{S + S_{g1}^z}}\, S_{g1}^+ S_{g+\Delta 1}^- \frac{c_{g+\Delta}^{(-)}(S_1^z)}{\sqrt{S + S_{g+\Delta 1}^z}}\,. \tag{3.20}$$

As a result of these and similar calculations, the equations (3.14) and (3.15) take the form (the index 1 of spin operators being dropped)

$$\left\{E - \frac{A(S+1)}{2} - \mathcal{H}_M\right\} c_g^{(-)}(S^z) =$$

$$= B \sum_{\Delta} \mathcal{F}^{(-)}(S_g, S_{g+\Delta})\, c_{g+\Delta}^{(-)}(S^z) + B \sum_{\Delta} G(S_g, S_{g+\Delta})\, c_{g+\Delta}^{(+)}(S^z)\,, \tag{3.21}$$

$$\left\{E + \frac{AS}{2} - \mathcal{H}_M\right\} c_g^{(+)}(S^z) =$$

$$= B \sum_{\Delta} \mathcal{F}^{(+)}(S_g, S_{g+\Delta})\, c_{g+\Delta}^{(+)}(S^z) + B \sum_{\Delta} G^*(S_{g+\Delta}, S_g)\, c_{g+\Delta}^{(-)}(S^z)\,, \tag{3.22}$$

where the following notations are used:

$$\left.\begin{aligned}
\mathcal{F}^{(-)}(S_g, S_{g+\Delta}) &= \frac{1}{2S+1}\left\{\sqrt{(S+S_g^z)(S+S_{g+\Delta}^z)}\, + \right.\\
&\quad \left. + \frac{1}{\sqrt{S+S_g^z}}\, S_g^+ S_{g+\Delta}^- \frac{1}{\sqrt{S+S_{g+\Delta}^z}}\right\},\\
\mathcal{F}^{(+)}(S_g, S_{g+\Delta}) &= \frac{1}{2S+1}\left\{\sqrt{(S+S_g^z+1)(S+S_{g+\Delta}^z+1)}\, + \right.\\
&\quad \left. + \frac{1}{\sqrt{S+S_g^z+1}}\, S_g^- S_{g+\Delta}^+ \frac{1}{\sqrt{S+S_{g+\Delta}^z+1}}\right\},\\
G(S_g, S_{g+\Delta}) &= \frac{1}{2S+1}\left\{\sqrt{S+S_g^z}\, S_{g+\Delta}^+ \frac{1}{\sqrt{S+S_{g+\Delta}^z+1}}\, - \right.\\
&\quad \left. - \frac{1}{\sqrt{S+S_g^z}}\, S_g^+ \sqrt{S+S_{g+\Delta}^z+1}\right\}.
\end{aligned}\right\} \quad (3.23)$$

In the equations (3.21) and (3.22), the terms which describe changing the exchange Hamiltonian $\mathcal{H}_M$ due to the conduction electron are omitted, since they are of the order of I/B relative to the terms entering (3.21) and (3.22). The dropped terms are presented in [19].

The form of the equations (3.22) and (3.23) makes it possible to write the Hamiltonian of the system under consideration as follows:

$$\left.\begin{aligned}
\mathcal{H} &= \mathcal{H}^{(+)} + \mathcal{H}^{(-)} + \mathcal{H}^{(+-)} + \mathcal{H}_M;\\
\mathcal{H}^{(+)} &= -\frac{AS}{2} \sum \alpha_{g(+)}^* \alpha_{g(+)} + B \sum \mathcal{F}^{(+)}(S_g, S_{g+\Delta})\, \alpha_{g(+)}^* \alpha_{g+\Delta(+)},\\
\mathcal{H}^{(-)} &= \frac{A(S+1)}{2} \sum \alpha_{g(-)}^* \alpha_{g(-)} + B \sum \mathcal{F}^{(-)}(S_g, S_{g+\Delta})\, \alpha_{g(-)}^* \alpha_{g+\Delta(-)},\\
\mathcal{H}^{(+-)} &= B \sum G(S_g, S_{g+\Delta})\, \alpha_{g(-)}^* \alpha_{g+\Delta(+)} + \text{c.c.}\,,
\end{aligned}\right\} \quad (3.24)$$

where $\alpha_{g(+)}^*$ and $\alpha_{g(+)}$ are the creation and annihilation operators of the spin polaron at atom g, when the total spin of the atom at which the electron is located, is $S + \frac{1}{2}$, while $\alpha_{g(-)}^*$ and $\alpha_{g(-)}$ refer to the case of total spin $S - \frac{1}{2}$. The commutation rules for the spin polaron operators can evidently be regarded as Fermi ones. This question is discussed in more detail in [19].

The equivalency of the Hamiltonian (3.24) to the Hamiltonian in the equations (3.22) and (3.23) is easily proved if one notes that the wave function of the system can be represented in terms of spin polaron operators by

$$\psi = \sum c_g^\mu(S^z)\, \alpha_{g\mu}^* \,|0\rangle\,, \qquad (3.25)$$

where $|0\rangle$ is the spin polaron vacuum state.

It should be emphasized that the "effective spin" S_g in (3.24) is the true spin of atom g only if the conduction electron is not located at it. In the opposite case, however, the "effective spin" has not any simple physical meaning.

3.3 Fermion elementary excitations

At low temperatures the structure of the Hamiltonian (3.24) simplifies strongly as a result of its linearization in the deviations of the "effective spins", introduced in the previous section. The deviation operators are introduced by means of the usual relations (3.2) now. In doing this we take into account that in the completely feromagnetic state at $T = 0$, all the "effective spin" projections as well as the real spin projections are maximum, which follows from the definition of the quantities S^z_{g0} given in Section 3.2.

To the first approximation in W/AS the term $\mathscr{H}^{(+-)}$ in the Hamiltonian (3.24) can be discarded so that the total spin of an atom and an electron at it may be regarded as an integral of motion. In the spin-wave approximation, the Hamiltonian $\mathscr{H}^{(+)}$ (3.24) coincides with the Hamiltonian (3.2 to 3.7), and the Hamiltonian $\mathscr{H}^{(-)}$ is given by

$$\mathscr{H}^{(-)} = \frac{A(S+1)}{2} \Sigma\, \alpha^*_{g(-)}\alpha_{g(-)} + \frac{2SB}{2S+1} \Sigma\, \alpha^*_{g(-)}\alpha_{g+\Delta(-)} +$$

$$+ \frac{B}{2S+1} \Sigma \left\{ b^*_{g+\Delta} b_g - 2S\left(1 - \sqrt{1 - \frac{1}{2S}}\right) (b^*_g b_g + b^*_{g+\Delta} b_{g+\Delta}) \right\} \alpha^*_{g(-)}\alpha_{g+\Delta(-)}\,, \tag{3.26}$$

where, like in Section 3.2, the operators of "effective spin" deviations mean those of real total spins with magnitude S or $S - \frac{1}{2}$ depending on whether or not the conduction electron is located at the atom.

In general we should add the term describing the "direct" exchange between atoms to $\mathscr{H}^{(+)}$ or $\mathscr{H}^{(-)}$:

$$\mathscr{H}_{\mathrm{M}} = IS\, \Sigma\, (b^*_g b_g - b^*_g b_{g+\Delta})\,. \tag{3.27}$$

From the expressions (3.2) to (3.7) and (3.26), (3.27) one can obtain the elementary excitation spectrum, making use of standard methods in solid state physics. In particular, at $T = 0$ we have

$$E^{(-)}(k) = \frac{A(S+1)}{2} + \frac{2S}{2S+1}\, zB\gamma_k \quad \text{for} \quad A < 0\,, \tag{3.28}$$

$$E^{(+)}(k) = -\frac{AS}{2} + zB\gamma_k \quad \text{for} \quad A > 0\,, \tag{3.29}$$

where the notation

$$\gamma_k = \frac{1}{z} \sum_{\Delta} \mathrm{e}^{ik\Delta}$$

is used, k being the spin polaron quasi-momentum.

The temperature shift in the spin polaron energy can also be easily found for large spins. In the main approximation with respect to $1/2S$, the spin polaron–magnon interaction leads only to a renormalization of the Bloch integral, leaving unchanged the type of dispersion law (3.28) or (3.29):

$$\tilde{B} = B\left\{1 + \frac{1}{2SN} \sum_q (\gamma_q - 1)\, m_q\right\}, \tag{3.30}$$

where m_q is the usual Bose distribution function for magnons with quasi-momentum q. The spin polaron damping appears only in higher orders of perturbation theory.

It is seen from (3.30) that the effective Bloch integral $\tilde{B}$ decreases with increasing temperature. At temperatures small compared with the maximum magnon frequency ω_m, the temperature shift in $\tilde{B}$ is proportional to $T^{5/2}$, i.e., it is weaker than the magnetization change ($\sim T^{3/2}$). In the temperature range $\omega_m < T < \omega_m S \sim T_C$ (T_C is the Curie point), the shift in $\tilde{B}$ is proportional to T, i.e. to the magnetization change. In the same manner the conduction band bottom shifts upwards, and the spin polaron effective mass being reciprocal to $\tilde{B}$ increases.

The obvious fact that in the limit of weak s–d coupling, the shift of the bottom of the conduction band should be proportional to the change of magnetization, was still indicated in [20] and [21]. But it seems to be nontrivial that the same result is also obtained in the opposite limit. Such a dependence of the band bottom shift on the change of magnetization was observed experimentally [22].

As is seen from (3.28), when a charge carrier is in the spin polaron state, the full gain is acquired in the largest of the typical energies of the system, i.e. the s–d exchange energy. But for $A < 0$ the spin polaron band is $1 + 1/(2S)$ times narrower than the electron band. These features of the spin polaron spectrum are similar to the polaron one, which is the reason to call the charge carriers in magnetic conductors spin polarons.

According to (3.29), for $A > 0$ the spin polaron spectrum at $T = 0$ coincides with the band electron spectrum if the electron spin is directed along the crystal magnetic moment, as is easily proved using the Hamiltonian (2.1). But taking into account the band bottom shift at finite temperatures, the considerations cited above for the case $A < 0$ are also valid for $A > 0$. Thus, the full gain in the s–d exchange energy is a peculiarity of spin polaron states, and this statement is true for arbitrary spin configurations according to (3.24).

It should be noted that attempts to treat carrier states for a strong antiferromagnetic s–d coupling were made earlier in [23] and [24]. However, the diagonalization of the Hamiltonian (2.1) in the zeroth approximation in W/AS was incorrect in these papers, and for this reason, the results of [23] and [24] are in error. This is already seen from the fact that the ground state energy found from (3.28) is sufficiently lower than that obtained in [23] and [24]. The correctness of the expression (3.28) is confirmed by the exact calculation for $T = 0$ valid in an arbitrary order in W/AS, which was carried out in [25].

3.4 Magnons in ferromagnetic conductors

In what follows an expression for magnon frequencies will be obtained when the spin polaron density is large enough to influence the magnetic properties of the crystal as a whole. As follows from (3.24), ferromagnetic ordering ensures the maximum width of the spin polaron band. For this reason, the indirect exchange via spin polarons tends to establish ferromagnetic ordering at least at spin polaron densities not too large.

First of all it should be pointed out that when the direct exchange can be neglected ($I = 0$), even at $T = 0$ magnons can be regarded as quasi-particles with well defined quasi-momenta only for large spins S. Only in that case the magnon–spin polaron interaction Hamiltonian has a small parameter, which makes it possible to develop the perturbation procedure and, in the main order, to separate the magnons from the spin polarons. This parameter is $1/2S$ (see, for instance, (3.26)). For small S only the total quasi-momentum of magnons and spin polarons is conserved. Henceforth only large spins will be considered.

Since the frequencies of spin motion are small compared with those of the spin polarons, the two subsystems can be separated making use of the Born-Oppenheimer approximation. To take account of the anharmonicity, we use the Holstein-Primakoff transformation [26] of a more general form than (3.1):

$$S_g^z = S - b_g^* b_g\,, \quad S_g^- = \sqrt{2S}\, b_g^* \left(1 - \frac{b_g^* b_g}{4S}\right), \quad S_g^+ = \sqrt{2S}\, b_g \left(1 - \frac{b_g^* b_g}{4S}\right). \tag{3.31}$$

We limit ourselves to the case $A > 0$, $I = 0$. After substituting (3.31) into the Hamiltonian (3.24) and changing from the site to the quasi-momentum representation, the Hamiltonian of the system up to the order of $1/4S^2$ takes the form

$$\begin{aligned}\mathcal{H}^{(+)} = &-\beta \sum \gamma_k \alpha_k^* \alpha_k + \frac{\beta}{4SN} \sum \Big[(\gamma_k + \gamma_{k'} - 2\gamma_{q+k}) - \\ &- \frac{1}{2S}\left(\frac{3}{4}\gamma_k + \frac{3}{4}\gamma_{k'} - 2\gamma_{k+q}\right)\Big] b_q^* b_{q'} \alpha_k^* \alpha_{k'} D\,(q + k, q' + k') - \\ &- \frac{\beta}{16S^2N^2} \sum \left(\gamma_{k+q-q'} - \frac{\gamma_k}{2} - \frac{\gamma_{k'}}{2}\right) \times \\ &\times b_q^* b_p^* b_{p'} b_{q'} \alpha_k^* \alpha_{k'} D\,(k + p + q, k' + p' + q')\,,\end{aligned} \tag{3.32}$$

where β denotes $z\,|B|$, $\alpha_k^* = 1/\sqrt{N} \sum \alpha_g^*\, e^{ikg}$, and so on. The symbol $D(k, k')$ expresses the law of the quasi-momentum conservation.

In the Born-Oppenheimer approximation, the fermion energy expressed in terms of magnon operators by means of the perturbation theory plays the role of the magnon Hamiltonian. To the second approximation in $1/2S$ we obtain the following expression for the magnon Hamiltonian using the Hamiltonian (3.32):

$$\mathcal{H}_S = \sum \omega_q b_q^* b_q + \sum_{q \neq q'} X_{qp|p'q'} b_q^* b_p^* b_{p'} b_{q'} D(p + q, p' + q')\,. \tag{3.33}$$

The magnon frequency ω_q and the anharmonicity parameters $X_{pq|p'q'}$ are given by

$$\omega_q = \omega_q^{(1)} + \omega_q^{(2)}\,, \tag{3.34}$$

$$\omega_q^{(1)} = J_{\text{eff}}(1 - \gamma_q)\,, \tag{3.35}$$

$$\omega_q^{(2)} = -\frac{J_{\text{eff}}}{4S}\left(\frac{3}{2} - 2\gamma_q\right) - \sum_p Z_{qp|qp}\,, \tag{3.36}$$

$$X_{pq|p'q'} = \frac{J_{\text{eff}}}{8SN}(1 - \gamma_{q-q'}) - Z_{qp|p'q'}\,, \tag{3.37}$$

where the notations

$$J_{\text{eff}} = \frac{\beta}{2SN} \sum_k n_k \gamma_k\,, \tag{3.38}$$

$$\begin{aligned}Z_{qp|p'q'} = &\frac{\beta}{16S^2N^2} \sum_{kk'} \frac{(\gamma_k + \gamma_{k'} - 2\gamma_{p+k})(\gamma_k + \gamma_{k'} - 2\gamma_{q+k})}{\gamma_k - \gamma_{k'} - i\eta} \times \\ &\times n_{k'}(1 - n_k)\, D(q + k, q' + k')\,; \quad \eta \to 0\end{aligned} \tag{3.39}$$

are used. n_k is Fermi distribution function for spin polarons.

In the first approximation in $1/2S$ the magnon frequency is given by (3.35).[5]) The quantity J_{eff} can be regarded as an effective indirect exchange integral. As follows from (3.38) this quantity is positive for small spin polaron densities as should be the case. It increases with density until the band is half filled, and then begins to decrease.

As follows from (3.34) to (3.39), if a magnon frequency is represented as a series in the magnon quasi-momentum, the leading term will be quadratic in the quasi-momentum. This result is quite natural since the spin wave with the zeroth quasi-momentum corresponds to the rotation of the total spin of the spin of the system as a whole in space. In the absence of a magnetic anisotropy such rotation cannot change the energy of the system. Thus the change in the magnetization obeys the usual law $T^{3/2}$ at low temperatures. It is also clear that the Curie temperature at $I = 0$ should be of order SJ_{eff}, i.e. $\nu\beta$, where $\nu = n/N$ is the spin polaron number per atom ($n = \sum\limits_k n_k$).

Attention should be paid to the difference between the result (3.34 to 3.39) obtained here and those of the Ruderman-Kittel-Kasuya-Yosida indirect exchange theory [26 to 28], which is valid in the opposite limit when the Fermi energy of carriers greatly exceeds the s–d exchange energy. At first sight it may seem to be strange that the magnon frequencies do not depend on the s–d exchange integral A, though in the case under consideration the ferromagnetic ordering itself is due to indirect exchange. But really this result is quite natural: the s–d exchange by itself could not establish ferromagnetic ordering if the conduction electrons were not able to move through the crystal going over from one atom to another. Thus a magnon frequency for $B = 0$ should vanish, i.e., it should be represented as a series in $W = 2\beta$ for small W/AS. The expression (3.34) corresponds to the first approximation in this parameter. To obtain the magnon frequencies in the second approximation, one should perform a canonical transformation of the Hamiltonian (3.24) which eliminates the Hamiltonian $\mathcal{H}^{(+-)}$ from it in the indicated approximation. The corresponding calculation is presented in [19].

Obviously the dispersion law (3.35) corresponds to the nearest-neighbour approximation which is easily seen from the relations

$$\left.\begin{aligned} \omega_q &= \sum_{h \neq 0} J_{\text{eff}}(h)\,[1 - e^{ihq}]\,, \\ J_{\text{eff}}(h) &= -\frac{1}{N} \sum_q \omega_q\, e^{-iqh}\,. \end{aligned}\right\} \tag{3.40}$$

Thus there are no oscillations of the effective exchange integral $J_{\text{eff}}(h)$ as a function of its coordinates h.

In the main approximation in $1/2S$ it is easy to write an expression for the Hamiltonian of the magnetic subsystem which is also valid for nonuniform distributions of the spin polaron density over the crystal (for instance, for spin polarons in localized states):

$$\mathcal{H}_S = \frac{|B|}{2S} \sum \langle \alpha_g^* \alpha_{g+\Delta} \rangle \left[\frac{1}{2} \left(b_g^* b_g + b_{g+\Delta}^* b_{g+\Delta} \right) - b_{g+\Delta}^* b_g \right], \tag{3.41}$$

where the angular brackets denote averaging over the ground state.

[5]) This expression is also valid for $W \gg AS$, if q is small enough [67].

The theory of the indirect exchange via the electrons of unionized donors (i.e. via localized spin polarons) was developed in [29]. It is shown that local or quasi-local magnons may arise due to such exchange. In [30] was pointed out that charging the semiconductor surface relative to its bulk changes the effective integral of the indirect exchange in a surface layer of the depth of the order of the screening radius or even more. For this reason specific surface magnons arise with frequencies depending on the surface charge. The latter can be changed by an external electric field.

3.5 Transport properties of spin polarons

The use of the spin polaron representation makes it possible to calculate the transport coefficients of magnetic semiconductors. In what follows, we shall restrict ourselves to calculate their conductivity [18, 31]. For this aim, we first need to construct the current operator in the spin polaron representation. We start from the standard expression for the current in terms of the electron operators in the nearest-neighbour approximation:

$$\boldsymbol{j} = e\boldsymbol{v} = ieB \sum_{\Delta} \Delta a^*_{g\sigma} a_{g+\Delta\sigma} \,. \tag{3.42}$$

The calculation of the matrix elements in (3.42) using the functions (3.12) is carried out in the same manner as in Section 3.2. In the principal approximation with respect to W/AS, it leads to the following result:

$$\boldsymbol{j}_\mu = ieB \sum_{\Delta} \mathcal{F}^\mu(S_g, S_{g+\Delta})\, \Delta \alpha^*_{g\mu} \alpha_{g+\Delta,\mu} \,. \tag{3.43}$$

In what follows, we shall consider the case $A < 0$ (the case $A > 0$ is treated in the same manner). Going over to the Fourier representation in the site number in the expression (3.43), we see at once that the standard relation between the spin polaron velocity and its energy, $v_k = \nabla_k E(k)$, is valid for $T = 0$, where the energy $E(k)$ is given by (3.28). Moreover, this relation remains valid at finite temperatures $\approx T_C/2S$ in the first approximation in $1/2S$.

The relation between the spin polaron velocity and its energy indicated above is sufficient for the usual Boltzmann equation to be applicable in the spin-wave region. This can be easily proved using the standard procedure for deriving the Boltzmann equation from the density matrix equation of motion (see, for instance, [31]). In the main approximation in $1/2S$, the Boltzmann equation is written in the form

$$e(\mathcal{E}v_k)\,\frac{\partial n_k^0}{\partial E_k} = 2\pi \sum M^2_{kq|k'q'}[n_{k'}(1 + m_q)\, m_{q'} - n_k(1 + m_{q'})\, m_q] \times \\ \times\, \delta(E_k + \omega_q - E_{k'} - \omega_{q'})\,; \tag{3.44}$$

$$\left.\begin{aligned} M_{kq|k'q'} &= \frac{3B}{2SN}[2\gamma_{q-k'} - \gamma_{k'} - \gamma_k]\, D(k + q, k' + q')\,, \\ \omega_q &= J(1 - \gamma_q);\quad J = J_{\text{eff}} + zIS\,, \end{aligned}\right\} \tag{3.45}$$

where n_k^0 is the equilibrium distribution function for spin polarons and m_q is that for magnons. Since $|B| \gg I$, the scattering of spin polarons by magnons is almost elastic for $n_k \ll 1$. In the degenerate case, this is true only if the condition

$$\left(\frac{JS}{|B|}\,\frac{\mu_F}{T}\right)^{1/2} \ll 1 \tag{3.46}$$

is satisfied [2], where μ_F is the Fermi energy and T the temperature in energy units.

Formula (3.44) leads to the following expression for the relaxation time when the scattering is elastic:

$$\frac{1}{\tau_k} = \frac{9\,|k|\,a\,|B|}{2\pi J^2 S^2 N} \sum \omega_q^2\, m_q(1 + m_q)\,. \tag{3.47}$$

It follows from (3.47) that the relaxation time τ_k is proportional to $T^{-7/2}$ for $T \ll J$. Thus the spin polaron mobility is proportional to T^{-4} in the nondegenerate case and to $T^{-7/2}$ in the degenerate case. When $J < T < JS$, the relaxation time is proportional to T^{-2}. Correspondingly the mobility is proportional to $T^{-5/2}$ in the nondegenerate case and to T^{-2} in the degenerate case. Thus the temperature dependence of mobility becomes weaker as the temperature increases.

One can also obtain an expression for the mobility in the high-temperature limit when $T > W \gg JS$. Certainly the states of the charge carriers cannot be described in terms of quasi-momentum: the degree of magnetic disordering is so large in the case under consideration that the system cannot be regarded as periodic at all. The mechanism of charge carrier motion in it should be of the same type as, for instance, in semiconducting glasses.

The case under consideration is interesting in that one is able to calculate the conductivity σ directly from the Kubo formula [32]:

$$\sigma = \int_0^\infty K(t)\,\mathrm{d}t\,; \qquad K(t) = \frac{1}{3T}\langle \mathrm{e}^{i\mathcal{H}t}\boldsymbol{j}\,\mathrm{e}^{-i\mathcal{H}t}\boldsymbol{j}\rangle \tag{3.48}$$

(angular brackets denote temperature average here). Such a possibility arises for all operators entering (3.48) depending on the only parameter — the Bloch integral B. Actually the statistical operator $\mathrm{e}^{-\mathcal{H}/T}$ can be replaced by the unit operator $\hat{1}$ in the zeroth approximation in W/T, and the terms $\sim JS$ can be omitted in the evolution operator $\mathrm{e}^{i\mathcal{H}t}$. The s–d exchange energy enters both $\mathcal{H}^{(+)}$ and $\mathcal{H}^{(-)}$ only as an additive constant.

For this reason the system considered has the only time scale $1/W$. This makes it possible to calculate the conductivity expanding the current correlator in powers of Wt and constructing an approximate analytic continuation of this series to the region $Wt \approx 1$ which gives the main contribution to the integral over t in (3.48). It should be emphasized that such a procedure cannot be used in cases when the usual Boltzmann equation is applicable, since then there are two time scales — the atomic τ_A and the relaxation τ_r, the latter greatly exceeding the former. The small parameter in powers of which the current correlator is expanded is t/τ_A. Such a series can be analytically continued only to $t \approx \tau_A$ but nor to $t \approx \tau_r$, whereas the main contribution to the integral (3.48) is given by $t \gtrsim \tau_r$.

It is natural to assume that in the right half-plane of the complex variable t^2, the function $K(t)$ has not any singularities at $|t^2| \to \infty$. Then it is convenient to construct an approximate analytic continuation to $t^2 \gtrsim W^{-2}$ by means of the Padé approximation method [33], in which $K(t)$ being even in t is approximated by

$$K(t) \approx \frac{P_L(t^2)}{Q_M(t^2)} = \frac{p_0 + \cdots + p_L t^{2L}}{q_0 + \cdots + q_M t^{2M}}\,. \tag{3.49}$$

This method is the most successful for $W^2t^2 \approx 1$ when using the diagonal approximants (3.49) with $L = M$.

Obviously when t tends to infinity, the diagnonal approximants (3.49) cannot represent $K(t)$ which should vanish at $t \to \infty$. For this reason the integration over t in (3.48) should be cut off at a finite time t_0. The latter is found taking into account the following considerations.

The current correlator is expressed in terms of eigenfunctions and eigenvalues of the Hamiltonian $\mathscr{H}$ by

$$K(t) = \sum |j_{mn}|^2 \, e^{i(\varepsilon_n - \varepsilon_m)t} . \tag{3.50}$$

If the periodicity of the crystal is perturbed weakly, the current matrix element is practically singular, being nonzero only for ε_n close to ε_m. But in the case under consideration, the periodicity of spin directions is broken at all, so that one can speak of charge carrier motion in an amorphous conductor. Thus the current matrix element certainly should be nonsingular here, i.e., it should vary slowly inside the energetical band. Replacing $|j_{mn}|^2$ in (3.50) by a constant, the quantity $K(t)$ is seen to be positive.

As the nominator of (3.49) goes to zero at some values t large enough, the approximation (3.49) with $L = M$ breaks for t exceeding the first root t_0 of the nominator. Keeping in mind that the contribution of large times $t \gg 1/W$ to the integral (3.49) is exponentially small, we can cut off the integration at $t = t_0$.

Generally speaking the expression (3.49) may also change its sign after going the denominator to zero. But certainly $K(t^2)$ has no poles on the real half-axis $t^2 > 0$. If L and M are large enough, it is not necessary to cut off the integral since "almost diagonal" approximants with $L = M - m$ $(m \ll L)$ can be used.

If we limit ourselves to the first diagonal approximation with $M = L = 1$, we obtain from (3.48) and (3.49)

$$\sigma = K(0) \left[\frac{1 - (p/q)}{\sqrt{q}} \arctan \sqrt{\frac{q}{|p|}} - \frac{\sqrt{|p|}}{q} \right], \tag{3.51}$$

where the notations

$$q = - \frac{1}{12} \left. \frac{\dfrac{d^4K(t)}{dt^4}}{\dfrac{d^2K(t)}{dt^2}} \right|_{t=0} \quad \text{and} \quad p = \left. \frac{\dfrac{d^2K(t)}{dt^2}}{K(t)} \right|_{t=0} - q$$

are used.

To estimate the accuracy of the method of approximate analytic continuation, it is convenient to put $K(t)$ equal to e^{-t^2}. Formula (3.51) gives a value of the integral (3.48) differing from the true value by 6%. In the absence of any small parameters, such an accuracy seems to be quite satisfactory.

A very cumbersome calculation with the aid of (3.24), (3.43), (3.48), and (3.51) leads to the following expression for the high-temperature mobility of spin polarons:

$$u = \frac{e\tilde{\tau}}{m^*}; \quad \tilde{\tau} = \frac{1}{2T}, \tag{3.52}$$

where $1/m^* = 2|B|a^2$ is the reciprocal effective mass of the conduction band electron. It should be noted that the expression (3.52), apart from a dimensionless factor of the order of 1, can be deduced directly from (3.48).

Thus the high-temperature mobility is given formally by an expression of the same type as in the standard band theory. But the quantity $\tilde{\tau}$, inversely proportional to the average termalh energy T, plays the role of the relaxation time. It is essential that the formula (3.52) is applicable only for conductors with small mobilities which certainly cannot be described within the framework of the band theory. So, for temperatures of ≈ 1000 °K our result is valid for conductors with mobilities less than 1 cm^2/Vs, independent of magnetic ordering type at low temperatures. Formula (3.51) agrees with experimental data for many transition element compounds [34].

Using the spin polaron representation makes it possible to find other transport coefficients, too. In [35] and [36] the spontaneous Hall and Nernst-Ettingshausen effects are investigated, in [37] the normal Hall effect. In the one-band approximation the high-temperature Hall constant should be proportional to T/W for the charge carrier density given. But it is not clear whether the one-band approximation is sufficient for such a calculation. It may turn out that a complete set of functions must be used for this aim.

The high-temperature thermo-power is given by the well-known Zener formula (in units k/e)

$$\alpha = \ln \frac{1-\nu}{\nu} \,. \tag{3.53}$$

3.6 Transition of a heavily doped ferromagnetic semiconductor into the insulating state

We shall now discuss specific features of the conductivity of doped ferromagnetic semiconductors. In semiconductors with small impurity density the impurity local levels are completely filled at $T = 0$. As has been pointed out when discussing formula (3.30), the spin polaron energy increases with temperature. At a given temperature, the energy shift is the less, the stronger the ferromagnetic coupling between the atoms is. When a spin polaron is localized at a defect, it enhances the ferromagnetic coupling in the vicinity of the defect. Thus the magnetic ordering degree in the region of maximum localization probability of the spin polaron is higher than that averaged over the crystal. This means that as the temperature increases, the local level shifts upwards less than the bottom of the conduction band, so that the activation energy should increase at first.

At temperatures above T_C the shift of the bottom of the conduction band practically ceases, but the shift of the donor levels increases since the local magnetic ordering near the donors begins to be destroyed. For this reason the activation energy of the conductivity should reach its maximum at a temperature $T \approx T_C$, and then begins to diminish as the temperature rises.

Let us assume now that at $T \to 0$ the donor concentration n exceeds its critical value n_c for which the impurity band appears, and the donor spin polarons get collectivized. Then, as the temperature increases, it may be thermodynamically favoured for spin polarons to go over from collectivized into localized states (one spin polaron at each donor like for $n < n_c$). Such a metal–insulator transition can be explained qualitatively as follows [37][6]:

If the indirect exchange via collectivized spin polarons is also taken into account, the magnetic ordering degree in the conducting state remains less

[6]) Only wide band semiconductors ($W \gg AS$) were treated quantitatively in [37].

than the local magnetic ordering degree in the vicinity of the unionized donors. This is due to the fact that the localized spin polaron density near the donor is certainly higher than the average spin polaron density in the conducting state. For this reason, the consideration cited above is also valid here according to which, as the temperature increases, the conduction band bottom shifts upwards stronger than the donor levels. If the energy difference between the conducting and insulating states is not to large for $T = 0$ ($n - n_c$ small enough), this may cause the metal–insulator transition.

One should compare the free energy of the conducting state with that of the insulating state when treating this problem consistently. Typical spin polaron energies (the Fermi energy in the conducting state or the local level depth in the insulating state) exceed the thermal energy by several orders of magnitude. Thus in both cases practically the whole temperature change in the free energy, $\delta F(T) = F(T) - F(0)$, is caused by the excitation of oscillations of the magnetic moment, and the contribution of spin polarons can be neglected.

Evidently the spin polaron density in the conducting state remains almost unchanged over a distance of the order of the lattice constant. As to the density of the localized spin polarons, it may change both smoothly and abruptly. It is advisible to begin with the former case when proving that the free energy of the insulating state diminishes with increasing temperature more rapidly than that of the conducting state.

For large spins the main contribution to the magnon free energy is due to magnons with wavelengths of the order of the lattice constant a for $T_C > T > T_C/S$. As the spin polaron density varies smoothly, the quasi-classical approximation can be used. The magnon free energy per unit volume is then given by

$$\delta F(T) = \frac{Ta^3}{(2\pi)^3} \int d^3r \int d^3k \ln\left[1 - \exp\left\{-\frac{\omega_k(r)}{T}\right\}\right]. \tag{3.54}$$

According to (3.27), (3.35), and (3.38), for $\nu \ll 1$ the local magnon frequency $\omega(r)$ is given by

$$\left.\begin{aligned} \omega_k(r) &= \overline{\omega}_k + \Delta\omega_k; \\ \overline{\omega}_k &= \left(ISz + \frac{\beta\nu}{2S}\right)(1 - \gamma_k) + O(\nu^{5/3}), \\ \Delta\omega_k(r) &= \frac{\beta}{2S}[\nu(r) - \nu](1 - \gamma_k), \end{aligned}\right\} \tag{3.55}$$

where $\nu(r)$ is the number of spin polarons per atom at the point r, and ν its mean value. Obviously the latter quantity is equal for both the conducting and insulating states (it is equal to the relative donor concentration). Thus, the mean magnon frequency ω_k is equal for both states.

The calculation of the magnon free energy (3.54) is carried out for the conditions $T > \overline{\omega}_k$ and $\overline{\omega}_k \gg \Delta\omega_k$. The latter is satisfied, for example, for arbitrary spin polaron densities when IS is large, or for arbitrary magnitudes of IS when the spin polaron density dispersion is small. Setting equal the free energies F_c and F_i of the conducting and the insulating states, we obtain the temperature T_M of the first-order phase transition from the conducting into the insulating state:

$$\left.\begin{aligned} T_{\mathrm{M}}(\nu) &= \frac{2E(\nu)}{D_{\mathrm{i}}^2 - D_{\mathrm{c}}^2}\left(\frac{2IS^2}{|B|\,\nu} + 1\right)^2; \\ D^2 &= \int \mathrm{d}^3r \left(\frac{\nu(r)}{\nu} - 1\right)^2, \end{aligned}\right\} \tag{3.56}$$

where $E(\nu)$ is the energy difference between the insulating and conducting states per magnetic atom at $T = 0$, D_{i}^2 and D_{c}^2 are the squared spin polaron density dispersions in the insulating and conducting states, respectively.

Obviously, the dispersion in the insulating state should exceed that in the conducting state. Thus, really, $\delta F(T)$ in the former case is less than in the latter, and the temperature of the insulator–metal transition defined by (3.56) is positive as should be the case. Further, as the temperature increases, the rate of free energy of the insulating state will diminish the more the larger is D_{i}^2. The latter increases with decreasing radius of the localized state. Thus, formula (3.56) ensures the transition temperature T_{M} to be positive (i.e. the possibility of the metal–insulator transition) also for the case of impurity states of small radii. It is clear, however, that (3.56) yields only a very rough estimate of T_{M} for small impurity radii.

It does not follow from (3.56) that the transition from the conduction into the insulating state does occur for any ν, since the applicability of (3.56) is limited to the case $T_{\mathrm{M}} < T_{\mathrm{C}}$. As the quantity $E(\nu)$ increases with $\nu - \nu_{\mathrm{c}}$, the temperature T_{M} determined from (3.56) should exceed T_{C} for $\nu - \nu_{\mathrm{c}}$ being large enough. In other words, formula (3.56) becomes inapplicable. Actually, T increasing above T_{C}, the difference $F_{\mathrm{c}}(T) - F_{\mathrm{i}}(T)$ reaches its maximum and then begins to decrease, while the local magnetic ordering in the vicinity of the non-ionized donors will be destroyed. For this reason, the metal–insulator transition does not occur for $\nu - \nu_{\mathrm{c}}$ being large enough. Strictly speaking, if the transition into the insulating state at $T_{\mathrm{M}} < T_{\mathrm{C}}$ is possible, the inverse transition into the conducting state may occur at $T_{\mathrm{I}} > T_{\mathrm{M}}$. However, for $\nu - \nu_{\mathrm{c}} \ll \ll \nu_{\mathrm{c}}$ the latter should take place in the high-temperature region or even be absent at all.[7])

So far the determination of the quantity $E(\nu)$ remains one of the most important unresolved problems of solid-state physics. It seems to be natural to assume

$$E(\nu) = \varepsilon\nu_{\mathrm{c}}\left(\frac{\nu}{\nu_{\mathrm{c}}} - 1\right)^y, \tag{3.57}$$

where ε is a constant of the order of a typical spin polaron energy and $y > 0$. Assuming $D_{\mathrm{i}}^2 - D_{\mathrm{c}}^2 \approx 1$, we obtain for $T_{\mathrm{C}} \gg \beta\nu_{\mathrm{c}}$:

$$T_{\mathrm{M}}(2\nu_{\mathrm{c}}) \sim \frac{\varepsilon T_{\mathrm{C}}^2}{W^2\nu_{\mathrm{c}}}. \tag{3.58}$$

A metal–insulator transition in heavily doped ferromagnetic semiconductors was observed experimentally in [63]. Besides, according to [63] and [49], heavily doped ferromagnetic semiconductors with ν large enough display a very high resistance peak near T_{C} which can be regarded as due to two successive transitions, the metal–insulator one and its inverse.

---

[7]) See Section 6.

4. Antiferromagnetic Conductors

4.1 Hamiltonian of an antiferromagnetic conductor

To begin with, it is advisable to reformulate the method developed in the preceding chapter in a way appropriate for the antiferromagnetic ordering [38]. We consider an antiferromagnet with two equivalent magnetic sublattices so that the direction of the magnetic moment of a given atom is opposite to those of its nearest neighbours. We introduce separate coordinate systems for each sublattice in such a way that the z-axis be directed along the moment of the sublattice. The symbol σ_g will denote that the z-axis onto which a conduction electron spin is projected, is directed along the moment of the sublattice to which atom g belongs. Taking the rules of spin projection transformation into account when going over from one coordinate system to another [39], the following relation for the electron operators can be shown to hold:

$$a(g + \Delta, \sigma_{g+\Delta}) = ia(g + \Delta, -\sigma_g)\, e^{ilg}\,; \tag{4.1}$$

$$a(g, \sigma) \equiv a_{g\sigma}\,, \qquad \boldsymbol{l} \equiv \left(\frac{\pi}{a}, \frac{\pi}{a}, \frac{\pi}{a}\right).$$

The sign factor $\exp\{ilg\}$ excludes the rotation of the coordinate system by 2π.

As before, the wave function of the system under consideration is sought for in the form (3.11, 3.12), with the only difference that the quantization axes are introduced in another manner. The Schrödinger equation formally takes the form (3.14, 3.15), but other matrix elements are nonzero here. Namely, it follows from the form of the functions (3.12) and the transformation rules of the electron operators (4.1) that the set of equations (3.14, 3.15) should connect the coefficients $c_g^{\mu}(\ldots, S_{g0}^z, S_{g+\Delta 0}^z, \ldots)$ with $c_{g+\Delta}^{\mu'}(\ldots, S_{g0}^z \pm 1, S_{g+\Delta 0}^z, \ldots)$ or $c_{g+\Delta}^{\mu'}(\ldots, S_{g0}^z, S_{g0}^z \pm 1, \ldots)$. The latter can be expressed in terms of the former by introducing the raising and lowering operators of the spin projection $S_{f0}^{\pm}$ which act in the space of functions of the indices S_{f0}^z. A calculation similar to that carried out in Section 3.2 makes it possible to write down the set of equations (3.14, 3.15) as follows:

$$\left.\begin{aligned}
&\left[E + \frac{AS}{2} - \mathscr{H}_{\mathrm{M}}\right] c_g^{(+)}(S^z) = -\frac{iB}{2S+1}\,\Sigma\, e^{ilg}\,[P^{(+)}(S_g^z, S_{g+\Delta}^z) + \\
&\quad + P^{(+)*}(S_{g+\Delta}^z, S_g^z)]\, c_{g+\Delta}^{(+)}(S^z) + \frac{iB}{2S+1}\,\Sigma\, R(S_g^z, S_{g+\Delta}^z)\, c_{g+\Delta}^{(-)}(S^z)\,, \\
&\left[E - \frac{A(S+1)}{2} - \mathscr{H}_{\mathrm{M}}\right] c_g^{(-)}(S^z) = \frac{iB}{2S+1}\,\Sigma\, e^{ilg}\,[P^{(-)}(S_g^z, S_{g+\Delta}^z) + \\
&\quad + P^{(-)*}(S_{g+\Delta}^z, S_g^z)]\, c_g^{(-)}(S^z) + \frac{iB}{2S+1}\,\Sigma\, R^*(S_{g+\Delta}^z, S_g^z)\, c_{g+\Delta}^{(+)}(S_g^z)\,; \\
&P^{(+)}(S_g^z, S_{g+\Delta}^z) = \frac{1}{\sqrt{S + S_g^z + 1}}\, S_g^-\, \sqrt{S + S_{g+\Delta}^z + 1}\,, \\
&P^{(-)}(S_g^z, S_{g+\Delta}^z) = \sqrt{S + S_g^z}\, S_{g+\Delta}^-\, \frac{1}{\sqrt{S + S_{g+\Delta}^z}}\,, \\
&R(S_g^z, S_{g+\Delta}^z) = e^{ilg}\left\{\sqrt{(S + S_g^z + 1)\,(S + S_{g+\Delta}^z)}\, - \right. \\
&\qquad \left. - \frac{1}{\sqrt{S + S_g^z + 1}}\, S_{g+\Delta}^-\, S_g^-\, \frac{1}{\sqrt{S + S_{g+\Delta}^z}}\right\},
\end{aligned}\right\} \tag{4.2}$$

where $c_g^\mu(S^z) \equiv c_g^\mu(\dots S_f^z \dots)$, the index "0" being left out from S_{f0} to simplify the formulae. As before, nonessential terms $\sim I$ describing a change in the Hamiltonian $\mathcal{H}_{\mathrm{M}}$ induced by s-electrons are omitted from (4.2).

Introducing the spin polaron operators $\alpha^*_{g\mu}$, $\alpha_{g\mu}$ makes it possible to construct a spin polaron Hamiltonian similar to (3.24). Using (3.21) we write it in a form linear in the spin deviation operators. To the first approximation in W/AS it divides into two independent Hamiltonians, respectively, for $A > 0$ and $A < 0$:

$$\mathcal{H}^{(+)} = \mathcal{H}_{\mathrm{M}} - \frac{AS}{2} \sum \alpha^*_{g(+)}\alpha_{g(+)} + \frac{iB}{\sqrt{2S+1}} \sum \mathrm{e}^{ilg} (b_{g+\Delta} + b_g^*)\, \alpha^*_{g(+)}\alpha_{g+\Delta(+)}, \tag{4.3}$$

$$\mathcal{H}^{(-)} = \mathcal{H}_{\mathrm{M}} + \frac{A(S+1)}{2} \sum \alpha^*_{g(-)}\alpha_{g(-)} + \frac{iB\sqrt{2S}}{2S+1} \sum \mathrm{e}^{ilg} (b_g + b^*_{g+\Delta})\, \alpha^*_{g(-)}\alpha_{g+\Delta(-)}; \tag{4.4}$$

$$\mathcal{H}_{\mathrm{M}} = \frac{|I|S}{2} \sum \left[b_g^* b_g + b^*_{g+\Delta} b_{g+\Delta} + b_g^* b^*_{g+\Delta} + b_g b_{g+\Delta}\right]. \tag{4.5}$$

In principle we should redefine the ground state of the system making it a vacuum for magnon operators by an appropriate transformation. Such a transformation would make it easily possible to take into account the zero-point spin oscillations. But according to [38] and [40] the zero-point oscillations weakly influence the charge carrier states. For this reason they will be disregarded in what follows.

Though the spin polaron–magnon Hamiltonian (4.3 and 4) seems to resemble the electron–phonon Hamiltonian, both actually differ very strongly: 1. The spin polaron–magnon Hamiltonian lacks a term corresponding to the free fermion field. 2. The spin polaron–magnon interaction is nonlocal. This leads to charge carrier states of quite another type than that of the polaron. For reasons which will be clarified further, such states will be called quasi-oscillator ones. The idea of them was introduced in [41] within the framework of a somewhat different model.

4.2 Quasi-oscillator state for antiferromagnetic s–d coupling

In what follows we shall limit ourselves to the case $T = 0$. We will consider states of a single charge carrier in a semiconductor in order to have a one-electron problem under consideration. In accordance with condition (2.5) the Born-Oppenheimer approximation can be used. To the zeroth approximation the magnon kinetic energy is left out, i.e., terms nondiagonal in the site label are omitted from $\mathcal{H}_{\mathrm{M}}$ (4.5).

It is convenient to use the method of the equations of motion for the two-time Green function $\langle\langle \alpha_g | \alpha_f^* \rangle\rangle$ and the higher-order Green functions connected with them. The notation

$$\langle\langle \hat{A} | \hat{B} \rangle\rangle = -i \langle \mathrm{e}^{i\mathcal{H}t} \hat{A}\, \mathrm{e}^{-i\mathcal{H}t} \hat{B} + \hat{B}\, \mathrm{e}^{i\mathcal{H}t} \hat{A}\, \mathrm{e}^{-i\mathcal{H}t} \rangle\, \theta(t)$$

is used where $\theta(t)$ is the unit step function ($\theta(t) = 1$ for $t \geqq 0$ and $\theta(t) = 0$ for $t < 0$); the angular brackets denote averaging over the magnon and spin polaron vacuum state.

Making use of the Hamiltonian (4.4), the following set of equations for the Fourier-transformed Green functions (with respect to time) can be obtained:

$$\varepsilon\langle\langle g|f\rangle\rangle = \delta(g, f) + i\beta_A \sum e^{ilg} \langle\langle \Delta g|f\rangle\rangle ,$$
$$(\varepsilon - n)\langle\langle \Delta_n \cdots \Delta_1 g|f\rangle\rangle = i\beta_A \exp\{il(g + \Delta_1 + \cdots + \Delta_n)\} \times$$
$$\times \Big[\sum_{\Delta_{n+1} \neq -\Delta_n} \langle\langle \Delta_{n+1} \cdots \Delta_1 g|f\rangle\rangle + \langle\langle \Delta \cdots \Delta_1 g|f\rangle\rangle\Big], \quad (4.6)$$

where the notations

$$\beta_A = \frac{\sqrt{2S}B}{(2S+1)|J|}, \qquad \varepsilon = \frac{E}{|J|}, \qquad J = zIS, \quad (4.7)$$
$$\langle\langle g|f\rangle\rangle = \langle\langle \alpha_g|\alpha_f^*\rangle\rangle ,$$
$$\langle\langle \Delta_n \cdots \Delta_1 g|f\rangle\rangle = \langle\langle \alpha_{g+\Delta_1+\cdots+\Delta_n} b_{g+\Delta_1+\cdots+\Delta_{n-1}} \cdots b_g|\alpha_f^*\rangle\rangle$$

are used.

Some simplifications were made when writing the equations (4.6). First the notation of the spin polaron trajectory should be introduced in order to explain its meaning and to substantiate its admissibility. The trajectory[8]) $(g/\Delta_1 \cdots \Delta_n)$ is defined as the whole set of atoms g, $g + \Delta_1$, ..., $g + \Delta_1 + \cdots + \Delta_n$ the spin polaron has passed inducing a diminishing of their spin projections. There are self-crossing trajectories among them for which the equality $\Delta_l + \cdots + \Delta_n = 0$ is fulfilled ($l < k \leqq n$). Trivial self-crossings are these with $\Delta_{m+1} = -\Delta_m$. When $S = 1/2$, such reverse spin polaron transitions are excluded automatically. When $S > 1/2$, trivial self-crossings can be neglected since their relative part with respect to all trajectories with the same step number n is $\approx 1/z \ll 1$ where z is the nearest-neighbour number.

Nontrivial self-crossings are possible only if the number of steps of a trajectory exceeds three. The relative part of self-crossed trajectories does not exceed $1/z^2$, i.e., it is very small. This makes it possible to neglect the difference between the true set of equations for the Green functions and (4.6), which is caused by the self-crossing and self-contacting trajectories (with $\Delta_l + \cdots + \Delta_k = \Delta$).

It should be noted that the linearized Hamiltonian (4.4 to 4.5) was used when deriving the equations (4.6). In principle it is necessary to go beyond this approximation to take more accurately into account the exchange interaction between neighbouring atoms along a trajectory. It is easily seen, however, that it would have caused changing $|J|$ to $|J|(1 - 2/z)$ in (4.7) for trajectories with $n > 1$. This goes formally beyond the accuracy of the employed approximation (neglecting of zero-point spin oscillations corresponds to the main approximation in $1/z$).

As follows from the structure of the equations (4.6), the longer is a trajectory, the more is the energy of disordering of the magnetic system, which is equal to $n|J|$. Thus a quasi-elastic force arises which tends to return the spin polaron to the initial atom g of the trajectory. As a result, the spin polaron oscillates relative to atom g, which plays the role of a centre of the autolocalized state. For this reason, such a fermion state is called a quasi-oscillator one.

The Green functions entering the equations (4.6) are functionals of the spin polaron trajectories. So far no mathematical methods are developed to solve such problems exactly. Solving the infinite set of (4.6) by its cut-off is possible practically if only trajectories with n close to 1 are important, since the number of

[8]) More usual is the name "path" for it.

trajectories increases exponentially with the number of steps n. For this reason, such a procedure is very inaccurate for $\beta \ll 1$. It is convenient to use the fact that (4.6) has solutions dependent only on the number of steps in the trajectory. Such solutions can be found exactly. After substituting

$$\langle\!\langle \Delta_n \cdots \Delta_1 g | f \rangle\!\rangle = \delta(g, f) \exp \{il (g + \Delta_1 + \cdots + \Delta_n)\} G_n , \tag{4.8}$$

the set of equations (4.6) takes the form

$$\left.\begin{aligned} - \varepsilon G_0 + \beta_A z G_1 &= 1 , \\ (n - \varepsilon) G_n + \beta_A [(z - 1) G_{n+1} + G_{n-1}] &= 0 . \end{aligned}\right\} \tag{4.9}$$

The quasi-oscillator energy spectrum is found from the poles of the Green function, i.e. as the totality of eigenvalues of the sets of homogeneous equations corresponding to (4.6) or (4.9).

Unfortunately, for arbitrary J we failed to prove rigorously that the minimum eigenvalues of both homogeneous sets of equations coincide with each other though for $J = 0$ their coincidence is obvious (both are equal to $-z|B|$). But even if to regard the minimum eigenvalue of (4.9) as the upper estimate for the true minimum energy of the system, such an estimate is much more precise than that obtained by cutting off (4.6).

The eigenvalues of the homogeneous set of equations that corresponds to (4.9) are real, since the matrix constructed of its coefficients is Jacobian with real elements. The latter are found from a transcendental equation obtained with the aid of the Laplace transformation of the solution G_n^0 of the homogeneous set of equations (4.9). After introducing the continuous index t instead of the discrete n, this homogeneous set of equations takes the form

$$(t - \varepsilon) G(t) + \beta_A [(z - 1) G(t + 1) + G(t - 1)] = - G_0^0 P(\varepsilon) \delta(t) ; \tag{4.10}$$

$$G(t) = \sum_n G_n^0 \delta(t - n) ; \tag{4.11}$$

$$t \equiv t - 0 , \qquad P(\varepsilon) = \frac{\varepsilon}{z} .$$

Using the rules of Laplace transformation, we obtain the following equation for the Laplace transform $G(s)$ of the quantity $G(t)$:

$$\left.\begin{aligned} \frac{dG(s)}{ds} + f'(s) G(s) &= G_0^0 [P - (z - 1) \beta_A e^s] ; \\ f(s) &= \varepsilon s - (z - 1) \beta_A e^s + \beta_A e^{-s} . \end{aligned}\right\} \tag{4.12}$$

The sought solution of (4.12) has the form

$$G(t) = \frac{G_0^0}{2\pi i} \int\limits_{x - i\infty}^{x + i\infty} ds \exp [ts - f(s)] \int\limits^{s} d\sigma \, e^{f(\sigma)} [P - (z - 1) \beta_A e^\sigma] . \tag{4.13}$$

The quasi-oscillator energy is found with the aid of (4.13) from the condition that the equality $G(t) = G_0^0 \delta(t)$ should be satisfied within the interval $0 < t < 1$. Expanding $\exp \{\pm f(\sigma)\}$ in (4.13) into a double series with respect to $e^{\pm\sigma}$, taking the inverse Laplace transform, and utilizing a representation for the

Bessel functions $I_n(x)$ as a series makes it possible to represent this equality in the form

$$\left.\begin{aligned} 1 + \frac{P}{\varepsilon} I_0^2(\tau) &= 2P\varepsilon \sum_{n=0}^{\infty} \frac{I_n^2(\tau)}{\varepsilon^2 - n^2} + \\ &\quad + \frac{\tau}{2} \sum_{n=0}^{\infty} \left\{ \frac{I_{n+1}(\tau)\, I_{n+2}(\tau)}{\varepsilon + n + 2} - \frac{I_{n-1}(\tau)\, I_n(\tau)}{\varepsilon - n + 1} \right\}; \\ \tau &= 2\sqrt{z-1}\,\beta_A \,. \end{aligned}\right\} \tag{4.14}$$

In the limit $|\beta_A| \ll 1$, equation (4.14) leads to the standard results of perturbation theory. Numerical calculations give $|\varepsilon/\beta_A|$ equal to 2.65, 3.0, 3.5, 3.67, and 3.93 for $|\beta_A|$ equal to 1.0, 1.5, 3.5, 5, and 10, respectively.

In [41] a method was indicated which makes it possible to estimate the quasi-oscillator energy from below. For this aim, the same spin deviation energy $|J|n_h$ was attributed to all trajectories connecting the equilibrium position with an atom h, where n_h is the least number of steps between these two points. When $|\beta_A| \gg 1$, the continuous medium approximation can be used. Thus the problem under consideration reduces to solving a one-particle Schrödinger equation with a spherically symmetric potential energy proportional to the distance between the particle and the centre. The lower limit for the quasi-oscillator energy is given by

$$\varepsilon = -\,6|\beta_A| + |\beta_A|^{1/3} \left(\frac{9\pi}{8}\right)^{2/3}; \;\; z = 6\,. \tag{4.15}$$

According to the approximation assumed so far, a quasi-oscillator equilibrium position remains fixed at a lattice site. If one takes into account more accurately the self-crossing trajectories as well as the zero-point spin oscillations, the motion of the quasi-oscillator equilibrium position through the crystal becomes possible. An estimation carried out in [38] shows that for typical cases the quasi-oscillator and magnon band widths are comparable with each other, i.e. effective mass of the quasi-oscillator exceeds that of the electron with the same value of the Bloch integral at least by one order.

4.3 *Quasi-oscillator state for ferromagnetic s–d coupling*

Though formally the Hamiltonian $\mathcal{H}^{(+)}$ (4.3) strongly resembles $\mathcal{H}^{(-)}$ (4.4), the charge carrier state for $A > 0$ differs very strongly from that for $A < 0$. This is essentially due to the difference in the multiplicity of states of the magnetic atom at which a conduction electron is located in both cases, as was indicated in Section 3.2. It becomes manifest itself in that the Hamiltonian $\mathcal{H}^{(-)}$ describes those transitions of a spin polaron from atom to atom where the spin deviation appears at the atom which the electron has left, while the Hamiltonian $\mathcal{H}^{(+)}$ describes those transitions where the spin deviation appears for the atom at which the electron has arrived. Correspondingly, in the first case the spin deviation is annihilated for the atom at which the spin polaron has arrived, but in the second case for the atom which the spin polaron has left.

Making use of the Hamiltonian $\mathcal{H}^{(+)}$ (4.3), we can write the equations of motion for the one-spin polaron Green function and those connected with it similar to the preceding section:

$$\varepsilon \langle\langle \alpha_g | \alpha_f^* \rangle\rangle = \delta(g, f) + i\beta_F \, e^{ilg} \sum \langle\langle b_{g+\Delta} \alpha_{g+\Delta} | \alpha_f^* \rangle\rangle \,, \tag{4.16}$$

$$\left.\begin{aligned} &(\varepsilon - 1) \langle\langle b_{g+\Delta_1} \alpha_{g+\Delta_1} | \alpha_f^* \rangle\rangle = i\beta_F \, e^{il(g+\Delta_1)} \times \\ &\times \sum [\langle\langle \alpha_{g+\Delta_1+\Delta_2} | \alpha_f^* \rangle\rangle + \langle\langle b_{g+\Delta_1} b_{g+\Delta_1+\Delta_2} \alpha_{g+\Delta_1+\Delta_2} | \alpha_f^* \rangle\rangle] \,, \end{aligned}\right\} \tag{4.17}$$

$$\begin{aligned} &(\varepsilon - 2) \langle\langle b_{g+\Delta_1} b_{g+\Delta_1+\Delta_2} \alpha_{g+\Delta_1+\Delta_2} | \alpha_f^* \rangle\rangle = \\ &= i\beta_F \exp \{il \cdot (g + \Delta_1 + \Delta_2)\} \sum [\langle\langle b_{g+\Delta_1} \alpha_{g+\Delta_1+\Delta_2+\Delta_3} | \alpha_f^* \rangle\rangle + \\ &+ \langle\langle b_{g+\Delta_1} b_{g+\Delta_1+\Delta_2} b_{g+\Delta_1+\Delta_2+\Delta_3} \alpha_{g+\Delta_1+\Delta_2+\Delta_3} | \alpha_f^* \rangle\rangle] \,, \end{aligned} \tag{4.18}$$

and so on, where the notation

$$\beta_F = \frac{\beta}{|J| \sqrt{2S + 1}}$$

is used.

As is seen from the structure of equation (4.16), if there are no spin deviations initially the spin polaron transition from atom g to atom $g + \Delta_1$ is accompanied by appearing of a spin deviation at atom $g + \Delta_1$. According to (4.17) the subsequent spin polaron transition can be realized in two different ways: either with annihilation of the spin deviation at atom $g + \Delta_1$, or with its conservation while a new spin deviation at atom $g + \Delta_1 + \Delta_2$ arises.

In the first case the spin polaron can continue its translational motion in the crystal without hindrance. However, transitions of the second kind are accompanied by a violation of the magnetic ordering and consequently by an increase in the interaction energy of the magnetic atoms with each other. It is quite clear that one conduction electron cannot destroy the magnetic ordering over an entire crystal of macroscopic dimensions. Therefore, in order to continue its motion in the crystal the spin polaron must return to the initial state without spin deviations. In other words, transitions of the second kind lead to the quasi-oscillator-type motion discussed in the preceding section.

Because there are no small parameters in the set of equations (4.16 to 4.18), it is convenient to use a method of energy calculation which is equivalent to the Ritz variational method (one can easily prove it by comparing the calculation presented below with a similar calculation based on the Ritz method [40]). We will use the continuous trajectory approximation when only the following Green functions are regarded as being nonzero:

$$\langle\langle g | f \rangle\rangle = \langle\langle \alpha_g | \alpha_f^* \rangle\rangle \,,$$

$$\langle\langle \Delta_n \cdots \Delta_1 0 g | f \rangle\rangle = \langle\langle \alpha_{g+\Delta_1+\cdots+\Delta_n} b_{g+\Delta_1+\cdots+\Delta_n} \cdots b_{g+\Delta_1} b_g | \alpha_f^* \rangle\rangle \,.$$

The Fourier transformation of the Green functions,

$$\left.\begin{aligned} \langle\langle g | f \rangle\rangle &= \frac{1}{N} \sum G(k) \exp \{ik(g - f) + ilg\} \,, \\ \langle\langle \Delta_n \cdots \Delta_1 0 g | f \rangle\rangle &= \frac{1}{N} \sum G(\Delta_n \cdots \Delta_1 0 | k) \exp \{ik(g - f) + \\ &\quad + il(g + \Delta_1 + \cdots + \Delta_n)\} \,, \end{aligned}\right\} \tag{4.19}$$

is used, where k is the quasi-particle quasi-momentum and N the total number of magnetic atoms. Inserting (4.19) into the equations (4.16) to (4.18), and so on, one easily obtains

$$\left.\begin{aligned}
&(-)E\,G(k) + \frac{zB\gamma_k}{\sqrt{2S+1}}\,G(0|k) = 1\,,\\
&(|J| - E)\,G(0|k) + \frac{B}{\sqrt{2S+1}}\Big[\gamma_k z G(k) + \sum_{\Delta_1} G(\Delta_1 0|k)\Big] = 0\,,\\
&[(1+n)\,|J| - E]\,G(\Delta_n \cdots \Delta_1 0|k) + \frac{B}{\sqrt{2S+1}} \times\\
&\times \Big[\sum_{\Delta_{n+1} \neq -\Delta_n} G(\Delta_{n+1} \cdots \Delta_1 0|k) + G(\Delta_{n-1} \cdots \Delta_1 0|k)\Big] = 0\,.
\end{aligned}\right\} \tag{4.20}$$

The same considerations as in the preceding section were used when writing the set of equations (4.20).

The energy can be found from (4.20) in the same manner as in the preceding section, i.e. by assuming $G(\Delta_n \cdots \Delta_1 0|k) = G_n$. Then the equation (4.14) holds where the quantities P, β_A, and ε should be replaced according to

$$\left.\begin{aligned}
&P(\varepsilon) \to \frac{1}{z}\left[\varepsilon - 1 + \frac{(z-1)\,(z\beta_F\gamma_k)^2}{\varepsilon}\right],\\
&\beta_A \to \beta_F\,, \qquad \varepsilon \to \varepsilon - 1\,.
\end{aligned}\right\} \tag{4.21}$$

For $|J| \to 0$ one succeeds to find the low-lying eigenvalues of the set of equations (4.20) more accurately since in this limit the equations (4.20) have a solution of the type

$$G(\Delta_n \cdots \Delta_1 0|k) = G(\Delta_n + \cdots + \Delta_1|k)\,. \tag{4.22}$$

As before the eigenvalues are determined by the solution of a homogeneous set of equations corresponding to (4.20), which due to (4.22) takes the form

$$\left[-\lambda + \frac{\gamma_k^2}{\lambda}\,\delta(f, 0)\right] G^0(f) + \frac{1}{z}\sum_{\Delta} G^0(f + \Delta) = 0 \tag{4.23}$$

with the notation

$$E = -\frac{6|B|\lambda}{\sqrt{2S+1}}\,; \quad z = 6\,. \tag{4.24}$$

Formally the problem (4.23) is equivalent to the local level problem in the case of δ-function-like impurity potentials [42], with the only difference that the "impurity potential" itself depends on the sought energy in the case under consideration. In the case of a simple cubic crystal, the energy is determined from the transcendental equation

$$\frac{a^3\gamma_k^2}{\pi^2}\int_0^{\pi/a} \frac{d^3q}{\lambda - \gamma_q} = \lambda \tag{4.25}$$

which is easily obtained by the Fourier transformation of G^0 in (4.23) with respect to the site number f.

If the equation (4.25) has a solution with $\lambda > 1$, then according to [42] the spin polaron oscillations with respect to site $f = 0$ are finite even if J will go to zero. (It should be noted that according to (4.22) site $f = 0$ is the origin of the coordinate system moving through the crystal with quasi-momentum k.) The radius of that state is given by

$$r_L = \frac{a}{\sqrt{\lambda - 1}}.$$

The mean number of deflected spins when the spin polaron is at the distance r_L from site $f = 0$ is of the order of $(r_L/a)^2$ (as for random walks).

Therefore neglecting the quantity $|J|$ is justified if the average energy of magnetic disordering due to spin polaron oscillations is small compared with the "local" level depth:

$$|J| \left(\frac{r_L}{a}\right)^2 = \frac{|J|}{\lambda - 1} \ll \frac{6|B|(\lambda - 1)}{\sqrt{2S + 1}}. \tag{4.26}$$

As follows from equations (4.25), (4.14), and (4.21), the energies of the states with quasi-momenta k and $l - k$ are equal, so that the energy spectrum is twofold degenerate. The quantity λ, as obtained by the solution of the equation (4.25), is equal to 1.1 for $k = 0$. This means that allowance for spin polaron oscillations lowers the ground-state energy of the quasi-partide by 10%. However, spin polaron oscillations affect the effective mass of the quasi-particle much more stronger: it exceeds its value m_0^* obtained without taking them into account by 40%.

As to the zero-point spin oscillations, they lower the bottom of the quasi-oscillator band and diminish its effective mass by about 15% [40]. Taking into account this effect, we obtain the quasi-oscillator effective mass:

$$m_q^* \approx 1.25 m_0^* = \frac{1.25\sqrt{2S + 1}}{2a^2|B|}. \tag{4.27}$$

Thus, for a ferromagnetic s–d coupling, in contrast to the antiferromagnetic one the quasi-oscillator effective mass has the same order as that of the band electron.

Let us now estimate the accuracy of the method employed above. First at all we note that the lowest eigenvalue of the Hamiltonian (2.1), E_m, exceeds the sum of the lowest eigenvalues E_{Bm}, E_{Am}, and E_{Mm} of the Hamiltonians $\mathcal{H}_B$, $\mathcal{H}_A$, and $\mathcal{H}_M$, respectively. Thus the accuracy of the approximation can be found by comparing the lower limit $E_{BM} = -6|B|$ for the quasi-oscillator energy with its value for $S = \frac{1}{2}$ obtained above. The latter according to (4.24) and with allowance for zero-point spin oscillations is equal to $-6|B| \times 1.25/\sqrt{2}$, i.e., it exceeds E_{Bm} by 12%. Evidently the true accuracy should be even more higher. This justifies the continuous trajectory approximation employed above.

4.4 Ferrons

As follows from (3.24), (4.14), (4.15), and (4.24), the ground-state energy of a charge carrier for antiferromagnetic ordering substantially exceeds that for ferromagnetic ordering. When $2S \gg 1$, the former is of the order of $-(z|B|/\sqrt{2S})$, i.e., it is $\approx \sqrt{2S}$ times more than the latter. For this reason it may be energetically favoured to change the antiferromagnetic ordering to a ferromagnetic one in a microregion. This microregion is a potential well for a charge carrier, its depth being $\approx z|B| (1 - 1/\sqrt{2S})$. Therefore, a spin polaron level can appear below the minimum energy of the charge carrier in the unperturbed antiferromagnet. The gain in charge carrier energy due to its transition to a local level may exceed the increase in energy due to replacing the antiferromagnetic ordering by a ferromagnetic one in the microregion mentioned above.

It is clear that the energy level in the potential well decreases with increasing radius R of the well. On the other hand, the energy of magnetic disordering increases with R. The equilibrium radius is to be found from the minimum condition for the total energy of the system. The spin polaron trapped by the ferromagnetic microregion produced by itself should move through the crystal together with it.

It should be noted that qualitative considerations were given earlier [14] concerning the possibility of energy gain resulting from deviation of only one localized spin in an antiferromagnet and subsequent autolocalization of a conduction electron at the deflected spin. In [43] and [44] the concept of the autolocalized state of the kind indicated above was formulated according to which the charge carrier is autolocalized rather in a ferromagnetic microregion consisting of many deflected spins. This implies that the equilibrium radius of the ferromagnetic microregion may well exceed the lattice constant. Actually it is the first example of quasi-particles in crystals characterized by an autolocalization of the charge carrier in a microregion of another phase. Further examples of quasi-particles of such a kind are given in [45], [46], and [6].

In papers [43] and [44] quasi-particles of the type considered here were called "magnetic polarons". This name turns out to be unhappy for two reasons: First, it was often used by other authors for quite different quasi-particles, and second, it can easily be confused with the name "spin polaron". In what follows, the name "ferron" will be used instead of "magnetic polaron". This name can also be used for the quasi-particles investigated in [6].

The nonlinearity of the interaction of the charge carrier with localized spins is of principal importance for the problem under consideration. The direct variational principle is used to calculate the ground-state energy of a ferron. It is assumed that the atomic spins of sublattice 2 in a microregion of the radius R are directed opposite to their sublattice moment, i.e., they are parallel to the momentum of sublattice 1 when constructing the trial function. For $A > 0$ the trial function is chosen as follows:

$$\Phi = \left\{\sum_{g_1} \psi(g_1)\, \alpha^*_{g_1} + \sum_{|g_2| \leq R} \psi(g_2)\, \alpha^*_{g_2} + \sum_{|g_2| > R} \psi(g_2)\; \alpha^*_{g_2} b^*_{g_2}\right\} |R\rangle\,, \tag{4.28}$$

where $|R\rangle$ is a wave function which is vacuum for the spin polaron and describes the localized spin alignment indicated above. The indices g_1 and g_2 denote atoms belonging to sublattices 1 and 2, respectively. The possibility to limit oneself to states with one magnon in the sublattice 2 follows from the results of the preceding section. (States with more magnons are unimportant, for neglecting the quasi-oscillator motion let the ground state energy almost unchanged.)

Inside the ferromagnetic microregion $|g| < R$ the system is described by the Hamiltonian (3.32), and outside it by the Hamiltonian (4.3 to 4.5). After substituting the wave function (4.28) in the Schrödinger equation we obtain the following equations for the unknown coefficients $\psi(g)$:

$$(-\, E_R)\, \psi(g) + B \sum_{\Delta} \psi(g + \Delta) = 0\,, \tag{4.29}$$

$$(-E_R)\,\psi(g) + \frac{B}{\sqrt{2S+1}} \sum_{\Delta} \psi(g+\Delta) = 0\,, \tag{4.30}$$

where E_R is the ferron energy for given R. The set of equations (4.29) corresponds to the ferromagnetic microregion and (4.30) to the antiferromagnetic region. The condition $|B| \gg |I|$ was taken into account when writing (4.30).

When $R \gg a$, the continuous medium approximation can be used. For this aim the amplitudes $\psi(g_1)$ and $\psi(g_2)$ are regarded as the components $\psi_1(r)$ and $\psi_2(r)$ of a spinor wave function. Their expansion in a Taylor series makes it possible to rewrite the sets of equations (4.29) and (4.30) in the form

$$\left.\begin{aligned}(-E_R)\,\psi_1(r) + B[z + a^2\nabla^2]\,\psi_2(r) &= 0\,,\\ (-E_R)\,\psi_2(r) + B[z + a^2\nabla^2]\,\psi_1(r) &= 0\,; \qquad r < R\,,\end{aligned}\right\} \tag{4.31}$$

$$\left.\begin{aligned}(-E_R)\,\psi_1(r) + \frac{B}{\sqrt{2S+1}}\,[z + a^2\nabla^2]\,\psi_2(r) &= 0\,,\\ (-E_R)\,\psi_2(r) + \frac{B}{\sqrt{2S+1}}\,[z + a^2\nabla^2]\,\psi_1(r) &= 0; \qquad r > R\,.\end{aligned}\right\} \tag{4.32}$$

The joining of the solutions of (4.31) and (4.32) at the boundary $r = R$ is achieved using the condition that an atom with $|g_1| = R$ belongs simultaneously to both the ferromagnetic and antiferromagnetic phases, i.e., the logarithmic derivative of $\psi_1(r)$ should be continuous at $r = R$. Representing $\psi_1(r)$ as a spherical wave with a real wave number k inside the ferromagnetic microregion and with an imaginary wave number $i\varkappa$ outside it, we obtain from (4.31) and (4.32) the following transcendental equation to determine the ferron energy:

$$\left.\begin{aligned}\frac{2}{\pi}\,kR_{\mathrm{M}} &= \sin kR\,; \qquad \frac{\pi}{2} < kR < \pi\,,\\ R_{\mathrm{M}} &= \frac{\pi}{2}\,a\left[z\left(1 - \frac{1}{\sqrt{2S+1}}\right)\right]^{-1/2},\end{aligned}\right\} \tag{4.33}$$

$$E_R = B[z - a^2k^2]\,. \tag{4.34}$$

As follows from (4.33), in accordance with the usual situation a discrete spin polaron level exists if only R exceeds R_{M}.

However, the existence of a discrete level in the potential well is not sufficient for the ferron state to be stable. It is also necessary that its depth should exceed the magnetic disordering energy

$$E_{\mathrm{d}}(R) = \frac{4\pi}{3}\left(\frac{R}{a}\right)^3 |J|\,S\,. \tag{4.35}$$

It is more convenient to treat the quantity k instead of R as a variational parameter keeping in mind their connection given by (4.33). Then the minimum condition for the total energy $E_{\mathrm{t}} = E_{\mathrm{d}} + E_R$ with the aid of (4.33) to (4.35) leads to the following equation for $x = kR_{\mathrm{M}}$:

$$\frac{1}{x}\,\frac{\mathrm{d}}{\mathrm{d}x}\left[\frac{\pi - \arcsin\,(2x/\pi)}{x}\right]^3 = -\frac{3a^5}{2\pi R_{\mathrm{M}}^5}\left|\frac{B}{JS}\right|; \tag{4.36}$$

$$0 < \arcsin\varphi < \frac{\pi}{2}\,.$$

It can be easily solved analytically in the limit $x \ll 1$:

$$\left.\begin{aligned} R &= a\left[\frac{\pi}{2}\left|\frac{B}{JS}\right|\right]^{1/5}, \\ E_t &= -z|B| + \frac{5}{3}\pi^{8/5}|B|^{3/5}|JS|^{2/5}. \end{aligned}\right\} \tag{4.37}$$

It follows from (4.36), (4.37) that for typical values of the parameters, $B \approx \approx 0.1$ eV and $JS \approx 0.001$ to 0.01 eV, the radius R of the ferromagnetic microregion is two to three times the lattice constant, which justifies the continuous medium approximation employed.

In cases, when JS/B is not small enough for the radius R to exceed substantially the lattice constant, another way of solving the problem is advisable [18]. If only the spin of the atom $g = 0$ is reversed, the wave function (4.28) can be used as before, but R must be taken to be zero. Then the set of equations analogous to (4.29) can be written in the form

$$\left.\begin{aligned} z\lambda\psi(g) &= \sum_{\Delta}\psi(g+\Delta) + \chi\sum_{\Delta}\psi(g+\Delta)\,[\delta(g,0)+\delta(g,-\Delta)]\,; \\ \chi &= \sqrt{2S+1}-1\,, \end{aligned}\right\} \tag{4.38}$$

where the notations (4.24) are used.

After a Fourier transformation of the amplitudes,

$$\psi(g) = \frac{1}{\sqrt{N}}\sum e^{-igk}\,\psi_k\,,$$

the equation (4.38) can be rewritten in the form

$$(\lambda - \gamma_k)\,\psi_k = \frac{\chi}{N}\sum\gamma_k\psi_k + \frac{\chi\gamma_p}{N}\sum\psi_k\,. \tag{4.39}$$

The condition that a solution of the integral equation (4.39) does exist is

$$[1 - \chi G_1(\lambda)]^2 = \chi^2 G_0(\lambda)\,G_2(\lambda)\,; \tag{4.40}$$

$$G_n(\lambda) = \frac{a^3}{\pi^3}\int_0^{\pi/a}\frac{\gamma_p^n\,d^3p}{\lambda - \gamma_p}\,. \tag{4.41}$$

The equality (4.41) makes it possible to express the Green functions G_1 and G_2 in terms of G_0:

$$G_1 = \lambda G_0 - 1\,; \qquad G_2 = \lambda G_1\,. \tag{4.42}$$

Then, using (4.42), equation (4.40) can be written in the form

$$1 + \frac{1}{2S} = \lambda G_0(\lambda)\,. \tag{4.43}$$

For $\lambda - 1$ small enough, the approximate solution of equation (4.43) is given by

$$\sqrt{\lambda - 1} \approx 0.4 \left(1 - \frac{1}{S}\right). \tag{4.44}$$

It is seen from (4.44) in particular that a small-radius ferron state is possible only for $S > 1$, and its energy diminishes with increasing S. In the limit $\sqrt{2S + 1} \gg 1$, we obtain from (4.39)

$$E_0 = - |B|\sqrt{z},$$

as should be the case. This result was first indicated in [14] as valid for all spin magnitudes. But actually it cannot be employed for real magnetics because $\sqrt{2S + 1}$ is of the order of unity for them.

The case of antiferromagnetic s–d coupling differs from that discussed above in that a spin polaron is in an autolocalized (quasi-oscillator) state even in the absence of a ferro-magnetic microregion (see Section 4.2). For this reason, a ferromagnetic microregion can be of arbitrary small radius when $A > 0$. A calculation of such a mixed ferron–quasi-oscillator state was carried out in [40]. In particular it was shown there that if only one atomic spin is reversed, the energy of the system is given by (4.14) with

$$P = \frac{z}{\varepsilon} \left[\frac{2SB}{(2S + 1) J}\right]^2.$$

The tendency to produce a ferromagnetic microregion becomes stronger as the atomic spin S increases in both cases $A > 0$ and $A < 0$.

So far nothing was said about the translational motion of a ferron through a crystal. Estimates presented in [18] show that the ferron effective mass is very large and exponentially increases with S.

It should be noted that unlike the quasi-oscillator states, the ferron states may also exists even if $AS \ll W$. Most probable is the existence of ferrons in rare-earth compounds which are characterized by low Néel points and large spins. Apparently their existence in antiferromagnetic europium chalcogenides is proved experimentally [47, 48].

Autolocalized ferron-like states are possible not only in antiferromagnets. They can also exists in paramagnets or in ferrites. One may hope that they are also possible in some ferromagnets at higher temperatures. This was first indicated in [49]. The quantitative investigation of this problem was carried out in [46, 70] in the wide-band limit.[9]) It should also be noted that the metal–insulator transition discussed in Section 3.6 can be regarded as a cooperative autolocalization of spin polarons, and the insulating state of heavily doped ferromagnetic semiconductors as a cooperative ferron state.

Ferromagnetic microregions can also appear in the vicinity of non-ionized donors. Their appearance can be explained in terms of the indirect exchange between magnetic atoms in the vicinity of donors via spin polarons trapped by these donors. These spin polarons can be regarded as localized ferrons. They

---

[9]) The calculations in [46, 70] are similar to those presented in [43, 44] for $W \gg AS$. But, as temperature is finite, the minimum condition for the free energy was used in [46, 70] instead of that of the ground state energy.

may strongly influence the magnetic suspectibility of an antiferromagnet at low temperatures. Their contribution to the transversal magnetic susceptibility is proportional to $(R/a)^5\nu$ where R is the radius of the ferromagnetic microregion and ν the relative concentration of defects [44]. Thus for $R \approx 2a$ the relative magnetic susceptibility of defects exceeds its relative concentration by two orders of magnitude. This result agrees with the experimental data [50].

If the defects are situated symmetrically relative to the magnetic sublattices, ferromagnetic ordering of their momenta is possible. This ordering may be caused both by the exchange and the dipolar spin–spin interaction between ferrons [51].

4.5 Noncollinear magnetic structures

As was already pointed out, the indirect exchange via spin polarons tends to establish ferromagnetic ordering. Actually this causes the existence of ferrons (a spin polaron establishes local ferromagnetic ordering). In what follows we shall consider such antiferromagnets where ferrons cannot exist, and charge carriers can move freely through the crystal (for example quasi-oscillators for $A > 0$). If the charge carrier density n is large enough, magnetic structures intermediate between ferromagnetic and antiferromagnetic can appear because of the competition between the direct antiferromagnetic and indirect ferromagnetic exchanges. First of all the noncollinear antiferromagnetic structure will be discussed, for which the angle 2θ between the magnetic moments of two equivalent sublattices differs from π.

It is not difficult to develop a phenomenological theory of noncollinear antiferromagnetic structures. For most crystal structures including those with a high symmetry, the thermodynamic potential $\Phi(n, T)$ of the system under consideration should be an even function of the spontaneous magnetic moment $\mathcal{M}$. Similar to the well-known Landau theory of phase transitions [52], the thermodynamic potential for small $\mathcal{M}$ at $T = 0$ (i.e. the ground state energy) should be represented as

$$\left.\begin{aligned} E(n, \theta) &= E_A(n) + A_2(n)\,\mathcal{M}^2 + A_4(n)\,\mathcal{M}^4\,; \\ \mathcal{M} &= S\cos\theta\,, \end{aligned}\right\} \tag{4.45}$$

where E_A is the ground-state energy in the antiferromagnetic state. The magnetic anisotropy is neglected.

Let us consider the behaviour of E as a function of n. At small concentrations n the crystal should be antiferromagnetic, so that the quantity A_2 should be positive. At high densities, however, the indirect exchange should dominate so that the quantity A_2 should be negative. Thus a critical density n_L should exist for which the coefficient $A_2(n)$ vanishes. If the coefficient $A_4(n_L)$ is positive, then it follows from the minimum condition for the thermodynamic potential that for $n > n_L$ the crystal has an spontaneous magnetic moment

$$\mathcal{M} = \sqrt{-\frac{A_2(n)}{A_4(n)}}\,. \tag{4.46}$$

Assuming that the quantity $A_2(n)$ is proportional to $n - n_L$ for n close to n_L, it follows from (4.46) that the moment $\mathcal{M}$ should depend on n as $\sqrt{n - n_L}$. The derivative of the Fermi energy with respect to the concentration should undergo a jump. Such behaviour can be regarded as an isothermal second-order phase transition, specifical in so far as is does not occur for a sample itself, but for a set of samples with different charge carrier densities [60]. Perhaps, it can also occur for a separate sample in a surface layer of the width of the order of the screening length, since within it the density can be changed by an external electric field (see [30]). The jump in $d\mu/dn$ should lead to singularities of electrical quantities as functions of the density (breaks in the work function and the conductivity curves, a jump in the thermo-power).

Emphasis should be laid on the principal difference between the noncollinear antiferromagnetism considered here and the Dzyaloshinskii weak ferromagnetism [53]. The latter is possible only in crystals with a symmetry which allows a term linear in $\mathcal{M}$, A_2 being assumed to be a large positive quantity. Since the Dzyaloshinskii noncollinearity is due to the relativistic interactions, it is small. Unlike that, no special symmetry conditions should be fulfilled for the structures under consideration, and the spontaneous magnetization can be arbitrarily large, being limited only by the saturation magnetization. However, an anomalous small coefficient A_2 is of principal importance.

For small angles θ the expansion of the ground-state energy in powers of "zero-point demagnetization", $\mathcal{N} = S - \mathcal{M} = S(1 - \cos 2\theta)$, should be carried out. As, besides the vector of demagnetization, the system is also characterized by another vector, that of the total moment, a scalar linear in $\mathcal{N}$ can be constructed. For this reason the expansion of the ground-state energy includes a term linear in $\mathcal{N}$:

$$E_F(n, \theta) = E_F(n) + B_1(n)\,\mathcal{N} + B_2(n)\,\mathcal{N}^2\,, \tag{4.47}$$

where E_F is the ground-state energy in the ferromagnetic state. The considerations concerning the concentrational phase transitions presented above can also be easily generalized for the case of small θ. In particular, if the coefficient B_1 is proportional to $n_R - n$ close to the phase transition point n_R, then the "demagnetization" $\mathcal{N}$ is also proportional to this difference.

It is easy to show that the noncolinear antiferromagnetism is principally impossible within the framework of the Heisenberg model. But, as was already pointed out, the influence of spin polarons on magnetic ordering cannot be described by the Heisenberg Hamiltonian. In what follows it will be shown that actually in the framework of the spin polaron model there is a density range $n_L < n < n_R$ in which the noncollinear antiferromagnetism is possible. The dependence of the spontaneous moment on the charge carrier concentrations agrees with that obtained above within the framework of a phenomenological treatment [54, 55].

First of all, the spin polaron energy spectrum for an arbitrary angles 2θ between the angular momenta of sublattices is to be found for the case $A > 0$. The wave function of the system is sought in a form similar to (4.28). Since the spin polaron operators were introduced only for $\theta = 0$ and $\pi/2$, it is advisable to return to the true operators of electrons and magnetic atoms using (2.6) and introducing the operators S_g^- according to the relation (3.9):

$$\psi = \sum \left\{\psi(g, +)\, a^*(g, \uparrow_g) + \frac{\psi(g, -)}{\sqrt{2S+1}}\,[a^*(g, \downarrow_g) + a^*(g, \uparrow_g)\, S_g^-]\right\} |0\rangle\,. \tag{4.48}$$

Here $|0\rangle$ is the vacuum state for the conduction electron and the spin deviations of magnetic atoms from the directions of the moments of the sublattices to which these atoms belong. (In other words, in the coordinate system in which the z-axis coincides with the moment of the sublattice to which an atom belongs, its spin projection S^z is equal to S.) Assuming the x-axes in both local coordinate systems to coincide with each other for the angle 2θ between their z-axes the following relations between the electron operators hold [39]:

$$\left.\begin{aligned} a^*(g+\Delta, \uparrow_{g+\Delta}) &= \cos\theta\, a^*(g+\Delta, \uparrow_g) + i\sin\theta\, a^*(g+\Delta, \downarrow_g)\,, \\ a^*(g+\Delta, \downarrow_{g+\Delta}) &= i\sin\theta\, a^*(g+\Delta, \uparrow_g) + \cos\theta\, a^*(g+\Delta, \downarrow_g)\,, \end{aligned}\right\} \quad (4.49)$$

where σ_g is the spin projection of the conduction electron in the coordinate system of atom g. We assume that the turning angle of the z-axis of the second sublattice relative to that of the first sublattice is $(+2\theta)$. Then the turning angle of the z-axis of the first sublattice relative to that of the second sublattice is (-2θ).

It is convenient to attribute an "isotopical spin" to the atoms. Its projection τ has the values ± 1 depending on whether an atom belongs to the first or to the second sublattice. The coefficients $\psi(g, \pm)$ of the wave function (4.48) are sought in the form

$$\psi(g, \pm) = \mathrm{e}^{ikg}\, \psi(\tau_g, \pm)\,. \quad (4.50)$$

Substituting the expressions (4.48) to (4.50) into the Hamiltonian (2.1) gives the following "bispinor" Schrödinger equation:

$$\left.\begin{aligned} E\psi(\tau, +) &= 6B\cos\theta\, \gamma_k \psi(-\tau, +) + \frac{i\tau 6B\gamma_k \sin\theta}{\sqrt{2S+1}}\, \psi(-\tau, -)\,, \\ E\psi(\tau, -) &= i\tau 6B\sin\theta\, \gamma_k \psi(-\tau, +) + \frac{6B\gamma_k \cos\theta}{2S+1}\, \psi(-\tau, -)\,. \end{aligned}\right\} \quad (4.51)$$

The condition $|B| \gg |I|$ was used when writing (4.51).

The solution of the equations (4.51) shows that the spin polaron energy spectrum consists of two bands with the dispersion laws

$$\varepsilon_\pm(\chi) = -6B_\pm \gamma_k\,; \quad (4.52)$$

$$B_\pm = \frac{|B|}{\sqrt{2S+1}} \left\{\sqrt{1+y^2} \pm y\right\}; \quad y = \frac{\mathcal{M}}{\sqrt{2S+1}}\,.$$

For complete antiferromagnetic ordering ($\theta = \pi/2$), both bands coincide with each other. But with increasing spontaneous momentum one of these bands becomes wider, and its bottom lowers. On the contrary, the other band becomes narrower, and its bottom raises. In the limit $\theta = 0$, the wider band corresponds to the state with maximum spin of the system, and the narrower one to the state with the spin less by a unit.

The ground-state energy is given by

$$\begin{aligned} E \approx {} & -3IS^2N\cos 2\theta - HSN\cos\theta\left(1 + \frac{\nu}{2S}\right) + \\ & + \sum \left[\varepsilon_+(k)\, n_+(k) + \varepsilon_-(k)\, n_-(k)\right]\,, \end{aligned} \quad (4.53)$$

where $n_\pm(k)$ is the Fermi distribution function for the charge carriers with energy $\varepsilon_\pm(k)$ for $T = 0$, $\nu = n/N$, and H is the external magnetic field in energy units directed parallel to the moment of the crystal.

At $T = 0$ the condition of independence of the number of charge carriers on the magnetic moment gives

$$\sum_{(\pm)k} \varepsilon_\pm(k) \frac{\mathrm{d}}{\mathrm{d}\theta}\, n_\pm(k) = \mu \frac{\mathrm{d}}{\mathrm{d}\theta} \sum_{(\pm)k} n_\pm(k) = \mu \frac{\mathrm{d}n}{\mathrm{d}\theta} = 0\,, \quad (4.54)$$

where μ is the Fermi energy. Using (4.54), the minimum condition for the ground state energy (4.53) with respect to the angle θ can be presented for $\nu \ll 1$ as

$$-12 I \mathcal{M} - H - 6 \left[\nu_+ \frac{d}{d\mathcal{M}} B_+ + \nu_- \frac{d}{d\mathcal{M}} B_-\right] = 0\,, \tag{4.55}$$

where the relative charge carrier concentrations ν_+ and ν_- in the wider and the narrower band are to be determined from the equality of the chemical potentials for both bands:

$$B_+[-6 + (6\pi^2\nu_+)]^{2/3} = B_-[-6 + (6\pi^2\nu_-)]^{2/3}\,. \tag{4.56}$$

For small $\mathcal{M}$ one obtains from (4.56) and (4.52) to the main order in the parameter $R \gg 1$:

$$\left.\begin{aligned} \nu_+ - \nu_- &\approx \nu \left[\frac{3}{4} Ry - \frac{1}{32} R^3y^3\right], \\ R(\nu) &= 12\,(3\pi^2\nu)^{-2/3}\,. \end{aligned}\right\} \tag{4.57}$$

It follows from (4.53), (4.55), and (4.57) that the noncollinear two-sublattice magnetic structure in the absence of an external magnetic field is possible for relative carrier densities ν exceeding the critical value ν_L:

$$\nu_L = \left(\frac{\pi}{3}\right)^4 \left[2 \left|\frac{I}{B}\right| (2S+1)^{3/2}\right]^3. \tag{4.58}$$

The angle θ is determined by

$$\mathcal{M} = S \cos\theta = \left[\frac{8\,(2S+1)}{R^2(\nu_L)\,\nu_L}\,(\nu - \nu_L)\right]^{1/2}.$$

This result makes more accurate a similar result obtained in [54].

As follows from (4.57), after reaching the critical value $\mathcal{M}_c$ of the magnetization, which according to (4.56) is equal to

$$\mathcal{M}_c = \frac{2^{2/3}\sqrt{2S+1}}{R(\nu)}, \tag{4.59}$$

the narrower spin polaron band becomes empty. Starting from this value of the moment, the relation between the moment and the carrier density with the aid of (4.55) is given by

$$\nu(\mathcal{M}) = 2\left|\frac{I}{B}\right| (2S+1) \left\{\frac{1}{\mathcal{M}} + \frac{1}{\sqrt{\mathcal{M}^2 + 2S + 1}}\right\}^{-1}, \tag{4.60}$$

$$\nu_R \equiv \nu(S) = \frac{2|I|\,S(S+1)}{|B|}. \tag{4.60a}$$

For densities ν exceeding ν_R, complete ferromagnetic ordering is stable. The value $\nu = \nu_R$ found from (4.60) is just that for which all magnon frequencies in the ferromagnetic state vanish, when $S \gg 1$. Indeed, according to (3.35) the magnon spectrum of a ferromagnet is given by

$$\omega_k = 6\left[\frac{|B|\nu}{2S} - |I|S\right](1 - \gamma_k)\,, \tag{4.61}$$

and formula (4.61) makes the above statement obvious.

Using formulae (4.55) and (4.56) it is also easy to find the magnetic susceptibility for both cases $\nu > \nu_L$ and $\nu < \nu_L$. For weak magnetic fields, when the field-induced moment is small compared with $\mathcal{M}_c$, the transverse susceptibility of a collinear antiferromagnet ($\nu < \nu_L$) is

$$\chi = S\left[2|J| - \frac{9}{2}\,|B|\,(2S+1)^{-3/2}\,\nu R(\nu)\right]^{-1}; \qquad J = ISz. \tag{4.62}$$

It is seen from (4.62) that the susceptibility increases as $(\nu_L - \nu)^{-1}$ with ν approaching ν_L (4.58).

For a noncollinear magnet, when its spontaneous moment $\mathcal{M}$ is smaller than $\mathcal{M}_c$, the weak-field magnetic susceptibility is reciprocal to $\mathcal{M}^2$:

$$\chi = \frac{8(2S+1)^{5/2}}{3R^3\mathcal{M}^2}. \tag{4.63}$$

If the moment in the presence of an external field exceeds $\mathcal{M}_c$, the susceptibility for both cases $\nu > \nu_L$ and $\nu < \nu_L$ is given by

$$\chi = S\,[2|J| - 6\nu|B|\,S^2(\mathcal{M}^2 + 2S + 1)^{-3/2}]^{-1}. \tag{4.64}$$

According to (4.52) and (4.55), the critical field H_s at which the complete saturation of the moment is achieved, is

$$H_s = 2|J| - \frac{6\nu|B|}{S+1}. \tag{4.65}$$

Setting it equal to zero gives the upper boundary ν_R of the density range in which the noncollinear antiferromagnetism is possible.

Finally, a singularity in the magnetic susceptibility at $\mathcal{M}_c$ should be pointed out. It follows from (4.52) and (4.56) that

$$\nu_- = \frac{\sqrt{6}}{\pi^2}\left(\frac{4}{2S+1}\right)^{3/4}(\mathcal{M}_c - \mathcal{M})^{3/2}\,\theta(\mathcal{M}_c - \mathcal{M}), \tag{4.66}$$

where as before $\theta(x)$ is the unit step function. As according to (4.59) $\mathcal{M}_c$ is small in comparison with S, the following expression for the magnetization is obtained from (4.55) and (4.64):

$$\mathcal{M} = \mathcal{M}_c + \frac{(H - H_c)S}{2|J'|} - \frac{12\sqrt{6}}{\pi^2}\left(\frac{4}{2S+1}\right)^{3/4} S^{5/2}|2J'|^{-5/2} \times$$
$$\times (H_c - H)^{3/2}\,\theta(H_c - H), \tag{4.67}$$

where the notations

$$J' = J + \frac{3\nu S|B|}{(2S+1)^{3/2}} \quad \text{and} \quad H_c = 2|J'|\frac{\mathcal{M}_c}{S} - \frac{6|B|\nu}{2S+1}$$

are used.

As is seen from (4.67), the quantity $d\chi/dH$ undergoes an infinite step when $H \to H_c$.

It should be pointed out that the considerations concerning the possibility of noncollinear antiferromagnetism in magnetic conductors with narrow bands were first presented in [14], but the analysis carried out there was not correct. In particular it was stated in [14] that noncollinear antiferromagnetism should be possible only in layer antiferromagnets and should arise at arbitrarily small charge carrier densities. These errors are due to the use of an inadequate Hamiltonian the form of which was postulated instead of being deduced from the basic principles (see Section 2.3).

The noncollinear antiferromagnetism of the type considered was observed experimentally in [56] to [59]. In [50] a nonlinear field dependence of the moment of a heavily doped antiferromagnetic semiconductor with $\nu < \nu_L$ was found, which is similar to that indicated above.

4.6 Unsaturated ferromagnetism

So far it was assumed that the direct exchange (more precisely the super-exchange) between magnetic atoms causes antiferromagnetic ordering. Meanwhile, it is well known that the negative sign of the exchange integral between nearest neighbours is only a necessary, but not a sufficient condition for antiferromagnetic ordering. In many cases Pauli paramagnetism does occur instead of antiferromagnetism. So far it remains unknown what conditions are sufficient for antiferromagnetic ordering. As a matter of fact, this problem is similar to that of crystallization of a quantum liquid. Further, even if a nearly perfect crystal is antiferromagnetic, an increase of the number of defects may lead to the disappearance of long-range order.

It is natural to suggest that a Pauli paramagnet can also undergo a concentrational phase transition into a state with a nonzero magnetic moment when increasing the charge carrier density. Such a nonsaturated ferromagnetism differs from noncollinear antiferromagnetism in that two magnetic sublattices are absent, although apparently there should exist a short-range antiferromagnetic order.

In the noncollinear antiferromagnetic state at $T = 0$, the average moment of each atom changes its direction relative to the total moment of the crystal, but does not change its magnitude when changing the carrier density. On the contrary, in the unsaturated ferromagnetic state a change in the carrier density leads to a change in the average moment of each atom, which is always directed along the total moment of the crystal.

All phenomenological considerations presented in the previous section for small $\mathcal{M}$ remain valid for the unsaturated ferromagnetism except for the connection between the moment and the angle between the moments of sublattices. In the opposite limit of small demagnetizations $\mathcal{N}$, the phenomenological treatment of the problem is far more complicated. At finite temperatures, the notion of the concentrational phase transition from the saturated to the unsaturated ferromagnetic state is meaningless as there is no order parameter which would be nonzero in only one of these two phases having the same symmetry. But for $T = 0$ one can formally introduce the notion of such a transition. The demagnetization $\mathcal{N}$ plays then the role of an order parameter, and the expansion (4.47) can be used.

To avoid misunderstandings, it is necessary to emphasize that the transition indicated is not a true second-order phase transition, so that the Landau prohibition for second-order phase transitions is not valid for the case under consideration. The principal possibility of singularities of the type of second-order phase transitions at $T = 0$ was pointed out in [40].

The quantum calculation of the lowest energy of an unsaturated ferromagnetic state for small $\mathcal{N}$ is based on the following considerations [61]: As seen from (4.61), all magnon frequencies become negative when diminishing ν to ν_R. Thus, if initially with $\nu > \nu_R$ the system was in the completely saturated ferromagnetic state, diminishing ν to ν_R should lead to magnon emission. The emission is limited by anharmonicity. It follows from the foregoing that for small $\mathcal{N}$ the ground state of the system can be regarded as the saturated ferromagnetic state with a macroscopically large number of excited magnons in it.

A variational method is used to calculate the energy. Since according to (4.61) all magnon frequencies go to zero simultaneously after reaching ν_R, the trial wave function is constructed in a form such that all states with equal total number of magnons, $\mathcal{N}$, are also of equal probability, the number $\mathcal{N}$ being regarded as a variational parameter:

$$\psi(\mathcal{N}) = \frac{1}{\sqrt{C_{N\mathcal{N}}}} \sum_{m_q=0}^{\infty} \prod_q \frac{(b_q^*)^{m_q}}{\sqrt{m_q!}} |0\rangle \exp\{i\varphi(\dots, m_q, \dots)\} \, \delta(\sum m_q, \mathcal{N}), \tag{4.68}$$

where $\delta(x, y)$ is the δ-function of a discrete argument with values 1 for $x = y$ and 0 for $x \neq y$, $C_{N\mathcal{N}}$ is a normalization factor. In the spirit of the random-phase approximation, the phases $\varphi(\dots, m_q, \dots)$ are assumed to be chosen in such a way that the mean value of the nondiagonal part of the Hamiltonian $\mathcal{H}_S + \mathcal{H}_M$ of the magnetic system vanishes. $\mathcal{H}_S$ is the indirect exchange Hamiltonian (3.33 to 3.39), $\mathcal{H}_M$ the direct exchange Hamiltonian (2.4) which with the aid of (3.31) takes the form

$$\mathcal{H}_M = J \sum (1 - \gamma_k) \, b_k^* b_k - \frac{J}{4SN} \sum (2\gamma_{q-q'} - \gamma_q - \gamma_{q'}) \, b_q^* b_p^* b_{p'} b_{q'} \times \\ \times D(q + p, q' + p') \,. \tag{4.69}$$

Explicit expressions for the phases are nonessential since they drop out when calculating the mean value of the diagonal part of the Hamiltonian.

Minimizing the ground-state energy with respect to $\mathcal{N}$ yields with the aid of (4.68), (4.69), and (3.33 to 3.39)

$$\left.\begin{aligned} E(\mathcal{N}) &= E_F + \frac{\mathcal{N}^2}{8SN} J[S - Z] \,, \\ \mathcal{N} &= \frac{4SN}{5 - Z}\left(1 + \frac{J_{\text{eff}}}{J}\right), \end{aligned}\right\} \tag{4.70}$$

where the notation

$$Z = \frac{16S^2}{\beta} \{\sum n_k \gamma_k\}^{-1} \sum Z_{qp|qp} \tag{4.71}$$

is used.

For small ν we succeeded in estimating the quantity $Z \approx 3.2$. It is seen from (4.70) the states with nonzero $\mathcal{N}$ really exist for $(-J) > J_{\text{eff}}$, which ensure the minimum energy of the system. The appropriate quantity $\mathcal{N}$ is proportional to $\nu_R - \nu$ where ν_R is the same as in the previous section. One can show that the energy (4.70) is close to the energy of the noncollinear antiferromagnet found above. But the use of the variational method in both cases makes it meaningless to elucidate the question which of these two energies is less.

Thus it is proved that nonsaturated ferromagnetism can occur in a system of localized spins if the exchange between them is non-Heisenbergian. Up to now it was believed that the nonsaturated ferromagnetism is restricted only to collectivized magnetic electrons (the Stoner model). Apparently the corresponding situation was observed experimentally in [62].

It should be noted that in perfect crystals there is no continuous transition between the saturated and unsaturated ferromagnetic states with density change. Actually, in general with diminishing ν the saturated ferromagnetic state becomes unstable as one of the magnetic frequencies goes to zero. Strictly speaking, the magnon frequency with quasi-momentum $(\pi/a, \pi/a, \pi/a)$ goes to zero first. For this reason, a magnetic two-sublattice structure would establish. However, this rule is not stringent for defect crystals such as heavily doped semiconductors and alloys, since they do not have translational symmetry. In particular, for the case of narrow bands the following considerations will be presented in order to elucidate the question which of these two structures should establish for $\nu < \nu_R$.

As was already indicated, according to (4.61) all magnon frequencies go to zero for $\nu = \nu_R$. If second-order terms in W/AS are taken into account, then according to [19] just the magnon frequency with quasi-momentum $(\pi/a, \pi/a, \pi/a)$ vanishes first with diminishing ν. If magnon attenuation due to their scattering by defects exceeds the second-order correction to the frequencies, the statement that one of the frequencies vanishes first is meaningless. In this case, the nonsaturated ferromagnetism is expected to occur. In the opposite case the noncollinear antiferromagnetism should take place at least for $(\nu_R - \nu) \ll \nu_R$.

In conclusion, it should be pointed out that the indirect exchange via spin polarons may not only lead to the structures with nonzero moment considered above in the range of instability of complete ferromagnetic and antiferromagnetic structures. Other structures, for example helical ones, can occur in this range in some cases.

The treatment presented above is based on the assumption that uniform state of the crystal is stable. But it can be shown that uniform states may be unstable in a density range $\nu < \nu_R$. In this case it is energetically favourable for a heavily doped antiferromagnetic semiconductor to divide into alternating layers with increased and decreased charge carrier densities, the former being ferromagnetic, the latter antiferromagnetic. Thus the state of the crystal should be the nonuniform ferro-antiferromagnetic one [64]. The randomness of the impurity distribution over the crystal promotes the nonuniform magnetic state of the crystal.

5. Conclusion

The theory presented above deals with narrow-band magnetic semiconductors. Its main point is that the conduction electron spin is always directed along the local magnetic moment in that region where the electron is located. The role of the local magnetic moment is played by the spin of the magnetic atom nearest to the electron. The question arises whether similar effects are possible in wide-band magnetic semiconductors.

The answer to this question would be negative if the electron energy were an analytic function of AS as is usually assumed. In the first approximation in this parameters, the s–d exchange would shift the electron energy by the amount $A\langle S\rangle/2$, where $\langle S\rangle$ is the average magnetic moment of the crystal per atom. This result corresponds to the conduction electron spin being fixed. But it can be easily shown that this is not true for the case of the local magnetic moment direction varying slowly in the crystal. In fact, according to [65] the electron energy $\tilde{\varepsilon}_k$ in a crystal with helical magnetic structure, its vector being λ, is

$$\tilde{\varepsilon}_k = \frac{1}{2} (\varepsilon_k + \varepsilon_{k+\lambda}) \pm \frac{1}{2} \sqrt{(\varepsilon_k - \varepsilon_{k+\lambda})^2 + (AS)^2}, \tag{5.1}$$

where ε_k is the electron energy in the absence of s–d exchange.

As is seen from this expression, for λ large enough, the s–d exchange correction to the electron energy is of the order $(AS)^2/W$, as it should be the case for $S = 0$. But for λ small enough but finite, so that $\langle S \rangle$ remains zero, this quantity is practically $AS/2$; that is the same as for ferromagnetic ordering. The last result is a clear indication of the existence of spin polaron states in the case of wide-band magnetic semiconductors. It means that at small λ, the conduction electron varies its spin direction on its way through the crystal in order to get its spin aligned along the local magnetic moment. It does not matter that, unlike the local magnetic moments, the total magnetic moment of the crystal is zero. In other words, the electron energy is determined not by the long-range magnetic ordering, but by the short-range one.

First at all, spin polaron effects should manifest themselves as an anomalous temperature dependence of the electron energy in wide-band magnetic semiconductors. It can be shown [66] that at $2S \gg 1$, the temperature shift of the conduction band bottom is given by an expression similar to (3.30) for temperatures low enough that only long-wave magnons are excited. This means that the shift is independent of the s–d exchange integral and proportional to $T^{5/2}$. The frequencies of long-wave magnons are also given by the expression (3.35) in the case under consideration [67].

Still more interesting is the situation in the vicinity of the Curie point T_C, where the direction of the local magnetic moment varies over distances of the order of the inverse correlation length. The latter tends to infinity with temperature approaching the Curie point. Thus, in the region $T > T_C$, the s–d exchange gain in the electron energy may be considerable, whereas the total magnetic moment is still zero. A detailed calculation [68] shows that at T_C the gain should be of the order $AS(AS/W)^{1/3}$, that is, the small parameter (AS/W) enters it only in the one-third power. At higher temperatures, the gain should be of the order $AS(AS/W)$. This means that the anomalously strong temperature shift of the conduction band bottom, which obviously takes place at $T < T_C$, should also continue after passing the Curie point. The result just indicated is confirmed by experiment [69].

The spin polaron effect providing an appreciable gain in the s–d exchange energy near the Curie point tends to prevent the conduction electron from autolocalization inside a fluctuation of the magnetic moment, the idea of which was advanced in [49].

As to other physical effects described in this paper, most of them are also possible in wide-band semiconductors (ferrons, canted magnetic structures, and so on). The exception is the quasi-oscillator states which are restricted only to narrow-band semiconductors.

6. Very Recent Advances in the Theory of Magnetic Semiconductors[10])

1. Since the review had been written a number of papers on the theory of magnetic semiconductors has appeared in which ideas presented in the review

[10]) This section bears the date of receipt June 21, 1974.

are developed. In [71, 72] problems of mathematical description of narrow-band magnetic semiconductors are considered. In the former Hamiltonian for the conduction electrons in the limit of large spins $2S \gg 1$ is deduced which is a generalization of (2.15) for a system of arbitrary number of magnetic atoms. It differs from that suggested in [14] (Section 2.2 of the review) in that the effective Bloch integrals describing electron transitions between various pairs of atoms are out of phase with respect to each other. The phase shift vanishes if spins of all atoms are in the same plane, and this is a condition of validity of De Gennes Hamiltonian [14]. In the limit of large spins Hamiltonian (3.24) goes over in that obtained in [71] by means of a canonical transformation shifting phases of electron operators. The accuracy of the latter is $1/2S$ when spin alignment differs strongly from the antiferromagnetic one, and $1/\sqrt{2S}$ in the opposite case.

In [72] the Hamiltonian of the system under consideration is expressed in terms of projection operators into states in which an electron spin is parallel or antiparallel to the spin of the atom it occupies. Such a treatment should lead to the same results as those presented in the review. For example, the result (3.28) was repeated in [72]. But it is less convenient virtually since the movements of spins and of electrons through the crystal are not separated from each other. This leads to some difficulties which manifest themselves, for example, when introducing magnons in a ferromagnet with allowance for the indirect exchange [72, 74]. They are treated there as oscillations of localized d-spins only without taking into account of electron spins taking part. For this reason the results of [72] in the spin-wave region are valid only in the main order in $1/(2S)$. The magnon spectrum obtained in [72, 74] coincides with (3.34 to 3.38) to this order. But an error was made in [72] when calculating scattering of charge carriers by magnons. When the parabolic dispersion law for magnons holds the resistivity of degenerate ferromagnetic semiconductors is proportional not to $T^{9/2}$ but to $T^{7/2}$ (Section 3.2).

In [71 to 73, 75] the spectrum of charge carriers in narrow-band semiconductors at temperatures well exceeding the Curie point T_C was investigated. It was shown in [71] by calculating the moments of electron density of states that though, on temperature increase, the electron band becomes narrower, its shape changes relatively weakly. This confirms the applicability of the coherent potential approximation used in [73, 75] for investigating its gross features. According to [75] the high-temperature band width is $[(S+1)/(2S+1)]^{1/2}$ times less than that at $T = 0$ which in the limit $S \to \infty$ coincides with the results [71].

2. The investigation of ferron states was continued. In [76] these are considered in an antiferromagnet. More accurate approximation for the ferromagnetic potential well than that in Section 4.4 was used in [76] when searching for the energy minimum, and numerical calculations were carried out for several values of parameters. For wide-band semiconductors the results of [76] are in agreement with those of [43, 44]. But for narrow-band semiconductors the calculation is valid only under the nonrealistic condition $\sqrt{2S} \gg 1$ since it is based on the De Gennes Hamiltonian [14]. Destruction of the ferron by the external magnetic field was investigated in [95]. In [77 to 79, 86, 87] attempts were made to investigate analogous states in ferromagnets at elevated temperatures.

Concerning the problem of high-temperature ferron states, it should be noted that in the paramagnetic region there is no difference in treating antiferromagnets and ferromagnets. A reasonable estimate of the ferron energy in a narrow-band semiconductor can be obtained, using the above mentioned results [75] for paramagnetic band width. The ferromagnetic well depth is given by the difference of ferromagnetic and paramagnetic band halfwidths, and the change in the magnetic free energy caused by the formation of ferromagnetic microregion is not $|J|S$ but $kT \ln (2S + 1)$ per magnetic atom. For T greatly exceeding critical T_C the former is nearly 3 times less, and the latter $\approx T/T_C$ more than these at $T = 0$. This means, of course, that ferrons in antiferromagnets disappear on temperature increase. Moreover, this leads to the conclusion that the existence of ferrons in ferromagnets, which unlike antiferromagnets is possible only at elevated temperatures, is considerably less probable than in antiferromagnets. Their energy and stability conditions are given as before by (4.33 to 4.36) with replacing $1/\sqrt{2S + 1}$ by $[(S + 1)/(2S + 1)]^{1/2}$ in (4.33) and $|J|S$ by $kT \ln (2S + 1)$ in (4.35).

Further, the correlations of spins of magnetic atoms should be taken into account when treating ferron states in a ferromagnet near Curie point. As follows from estimates for band width at T_C and $T \gg T_C$ [71] even with allowance for correlations between only nearest neighbours the ferromagnetic potential well depth at T_C is 1.5 times less than at $T \gg T_C$. The long-range correlations should make it still more shallow. At any case, for wide-band ferromagnetic semiconductor they have been shown to diminish considerably the energy of a free conduction electron, i.e. tend to prevent it from going over to the ferron state [68].

In [88 to 90] problems similar to the quasi-oscillator problem (Section 4.2) are investigated within the framework of the Hubbard model which in the semiconductor limit is equivalent to a particular case of the s–d model. But in them only moments of electron density of states are calculated even without taking into account of the d–d exchange which causes the appearance of the quasi-elastic force discussed above.

3. A number of papers on cooperative ferron states has been published since. Though in most of them wide-band magnetic semiconductors were treated, many qualitative results of them remain valid for narrow-band ones. In [66] a further development of the theory of the metal–insulator transition in degenerate semiconductors (Section 3.6) is presented. In particular, it is proved there that, if the donor density is close to the critical ν_c, temperature increasing causes only transition from conducting to insulating state, so that the opposite transition to the conducting state does not occur with its further raising. The origin for this is that the electron energy is the lower the higher is the magnetic moment per atom of the region allowed for the electron. For $T \gg T_C$ the moment is $\sim 1/\sqrt{V}$ where V is volume of the region. This means that the gain in the electron energy caused by the local magnetic ordering is higher for the localized electron than for the conducting one.

A theory of the same effect was also presented in [91]. The basic physical ideas of [91] are similar to those indicated in Section 3.6, but the calculation was carried out inconsistently. The condition of the minimum of the energy of the conduction electrons was used as the starting point whereas that of the free energy of the whole system should be used instead of. The calculation of the static dielectric function is not very convincing, too, since the representation of the magnetic free energy in the form appropriate to the vicinity of the Curie point does not match the assumption of the complete polarization of electron spins.

In [80, 81] a new type of cooperative ferron states was investigated. It was shown that some of the degenerate antiferromagnetic semiconductors within a certain range of electron density may decay into ferromagnetic and antiferromagnetic domains with dimensions well exceeding the mean distance between donors. There is an enhanced electron density inside the former and a reduced one inside the latter. The origin for this is the same as for individual ferrons in antiferromagnets (Section 4.4). At lower impurity densities the nonuniform state of the crystal is as follows: Ferromagnetic droplets embedded in the antiferromagnetic host crystal. After reaching a critical impurity density the topology of the nonuniform state changes: the ferromagnetic part of the crystal becomes single-connected and the antiferromagnetic one multiconnected (antiferromagnetic droplets in the ferromagnetic crystal). At this density a sharp rise in the conductivity should occur.

A nonuniform noncollinear antiferromagnetic ordering is also possible in degenerate antiferromagnetic semiconductors under certain conditions. It differs from that discussed in Section 4.5 in that the angle between moments of sublattices changes periodically in the crystal, and so does the electron density [81].[11])

Existence of ferromagnetic domains in such systems was proved experimentally [84].

Some of the degenerate ferromagnetic semiconductors at elevated temperatures may also go over to a similar nonuniform state in which the regions of enhanced both magnetization and charge carrier density and of reduced ones alternate [80, 82, 83, 94]. They should realize at densities well exceeding those at which the metal–insulator transition occurs. The phase transition from the uniform state to the nonuniform one should be of the first kind. Though as a result of it the resistivity increases, the crystal does not get insulating. Below the transition point the nonuniform state may exist as a metastable one which explains experimental results [85].

Attempts to investigate the transition of magnetic semiconductors from the uniform state to a nonuniform one were made in [92, 93], too.

4. New results were obtained for the temperature dependence of charge carrier density and their mobility in ferromagnetic semiconductors [66, 83]. In particular, it is shown in [83] for heavily doped semiconductors that the resistivity increase with temperature is due mainly to the scattering of electrons not by magnons but by the static spatial fluctuations of magnetization. The latter are caused by similar fluctuations of electron density due to the fact that the positive charge of donors is distributed nonuniformly over the crystal. The magnetization depends on the charge carrier density since these realize indirect exchange between magnetic atoms tending to support ferromagnetic ordering. Thus, on temperature increase, the magnetization of regions of enhanced electron density diminishes slower than that of reduced density, i.e. spatial fluctuations of magnetization grow.

---

[11]) To avoid misunderstandings: this is caused not by the randomness of impurity distribution, but only by competition of the electrostatic, kinetic, and exchange energies.

After passing T_C they diminish and so does the scattering by them. This leads to a minimum of the mobility in the vicinity of T_C. Its depth increases first and then after reaching its maximum diminishes during the charge carrier density increase. The scattering by the critical fluctuations cannot cause such a maximum. Numerous experimental data on degenerate ferromagnetic semiconductors confirm the existence of the maximum and, thus, indicate that the critical scattering does not play an important role in them.

According to [83, 94] the screening length first diminishes with temperature increasing in degenerate ferromagnetic semiconductors. After reaching its minimum it begins to grow, and the screened potential becomes oscillating. At the point of loss of absolute stability for the uniform state, if that exists, the amplitude of oscillations and the screening length should get infinite. But still earlier the transition to a nonuniform state should occur under the condition just indicated.

The activation energy for impurity nondegenerate ferromagnetic semiconductors should reach its maximum at $T \approx T_C$, but for reasons presented in the previous point its high-temperature value is higher than that for $T = 0$ [66].

5. Some progress was achieved in investigating magnetic properties of magnetic semiconductors. A new type of magnons in degenerate ferromagnets was investigated which are caused by the surface being charged with respect to the bulk of the crystal [96]: the electron density in the surface charge layer differs from that in the bulk and this results in changing indirect exchange close to the surface (see Section 3.2). It suffices for the surface magnetization to differ from the bulk one, and this may explain the experimentally observed paramagnetic sheet on the surface of ferromagnetic semiconductors [97]. Conditions at the surface of heavily doped antiferromagnetic semiconductors may be created when the ferromagnetic sheet would exist on their surfaces (an external electric field, chemisorption, and so on).

One should keep in mind that the indirect exchange in magnetic semiconductors irrespective of the band width is essentially non-Heisenbergian since the conditions for applicability of the Ruderman-Kittel theory are not met in them. This manifests itself not only in the form of the dispersion law for magnons established in [67] but also in the magnetic susceptibility. For example, that for degenerate antiferromagnets is strongly field-dependent, imitating the existence of ferromagnetic clusters inside the crystal even if the conditions of their existence are not satisfied [55].

Another important point is that, strictly speaking, impurity magnetic semiconductors are disordered magnetic systems. If the impurity density is small enough, local or quasi-local magnons appear in the vicinity of unionized donors due to the indirect exchange via the electrons localized at donors [66]. In heavily doped ferromagnetic semiconductors there appears a tail of the magnon band directed to the higher frequencies which is an analogue of the electron density of states tail inside the gap in heavily doped semiconductors. Both are caused by fluctuations of impurity potential and grow with temperature [83]

References

[1] S. V. Vonsovskii, Magnetizm, Izd. Nauka, Moscow 1971 (p. 553) (in Russian).
[2] E. L. Nagaev, Fiz. tverd. Tela **13**, 1163 (1971).
[3] E. L. Nagaev, Zh. eksper. teor. Fiz., Pisma **8**, 596 (1968).
[4] E. L. Nagaev, Zh. eksper. teor. Fiz. **57**, 469 (1969).

[5] E. L. Nagaev, Fiz. tverd. Tela **12**, 607 (1970).
[6] E. L. Nagaev, Zh. eksper. teor. Fiz. **59**, 1215 (1970).
[7] E. L. Nagaev, Fiz. tverd. Tela **13**, 41 (1971).
[8] E. L. Nagaev, Teoret. i matem. Fiz. **13**, 91 (1973).
[9] V. M. Matveev and E. L. Nagaev, Phys. Letters, to be published.
[10] D. C. Mattis, The Theory of Magnetism, Harper & Row Publishers, New York/Evanston/London 1965.
[11] C. Zener, Phys. Rev. **82**, 403 (1951).
[12] P. W. Anderson and H. Hasegawa, Phys. Rev. **100**, 675 (1955).
[13] L. D. Landau and E. M. Lifshits, Kvantovaya mekhanika, Fizmatgiz, Moskva 1963 (p. 468).
L. D. Landau and E. M. Lifshits, Lehrbuch der theoretischen Physik, Vol. III: Quantenmechanik, Akademie-Verlag, Berlin 1970.
[14] P. de Gennes, Phys. Rev. **118**, 141 (1960).
[15] S. V. Tyablikov, Zh. eksper. teor. Fiz. **21**, 377 (1951).
[16] E. L. Nagaev, Fiz. tverd. Tela **4**, 2201 (1962).
[17] I. G. Lang and Yu. A. Firsov, Zh. eksper. teor. Fiz. **43**, 1843 (1962).
[18] E. L. Nagaev, Zh. eksper. teor. Fiz. **56**, 1013 (1969).
[19] E. L. Nagaev, Fiz. Metallov i Metallovedenie **29**, 905 (1970).
[20] Yu. P. Irkhin and E. A. Turov, Fiz. Metallov i Metallovedenie **4**, 9 (1957).
[21] F. Rys, J. Helman, and W. Baltensperger, Phys. kondens. Materie **6**, 105 (1967).
[22] G. Busch, P. Junod, and P. Wachter, Phys. Letters **12**, 11 (1969).
[23] J. Callaway and T. Wolfram, Phys. Rev. **127**, 1606 (1962).
[24] J. Callaway and T. Wolfram, Phys. Rev. **130**, 46 (1963).
[25] Yu. A. Izyumov and M. V. Medvedev, Zh. eksper. teor. Fiz. **59**, 553 (1970).
[26] M. Ruderman and C. Kittel, Phys. Rev. **96**, 99 (1954).
[27] K. Yosida, Phys. Rev. **106**, 893 (1957).
[28] T. Kasuya, Progr. theor. Phys. (Kyoto) **16**, 45 (1956).
[29] E. L. Nagaev, Fiz. tverd. Tela **11**, 3438 (1969).
[30] E. L. Nagaev, Fiz. tverd. Tela **11**, 2779 (1969).
[31] D. N. Zubarev, Uspekhi fiz. Nauk **71**, 71 (1960).
[32] G. L. Lazarev and E. L. Nagaev, to be published.
[33] G. A. Baker and J. L. Gammel, J. math. Anal. Applic. **2**, 21 (1961).
[34] K. P. Belov and E. P. Svirina, Uspekhi fiz. Nauk **96**, 21 (1968).
[35] G. L. Lazarev, Fiz. tverd. Tela **14**, 1 (1972).
[36] G. L. Lazarev, to be published.
[37] E. L. Nagaev and A. P. Grigin, Phys. Letters A **38**, 469 (1972).
[38] E. L. Nagaev, Zh. eksper. teor. Fiz. **58**, 1269 (1970).
[39] L. D. Landau and E. M. Lifshits, Kvantovaya mekhanika, Fizmatgiz, Moskva 1963 (p. 246).
L. D. Landau and E. M. Lifshits, Lehrbuch der theoretischen Physik, Vol. III: Quantenmechanik, Akademie-Verlag, Berlin 1970.
[40] E. L. Nagaev, Fiz. tverd. Tela **13**, 958 (1971).
[41] L. N. Bulaevskii, E. L. Nagaev, and D. I. Khomskii, Zh. eksper. teor. Fiz. **54**, 1562 (1968).
[42] E. M. Lifshits, Zh. eksper. teor. Fiz. **17**, 1017 (1947); **17**, 1079 (1947).
[43] E. L. Nagaev, Zh. eksper. teor. Fiz., Pisma **6**, 485 (1967).
]44] E. L. Nagaev, Zh. eksper. teor. Fiz. **54**, 228 (1968).
[45] M. A. Krivoglaz, Fiz. tverd. Tela **11**, 2230 (1969).
[46] M. A. Krivoglaz and A. A. Trutshenko, Fiz. tverd. Tela **11**, 3119 (1969).
[47] S. Streit and P. Wachter, Phys. kondens. Materie **11**, 231 (1970).
[48] G. Busch, P. Streit, and P. Wachter, Solid State Commun. **8**, 1759 (1970).
[49] S. von Molnar and S. Methfessel, J. appl. Phys. **38**, 959 (1967).
[50] F. Holtzberg, T. McGuire, and S. Methfessel, J. appl. Phys. **37**, 976 (1966).
[51] E. L. Nagaev, Fiz. tverd. Tela **13**, 891 (1971).

[52] L. D. LANDAU and E. M. LIFSHITS, Statisticheskaya fizika, Izd. Nauka, Moskva 1964 (p. 501).
L. D. LANDAU and E. M. LIFSHITS, Lehrbuch der theoretischen Physik, Vol. V: Statistische Physik, Akademie-Verlag, Berlin 1971.
[53] I. E. DZYALOSHINSKII, Zh. eksper. teor. Fiz. **32**, 1547 (1957).
[54] E. L. NAGAEV, Zh. eksper. teor. Fiz. **57**, 1274 (1969).
[55] E. L. NAGAEV, Fiz. tverd. Tela **14**, 773 (1972).
[56] E. WOLLAN and W. KOEHLER, Phys. Rev. **100**, 545 (1955).
[57] J. KOUVEL and J. FORSYTH, J. appl. Phys. **40**, 1359 (1969).
[58] W. TAKEI, D. COX, and G. SHIRANE, Phys. Rev. **129**, 2008 (1963).
[59] W. TAKEI, D. COX, and G. SHIRANE, J. appl. Phys. **37**, 973 (1966).
[60] E. M. LIFSHITS and M. I. KAGANOV, Uspekhi fiz. Nauk **78**, 411 (1962).
[61] E. L. NAGAEV, Fiz. tverd. Tela **12**, 1109 (1970).
[62] H. JARRETT, W. CLOUD, R. BOUCHARD, S. BUTLER, C. FREDERICK, and J. GILLSON, Phys. Rev. Letters **21**, 517 (1968).
[63] M. OLIVER, J. KAFALAS, J. DIMMOCK, and T. REED, Phys. Rev. Letters **24**, 1064 (1970).
[64] E. L. NAGAEV, Zh. eksper. teor. Fiz., Pisma **16**, 558 (1972).
[65] I. E. DZYALOSHINSKII, Zh. eksper. teor. Fiz. **47**, 336 (1964).
[66] A. P. GRIGIN and E. L. NAGAEV, phys. stat. sol. (b) **61**, 65 (1974).
[67] E. L. NAGAEV, Fiz. tverd. Tela **11**, 2779 (1969).
[68] E. L. NAGAEV, Solid State Commun., in the press.
[69] S. METHFESSEL and D. C. MATTIS, Hdb. Phys., Vol. XVIII/1, Springer-Verlag, Berlin/New York 1968.
[70] T. KASUYA, A. YANASE, and T. TAKEDO, Solid State Commun. **8**, 1543 (1970); **8**, 1551 (1970).
[71] E. L. NAGAEV and E. B. SOKOLOVA, Fiz. tverd. Tela **16**, 1293 (1974).
[72] K. KUBO and N. OHATA, J. Phys. Soc. Japan **33**, 21 (1972).
[73] K. KUBO, J. Phys. Soc. Japan **33**, 929 (1972).
[74] N. OHATA, J. Phys. Soc. Japan **34**, 343 (1973).
[75] A. RANGETTE, A. YANASE, and J. KÜBLER, Solid State Commun. **12**, 171 (1973).
[76] M. UMEHARA and T. KASUYA, J. Phys. Soc. Japan **33**, 602 (1972).
[77] A. YANASE, Internat. J. Magnetism **2**, 99 (1972).
[78] M. A. KRIVOGLAZ, Zh. eksper. teor. Fiz., Pisma **17**, 262 (1973); Fiz. tverd. Tela **15**, 1765 (1973).
[79] A. YANASE, Solid State Commun. **9**, 2111 (1971).
[80] E. L. NAGAEV, Zh. eksper. teor. Fiz., Pisma **16**, 558 (1972).
[81] V. A. KASHIN and E. L. NAGAEV, Zh. eksper. teor. Fiz. **66**, No. 6 (1974).
[82] E. L. NAGAEV and A. P. GRIGIN, Solid State Commun. in the press.
[83] E. L. NAGAEV and A. P. GRIGIN, Zh. eksper. teor. Fiz., in the press.
[84] J. VITINS and P. WACHTER, Solid State Commun. **13**, 1273 (1973).
[85] Y. SHAPIRA and T. REED, Phys. Letters A **36**, 105 (1971).
[86] J. KÜBLER, Z. Phys. **250**, 324 (1972).
[87] D. VIGREN, J. Physique **60**, 967 (1973).
[88] W. BRINKMAN and T. RICE, Phys. Rev. B **2**, 1324 (1970).
[89] J. SOKOLOFF, Phys. Rev. B **2**, 3707 (1970); B **3**, 3826 (1971).
[90] B. ROULET, M. FISHER, and S. DONIACH, Phys. Rev. B **7**, 403 (1973).
[91] P. LEROUX-HUGON, Phys. Rev. Letters **29**, 939 (1972).
[92] M. A. KRIVOGLAZ, Zh. eksper. teor. Fiz. **63**, 671 (1972).
[93] YU. I. BALKAREI, V. G. BARU, and A. A. SUKHANOV, Fiz. tverd. Tela **15**, 2223 (1973).
[94] V. E. ZILBERVARG and E. L. NAGAEV, Fiz. tverd. Tela **16**, No. 10 (1974).
[95] G. L. LAZAREV and E. L. NAGAEV, Fiz. tverd. Tela **15**, 1635 (1973).
[96] V. E. ZILBERVARG and E. L. NAGAEV, Fiz. tverd. Tela **15**, 1585 (1973).
[97] K. SATTLER and H. SIEGMANN, Phys. Rev. Letters **29**, 1565 (1972).

(Received September 20, 1973)

Original Papers

phys. stat. sol. (b) **65**, 61 (1974)

Subject classification: 14.3; 22.2.1

Istituto di Fisica della Università di Modena

Generalization of Fick's Law for Non-Local Complex Diffusion in Semiconductors[1])

By

C. Jacoboni

Electron diffusion in semiconductors is studied at short distances and/or times. An extension of Fick's law with higher derivatives is introduced, and a numerical method determines how many terms are to be kept and the values of the corresponding coefficients. At high applied electric fields the asymmetric diffusion introduces an imaginary part of the Fourier-transformed diffusion coefficient, that is, even order derivatives in the diffusion equation.

Es wird die Diffusion in Halbleitern bei kurzen Distanzen und/oder Zeiten untersucht. Eine Erweiterung des Fickschen Gesetzes mit höheren Ableitungen wird eingeführt und eine numerische Methode ergibt, wieviele Terme berücksichtigt werden müssen, sowie die Werte der entsprechenden Koeffizienten. Bei hohen äußeren elektrischen Feldern ergibt die asymmetrische Diffusion einen imaginären Anteil des Fourier-transformierten Diffusionskoeffizienten, d. h. gerade Ableitungen in der Diffusionsgleichung.

1. Introduction

Although most of the effort in studying the transport properties of semiconductors is devoted to determination of the charge-carrier mobility, diffusion does not seem to be less important either from a fundamental point of view or for device applications. At low fields the analyses of the mobility μ and of the diffusion constant D coincide due to the Einstein relation; at high fields, however, these two quantities must be determined independently [1]. Furthermore, at short times or at small distances the diffusion phenomenon is not yet described by the phenomenological Fick equation. However, diffusion can in general be analysed by direct Monte Carlo simulation [2, 3].

In this paper we report a new, deeper approach, both analytical and numerical, to a study of the diffusion phenomenon at high fields and/or small distances and/or short times.

2. Theory

Let us start from the one-dimensional classical diffusion equation

$$j(x) = n(x)\, v_\mathrm{d} - D\, \frac{\partial n}{\partial x} \tag{1}$$

or its Fourier-transformed

$$j_q = n_q v_\mathrm{d} - iqDn_q\,, \tag{2}$$

where $j(x)$ is the current density, $n(x)$ the particle density, v_d their drift velocity,

[1]) Partially supported by Consiglio Nazionale delle Ricerche, Italy.

assumed constant, and D the diffusion coefficient; j_q and n_q are the Fourier transforms of $j(x)$ and $n(x)$.

If we now start from (2) and antitransform, we obtain back (1) if D is q-independent; in this case D must be real. However, if a q-dependent D is inserted in (2), real physical quantities may be obtained with complex $D(q)$. An imaginary part of D means that a component of the diffusion current in phase with n is introduced (which is different from zero when $\partial n/\partial x$ is zero). A physical situation of this kind may come about when the relaxation processes of electrons coming from "the right" or from "the left" of the point x are different. A greatly distorted distribution function may cause such an effect if ql is not much less than unity (where l is the average electron mean free path). This is essentially the situation discussed in [3] with regard to asymmetrical initial diffusion. In such a case electrons contributing to the diffusion current in x arrive with "free paths" from distances such that (1) does not hold anymore. However, for n so small that interparticle collisions do not contribute to the transport phenomenon, we may still assume a diffusion current linear with n; then we extend the starting equation as

$$j(x) = nv_\mathrm{d} + \mathcal{D}_1 \frac{\partial n}{\partial x} + \mathcal{D}_2 \frac{\partial^2 n}{\partial x^2} + \cdots . \tag{3}$$

Below we present a simulation method from which the information is obtained on how many terms are to be kept in (3) and what values to give to the corresponding coefficients in order to correctly describe the diffusion process when short distances are involved.

By Fourier-transforming (3) we obtain

$$j_q = n_q v_\mathrm{d} - iqD(q)\, n_q = \mathcal{D}(q)\, n_q \tag{4}$$

with

$$\mathcal{D}(q) = \sum_{m=0}^{\infty} \mathcal{D}_m (iq)^m ; \qquad \mathcal{D}_0 = v_\mathrm{d} , \tag{5}$$

$$D(q) = -\sum_{m=1}^{\infty} \mathcal{D}_m (iq)^{m-1} . \tag{6}$$

In order to have real physical quantities the expansion of $D(q)$ in powers of q must then contain real even terms and imaginary odd terms. When the distribution function is an even function of v, we expect even terms higher than the zero-order term to vanish.

Antitransforming (4) we may write, as in the more familiar formalism of linear response theories,

$$j(x) = nv_\mathrm{d} + \int_{-\infty}^{\infty} \tilde{D}(x - x') \left.\frac{\partial n}{\partial x}\right|_{x'} \mathrm{d}x' , \tag{7}$$

where

$$\tilde{D}(\xi) = \frac{1}{2\pi} \int_{-\infty}^{\infty} D(q)\, \mathrm{e}^{iq\xi} \mathrm{d}q . \tag{8}$$

Equation (7) is equivalent to (3) and may be taken as the starting equation.

3. Monte Carlo Simulation

Let us now consider the simulation process devised to obtain the generalized diffusion coefficient $D(q)$ of (4). In order to introduce also a time variable to analyse diffusion at short times we consider particles which have a constant probability τ_d^{-1} of being trapped, or decaying.

Let $\mathcal{N}(x, t)$ be the probability that a particle in $x = 0$ at $t = 0$ will be in x at t when $\tau_d^{-1} = 0$. Then

$$Q(x, \tau_d) = \int_0^\infty \mathcal{N}(x, t)\, e^{-t/\tau_d} \frac{dt}{\tau_d} \tag{9}$$

is the probability that the particle with lifetime τ_d will decay in x. After integration by parts of (9), from the continuity and diffusion equations we obtain the equation satisfied by $Q(x, \tau_d)$:

$$Q(x, \tau_d) = \delta(x) - v_d\tau_d \frac{\partial Q}{\partial x} - \mathcal{D}_1\tau_d \frac{\partial^2 Q}{\partial x^2} - \mathcal{D}_2\tau_d \frac{\partial^3 Q}{\partial x^3} - \dots \tag{10}$$

Let

$$R(q, \tau_d) = \int_{-\infty}^{\infty} Q(x, \tau_d)\, e^{-iqx}\, dx\,. \tag{11}$$

Then by substitution (10) into (11) we obtain

$$\mathcal{D}(q) = \frac{1 - R}{iq\tau_d R}$$

or

$$D(q) = \frac{1 - R - iq\tau_d v_d R}{q^2\tau_d R}. \tag{12}$$

We then consider a normal Monte Carlo simulation in which a constant probability τ_d^{-1} for a decay process is added to the normal scattering processes (Rees' included). If the particle decays after covering a longitudinal distance X, the quantity $\exp(-iqX)$ is recorded. The particle then starts again at $x = 0$ with the velocity it had when it decayed.[2]) Then R is obtained as the mean value of these $\exp(-iqX)$ values.

4. Numerical Application to GaAs

Numerical calculations have been performed for GaAs at room temperature with the same physical model and constants as in [4].

Fig. 1 shows the results obtained for an Ohmic field (200 V/cm). For long τ_d and long wavelengths we obtain the value of the diffusion constant (235 cm^2/s) given by the Ohmic mobility with the Einstein relation. At higher q, $D(q)$ decreases since the electrons begin to feel the periodic character of the disturbance. At low fields, an analytic expression can be obtained for $D(q)$ if the electrons have a constant momentum relaxation time τ_0, by means of Chambers'

---

[2]) Here the physical assumption is made that at $t = 0$ all particles have the stationary-state velocity distribution. In physical situations with short-distance diffusion, as for example in small devices, this assumption must be checked and, if necessary, substituted with a different one which takes into account the true physical situation.

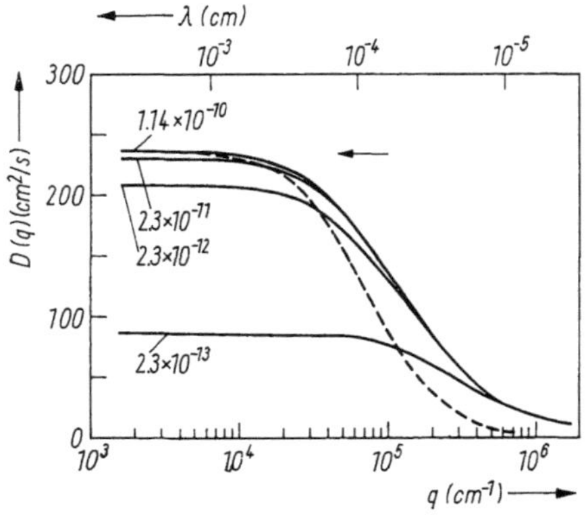

Fig. 1. Diffusion coefficient as a function of wavenumber for an Ohmic applied electric field (200 V/cm) in GaAs at 300 °K. The numbers on the curves indicate particle lifetimes in seconds. The arrow indicates the value of D as obtained by the Ohmic mobility through the Einstein relation. The dashed line represents the analytical solution obtained for a constant momentum relaxation time

method. The result is contained as a particular case in [5] and is given by

$$D(q) = D(0)\, 2y^2\, [1 - \sqrt{\pi}\, y\, e^{y^2} \operatorname{erfc}(y)]\,, \tag{13}$$

where $y = 1/|q|\, l$, $l = v_0 \tau_0 = \tau_0\, (2\, kT/m)^{1/2}$. In agreement with the discussion above, this expression is real and even in q. For comparison, the result obtained from this expression (normalized to the same mobility) is reported in Fig. 1. It can be seen that it has the same qualitative behaviour as the curve obtained with direct simulation.

When τ_d is reduced below about 10^{-10} s the diffusion constant at the long-wavelength limit is also reduced since, like what happens at high q, the contribution of electrons coming from far away is missing. For high q, however, all curves coincide, since the contribution of electrons from far away is missing anyway.

No imaginary part of $D(q)$ is obtained at Ohmic fields, within the statistical uncertainty, this being below 2% in the real part of D.[3])

Fig. 2 shows the results obtained at an applied field $E = 10$ kV/cm, well above the threshold value for the onset of negative differential mobility. For $\tau_d > 8.6 \times 10^{-12}$ s the curve for τ_d equal to this value is reproduced within few percent. At long lifetimes and intermediate q, D passes through a maximum.

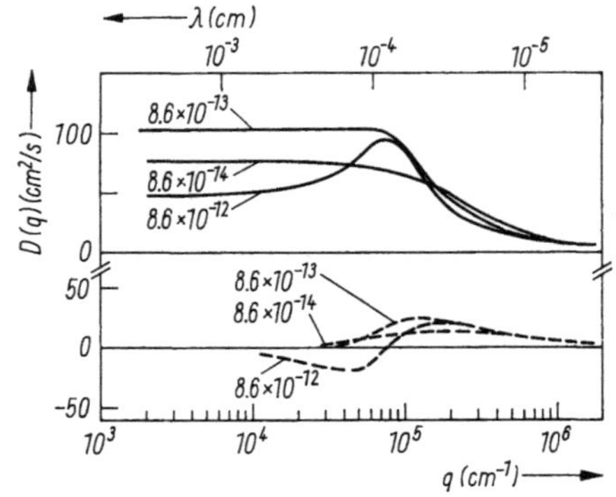

Fig. 2. Real (———) and imaginary (— — —) part of the diffusion coefficient as function of wavenumber for an electric field $E =$ $= 10$ kV/cm in GaAs at 300 °K. The numbers on the curves indicate particle lifetimes in seconds

[3]) This statistical error has been determined by performing many independent determinations of R and v_d, assuming their standard deviations as their uncertainty, and then applying the error propagation theory to (12).

This behaviour is related to the high value of the average velocity which brings about a high diffusion at intermediate wavelengths. For longer wavelengths the oscillatory character of the longitudinal motion of the electrons discussed by Fawcett and Rees [2] reduces the longitudinal diffusion. For the same reason in the limit of small q, $D(q)$ has non-monotonic behaviour as a function of τ_d.

In agreement with the above discussion an imaginary part of the diffusion constant is found with the simulation at high fields, as shown in Fig. 2. It may be seen there that at intermediate q, Im D is of the same order of magnitude as Re D, and in the case of long lifetimes it changes sign in correspondence with the maximum of Re D.

At fields around the threshold value for the onset of negative differential mobility, similar results are obtained, although, in agreement with the discussion in [2], longer lifetimes are necessary to obtain the limiting curve, which has a higher maximum, shifted toward longer wavelengths.

From the results shown in Fig. 1 and 2, and the theory discussed above it may be concluded that for a diffusion phenomenon in GaAs at 300 °K:

a) For $n(x)$ with structure within about 10^{-3} cm it is necessary to consider derivatives higher than first order in the diffusion equation.

b) At low fields only odd-order derivatives can be taken into account, while at high fields, because of asymmetry of the distribution function, complex diffusion comes about and even derivatives must be included.

c) How many terms must be kept in the diffusion equation (3) and the numerical values of the corresponding constants can be obtained from $D(q)$, as given by the direct simulation, in the following way: $D(q)$ must be expanded in powers of q (see equation (6)). The number of terms to be kept in this expansion must be such as to approximate $D(q)$ up to the wavenumbers involved in the Fourier expansion of the profile $n(x)$ of interest in each particular case. The coefficients $\mathcal{D}_m$ obtained in this way are those to be used in (3).

d) From the simulation it is also possible to know the time scale necessary for the applicability of the diffusion equation (3). It may be interesting to note that this time scale is of the same order as the time necessary for a Gunn domain to travel a distance comparable to its width.

Acknowledgements

I wish to thank Dr. P. Price for stimulating discussions which greatly contributed to this research. I am also grateful to Dr. P. Lebwohl for making available his original Monte Carlo program and to Prof. A. Alberigi-Quaranta for his continuous interest in the present work. I am also indebted to IBM Co. for an IBM World Trade Summer Fellowship and the hospitality at the T. J. Watson Research Center in summer 1972, where the program was developed and the initial results were obtained.

References

[1] P. J. Price, Fluctuation Phenomena in Solids, Chap. 8, Ed. R. E. Burgess, Academic Press, New York/London 1965.

[2] W. Fawcett and H. D. Rees, Phys. Letters A **29**, 578 (1969).

[3] A. Alberigi-Quaranta, V. Borsari, C. Jacoboni, and G. Zanarini, Appl. Phys. Letters **22**, 103 (1973).

[4] P. A. Lebwohl, J. appl. Phys. **44**, 1744 (1973).

[5] C. Jacoboni and E. W. Prohofsky, J. appl. Phys. **40**, 454 (1969).

(Received March 18, 1974)

phys. stat. sol. (b) **65**, 67 (1974)

Subject classification: 13.4 and 20; 8; 18; 22.8.1

Department of Physics, D.B.S. College, Dehra Dun, India (a) and
Department of Physics, University of Roorkee, Roorkee, India (b)

Crystal Field Calculations of Nd^{3+} Ions in Scheelites with D_{2d} Symmetry

By

V. KUMAR (a), N. DASS (b), and K. CHANDRA (b)[1])

Crystal field calculations of the temperature dependent principal and mean magnetic susceptibilities, μ_{eff}, and the Schottky specific heat have been considered over the temperature range 10 to 400 °K in $Nd^{3+}:PbMoO_4$ and $Nd^{3+}:CaWO_4$ systems with D_{2d} point symmetry at the crystal ion site. The spectroscopic splitting factors, Curie constants, and the extra entropy associated with Schottky anomaly are compared with available experimental data.

Kristallfeldberechnungen der temperaturabhängigen Haupt- und Mittelwerte der magnetischen Suszeptibilität μ_{eff} und der Schottkyschen spezifischen Wärme wurden im Temperaturbereich 10 bis 400 °K für $Nd^{3+}:PbMoO_4$- und $Nd^{3+}:CaWO_4$-Systeme mit D_{2d}-Punktsymmetrie der Kristallionenplätze untersucht. Die spektroskopischen Aufspaltungsfaktoren, Curiekonstanten und die zusätzlich mit der Schottky-Anomalie verknüpfte Entropie werden mit verfügbaren experimentellen Werten verglichen.

1. Introduction

The optical properties of rare-earth ions in a single crystal of scheelite structure have been of much importance in laser technology. Specially trivalent neodymium (Nd^{3+}) has been studied by many workers [1, 2], because it exhibits good laser action in materials. It is used in single crystals such as YAG which has low threshold and in other materials including glass. When Nd^{3+} is doped in various laser host materials, its tendency is to modify some features like life-time, linewidth, and extra lines, etc. Here in fact we are interested in studying the thermal and magnetic properties like Schottky specific heat, extra entropy associated with Schottky anomaly, principal magnetic susceptibilities and effective magnetic moments in $Nd^{3+}:PbMoO_4$ and $Nd^{3+}: CaWO_4$ systems.

2. Theory

$CaWO_4$ and $PbMoO_4$ are representatives of the scheelites, and crystallize in tetragonal (caa) symmetry corresponding to space group C_{4h}^6. The Ca (or Pb) and W (or Mo) positions are fixed in terms of lattice parameters c and a. The oxygen sites require additional special position parameters (x, y, z) for specification. The rare-earth ions are supposed to be substituted at Ca^{2+} (or Pb^{2+}) sites thus forming optically active magnetic centres. At present it is believed that Coulomb and spin–orbital interactions within the $4f^n$ shall of rare-earth ion is not much affected in different host crystals. Further by substitution of Nd^{3+} ion at the Ca^{2+} (or Pb^{2+}) site, the crystalline field at rare-earth ion remains to be S_4.

[1]) Present address: Institute of Physics, University of Uppsala, Uppsala, Sweden.

In the present calculations we have followed Eremin et al. [3] in the sense that we will consider the crystalline potential in D_{2d} point symmetry. In considering this point symmetry we are justified due to the fact that the S_4 point symmetry is very near to D_{2d} and is attainable only by a very slight displacement of oxygen ions surrounding the Ca^{2+} (or Pb^{2+}) ions. The advantage of D_{2d} symmetry is that the imaginary CEF parameters automatically vanish [4] making the computations simpler.

The complete Hamiltonian of the Scheelite system with D_{2d} point symmetry at the rare-earth ion site may be written as

$$H = H_0 + V_h + V\,, \tag{1}$$

where H_0 is free-ion Hamiltonian. $V_h = g_L \mu_B \bar{H} \cdot \bar{J}$ is the magnetic field operator in which g_L is the Landé splitting factor.

$$V = A_2^0\langle r^2\rangle\, V_2^0 + A_4^0\langle r^4\rangle\, V_4^0 + A_6^0\langle r^6\rangle\, V_6^0 + A_4^4\langle r^4\rangle\, V_4^4 + A_6^4\langle r^6\rangle\, V_6^4\,.$$

The $A_n^m\langle r^n\rangle$ are empirical crystal electric field parameters and V_n^m are the operator equivalents multiplied by coefficients α, β, and γ.

Using the standard operator equivalent technique [5 to 8] and omitting the magnetic field term, the matrices are diagonalized to give eigenvalues (E_i^0) along with eigenvectors. Thus the gram atomic Schottky specific heat is calculated from the relation

$$C_s = \frac{Nk}{Z^2}\left\{Z \sum_{i=1}^{n} \left(\frac{E_i^0}{kT}\right)^2 \exp\left(-\frac{E_i^0}{kT}\right) - \left[\sum_{i=1}^{n} \left(\frac{E_i^0}{kT}\right) \exp\left(-\frac{E_i^0}{kT}\right)\right]^2\right\}, \tag{2}$$

where

$$Z = \sum_{i=1}^{n} \exp\left(-\frac{E_i^0}{kT}\right),$$

and E_i^0 is the zeroth-order energy eigenvalue of the i-th level.

The standard perturbation technique is used to obtain the Zeeman splitting of Stark levels to get E_i^1 and E_i^2 which refer to first- and second-order perturbation energies, respectively. Thus knowing the values of E_i^1 and E_i^2, along with E_i^0 the gram atomic magnetic susceptibility (K_j) is obtained by the van Vleck formula

$$K_j = \frac{N}{Z} \sum_i \left[\frac{(E_i^1)^2}{kT} - 2E_i^2\right]_j \exp\left(-\frac{E_i^0}{kT}\right). \tag{3}$$

The principal magnetic susceptibility K_c ($j = c$) and K_a ($j = a$) and the anisotropy $\Delta K = K_c - K_a$ are obtained from this formula. The effective value of the magnetic dipole moment μ_{eff} is calculated from the relation

$$\mu_{eff} = \left(\frac{3kT\bar{K}}{N}\right)^{1/2},$$

where $\bar{K} = \frac{1}{3}(K_c + 2K_a)$ is the mean gram atomic susceptibility.

3. Calculations

The ground multiplet of the Nd^{3+} ($4f^3$) ion is $^4I_{9/2}$ and the first excited level $^4I_{11/2}$ is about 1800 cm^{-1} higher [5] and therefore the effect of higher multiplets on the ground level may be neglected. The D_{2d} crystalline field splits the tenfold

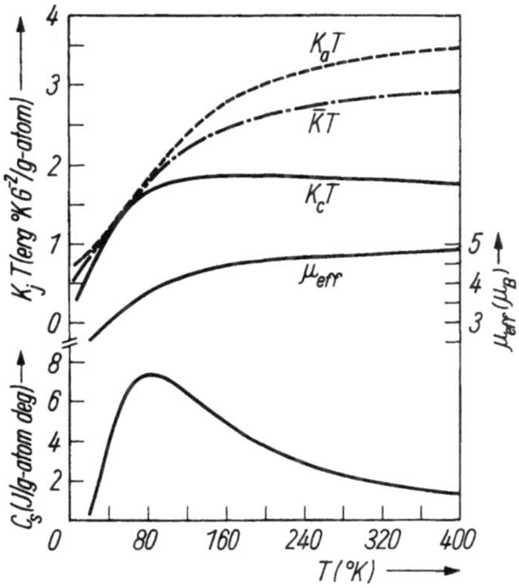

Fig. 1. Calculated Schottky specific heat, effective magnetic moment, principal and mean gram atomic susceptibility curves for $Nd^{3+}:PbMoO_4$

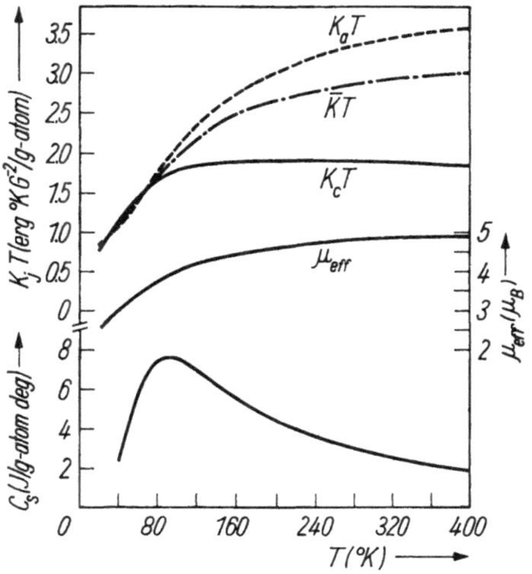

Fig. 2. Calculated Schottky specific heat, effective magnetic moment, principal and mean gram atomic susceptibility curves for $Nd^{3+}:CaWO_4$

ground level into five Kramers doublets, three with crystal quantum number $\pm 1/2$ and two with $\pm 3/2$. The relevant crystal field parameters [9, 10] used in calculations are listed in Table 1.

Table 1

Crystal field parameters for scheelites (in cm^{-1})

system	$A_2^0\langle r^2\rangle$	$A_4^0\langle r^4\rangle$	$A_6^0\langle r^6\rangle$	$A_4^4\langle r^4\rangle$	$A_6^4\langle r^6\rangle$	ref.
$Nd^{3+}:PbMoO_4$	260.0	− 83.0	−10.0	−750.0	−500.0	[8]
$Nd^{3+}:CaWO_4$*)	405.5	−131.12	5.56	1107.53	646.14	[9]

*) The crystal field parameters are converted by using the table given in the Appendix.

In view of the large separation of the excited state from the lower state it seems plausible that the influence of higher multiplets on the ground level may be neglected and the calculations were carried out within the ground manifold J using the mentioned CF parameters. The Stark splittings are obtained as listed in Table 2. Further from these the temperature dependence of the Schottky specific heat and principal magnetic susceptibilities are calculated as reported in Fig. 1 and 2.

4. Discussion

4.1 Thermal properties

From Fig. 1 and Fig. 2 it clearly appears that the Schottky specific heat is maximum around 80 °K with the value 7.43 J $(g\text{-atom})^{-1}$ deg^{-1} for $Nd^{3+}:PbMoO_4$ and at 90 °K with the value 7.61 J $(g\text{-atom})^{-1}$ deg^{-1} for $Nd^{3+}:CaWO_4$. The extra

Table 2

Calculated Stark levels and their corresponding Zeeman level along with spectroscopic splitting factors

ground term	μ	E_i^0 (cm^{-1})	$H \parallel c$		$H \parallel a$		g_c	g_a
			E_i^1*)	E_i^2**)	E_i^1*)	E_i^2**)		
	±1/2	−169.120	±0.8903	−0.0966	±1.8358	−0.0447	1.295	2.670
	±1/2	− 69.712	±1.0559	+0.0800	±1.2519	−0.1867	1.536	1.821
Nd^{3+}:$PbMoO_4$	±3/2	− 14.269	±1.9964	−0.0249	±1.5215	+0.0461	2.902	2.213
	±3/2	56.598	±0.9953	+0.0249	±1.5219	+0.1063	1.447	2.213
$^4I_{9/2}$	±1/2	196.503	±1.6656	+0.0166	±1.9161	+0.0790	2.423	2.787
	±1/2	−212.859	±1.3904	−0.0705	±1.7583	−0.0485	2.022	2.558
Nd^{3+}:$CaWO_4$	±1/2	− 78.009	±1.2682	+0.0592	±1.3061	−0.3719	1.845	1.900
	±3/2	− 53.139	±2.3646	−0.0049	±0.8288	+0.2724	3.439	1.205
$^4I_{9/2}$	±3/2	53.766	±1.3646	+0.0049	±0.8288	+0.0940	1.985	1.205
	±1/2	290.241	±1.3777	+0.0113	±2.0478	+0.0540	2.004	2.979

*) In terms of $\lambda = \mu_B g_L H$.
**) In terms of λ^2.

entropy associated with the Schottky anomaly estimated from the specific heat curve up to 400 °K is 13.78 J (g-atom)$^{-1}$ deg^{-1} for Nd^{3+}:$PbMoO_4$ and 12.61 J (g-atom)$^{-1}$ deg^{-1} for Nd^{3+}:$CaWO_4$. These values are in good agreement with the calculated value 13.4 for the Nd^{3+} free ion, obtained from the entropy expression for a system of doublet ground state, $S = R \ln (J + 1/2)$.

4.2 Magnetic properties

Nd^{3+}:$PbMoO_4$

From Fig. 1 it is clear that $K_a T$ decreases up to 60 °K and there after still decreases but with opposite trend, whereas $K_c T$ slowly increases with the fall of temperature up to 160 °K, and thereafter decreases sharply with decrease of temperature. In the whole temperature range $\bar{K} T$ decreases with fall of temperature whereas $K_a T > K_c T$.

Nd^{3+}:$CaWO_4$

In this system, from Fig. 2 it is clear that $K_a T$ decreases up to 50 °K and thereafter still decreases but with opposite trend. On the other hand $K_c T$ increases with decrease of temperature up to 160 °K and thereafter decreases fastly and continuously with fall of temperature. However, it is interesting to note that in the temperature region 35 to 65 °K, $K_a T < K_c T$ and this may reflect that the magnetic behaviour of Nd^{3+} ion in $CaWO_4$ scheelite is modified as compared with $PbMoO_4$.

The experimental values for the system Nd^{3+}: $PbMoO_4$ for the lowest Stark doublet $g_c = 1.351$ and $g_a = 2.592$ [11] are in good agreement with the calculated values $g_c = 1.295$ and $g_a = 2.670$. Also the experimental values for Nd^{3+}:$CaWO_4$ for the lowest Stark doublet $g_c = 2.034$ and $g_a = 2.535$ [12] are in good agreement with the calculated values $g_c = 2.022$ and $g_a = 2.558$. Further the g-values for both systems corresponding to various Stark doublets are listed in Table 2.

The calculated value of μ_{eff} at room temperature for Nd^{3+}:$PbMoO_4$ is 4.76 μ_B and for Nd^{3+}:$CaWO_4$ is 4.82 μ_B which is in reasonable agreement with the free-ion value 3.62 μ_B as well as 3.49 μ_B for neodymium metal [13]. The magnetic susceptibilities are fitted with the Curie-Weiss law $K_j = C_j/(T + \theta_j)$ and the values of constants so obtained by a least-squares fit in the temperature range $T >$ > 120 °K are given in Table 3. The experimental value of $C = 1.72$ for neodymium metal [14] is in good agreement with $C_c = 1.810$ for Nd^{3+}:$PbMoO_4$ and $C_c = 1.905$ for Nd^{3+}:$CaWO_4$. However, the agreement for $\bar{C}$ is not much satisfactory.

Table 3

Curie-Weiss constants in the temperature range $T > 120$ °K

system	C_a	θ_a	C_c	θ_c	$\bar{C}$	$\bar{\theta}$
Nd^{3+}: $PbMoO_4$	4.455	96.783	1.810	−4.306	3.461	63.909
Nd^{3+}: $CaWO_4$	4.801	118.823	1.905	−0.322	3.684	78.120

It is to be noted that the extra entropy associated with the Schottky specific heat and the spectroscopic splitting factors g_c and g_a are in good agreement with that of the experimental values for both systems as remarked above. However, a slight deviation from experimental values of μ_{eff} and $\bar{C}$ is observed in them. The basic possible reason for such discrepancy may be due to the fact that we have considered the Nd^{3+} site symmetry to be D_{2d} rather than S_4.

Appendix

Conversion table for crystal field parameters [15] used in present study:

$$B_0^2 = 2A_2^0\langle r^2\rangle, \qquad B_4^4 = -\frac{8}{\sqrt{70}} A_4^4\langle r^4\rangle,$$

$$B_0^4 = 8A_4^0\langle r^4\rangle, \qquad B_4^6 = -\frac{8\sqrt{14}}{21} A_6^4\langle r^6\rangle,$$

$$B_0^6 = 16A_6^0\langle r^6\rangle.$$

References

[1] L. F. JOHNSON and R. A. THOMAS, Phys. Rev. **131**, 2038 (1963).
[2] YA. E. KARISS and P. P. FEOFILOV, Optika i Spectroskopiya **17**, 387 (1964).
[3] M. V. EREMIN et al. Soviet Phys. — Solid State **11**, 1967 (1970).
[4] J. L. PARTHER, Nat. Bur. Standards U.S. Monograph (Table IV) **19**, 9 (1961).
[5] K. W. H. STEVENS, Proc. Phys. Soc. **65**, 209 (1952).
[6] R. J. ELLIOT and K. W. H. STEVENS, Proc. Roy. Soc. **A218**, 553 (1953).
[7] R. J. ELLIOT and K. W. H. STEVENS, Proc. Roy. Soc. **A219**, 387 (1953).
[8] B. R. JUDD, Proc. Roy. Soc. **A227**, 552 (1955).
[9] L. YA. SHEKUN, Soviet Phys. — Solid State **6**, 1144 (1964).
[10] L. YA. SHEKUN, Soviet Phys. — Solid State **8**, 1370 (1966).
[11] J. G. GUALTIERI and G. P. DELHERY, J. chem. Phys. **55**, 1541 (1971).
[12] I. N. KURKIN and L. YA. SHEKUN, Soviet Phys. — Solid State **7**, 2308 (1966).
[13] M. S. ABU ELAMAYAN, J. Chem. (U.A.R.) **3**, 75 (1960).
[14] D. R. BEHRENDT, S. LEGVOLD, and F. H. SPEDDING, Phys. Rev. **106**, 723 (1957).
[15] N. KARAYANIS, J. Phys. Chem. Solids **32**, 2385 (1971).

(Received January 7, 1974)

phys. stat. sol. (b) **65**, 73 (1974)

Subject classification: 13; 2; 21.2; 21.3

Institute of Physical Metallurgy and Metal Physics, I. P. Bardin, Central Scientific Research Institute of Ferrous Metallurgy, Moscow

Quasi-Localized Electronic States in the Model Pseudopotential Theory of Disordered Alloys of Simple Metals

By

A. Ya. Belenkii and G. L. Krasko

The energy of disordered alloys with different ion valencies apparently cannot be described in the framework of the pseudopotential perturbation theory because of the strong electron scattering on the long wavelength concentration fluctuations. A model is suggested which provides for the existence of quasi-localized (virtual) electronic states in alloys. A variational procedure of energy calculation is considered to modify the pseudopotential method. The concrete calculations are performed for the Li–Mg system. The two-phase equilibrium, predicted by the theory is discussed.

Энергия неупорядоченного сплава, ионы-компоненты которого имеют разную валентность, повидимому не может быть описана в рамках теории возмущении по псевдопотенциалу из-за сильного рассеяния электронов на длинноволновых флуктуациях концентрации. Рассматривается модель, предполагающая существование в сплаве квазилокализованных (виртуальных) электронных состояний. Предлагается вариационная процедура вычисления энергии, представляющая собой модификацию метода псевдопотенциалов. Выполнены конкретные расчёты для системы Li–Mg. Обуждается предсказываемое теорией двухфазное равновесие в сплаве.

1. Introduction

As is well known, the pseudopotential theory originally formulated for pure non-transition metals, formally may be easily generalized to the case of metallic solid solutions [1, 2]. But when attempting to describe a disordered alloy the theory meets the essential difficulty arising from the use of perturbation theory in the calculation of the contribution to the total energy due to concentration fluctuations [2]. In a completely disordered alloy the total energy may be written as a sum of two terms [1 to 4]:

$$E = \overline{E} + E_{\mathrm{bs}}^{\mathrm{dif}}\,, \tag{1}$$

where $\overline{E}$ the energy of the "average crystal" — a regular crystal lattice with the sites occupied by the "average" ions with average valency $Z = (1 - c)\, Z_{\mathrm{A}} + cZ_{\mathrm{B}}$ and average pseudopotential

$$\overline{w} = (1 - c)\, w_{\mathrm{A}} + cw_{\mathrm{B}} \tag{2}$$

($Z_{\mathrm{A,B}}$ and $w_{\mathrm{A,B}}$ the valencies and the screened pseudopotentials form factors of ions A and B, respectively), and

$$E_{\mathrm{bs}}^{\mathrm{dif}} = c(1 - c) \int \mathrm{d}^3\boldsymbol{q}\, |\Delta w(\boldsymbol{q})|^2\, f(\boldsymbol{q}) \tag{3}$$

is the contribution to the band structure energy of the component concentration fluctuations; if there is no short-range order the similar contribution to the electrostatic energy $\sim\Delta Z^2$ ($\Delta Z = Z_A - Z_B$) exactly vanishes [5] (in (3) $\Delta w = w_A - w_B$ and $f(q)$ the function characterizing the self-consistent electronic interaction; see e.g. [1, 2]).

The integration in (3) goes over all the wave-number space, including the region of small $\boldsymbol{q}$, where $\Delta w/\varepsilon_F$ (ε_F the Fermi energy of the alloy) — the perturbation theory quantity — generally is not small: at $\boldsymbol{q} \to 0$ $\Delta w \to (-2/3)\, \varepsilon_F\, \Delta Z/Z$ and the integral (3) may happen to be of the order of ε_F. The latter means that in this case the pertubation series most probably would diverge.

If the valencies of ions A and B are equal ($\Delta Z = 0$) no difficulties actually appear and the contribution to the energy (3) is correct. In [2] the supposition was claimed that if $\Delta Z = 1$ the perturbation theory was still valid. However, perhaps even this case is doubtful. In [4] the system Li–Mg has been considered. It has been shown that the quantity (2) at about all the concentrations, though much smaller than ε_F, is of the same order as the band structure energy of the "average" crystal. It is clear from the physical arguments that the energy of fluctuations in a stable system can hardly be comparable to the average energy. Indeed, the energy contribution (3) is overestimated, giving rise to a significant deviation of calculated and experimental equilibrium lattice spacings [4].

From the physical point of view the main cause of the difficulty lies in the fact that in zero approximation the conduction electron density distribution in disordered alloy is assumed to be a uniform one. Indeed, though the electroneutrality condition in an alloy does hold "on the average", but each of the crystal cells occupied by ion A or B is not electrically neutral. The charge excess and lack equal respectively to $Z_B - Z$ and $Z - Z_A$ ($Z_B > Z_A$), with the loss in electrostatic energy (compared to the case of all cells electrically neutral) being equal approximately to $1.8c\,(1 - c)\,\Delta Z^2/r_0$ (r_0 the Wigner-Seitz cell radius: $4/3\pi r_0^3 = \Omega_0$; Ω_0 atomic volume)[1]). In reality an exceeding charge must be screened up by electron gas; as a result most of the energy loss mentioned would be eliminated, but the electron density distribution would become strongly non-uniform. The charge changes of each cell would not be small (actually of the order of an ion charge), that is it would hardly be obtained in finite order perturbation theory.[2])

To avoid the difficulty one may assume — already in the zeroth approximation — the existence in the alloy of some non-uniform electron distribution. Such a distribution may, in principle, arise when adding to the solvent the atoms with valency greater than that of the host atoms. Part (equal to ΔZ) of electrons of the atom solvent does not go to the conduction band but stays localized in its own cell. If the valency of host atoms is greater some non-occupied localized electronic states ("the holes") — positively charged — may arise in the impurity cells.

The possibility that such localized (precisely speaking quasi-localized) states in impurity cells may appear has been widely discussed in literatur (see e.g. [9]).

---

1) We use the atomic units with Rydbergs for energy.

2) In ordered alloys (as well as intermetallic compounds) the screening also leads to a non-uniformity in the charge distribution, i.e. of "charge on bond" type (see [6]), but those inhomogeneities are "ordered" ones (the fluctuations in their distribution are small) and may be described when using the third and some terms of the fourth order perturbation theory [7, 8].

Particularly the detailed analysis was performed by Friedel [10]. The peculiarities in the density of states curves characteristic of quasi-localized electronic states seem to be observed experimentally [11, 12].

Thus it seems reasonable to assume the existence of some quasi-localized, virtual energy levels inside the conduction band of the alloy. Strictly speaking the electron energy spectrum should be found by means of a self-consistent calculation rather than postulated a priori. In that case, however, we would inevitably meet all the difficulties of many-electron theory at metallic densities. Moreover, as has been pointed out by Friedel [10] it is just in this case that the self-consistent field theory becomes most doubtful. Therefore the purpose of the present paper is to consider a model problem of calculating the total energy of the disordered alloy (provided the valencies of atom-components are different) in the light of what was said above about the essential inhomogeneity in the electron density distribution. One may hope that the pseudopotential method properly modified would still be a convenient means of investigations.

Of course the pseudopotential formulation of the alloy problem should be changed now. Particularly, the pseudopotential acting on a collectivized electron should be constructed, for instance, with the quasi-localized (virtual) states treated like the states of the ion core.

Thus, say the Mg ion dissolved in Li should be considered as "quasi-one-valent"; as a result the alloy problem of different ion valencies would be reduced to the equal-valencies one. In the latter case, as was mentioned above, the "difference" term $E_{\text{bs}}^{\text{dif}}$ in (1) is rather small and the alloy may be satisfactorily described already in the "average crystal" approximation.

It should be noted, however, for that a quasi-localized virtual state not being a "genuine" impurity one, its energy is located inside the conduction band and plane wave-like wave functions do admix to the state, the energy level itself being widened. The quasi-localized state in turn admix to the plane-wave states giving a contribution to the density of states of conduction electrons. In other words, the "hybridization" of the states take place. In the present paper this effect will be neglected. At the same time, as it will be seen, even such a simple approximation enables us to rather reasonably describe a disordered alloy.

In Section 2 we present the variational procedure of energy calculation in the framework of local model pseudopotentials theory. As an illustration a concrete disordered system — the Li–Mg one — is considered in Section 3. In Section 4 the phase equilibrium of an alloy in the quasi-localized states model is briefly discussed.

2. The Variational Approach

As was shown in [15] the assumption that a virtual level is energetically favourable leads to a rather reasonable estimation of its location in the conduction band. This may be easily done in the so-called "zero-order model" [14]. Now we will try to more rigorously compute the alloy energy on the basis of pseudopotential theory.

Let us assume again that there are $\Delta Z = Z_{\text{B}} - Z_{\text{A}}$ electrons localized on virtual levels near a B ion with greater valency (we will call these "l-electrons"). We suppose for simplicity $\Delta Z = 1$ and assume the wave function of l-electron to be $\psi_{\text{l}}(r)$. With that trial function we will compute the total energy of the system and then minimize it over the variational parameters of the function $\psi_l(r)$.

Obviously the remainder Z_A electrons per atom B are almost free and may fill the plane-wave states. The energy of the homogeneous electrons gas of density Z_A/Ω_0 including the exchange-correlation effects (E_0) may be easily written (see [1, 2]).

First we find E_{es}, the energy of direct electrostatic interaction of ions (more precisely, the energy of cells, containing the ions and l-electrons). Let the charge densities of ions A, B, and l-electrons be respectively $\varrho_A(\boldsymbol{r})$, $\varrho_B(\boldsymbol{r})$, and $\varrho_l(\boldsymbol{r}) = |\psi_l(\boldsymbol{r})|^2$. Then the total charge density in the alloy is

$$\varrho(\boldsymbol{r}) = \sum_{\boldsymbol{R}} \varrho_A(\boldsymbol{r}-\boldsymbol{R})\,(1-C_{\boldsymbol{R}}) + \sum_{\boldsymbol{R}} [\varrho_B(\boldsymbol{r}-\boldsymbol{R}) - \varrho_l(\boldsymbol{r}-\boldsymbol{R})]\,C_{\boldsymbol{R}} - \frac{Z_A}{\Omega_0} \tag{4}$$

($C_{\boldsymbol{R}}$ is the random variable, equal 0 or 1 depending on whether the site $\boldsymbol{R}$ is occupied by an ion A or B), and one may easily find:

$$E_{es} = E_{es}^A + \frac{1}{\Omega_0} \sum_{\boldsymbol{q}}' \frac{4\pi}{q^2} \varrho_A^*(\boldsymbol{q})\,[\Delta\varrho(\boldsymbol{q}) - \varrho_l(\boldsymbol{q})]\,\frac{1}{N^2} \sum_{\boldsymbol{R},\boldsymbol{R}'} C_{\boldsymbol{R}}\, e^{i\boldsymbol{q}(\boldsymbol{R}-\boldsymbol{R}')} + $$
$$+ \frac{1}{2\Omega_0} \sum_{\boldsymbol{q}}' \frac{4\pi}{q^2} [\Delta\varrho(\boldsymbol{q}) - \varrho_l(\boldsymbol{q})]^2\, \frac{1}{N^2} \sum_{\boldsymbol{R},\boldsymbol{R}'} C_{\boldsymbol{R}} C_{\boldsymbol{R}'}\, e^{i\boldsymbol{q}(\boldsymbol{R}-\boldsymbol{R}')} \tag{5}$$

$$\Delta\varrho = \varrho_B - \varrho_A$$

with E_{es}^A being the electrostatic energy of pure A metal, N the total number of ions in the crystal, and $\varrho_{A,B,l}(\boldsymbol{q})$ the Fourier transforms of the $\varrho_{A,B,l}(\boldsymbol{r})$. Note that in all the contributions in (5) but those corresponding to l-electron–ion interaction the terms with $\boldsymbol{R} = \boldsymbol{R}'$ must be omitted.

For an absolutely disordered ion configuration (with no long or short range order) the averaging of (5) is straightforward (see, e.g. [1, 2, 5]).

Before writing down the final expression we note that in the framework of the traditional pseudopotential theory [1, 2] we should regard the A and B ions as point ones, i.e. $\varrho_A(\boldsymbol{q}) = Z_A$, $\varrho_B(\boldsymbol{q}) = Z_B$ when calculating their interaction with each other and the negative uniform background; at the same time one has to take into consideration the fact that an l-electron in its own cell does interact with the non-point pseudo-ion B. Consequently in the corresponding term of (5) one should substitute $\varrho_B(\boldsymbol{q}) = (\Omega_0/4\pi)\, q^2 w_B^0(\boldsymbol{q})$, $w_B^0(\boldsymbol{q})$ being a "bare" ion B pseudopotential form factor.

Now, one may write down the final expression for the electrostatic energy of a disordered alloy:

$$E_{es} = E_{es}^A + \frac{c}{\Omega_0} \sum_{\boldsymbol{q}}' \frac{4\pi}{q^2} \varrho_A(\boldsymbol{q})\,[\Delta\varrho(\boldsymbol{q}) - \varrho_l(\boldsymbol{q})]\,|S_{\boldsymbol{q}}|^2 -$$
$$- \frac{c}{(2\pi)^3} \int d^3\boldsymbol{q}\, \frac{4\pi}{q^2} \varrho_A(\boldsymbol{q})\,[\Delta\varrho(\boldsymbol{q}) - \varrho_l(\boldsymbol{q})] +$$
$$+ \frac{c^2}{2\Omega_0} \sum_{\boldsymbol{q}}' \frac{4\pi}{q^2} [\Delta\varrho(\boldsymbol{q}) - \varrho_l(\boldsymbol{q})]^2\,|S_{\boldsymbol{q}}|^2 - \frac{c^2}{2(2\pi)^3} \int d^3\boldsymbol{q}\, \frac{4\pi}{q^2} [\Delta\varrho(\boldsymbol{q}) - \varrho_l(\boldsymbol{q})]^2 \tag{6}$$

(here $S_{\boldsymbol{q}} = (1/N) \sum_{\boldsymbol{R}} \exp(i\boldsymbol{q}\boldsymbol{R})$ the structure factor of the lattice).

The next step is to calculate the energy of "indirect" interaction, i.e. the band structure energy of conduction electrons in the field of the effective (including l-electrons) pseudopotentials. The effective pseudopotential of a B cell may be written as

$$\tilde{w}_{\rm B}^0(\boldsymbol{q}) = w_{\rm B}^0(\boldsymbol{q}) - \frac{4\pi}{\Omega_0 q^2}\, \varrho_{\rm l}(\boldsymbol{q}) \,. \tag{7}$$

Then like in the usual pseudopotential theory

$$E_{\rm bs} = \bar{E}_{\rm bs} + E_{\rm bs}^{\rm dif} , \tag{8}$$

where $\bar{E}_{\rm bs}$ is defined through the renormalized average pseudopotential

$$\bar{\tilde{w}}^0 = w_{\rm A}^0(1-c) + \tilde{w}_{\rm B}^0 c \tag{9}$$

and $E_{\rm bs}^{\rm dif}$ through

$$\Delta\tilde{w}^0 = \tilde{w}_{\rm B}^0 - w_{\rm A}^0 \,. \tag{10}$$

Note that at $\boldsymbol{q} \to 0$ the quantity (10) vanishes; therefore the difficulty with the integration in (3) over all the reciprocal space (pointed out in Section 1) does not arise. Thus the band structure energy $E_{\rm bs}$ in this model is described by the usual relations of the second order perturbation theory for a disordered alloy [1, 2] with pseudopotentials (9), (10). As for the first order contribution — the non-Coulombic part of the electron–ion interaction — it is determined like in the usual theory:

$$E_1 = 2Z_{\rm A} \lim_{\boldsymbol{q}\to 0} \left[(1-c)\, w_{\rm A}^0(\boldsymbol{q}) + c w_{\rm B}^0(\boldsymbol{q}) + \frac{4\pi}{\Omega_0}\frac{Z}{q^2} \right]. \tag{11}$$

Now the only contribution left is one of the kinetic ($E_{\rm l}^{\rm kin}$) and conduction–l-electrons exchange ($E_{\rm l}^{\rm ex}$) energies. The former equals to

$$E_{\rm l}^{\rm kin} = c \int {\rm d}^3\boldsymbol{r}\, |\nabla\psi_{\rm l}(\boldsymbol{r})|^2 . \tag{12}$$

Equation (12) is valid only if the overlapping of the function $\psi_{\rm l}(\boldsymbol{r})$ on different B ions may be neglected, i.e. at $c \ll 1$. Below (in Section 4) one will see this approximation might be reasonable.

The exchange energy will be determined following [16] on approximation similar to the Slater one:

$$E_{\rm l}^{\rm ex} = -2Z_{\rm A} \int {\rm d}^3\boldsymbol{r} \left\{ \left[\frac{3}{8\pi} \left(\varrho_{\rm l}(\boldsymbol{r}) + \frac{Z_{\rm A}}{\Omega_0} \right) \right]^{1/3} - \left[\frac{3}{8\pi} \frac{Z_{\rm A}}{\Omega_0} \right]^{1/3} \right\}. \tag{13}$$

Finally the total energy of the disordered alloy may be written as

$$E = E_0 + E_1 + E_{\rm bs} + E_{\rm es} + E_{\rm l}^{\rm kin} + E_{\rm l}^{\rm ex}. \tag{14}$$

If $\Delta Z > 1$ the Coulomb and exchange energies of l-electrons localized on the same B ion must be added.

Now the energy (14) should be minimized over the variational parameters to determine the equilibrium energies and corresponding atomic volumes of different crystal structures.

3. The System Li–Mg

In this section we present the results of a variational calculation for the alloys Li–Mg. As a pseudopotential of "bare" ions we used the local model pseudopotential suggested in [17] with the form factor

$$w^0(q) = \frac{4\pi Z}{\Omega_0 q^2} \frac{(2a-1)(qr_c)^2 - 1}{[(qr_c)^2 + 1]^2}. \tag{15}$$

The pseudopotential (15) was earlier successfully used in the analysis of Vegard's rule and the deviations from it in the disordered alloys of alkali metals [3] as well as in the investigations of the dependence of atomic volume on concentration and the singularities (similar to Kohn's ones in phonon spectra) in alloys with different ion valencies (using the "average crystal" approximation) [4, 18]. The values of the parameters a and r_c were chosen to satisfy on one hand the Schrödinger equation for the ground state of an electron in the field of a single ion provided the self-energy is equal to its experimental value (see [18]) and, on the other hand, to satisfy the equation of state at $T = 0$, i.e. the requirement of equality the experimental and theoretical values of equilibrium atomic volumes of pure metals. Table 1 presents the values of Z, a, r_c, and Ω_0 for pure Li and Mg metals.

Table 1

ion	Z	a	r_c	Ω_0
Li	1	5.954	0.365	143.9
Mg	2	2.933	0.393	154.0

For the description of an l-electron in the vicinity of Mg ion we have chosen the trial function:

$$\psi_l(r) = Ar\, e^{-r/2r_l}, \tag{16}$$

where r_l is the variational parameter (it characterizes the size of the region of electron localization) and A the normalization factor. The charge density and its Fourier transform are

$$\varrho_l(r) = A^2 r^2\, e^{-r/r_l}; \qquad \varrho_l(\boldsymbol{q}) = \Delta Z \frac{1 - (qr_l)^2}{[1 + (qr_l)^2]^4}. \tag{17}$$

As one may see from (16) and (17) the probability for an l-electron to be found at small r goes to 0 with the maximum at $r = 2r_l$. Note the distribution (17) is normalized: $\varrho_l(q = 0) = \Delta Z$.

Following the procedure described in Section 2 we minimized the total energy (14) with respect to r_l and Ω_0 at fixed c. As a result for a given crystallographic modification (b.c.c., f.c.c., and h.c.p.) the equilibrium values $\Omega_0(c)$ and corresponding energy (14) were found. In a close analogy the calculations were performed for the alloys with low Li concentrations, i.e. those with localized holes states (the trial function for the latter was also chosen as in (16)).

As the calculations showed the variational parameter r_l for both l-electrons and l-holes appeared to depend on c rather weakly, the corresponding values being 0.62 and 0.41. These quantities seem to be quite reasonable ones (though, perhaps somewhat lowered): $2r_l$, the distance where maximum (minimum) of l-charge densities takes place, is greater than the characteristic core size ($r_c \approx 0.4$) at the same time being much smaller than the atomic radius ($r_0 \approx 3.0$).

Fig. 1. The reduced energy $E^* = E - E_{\mathrm{Mg}} + E_{\mathrm{Li}}$ versus concentration curves for the Li–Mg system. (1) the usual pseudopotential theory, "average crystal" approximation (b.c.c. lattice); (2) l-electrons present (b.c.c. lattice); (3) l-holes present (h.c.p. lattice)

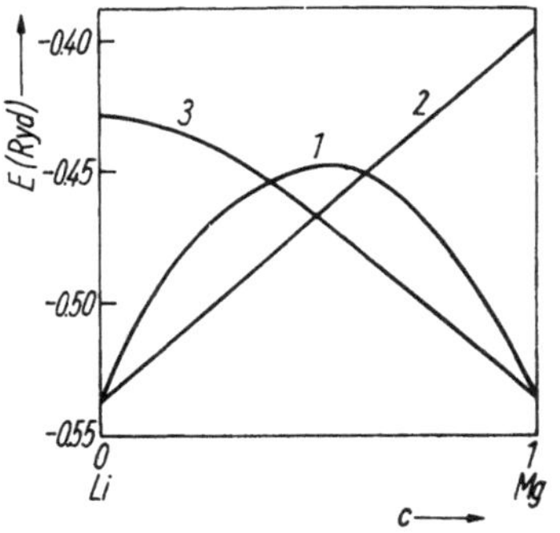

Fig. 1 shows the dependence of the reduced total energies of the Li–Mg alloys

$$E^* = E(c) - E_{\mathrm{Mg}} + E_{\mathrm{Li}}$$

($E_{\mathrm{Mg,Li}}$ energies of corresponding pure metals) on Mg concentration c. Curve 1 is the energy of the average crystal $\bar{E}(c)$ (b.c.c. structure) calculated using the usual pseudopotential scheme; two other curves correspond to alloys with quasi-localized states in the crystallographic modifications of the lowest energy: (2) l-electrons (b.c.c. structure), (3) l-holes (h.c.p. structure). As one may see in the regions of small ($c \to 0$) and large ($c \to 1$) concentration most energetically favourable are l-electrons and l-holes containing alloys, respectively; on the other hand, as was assumed in Section 2, the quasi-local states are not advantageous in pure Li and Mg metals. The "average crystal" loses in energy at all the concentrations.

4. The Model of Quasi-Local States and Phase Equilibrium in Alloys

As can already be seen from the simplified "zero-order" model [15] the free-energy curves of the alloys with l-electrons and l-holes must intersect. Really, at $c \to 0$ $\partial E_{\mathrm{e}}/\partial c < \partial \bar{E}/\partial c$; $E_{\mathrm{h}} > \bar{E}$; on the other hand, at $c \to 1$ $E_{\mathrm{e}} > \bar{E}$; $\partial E_{\mathrm{h}}/\partial c < \partial \bar{E}/\partial c$ ($E_{\mathrm{e}}(c)$ and $E_{\mathrm{h}}(c)$ are the energies of l-electrons and l-hole alloys, respectively). Now considering the free energy $F = E - TS$ one may see that a common tangent may be drawn between the curves $F_{\mathrm{e}}(c)$ and $F_{\mathrm{h}}(c)$. The latter means that at some concentrations c_α and c_β a one-phase solid solution becomes thermodynamically non-stable: at $c_\alpha < c < c_\beta$ a two-phase region $(\alpha + \beta)$ arises with l-electrons in the B atom cells in α-phase and l-holes in A atom cells in β-phase.

One may write the simplest approximation for the configurational entropy

$$S = -[c \ln c + (1 - c) \ln (1 - c)]$$

and easily obtain for low temperatures the expressions for the phase boundaries:

$$c_\alpha = \exp\left[-\frac{1}{T}\left(E_{\mathrm{e}}(1) - E_{\mathrm{h}}(1)\right)\right], \quad 1 - c_\beta = \exp\left[-\frac{1}{T}\left(E_{\mathrm{h}}(0) - E_{\mathrm{e}}(0)\right)\right], \quad (18)$$

where T is the absolute temperature in energy units.

Fig. 2 shows the free energy and corresponding "phase diagram". As already follows from (18) at $T \to 0$ $c_\alpha \to 0$ and $c_\beta \to 1$. At a finite temperature c_α is finite and c_β is close to 1. This justifies the approximation (12) taken in Section 2,

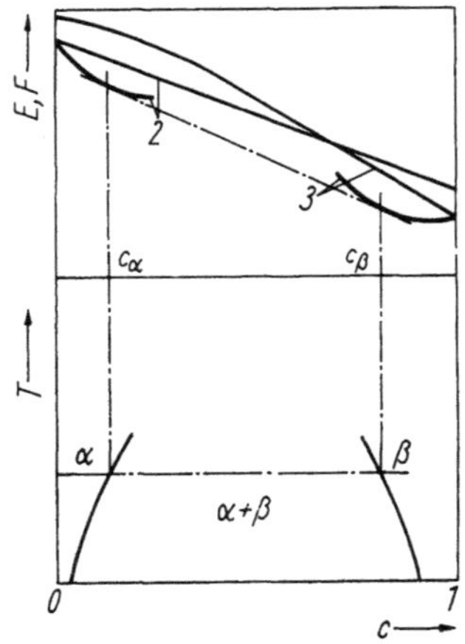

Fig. 2. Above: the internal (thick lines) and free energies (thin lines) versus concentration at a temperature T (schematic); the marks 2 and 3 meen the same as on Fig. 1. Below: the corresponding phase-equilibrium diagram; c_α and c_β are limiting concentrations of α and β phases

i.e. the fact that the overlapping of l-states at different ions may be considered as negligible, since in the region $c_\alpha < c < c_\beta$ the energy is a weighted sum of the energies at the borders c_α and c_β.

Note that the appearance in the alloy with quasi-localized states of a two-phase equilibrium means on alloy decomposition with the phases α and β being enriched by components A and B, respectively. As for the Li–Mg system it really contains the two-phase region [20].

5. Conclusion

We have considered one of the possible models of electronic structure in disordered alloy which allows to overcome the difficulty mentioned in Section 1. The most essential defect of the model is perhaps that the hybridization of plane wave-like and quasi-localized states is not taken into account. Actually the two types of states were being considered separately, particularly, the density of near-free electrons was taken independent of concentration and equal to Z_A/Ω_0. As a result, for example, the Kohn singularities on the $\Omega_0(c)$ curve vanish, while in the usual theory one may get them already in the "average crystal" approximation [4, 18].

The theory of alloys might in principle be built in close analogy with the pseudopotential theory for transition metals [13], the role of d-states being played by quasi-localized states. At low B ions concentration the latter are "similar" to localized sp-states in a single B ion of valency ΔZ, so these are the ion states that may be used (together with the core states) in the orthogonalization procedure of plane waves. The expansion of $\psi_l(r)$ in the thus obtained "overfilled" system of functions might be done just as in [13]. The energy calculation (including the hybridization effects) would be performed likewise, the corresponding ionization potential of a single B ion playing the role of the quantity E_d^0 in [13].

Less obvious is the use of theory [13] to describe quasi-localized "holes" on the ions A, since in pure A metal there are no similar states.

At the same time, as may be seen from Sections 2 and 3, even the simplified model suggested enables us to describe in a reasonable way disordered alloys; particularly, the correct crystal structures were found for the Li–Mg system

In conclusion it should be noted that the creation of the two-phase l-electron–hole region in the electron subsystem resulting in a real two-phase equilibrium may have some close relation to the processes of phase boundaries formation in the real alloys of simple metals. Of great interest is the fact that the so-called "primary solid solutions" homogeneity regions in the latter are as a rule rather narrow ($c_\alpha \to 0$, $c_\beta \to 1$) [20] which matches the concept of l-electron–hole decomposition considered above. The question however needs further investigations.

Acknowledgement

The authors are grateful to Prof. Yu. Kagan for fruitful discussion.

References

[1] W. A. Harrison, Pseudopotentials in the Theory of Metals, W. A. Benjamin, Inc., New York 1966.
[2] V. Heine and D. Weaire, Solid State Phys., **24**, 249 (1970).
[3] G. L. Krasko and Z. A. Gurskii, Fiz. tverd Tela **14**, 321 (1972).
[4] A. Ya. Belenkii, Z. A. Gurskii, and G. L. Krasko, Fiz. tverd Tela **15**, 3473 (1973).
[5] G. L. Krasko, Zh. eksper. teor. Fiz., Pisma **13**, 218 (1971).
[6] J. C. Phillips, Covalent Bonding in Crystals, Molecules and Polymers, University Press, Chicago 1970.
[7] J. Inglesfield, J. Phys. C **4**, 1003 (1971).
[8] G. L. Krasko and A. B. Makhnovetskii, Fiz. tverd. Tela **15**, 3114 (1973).
[9] C. Kittel, Quantum Theory of Solids, Wiley, New York 1963.
[10] J. Friedel, Phil. Mag., Suppl. **3**, 446 (1954).
[11] G. A. Rooke, in: Soft X-Ray Band Spectra, Ed. D. Fabian, Academic Press, New York 1968.
[12] A. Appelton and C. Curry, Phil. Mag. **12**, 245 (1965).
[13] W. A. Harrison, Phys. Rev. **181**, 1036 (1969).
[14] E. G. Brovman and Yu. Kagan, Uspekhi fiz. Nauk **112**, 369 (1974).
E. G. Brovman and Yu. Kagan, in: Lattice Dynamics, Ed. A. A. Maradudin and G. K. Horten, North-Holland Publ. Co., Amsterdam 1974.
[15] A. Ya. Belenkii and G. L. Krasko, Dokl. Akad. Nauk SSSR **217**, 48 (1974).
[16] J. Moriarty, Phys. Rev. B **8**, 1338 (1973).
[17] G. L. Krasko and Z. A. Gurskii, Zh. eksper. teor. Fiz., Pisma **9**, 596 (1969).
[18] A. Ya. Belenkii and G. L. Krasko, Fiz. tverd. Tela **16**, 1248 (1974).
[19] Z. A. Gurskii and G. L. Krasko, Dokl. Akad. Nauk SSSR **197**, 810 (1972).
[20] M. Hansen and K. Anderko, Constitution of Binary Alloys, McGraw-Hill, New York 1958.

(Received June 26, 1974)

phys. stat. sol. (b) **65**, 83 (1974)

Subject classification: 6 and 8

Institute of Physics, Academy of Sciences of the Estonian SSR, Tartu

Theory of Lattice Thermal Conductivity of Anharmonic Crystals with Impurities

II. Interference Effects

By

V. I. ALTUKHOV and G. S. ZAVT

The interference of the anharmonic and impurity scattering of phonons is investigated. In particular, a general formula is obtained for the resonant scattering of phonons by impurities taking into account the inelastic channel which is responsible for the temperature dependence of the total cross-section of impurity scattering. It is found that the anharmonic scattering is decreased by interference effects. The latter is connected with the finite lifetime of the phonons scattered by impurities. The case of resonant scattering of phonons by isotopic defects is considered in detail.

Рассмотрена интерференция ангармонического и примесного рассеяния фононов. В частности, получена общая формула резонансного рассеяния фононов на примесях с учетом неупругого канала рассеяния, который отвечает за температурную зависимость полного сечения примесного рассеяния. Найдено, что ангармоническое рассеяние убывает из-за эффектов интерференции. Последнее связано с конечным временем жизни фононов, релаксирующих на примесях. Подробно рассматривается случай резонансного рассеяния фононов на изотопическом дефекте.

1. Introduction

The scattering of phonons by impurities is commonly considered as independent of other scattering processes, in particular, of phonon–phonon collisions. On the other hand, treating the latter processes, the presence of impurities is usually ignored. The inverse relaxation rates, obtained in such a way, are then summed up (Matthiessen rule). This means that the mutual influence of these processes is absent except for the change of their relative strength with the change of phonon frequency and temperature.

One can point out the following defects of this simple picture. The first one is connected with the special role of phonon–phonon normal (N) processes. As it has been first shown by Peierls [1], these processes themselves do not lead to finite thermal resistance but play a significant role in the redistribution of the phonon momentum and make an essential contribution to the thermal resistance in combination with the resistive (R) processes [2 to 6].

The second defect is connected with the fact that the summing of the inverse relaxation times means the summing of the squared matrix elements of the collision integral describing the separate processes, whereas one ought actually to sum the matrix elements themselves. In terms of perturbation theory the interference terms appear in the orders higher than the first one. In such an approach Carruthers [7] has studied the corrections to the three-phonon processes due to the mass-defect impurities in the second order of the perturbation theory. The modification of three-phonon processes in the Carruthers theory is caused by the impurity-conditioned violation of the quasi-momentum conservation rule. It is necessary to mention also the paper by Pomeranchuk [8] who has studied the phonon–phonon collision processes in the vicinity of an impurity, and the paper by Ivanov and Krivoglaz [9] which is devoted to the study of the phonon attenuation caused by the change of anharmonic force constants by impurities.

In paper [10] (hereafter cited as I), the general theory of the lattice thermal conductivity for the disordered anharmonic crystal was developed and the general formula which describes the simultaneous action of anharmonic and impurity mechanisms of phonon scattering was obtained. Proceeding from this formula we consider here the interference processes connected with the fact that phonons, scattered by impurities, have anharmonic widths and shifts and, on the other hand, with the fact that the phonon–phonon interaction is modified by impurity scattering.

2. τ-Approximation

Starting from expressions (40) and (41) in I and taking into account that $\langle \operatorname{Im} D(\omega)\rangle_c$ has a maximum in the vicinity of $\omega_c^2 = E_\nu^2(\omega_c)$, we obtain

$$\left.\begin{aligned} \langle K(T)\rangle_c &= \frac{1}{3\Omega} \sum_\nu v_\nu^2 C(\omega_c)\, \tau_\nu(\omega_c)\,, \\ \tau_\nu^{-1}(\omega_c) &= 2Z_\nu^{-1}(\omega_c)\,[\gamma_\nu(\omega_c) + \Gamma_\nu(\omega_c)]\,, \\ Z_\nu^{-1}(\omega_c) &= \frac{\omega_\nu}{\omega_c}\left(1 - \left.\frac{\partial E_\nu(\omega)}{\partial \omega}\right|_{\omega=\omega_c}\right), \end{aligned}\right\} \tag{1}$$

where

$$2\omega_\nu \Gamma_\nu(\omega) = \frac{\hbar}{2\pi} \sum_{\nu_1\ldots\nu_4} \int\!\!\int_{-\infty}^{\infty} d\omega'\, d\omega''\, (n' + n'' + 1)\, \delta(\omega' + \omega'' - \omega) \times$$

$$\times \langle \Phi_{\nu\nu_1\nu_4} \Phi_{\nu_2\nu_3\nu} \operatorname{Im} G_{\nu_1\nu_2}(\omega') \operatorname{Im} G_{\nu_3\nu_4}(\omega'')\rangle_c \tag{2}$$

and

$$2\omega_\nu\gamma_\nu(\omega) = c \operatorname{Im} \lambda_\nu(\omega) = 3N \operatorname{Im} \sum_{LL_1L_2} \overline{w}_\nu(L)\, V(LL_1)\, (T - 1)^{-1}\, (L_1L_2)\, w_\nu(L_2) \tag{3}$$

(see notations in I).

The formulas (2) and (3) describe the interference of the anharmonic and impurity scatterings.

3. Three-Phonon Interaction

In the one-particle approximation (see I)

$$\langle \Phi\Phi \operatorname{Im} G \operatorname{Im} G\rangle_c \approx \langle \Phi\Phi\rangle_c \langle \operatorname{Im} G\rangle_c \langle \operatorname{Im} G\rangle_c\,,$$

and according to [11] in harmonic approximation

$$\langle \operatorname{Im} G_{\nu\nu'}(\omega)\rangle_c = \frac{\delta_{\nu\nu'} Z_\nu(\omega_i)}{2\omega_\nu}\left[\frac{Z_\nu\gamma_\nu^0(\omega_i)}{(\omega - \omega_i)^2 + (Z_\nu\gamma_\nu^0(\omega_i))^2} - \frac{Z_\nu\gamma_\nu^0(\omega_i)}{(\omega + \omega_i)^2 + (Z_\nu\gamma_\nu^0(\omega_i))^2}\right], \tag{4}$$

where

$$\omega_i^2 = \varepsilon_\nu^2(\omega_i) = \omega_\nu^2 - c \operatorname{Re} \lambda_\nu^0(\omega_i) ,$$

$$Z_\nu^{-1}(\omega_i) = \frac{\omega_\nu}{\omega_i}\left(1 - \frac{\partial \varepsilon_\nu(\omega)}{\partial \omega}\bigg|_{\omega=\omega_i}\right).$$

Substituting (4) into (2) and after integrating over ω' and ω'' [12], we obtain

$$\Gamma_\nu(\omega) = \frac{\hbar}{2} \sum_{\nu_1\nu_2} \frac{\langle|\Phi_{\nu\nu_1\nu_2}|^2\rangle_c Z_1 Z_2}{2\omega_\nu\omega_1\omega_2} \left[\frac{2(n_1 - n_2)\,\gamma_{12}}{(\omega_1 - \omega_2 + \omega)^2 + \gamma_{12}^2} \pm \frac{(n_1 + n_2 + 1)\,\gamma_{12}}{(\omega_1 + \omega_2 \mp \omega)^2 + \gamma_{12}^2}\right]. \tag{5}$$

Here ω_1 denotes ω_{ν_1}, $\omega_2 = \omega_{\nu_2}$; $n_1 = n(\omega_1)$, $Z_1 = Z_{\nu_1}$; $\gamma_{12} = Z_1\gamma_1^0 + Z_2\gamma_2^0$. The first term with $(\omega_1 - \omega_2 + \omega)$ describes the collision of the phonons ω and ω_1 with the creation of the phonon ω_2; the second term with $(\omega_1 + \omega_2 - \omega)$ describes the decay of the phonon ω into the phonons ω_1 and ω_2.

The average $\langle|\Phi_{\nu\nu_1\nu_2}|^2\rangle_c$ may be written as

$$|\Phi^0_{\nu\nu_1\nu_2}|^2\, \Delta(\boldsymbol{k} + \boldsymbol{k}_1 + \boldsymbol{k}_2) + [\langle|\Phi_{\nu\nu_1\nu_2}|^2\rangle_c - |\Phi^0_{\nu\nu_1\nu_2}|^2\, \Delta(\boldsymbol{k} + \boldsymbol{k}_1 + \boldsymbol{k}_2)] .$$

Then $\Gamma_\nu(\omega)$ is the sum of two terms. The second term describes the scattering of phonons due to $\Delta\Phi_{\nu\nu_1\nu_2}$ and gives the formula obtained in [9], if $\gamma_{12} \to 0$. The first term describes the collision of the phonons ω_1 and ω_2 with the account of the finite lifetimes of those phonons, $(2\gamma_1)^{-1}$ and $(2\gamma_2)^{-1}$, caused by their scattering by impurities. This term has a complicated dependence on ω, T, and through γ_{12}, also on c, ε, and $\Delta\Phi(LL')$. In spite of this one can obtain a number of general regularities in the behaviour of $\Gamma_\nu(\omega)$ using the model of an isotropic solid with dispersion.

Let us first examine N-processes $(\boldsymbol{k} + \boldsymbol{k}_1 + \boldsymbol{k}_2 = 0)$. For a crystal with defects the most actual is the case $T \ll \theta$. At the same time for $\hbar\omega \gg k_B T$, as follows from (5), collision processes $\omega + \omega_1 \rightleftarrows \omega_2$ get frozen: their contribution decreases exponentially with temperature (see also [13]). At $\hbar\omega \ll k_B T$ and $\hbar\omega_1 \approx \hbar\omega_2 \approx k_B T$, we have $n_1 - n_2 \approx -(\partial n_1/\partial\omega_1)(\omega_2 - \omega_1)$ and $(\omega_2 - \omega_1) = (\partial\omega_1/\partial k_1)\, k \cos(\widehat{\boldsymbol{k}_1\boldsymbol{k}})$. Expanding $\partial\omega_1/\partial k_1$ in a power series with accuracy up to k_1^2, we get $\omega_2 - \omega_1 = \omega r \sigma \xi$, where $r = v_1/v$ is the ratio of the velocities of the phonons ω_1 and ω; $\xi = \cos(\widehat{\boldsymbol{k}_1\boldsymbol{k}})$, $\sigma = 1 - \beta a^2\omega_1^2 v_1^{-2}$, a is the lattice constant, β the dispersion parameter $(\beta \approx \pi^{-1})$. Now, summing in (5) over k_2 and then integrating over ξ, we get for collision processes

$$\Gamma_\nu^c(\omega) = \frac{\hbar\omega\, |A_{\nu 11}|^2}{32\pi^2\varrho^3 v_1^7 v^2 r\sigma} \int_0^{\omega_D} d\omega_1\, \omega_1^4 Z_1^2 \left(-\frac{\partial n_1}{\partial\omega_1}\right) \times$$
$$\times \left[\tan^{-1}\frac{(r\sigma - 1)\,\omega}{2Z_1\gamma_1^0} + \tan^{-1}\frac{(r\sigma + 1)\,\omega}{2Z_1\gamma_1^0}\right]. \tag{6}$$

Here $\omega_D = k_B\theta/\hbar$ and it is taken into consideration that for acoustic phonons in the case of small $\boldsymbol{k}$, $\boldsymbol{k}_1$, $\boldsymbol{k}_2$ (see [14])

$$|\Phi^0_{\nu\nu_1\nu_2}|^2 = \frac{|A_{\nu 12}|^2\, \omega_\nu^2\omega_1^2\omega_2^2}{\Omega\varrho^3 v_\nu^2 v_1^2 v_2^2} ,$$

where ϱ is the crystal density. $A_{\nu 12}$ does not depend on the values of k, k_1, and k_2 and at $T \ll \theta$ it only weakly depends on the directions of $\boldsymbol{k}$, $\boldsymbol{k}_1$, and $\boldsymbol{k}_2$ so that in integrating over ξ, $A_{\nu 12}$ is considered to be constant.

The formulas of type (6) were earlier obtained in [15, 16] and used in the theory of ultrasonic attenuation [17, 18], while only such cases were studied when the damping $\gamma_1^0(\omega_1)$ is caused by the anharmonicity or the Rayleigh scattering by impurities without considering $\Delta\Phi(LL')$ and only for small ε, i.e. in the absence of localized (LM) and quasi-localized modes (QM). At $\gamma_{12} \to 0$ in (5) $\delta(\omega + \omega_1 - \omega_2)$ arises whence it follows that $v < v_1$ and $v < v_2$; for the elastic continuum this means that only the absorption of a transverse phonon by the longitudinal one $\mathrm{t} + \mathrm{l} \rightleftarrows \mathrm{l}$ is possible. Consequently, at $\gamma_1^0(\omega_1) \to 0$ expression (6) gives the Landau-Rumer law [19] ($Z_1 \approx 1$)

$$\Gamma_{\mathrm{L}}(\omega) = \frac{\hbar\, |A_{\mathrm{tll}}|^2}{32\pi^2\varrho^3 v_1^8 v_{\mathrm{t}}\sigma} \left(\frac{k_{\mathrm{B}}T}{\hbar}\right)^4 \frac{4\pi}{15}.$$

In case $\omega \rightleftarrows \omega_1 + \omega_2$ at $T \ll \theta$, $\hbar\omega < k_{\mathrm{B}}T$ for the decay $\mathrm{l} \rightleftarrows \mathrm{t} + \mathrm{t}$, analogously to (6) we get

$$\Gamma_1^{\mathrm{r}}(\omega) = \frac{\hbar\omega\, |A_{\mathrm{ltt}}|^2}{64\pi^2\varrho^3 v_1 v_{\mathrm{t}}^8} \int\limits_0^{\omega} \mathrm{d}\omega_1\, Z_1^2 \omega_1^3\, (\omega_1 - \omega) \left(-\frac{\partial n_1}{\partial \omega_1}\right) \times$$
$$\times \left[\tan^{-1} \frac{(1 - r^{-1})\,\omega - 2\omega_1}{2Z_1\gamma_1^0} - \tan^{-1} \frac{(1 - r^{-1})\,\omega}{2Z_1\gamma_1^0}\right], \tag{7}$$

at $\gamma_1^0(\omega_1) \to 0$ (7) gives the Pomeranchuk damping for longitudinal phonons [20]

$$\Gamma_{\mathrm{P}}(\omega) = \frac{\hbar\omega^4\, |A_{\mathrm{ltt}}|^2}{64\pi^2\varrho^3 v_1 v_{\mathrm{t}}^8} \frac{k_{\mathrm{B}}T}{\hbar} \frac{\pi}{6}.$$

Crude estimations about the influence of impurities on $\Gamma_\nu(\omega)$ can be made if one takes into account that the factor $\partial n_1/\partial\omega_1$ in (6) and (7) cuts off a certain region near $\omega_1 = \overline{\omega} \sim k_{\mathrm{B}}T/\hbar$ which makes the main contribution to the integrals in (6) and (7). If the concentration is sufficiently small, so that $\omega/\gamma_1^{\mathrm{c}}(\overline{\omega}) \gg 1$ (except the small region near $\omega = 0$), we obtain from (6) for $\Gamma_{\mathrm{t}}^{\mathrm{c}}(\omega)$ (when $\sigma \approx 1$, $r > 1$) and for $\Gamma_1^{\mathrm{c}}(\omega)$ (when $r \approx 1$, $\sigma \neq 1$, $\beta a^2/v_1^2 \equiv \alpha$) and from (7) for $\Gamma_1^{\mathrm{r}}(\omega)$ (when $r > 1$)

$$\left.\begin{aligned} \Gamma_{\mathrm{t}}^{\mathrm{c}}(\omega) &= \Gamma_{\mathrm{L}}(\omega)\left(1 - \frac{2\gamma_1^0(\overline{\omega})}{\pi\omega}\,\frac{2r\sigma}{r^2\sigma^2 - 1}\right), \\ \Gamma_1^{\mathrm{c}}(\omega) &= \Gamma_{\mathrm{L}}(\omega)\left(\frac{\hbar}{k_{\mathrm{B}}T}\right)^2 \frac{2\gamma_1^0(\overline{\omega})}{r\alpha\omega}\,\frac{5\pi}{4}, \\ \Gamma_1^{\mathrm{r}}(\omega) &= \Gamma_{\mathrm{P}}(\omega)\left(1 - \frac{2\gamma_{\mathrm{t}}^0(\overline{\omega})}{\pi\omega}\,\frac{r}{r - 1}\right). \end{aligned}\right\} \tag{8}$$

In another case, when $\omega/\gamma_1^0(\overline{\omega}) \ll 1$, i.e. the concentration is not too small, we get

$$\left.\begin{aligned} \Gamma_{\mathrm{t}}^{\mathrm{c}}(\omega) &= \Gamma_{\mathrm{L}}(\omega)\,\frac{r\sigma}{\pi\gamma_1^0(\overline{\omega})}, \\ \Gamma_1^{\mathrm{c}}(\omega) &= \Gamma_{\mathrm{L}}(\omega)\,\frac{\omega}{\pi r\gamma_1^0(\overline{\omega})}, \\ \Gamma_1^{\mathrm{r}}(\omega) &= \Gamma_{\mathrm{P}}(\omega)\,\frac{\omega}{2\pi\gamma_{\mathrm{t}}^0(\overline{\omega})}. \end{aligned}\right\} \tag{9}$$

At $\gamma^0(\overline{\omega}) \to 0$, as was first shown in [19, 20], the processes $\mathrm{l} + \mathrm{l} \rightleftarrows \mathrm{l}$ are forbidden and $\Gamma_1^{\mathrm{c}}(\omega)$ is equal to zero. However, the anisotropy [21] and the finite lifetime of longitudinal phonons, caused by their decay processes [18], lead to the circumstance that for the damping $\Gamma_1^c(\omega)$ in an anharmonic perfect crystal for not too low T and not too high ω follows the same dependence as the damping of transverse phonons.

Formulas (8) allow one to draw a general conclusion that nonvanishing $\gamma_1^0(\omega_1)$ leads to a decrease of the anharmonic damping of phonons as compared with their damping in a pure crystal, which is connected with the decrease of the mean lifetime of phonons due to the impurity scattering.

For the U-processes ($\boldsymbol{k} + \boldsymbol{k}_1 + \boldsymbol{k}_2 = \boldsymbol{K} \neq 0$) taking place if $\omega_1 \approx \omega_2 \approx \omega_D/2$, at $\hbar\omega \approx k_B T \ll k_B\theta$, the terms of the type $\omega_1 + \omega_2 \rightleftarrows \omega$ are actual. In this case in (5) we may put $(1 + n_1 + n_2) \approx -(\hbar\omega/k_B T)\, e^{-\theta/\varkappa T}$, ($\varkappa \approx 2$) and at $\gamma_1^0(\omega_1) \to 0$ we get

$$\Gamma_\nu^U(\omega) = \frac{\hbar\omega_\nu\, |A_{\nu 11}|^2}{\Omega\varrho^3 v^6} \frac{9N\omega_D^2}{320} \frac{\pi}{16} \frac{\hbar\omega}{k_B T} e^{-\theta/\varkappa T} . \tag{10}$$

At $\gamma_1^0(\omega_1) \neq 0$ the pre-exponential factor in $\Gamma_\nu^U(\omega)$ is changed but at $T \ll \theta$ the U-processes are not significant as before.

4. Resonant Scattering of Phonons

Formula (3) determines the polarization operator $\lambda_\nu(\omega)$ describing the scattering of phonons by impurities in anharmonic crystal. Averaging $\lambda_\nu(\omega)$ over phonons with $\omega_\nu = \omega$ we get (for details see [11])

$$\lambda(\omega) = \frac{\operatorname{Sp} V(T-1)^{-1} \operatorname{Im} G^0(\omega)}{\pi p N \varrho(\omega^2)}, \tag{11}$$

where $\varrho(\omega^2)$ is the density of squared frequencies, normalized to unity. If the symmetry group of the matrix $T = -G^0(V + P)$ is not lower than the point group of the crystal then in the representation of the right $-\xi_{\mu i}^\sigma(L)$ and left $-\eta_{\mu i}^\sigma(L)$ eigenvectors of the matrix T (μ is the irreducible representation of the symmetry group of the T-matrix, σ are repeated representations, i are lines of the degenerate representations) the operators P, V, and G^0 are diagonal in μ and i. In this case for the trace in (11) in the representation of the vectors ξ and η, considering that the components of the degenerate state make an equal contribution, we obtain

$$\lambda(\omega) = \frac{1}{\pi p N \varrho(\omega^2)} \sum_\mu n_\mu \operatorname{Sp} V_\mu (T_\mu - 1)^{-1} \operatorname{Im} G_\mu^0(\omega) , \tag{12}$$

where $T_\mu = T_{\mu i}(\sigma\sigma') = -\sum_{\sigma''} G_{\mu i}^0(\sigma\sigma'')\,(V + P)_{\mu i}\,(\sigma''\sigma')$, n_μ is the degree of degeneracy of the representation μ. Then, according to (3)

$$\left.\begin{aligned} \gamma(\omega) &= \frac{c}{2\pi p \omega \varrho(\omega^2)} \sum_\mu n_\mu g_\mu(\omega) , \\ g_\mu(\omega) &= \operatorname{Sp} V_\mu \operatorname{Im} (T_\mu - 1)^{-1} \operatorname{Im} G_\mu^0(\omega) . \end{aligned}\right\} \tag{13}$$

$g_\mu(\omega)$ has an especially simple form in the cases when the representation μ occurs only once ($\sigma = 1$). For example, for a crystal of the NaCl type, if the change of central forces with nearest neighbours only is taken into account, this is fulfilled for A_{1g} and E_g representations of the three types of scattering (A_{1g}, E_g, $2F_{1u}$).

For the F_{1u} representation $\sigma = 1$ in the case of an isotropic defect. If $\sigma = 1$,

$$g_\mu(\omega) = \frac{-V_\mu \operatorname{Im} T_\mu(\omega) \operatorname{Im} G^0_\mu(\omega)}{(1 - \operatorname{Re} T_\mu(\omega))^2 + (\operatorname{Im} T_\mu(\omega))^2}. \tag{14}$$

The condition $1 - \operatorname{Re} T_\mu(\omega_0) = 0$ determines the frequencies of local and quasi-local modes in μ-th representation. Using the Lorentz approximation for the frequencies ω close to ω_0 we get

$$\left.\begin{aligned} g_\mu(\omega) &= \frac{Z_\mu \gamma^e_\mu(\omega)\, \Gamma_\mu(\omega)}{(\omega_0^2 - \omega^2)^2 + \Gamma^2_\mu(\omega)}, \\ \Gamma_\mu(\omega) &= Z_\mu(\omega_0)\,[\gamma^e_\mu(\omega) + \Gamma^r_\mu(\omega)]; \qquad Z^{-1}_\mu(\omega_0) = \left.\frac{\partial \operatorname{Re} T_\mu(\omega)}{\partial \omega^2}\right|_{\omega^2=\omega_0^2}, \\ \Gamma^r_\mu(\omega) &= \operatorname{Re} P_\mu \operatorname{Im} G^0_\mu(\omega) + \operatorname{Im} P_\mu \operatorname{Re} G^0_\mu(\omega), \\ \gamma_\mu(\omega) &= V_\mu \operatorname{Im} G^0_\mu(\omega). \end{aligned}\right\} \tag{15}$$

Formula (15) is well-known in the general theory of resonant scattering [22]. Here it describes the total cross-section of elastic and inelastic scattering of phonons by impurities, $Z_\mu(\omega_0)$ gives an additional dependence of the total scattering cross-section on temperature (as in the theory of infrared absorption [23], it is possible to show that Z_μ has the form of the Debye-Waller factor), $Z_\mu \gamma^e_\mu(\omega_0)$ is the "elastic width" and $Z_\mu \Gamma^r_\mu(\omega_0)$ the full "inelastic width" of the level ω_0.

5. Resonant Scattering by Isotopic Defect in an Anharmonic Crystal

In calculating $g_\mu(\omega)$ according to formula (15) most difficulties face the computation of the matrix

$$\left.\begin{aligned} P_\mu(\omega) &= \sum_{LL'} \eta_\mu(L)\, P(LL')\, \xi_\mu(L'), \\ P(LL') &= \frac{\hbar}{2\pi^2 M^3} \iint_{-\infty}^{\infty} d\omega'\, d\omega'' \sum_{ll'} \sum_{l_1 l_1'} \left[\Phi(Lll')\Phi(L'l_1l_1') \frac{n' + n'' + 1}{\omega' + \omega'' - \omega} + \right. \\ &\quad \left. + \Phi(LL'l)\, \Phi(l'l_1l_1') \frac{n''}{\omega'}\right] \operatorname{Im} G(ll_1; \omega') \operatorname{Im} G(l'l_1'; \omega'') + \\ &\quad + \frac{\hbar}{2\pi M^2} \int_{-\infty}^{\infty} d\omega'\, n' \sum_{ll'} \Phi(LL'll') \operatorname{Im} G(ll'; \omega'). \end{aligned}\right\} \tag{16}$$

In the following we shall calculate the first and most essential term in $P_\mu = P^1_\mu + P^2_\mu + P^3_\mu$ in the simplest case of an isotopic defect. As $\Phi(Lll')$ and $\Phi(L'l_1l_1')$ decrease rapidly when moving away from the defect, the summation over ll' and l_1l_1' may be restricted to the region of the nearest neighbours of the defect site. Thus we can write

$$\left.\begin{aligned} P^1_\mu(\omega) &= \frac{\hbar}{2\pi^2} \iint_{-\infty}^{\infty} d\omega'\, d\omega'' \frac{n' + n'' + 1}{\omega' + \omega'' - \omega} \sum_{\mu_1\mu_2} |\Phi_{\mu\mu_1\mu_2}|^2 \operatorname{Im} U_{\mu_1}(\omega') \operatorname{Im} U_{\mu_2}(\omega''), \\ U_\mu(\omega) &= \sum_{LL'} \eta_\mu(L)\, G(LL')\, \xi_\mu(L'), \\ \Phi_{\mu\mu_1\mu_2} &= \sum_{LL_1L_2} M^{-3/2} \Phi(LL_1L_2)\, \xi_\mu(L)\, \xi_{\mu_1}(L_1)\, \xi_{\mu_2}(L_2). \end{aligned}\right\} \tag{17}$$

By transformations in the symmetry group of the T-matrix, $\Phi_{\mu\mu_1\mu_2}$ are transformed according to the direct product of irreducible representations Γ_μ, Γ_{μ_1}, and Γ_{μ_2} [23]. For $\Phi_{\mu\mu_1\mu_2}$ we obtain the selection rules $\Gamma_\mu \otimes \Gamma_{\mu_1} \otimes \Gamma_{\mu_2} \in \mathrm{I}$ (I is the unitary representation). In the case of central forces for the group O_h, when $\Gamma = A_{1g} + E_g + 2F_{1u}$, out of the 27 elements $\Phi_{\mu\mu_1\mu_2}$ only 11 are different from zero. For the isotopic defect 4 elements are left: $F_{1u}\,(2A_{1g} + 2E_g)\,F_{1u}$. In this case, e.g. for the cation site

$$\eta^\alpha_{F_{1u}}\begin{pmatrix} s \\ L \end{pmatrix} \xi^\beta_{F_{1u}}\begin{pmatrix} s' \\ L' \end{pmatrix} = \delta_{\alpha x}\delta_{\beta x}\delta_{L0}\delta_{L'0}\delta_{s+}\delta_{s'+} \tag{18}$$

and

$$|\Phi_{F_{1u}F_{1u}\mu}|^2 \equiv \Phi_\mu = \sum_{\substack{\alpha\beta LL'\\ \gamma\varrho}} \frac{\Phi_{\alpha\gamma\varrho}\begin{pmatrix} - & + & + \\ L & 0 & 0 \end{pmatrix} \Phi_{\beta\gamma\varrho}\begin{pmatrix} - & + & + \\ L' & 0 & 0 \end{pmatrix}}{m_+ m_+ m_-} \eta^\alpha_\mu\begin{pmatrix} - \\ L \end{pmatrix} \xi^\beta_\mu\begin{pmatrix} - \\ L' \end{pmatrix},$$

where L and L' take the values (001), (010), (100), (00$\bar{1}$), (0$\bar{1}$0), ($\bar{1}$00). In the case of central forces [24]

$$\left.\begin{aligned} \Phi_{\alpha\gamma\varrho}(L00) &= A R^0_\alpha(L)\, R^0_\gamma(L)\, R^0_\varrho(L) + B[\delta_{\alpha\gamma} R^0_\varrho(L) + \delta_{\gamma\varrho} R^0_\alpha(L) + \delta_{\varrho\alpha} R^0_\gamma(L)]\,,\\ A &= a^{-3}\left[\varphi'''(a) - \frac{3}{a}\varphi''(a) + \frac{3}{a^2}\varphi'(a)\right],\\ B &= a^{-2}\left[\varphi''(a) - \frac{1}{a}\varphi'(a)\right],\\ \varphi &= \varphi(|X(L) - X(L')|) = \varphi(r)\,. \end{aligned}\right\} \tag{19}$$

$\varphi(r)$ is the repulsive pair interaction of neighbouring atoms (ions), $X(l) = R^0(l) + x(l)$. As a result

$$\left.\begin{aligned} P^1_{F_{1u}}(\omega) &= \frac{\hbar}{\pi^2}\int\!\!\int_{-\infty}^{\infty} d\omega'\, d\omega''\, \frac{n' + n'' + 1}{\omega' + \omega'' - \omega} \operatorname{Im} U_{F_{1u}}(\omega') \times\\ &\quad\times [\Phi_{A_{1g}} \operatorname{Im} U_{A_{1g}}(\omega'') + \Phi_{E_g} \operatorname{Im} U_{E_g}(\omega'')]\,,\\ \operatorname{Im} U_{F_{1u}}(\omega) &= G^0_{F_{1u}}(\omega)\,(1 - \varepsilon_+\omega^2 G^0_{F_{1u}}(\omega))^{-1}\,, \qquad G^0_{F_{1u}}(\omega) = G^{++}_{xx}(000)\,;\\ \operatorname{Im} U_\mu(\omega) &= \pi \sum_\nu |Q_\mu(\nu)|^2\, \delta(\omega^2_\nu - \omega^2)\,, \qquad Q_\mu(\nu) = \sum_L \eta_\mu(L)\, w_\nu(L)\,,\\ &\qquad \mu \in A_{1g}, E_g;\\ \Phi_{A_{1g}} &= \frac{2a^2}{m^2_+ m_-}(Aa^2 + 5B)^2\,, \qquad \Phi_{E_g} = \frac{4a^2}{m^2_+ m_-}(Aa^2 + 2B)^2\,. \end{aligned}\right\} \tag{20}$$

Since $P^1(\omega) = \Delta(\omega) + i\Gamma(\omega)$ and $G^0_{F_{1u}}(\omega) = R(\omega) + iI(\omega)$, according to (13) and (14) for $\gamma(\omega)$ we obtain ($\varepsilon_+ = \varepsilon$)

$$\gamma(\omega) = \frac{3}{4} c\varepsilon\omega \times$$
$$\times \frac{I(\omega)\,[\varepsilon\omega^2 - \Delta(\omega)] + R(\omega)\,\Gamma(\omega)}{[1 - R(\omega)\,(\varepsilon\omega^2 - \Delta(\omega)) + I(\omega)\,\Gamma(\omega)]^2 + [I(\omega)\,(\varepsilon\omega^2 - \Delta(\omega)) + R(\omega)\,\Gamma(\omega)]^2}\,. \tag{21}$$

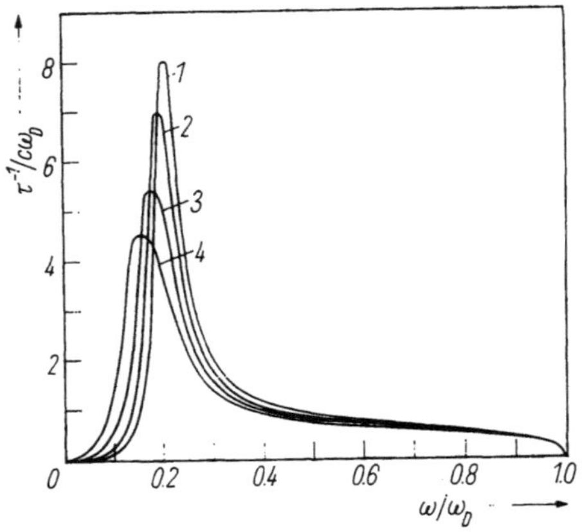

Fig. 1. Inverse relaxation rate for isotopic impurity ($\varepsilon = 8$); $T = 0$ (curve 1), 30 (2), 90 (3), 150 (4) (in °K)

In Fig. 1 the form of $\tau_d^{-1}(\omega) = 2\gamma(\omega)$ is shown for several temperatures. The calculation of $\gamma(\omega)$ is made for the resonant peak with the frequency $\omega_0 = 0.208\omega_D$ ($\varepsilon = 8$). The Born-Mayer potential and Debye approximation with the parameters of KCl crystals were used in these calculations. It is evident from Fig 1 that the peak value of $\tau_d^{-1}(\omega_0)$ essentially decreases with the rise of temperature. The peak is broadened, being shifted to the low-frequency region. Such a behaviour of $\tau_d^{-1}(\omega)$ may lead to the lowering of the effectiveness of resonant scattering at not too low temperatures, and thus to the change of the $K(T)$ curve immediately after the dip. Indeed, the remarkable exclusion of impurity scattering with the rise of temperature is clearly revealed in a number of experiments [25, 26] (see also [11]).

It should be noted that expression (21) is applicable also for the localized modes: in this case the term describing the collision processes coincides with the result obtained by Wagner [27].

6. Discussion

Formula (2) for $\Gamma_\nu(\omega)$ is obtained in the one-particle approximation (see I) and it represents the inverse lifetime of the phonon. It is natural because the N- and U-processes enter it in the same way. Thus, expression (1) for $\langle K(T)\rangle_c$ corresponds to the first approximation of the Callaway theory [4], which does not take account of the circumstance that N-processes themselves do not give finite thermal conductivity. Consideration of the special role of N-processes gives the correction [4, 5] to expression (1). This correction, however, is often very small, so that the one-particle approximation is sufficiently justified.

N-processes may be taken into consideration in a more correct way if one estimates the contribution of $\Sigma(\omega)$. Such an estimate for ideal anharmonic crystals was obtained by Ranninger [28] who demonstrated that the account of $\Sigma(\omega)$ leads to the appearance of the term $\left(1 - \frac{\boldsymbol{v}_1\boldsymbol{v}_2\,\omega_2\Gamma_1}{v_1^2\,\omega_1\Gamma_2} \pm \frac{\boldsymbol{v}_1\boldsymbol{v}_3\,\omega_3\Gamma_1}{v_1^2\,\omega_1\Gamma_3}\right)$ under the sum in (5). In the case of the Debye spectrum this term gives $\boldsymbol{k}_1/k_1^2\,(\boldsymbol{k}_1 - \boldsymbol{k}_2 \pm \boldsymbol{k}_3)$ and thus completely excludes N-processes. The latter fact together with the Peierls conclusion [1] represents another extremity in comparison with the one-particle approximation.

The existence of defects makes the task still more complicated. The fact is that the behaviour of $K(T)$ in every concrete case depends on the real ratio of the mean free path of phonons, due to the anharmonicity and defects. This ratio is different in different spectral regions, depends on temperature and the concentration of impurities [29].

The account of N-processes equally with U-processes in the one-particle approximation, as well as the Callaway theory, leads to the convergence of expression (1) already in an ideal anharmonic crystal.This sum also converges in case only the $\tau_d^{-1}(\omega)$ are left, corresponding to the scattering by a defect. That is caused by the function $\omega^2\Delta(\omega)$ in the numerator of (21) due to which at $\omega \to 0$, $\gamma \sim \omega^2$ guaranteeing the convergent expression for $K(T)$; this term changes the Rayleigh scattering ($\sim\omega^4$). The change of the Rayleigh scattering with the rise of T on the account of $\omega^2\Delta$ is shown in Fig. 1. At large Δ this may yield a too much lowered value of heat conductivity in the region of lowest temperatures. Thus, in this region of temperatures the one-particle approximation may give a wrong result. On the other hand, in the case of higher temperatures formula (21), obtained in the one-particle approximation, correctly describes the change of the role of resonant scattering with the rise of temperature which becomes evident in the behaviour of the curve $K(T)$ for a number of systems where a considerable exclusion of impurity scattering immediately after the dip is observed.

Therefore, it seems that the approximations used for calculation of τ_d^{-1} are justified only if the phonon mean free path with respect to anharmonic processes is shorter than that with respect to impurity scattering.

References

[1] R. E. Peierls, Ann. Phys. (Germany) **3**, 1055 (1929).
[2] P. G. Klemens, Solid State Phys. **7**, 1 (1958).
[3] R. Berman, P. T. Nettley, F. W. Sherard, A. H. Spencer, R. W. H. Stevenson, and J. M. Ziman, Proc. Roy. Soc. **A253**, 403 (1959).
[4] J. Callaway, Phys. Rev. **113**, 1046 (1959).
[5] J. A. Krumhansl, Proc. Phys. Soc. **85**, 921 (1965).
[6] V. S. Oskotskii, and I. A. Smirnov, Defects in Crystals and Thermal Conductivity, Nauka, Moscow 1972.
[7] P. Carruthers, Phys. Rev. **126**, 1449 (1962).
[8] I. Pomeranchuk, J. Phys. (USSR) **6**, 237 (1942).
[9] M. A. Ivanov and M. A. Krivoglaz, Fiz. tverd. Tela **6**, 200 (1964).
[10] V. I. Altukhov, phys. stat. sol. (b) **64**, 403 (1974).
[11] V. I. Altukhov and G. S. Zavt, Fiz. tverd. Tela **13**, 2438 (1971).
[12] H. I. Maris, Phil. Mag. **12**, 89 (1965).
[13] V. N. Kashcheev and M. A. Krivoglaz, Fiz. tverd. Tela **3**, 1528 (1961).
[14] R. E. Peierls, Quantum Theory of Solids, Clarendon Press, Oxford 1955.
[15] S. Simons, Proc. Phys. Soc. **82**, 401 (1963).
[16] P. B. Miller, Phys. Rev. **137**, A1937 (1965).
[17] P. C. Kwok, Solid State Phys. **20**, 213 (1967).
[18] L. V. Gurevich and B. I. Shklovskii, Fiz. tverd. Tela **9**, 526 (1967).
[19] L. D. Landau and G. Rumer, Phys. Z. SU **11**, 18 (1937).
[20] I. Pomeranchuk, J. Phys. (USSR) **4**, 259 (1941).
[21] C. Herring, Phys. Rev. **95**, 954 (1954).
[22] L. D. Landau and E. M. Lifshits, Quantum Mechanics, Fizmatgiz, Moscow 1963.
[23] I. P. Ipatova, A. V. Subashiev, and A. A. Maradudin, Ann. Phys. (U.S.A.) **53**, 376 (1969).

[24] G. LEIBFRIED and W. LUDWIG, Theory of Anharmonic Effects in Crystals, Academic Press, New York/London 1961.
[25] F. C. BAUMANN and R. O. POHL, Phys. Rev. **163**, 843 (1967).
[26] R. F. CALDWELL and M. V. KLEIN, Phys. Rev. **158**, 851 (1967).
[27] M. WAGNER, Phys. Rev. **131**, 1443 (1963).
[28] J. RANNINGER, Ann. Phys. (U.S.A.) **45**, 452 (1967).
[29] G. S. ZAVT and V. I. ALTUKHOV, Fiz. tverd. Tela **15**, 766 (1973).

(Received May 8, 1974)

phys. stat. sol. (b) **65**, 93 (1974)

Subject classification: 6; 4; 10.1; 21.1

A. F. Ioffe Physico-Technical Institute, Academy of Sciences of the USSR, Leningrad

Phenomenological Force Model for a Lattice from Experimental Phonon Spectra Appropriate to Lattice Distortion by Point Defects

By

A. D. CHEVYCHELOV and I. L. LIFSHITS

A general formal force model is constructed for a lattice with pair interactions between the atoms and appropriate to the problem of the point defects using group theory analysis. The pair interactions include bend and mixed parts. The concrete analysis is performed for cubic lattices. Numerical constants are chosen from experimental phonon dispersion curves. As an example of the application of the model the lattice distortion around a vacancy is calculated for copper.

Un général force modèle formel du réseau crystalline, dont les atomes sont joints par des interactions non-centrales, est construit à l'aide de la théorie de la symétrie. Le modèle est approprié à la problème des défauts ponctuels. L'interaction interatomique contient des parties angulaires et mixtes. L'analyse concrète est effectuée pour un réseau appartenant à la groupe cubique ponctuelle. Les constantes sont choisies d'après des spectres de vibration experimentaux du réseau. Comme un exemple de l'application du modèle la distortion du réseau autour d'une lacune est calculée pour cuivre.

1. Introduction

At present some phenomenological lattice models are used in the theory of the lattice vibration [1 to 2]. The aim of the present paper is to construct a general phenomenological force model for a lattice completely compatible with the whole symmetry of the lattice and appropriate to calculate the lattice distortion around point defects in the case of small displacements of the atoms in the vicinity of the defect. As it is remarked below, the existing models elaborated for phonon spectra calculations, are not sufficient for the problem of a defect. The idea of our investigation is taken from a series of papers concerning the properties of electronic spectra [3 to 5]. In these papers the correct symmetry relations are directly incorporated into the theory, and the arbitrary parameters reserved in such an analysis are calculated from the available experiments. In our case, these experiments are the complete dispersion phonon curves along the symmetry directions.

The consideration of the present paper is restricted to the case of a lattice model with pair (but in general non-central) interactions between the atoms. An actual analysis is performed for the case of f.c.c. lattice. The lattice distortion around a vacancy in copper is calculated as an example of application of the theory.

2. Construction of a Force Model for a Lattice

We suppose that the interactions between the atoms are pairwise. The generalization of the method to the case of three-center interactions is obvious but is not discussed here,

$$U(\boldsymbol{r} - \boldsymbol{r}') = U_\mathrm{r}(\boldsymbol{r} - \boldsymbol{r}') + U_\mathrm{b}(\boldsymbol{r} - \boldsymbol{r}') + U_\mathrm{m}(\boldsymbol{r} - \boldsymbol{r}') . \tag{1}$$

Here U_r is the spherically symmetric interaction between the atoms placed at $\boldsymbol{r}$ and $\boldsymbol{r}'$, U_b is the bend energy, and U_m is the mixed part (the meaning of the terms is precised below). It is assumed that all energies are expanded in Taylor series and only the harmonic terms are retained

$$\left.\begin{aligned} \boldsymbol{r} - \boldsymbol{r}' &= \boldsymbol{X} - \boldsymbol{X}' + \boldsymbol{u}(\boldsymbol{X}) - \boldsymbol{u}(\boldsymbol{X}') = \boldsymbol{Y} + \boldsymbol{v}(\boldsymbol{X}, \boldsymbol{X}') , \\ \boldsymbol{Y} &= \boldsymbol{X} - \boldsymbol{X}' , \qquad \boldsymbol{v}(\boldsymbol{X}, \boldsymbol{X}') = \boldsymbol{u}(\boldsymbol{X}) - \boldsymbol{u}(\boldsymbol{X}') , \end{aligned}\right\} \tag{2}$$

$\boldsymbol{X}$ and $\boldsymbol{X}'$ being the lattice vectors, $\boldsymbol{u}(\boldsymbol{X})$ and $\boldsymbol{u}(\boldsymbol{X}')$ the displacements of the corresponding atoms.

The expansion for the radial part is the following:

$$U_\mathrm{r}(\boldsymbol{Y} + \boldsymbol{v}) = U_\mathrm{r}(\boldsymbol{Y}) + c_1(\boldsymbol{Y})\, \boldsymbol{e} \cdot \boldsymbol{v} + c_2(\boldsymbol{Y})\, (\boldsymbol{e} \cdot \boldsymbol{v})^2 + c_3(\boldsymbol{Y})\, \boldsymbol{v}^2 . \tag{3}$$

Here the notations

$$\boldsymbol{e} = \boldsymbol{Y}/|\boldsymbol{Y}| , \quad c_1 = U_\mathrm{r}'(|\boldsymbol{Y}|) , \quad c_2 = \tfrac{1}{2} U_\mathrm{r}''(|\boldsymbol{Y}|) - c_3 , \quad c_3 = c_1/2\, |\boldsymbol{Y}| \tag{4}$$

are introduced, U_r' and U_r'' being the first and the second radial derivatives of the scalar function $U_\mathrm{r}(|\boldsymbol{Y}|)$ of the argument $|\boldsymbol{Y}|$, respectively.

As an expansion for $U_\mathrm{b}(\boldsymbol{Y} + \boldsymbol{v})$ we take the quadratic form

$$U_\mathrm{b}(\boldsymbol{Y} + \boldsymbol{v}) = \tfrac{1}{2} B(\boldsymbol{Y}) : \boldsymbol{v}_\mathrm{t} \boldsymbol{v}_\mathrm{t} . \tag{5}$$

$B(\boldsymbol{Y})$ is a constant three-dimensional matrix and $\boldsymbol{v}_\mathrm{t}$ the tangential part of the whole displacement $\boldsymbol{v}$:

$$\boldsymbol{v}_\mathrm{t} = \boldsymbol{e} \times \boldsymbol{v} \times \boldsymbol{e} = \boldsymbol{v} - (\boldsymbol{e} \cdot \boldsymbol{v})\, \boldsymbol{e} . \tag{6}$$

The expansion of the mixed part $U_\mathrm{m}(\boldsymbol{Y} + \boldsymbol{v})$ is chosen as

$$U_\mathrm{m}(\boldsymbol{Y} + \boldsymbol{v}) = \tfrac{1}{2} M(\boldsymbol{Y}) : (\boldsymbol{v} \boldsymbol{v}_\mathrm{t} + \boldsymbol{v}_\mathrm{t} \boldsymbol{v}) . \tag{7}$$

$M(\boldsymbol{Y})$ being a constant three-dimensional matrix.

The whole energy (1) and every part (2), (5), and (7) are invariant with respect to an arbitrary translation and rotation of the lattice.

For the calculation of the vectorial derivatives of U_b and U_m, it is necessary to express the quadratic forms (5) and (7) as the quadratic forms of the whole displacement $\boldsymbol{v}$. The bend part may be written as follows:

$$U_\mathrm{b}(\boldsymbol{Y} + \boldsymbol{v}) = \tfrac{1}{2} (\boldsymbol{e} \times \boldsymbol{v} \times \boldsymbol{e}) \cdot B(\boldsymbol{Y}) \cdot (\boldsymbol{e} \times \boldsymbol{v} \times \boldsymbol{e}) = \tfrac{1}{2} \boldsymbol{v} \cdot b(\boldsymbol{Y}) \cdot \boldsymbol{v} . \tag{8}$$

Here the matrix $b(\boldsymbol{Y})$ has the form

$$b(\boldsymbol{Y}) = \boldsymbol{e} \times \big(\boldsymbol{e} \times B(\boldsymbol{Y}) \times \boldsymbol{e}\big) \times \boldsymbol{e} . \tag{9}$$

The matrix $\boldsymbol{e} \times B$ is the cross product [6] of a vector and a matrix, and the matrix $B \times \boldsymbol{e}$ is the cross product of a matrix and a vector. For obtaining (9) we employ the following properties of cross products [6]

$$(\boldsymbol{c}\times\boldsymbol{d})\cdot B = \boldsymbol{c}\cdot(\boldsymbol{d}\times B)\,,\quad B\cdot(\boldsymbol{c}\times\boldsymbol{d}) = (B\times\boldsymbol{c})\cdot\boldsymbol{d}\,, \tag{10}$$

$\boldsymbol{c}$ and $\boldsymbol{d}$ being any vectors.

Similarly

$$U_{\mathrm{m}}(\boldsymbol{Y}+\boldsymbol{v}) = \tfrac{1}{2}\boldsymbol{v}\cdot m(\boldsymbol{Y})\cdot\boldsymbol{v}\,. \tag{11}$$

Here we employ the notation $m(\boldsymbol{Y})$ for the matrix

$$m(\boldsymbol{Y}) = -\boldsymbol{e}\times\big(\boldsymbol{e}\times M(\boldsymbol{Y})\big) - \big(M(\boldsymbol{Y})\times\boldsymbol{e}\big)\times\boldsymbol{e}\,. \tag{12}$$

The mixed part U_{m} may be determined in some different manner, for example

$$U_{\mathrm{m}}(\boldsymbol{Y}+\boldsymbol{v}) = \tfrac{1}{2}M(\boldsymbol{Y}):(\boldsymbol{v}_{\mathrm{t}}\boldsymbol{v}_{\mathrm{r}} + \boldsymbol{v}_{\mathrm{r}}\boldsymbol{v}_{\mathrm{t}})\,, \tag{13}$$

$$\boldsymbol{v}_{\mathrm{r}} = \boldsymbol{v} - \boldsymbol{v}_{\mathrm{t}}\,. \tag{14}$$

The expressions (7) and (13) differ on the sum of the structure (5). We can include this sum into (5).

3. Analysis of the Symmetric Properties of the Matrices $\boldsymbol{b}(\boldsymbol{Y})$ and $\boldsymbol{m}(\boldsymbol{Y})$ for Cubic Lattices

The symmetry properties of the lattice impose some restrictions on the structure of the matrices $b(\boldsymbol{Y})$ and $m(\boldsymbol{Y})$. As an actual example we take the lattices of the cubic point group m3m. The simple cubic lattice as well as f.c.c. and b.c.c. lattices belong to this point group. The matrices of the operations of m3m point group are reported in [7].

The formal analysis is most conveniently performed in the following manner. For every lattice vector $\boldsymbol{Y}$ under consideration, we choose some coordinate system in which the vector $\boldsymbol{e}$ has the form (1, 0, 0). We take the matrices $B(\boldsymbol{Y})$ and $M(\boldsymbol{Y})$ with all its arbitrary elements and perform the operations (9) and (12). At this stage the results do not depend on the vector $\boldsymbol{Y}$ and the calculations may be realized once and for all. Then we take the group of vector $\boldsymbol{Y}$, i.e. the subgroup of the whole cubic group that does not change the vector $\boldsymbol{Y}$, and transform the matrices of the group operations onto the above coordinate system by the appropriate rotations. It is the most convenient to find the matrix T of the rotation of a vector around any vector $\boldsymbol{c}$ by means of the equation ([6] p. 147):

$$T = I\cos\varphi - \mathbf{P}\sin\varphi + \boldsymbol{pp}(1-\cos\varphi)\,,\qquad \boldsymbol{p} = \frac{\boldsymbol{c}}{|\boldsymbol{c}|}\,. \tag{15}$$

I is the unit matrix, φ the angle of rotation and the tensor $\mathbf{P}$ has the form

$$\mathbf{P} = \begin{Vmatrix} 0 & +p_z & -p_y \\ -p_z & 0 & +p_x \\ +p_y & -p_x & 0 \end{Vmatrix}. \tag{16}$$

In the new coordinate system, we find the relations among the elements for the matrices $b(\boldsymbol{Y})$ and $m(\boldsymbol{Y})$ by the ordinary procedure [8] and then transform the matrices obtained onto the basic coordinate system. If the structures of the matrices $b(\boldsymbol{Y})$ and $m(\boldsymbol{Y})$ are identical, we formally add the matrix m to b. The matrix m is assumed in this case being equal to zero.

The above analysis may be similarly performed for any lattice.

4. Solution of the Equations for the Parameters

At the present stage we must find the matrix $U''(\boldsymbol{Y})$ of the second vectorial derivative of the whole potential U at $\boldsymbol{v} = 0$,

$$U''(\boldsymbol{Y}) = \frac{\partial^2 U_\mathrm{r}}{\partial \boldsymbol{v}\, \partial \boldsymbol{v}} + \frac{\partial^2 U_\mathrm{b}}{\partial \boldsymbol{v}\, \partial \boldsymbol{v}} + \frac{\partial^2 U_\mathrm{m}}{\partial \boldsymbol{v}\, \partial \boldsymbol{v}}. \tag{17}$$

The expressions in the above relation are

$$\frac{\partial^2 U_\mathrm{r}}{\partial \boldsymbol{v}\, \partial \boldsymbol{v}} = 2c_2(\boldsymbol{Y})\, \boldsymbol{ee} + 2c_3(\boldsymbol{Y})\, I\,, \quad \frac{\partial^2 U_\mathrm{b}}{\partial \boldsymbol{v}\, \partial \boldsymbol{v}} = b(\boldsymbol{Y})\,, \quad \frac{\partial^2 U_\mathrm{m}}{\partial \boldsymbol{v}\, \partial \boldsymbol{v}} = m(\boldsymbol{Y})\,. \tag{18}$$

For every $\boldsymbol{Y}$ under consideration the matrix (17) is assumed to be equal to the corresponding matrix $\Phi''(\boldsymbol{Y})$ obtained from experiments of neutron diffraction [9]

$$U''(\boldsymbol{Y}) = \Phi''(\boldsymbol{Y})\,. \tag{19}$$

We have an inhomogeneous system of linear equations to determine the elements of the matrices $b(\boldsymbol{Y})$ and $m(\boldsymbol{Y})$ and the constants $c_2(\boldsymbol{Y})$ and $c_3(\boldsymbol{Y})$.

In Table 1 the results of the above analysis are listed together with the matrices $\Phi''(\boldsymbol{Y})$ and $U_\mathrm{r}''(\boldsymbol{Y})$ for some lattice vectors. These vectors represent the irreducible vectors of the f.c.c. lattice up to sixth coordination shell. The combination of these matrices for every vector $\boldsymbol{Y}$ form the equation (19).

In (19) we cannot in general take the part U_b as isotropic. For example in the first coordination sphere of the f.c.c. lattice (see Table 1) the mixed part U_m is zero. If we suppose in U_b the relation $b_{22} = b_{33}$ (this is the condition of isotropy for U_b), the resulting matrix $U''(\boldsymbol{Y})$ has higher symmetry then has the matrix $\Phi''(\boldsymbol{Y})$. The conditions of the high symmetry impose some supplementary relations among the elements of the matrix $U''(\boldsymbol{Y})$ that make the equation (19) incompatible. This means that the angular part U_b must be sufficiently anisotropic. This is the reason for the well-known de Launay model [10] of angular forces to be insufficient for a complete formal description of the lattice force model.

In some cases (third and fifth coordination shells in Table 1) even an anisotropic angular part does not guarantee the minimal symmetry imposed on the matrix $U''(\boldsymbol{Y})$ by the symmetry of the lattice. The introduction of the mixed part (7) becomes then indispensable.

From (19) we obtain the numerical values of c_2, c_3 and the elements of the matrices b and m. Then from (4) we have the value of c_1 that is necessary for the calculation of the first vectorial derivative of U, i.e. the force of the interatomic interaction. In general the result is that $c_1 \neq 0$. This is a serious difference between our force model and the models [1, 2] in which the radial part has the form

$$U_\mathrm{r}(\boldsymbol{Y} + \boldsymbol{v}(\boldsymbol{Y})) = c v_\mathrm{r}^2\,. \tag{20}$$

The first vectorial derivative of the radial part (20) is zero at $\boldsymbol{v} = 0$ for every coordination shell. In such a model it is possible to calculate the phonon spectra for which only the matrix U'' is necessary but it is impossible to calculate the lattice distortion around a vacancy when we are obliged to calculate the forces. The force acting on an atom in a crystal with a vacancy is zero in such a model if the atoms are not displaced from the lattice sites. Our force model is exempt from this unphysical assumption.

Table 1

Structure of the matrices $\Phi''(\boldsymbol{Y})$, $\frac{\partial^2 U_r}{\partial \boldsymbol{v}\, \partial \boldsymbol{v}}$, $b(\boldsymbol{Y})$, and $m(\boldsymbol{Y})$

$\boldsymbol{Y}$	$\Phi''(\boldsymbol{Y})$	$\frac{\partial^2 U_r}{\partial \boldsymbol{v}\, \partial \boldsymbol{v}}$	$b(\boldsymbol{Y})$	$m(\boldsymbol{Y})$
1 1 0	$\begin{matrix} \Phi_{11} & \Phi_{12} & 0 \\ \Phi_{12} & \Phi_{11} & 0 \\ 0 & 0 & \Phi_{33} \end{matrix}$	$\begin{matrix} c_2 + 2c_3 & c_2 & 0 \\ c_2 & c_2 + 2c_3 & 0 \\ 0 & 0 & 2c_3 \end{matrix}$	$\begin{matrix} \frac{1}{2} b_{22} & -\frac{1}{2} b_{22} & 0 \\ -\frac{1}{2} b_{22} & \frac{1}{2} b_{22} & 0 \\ 0 & 0 & b_{33} \end{matrix}$	
2 0 0	$\begin{matrix} \Phi_{11} & 0 & 0 \\ 0 & \Phi_{22} & 0 \\ 0 & 0 & \Phi_{22} \end{matrix}$	$\begin{matrix} 2(c_2 + c_3) & 0 & 0 \\ 0 & 2c_3 & 0 \\ 0 & 0 & 2c_3 \end{matrix}$	$\begin{matrix} 0 & 0 & 0 \\ 0 & b & 0 \\ 0 & 0 & b \end{matrix}$	
2 1 1	$\begin{matrix} \Phi_{11} & \Phi_{12} & \Phi_{12} \\ \Phi_{12} & \Phi_{22} & \Phi_{23} \\ \Phi_{12} & \Phi_{23} & \Phi_{22} \end{matrix}$	$\begin{matrix} \frac{4}{3} c_2 + 2c_3 & \frac{2}{3} c_2 & \frac{2}{3} c_2 \\ \frac{2}{3} c_2 & \frac{1}{3} c_2 + 2c_3 & \frac{1}{3} c_2 \\ \frac{2}{3} c_2 & \frac{1}{3} c_2 & \frac{1}{3} c_2 + 2c_3 \end{matrix}$	$\begin{matrix} \frac{1}{3} b_{33} & -\frac{1}{3} b_{33} & -\frac{1}{3} b_{33} \\ -\frac{1}{3} b_{33} & \frac{1}{3} b_{33} + \frac{1}{2} b_{22} & \frac{1}{3} b_{33} - \frac{1}{2} b_{22} \\ -\frac{1}{3} b_{33} & \frac{1}{3} b_{33} - \frac{1}{2} b_{22} & \frac{1}{3} b_{33} + \frac{1}{2} b_{22} \end{matrix}$	$m \begin{Vmatrix} -4 & 1 & 1 \\ 1 & 2 & 2 \\ 1 & 2 & 2 \end{Vmatrix}$
2 2 0	$\begin{matrix} \Phi_{11} & \Phi_{12} & 0 \\ \Phi_{12} & \Phi_{11} & 0 \\ 0 & 0 & \Phi_{33} \end{matrix}$	$\begin{matrix} c_2 + 2c_3 & c_2 & 0 \\ c_2 & c_2 + 2c_3 & 0 \\ 0 & 0 & 2c_3 \end{matrix}$	$\begin{matrix} \frac{1}{2} b_{22} & -\frac{1}{2} b_{22} & 0 \\ -\frac{1}{2} b_{22} & \frac{1}{2} b_{22} & 0 \\ 0 & 0 & b_{33} \end{matrix}$	
3 1 0	$\begin{matrix} \Phi_{11} & \Phi_{12} & 0 \\ \Phi_{12} & \Phi_{22} & 0 \\ 0 & 0 & \Phi_{33} \end{matrix}$	$\begin{matrix} \frac{9}{5} c_2 + 2c_3 & \frac{3}{5} c_2 & 0 \\ \frac{3}{5} c_2 & \frac{1}{5} c_2 + 2c_3 & 0 \\ 0 & 0 & 2c_3 \end{matrix}$	$\begin{matrix} \frac{1}{10} b_{22} & -\frac{3}{10} b_{22} & 0 \\ -\frac{3}{10} b_{22} & \frac{9}{10} b_{22} & 0 \\ 0 & 0 & b_{33} \end{matrix}$	$m \begin{Vmatrix} -3 & 4 & 0 \\ 4 & 3 & 0 \\ 0 & 0 & 0 \end{Vmatrix}$
2 2 2	$\begin{matrix} \Phi_{11} & \Phi_{12} & \Phi_{12} \\ \Phi_{12} & \Phi_{11} & \Phi_{12} \\ \Phi_{12} & \Phi_{12} & \Phi_{11} \end{matrix}$	$\begin{matrix} \frac{2}{3} c_2 + 2c_3 & \frac{2}{3} c_2 & \frac{2}{3} c_2 \\ \frac{2}{3} c_2 & \frac{2}{3} c_2 + 2c_3 & \frac{2}{3} c_2 \\ \frac{2}{3} c_2 & \frac{2}{3} c_2 & \frac{2}{3} c_2 + 2c_3 \end{matrix}$	$b \begin{Vmatrix} 2 & -1 & -1 \\ -1 & 2 & -1 \\ -1 & -1 & 2 \end{Vmatrix}$	

In every coordination shell under consideration the number of the independent variables in equation (19) equals the number of equations plus 1. A unique solution of this system does not exist, although every possible solution is completely compatible with the experimental phonon curves determined by the set of the matrices $\Phi''(\boldsymbol{Y})$. For making (19) close we must take into account a supplementary information. This information may be the volume change of the crystal due to a vacancy or any other physical properties due to the presence of the vacancies in the crystal.

Here we choose the parameters quite arbitrary in the framework of the equation (19). In the second and the sixth coordination shells the number of the equations becomes equal the number of the variables if we put $U_b = 0$. In the other coordination spheres we can put either $b_{22} = 0$ or $b_{33} = 0$. Then the solutions of (19) are completely determined.

Table 2

Parameters of the force models in atomic units for dimensionless displacements of the atoms in units of the cubic lattice constant *a*

$\boldsymbol{Y}$		variant 1	variant 2	variant 3	variant 4
1 1 0	c_1	−0.04087	−0.02787	−0.03188	−0.02575
	c_2	+0.45078	+0.44159	+0.44913	+0.44480
	c_3	−0.02890	−0.01971	−0.02254	−0.01821
	b_{22}	0	−0.01838	0	−0.00866
	b_{33}	+0.01838	0	+0.00866	0
2 0 0	c_1	−0.00752	−0.00752	−0.00144	−0.00144
	c_2	+0.00698	+0.00698	+0.00099	+0.00099
	c_3	−0.00376	−0.00376	−0.00072	−0.00072
	b	0	0	0	0
2 1 1	c_1	+0.00973	+0.00593	+0.00294	−0.00256
	c_2	+0.01077	+0.01717	+0.01341	+0.01565
	c_3	+0.00397	−0.00242	+0.00120	−0.00104
	b_{22}	0	+0.01279	0	+0.00449
	b_{33}	−0.01279	0	−0.00449	0
	m	+0.00055	+0.00055	+0.00090	+0.00090
2 2 0	c_1	+0.00081	−0.01314	+0.01285	−0.00136
	c_2	−0.00129	+0.00364	−0.00108	+0.00394
	c_3	+0.00028	−0.00465	+0.00454	−0.00048
	b_{22}	0	+0.00986	0	+0.01004
	b_{33}	−0.00986	0	−0.01004	0
3 1 0	c_1	0	0	−0.00969	+0.00175
	c_2	0	0	+0.00144	−0.00218
	c_3	0	0	−0.00307	+0.00055
	b_{22}	0	0	0	−0.00724
	b_{33}	0	0	+0.00724	0
	m	0	0	−0.00008	−0.00008
2 2 2	c_1	0	0	−0.00514	−0.00514
	c_2	0	0	−0.00261	−0.00261
	c_3	0	0	−0.00148	−0.00148
	b	0	0	0	0

5. Choice of the Parameters and the Calculation of the Lattice Distortion around a Vacancy in Copper

For the numerical calculations we take a vacancy in copper. Some sets of the matrices $\Phi''(Y)$ for Cu are reported in [9] and [11]. We take the parameters of Sinha [9] (S) and the second set (set M_2) of the parameters of Swensson, Brockhouse, and Rowe [11] (SBR). For every set of the parameters we take two variants:

Variant 1: SBR, $b_{22} = 0$, variant 2: SBR, $b_{33} = 0$, variant 3: S, $b_{22} = 0$, variant 4: S, $b_{33} = 0$.

The calculation of the lattice distortion around a vacancy is performed by means of the Matsubara method [12] in the lattice-site representation [13]. We have realized this method computationally for an arbitrary Bravais lattice, arbitrary symmetry of the point defects, and a potential influenced up to arbitrary number of coordination spheres. We obtain from (1), (3), (5), and (7) the following expression for the interaction force between the atoms:

$$-\boldsymbol{F}(\boldsymbol{Y} + \boldsymbol{v}(\boldsymbol{Y})) = (c_1(\boldsymbol{Y}) + 2c_2(\boldsymbol{Y})(\boldsymbol{e} \cdot \boldsymbol{v}))\,\boldsymbol{e} + 2c_3\boldsymbol{v} + (b(\boldsymbol{Y}) + m(\boldsymbol{Y}))\,\boldsymbol{v}\,. \quad (21)$$

The matrices of the Green function do not depend on the details of the force model. They depend only on the set of matrices $\Phi''(Y)$ and can be calculated once for the variants 1 and 2 and once for the variants 3 and 4. The calculation of the Green function matrices is performed for 63439 points in the Brillouin zone with 1637 points in the irreducible part of the Brillouin zone. The four sets of parameters of the force model, the displacements of the atoms around a vacancy for six coordination shells, values of the volume change due to a vacancy and the relaxation energy are represented in Tables 2 to 4.

Table 3

Dimensionless displacements of the atoms around a vacancy in units of the cubic lattice constant a

Γ	variant 1			variant 2		
	x	y	z	x	y	z
1 1 0	−0.0089	−0.0089	0	−0.0074	−0.0074	0
2 0 0	−0.0019	0	0	−0.0022	0	0
2 1 1	+0.0001	+0.0001	+0.0001	−0.0005	−0.0004	−0.0004
2 2 0	−0.0028	−0.0028	0	−0.0060	−0.0060	0
3 1 0	−0.0003	−0.0004	0	−0.0004	−0.0006	0
2 2 2	−0.0002	−0.0002	−0.0002	−0.0007	−0.0007	−0.0007
Γ	variant 3			variant 4		
	x	y	z	x	y	z
1 1 0	−0.0068	−0.0068	0	−0.0072	−0.0072	0
2 0 0	−0.0028	0	0	+0.0003	0	0
2 1 1	−0.0022	−0.0006	−0.0006	−0.0028	−0.0024	−0.0024
2 2 0	+0.0008	+0.0008	0	−0.0031	−0.0031	0
3 1 0	−0.0048	−0.0013	0	+0.0005	−0.0004	0
2 2 2	−0.0017	−0.0017	−0.0017	−0.0030	−0.0030	−0.0030

Table 4

The relaxation energy (in eV) and the volume change due to a vacancy in units of atomic volume $a^3/4$

	variants			
	1	2	3	4
relaxation energy	−0.086	−0.048	−0.050	−0.044
volume change	−0.32	−0.25	−0.17	−0.30

6. Discussion

From the tables of the results it is seen that the results are not very sensitive to the arbitrariness in the choice of parameters. This arbitrariness most strongly influences the angular and mixed parts of the energy, but these parts are small compared to the spherically symmetric part.

There are two types of uncertainty in the choice of the force model in the framework of our approach. The uncertainty of the set of matrices $\Phi''(\boldsymbol{Y})$ is discussed in [9, 11, 13]. The source of the second type of uncertainty is the insufficiency of the orthogonal information contained in the given set of matrices $\Phi''(\boldsymbol{Y})$ for the construction of a unique general force model with the complete transformation properties. In principle the complementary information may be incorporated into the theory and then the choice of parameters may be refined.

We believe however that our approach has some advantage. Our force model is completely compatible with the correct experimental phonon spectra. The possibility to obtain good phonon curves in a lattice model is a necessary (although not sufficient) condition imposed on the model representing adequately the interactions between the atoms. In the theory of point defects, sometimes models are used that do not satisfy this necessary condition.

The sensitivity of the results concerning the defects on the model (even if the models give good phonon spectra) is a common feature of the existent theories (see the discussion in [14]).

We consider it useful to develop a similar theory for crystals with strong three-center angular forces.

Acknowledgement

We are indebted to A. N. Orlov for valuable discussions.

References

[1] M. J. P. Musgrave and J. A. Pople, Proc. Roy. Soc. **A268**, 474 (1962).
[2] T. Wolfram and G. W. Lehman, Phys. Rev. **129**, 2483 (1963).
[3] T. W. McClure, Phys. Rev. **108**, 612 (1957).
[4] J. C. Slonczewski and P. R. Weiss, Phys. Rev. **109**, 272 (1958).
[5] G. Dresselhaus and M. S. Dresselhaus, Phys. Rev. **160**, 649 (1967).
[6] E. A. Milne, Vectorial Mechanics, Methuen and Co., London 1948.
[7] H. W. Streitwolf, Gruppentheorie in der Festkörperphysik, Akademische Verlagsschaft Geest & Portig, Leipzig 1967.
[8] G. Leibfried, Hdb. Phys., Vol. VII/1, Ed. S. Flügge, Springer Verlag, Berlin/Göttingen/Heidelberg 1955 (p. 148).

[9] S. K. Sinha, Phys. Rev. **143**, 422 (1966).
[10] J. de Launay, Solid State Phys. **2**, 249 (1956).
[11] E. C. Svensson, B. N. Brockhouse, and J. M. Rowe, Phys. Rev. **155**, 619 (1967).
[12] T. J. Matsubara, J. Phys. Soc. Japan **7**, 270 (1952).
[13] V. K. Tewary, A.E.R.E. Rep. theor. Phys. **388**, Berks. 1969.
[14] J. W. Flocken and J. R. Hardy, Phys. Rev. **177**, 1054 (1969).

(Received April 22, 1974)

phys. stat. sol. (b) 65, 103 (1974)

Subject classification: 20.1; 14.4.1; 22.8

A. F. Ioffe Physico-Technical Institute, Academy of Sciences of the USSR, Leningrad

Optical Absorption of Ferroelectrical Copper and Chromium Boracites

By

N. N. NESTEROVA, R. V. PISAREV, and G. T. ANDREEVA

Optical absorption spectra of ferroelectrical boracites $Cr_3B_7O_{13}Cl$, $Cu_3B_7O_{13}Cl$, $Cu_3B_7O_{13}Br$, and also borate CuB_2O_4 were studied in the spectral range from 2000 to 30000 cm^{-1} in the temperature interval 77 to 450 °K. The interpretation of the spectra are given on the basis of the crystal field theory including tetragonal distortion of the local environment of 3d-ions in the lattice. The temperature dependence of the oscillator strength of the absorption band and absorption coefficient at a fixed frequency were investigated around the temperature of the ferroelectric phase transition in the copper boracites. The results show that the absorption is very sensitive to the displacement of the ions in the crystal lattice at the phase transition. For Cu–Cl boracite a change of the slope of the temperature dependence of the absorption near the fundamental absorption edge was observed.

Im Temperaturbereich von 77 bis 450 °K wurden die optischen Absorptionsspektren der ferroelektrischen $Cr_3B_7O_{13}Cl$-, $Cu_3B_7O_{13}Cl$-, $Cu_3B_7O_{13}Br$-Borazite und auch des CuB_2O_4-Borats im Spektralgebiet von 2000 bis 30000 cm^{-1} untersucht. Die Interpretation der Spektren wird auf der Grundlage der Kristallfeldtheorie unter Einschluß einer Störung der Umgebung der 3d-Ionen im Gitter gegeben. Die Temperaturabhängigkeit der Oszillatorstärke der Absorptionsbande und des Absorptionskoeffizienten bei einer festen Frequenz wurden in der Nähe der Temperatur des ferroelektrischen Phasenübergangs in den Kupferboraziten untersucht. Die Ergebnisse zeigen, daß die Absorption sehr empfindlich gegenüber einer Verschiebung der Ionen im Kristallgitter am Phasenübergang ist. Für Cu–Cl-Borazite wurde eine Änderung des Anstiegs der Temperaturabhängigkeit der Absorption in der Nähe der Grundgitterabsorptionskante beobachtet.

1. Introduction

Boracites are compounds described by the general formula $Me_3B_7O_{13}X$, where Me stays for divalent 3d-ions and X stays for Cl^-, Br^-, or I^-. All the boracites studied so far are ferroelectrics in the low temperature phase [1]. A magnetic ordering of 3d-ions is also found at low temperatures and magnetic Curie points are lying below ferroelectric Curie points in known boracites [2]. X-ray structure analysis shows [3] that paramagnetic 3d-ions occupy the sites surrounded by four oxygens and two halogen ions. The symmetry of this allongated octahedron is low — C_1, but, as it was already discussed earlier [4], in first approximation it can be regarded as D_{4h} and C_{4v} in paraelectric and ferroelectric phases, respectively.

In recent publications [4 to 6] optical absorption spectra of iron, cobalt, and nickel boracites were reported. In [4] the crystal field theory was developed in order to describe the main features of the absorption. It was found that the spectra may be interpreted taking into account a strong tetragonal component of the crystal field (parameters *Ds* and *Dt*). Pronounced changes of the cubic field parameter *Dq* and tetragonal parameters *Ds* and *Dt* were found at the temperature of transition from the ferroelectric to the parelectric phase. These changes were in accordance with the developed theory.

An unusual temperature behaviour was observed for the intensity of the absorption. In different crystals both an increase and decrease of the intensity was depicted while raising the temperature. From this behaviour it was concluded that different mechanisms due to electron–phonon interaction or to the displacement of 3d-ions in the lattice give a contribution to the observed intensity.

In the present paper we report on the absorption spectra of chromium and copper boracites and their analysis on the basis of the crystal field theory. The paper includes also the study of the temperature behaviour of absorption spectra undertaken for a better understanding of the influence of the ferroelectric transition on the position, intensity, and some other features of the absorption.

2. Physical Properties of Chromium and Copper Boracites

The single crystals of boracites $Cr_3B_7O_{13}Cl$, $Cu_3B_7O_{13}Cl$, and $Cu_3B_7O_{13}Br$ were grown by the transport reaction method [1]. Crystals having edges as large as 1 mm for Cr–Cl, 3 mm for Cu–Cl, and 6 mm for Cu–Br were obtained. The colour of Cr–Cl boracite was bright-blue. The crystals of Cu-boracites were grey. The developed faces of crystals were typically (111), (110), and (100). The structure and composition of crystals were verified by X-ray and chemical analyses. Some properties of the studied boracites are shown in Table 1.

Table 1

Lattice parameters at $T = 295$ °K, magnetic Néel temperature T_N, and ferroelectric Curie temperature T_c

boracite	a (Å) [1]	T_N (°K) [2]	T_c (°K) [7]
$Cr_3B_7O_{13}Cl$	12.121	25	251 to 257
$Cu_3B_7O_{13}Cl$	11.935	20	365
$Cu_3B_7O_{13}Br$	11.955	24	222 to 226

3. Energy Level Scheme of Cr^{2+} and Cu^{2+} Ions in Cubic and Tetragonal Crystal Fields

3.1 The case of Cr^{2+} ions

Divalent chromium ion Cr^{2+} has four electrons in unfilled 3d-shell and the ground state of the free ion is ^5D. In cubic field of octahedral symmetry three 3d-electrons occupy three t_{2g}-orbitals, the fourth one can be in both e_g- and t_{2g}-orbitals depending on the values of the Slater integrals F_2 and F_4 and the value of the cubic crystal field Dq. The energy of these two different state are expressed as [8]

$$\left.\begin{aligned} E(^5E_g, t_{2g}^3e_g) &= -6Dq - 21F_2 - 189F_4\,,\\ E(^3T_{1g}, t_{2g}^4) &= -16Dq - 15F_2 - 44F_4\,. \end{aligned}\right\} \quad (1)$$

The estimation of these two energies by using the known values of $F_2 = 1130\ cm^{-1}$ and $F_4 = 147\ cm^{-1}$ for the isoelectronic Mn^{3+} ion shows that the low-spin state $^3T_{1g}$ is realized when $Dq \approx 2800\ cm^{-1}$. From the experimental results for $Cr_3B_7O_{13}Cl$ we have $Dq \approx 1300\ cm^{-1}$. So we conclude that the ground state is the quintet $^5E_g(t_{2g}^3, e_g)$. The excited state is the quintet $^5T_{2g}(t_{2g}^2, e_g^2)$. These quintet states are shown in Fig. 1, the triplet states are not shown because we did not study them in absorption.

The wave functions of the 3d-states in cubic crystal field can now be used for the tetragonal splitting. As it was already discussed, the crystal field for 3d-ions in boracites may be taken as C_{4v} and D_{4h} in ferroelectric and paraelectric phases, respectively.

For the doubly degenerate ground state ^5E_g in cubic field the wave functions are

$$\left.\begin{aligned} \psi_1 &= |\overset{+}{(xy)}\,\overset{+}{(xz)}\,\overset{+}{(yz)}\,\overset{+}{(x^2 - y^2)}| \,, \\ \psi_2 &= |\overset{+}{(xy)}\,\overset{+}{(xz)}\,\overset{+}{(yz)}\,\overset{+}{(z^2)}| \,. \end{aligned}\right\} \tag{2}$$

By applying to these functions the symmetry operations of group C_{4v} we see that ψ_1 and ψ_2 transform as the irreducible representations B_1 and B_2. The splitting of the ground state now may be expresed in terms of tetragonal parameters Ds and Dt [8]. The energy of the ψ_1 state is changed by $(3Ds + 6Dt)$, and of the ψ_2 state by $(-2Ds + Dt)$ resulting in the total splitting of the ^5E_g state:

$$\Delta E = E(^5B_1) - E(^5A_1) = 4Ds' + 5Dt \,. \tag{3}$$

Now we proceed to the splitting of the excited state $^5T_{2g}(t_{2g}^2, e_g^2)$ by a tetragonal field. We have three wave functions:

$$\left.\begin{aligned} \varphi_1 &= |\overset{+}{(xy)}\,\overset{+}{(yz)}\,\overset{+}{(z^2)}\,\overset{+}{(x^2 - y^2)}| \,, \\ \varphi_2 &= |\overset{+}{(xy)}\,\overset{+}{(xz)}\,\overset{+}{(z^2)}\,\overset{+}{(x^2 - y^2)}| \,, \\ \varphi_3 &= |\overset{+}{(xz)}\,\overset{+}{(yz)}\,\overset{+}{(z^2)}\,\overset{+}{(x^2 - y^2)}| \,. \end{aligned}\right\} \quad . \tag{4}$$

In C_{4v} φ_3 transforms as the irreducible representation B_2, φ_1 and φ_2 transform one into another and correspond to the doubly degenerate irreducible representation E. The displacements of these levels are $E(^5B_2) = (-2Ds + Dt)$ and $E(^5E) = (Ds - 4Dt)$, the total splitting being equal to

$$\Delta E = E(^5B_2) - E(^5E) = 3Ds - 5Dt \,. \tag{5}$$

From the energy level scheme shown in Fig. 1 we see that the rule of the centre of gravity fulfils for Ds within e_g- and t_{2g}-levels and as for Dt within the whole configuration.

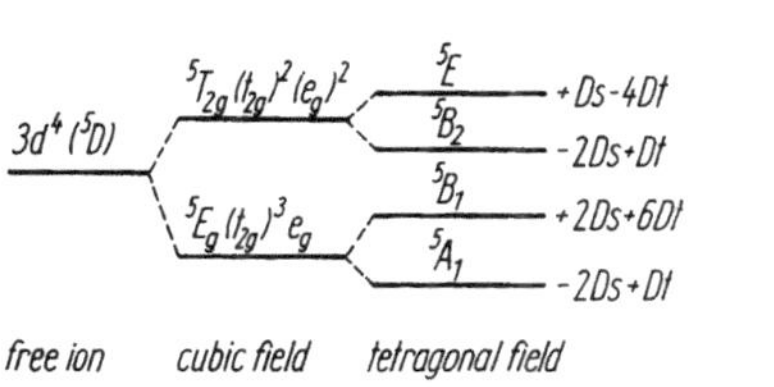

Fig. 1. Energy level diagram of Cr^{2+} in cubic and tetragonal fields

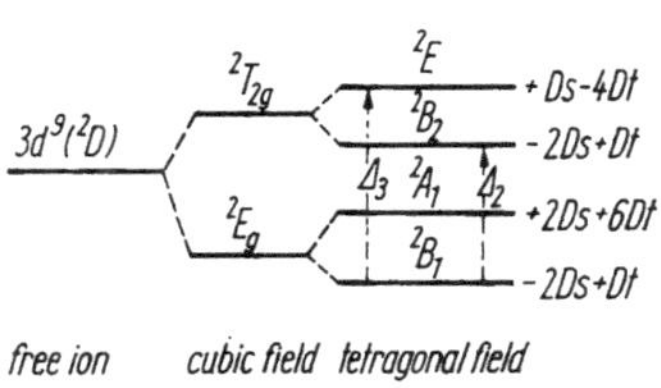

Fig. 2. Energy level diagram of Cu^{2+} in cubic and tetragonal fields

3.2 The case of Cu^{2+} ions

The divalent copper ion Cu^{2+} has nine electrons in the 3d-shell and the ground state of the free ion is ^2D. In the cubic field of octahedral symmetry the ^2D state is split into ^2E_g (ground) and $^2T_{2g}$ (excited) ones. The wave functions are

$$\left.\begin{aligned} &{}^2E_g(t_{2g}^6, e_g^3): \quad \alpha_1 = |(x^2 \overset{+}{-} y^2)|\,; \quad \alpha_2 = |(\overset{+}{z^2})|\,, \\ &{}^2T_{2g}(t_{2g}^5, e_g^4): \quad \beta_1 = |(\overset{+}{xy})|\,; \quad \beta_2 = |(\overset{+}{xz})|\,; \quad \beta_3 = |(\overset{+}{yz})|\,. \end{aligned}\right\} \tag{6}$$

As in the case of the Cr^{2+} ion we shall express the splitting of these states by the tetragonal field in terms of parameters Ds and Dt. For the ground state ^2E_g we have

$$\begin{aligned} &{}^2B_1(\alpha_1)\,; \quad E(^2B_1) = -\,2Ds + Dt\,, \\ &{}^2A_1(\alpha_2)\,; \quad E(^2A_1) = 2Ds + 6Dt\,, \end{aligned}$$

that gives for the resulting splitting

$$\Delta E = E(^2B_1) - E(^2A_1) = 4Ds + 5Dt\,. \tag{7}$$

For the excited state $^2T_{2g}$ one obtains

$$\begin{aligned} &{}^2E(\beta_2, \beta_3)\,; \quad E(^2E) = Ds - 4Dt\,, \\ &{}^2B_2(\beta_1)\,; \quad E(^2B_2) = -\,2Ds + Dt\,, \end{aligned}$$

and the splitting is

$$\Delta E = E(^2E) - E(^2B_2) = 3Ds - 5Dt\,. \tag{8}$$

The energy level scheme of the Cu^{2+} ion in cubic and tetragonal fields is shown in Fig. 2.

4. Optical Absorption Spectra of Copper and Chromium Boracites

Despite the fact that the energy levels scheme of the Cu^{2+} ion is rather simple as depicted in Fig. 2, different authors propose different explanations of the observed absorption spectra. To our opinion, in all absorption spectra of the Cu^{2+} ion in a cubic crystal field a broad and rather intense band is observed in the region 8000 to 16000 cm^{-1} depending on the ligand ions. This band can be interpreted as $^2E_g \rightarrow {}^2T_{2g}$ transition.

In a tetragonal crystal field both the ground ^2E_g and the excited $^2T_{2g}$ states are split. As a matter of fact the broad absorption band usually has one or more shoulders in those compounds where an octahedron of ligands is distorted [9] or where an octahedron is composed of different ligands and consequently has a pronounced tetragonality [10 to 13]. So it was interesting to study the absorption spectra of copper boracites in which the tetragonal component of the crystal field is rather important.

Absorption spectra of boracites $Cu_3B_7O_{13}Cl$ and $Cu_3B_7O_{13}Br$ are shown in Fig. 3 a, b. The absorption consists of a broad band in both crystals in the region 8000 to 12000 cm^{-1}, the absorption coefficient in the maximum is approximately 3000 cm^{-1}. In both crystals the absorption band is split, there are two peaks in Cu–Cl boracite and three peaks in Cu–Br boracite. The splitting is about 2000 cm^{-1}. It is reasonable to suggest that these peaks are due to the tetragonal component of the crystal field, i.e. the splitting of the $^2T_{2g}$ state by the tetragonal component of the crystal field or even by the crystal field components of lower symmetry. So we associate these peaks with the $^2B_1 \rightarrow {}^2B_2$, ^2E transitions (Fig. 2).

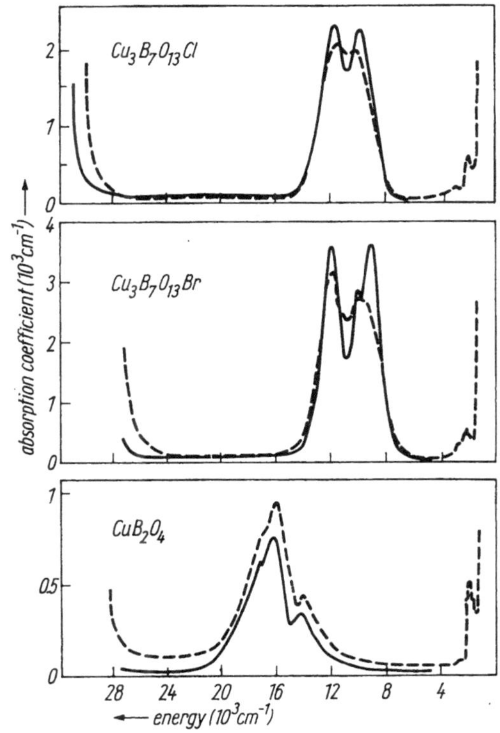

Fig. 3. Optical absorption of a) $Cu_3B_7O_{13}Cl$, b) $Cu_3B_7O_{13}Br$, c) CuB_2O_4. Solid line: 77 °K; dashed line: 295 °K

The ground state $^2E_g(O_h)$ is also split in the tetragonal field. According to calculations of [14] the splitting is about ≈ 1300 cm^{-1}. We did not observe any pronounced absorption in this spectral range and probably this transition is overlapped with the strong infrared lattice absorption. If we suppose that this transition is shifted to the visible spectral range we find that the tetragonal field parameters Ds and Dt are larger than Dq.

Actually we observed two bands at 2000 and 2700 cm^{-1} in the infrared region, but they cannot be identified with Cu^{2+}-ion transitions because they are also observed in boracites of iron [5], cobalt [4], and cadmium [6]. We found these two bands equally in Zn–Cl boracite. These bands are typical of boron–oxygen compounds and were observed in some other borates [15, 16]. We did not find any other bands in the infrared region though samples as thick as 0.8 mm were used.

The proposed identification of the 8000 to 12000 cm^{-1} bands is also supported by the calculation of the g-factor of the ground state in $Cu_3B_7O_{13}Cl$. According to [17] the two values of the g-factor are expressed as

$$\left.\begin{aligned} g_{\parallel} &= 2\,(1 - 4\lambda/\Delta_2)\,, \\ g_{\perp} &= 2\,(1 - \lambda/\Delta_3)\,, \end{aligned}\right\} \tag{9}$$

where $\Delta_2 = E(^2B_2) - E(^2B_1) = 10250$ cm^{-1} and $\Delta_3 = E(^2E) - E(^2B_1) = 11600$ cm^{-1}. The spin–orbit constant $\lambda = -852$ cm^{-1} is for the free ion of Cu^{2+}. Using these numerical values we obtain $g_{\parallel} = 2.68$ and $g_{\perp} = 2.14$.

The g-factors of Cu^{2+} ions were measured by the electron paramagnetic resonance method for $Zn_3B_7O_{13}Cl$ boracite doped with Cu^{2+} (0.2%) [18]. At room temperature $g_{\parallel} = 2.44$ and $g_{\perp} = 2.09$ were found, that are in good agreement with our estimations. The small discrepancy may be easily understood taking into account the usual lowering of the spin–orbit constants λ by 15 to 20% in crystals comparing to the free ion value.

Besides the copper boracites, we studied also the optical absorption spectrum of copper borate CuB_2O_4 that is shown in Fig. 3c. In this crystal the Cu^{2+} ions are situated in the oxygen octahedron. As the oxygen ion gives a greater con-

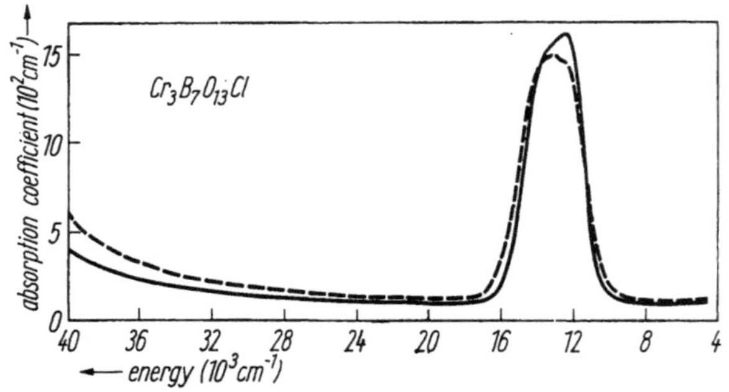

Fig. 4. Optical absorption of $Cr_3B_7O_{13}Cl$. Solid line: 77 °K, dashed line: 295 °K

tribution to the crystal field than halogen ions, the crystal field in borates is greater than in boracites. A strong absorption band is observed in the visible region of the spectrum at $\approx$16000 cm^{-1} and the cubic crystal field parameter is $Dq \approx$ 1600 cm^{-1}. The shoulders in the absorption are interpreted as due to a low-symmetry crystal field of the distorted octahedron.

As it was discussed in Section 3 of this paper for the Cr^{2+} ion in the tetragonal field two transitions $^5A_1 \rightarrow {}^5B_2$, ^5E are expected in the visible region (Fig. 4). As a matter of fact a strong absorption ($\alpha \approx$ 1500 cm^{-1}) centered at $\nu \approx$ $\approx$13000 cm^{-1} is observed. The splitting of the band is not well resolved and is roughly estimated to be $\approx$1000 cm^{-1}. The splitting of the ground state must be of the same order of magnitude.

No spin-forbidden bands were observed in this study because of the unsufficient thickness of chromium boracite samples.

5. Influence of Ferroelectric Phase Transition on the Absorption Spectra of Copper Boracites

The ferroelectric phase transition in boracites changes the crystallographic unit cell and the local symmetry of the ions in the lattice. As the positions of

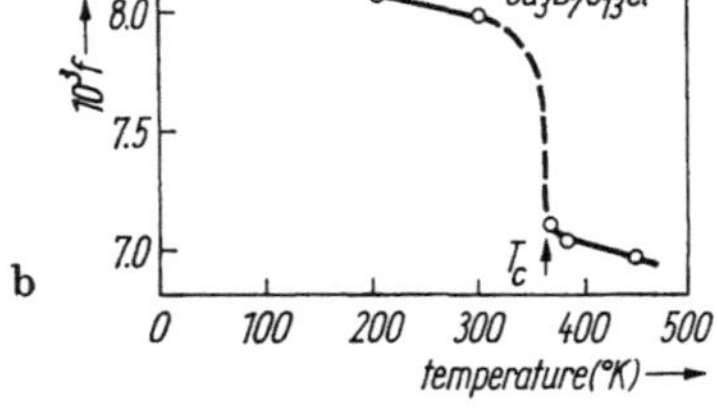

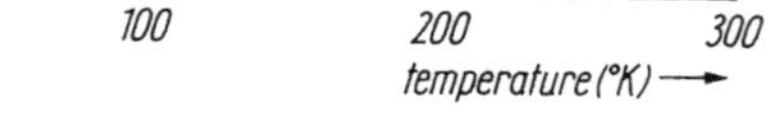

Fig. 5 Fig. 6

Fig. 5. Temperature dependence of a) the splitting of the $^2T_{2g}$ state by the tetragonal component of the crystal field in $Cu_3B_7O_{13}Cl$, b) the oscillator strength of the $^2B_1 \rightarrow {}^2B_2 + {}^2E$ transition in $Cu_3B_7O_{13}Cl$

Fig. 6. Temperature dependence of the absorption coefficient at fixed frequency ν = 12000 cm^{-1} of $Cu_3B_7O_{13}Br$

the energy levels of 3d-ions depend on the strength and symmetry of the crystal field changes in the optical spectra of boracites may be expected.

In Fig. 5a we show the temperature dependence of the splitting $E(^2E) - E(^2B_2)$ of boracite $Cu_3B_7O_{13}Cl$. This splitting depends on the tetragonal component of the crystal field. There is a rapid variation of this splitting near $T_c = 365$ °K.

The temperature dependence of the oscillator strength of the same transition is shown in Fig. 5b. As two components ^2E and $^2B_2$ of the excited state are not well resolved Fig. 5b shows the variation of the integrated intensity. Near the temperature of phase transition the intensity is changed by $\approx 10\%$.

Besides the integrated intensity of the transition $^2B_1 \rightarrow {}^2B_2 + {}^2E$ we measured the absorption coefficient at fixed frequency $\nu = 12000$ cm^{-1} on the blue side of the absorption band. As it was earlier observed in iron boracites [5] near the Curie temperature there is an abrupt change of the absorption coefficient (Fig. 6). This variation is caused both by the change of the intensity of the transition and by the change of the splitting.

We studied also the variation of the absorption near the edge of the fundamental absorption in the ultraviolet part of the spectral range. In this region there are no 3d-transitions and the absorption coefficient is changed due to the displacement of the fundamental absorption edge. Fig. 7 shows that near T_c the slope of the curve $\alpha(T)$ is markedly changed.

In order to study the domain structure formation of Cu–Cl boracite by optical methods the single domain crystal was placed between crossed polarizers. The sample of $1.5 \times 1 \times 0.15$ mm^3 size was cut from the (as-grown) natural (100)-face and the direction [110] (one of the main axes of the refraction index ellipsoid) was parallel to the polarization direction of the polarizer. The wavelength $\lambda = 6328$ Å ($\nu = 15802$ cm^{-1}) was used. The absorption coefficient at this wavelength was about 50 cm^{-1}.

In this geometry the signal of the photomultiplier was zero. When the crystal was heated the light signal due to the breaking up of the single domain crystal into a polydomain crystal was observed at about 343°. The main indexes of refraction of the domains are parallel to [100] type directions of the cubic lattice and form the angle of 45° with the polarization plane of the light. The temperature dependence of the light intensity transmitted through the polarizer, the sample, and the crossed analyzer is shown in Fig. 8. The maximum signal was

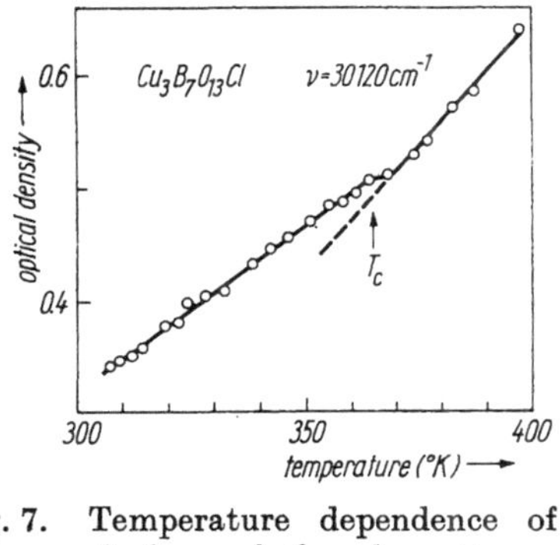

Fig. 7. Temperature dependence of the change of slope of the absorption coefficient in $Cu_3B_7O_{13}Cl$

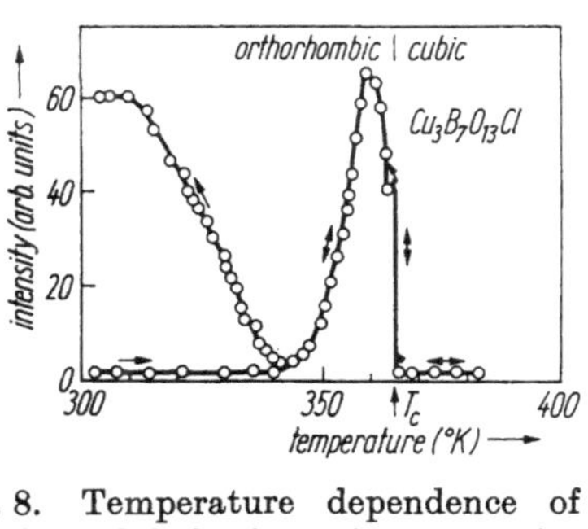

Fig. 8. Temperature dependence of the polarized light intensity transmitted through a sample of $Cu_3B_7O_{13}Cl$

observed at $T = 360$ °K, the decrease of the signal above this temperature may be caused by a change of the difference $n_\gamma - n_\beta$ near T_c, as it was early observed in Ni–Br [7]. No signal was detected in the paraelectric cubic phase above $T_c = 365$ °K. The cooling of the crystal from the paraelectric phase is accompanied by the same variation of the intensity of transmitted light until 343 °K (there is also a small temperature hysteresis at $\approx$365 °K), but below this temperature the single crystal again was divided into domains.

References

[1] H. Schmid, J. Phys. Chem. Solids **26**, 973 (1965).
[2] G. Qúezel and H. Schmid, Solid State Commun. **6**, 447 (1968).
[3] T. Ito, N. Morimoto, and R. Sadanaga, Acta Cryst. **4**, 310 (1951).
[4] R. V. Pisarev, V. V. Druzhinin, S. D. Prochorova, N. N. Nesterova, and G. T. Andreeva, phys. stat. sol. **35**, 145 (1969).
[5] R. V. Pisarev, V. V. Druzhinin, N. N. Nesterova, S. D. Prochorova, and G. T. Andreeva, phys. stat. sol. **40**, 503 (1970).
[6] E. Dormann, J. Phys. Chem. Solids **31**, 199 (1970).
[7] E. Ascher, H. Schmid, and D. Tar, Solid State Commun. **2**, 45 (1964).
[8] C. J. Ballhausen, Introduction to Ligand Field Theory, McGraw-Hill Publ. Co., New York/London 1962.
[9] O. von Schmitz-DuMont and D. Grimm, Z. anorg. allg. Chem. **355**, 280 (1967).
[10] O. von Schmitz-DuMont und H. Fendel, Monatshefte für Chemie **96**, 495 (1965).
[11] O. G. Holmes and D. S. McClure, J. chem. Phys. **26**, 1686 (1957).
[12] J. Bjerrum, C. J. Ballhausen, and C. K. Jørgensen, Acta Chem. Scand. **8**, 1275 (1954).
[13] H. Elliott and B. J. Hathaway, Inorganic Chemistry **5** (1966).
[14] A. D. Liehr and C. J. Ballhausen, Ann. Phys. (U.S.A.) **3**, 304 (1958).
[15] A. Sommer, D. White, M. J. Linevsky, and D. E. Mann, J. chem. Phys. **38**, 87 (1963).
[16] K. E. Lawson, Infrared Absorption of Inorganic Substances, Reinhold Publ. Co., New York 1961.
[17] A. Abragham and M. H. L. Pryce, Proc. Roy. Soc. **A206**, 164 (1951).
[18] M. P. Petrov, S. A. Kizhaev, G. T. Andreeva, and G. A. Smolensky, J. Phys. Soc. Japan Suppl. **28**, 128 (1970).

(Received April 22, 1974)

phys. stat. sol (b) 65, 111 (1974)

Subject classification: 18.2

Institute for Metal Physics, Academy of Sciences of the USSR, Sverdlovsk

Parametric Excitation of Spin Waves in Ferromagnets with Domain Structure

By

B. N. Filippov, V. P. Lukomskii, and Yu. G. Lebedev

An investigation is made of the mechanism of the influence of an rf field on ferromagnets with different domain structures leading to a modulation of the internal effective magnetic field, in which the magnetization vector precesses. The calculations are made for the case when external polarizing magnetic field $H_0 = 0$. It is shown that the mechanism of the parametric excitation investigated is strongly sensitive to the type of the domain structure. It turns out that the parametric excitation of the spin waves in a ferromagnet with a Shirobokov domain structure is impossible. However, in a ferromagnet with a striped domain structure the investigated mechanism leads to the excitation of wall and intra-domain modes and to the simultaneous excitation of both the modes as well. The excitation thresholds are found of all the modes under consideration and analysed.

Рассматривается механизм влияния высокочастотного поля на ферромагнетики с разными доменными структурами, приводящий к модуляции внутреннего эффективного магнитного поля, в котором прецессирует намагниченность ферромагнетика. Расчет проводится для случая, когда внешнее поляризующее магнитное поле $H_0 = 0$. Показывается, что рассматриваемый механизм параметрического возбуждения спиновых волн оказывается весьма чувствительным к типу доменной структуры. Так оказывается, что параметрическое возбуждение спиновых волн в ферромагнетике с доменной структурой широбоковского типа невозможно. Однако в ферромагнетике с полосовой доменной структурой рассматриваемый механизм приводит к возбуждению внутриграничных и внутридоменных спиновых волн, а так же к одновременному возбуждению тех и других. Найдены и проанализированы пороги всех этих возбуждений.

1. Introduction

The investigation of parametric excitation of the spin-wave modes in the saturated ferromagnet [1, 2] has permitted to give a consistent explanation of known experimental data and put a number of new problems connected with the measurements of the ferromagnetic material characteristics and with application of the instability effects in practice. However, ferromagnets used in technology sometimes work in the static magnetic field H_0 below saturation, therefore the parametric effects must have peculiarities in them.

The initial stable state is now essentially inhomogeneous in the presence of the domain structure (DS) and, as a result of this, the spin-wave spectrum differs considerably from that of the homogeneously magnetized ferromagnet. At present the rigorous spin–wave theory in the presence of some types of DS is worked out under the assumption that $H_0 = 0$ only. The main features of the spin-wave spectrum in the ferromagnet with a single domain wall and in the presence of the periodical DS [3] were stated by Winter [4] and Farztdinov and Turov [5, 6]. In [6], in particular, it was shown that in the presence of the periodical DS there exist two spin-wave energy bands. One of these bands actually corresponds to the usual spin-wave modes strongly altered, however, due to magnetic inhomogeneity. The other one characterizes the spin-wave modes of a new type with the maximum of the amplitude in the centre of the domain walls. For the first time this periodical structure was studied by Shirobokov [3], so we call it Shirobokov DS. Erukhimov [7] has recently investigated the spin-wave spectrum of so-called films with striped DS, where the magnetization does not rotate, but undergoes a nutation in some angular interval less then π relative to the hard-magnetized plane.

For the spin-wave spectrum the eigenfunctions and the orthogonality relations are known, therefore it is possible to calculate strictly all the parametric effects arising from the influence of the rf magnetic field upon the ferromagnet.

If one considers the domain walls as certain two-dimensional imperfection we would come to the conclusion that the spin waves must undergo additional scattering, hence, the presence of the domains must result in an increase of the parametric instability threshold for these waves. In [8], however, it was found by an experimental study of the threshold $h_{\text{thr}}(H_0)$ that at given magnetization and pump frequency and in the fields H_0 immediately below saturation, h_{thr} reached its maximum value and decreased with decreasing H_0 reaching sometimes values less than $(h_{\text{thr}})_{\min}$ in the saturated sample. It is possible that the perfect periodical DS as a matter of fact cannot lead to scattering of spin waves (for example, in full analogy with the Bloch waves in a perfect crystal lattice), but stipulates the existence of a certain spin-wave spectrum only. As for damping it will depend, for example, on the interaction between the spin waves.

The calculation of the spin-wave instability under parallel pumping in the presence of striped DS [9] was carried out only under the assumption that the spin-wave instability conditions in each domain were just the same as in the saturated single crystal. This approach, however, wholly ignores the possible excitation of the other types of the spin-wave modes existing in the ferromagnet with DS, though, in principle, the instability threshold of these modes can be the lowest.

In this paper the calculation of the parametric excitation thresholds was carried out for two different DS [3, 10] at $H_0 = 0$ both for the above-mentioned types of the spin-wave modes and for the simultaneous excitation of both types of spin-wave modes in the presence of the rf pumping field applied parallel to the domain wall. It should be noted that in the presence of DS the terms parallel or perpendicular pumping have lost their usual meaning if they were connected with the geometry of the problem. However, both ways of pumping can be characterized by the influence of the homogeneous rf magnetic field on the crystal. As a rule, the rf field gives rise to modulation of the internal field under parallel pumping, while under perpendicular pumping the rf field is connected only with the magnetization changing its direction. In the present paper the spin-wave stability is examined relative to the first way of excitation only. It is shown that the finite instability thresholds occur only in films with striped DS of the nutation type.

2. General Expression for the Parametric Excitation Threshold

Let us consider a plate-like specimen of a uniaxial ferromagnetic single crystal with dimensions $2d \times L$ placed in the xy plane of the laboratory coordinate system x, y, z (Fig. 1). We assume that in the plate there exists a striped DS

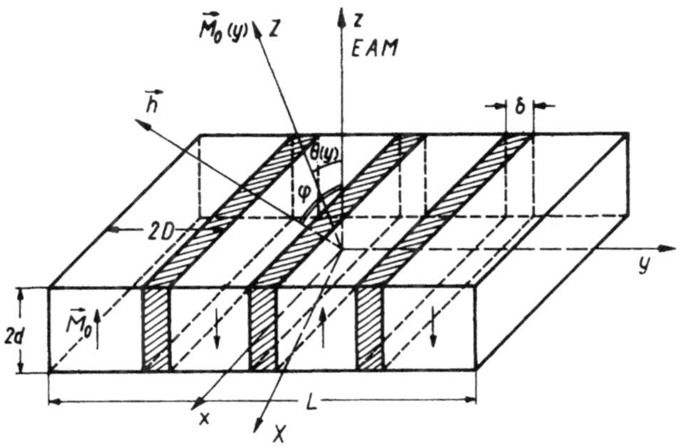

Fig. 1. Depiction of the domain structure and geometry of the problem

with domain walls perpendicular to the y-axis. The easy axis of magnetization (EAM) is considered to be parallel to the z-axis. Let us assume following Winter [4] the local coordinate system XyZ to be turned at the angle $\theta(y)$ relative to the x, y, z coordinate system and connected with the direction of the local stable magnetization $M_0(y)$. $\theta(y)$ is counted from the z-axis. Both origins coincide and are placed in the centre of one of the walls.

In the local coordinate system the Hamiltonian of the ferromagnet can be written as [11]

$$\mathcal{H} = \int\limits_V \left\{\frac{Ja_0^2}{M_0^2}\left[(\nabla M_X)^2 + (\nabla M_y)^2 + \left(\frac{d\theta}{dy}\right)^2 (M_X^2 + M_Z^2) + 2\,\frac{d^2\theta}{dy^2}\,M_X M_Z\right] + \right.$$
$$+ \frac{K_A}{M_0^2}\,[M_Z^2 \sin^2\theta - M_X M_Z \sin 2\theta + M_X^2 \cos^2\theta + M_y^2] + \frac{1}{2}\,N_y M_y^2 +$$
$$+ \frac{1}{2}\,[M_X^2 \sin^2\theta + M_Z^2 \cos^2\theta + M_X M_Z \sin 2\theta] +$$
$$\left. + \frac{K'_A}{M_0^2}\,M_X^2 - h_1(t)\,M_X + h_2(t)\,M_Z\right\} d\boldsymbol{r}\,. \tag{1}$$

Here V is the volume of the ferromagnet, J is the exchange integral, a_0 denotes the crystal lattice constant, K_A is the crystallographic magnetic anisotropy constant, K'_A is a constant in the expression for the quasi-elastic energy $(K'_A/M_0^2)\,M_X^2$ characterizing the domain wall stability relative to their displacement (see [4, 12]). M_X, M_y, M_Z are the Cartesian components of M in the X,y,Z coordinate system. M_0 is the saturation magnetization. The rigorous evaluation of the demagnetizing field energy is a fairly complicated problem. We took it into account making use of the rough approximation of Erukhimov [7] assuming the demagnetizing factors $N_y = 4\pi - N_Z$,

$$N_Z = \frac{2D}{d}\left[1 - \exp\left(-\frac{2\pi d}{D}\right)\right], \tag{2}$$

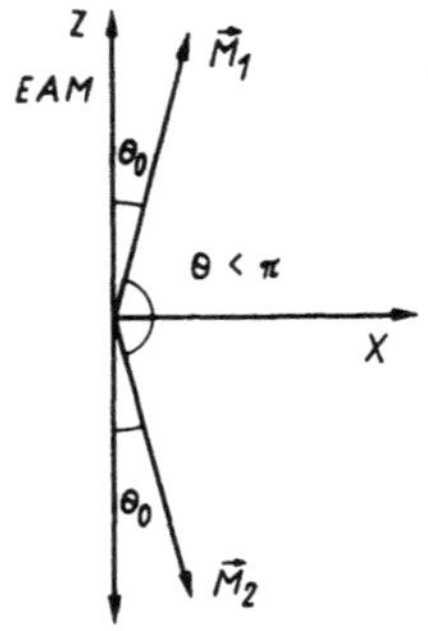

Fig. 2. Plane section of the nutating domain structure perpendicular to the y-axis and parallel to the plane in which the magnetization vector $\boldsymbol{M}$ rotates. $\boldsymbol{M}_1$ and $\boldsymbol{M}_2$ are boundary orientations of the magnetization vectors

where D is the stable width of a domain. The approximation made is equivalent to that of Winter [4], provided $D/d \ll 1$,

$$h_1(y, t) = h_x(t) \cos \theta + h_z(t) \sin \theta \ , \ h_2(y, t) = h_x(t) \sin \theta - h_z(t) \cos \theta \ . \qquad (3)$$

The first and second terms in (1) take account of the exchange and anisotropy energies, respectively, the last two terms allow for the energy of the system in the external field h. In the stable state the components $M_X = 0$, $M_y = 0$, and $M_Z = M_0$ provided $h = 0$. Then equating the variation of $\mathcal{H}$ with respect to θ to zero one can readily obtain the following equation describing the stable magnetization distribution:

$$\delta^2 \left(\frac{\mathrm{d}\theta}{\mathrm{d}y}\right)^2 - \sin^2 \theta = C \ , \qquad (4)$$

where $\delta = a_0(J/K_{1\mathrm{A}})^{1/2}$, $K_{1\mathrm{A}} = K_\mathrm{A} - \frac{1}{2} N_Z M_0^2$, and C is a constant which can be determined by minimization of the energy of the domain walls and the demagnetizing energy. With $C = 0$ there is a distribution corresponding to a single domain wall [13]. $C > 0$ corresponds to the Shirobokov DS. At last, if $C < 0$ there is a striped DS in which M_0 undergoes a nutation relative to the film plane (plane of hard magnetization) in the angular interval less than π (see Fig. 2). For the sake of convenience we shall call this type of DS the nutating DS. Available experimental data indicate that the nutating DS is likely realizing in the so-called films with striped DS. We assume that this type of DS can be realized also in sufficiently thick films. In such a case the magnetization vector can approach EAM up to very small angles θ_0, but nevertheless the fact that θ is nutating in the angular interval less than 2π exerts a decisive influence upon the parametric effects under consideration.

Assuming $M_X, M_y \ll M_0$, $M_Z \approx M_0 [1 - (1/2M_0^2) (M_X^2 + M_y^2)]$ we perform the Fourier transformation of M_X, M_y with respect to X and Z coordinates and introduce the creation and destruction operators

$$M_X = \frac{1}{2} \left(\frac{2\mu_\mathrm{e} M_0}{S}\right)^{1/2} \sum_\varkappa [a_\varkappa(y) \, \mathrm{e}^{i(\varkappa_X X + \varkappa_Z Z)} + a_\varkappa^+(y) \, \mathrm{e}^{-i(\varkappa_X X + \varkappa_Z Z)}] \ ,$$

$$M_y = \frac{1}{2i} \left(\frac{2\mu_\mathrm{e} M_0}{S}\right)^{1/2} \sum_\varkappa [a_\varkappa(y) \, \mathrm{e}^{i(\varkappa_X X + \varkappa_Z Z)} - a_\varkappa^+(y) \, \mathrm{e}^{-i(\varkappa_X X + \varkappa_Z Z)}] \ ,$$

$\varkappa = \{\varkappa_X, 0, \varkappa_Z\}$ being the wave vector lying in the domain wall plane, $\mu_\mathrm{e} = \gamma \hbar$ is the Bohr magneton, γ the gyromagnetic ratio, S the domain wall area, $a_\varkappa$ and $a_\varkappa^+$ obey the following commutation relations:

$$[a_\varkappa(y) \, a_{\varkappa'}^+(y')] = \delta_{\varkappa\varkappa'} \delta(y - y') \ .$$

The mentioned operation being performed, the Hamiltonian (1) can be written as

$$\mathcal{H} = \sum_{\varkappa} \int_V \Big\{ \Phi_\varkappa(y)\, a_\varkappa^+(y)\, a_\varkappa(y) + \alpha \frac{\partial a_\varkappa^+(y)}{\partial y} \frac{\partial a_\varkappa(y)}{\partial y} - \frac{1}{2} B[a_\varkappa(y)\, a_{-\varkappa}(y) + a_\varkappa^+(y)\, a_{-\varkappa}^+(y)] - \frac{1}{2} (2\mu_e M_0)^{1/2}\, h_1(y, t)\, (a_0^+ + a_0) \Big\} \, \mathrm{d}y \,, \tag{5}$$

$$\Phi_\varkappa(y) = \alpha \varkappa^2 + \frac{2\mu_e}{M_0} K_{1A} \cos 2\theta(y) + \frac{\mu_e}{M_0} K_A' + N_A \mu_e M_0 - \mu_e h_2(y, t) \,, \tag{6}$$

where

$$B = N_A \mu_e M_0 - \frac{\mu_e}{M_0} K_A' \,, \qquad \alpha = 2\mu_e J a_0 \,.$$

We omitted the terms without $a_\varkappa^+$ and $a_\varkappa$. From (5) and (6) it follows that h_2 leads to the modulation of the internal effective field and $h_1(y, t)$ gives only linear excitation of the spin-wave modes with $\varkappa = 0$. Assuming that the pumping field frequency does not coincide with the frequencies of these modes we omitted the term with h_1 and later on consider the phenomena connected with the modulation of the internal effective field. Making use of (5) the equations of motion for $a_\varkappa(y)$ and $a_{-\varkappa}^+(y)$ can be easily obtained:

$$\left.\begin{aligned} i\hbar \dot{a}_\varkappa(y) &= \Phi_\varkappa(y)\, a_\varkappa(y) - \alpha \frac{\mathrm{d}^2 a_\varkappa(y)}{\mathrm{d}y^2} - B a_{-\varkappa}^+(y) \,, \\ i\hbar \dot{a}_{-\varkappa}^+(y) &= -\Phi_\varkappa(y)\, a_{-\varkappa}^+(y) + \alpha \frac{\mathrm{d}^2 a_{-\varkappa}^+(y)}{\mathrm{d}y^2} + B a_\varkappa(y) \,. \end{aligned}\right\} \tag{7}$$

The solution of (7) can be represented in terms of $b_{\varkappa f}$ and $b_{\varkappa f}^+$ operators

$$\left.\begin{aligned} a_\varkappa(y, t) &= \sum_f [u_{\varkappa f}(y)\, b_{\varkappa f}(t) + v_{-\varkappa f}^*(y)\, b_{-\varkappa f}^+(t)] \,, \\ a_{-\varkappa}^+(y, t) &= \sum_f [u_{-\varkappa f}^*(y)\, b_{-\varkappa f}^+(t) + v_{\varkappa f}(y)\, b_{\varkappa f}(t)] \,, \end{aligned}\right\} \tag{8}$$

where $b_{\varkappa f}$ and $b_{\varkappa f}^+$ are the Bose operators. Index f stands for both the possible band branches and the modes of every band. $u_{\varkappa f}$ and $v_{\varkappa f}$ are unknown and chosen so that the Hamiltonian (3) would be diagonal in terms of $b_{\varkappa f}$ and $b_{\varkappa f}^+$. This being performed one can readily obtain that $u_{\varkappa f}$ and $v_{\varkappa f}$ obey the following differential equations:

$$\left.\begin{aligned} \alpha \frac{\mathrm{d}^2 u_{\varkappa f}}{\mathrm{d}y^2} + (p \sin^2\theta + \varepsilon_{\varkappa f} - \mathcal{D}_\varkappa)\, u_{\varkappa f} + B v_{\varkappa f} &= 0 \,, \\ \alpha \frac{\mathrm{d}^2 v_{\varkappa f}}{\mathrm{d}y^2} + (p \sin^2\theta - \varepsilon_{\varkappa f} - \mathcal{D}_\varkappa)\, v_{\varkappa f} + B u_{\varkappa f} &= 0 \,, \end{aligned}\right\} \tag{9}$$

and the orthonormalization relations

$$\int (u_{\varkappa f} u_{\varkappa f'}^* - v_{\varkappa f} v_{\varkappa f'}^*)\, \mathrm{d}y = \delta_{ff'} \,, \qquad \int (u_{\varkappa f} v_{-\varkappa f'} - u_{-\varkappa f} v_{\varkappa f'})\, \mathrm{d}y = 0 \,. \tag{10}$$

Here $p = 2(2\mu_e K_{1A}/M_0)$, $\varepsilon_{\varkappa f}$ is the spectrum of the spin waves and

$$\mathcal{D}_\varkappa = \frac{2\mu_e}{M_0} J (a_0 \varkappa)^2 + \frac{2\mu_e}{M_0} K_{1A} + \frac{\mu_e K_A'}{M_0} + \mu_e M_0 N_A \,.$$

Substituting (8) into (7) and taking into account (9) and (10) one can obtain

$$\left.\begin{aligned} i\hbar\dot{b}_{\varkappa f} &= \varepsilon_{\varkappa f} b_{\varkappa f} - \mu_e h_0 \cos\omega t \sum_{f'} \Lambda^*_{ff'}(\varkappa)\, b^+_{-\varkappa f'}\,, \\ -i\hbar\dot{b}_{-\varkappa f} &= \varepsilon_{\varkappa f} b^+_{-\varkappa f} - \mu_e h_0 \cos\omega t \sum_{f'} \Lambda_{ff'}(-\varkappa)\, b_{\varkappa f'}\,, \end{aligned}\right\} \tag{11}$$

where

$$\Lambda_{ff'}(\varkappa) = \mu_{ff'}(\varkappa)\sin\varphi - \lambda_{ff'}(\varkappa)\cos\varphi\,, \tag{12}$$

$$\mu_{ff'}(\varkappa) = \int_V \sin\theta(y)\,[v_{\varkappa f'} u_{-\varkappa f} + v_{-\varkappa f} u_{\varkappa f'}]\,dy\,, \tag{13}$$

$$\lambda_{ff'}(\varkappa) = \int_V \cos\theta(y)\,[v_{\varkappa f'} u_{-\varkappa f} + v_{-\varkappa f} u_{\varkappa f'}]\,dy\,. \tag{14}$$

Here $h_x(t) = h_0\cos\omega t\sin\varphi$, $h_z(t) = h_0\cos\omega t\cos\varphi$, ω is the frequency of the external rf field (the pumping field). In (11) the term corresponding to the modulation of $\varepsilon_{\varkappa f}$ with the frequency ω was omitted for it does not lead to instability [1]. We have taken dissipation into account assuming

$$\varepsilon_{\varkappa f} \to \varepsilon_{\varkappa f} + \frac{i}{2}\mu_e \Delta H_{\varkappa f}\,, \tag{15}$$

where $\Delta H_{\varkappa f}$ is the equivalent linewidth of spin waves under consideration and consider the solution of the system (11) in the form $\exp(i\omega_{\varkappa f} t)$. Then for the waves satisfying the condition

$$\omega_{\varkappa f} + \omega_{\varkappa f'} = \omega \tag{16}$$

one can obtain the following expression for the instability threshold:

$$h_{0,\,\mathrm{thr}} = \left(\frac{\Delta H_{\varkappa f}\,\Delta H_{-\varkappa f'}}{\Lambda_{f'f}\Lambda^*_{ff'}}\right)^{1/2}. \tag{17}$$

While obtaining (17) we assumed the existence of a spin-wave spectrum with eigenfunctions satisfying the orthogonality relations (10), but did not define concretely these functions. In the next sections we shall calculate $\Lambda_{ff'}(\varkappa)$ for the Shirobokov and nutating DS.

3. Homogeneous Magnetizing

If one considers $2d \gg D > L_y$, practically $d\theta/dy = \theta(y) = 0$. Therefore there is a long plate in the z-direction magnetized practically homogeneously and $\mu_{ff'} = 0$. Assuming further $K'_A = 0$, $C = 0$ one obtains that $N_A = 4\pi$. Strictly speaking this result is correct provided $\varkappa = 0$. Then taking (11) into account we have the following solution of (9):

$$u_{0,k_y} = \left(\frac{A_{k_y} + \varepsilon_{k_y}}{2L\varepsilon_{k_y}}\right)^{1/2} e^{\pm ik_y y}\,, \qquad v_{0,k_y} = \left(\frac{A_{k_y} - \varepsilon_{k_y}}{2L\varepsilon_{k_y}}\right)^{1/2} e^{\pm ik_y y}\,, \tag{18}$$

$$\varepsilon^2_{k_y} = \left(\alpha k_y^2 + \frac{2\mu_e K_A}{M_0}\right)\left(\alpha k_y^2 + \frac{2\mu_e K_A}{M_0} + 4\pi\mu_e M_0\right), \tag{19}$$

where $A_{k_y} = \alpha k_y^2 + 2\mu_e K_A/M_0 + 2\pi\mu_e M_0$, L being the specimen length in the y-direction. Calculating $\lambda_{ff'}(0)$ and substituting it into (17) one obtains

$$h_{0,\,\mathrm{thr}} = \frac{\varepsilon_{k_y}\,\Delta H_{k_y}}{B\,|\cos\varphi|}\,. \tag{20}$$

The expression (20) is just the same as that from [1] for the parallel pumping instability threshold in a single domain specimen, provided $H_0 = 0$ and $\varphi = 0$ when the excited waves propagate perpendicular to M_0.

4. Spin-Wave Instability Threshold in a Ferromagnet with Shirobokov DS

Now the stable magnetization distribution is described by the following expression:

$$\cos \theta = - \operatorname{sn}\left(\frac{y}{\delta k}\right), \tag{21}$$

where $\operatorname{sn} u$ is the elliptic sine (see e.g. [14]). Here $k = (C + 1)^{-1/2}$, $(C > 0)$, $D = 2k\delta K(k)$, k is the parameter of the elliptic integral, and $K(k) \equiv K$ is the complete elliptic integral of the first kind.

For the Shirobokov DS the spin-wave spectrum was studied by Farztdinov and Turov [5, 6]. Following [6] we shall call the first type of elementary excitations as spin-wave (intradomain), the second one as translational (wall). Making use of the functions $u_\varkappa(f, y)$, $v_\varkappa(f, y)$ determined in [5, 6] we are able now to find $\Lambda_{ff'}(\varkappa)$ involved in the expression for the excitation thresholds for spin-wave and translational excitation, and for the simultaneous excitation of both modes. The elaborated calculation gives the following expression for all cases:

$$\Lambda_{ff'}(\varkappa) = 0\,. \tag{22}$$

Thus in accordance with (22) and (17) one can come to the conclusion that the excitation thresholds for the wall and intradomain modes and the threshold of the simultaneous excitation of both modes in the ferromagnet with periodical Shirobokov DS due to modulation of the internal effective field by the rf field h are all infinite. Therefore, in this case the parametric excitation of the mentioned modes is impossible. The physical essence of this result can be elucidated in the following way. The effect of the parametric excitation arises from taking into account the interaction of the magnetization with the rf field. Within the limits of the model the energy density E_h of this interaction is proportional to $(h_x \sin \theta(y) - h_z \cos \theta(y))\,(M_X^2 + M_y^2)$. The alternative part of the magnetization (the components M_X and M_y) at the pumping frequency induces a complete revolution following M_0. As $M_0(y)$ is oriented at all possible angles relative to h, E_h is positive for some values of y and E_h is negative for the other values of y, and as a result of the symmetry of the magnetization distribution the integral contribution of the negative part of the interaction energy will be compensated by the integral contribution of the positive part of this energy.

This allows to understand in what cases the energy under consideration must be different from zero. For this it is necessary that the angle θ should nutate in the angular interval less than 2π, being essentially less than 2π; otherwise the excitation threshold different from zero may be too large. This situation occurs in the presence of the nutating DS where θ lies in the angular interval less than π relative to the direction $\theta = \pi/2$. From the mentioned considerations, however, it is clear that in this case the integral contribution arising from the term $h_z \cos \theta(y)\,(M_X^2 + M_y^2)$ must be equal to zero too and, hence, the contribution to the excitation threshold of the spin waves will arise only from the term $\sin \theta(y)\,(M_X^2 + M_y^2)\,h \sin \varphi$. In the next sections we shall show it by detailed calculations.

5. Nutating DS

In this case the stable magnetization distribution can be expressed as

$$\cos\theta = -k \operatorname{sn} \xi\,, \tag{23}$$

where $\xi = y/\delta$. In this case $k = \sqrt{C+1}$, $(-1 \leqq C < 0)$. The angle $\pi/2 - \theta_0 = \arccos k$, hence, it is defined in terms of the parameter k of the elliptic integral. In contrast to the former distribution $D = 2K\delta$ now. From (9) and (10) one can see that two types of waves are possible [7], which similarly to the case of the Shirobokov DS were called spin-wave (intradomain) and translational (wall). All the quantities relating to the first type of the elementary excitations we shall label by the index 1 while relating to the second one by the index 2 in parentheses. As in [7] $u_{\varkappa g}^{(1,2)}$ and $v_{\varkappa g}^{(1,2)}$ were given without normalization, we put them down here having taken into account the orthogonality relations (10). Thus, from (9) and (10) by full analogy with [7] one can obtain:

$$u_{\varkappa g}^{(1,2)}(y) = c_{1\varkappa g}^{(1,2)}\psi_{\varkappa g}^{(1,2)}\,, \qquad v_{\varkappa g}^{(1,2)}(y) = c_{2\varkappa g}^{(1,2)}\psi_{\varkappa g}^{(1,2)}(y)\,, \tag{24}$$

$$\psi_{\varkappa g}^{(1,2)} = e^{iq_y\delta\xi}\,F_{\varkappa g}^{(1,2)}(\xi); \qquad F_{\varkappa g}^{(1,2)} = e^{-i\pi\xi/2K}\,\frac{\vartheta_1[(\pi/2K)\,(\xi + is_g^{(1,2)} + \xi_n^{(1,2)})]}{\vartheta_4(\pi\xi/2K)}, \tag{25}$$

$$c_{1\varkappa g}^{(1,2)} = -\frac{B_{\varkappa g}}{|B_{\varkappa g}|}\sqrt{\frac{A_{\varkappa g}^{(1,2)} + \varepsilon_{\varkappa g}^{(1,2)}}{2\varepsilon_{\varkappa g}^{(1,2)}R_g^{(1,2)}}}\,\chi_{\varkappa g}^{(1,2)}; \quad c_{2\varkappa g}^{(1,2)} = \sqrt{\frac{A_{\varkappa g}^{(1,2)} - \varepsilon_{\varkappa g}^{(1,2)}}{2\varepsilon_{\varkappa g}^{(1,2)}R_g^{(1,2)}}}\,\chi_{\varkappa g}^{(1,2)}\,, \tag{26}$$

$$\chi_{\varkappa g}^{(1,2)} = \nu^{(1,2)}\left(\frac{K'}{K}\right)^{1/2}\frac{\vartheta_4(0|\tau)}{\vartheta_2\left(\dfrac{\pi s_g^{(1,2)}}{2K'}\middle|\,\tau'\right)}\exp\left(-\frac{\pi(s_g^{(1,2)})^2}{4KK'}\right), \tag{27}$$

$$\left.\begin{aligned}
&\nu^{(1)} = 1\,, \quad \nu^{(2)} = \sqrt{k'}\,, \quad \xi_n^{(1)} = 2nK\,, \quad \xi_n^{(2)} = (2n+1)\,K\,,\\
&R_g^{(1)} = 2N\delta\left[\frac{K-E}{k^2} + K\operatorname{sc}^2(s_g^{(1)}, k')\right],\\
&R_g^{(2)} = 2N\delta\left[\frac{E-k'^2K}{k^2} + E\operatorname{sc}^2(s_g^{(2)}, k')\right],\\
&\varepsilon_{\varkappa g}^{(1,2)} = \sqrt{(A_{\varkappa g}^{(1,2)})^2 + B_{\varkappa g}^2}\,,
\end{aligned}\right\} \tag{28}$$

$$A_{\varkappa g}^{(1,2)} = \frac{2\mu_e}{M_0}\left[J(a_0\varkappa)^2 + \frac{1}{2}\left(K'_A + \frac{1}{2}N_yM_0^2 - (k^2+1)\,K_{1A}\right) - K_{1A}C_{\varkappa g}^{(1,2)}\right], \tag{29}$$

$$B_{\varkappa g} = \mu_e M_0\left[\frac{1}{2}N_y - \frac{(k^2-1)\,K_{1A} + K'_A}{M_0^2}\right] \equiv B\,, \tag{30}$$

$$\left.\begin{aligned}
&C_{\varkappa g}^{(1)} = \frac{1 - \operatorname{nc}^2(s_g^{(1)}, k')\operatorname{dc}^2(s_g^{(1)}, k')}{\operatorname{sc}^2(s_g^{(1)}, k')},\\
&C_{\varkappa g}^{(2)} = -\frac{1 + (k')^4\operatorname{sd}^2(s_g^{(2)}, k')\operatorname{cd}^2(s_g^{(2)}, k')}{\operatorname{nd}^2(s_g^2, k')}.
\end{aligned}\right\} \tag{31}$$

The quantities $s_g^{(1,2)}$ can be obtained from the following equations:

$$s_g^{(1)} + \operatorname{dn}(s_g^{(1)}, k') \operatorname{sc}(s_g^{(1)}, k') - E(s_g^{(1)}, k') - s_g^{(1)} \frac{E}{K} - \frac{\pi}{2K} = -\frac{\pi g}{KN}, \quad (32)$$

$$s_g^{(2)} + \operatorname{dn}(s_g^{(2)}, k') \operatorname{sc}(s_g^{(2)}, k') - E(s_g^{(2)}, k') - \frac{s_g^{(2)} E}{K} - \frac{\pi}{2K} -$$

$$- k^2 \operatorname{sc}(s_g^{(2)}, k') \operatorname{nc}(s_g^{(2)}, k') \operatorname{dc}(s_g^{(2)}, k') [1 + k^2 \operatorname{sc}^2(s_g^{(2)}, k')]^{-1} = -\frac{\pi g}{KN}. \quad (33)$$

$\operatorname{cn} u$, $\operatorname{dn} u$ are elliptic cosine and delta of the amplitude. $\operatorname{nc} u$, $\operatorname{dc} u$ and other functions of this type in (29) to (33) are elliptic functions connected with $\operatorname{sn} u$, $\operatorname{cn} u$, and $\operatorname{dn} u$ (see, for example, [14]); $E(\xi) = \int_0^\xi \operatorname{dn}^2 \xi \, d\xi$; E is the complete elliptic integral of the second kind; $K' \equiv K(k')$, $k' = (1 - k'^2)^{1/2}$, $\tau = iK'/K$, $\tau' = -\tau^{-1}$, $N = L/D$. In (25) $\vartheta_1(t)$ and $\vartheta_4(t)$ are Jacobian ϑ-functions,

$$q_y = \frac{2\pi}{L} g \equiv \frac{\pi g}{\delta K N}; \qquad g = 0, \pm 1, \pm 2, \dots . \quad (34)$$

For defining the allowed values of q_y we make use of the Born-Karman condition assuming $\psi_{\varkappa g}^{(1,2)}(y) = \psi_{\varkappa g}^{(1,2)}(y + L)$, and take also into account that $F_{\varkappa g}^{(1,2)}(y/\delta)$ is periodical along the y-axis with the period $a_n = nD$ (or $2nK$ along ξ, for $D = 2\delta K$) that is an arbitrary vector of the crystal lattice connected with the periodical DS. In this sense (25) is analogous to the Bloch function. The fact that the primitive period of $F_{\varkappa g}^{(1,2)}$ is equal to D, but not to $2D$, is connected with the symmetry of the equation (9). The quantity q_y (hence, g) is limited by the size of the first Brillouin zone for the translational modes. For the intradomain modes there is no restriction upon g which is a consequence of the continuous-medium approximation. Index f is substituted here by the indices 1 and 2, which characterized the spin-wave bands and index g characterizing the number of the modes in the band.

Making use of (24) to (31) one can obtain

$$\mu_{g,-g+Nm}^{(1)} = -\frac{\pi B k^2}{4\varepsilon_{\varkappa g}^{(1)}} [1 + 2 \operatorname{sc}^2(s_g^{(1)}, k')] [K - E + k^2 \operatorname{sc}^2(s_g^{(1)}, k')]^{-1}, \quad (35)$$

$$\mu_{g,-g+Nm}^{(2)} = -\frac{\pi B}{4\varepsilon_{\varkappa g}^{(2)}} k^2 [1 + (2 - k^2) \operatorname{sc}^2(s_g^{(2)}, k')] \times$$

$$\times [E - k'^2 K + k^2 E \operatorname{sc}^2(s_g^{(2)}, k')]^{-1}, \quad (36)$$

$$\mu_{g,-g+Nm}^{(0)} = (-1)^m \frac{\pi^3 k' K' \delta}{4 k^{5/2} K^3} B Q_1 \frac{\exp(-3\pi K'/4K)}{\nu_{\varkappa g} \vartheta_4(0|\tau)} \frac{1 + \nu_{\varkappa g}^2}{\sqrt{\varepsilon_{\varkappa g}^{(1)} \varepsilon_{\varkappa g}^{(2)} (R_g^{(1)} R_g^{(2)} / N^2)}} \times$$

$$\times \frac{\exp(\varepsilon_g^{(2)} - \varepsilon_g^{(1)})}{\sinh(\varepsilon_g^{(1)} - \varepsilon_g^{(2)})} \frac{\exp\left\{-\frac{\pi}{4KK'} [(s_g^{(1)})^2 + (s_g^{(2)})^2]\right\}}{\vartheta_2\left(\frac{\pi s_g^{(1)}}{2K'} \middle| \tau'\right) \vartheta_2\left(\frac{\pi s_g^{(2)}}{2K'} \middle| \tau'\right)}, \quad (37)$$

$$\lambda_{g,-g+Nm}^{(1)} = \lambda_{g,-g+Nm}^{(2)} = \lambda_{g,-g+Nm}^{(0)} = 0. \quad (38)$$

Here $m = 0, 1, 2, \dots$ Index 0 in parentheses labels the quantities corresponding to the threshold of the simultaneous excitation of the intradomain and wall modes.

$$Q_1 = \frac{\mathrm{d}^2}{\mathrm{d}t^2} \{\vartheta_3(t) [\vartheta_1(t + i\varepsilon_g^{(1)}) \vartheta_2(t - i\varepsilon_g^{(2)})]\}|_{t=t_1},$$

$$t_1 = \frac{\pi\tau}{2}, \qquad \varepsilon_g^{(1,2)} = \frac{\pi s_g^{(1,2)}}{2K}, \qquad \nu_{\varkappa g} = \left(\frac{A_{\varkappa g}^{(2)} + \varepsilon_{\varkappa g}^{(2)}}{A_{\varkappa g}^{(1)} + \varepsilon_{\varkappa g}^{(1)}}\right)^{1/2}.$$

The thresholds for the intradomain and wall modes can be written as

$$h_{0,\,\mathrm{thr}}^{(1,2)} = \frac{\Delta H_{\varkappa g}^{(1,2)}}{|\mu_{g,\,-g+Nm}^{(1,2)}(\varkappa)|\,|\sin\varphi|}. \tag{39}$$

The threshold of the simultaneous excitation of the intradomain and wall modes according to (37), (38), and (17) can be represented in the following way:

$$h_{0,\,\mathrm{thr}}^{(0)} = \frac{(\Delta H_{\varkappa g}^{(1)}\,\Delta H_{\varkappa g}^{(2)})^{1/2}}{|\sin\varphi|} \frac{4k^{5/2}K^3\nu_{\varkappa g}}{\pi^3 k' K' B} \frac{\vartheta_4(0|\tau)\,\mathrm{e}^{3\pi K'/2K}}{(1+\nu_{\varkappa g}^2)\,|Q_1|} \sqrt{\varepsilon_{\varkappa g}^{(1)}\varepsilon_{\varkappa g}^{(2)}(R_g^{(1)}R_g^{(2)}/N^2)}\times$$
$$\times \frac{\sinh(\varepsilon_g^{(2)} - \varepsilon_g^{(1)})}{\exp(\varepsilon_g^{(2)} - \varepsilon_g^{(1)})} \frac{\vartheta_2\left(\dfrac{\pi s_g^{(1)}}{2K'}\middle|\tau'\right)\vartheta_2\left(\dfrac{\pi s_g^{(2)}}{2K'}\middle|\tau'\right)}{\exp\left\{-\dfrac{\pi}{4KK'}[(s_g^{(1)})^2 + (s_g^{(2)})^2]\right\}}. \tag{40}$$

In these expressions $\Delta H_{\varkappa g}^{(1)}$ and $\Delta H_{\varkappa g}^{(2)}$ are equivalent linewidths of the intradomain and wall modes, respectively.

Thus, the lowest excitation thresholds correspond to the field polarization perpendicular to EAM. This was qualitatively explained at the end of Section 4. It should be noted that in [8] an increasing of the contribution of the perpendicular component of h (relative to EAM) was observed if there was DS in the specimen, not homogeneity. Formulas (39), (40) correspond to the lowest excitation threshold for the wall modes provided $\Delta H_{\varkappa g}^{(1)} \gtrapprox \Delta H_{\varkappa g}^{(2)}$. This can be seen fairly clear when $k \to 1$. In this limit

$$h_{0,\,\mathrm{thr}}^{(1,2)} = \frac{8\varepsilon_{\varkappa g}^{(1,2)}\,\Delta H_{\varkappa g}^{(1,2)}}{\hbar\omega_\mathrm{M}\,|\zeta|\,|\sin\varphi|}\,\Gamma^{(1,2)}, \tag{41}$$

where $\Gamma^{(1)} = \ln(4/k')$, $\Gamma^{(2)} = 1$,

$$h_{0,\,\mathrm{thr}}^{(0)} = \frac{\pi}{2}\left|\frac{1 - \dfrac{2g}{N} + \pi m}{\sin\dfrac{\pi g}{N}}\right| \frac{(\Delta H_{\varkappa g}^{(1)}\,\Delta H_{\varkappa g}^{(2)})^{1/2}}{|\sin\varphi|} \frac{\nu_{\varkappa g}(\varepsilon_{\varkappa g}^{(1)}\varepsilon_{\varkappa g}^{(2)})^{1/2}}{\omega_\mathrm{M}\,|\zeta|\,(1+\nu_{\varkappa g}^2)}\left(\ln\frac{4}{k'}\right)^{1/2}, \quad m = 1, 2, \dots. \tag{42}$$

Then from (28) to (30) and at $k \to 1$, and taking into account that $D/d \ll 1$ in this case (see [7]) we can obtain

$$\left(\frac{\varepsilon_{\varkappa g}^{(1)}}{\hbar}\right)^2 = [\omega_\mathrm{a}(a_0q)^2 + \omega_\mathrm{A} + \omega_\mathrm{A}']\,[\omega_\mathrm{a}(a_0q)^2 + \omega_\mathrm{A} + \omega_\mathrm{M}], \tag{43}$$

$$\left(\frac{\varepsilon_{\varkappa g}^{(2)}}{\hbar}\right)^2 = [\omega_\mathrm{a}(a_0q)^2 + \omega_\mathrm{A}']\,[\omega_\mathrm{a}(a_0q)^2 + \omega_\mathrm{M}]. \tag{44}$$

In (41) to (44)

$$\omega_a = \frac{2I}{M_0}\gamma; \qquad \omega_A = \frac{2K_A}{M_0}\gamma; \qquad \omega_A' = \frac{2K_A'}{M_0}\gamma; \qquad \omega_M = 4\pi M_0\gamma;$$

$$\zeta = 1 - \frac{\omega_A'}{\omega_M}; \qquad q^2 = \varkappa^2 + q_y^2 .$$

Formulas (43) and (44) are just the same as those found by Winter [4] for a single domain wall. Therefore (41) and (42) correspond to the excitation thresholds for the Winter modes. The lowest one corresponds to the excitation of the wall modes (44). The excitation wave vectors of these modes at the pumping frequencies $\omega \geqq \omega_0 = 2(\omega_A'\omega_M)^{1/2}$ can be found from the equation

$$\varkappa = \frac{1}{a_0}\left[\frac{1}{2\omega_a}\left(\sqrt{(\omega_A - \omega_A')^2 + \omega^2} - \omega_M - \omega_A'\right)\right]^{1/2}. \tag{45}$$

In (44) the quantity K_A' is stipulated by different imperfections or mechanical stresses. As a rule $K_A' \ll K_A$ and $K_A' \ll 4\pi M_0$. Therefore the energy of the wall modes can be lower than the energy of the intradomain modes and spin-wave modes in the homogeneously magnetized sample as well. From here it follows that the frequency band for the excitation of the wall modes can lie lower than that used for the homogeneously magnetized samples. Therefore though there is the factor $|\zeta| < 1$ in the threshold of the wall modes, this threshold can lie lower than that of the homogeneously magnetized sample where $|\zeta| = 1$.

Returning to the study of the threshold of the wall modes, which is defined by (39), one can see that the lowest value of this threshold is being reached when $s_g^{(2)} = 0, \varphi = \pi/2$. This corresponds to excited waves which propogate along the domain walls. In this case the threshold is

$$(h_{0,\,\mathrm{thr}}^{(2)})_{\min} = \frac{2}{\pi}\frac{|\Delta H_\varkappa|}{k^2\,|B|}\hbar\omega\,|E - k'^2 K| . \tag{46}$$

Here B is a function of k and crystal thickness $2d$. Thus, the threshold $(h_{0,\,\mathrm{thr}}^{(2)})_{\min}$ depends also on $2d$. However, this dependence is a weak one. According to [7] on decreasing $2d$ when the plate becomes a thin magnetic film with thickness $2d = 2d_c$, where $d_c = (3\pi M_0^2/K_A)\,(6J/K_A)^{1/2}a_0$ and $K_A < 2\pi M_0^2$ there occurs a phase transition of the second kind and the film becomes homogeneously magnetized in the film plane. According to [7], if $d = d_c$, so $D = D_c = \pi(6J/K_A)^{1/2}\,a_0$ and $N_Z = 2K_A/3M_0$. $k = 0$ in the vicinity of the critical point and the threshold (46) is

$$(h_{0,\,\mathrm{thr}}^{(2)})_{\min} = \frac{\omega\,|\Delta H_\varkappa|}{\omega_M}\tau , \tag{47}$$

where $\tau = 1 + 4\omega_A/3\omega_M - \omega_A'/\omega_M$. If $\omega_A \ll \omega_M \gg \omega_A'$ so $\tau \approx 1$ and the threshold (47) coincides with that for spin-wave modes in the homogeneously magnetized film (see, for example [15]) but corresponds to the excitation of waves of different character and length.

6. Conclusions

As the result of the calculation of the parametric effects connected with the modulation of the internal effective field by the rf field in the ferromagnets with some types of DS at $H_0 = 0$ it turned out that:

1. The parametric excitation of spin waves with frequency $\omega/2$ in the presence of the Shirobokov DS in the framework of the approximations made is impossible.

2. In the presence of the nutating DS the excitation of both the wall and intradomain modes is possible at the pumping frequency ω divided by the factor 2. The simultaneous excitation of both modes is also possible. The lowest threshold corresponds to the excitation of the wall modes propagating along the domain walls. In sufficiently thick plates and when $k \to 1$ the excitation thresholds of all the modes under consideration are increasing infinitely. The excitation threshold of the wall modes is, nevertheless, finite. Thus, the mechanism of parametric excitation under consideration proved to be strongly sensitive to the type of DS.

These results cannot be obtained by the usual technique [9] which starts from the assumption that the excitation conditions in every domain are just the same as that in the homogeneously magnetized specimen, however, at present we have no available data to give evidence to the calculations.

References

[1] E. Schlömann and R. Joseph, J. appl. Phys. **32**, 1006 (1961).
[2] H. Suhl, J. Phys. Chem. Solids **1**, 209 (1957).
[3] M. Shirobokov, Zh. eksper. teor. Fiz. **15**, 57 (1945).
[4] J. M. Winter, Phys. Rev. **124**, 452 (1961).
[5] M. M. Farztdinov and E. A. Turov, Fiz. Metallov i Metallovedenie **29**, 458 (1970).
[6] M. M. Farztdinov and E. A. Turov, Fiz. Metallov i Metallovedenie **30**, 1064 (1970).
[7] M. S. Erukhimov, Izv. Akad. Nauk SSSR, Ser. fiz. **36**, 1485 (1972).
[8] E. V. Lebedeva, A. I. Pilshchikov, and N. S. Sedletskaya, Fiz. tverd. Tela **8**, 3204 (1966).
[9] W. E. Courtney, Electronics Letters **1**, 21 (1965).
[10] J. Kaczér, M. Zeleny, and P. Suda, Czech. J. Phys. **13**, 579 (1963).
[11] A. P. Tankeev, Kandidatskaya dissertatsiya, Sverdlovsk 1970.
[12] E. A. Turov and M. P. Petrov, Yaderny magnitny resonans v ferro- i antiferromagnetikakh, Nauka, Moscow 1969 (p. 168).
[13] L. D. Landau and E. M. Lifshits, Phys. Z. SU **8**, 153 (1935).
[14] E. T. Whittaker and G. N. Watson, A Course of Modern Analysis, Cambridge University Press 1927.
[15] V. P. Lukomskii and A. V. Kuzko, Fiz. tverd. Tela **11**, 2951 (1969).

(Received February 27, 1974)

phys. stat. sol. (b) 65, 123 (1974)

Subject classification: 13.5 and 20.1; 22.6

Sektion Physik der Universität München, Lehrstuhl Prof. Brandmüller (a), Fachbereich Physik der Universität Münster, Lehrstuhl Prof. Merten (b), and Sektion Physik der Universität München (c)

Dispersion of Generalized Surface Phonon Polaritons on α-Quartz Observed by Attenuated Total Reflection

By

H. J. FALGE (a), G. BORSTEL (b), and A. OTTO (c)

Real and virtual excitation generalized surface polaritons are investigated on a 45° cut of an α-quartz crystal. For the wave vector of the surface polariton coplanar with the normal to the surface and the optical axis they are transverse magnetic polarized extraordinary waves. The experimental results agree with calculated dispersion curves. By attenuated total reflection for transverse magnetic polarized radiation it is shown that virtual excitation generalized surface polaritons exist only for certain ranges of the angle between the surface and the optical axis.

Es werden reell und virtuell angeregte, verallgemeinerte Oberflächenpolaritonen für den 45°-Schnitt eines α-Quarz-Kristalls untersucht. Wenn der Wellenvektor des Oberflächenpolaritons koplanar mit der Normalen zur Oberfläche und der optischen Achse ist, sind sie transversal magnetisch polarisierte, außerordentliche Wellen. Die experimentellen Ergebnisse stimmen mit berechneten Dispersionskurven überein. Durch verhinderte Totalreflektion für transversal magnetisch polarisierte Strahlung wird gezeigt, daß virtuell angeregte, verallgemeinerte Oberflächenpolaritonen nur innerhalb von bestimmten Bereichen des Winkels zwischen Oberfläche und optischer Achse existieren.

1. Introduction

In [1] ordinary and extraordinary surface polaritons on α-quartz have been observed. As in [2, 3] the measurements were limited to regular surface polaritons. These studies of uniaxial crystals are based on discussions of special crystal cuts [4 to 7] and special directions of the wave vector which lead to regular waves. Generalized surface polaritons of the Fano type[1]) recently have been treated by one of the authors [8]. The results in [8] allow the calculation of the dispersion curve of Fano surface modes on arbitrary crystal cuts. Here we give a short review of the theory and show the necessary conditions for the existence of Fano surface modes on biaxial and uniaxial crystals. We calculated their dispersion for a 45° cut of an α-quartz crystal. The theoretical curves are compared with experimental values obtained by attenuated total reflection (ATR) [9]. Both results are in good agreement. Especially they confirm that virtual excitation surface polaritons exist only within certain ranges of the angle ϑ (Fig. 1).

2. Theory

To investigate polaritons propagating along a plane interface between a biaxial semi-infinite crystal and an isotropic semi-infinite medium we choose the geometry shown in Fig. 2. An orthorhombic crystal characterized by a fre-

[1]) Following Burstein et al. [18], we call transverse magnetic (TM) polarized surface polaritons Fano modes.

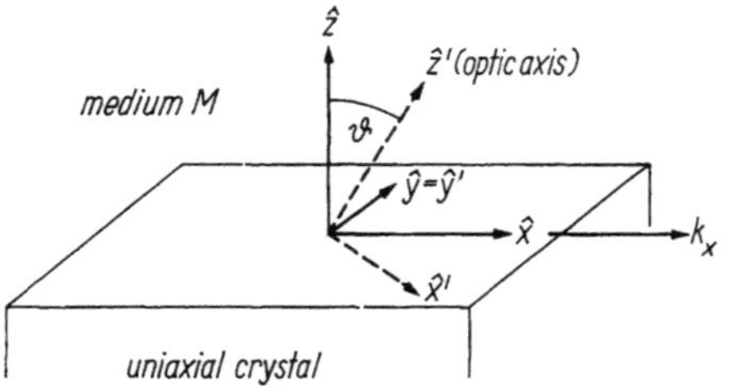

Fig. 1. Geometry for surface polaritons in uniaxial crystals

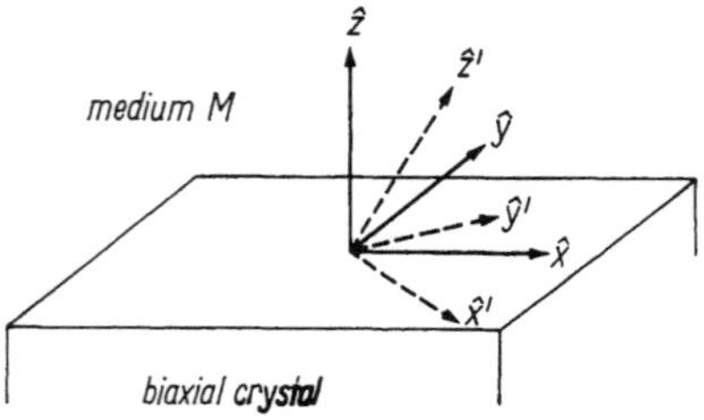

Fig. 2. Geometry for surface polaritons in biaxial crystals

quency dependent dielectric tensor $\boldsymbol{\epsilon}(\omega)$ fills the lower half space $z < 0$, the upper half space $z > 0$ consists of an isotropic medium M with dielectric constant $\varepsilon_M \geqq 1$.

The unit vectors $\hat{\boldsymbol{x}}$, $\hat{\boldsymbol{y}}$, $\hat{\boldsymbol{z}}$ build up an orthonormal basis B of the interface at $z = 0$. $\hat{\boldsymbol{x}}'$, $\hat{\boldsymbol{y}}'$, $\hat{\boldsymbol{z}}'$ denote the principal axes of the dielectric tensor $\boldsymbol{\epsilon}(\omega)$ of the crystal. They form a second orthonormal system B′ and may be oriented in arbitrary directions with respect to B.

The connection of B and B′ is described by an orthogonal transformation **T**:

$$\mathbf{T}: \mathrm{B} \to \mathrm{B}' , \quad \mathbf{T}^+: \mathrm{B}' \to \mathrm{B} .$$

The operator **T** is completely determined by the crystal cut and its components therefore may be regarded as known quantities.

Particular nonradiative solutions (ω real) of Maxwell's equations for the crystal and the isotropic medium M now may be chosen in the form

$$\boldsymbol{E}(\boldsymbol{r}, t) = \boldsymbol{E} \exp (i\boldsymbol{k} \cdot \boldsymbol{r} - i\omega t) , \tag{1a}$$

$$\boldsymbol{E}_M(\boldsymbol{r}, t) = \boldsymbol{E}_M \exp (i\boldsymbol{k}_M \cdot \boldsymbol{r} - i\omega t) . \tag{1b}$$

Herein the amplitudes $\boldsymbol{E}$, $\boldsymbol{E}_M$ and the wave vectors $\boldsymbol{k}$, $\boldsymbol{k}_M$ are complex quantities, i.e. we deal with inhomogeneous waves which in general are elliptically polarized.

Because of the invariance of the whole system with respect to $\hat{\boldsymbol{x}}$ and $\hat{\boldsymbol{y}}$ it follows

$$k_x = k_{Mx} , \quad k_y = k_{My} . \tag{2}$$

As shown in [8] the characteristic equations to determine the wave vector and the frequency in both media are

$$F(k_x, k_y, k_z, \omega) \equiv \sum_\alpha \frac{\varepsilon'_\alpha \left(\sum_\beta T_{\alpha\beta} k_\beta\right)^2}{n^2 - \varepsilon'_\alpha} = 0 , \qquad (z < 0) \tag{3a}$$

$$k_x^2 + k_y^2 + k_{Mz}^2 = \frac{\omega^2}{c^2} \varepsilon_M , \quad (z > 0) \tag{3b}$$

where

$$n^2 \equiv \frac{c^2 \boldsymbol{k} \cdot \boldsymbol{k}}{\omega^2} .$$

Equations (3a, b) must be solved with regard to the electromagnetic boundary conditions at $z = 0$:

$$\hat{z} \cdot (\boldsymbol{D} - \boldsymbol{D}_{\mathrm{M}}) = 0\,, \tag{4a}$$

$$\hat{z} \times (\boldsymbol{E} - \boldsymbol{E}_{\mathrm{M}}) = 0\,, \tag{4b}$$

$$\hat{z} \cdot (\boldsymbol{H} - \boldsymbol{H}_{\mathrm{M}}) = 0\,, \tag{4c}$$

$$\hat{z} \times (\boldsymbol{H} - \boldsymbol{H}_{\mathrm{M}}) = 0\,, \tag{4d}$$

where $\boldsymbol{D} = \boldsymbol{\varepsilon}(\omega)\,\boldsymbol{E}$ denotes the dielectric displacement field, $\boldsymbol{H} = (c/\omega)\,\boldsymbol{k} \times \boldsymbol{E}$ the magnetic field.

As shown in [5] surface polaritons on anisotropic dielectric media in general are combinational waves which involve two electromagnetic waves with different attenuation constants in the anisotropic medium. In the following we restrict our discussion to Fano waves which involve only one single electromagnetic wave in each medium.

In this case the general solution in both media is given by (1) and the boundary conditions (4b, d) for the field components parallel to the surface reduce to

$$E_x = E_{\mathrm{M}x}\,, \tag{5a}$$

$$E_y = E_{\mathrm{M}y}\,, \tag{5b}$$

$$k_y E_z - k_z E_y = k_y E_{\mathrm{M}z} - k_{\mathrm{M}z} E_{\mathrm{M}y}\,, \tag{5c}$$

$$k_z E_x - k_x E_z = k_{\mathrm{M}z} E_{\mathrm{M}x} - k_x E_{\mathrm{M}z}\,. \tag{5d}$$

Boundary conditions (4a) and (4c) follow from (4b) and (4d). Rotating the coordinate system B in such a way that $k_y = 0$ we get from (5c) with (5b)

$$k_z E_y = k_{\mathrm{M}z} E_y\,. \tag{6}$$

Excluding the solution $k_z = k_{\mathrm{M}z}$ of (6) which is not interesting for our problem we have therefore

$$E_y = 0\,. \tag{7}$$

From the definition of $\boldsymbol{H}$ and the boundary conditions (5) it is now easily verified that the only nonzero component of $\boldsymbol{H}$ is H_y, i.e. apart from singular cases surface polaritons with only one electromagnetic wave in the orthorhombic crystal (according to (1a), (1b)) are necessarily transverse magnetically (TM) polarized.

In order to satisfy (7) a certain form of the operator $\mathbf{T}$ is necessary which we will now derive.

According to [8] the different components of the eigenvector $\boldsymbol{E}'$ in the coordinate system B′ are given by

$$E'_\alpha = \frac{k'_\alpha}{k'_\beta} \frac{n^2 - \varepsilon'_\beta}{n^2 - \varepsilon'_\alpha} E'_\beta\,. \tag{8}$$

From (7), (8) it follows:

$$E_y = \left(T_{xy} \frac{k'_x}{k'_z} \frac{n^2 - \varepsilon'_z}{n^2 - \varepsilon'_x} + T_{yy} \frac{k'_y}{k'_z} \frac{n^2 - \varepsilon'_z}{n^2 - \varepsilon'_y} + T_{zy} \right) E'_z = 0\,. \tag{9}$$

Apart from singular cases (9) may be fulfilled only for $T_{xy} = T_{zy} = 0$, $k'_y = 0$. Obviously the operator $\mathbf{T}$ must have the form

$$\mathbf{T} = \begin{bmatrix} T_{xx} & 0 & T_{xz} \\ 0 & 1 & 0 \\ T_{zx} & 0 & T_{zz} \end{bmatrix}, \tag{10}$$

i.e. the coordinate systems B and B′ for Fano waves are necessarily connected by a rotation of the three-dimensional space around $\hat{\boldsymbol{y}}$.

This means that Fano surface modes in orthorhombic crystals may exist only if

1. at least one principal axis of $\boldsymbol{\varepsilon}(\omega)$ lies in the surface of the crystal, and
2. the polariton propagates perpendicular to that principal axis.

Since for Fano modes propagating along $\hat{\boldsymbol{x}}$ and fulfilling (7) the validity of the electric boundary conditions implies that of the magnetic ones only (4a, b) must be satisfied. By means of the operator **T** given by (10) we get from (4a, b), (7), (8), and div $\boldsymbol{E}_{\mathrm{M}} = 0$ for $k_y = 0$

$$G(k_x, k_z, k_{\mathrm{M}z}, \omega) \equiv \sum_{\alpha \neq y} \frac{T_{\alpha x} k_x + T_{\alpha z} k_z}{n^2 - \varepsilon_\alpha'} (\varepsilon_{\mathrm{M}} T_{\alpha x} k_x + \varepsilon_\alpha' T_{\alpha z} k_{\mathrm{M}z}) = 0 . \tag{11}$$

Equation (11) together with (3a, b) completely determines the three unknown eigenvalues ω, k_z, $k_{\mathrm{M}z}$ as a function of the two-dimensional wave vector $\boldsymbol{\varphi} = (k_x, 0)$. The fact that Fano modes in orthorhombic crystals may exist only in special configurations has been overlooked in [8]. As a consequence (6) in [8] are valid only if the operator **T** is assumed to have the form (10) and in addition $\boldsymbol{\varphi} = (k_x, 0)$ holds.

In general it is not possible to combine (3) and (11) to an explicit dispersion relation. Only in the case **T** = **I** (**I** unit tensor) we get

$$\frac{k_x^2 c^2}{\omega^2} = \varepsilon_{\mathrm{M}} \varepsilon_z' \frac{\varepsilon_x' - \varepsilon_{\mathrm{M}}}{\varepsilon_x' \varepsilon_z' - \varepsilon_{\mathrm{M}}^2} \tag{12}$$

which has been derived previously by Hartstein and Burstein [6].

It easily follows from (3a) that regular surface excitations (k_z pure imaginary) may exist only if the direction of propagation $\boldsymbol{\varphi} = (k_x, 0)$ coincides with a principal axis of $\boldsymbol{\varepsilon}(\omega)$. In all other cases the excitation is generalized i.e. inside the crystal the corresponding fields decay sinusoidally with increasing distance from the surface.

The analysis shows that all dispersion branches of TM surface polaritons start at the light line in the dielectric $k = (\omega/c)\sqrt{\varepsilon_{\mathrm{M}}}$ at frequencies which are either transverse phonon frequencies or solutions of

$$\varepsilon_{\mathrm{M}} = \varepsilon_x' T_{xz}^2 + \varepsilon_z' T_{xx}^2 \tag{13}$$

with the additional condition $\varepsilon_x' \varepsilon_z' < 0$. For $\vartheta = 0$ and 90°, special cases are depicted in Fig. 1 of [15].

In the electrostatic limiting case $k_x \to \infty$ the limiting frequencies ω_j^{S} are given by

$$\varepsilon_x'(\omega_j^{\mathrm{S}})\, \varepsilon_z'(\omega_j^{\mathrm{S}}) = \varepsilon_{\mathrm{M}}^2 , \tag{14}$$

where $\varepsilon_x' < 0$, $\varepsilon_z' < 0$.

Introducing the commonly used Kurosawa-relation [10] for ε_x' and ε_z',

$$\varepsilon_x' = \varepsilon_x^\infty \prod_{j=1}^{n_x} \frac{(\omega_{xj}^{\mathrm{LO}})^2 - \omega^2}{(\omega_{xj}^{\mathrm{TO}})^2 - \omega^2} ; \qquad \varepsilon_z' = \varepsilon_z^\infty \prod_{j=1}^{n_z} \frac{(\omega_{zj}^{\mathrm{LO}})^2 - \omega^2}{(\omega_{zj}^{\mathrm{TO}})^2 - \omega^2} , \tag{15}$$

it is found from (13) and (14) that the maximum possible number of Fano modes is just $n_x + n_z$.

Some dispersion branches may exist only for finite values of k_x. According to [6] such excitations are called virtual excitation modes whereas all solutions with limit $k_x \to \infty$ are called real excitation modes.

In regions of frequency where $\varepsilon_x' < 0$, $\varepsilon_z' < 0$ the product $\varepsilon_x' \varepsilon_z'$ may take every value between zero and infinity. Consequently in such regions (14) may be satisfied for every ε_M. Thus the number of real surface excitations is a characteristic quantity of the crystal. On the other hand according to (13) the total number of surface excitations depends on the crystal cut and the dielectric constant ε_M of the medium M. That means that the number of virtual excitations may be varied by the experimental conditions. It is possible that no Fano surface modes do exist. This case occurs if the crystal shows no overlap in the negative regions of ε_x' and ε_z' and in addition ε_M is chosen sufficiently large.

Our results for biaxial crystals may be easily specialized to uniaxial crystals. Choosing the geometry in Fig. 1 where ϑ denotes the angle between the optic axis $\hat{z}'$ and the normal vector $\hat{z}$ of the crystal surface we have

$$T_{xx} = T_{zz} = \cos \vartheta \,, \quad T_{yy} = 1 \,, \quad T_{zx} = -T_{xz} = \sin \vartheta \,.$$

With the notation $\varepsilon_{||} \equiv \varepsilon_z'$, $\varepsilon_\perp \equiv \varepsilon_x' = \varepsilon_y'$ we get from (3a) the dispersion of the bulk polairtons, which is given by (16a) for the ordinary one and by (16b) for the extraordinary one:

$$\text{(A)} \qquad n^2 - \varepsilon_\perp = 0 \,, \tag{16a}$$

$$\text{(B)} \qquad \varepsilon_\perp (n^2 - \varepsilon_{||}) \, (k_x'^2 + k_y'^2) + \varepsilon_{||} (n^2 - \varepsilon_\perp) \, k_z'^2 = 0 \,. \tag{16b}$$

The most general surface mode on uniaxial crystals involves two electromagnetic waves in the crystal which have different attenuation constants [5].

The conditions for the existence of TM surface polaritons in uniaxial crystals may be directly won from the respective conclusions for orthorhombic crystals: Either the wave vector of the surface polariton is $\boldsymbol{\varphi} = (0, k_y)$ and $\vartheta = \pi/2$, or $\boldsymbol{\varphi} = (k_x, 0)$ and ϑ may have any value.

For the first case, the surface polariton involves fields of type A in the crystal polariton. Surface polaritons of the second case involve fields B and are hence extraordinary.

For the first case ($\boldsymbol{\varphi} = (0, k_y)$, $\vartheta = \pi/2$) the explicit dispersion relation

$$k_y^2 = \frac{\omega^2}{c^2} \frac{\varepsilon_\perp \varepsilon_M}{\varepsilon_\perp + \varepsilon_M} \tag{17}$$

is obtained which corresponds to (16) in [8] for $k_x = 0$.

Equation (17) has the same structure as the dispersion relation of surface polaritons on isotropic crystals but differs in the fact that the existence of ordinary Fano modes is restricted to one crystal cut and one direction of propagation.

Fano modes of type B in general are generalized extraordinary modes. Applying the general conditions for the occurrence of Fano waves it is found that extraordinary Fano modes may exist only in the following three cases:

1. For $\vartheta = 0$ the crystal is invariant for rotations around the optic axis. Using (12) the dispersion relation is given by

$$k_x^2 + k_y^2 = \frac{\omega^2}{c^2} \varepsilon_M \frac{\varepsilon_{||}(\varepsilon_\perp - \varepsilon_M)}{\varepsilon_{||}\varepsilon_\perp - \varepsilon_M^2} \,. \tag{18a}$$

2. For $\vartheta = \pi/2$ extraordinary Fano modes may exist only for $k_y = 0$. We get from (12)

$$k_x^2 = \frac{\omega^2}{c^2} \varepsilon_M \frac{\varepsilon_\perp(\varepsilon_\parallel - \varepsilon_M)}{\varepsilon_\parallel \varepsilon_\perp - \varepsilon_M^2} . \tag{18b}$$

In both cases of (18) the surface polariton is regular.

3. In the region $0 < \vartheta < \pi/2$ extraordinary Fano modes occur only for $k_y = 0$. In this case (3b), (16b), (11) reduce to

$$k_x^2 + k_{Mz}^2 = \frac{\omega^2}{c^2} \varepsilon_M , \tag{19a}$$

$$k_z^2(\varepsilon_\perp \sin^2 \vartheta + \varepsilon_\parallel \cos^2 \vartheta) + k_z [2 \cos \vartheta \sin \vartheta k_x (\varepsilon_\parallel - \varepsilon_\perp)] + \\ + k_x^2(\varepsilon_\perp \cos^2 \vartheta + \varepsilon_\parallel \sin^2 \vartheta) - \frac{\omega^2}{c^2} \varepsilon_\perp \varepsilon_\parallel = 0 , \tag{19b}$$

$$\left(k_x^2 + k_z^2 - \frac{\omega^2}{c^2} \varepsilon_\parallel\right) (\cos \vartheta k_x - \sin \vartheta k_z) (\varepsilon_M \cos \vartheta k_x - \varepsilon_\perp \sin \vartheta k_{Mz}) + \\ + \left(k_x^2 + k_z^2 - \frac{\omega^2}{c^2} \varepsilon_\perp\right) (\sin \vartheta k_x + \cos \vartheta k_z) (\varepsilon_M \sin \vartheta k_x + \varepsilon_\parallel \cos \vartheta k_{Mz}) = 0 \tag{19c}$$

which correspond to (6b), (17a), (17b) in [8] for $k_y = 0$. This result restricts the validity of the dispersion relation (18c) in [8] to the case $\vartheta = 0$.

The properties of the dispersion curves which correspond to the generalized TM surface polaritons described by (19) are discussed in [8]. Especially in [8] the behaviour of real and virtual surface excitations for $0 < \vartheta < \pi/2$ is demonstrated. It is shown that virtual surface polaritons should exist only for certain ranges of ϑ.

Theoretical details will be reviewed in comprehensive form in [19].

3. Results and Discussion

The experimental method has been described in [1]. Here we investigated a crystal of α-quartz with an angle of the optical axis to the surface of 45°. It was cut and optically polished by the Steinheil company. The geometry used is shown in Fig. 1. Our measurements are limited by the optical constants of the ATR hemicylinder which has been KRS 5. Its transmittance decreases at 270 cm^{-1}. Therefore the two lowest branches in Fig. 3 could not be investigated. The observable k_x range is approximately given by line B in Fig. 3. It corresponds to a maximum angle of 65° and $n = 2.4$ for KRS 5 in (20). The values of k_x are calculated by the equation

$$k_x = \frac{\omega}{c} n(\omega) \sin \alpha . \tag{20}$$

α is the angle of incidence, ω the frequency of maximum absorption, $n(\omega)$ the refractive index of KRS 5 [11].

The input data for calculating the dispersion by (19) have been the same as in [1, 12]. The curves C_i (Im $k_z = 0$) are defined in [8]. We assign the dispersion branches of the surface polaritons with increasing frequency by e.o. $SP_i^{(\infty)}(45°)$. For $\vartheta = 0°$ and $\vartheta = 90°$ they coincide with the e.o. $SP^{\infty}_{\parallel,\perp j}$ given in [1]. The correspondence is shown in Table 1.

Table 1

Ranges of ϑ for the generalized surface polaritons with $k_y = 0$

$\vartheta = 0°$	$\alpha < \vartheta < \beta$		$\vartheta = 90°$
e.o. $SP_{\perp 1}$	e.o. SP_1	(0° to 62°)	
e.o. $SP_{\perp 2}$	e.o. SP_2	(0° to 65°)	
	e.o. SP_3	(15° to 90°)	e.o. $SP_{\parallel 1}$
e.o. $SP_{\perp 3}$	e.o. SP_4	(0° to 47°)	
e.o. $SP^{\infty}_{\perp 4}$	e.o. SP^{∞}_5	(0° to 90°)	e.o. $SP^{\infty}_{\parallel 2}$
	e.o. SP_6	(59° to 90°)	e.o. $SP_{\parallel 2}$
e.o. $SP_{\perp 5}$	e.o. SP_7	(0° to 55°)	
	e.o. SP_8	(20° to 90°)	e.o. $SP_{\parallel 3}$
e.o. $SP_{\perp 6}$	e.o. SP_9	(0° to 47°)	
e.o. $SP^{\infty}_{\perp 7}$	e.o. SP^{∞}_{10}	(0° to 90°)	e.o. $SP^{\infty}_{\parallel 4}$ (ω_{S47})
e.o. $SP^{\infty}_{\perp 8}$	e.o. SP^{∞}_{11}	(0° to 90°)	e.o. $SP^{\infty}_{\parallel 4}$ (ω_{S48})

α and β are the limiting angles for ϑ. The assignments for $\vartheta = 0°$ and $\vartheta = 90°$ are given in [1].

There are three branches of real excitation (e.o. SP_i^{∞}) and eight branches of virtual excitation (e.o. SP_i) surface polaritons. All branches of real excitation could be observed. The limiting frequencies of $SP_5^{\infty}(45°)$, $SP_{10}^{\infty}(45°)$, and $SP_{11}^{\infty}(45°)$ are in agreement with (14) and equation (4) in [1]. As is shown in [8] the equations (19) can be solved only for certain ranges of ϑ with respect to virtual excitation surface polaritons. The limiting angles are given in Table 1. At the 45° cut seven branches of virtual excitation surface polaritons occur, five of which lie in the observable range. Two of these were detected. The other, SP_4 (45°) SP_7(45°), SP_9(45°) exist in a frequency splitting range $\Delta\omega < 1\text{cm}^{-1}$ and in a very small k_x range. For angles α a little off 45° they would not exist at all (see Table 1). The limit of detection is given by a frequency splitting of about 2 cm^{-1} as it is the case in Fig. 4. This is the reason, why these SP have not been observed.

The deviations of SP_{11}^{∞} (45°) from the calculated curve is caused by the application of spacers with a thickness of $d \approx 1.4$ µm between the α-quartz crystal and the ATR hemicylinder. In this case the frequencies of surface polaritons are shifted to lower frequencies [1, 13, 14]. Because the dispersion curves are calculated without any damping this effect is not taken into account. As is shown in [15] for $\vartheta = 0°$ and $\vartheta = 90°$ damping can be included by calculating the reflectivity with a formula given in [16]. The dispersion curves then are defined by the minima of reflection [17]. But as yet this method has not been used for arbitrary crystal cuts.

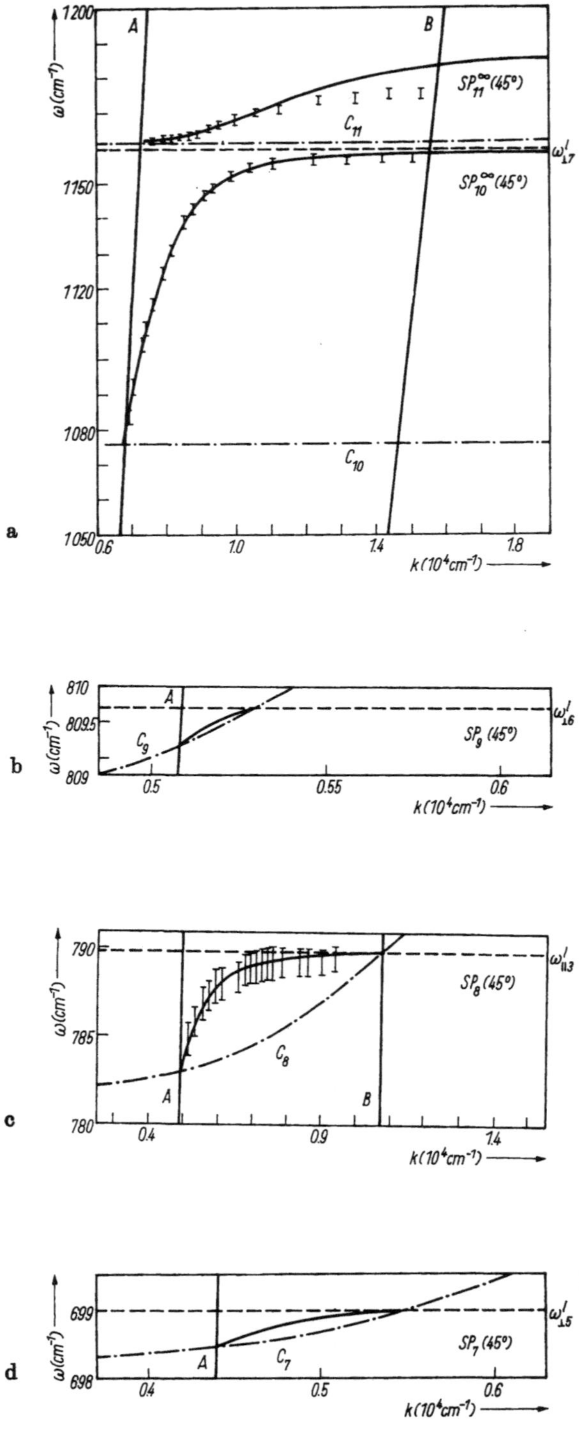

Fig. 3a to i. Calculated and observed dispersion of generalized surface polaritons of α-quartz with $k_y = 0$. ϑ is 45°. A denotes $k_x = \omega/c$. The left range of B is the accessible range for the experiment. —·—·—: C_i calculated by (22a) in [8]. – – –: frequencies of the longitudinal optical phonons $\omega^{l}_{\|, \perp i}$

Fig. 3

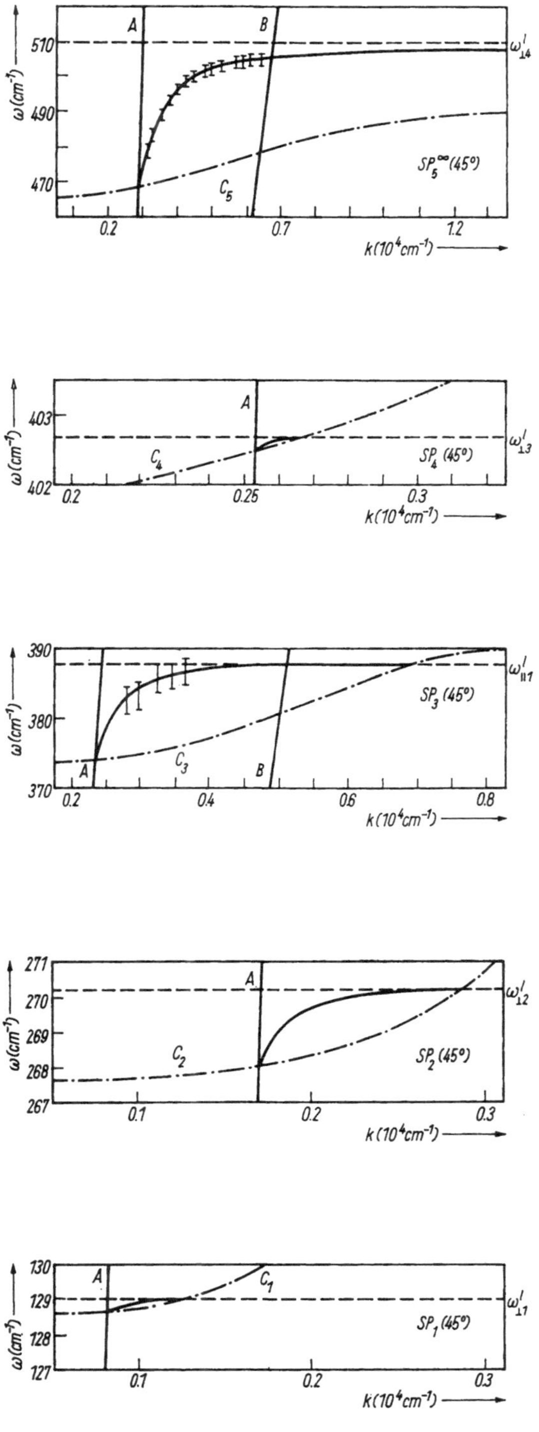

A
B
510
490
470
ω (cm⁻¹)
SP₅^∞ (45°)
C₅
0.2
0.7
1.2
k(10⁴cm⁻¹)
403
402
C₄
SP₄ (45°)
0.25
0.3
390
380
370
SP₃ (45°)
C₃
0.4
0.6
0.8
271
270
269
268
267
C₂
SP₂ (45°)
0.1
130
129
128
127
C₁
SP₁ (45°)

e

f

g

h

i

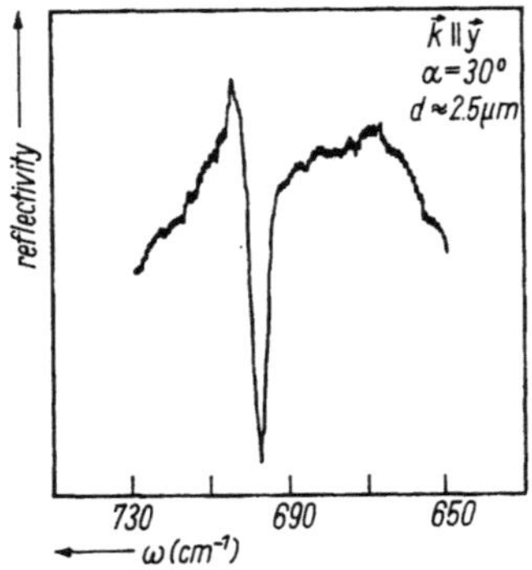

Fig. 4. Spectrum of surface phonon polaritons propagating parallel to y for a 45° cut. d thickness of the spacer; α angle of incidence

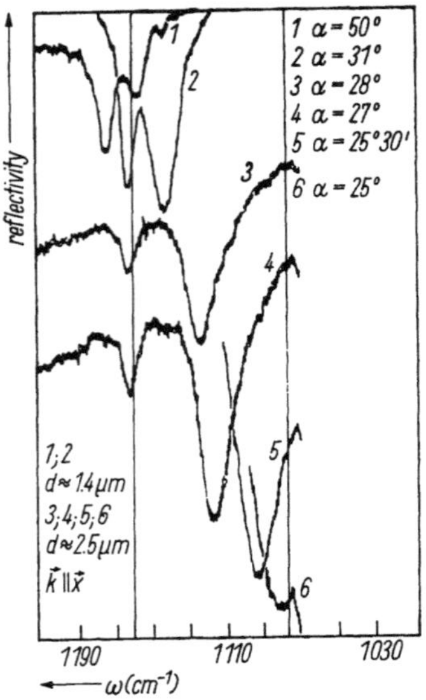

Fig. 5. Spectra of $SP_{10}^{\infty}(45°)$ and $SP_{11}^{\infty}(45°)$ for different angles of incidence. The vertical lines correspond to the frequencies of C_{10} and C_{11} at the crosspoint with A (Fig. 3a). d, α see Fig. 4

Acknowledgement

The authors would like to thank the Deutsche Forschungsgemeinschaft for financial support.

References

[1] H. J. FALGE and A. OTTO, phys. stat. sol. (b) **56**, 523 (1973).
[2] V. V. BRYKSIN, D. N. MIRLIN, and I. I. RESHINA, Soviet Phys. — Solid State **15**, 760 (1973).
[3] C. H. PERRY, B. FISCHER, and W. BUCKEL, Solid State Commun. **13**, 1261 (1973).
[4] O. A. DUBOVSKII, Soviet Phys. — Solid State **12**, 2471 (1971).
[5] V. N. LYUBIMOV and D. G. SANNIKOV, Soviet Phys. — Solid State **14**, 574 (1972).
[6] A. HARTSTEIN and E. BURSTEIN, Surface Sci. **34**, 81 (1973).
[7] A. HARTSTEIN and E. BURSTEIN, Solid State Commun. **12**, 1083 (1973).
[8] G. BORSTEL, phys. stat. sol. (b) **60**, 427 (1973).
[9] A. OTTO, Z. Phys. **216**, 398 (1968).
[10] T. KUROSAWA, J. Phys. Soc. Japan **16**, 1298 (1961).
[11] R. A. SMITH, F. E. JONES, and R. P. CHASMAR, The Detection and Measurement of Infrared Radiation, Clarendon Press, Oxford 1968.
[12] L. MERTEN, Z. Naturf. **24a**, 1878 (1969).
[13] N. MARSCHALL and B. FISCHER, Phys. Rev. Letters **28**, 811 (1972).
[14] A. OTTO, Proc. Taormina Conf. on Polaritons, Ed. E. BURSTEIN, Pergamon Press, New York 1974.
[15] H. J. FALGE, A. OTTO, and W. SOHLER, phys. stat. sol. (b) **63**, 259 (1974).
[16] W. SOHLER, Optics Commun. **10**, 203 (1974).
[17] R. W. ALEXANDER, G. S. KOVENER, and R. J. BELL, Phys. Rev. Letters **32**, 154 (1974).
[18] E. BURSTEIN et al., Proc. Taormina Conf. on Polaritons, Ed. E. BURSTEIN, Pergamon Press, New York 1974.
[19] G. BORSTEL, H. J. FALGE, and A. OTTO, Springer Tracts mod. Phys., to be published.

(Received May 27, 1974)

phys. stat. sol. (b) **65**, 133 (1974)

Subject classification: 6 and 20.1; 22.8.1

Sektion Physik der Universität München, Lehrstuhl Prof. Brandmüller

The Transferability of Raman Tensors of Isolated Molecules to Crystals — Application to the Computation of Relative Intensities of the Internal Modes of SO_4^{--} in K_2SO_4 Crystal

By

S. MONTERO[1])

A general group theoretical method for the construction of the Raman tensors of crystals as a function of the crystallographic data and the Raman tensors of the 'free' molecules or ions of the crystal is presented. In a first application of this method, the Raman tensors of all internal modes of K_2SO_4 were calculated. The relative intensities derived from the computed tensors showed a satisfactory agreement with the measured integrated intensities of the Raman spectra as obtained on oriented K_2SO_4 crystals.

Eine allgemeine Methode zum Aufbau der Raman-Tensoren von Kristallen als Funktion der kristallographischen Daten und der Raman-Tensoren der ‚freien' Moleküle oder Ionen des Kristalls wird vorgestellt. Als erste Anwendung der Methode wurden die Raman-Tensoren von allen inneren Schwingungen des K_2SO_4-Kristalls ausgerechnet. Die von diesen Raman-Tensoren abgeleiteten Intensitäten zeigten eine gute Übereinstimmung mit den gemessenen integrierten Intensitäten der Ramanspektren, die an orientierten K_2SO_4-Kristallen gemessen wurden.

1. Introduction

Several authors [1 to 3] have treated the evaluation of the relative intensities of the different Raman bands originating from the internal vibrations of a molecule or a molecular group which are split in the crystal into several components due to crystal field effects. The considerations and theoretical results in [3], where the distribution of relative intensities for some symmetry species of a crystal were calculated, are correct. However, these results are restricted to a particular crystal structure.

In the present paper a systematic group theoretical model is developed which gives expressions for the Raman tensors of a crystal as a function of the crystallographic data for all Raman active symmetry species. In this way, the relative intensities of the bands can be computed for all possible $a(bc)\,d$ orientations of the crystal. As a first application of this method, we have calculated the 30 tensor elements corresponding to the 18 internal vibrations of the crystal K_2SO_4. The agreement with the spectra recorded by us of a single crystal is good for those bands originating from the ν_1, ν_2, and ν_4 vibrations of the free ion (designation after Herzberg). For the ν_3 components there are however a few deviations.

The frequency assignments of the Raman spectrum of oriented K_2SO_4 have been investigated by Debeau [4] with good results for 17 of the 18 internal vibrations. The remaining band was observed by us at 1164 cm^{-1}. We could also

[1]) Permanent address: Instituto de Quimica Fisica Rocasolano, Serrano 119, Madrid (6).

observe that the A_g-band at 1092 cm^{-1} is in reality a doublet with a component at 1086 cm^{-1}, the intensity of which is approximately 30% of the 1092 cm^{-1} component at 80 K. Debeau has attempted an evaluation of the intensity distribution caused by the crystal field splitting of the ν_1 and ν_2 non-polar vibrations of the free SO_4^{--} ion using the formulas given in [3]. The computed values are for both modes in good agreement with our calculated and measured values, but show considerable deviations from his own experimental results. The interpretation of these deviations given by the author as a result of a strong perturbation of the SO_4^{--} ion in the crystal is no longer convincing because recentlyt it has been shown [5] that the perturbation of the SO_4^{--} ion in the K_2SO_4 crystal is much smaller than supposed earlier.

The spectra used for the evaluation of the relative intensities in the present paper were recorded with a Coderg Phl Raman spectrometer at room temperature. The exciting source was a Kr^+ laser of Coherent Radiation with a power of about 300 mW at 647.1 nm. The mean resolution was 2 cm^{-1}.

2. Theory

In a single crystal let be $N_{M_s}^{PC}$ molecules or molecule-like groups M_s (e.g. SO_4^{--}) per primitive cell (PC). P_{M_s} is the point group of the molecules M_s, S the space group of the crystal, P′ the factor group, and P the crystal class group (or direction group). Every free molecule M_s may have $m = m_1 + m_2$ Raman active modes, where m_1 are the non-polar and m_2 the polar modes. Then there are in the crystal $r = r_1 + r_2$ Raman active internal vibrations of two different kinds. Firstly, there are r_1-bands which originate from the m_1 non-polar vibrations of the free molecule and secondly there are r_2-bands arising from the m_2 polar vibrations. For the r_1- and r_2-bands we form $S_1, S_2, \dots, S_{r_1+r_2}$ vibrational symmetry coordinates in the factor group P′, which can be classified after the $\Gamma^{(p)}$ symmetry species of P′. To every symmetry coordinate $S^{(p)}$ of species $\Gamma^{(p)}$ in P′ corresponds uniquely a second rank Raman tensor $\boldsymbol{\alpha}_0(S_l^{(p)})$ of species $\Gamma^{(p)}$ in P, from which the Raman intensity can be calculated. The correspondence

$$S_l^{(p)} \to \boldsymbol{\alpha}_0(S_l^{(p)}) \tag{1}$$

is possible, because $S_l^{(p)}$ and $\boldsymbol{\alpha}_0(S_l^{(p)})$ have the same symmetry properties in P′ and in P, respectively. The symmetry coordinates $S_l^{(p)}$ can be found without great difficulties with the help of the projection operators in P′. The necessary symmetry operators of P′ can be found in [10]. The basis on which such operators act are:

a) The m_1 non-polar normal coordinates Q_{j1} ($j = 1, 2, \dots, m_1$) of the molecule carrying the index 1 in the crystal. The form of the corresponding r_1 symmetry coordinates is

$$S_l^{(p)} = \sum_{k=1}^{N_{M_s}^{PC}} a_{jkl} Q_{jk}\,, \qquad j = 1, \dots, m_1;\; l = 1, \dots, r_1\,. \tag{2}$$

b) The set of orthonormalized polar vectors ξ_{j1}^i associated with the j-th polar normal coordinate of the molecule carrying index 1 in the crystal. The generated symmetry coordinates are of the form

$$S_l^{(p)} = \sum_{k=1}^{N_{M_s}^{PC}} a_{jkl} \xi_{jk}^i\,, \qquad i = 1, \dots, g_{gj};\; j = 1, \dots, m_2;\; l = 1, \dots, r_2\,. \tag{3}$$

g_j is the degree of degeneration of mode j in the free molecule.

Due to the correspondences

$$Q_{jk}(\text{of } \mathrm{P}_{M_s}) \rightarrow \boldsymbol{\alpha}(Q_{jk}) \; (\text{of } \mathrm{P}_{M_s}) \,, \tag{4}$$

$$\xi^i_{jk}(\text{of } \mathrm{P}_{M_s}) \rightarrow \boldsymbol{\alpha}(\xi^i_{jk}) \; (\text{of } \mathrm{P}_{M_s}) \tag{5}$$

are also uniques, the construction of the Raman tensors of the crystal in the zero order approximation (i.e. neglecting all kinds of interactions between the ions in the crystal) is possible. The expressions for the Raman tensors in the crystal are

a) Modes of non-polar origin:

$$\boldsymbol{\alpha}_0(S_l^{(\mathrm{p})}) = \sum_{k=1}^{N^{\mathrm{PC}}_{M_s}} \left(a_{jlk} T^+_k \boldsymbol{\alpha}(Q_{jk}) \; T_k\right) . \tag{6}$$

b) Modes of polar origin:

$$\boldsymbol{\alpha}_0(S_l^{(\mathrm{p})}) = \sum_{k=1}^{N^{\mathrm{PC}}_{M_s}} \left(a_{jlk} T^+_k \, \{\mathbf{R}_x \mathbf{R}_y \mathbf{R}_z\}_{jk} \, \xi^i_{jk} T_k\right) . \tag{7}$$

T_k is the unitary transformation matrix which orients the system of axes (x, y, z) of the molecule k parallel to the system of crystal axes (X, Y, Z). $\boldsymbol{\alpha}(Q_{jk})$ is the Raman tensor of the non-polar vibrational mode j in the point group P_{M_s} of the molecule k in the crystal. $\{\mathbf{R}_x \mathbf{R}_y \mathbf{R}_z\}_{jk}$ are the Raman tensors of the polar vibrational mode j in the same conditions. If the mode j is not degenerate, two of the three $\mathbf{R}_i$ tensors are zero. If the mode is double degenerate only one of the $\mathbf{R}_i$ is zero and in the case of a threefold degeneracy the three $\mathbf{R}_i$ tensors are different from zero. Sometimes the ξ^i_{jk} vectors are not fully determined by the symmetry of the crystal. In this case a fitting of one or more parameters is necessary in order to obtain a good agreement between observed and computed distribution of intensities. The vectors ξ^i_{jk} describe the breakdown of the degeneracy of polar degenerate vibrations (of the free molecule M_s) within the crystal. ξ^i_{jk} must be chosen with the highest symmetry with respect to the symmetry elements of the site group of the molecule in the crystal.

Through $\boldsymbol{\alpha}_0(S_l^{(\mathrm{p})})$ is now possible to compute the relative intensities using the usual formula [7, 8] for the scattered intensity in crystals.

Finally, we want to remark that from (6) and (7) so many Raman tensors are obtained for each symmetry species as predicted by the factor group analysis for the internal vibrations in the crystal. The computed tensors show exactly the same structure of the Raman tensors of group P which can be found in tables [7, 8].

3. Application to K_2SO_4

3.1 Structure of K_2SO_4 and factor group analysis

At room temperature K_2SO_4 is an orthorhombic crystal of space group D^{16}_{2h} with four K_2SO_4 units in the primitive cell. The sites of the sulfate groups in the crystal and the selected system of axes are shown in Fig. 1 and 2. The standard setting Pnma [9, 11] has been used here. Using the conventional factor group method [6] the distribution of vibrations shown in Table 1 are to be expected in the crystal class group D_{2h} within the $k = 0$ approximation. Since

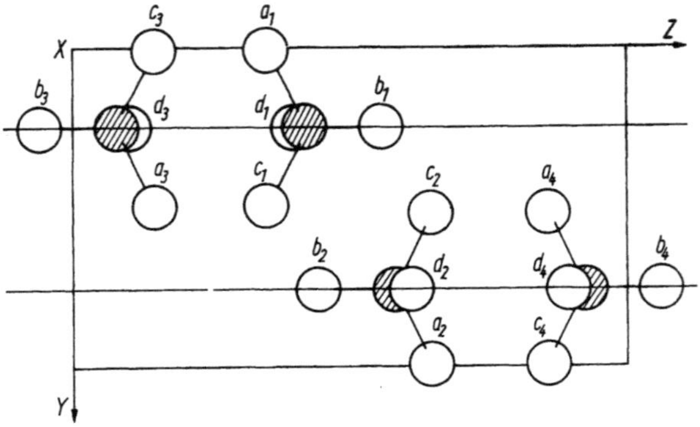

Fig. 1. Labelling of the oxygen atoms in the primitive cell of K_2SO_4. Potassium atoms are omitted in the figure

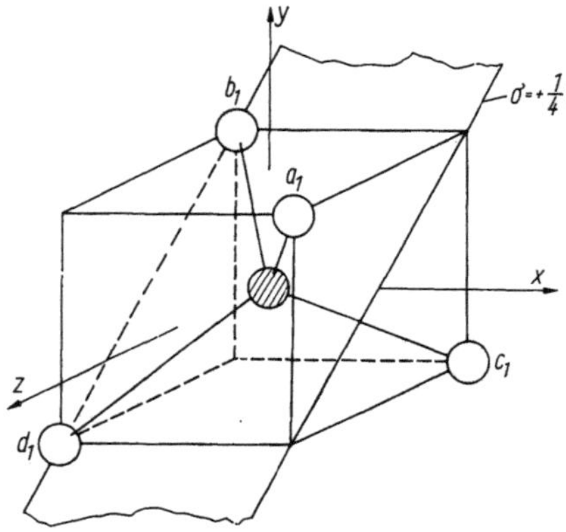

Fig. 2. Relative orientation of the system of axes in crystal (X, Y, Z) and in molecule 1 (x, y, z)

Table 1

Modes distribution of the K_2SO_4 crystal

	A_g	B_{1g}	B_{2g}	B_{3g}	A_u	B_{1u}	B_{2u}	B_{3u}	total
internal modes	6	3	6	3	3	6	3	6	36
translational modes	6	3	6	3	3	5	2	5	33
librational modes	1	2	1	2	2	1	2	1	12
acoustic branch						1	1	1	3

the primitive cell of this crystal has a centre of symmetry only the "gerade" species are Raman active. Consequently, the following internal Raman active modes are to be expected:

$$6A_g + 3B_{1g} + 6B_{2g} + 3B_{3g} \,. \tag{8}$$

The crystal field splitting predicted by the correlation of the point group T_d with the site group C_s of the SO_4^{--} ion in the crystal into the direction group D_{2h} is given in Table 2 for the 18 Raman active internal vibrations.

Table 2

Correlation diagram for factor group D_{2h}^{16}.
(Only Raman active species)

free SO_4^{--}	SO_4^{--} in crystal	$4SO_4^{--}$ in crystal
T_d	C_s	D_{2h}^{16}
A_1 (unpolar) →	A' →	$A_g + B_{2g}$
E (unpolar) →	$A' + A''$ →	$A_g + B_{2g} + B_{1g} + B_{3g}$
$2F_2$ (polar) →	$2(2A' + A'')$ →	$2(2A_g + B_{1g} + 2B_{2g} + B_{3g})$

3.2 The internal Raman tensors of the crystal

For the determination of the internal Raman tensors of the crystal, equations (2), (3), (6), and (7) are used. For an accurate definition of T_k let us call r the (4×3) matrix with the coordinates of the atoms a, b, c, d in the free ideal molecule (Fig. 2). R_k is the (4×3) matrix with the coordinates of the atoms a_k, b_k, c_k, d_k (Fig. 1) of the k-th SO_4^{--} ion, referred to the corresponding sulphur atom k. The application of T_k to matrix r (orientation) is

$$\underset{(4\times 3)}{r} \xrightarrow{T_k} \underset{(4\times 3)}{R_k} \tag{9}$$

or in its matrix formulation:

$$rT_k = cR_k\,. \tag{10}$$

Table 3

Internal symmetry coordinates for K_2SO_4 crystal

$$S_1^{(A_g)} = Q_{1,1} + Q_{1,2} + Q_{1,3} + Q_{1,4}$$
$$S_2^{(B_{2g})} = Q_{1,1} + Q_{1,2} - Q_{1,3} - Q_{1,4}$$

$$S_3^{(A_g)} = Q_{2a,1} + Q_{2a,2} + Q_{2a,3} + Q_{2a,4}$$
$$S_4^{(B_{2g})} = Q_{2a,1} + Q_{2a,2} - Q_{2a,3} - Q_{2a,4}$$
$$S_5^{(B_{3g})} = Q_{2b,1} + Q_{2b,2} + Q_{2b,3} + Q_{2b,4}$$
$$S_6^{(B_{1g})} = Q_{2b,1} + Q_{2b,2} - Q_{2b,3} - Q_{2b,4}$$

$$S_7^{(A_g)} = \cos\varphi_4(\xi_1^x - \xi_2^x + \xi_3^x - \xi_4^x) + \sin\varphi_4(\xi_1^z - \xi_2^z - \xi_3^z + \xi_4^z)$$
$$S_8^{(B_{2g})} = \cos\varphi_4(\xi_1^x - \xi_2^x - \xi_3^x + \xi_4^x) + \sin\varphi_4(\xi_1^z - \xi_2^z + \xi_3^z - \xi_4^z)$$
$$S_9^{(B_{1g})} = \xi_1^y - \xi_2^y + \xi_3^y - \xi_4^y$$
$$S_{10}^{(B_{3g})} = \xi_1^y - \xi_2^y - \xi_3^y + \xi_4^y$$
$$S_{11}^{(A_g)} = \sin\varphi_4(\xi_1^x - \xi_2^x + \xi_3^x - \xi_4^x) - \cos\varphi_4(\xi_1^z - \xi_2^z - \xi_3^z + \xi_4^z)$$
$$S_{12}^{(B_{2g})} = \sin\varphi_4(\xi_1^x - \xi_2^x - \xi_3^x + \xi_4^x) - \cos\varphi_4(\xi_1^z - \xi_2^z + \xi_3^z - \xi_4^z)$$

The remaining $S_{13}^{(A_g)}$ to $S_{18}^{(B_{2g})}$ coordinates can be obtained from $S_7^{(A_g)}$ to $S_{12}^{(B_{2g})}$ by substitution of φ_4 by φ_3.

c is a proportionality factor. The T_k matrices obtained according to this definition are

$$T_1 = \begin{pmatrix} a & 0 & c \\ d & e & f \\ -d & e & -f \end{pmatrix}, \quad T_2 = -T_1, \quad T_3 = \begin{pmatrix} a & 0 & -c \\ d & -e & -f \\ -d & -e & f \end{pmatrix}, \quad T_4 = -T_3, \tag{11}$$

with $a = 0.590$, $c = -0.803$, $d = 0.569$, $e = -0.710$, $f = 0.420$. The symmetry coordinates given in Table 3 were obtained using the basis described in Section 2.

The Raman tensors of the free SO_4^{--} ion which have to be combined according to (6) and (7) are

$$\left.\begin{array}{l} \boldsymbol{\alpha}(Q_{1j}) \sim \begin{pmatrix} 1 & 0 & 0 \\ 0 & 1 & 0 \\ 0 & 0 & 1 \end{pmatrix}, \quad \boldsymbol{\alpha}(Q_{2aj}) \sim \begin{pmatrix} -2 & 0 & 0 \\ 0 & 1 & 0 \\ 0 & 0 & 1 \end{pmatrix}, \quad \boldsymbol{\alpha}(Q_{2bj}) \sim \begin{pmatrix} 0 & 0 & 0 \\ 0 & -\sqrt{3} & 0 \\ 0 & 0 & -\sqrt{3} \end{pmatrix}, \\ \mathbf{R}_x \sim \begin{pmatrix} 0 & 0 & 0 \\ 0 & 0 & 1 \\ 0 & 1 & 0 \end{pmatrix}, \quad \mathbf{R}_y \sim \begin{pmatrix} 0 & 0 & 1 \\ 0 & 0 & 0 \\ 1 & 0 & 0 \end{pmatrix}, \quad \mathbf{R}_z \sim \begin{pmatrix} 0 & 1 & 0 \\ 1 & 0 & 0 \\ 0 & 0 & 0 \end{pmatrix}. \end{array}\right\} \tag{12}$$

For the $\boldsymbol{\alpha}(Q_{2aj})$ and $\boldsymbol{\alpha}(Q_{2bj})$ one has to pay attention to the orientation of the system of molecular axes (x, y, z). The calculated Raman tensors for the $r_1 = 6$ modes of non-polar origin, according to equation (6) and Table 3, are

$$\left.\begin{array}{l} \boldsymbol{\alpha}_0(S_1^{A_g}) = k_1 \begin{pmatrix} 1 & 0 & 0 \\ 0 & 1 & 0 \\ 0 & 0 & 1 \end{pmatrix}, \quad \boldsymbol{\alpha}_0(S_2^{B_{2g}}) = k_1 \begin{pmatrix} 0 & 0 & 0 \\ 0 & 0 & 0 \\ 0 & 0 & 0 \end{pmatrix}, \\ \boldsymbol{\alpha}_0(S_3^{A_g}) = k_2 \begin{pmatrix} 1-3a^2 & 0 & 0 \\ 0 & 1 & 0 \\ 0 & 0 & 1-3c^2 \end{pmatrix}, \quad \boldsymbol{\alpha}_0(S_4^{B_{2g}}) = k_2 \begin{pmatrix} 0 & 0 & -3ac \\ 0 & 0 & 0 \\ -3ac & 0 & 0 \end{pmatrix}, \\ \boldsymbol{\alpha}_0(S_5^{B_{3g}}) = k_2 \begin{pmatrix} 0 & 0 & 0 \\ 0 & 0 & -2\sqrt{3}\,ef \\ 0 & -2\sqrt{3}\,ef & 0 \end{pmatrix}, \quad \boldsymbol{\alpha}_0(S_6^{B_{1g}}) = k_2 \begin{pmatrix} 0 & -2\sqrt{3}\,de & 0 \\ -2\sqrt{3}\,de & 0 & 0 \\ 0 & 0 & 0 \end{pmatrix}. \end{array}\right\} \tag{13}$$

For the $r_2 = 12$ modes of polar origin according to equation (7) and Table 3, we obtain

$$\boldsymbol{\alpha}_0(S_7^{A_g}) = k_4 \left[\cos\varphi_4 \begin{pmatrix} -3d^2a & & \\ & e^2a & \\ & & -f(2cd+af) \end{pmatrix} + \sin\varphi_4 \begin{pmatrix} -d(2af+cd) & & \\ & e^2c & \\ & & -3f^2c \end{pmatrix} \right],$$

$$\boldsymbol{\alpha}_0(S_8^{B_{2g}}) = k_4 \left[\cos\varphi_4 \begin{pmatrix} & & -d(2af+cd) \\ & 0 & \\ -d(2af+cd) & & \end{pmatrix} + \right.$$

$$\left. + \sin\varphi_4 \begin{pmatrix} & & -f(2cd+af) \\ & 0 & \\ -f(2cd+af) & & \end{pmatrix} \right],$$

$$\left.\begin{aligned}
&\boldsymbol{\alpha}(S_9^{\mathrm{B_{1g}}}) = k_4\begin{pmatrix} 0 & ae^2 & 0 \\ ae^2 & 0 & 0 \\ 0 & 0 & 0 \end{pmatrix}, \qquad \boldsymbol{\alpha}({}_0S_{10}^{\mathrm{B_{3g}}}) = k_4\begin{pmatrix} 0 & 0 & 0 \\ 0 & 0 & e^2c \\ 0 & e^2c & 0 \end{pmatrix}, \\
&\boldsymbol{\alpha}_0(S_{11}^{\mathrm{A_g}}) = k_4\left[\sin\varphi_4\begin{pmatrix} -3d^2a & & \\ & e^2a & \\ & & -f(2cd+af) \end{pmatrix} - \cos\varphi_4\begin{pmatrix} -d(2af+cd) & & \\ & e^2c & \\ & & -3f^2c \end{pmatrix}\right], \\
&\boldsymbol{\alpha}_0(S_{12}^{\mathrm{B_{2g}}}) = k_4\left[\sin\varphi_4\begin{pmatrix} & & -d(2af+cd) \\ & 0 & \\ -d(2af+dc) & & \end{pmatrix} - \right. \\
&\qquad \left. - \cos\varphi_4\begin{pmatrix} & & -f(2cd+af) \\ & 0 & \\ -f(2cd+af) & & \end{pmatrix}\right].
\end{aligned}\right\} \quad (14),$$

The remaining $\boldsymbol{\alpha}_0(S_{13}^{\mathrm{A_g}})$ to $\boldsymbol{\alpha}_0(S_{18}^{\mathrm{B_{2g}}})$ coming from the degenerate mode ν_3 are obtained changing the constant k_4 and the parameter φ_4 by k_3 and φ_3 respectively. From (13) and (14) there are obtained the relative intensities within each family of bands with a common vibration of the SO_4^{--} ion as origin. The results are summarized in Table 4. Observed and computed intensities in Table 4 are normalized to 100 for the sum of intensities of each ν_i group.

The Raman tensors (13) and (14) can in principle be used for every crystal of space group D_{2h}^{16} isomorphic with K_2SO_4. However, one has to consider the possibility of interactions between ions, which can change the theoretical distribution of intensities appreciably. As an indication of such interactions one can regard the geometry of ions or molecules within the crystal, because strong interactions often cause a change in geometry with respect to the free ion or molecule.

4. Summary

The results of Table 4 show a good agreement between calculated and observed intensity distributions, especially for the groups of ν_1- and ν_2-bands. The group of ν_3- and ν_4-bands of polar origin shows a few discrepancies which in our opinion are to be considered mainly as a consequence of the isotropic approximation since the Raman tensor of SO_4^{--} ions in the crystal are in reality anisotropic, showing C_s instead of T_d symmetry.

The present method of calculation clearly shows that the Raman tensors of the SO_4^{--} ion can be transfered into the K_2SO_4 crystal in the zero order approximation with satisfactory results. In any case, K_2SO_4 is a crystal well suited for our purposes because it is almost optically isotropic with only a slight distortion of the SO_4^{--} ion.

Finally we want to emphasize that the calculated intensity of the B_{2g}-band originating from ν_1 is exactly zero and that this corresponds to a kind of selection rule which is not considered in the factor group analysis.

Table 4
Assignment of the Raman active internal vibrations of K_2SO_4 crystal with the computed and observed intensities distribution

K_2SO_4 (T_d) aqueous sol.	K_2SO_4 (D_{2h}^{16}) crystal	$S_j^{(p)}$	obs. intens.	comp. intens.
$\nu_2(E)$, 451 cm^{-1}	A_g 447 cm^{-1}	S_3	XX 0	XX 0.033
			YY 15.6	YY 14.6
			ZZ 9.9	ZZ 12.6
	B_{1g} 456 cm^{-1}	S_6	26.8	29.1
	B_{2g} 453 cm^{-1}	S_4	29.6	28.2
	B_{3g} 457 cm^{-1}	S_5	14.1	15.4
$\nu_4(F_2)$, 615 cm^{-1}	for $\varphi_4 = 0°$			
	A_g 617 cm^{-1}	S_{11}	XX 1.4	XX 0.04
			YY 17.4	YY 13.9
			ZZ 12.7	ZZ 15.4
	A_g 627 cm^{-1}	S_7	XX 28.0	XX 28.1
			YY 2.8	YY 7.5
			ZZ 7.7	ZZ 6.7
	B_{1g} 622 cm^{-1}	S_9	7.7	6.7
	B_{2g} 619.5 cm^{-1}	S_{12}	7.4	7.5
	B_{2g} 633.5 cm^{-1}	S_8	0.3	0.04
	B_{3g} 620.5 cm^{-1}	S_{10}	14.6	13.9
$\nu_1(A_1)$, 981 cm^{-1}	A_g 983 cm^{-1}	S_1	XX 32.6	XX 33.3
			YY 32.3	YY 33.3
			ZZ 35.1	ZZ 33.3
	B_{2g} not obs.	S_2	0	0
$\nu_3(F_2)$, 1105 cm^{-1}	for $\varphi_3 = 0°$			
	A_g 1087 cm^{-1} 1093 cm^{-1}	S_{17}	XX 0	XX 0.04
			YY 14.6	YY 13.9
			ZZ 9.7	ZZ 15.4
	A_g 1145 cm^{-1}	S_{13}	XX 41.0	XX 28.1
			YY 3.3	YY 7.5
			ZZ 4.5	ZZ 6.7
	B_{1g} 1109 cm^{-1}	S_{15}	7.1	6.7
	B_{2g} 1110.5 cm^{-1}	S_{18}	6.2	7.5
	B_{2g} 1164 cm^{-1}	S_{14}	0.3	0.04
	B_{3g} 1104.5 cm^{-1}	S_{16}	12.3	13.9

Acknowledgements

I want to thank to J. Brandmüller, J. Behringer, and H. W. Schrötter for stimulating discussions and interest in this work and Prof. S. Haussühl, Köln, for providing crystal samples. Finally thanks are due to the Max-Planck-Gesellschaft for financial support and the Deutsche Forschungsgemeinschaft for supplying the instruments.

References

[1] J. Cabannes and R. Aynard, J. Physique **3**, 137 (1942).
[2] L. Couture and J. P. Mathieu, J. Physique **6**, 314 (1945).
[3] L. Couture, Ann. Phys. (France) **2**, 5 (1947).

[4] M. Debeau, Rev. Phys. appl. 7, 49 (1971).
[5] J. A. McGuinnety, Acta cryst. **B28**, 2845 (1972).
[6] J. Behringer, Springer Tracts in Modern Physics, Vol. 68, Springer-Verlag, Berlin/Heidelberg/New York 1973.
[7] R. Loudon, Adv. Phys. **13**, 423 (1964).
[8] H. Poulet and J. P. Mathieu, Spectres de Vibration et Symetrie des Cristaux, Gordon and Breach, New York 1970.
[9] R. W. G. Wykoff, Crystal Structures, Vol. III, 2nd edition, Wiley, New York 1963.
[10] J. Zak, The Irreducible Representations of Space Groups, W. A. Benjamin Inc., New York 1969.
[11] N. F. M. Henry and K. Lonsdale, International Tables for X-Ray Crystallography, Vol. I, The Kynoch Press, Birmingham 1969.

(Received May 30, 1974)

phys. stat. sol. (b) **65**, 143 (1974)

Subject classification: 13; 6; 14

A. F. Ioffe Physico-Technical Institute, Academy of Sciences of the USSR, Leningrad

A Theory Accounting for Electron–Phonon Interaction in a Compositionally Disordered System

By

H. Böttger[1])

An incorporation of the electron–phonon interaction into the current theoretical concept of electronic states in disordered solids is investigated on the basis of a simple model of a compositionally disordered solid. In particular a single-site approximation including contributions of the electron–phonon interaction is discussed.

Es wird versucht, auf der Grundlage des gegenwärtigen theoretischen Konzepts zur Beschreibung der Elektronenzustände in nichtgeordneten Festkörpern auch die Elektron-Phonon-Wechselwirkung zu berücksichtigen. Dabei wird ein einfaches Modell eines Festkörpers mit einer Unordnung in der chemischen Zusammensetzung betrachtet. Speziell wird eine Einzentrenapproximation, die Beiträge der Elektron–Phonon-Wechselwirkung einschließt, diskutiert.

1. Introduction

Electronic properties of compositionally disordered solids, especially of disordered binary alloys, have been of great theoretical interest in recent years. Most of the investigations dealing with such systems start from a more or less simplified one-electron Hamiltonian and assume the atomic positions to be fixed. Within this concept single-site approximations — regarding each atom embedded in the same effective medium — have turned out to give some valuable insight into the electronic behaviour of compositionally disordered systems [1, 2]. A deeper understanding of this behaviour, however, requires to go beyond such an approximation by accounting for atomic clusters (see e.g. [3, 4]) and to extend the above concept by including electron–phonon interaction [5] and, if necessary, electron–electron interaction [6], too.

Electron–phonon interaction is expected to lead in the systems under consideration to interesting effects resulting from the interplay between the compositionally disorder and the thermal disorder. Some of these effects have recently been discussed by Chen et al. [5] by treating the lattice vibrations classically.

The aim of the paper presented here is an incorporation of electron–phonon interaction into a theory of electronic states in compositionally disordered systems without a limitation to a classical atomic motion. For this reason physical quantities — such as the density of electronic states and the dc conductivity — are expressed by Green's functions (Section 2) and the equations of motion of the needed Green's functions are set up and discussed (Section 3). In particular a single-site approximation including contributions of electron–phonon interaction is discussed (Sections 4 and 5).

---

[1]) Permanent address: Zentralinstitut für Elektronenphysik der Akademie der Wissenschaften der DDR, Berlin.

2. Formulation of the Problem

In this paper the model under consideration is the following one: two types of atoms — called A and B, respectively — are randomly placed on the sites of a regular lattice. Each atom is assumed to have only one level. The corresponding wave functions are assumed to be very localized and orthogonal, so that a tight binding model is adequate to describe the electron motion.

The Hamiltonian, H, is taken to be

$$\left.\begin{aligned} H &= H_\mathrm{e} + H_\mathrm{p} + H_\mathrm{ep}\,, \\ \text{where} \\ H_\mathrm{e} &= \sum_n \varepsilon_n a_n^+ a_n + \sum_{n \neq m} V_{nm} a_n^+ a_m\,, \\ H_\mathrm{p} &= \sum_s \hbar\omega_s \left(\tfrac{1}{2} + b_s^+ b_s\right), \\ H_\mathrm{ep} &= \sum_{ns} \left[\gamma_s(n)\, b_s + \gamma_s^*(n)\, b_s^+\right] a_n^+ a_n \\ \text{with} \\ [a_n, a_m^+]_+ &= \delta_{nm}\,, \qquad [a_n, a_m]_+ = [a_n^+, a_m^+]_+ = 0\,, \\ [b_s, b_{s'}^+]_- &= \delta_{ss'}\,, \qquad [b_s, b_{s'}]_- = [b_s^+, b_{s'}^+]_- = 0\,. \end{aligned}\right\} \tag{2.1}$$

The (one-)electron Hamiltonian for the static lattice, H_e, is identical with that of the model investigated by Velický et al. [1]. The first part of H_e represents the "impurity" Hamiltonian with ε_n either ε^A or ε^B according to whether an A or B atom is on site n. The second part of H_e, being off-diagonal in site indices, is assumed to be independent of atomic configuration.

H_p describes the lattice vibrations in harmonic approximation. In this paper the lattice vibrations are for simplicity assumed to be not affected by compositional disorder. The normal mode index s is therefore determined by the phonon wave vector $\boldsymbol{q}$ and the vibrational branch index j.

For H_ep, the electron–phonon part of H, a Fröhlich-type Hamiltonian is chosen, i.e. only terms being diagonal in site index n are taken into account in H_ep. $|\gamma_s(n)|^2$ is admitted to depend on the type of atom placed on site n.

Green's functions are introduced by

$$\left.\begin{aligned} G(11') &= \mathrm{Sp}\left(WT\,\{a(1)\,a^+(1')\}\right), \\ G(121'2') &= \mathrm{Sp}\left(WT\,\{a(1)\,a(2)\,a^+(2')\,a^+(1')\}\right), \\ L(11'22') &= G(121'2') - G(11')\,G(22')\,, \end{aligned}\right\} \tag{2.2}$$

where $a(1) \equiv a_{m_1}(t_1)$ is a_{m_1} in Heisenberg representation; W is the statistical operator $W = \exp\{-\beta(H - \mu N)\}/\mathrm{Sp}\,[\exp\{-\beta(H - \mu N)\}]$ with the electron number operator, N, and the chemical potential, μ; T represents the time ordering operation. The time variable is assumed to be restricted to the interval $0 \leqq it \leqq \hbar\beta$.

Adopting the customary concept of replacing physical properties of a single disordered system by an ensemble average of corresponding properties of systems with well-defined configurations, the following ensemble-averaged Green's functions are needed to calculate physical properties such as the electronic density of states and the dc electrical conductivity:

$$\mathcal{G}(11') = \langle G(11')\rangle\,, \quad \mathcal{G}(121'2') = \langle G(121'2')\rangle\,, \quad \mathcal{L}(11'22') = \langle L(11'22')\rangle\,, \tag{2.3}$$

where $\langle\ldots\rangle$ denotes the averaging procedure.

The averaged density of electronic states at energy $\hbar\omega$, $\mathcal{N}(\hbar\omega)$, may be related to $\mathcal{A}(mm', \omega)$ — the spectral density of $\mathcal{G}(11')$ — by

$$\mathcal{N}(\hbar\omega) = \frac{1}{2\pi\hbar V} \sum_m \mathcal{A}(mm, \omega) , \tag{2.4}$$

where V is the volume of the system. $\mathcal{A}(mm', \omega)$ results from

$$\mathcal{A}(mm', \omega) = \lim_{\varepsilon\to 0} \{\mathcal{G}(mm', \omega + i\varepsilon) - \mathcal{G}(mm', \omega - i\varepsilon)\} , \tag{2.5}$$

where the quantities on the right of (2.5) are the analytical continuation of the Fourier coefficients of $\mathcal{G}(11')$ into the complex z-plane taken at $z = \omega \pm i\varepsilon$.

The averaged dc conductivity tensor, σ_{ij}, may be written

$$\sigma_{ij} = -\frac{e^2}{2\hbar^3 V} \sum_{m_{1'}m_1m_{2'}m_2} V_{m_{1'}m_2} V_{m_{2'}m_2} R^i_{m_{1'}m_1} R^j_{m_{2'}m_2} \frac{\partial}{\partial\omega} \mathcal{B}(m_1m_{1'}m_2m_{2'}, \omega)|_{\omega=0} . \tag{2.6}$$

This relation results from Kubo's conductivity formula using the following expression for the current $\boldsymbol{J}$:

$$\boldsymbol{J} = i\,\frac{e}{\hbar} \sum_{nm} V_{nm} \boldsymbol{R}_{nm} a_n^+ a_m , \qquad \boldsymbol{R}_{nm} = \boldsymbol{R}_m - \boldsymbol{R}_n , \tag{2.7}$$

where the position of the m-th atom is described by $\boldsymbol{R}_m$. The quantity $\mathcal{B}(m_1m_{1'}m_2m_{2'}, \omega)$ involved in (2.6) results from

$$\mathcal{B}(m_1m_{1'}m_2m_{2'}, \omega) = \lim_{\varepsilon\to 0} \{\mathcal{L}(m_1m_{1'}m_2m_{2'}, \omega + i\varepsilon) - \mathcal{L}(m_1m_{1'}m_2m_{2'}, \omega - i\varepsilon)\} , \tag{2.8}$$

where the quantitites on the right of (2.8) are the analytical continuation of the Fourier coefficients of $\mathcal{L}(11'22')|_{t_{1'}=t_1^+, t_{2'}=t_2^+}$ into the complex z-plane taken at $z = \omega \pm i\varepsilon$.

The investigation of the averaged Green's functions (2.3) starting from the Hamiltonian (2.1) is the subject of the following sections.

3. Formalism of Green's Functions

According to the usual procedure [7] the following more general one-particle Green's function is introduced:

$$G(11', U) = \frac{\mathrm{Sp}\,(W T\, \{S(-i\hbar\beta, 0)\, a(1)\, a^+(1')\})}{\mathrm{Sp}\, \{W S\, (-i\hbar\beta, 0)\}} , \tag{3.1}$$

where $S(t, t')$ is defined by

$$\frac{\partial}{\partial t} S(t, t') = -\frac{i}{\hbar} U(t)\, S(t, t') , \qquad S(t, t) = 1 , \tag{3.2}$$

and $U(t)$ is taken to be

$$\left.\begin{aligned} U(t) &= U^{(1)}(t) + U^{(2)}(t) , \\ U^{(1)}(t) &= \sum_{ms} U^{(1)}_{ms}(t)\, [\gamma_s(m)\, b_s(t) + \gamma_s^*(m)\, b_s^+(t)] , \\ U^{(2)}(t) &= \sum_{m_1m_2} U^{(2)}_{m_1m_2}(t)\, a^+_{m_1}(t)\, a_{m_2}(t) \\ \text{with}\qquad & \\ [U^{(1)}_{ms}(t)]^* &= U^{(1)}_{ms}(t) , \qquad [U^{(2)}_{m_1m_2}(t)]^* = U_{m_2m_1}(t) . \end{aligned}\right\} \tag{3.3}$$

$U^{(1)}(t)$ and $U^{(2)}(t)$ are auxiliary potentials. They are assumed to be independent of a special atomic configuration.

The following Dyson equation may be written down for the Green's function (3.1):

$$\int d2\,\{G_0^{-1}(12) - U^{(2)}(12)\}\,G(21', U) = i\hbar\delta(1-1') + \int d2\,\Sigma\,(12, U)\,G(21', U)\,, \tag{3.4}$$

where the following definitions and abbreviations have been used:

$$\left.\begin{aligned} G_0^{-1}(12) &= \delta(t_1 - t_2)\left[i\hbar\delta_{m_1m_2}\frac{\partial}{\partial t_2} - \delta_{m_1m_2}\varepsilon_{m_2} - V_{m_1m_2}\right], \\ U^{(2)}(12) &= U^{(2)}_{m_1m_2}(t_1)\,\delta(t_1 - t_2)\,, \\ \delta(1-1') &= \delta(t_1 - t_{1'})\,\delta_{m_1m_{1'}}\,,\quad \int d2 \cdots = \sum_{m_2}\int_0^{-i\hbar\beta} dt_2 \ldots . \end{aligned}\right\} \tag{3.5}$$

The self-energy Σ occurring in (3.4) is defined by

$$\int d2\,\Sigma(12, U)\,G(21', U) = $$
$$= \frac{\operatorname{Sp}\left(WT\left\{S(-i\hbar\beta, 0)\left[\sum_s \left(\gamma_s(m_1)\,b_s(t_1) + \gamma_s^*(m_1)\,b_s^+(t_1)\right)\right]a(1)\,a^+(1')\right\}\right)}{\operatorname{Sp}\{WS\,(-i\hbar\beta, 0)\}}. \tag{3.6}$$

Using

$$\frac{\delta G(11', U)}{\delta U^{(1)}_{ms}(t)} = \frac{1}{i\hbar}\int d2\,d2'\,G(12, U)\,\frac{\delta\Sigma(22', U)}{\delta U^{(1)}_{ms}(t)}\,G(2'1', U) \tag{3.7}$$

the auxiliary potential $U^{(1)}(t)$ allows to derive the following functional differential equation for Σ:

$$\Sigma(11', U) = \delta(1-1')\,\frac{\operatorname{Sp}\left(WT\left\{S(-i\hbar\beta, 0)\left[\sum_s \left(\gamma_s(m_1)\,b_s(t_1) + \gamma_s^*(m_1)\,b_s^+(t_1)\right)\right]\right\}\right)}{\operatorname{Sp}\{WS(-i\hbar\beta, 0)\}} +$$
$$+ \sum_s \int d2\,G(12, U)\,\frac{\delta\Sigma(21', U)}{\delta U^{(1)}_{sm_1}(t_1)}. \tag{3.8}$$

Introducing the two-particle Green's function

$$L(11'22', U) = G(121'2', U) - G(11', U)\,G(22', U)\,, \tag{3.9}$$

where

$$G(121'2', U) = \frac{\operatorname{Sp}\left(WT\{S(-i\hbar\beta, 0)\,a(1)\,a(2)\,a^+(2')\,a^+(1')\}\right)}{\operatorname{Sp}\{WS(-i\hbar\beta, 0)\}}, \tag{3.10}$$

the auxilary potential $U^{(2)}(t)$ allows to derive the Bethe-Salpeter equation

$$L(11'22', U) = L_0(11'22', U) +$$
$$+ \int d3\,d4\,d5\,d6\,L_0(11'43, U)\,\Lambda(3456, U)\,L(5622', U) \text{ for } t_{2'} = t_2^+\,, \tag{3.11}$$

where the following identifications have been made:

$$\left.\begin{aligned} L(11'22', U)\big|_{t_{2'}=t_2^+} &= -\,i\hbar\,\frac{\delta G(11', U)}{\delta U^{(2)}_{m_{2'}m_2}(t_2)}, \\ L_0(11'22', U) &= -\,G(12', U)\,G(21', U)\,, \\ \Lambda(11'22', U) &= \frac{i}{\hbar}\,\frac{\delta\Sigma(11', U)}{\delta G(22', U)}. \end{aligned}\right\} \tag{3.12}$$

Deriving (3.11) the fact has been used that Σ depends on $U^{(2)}$ only as $\Sigma = \Sigma(G(U^{(2)}))$.

Analogously as done in [3, 8, 9] now a further auxiliary quantity is introduced by

$$\Lambda = \int \mathrm{d}1\, \mathrm{d}2\, \lambda(12)\, v(12, U)\,, \tag{3.13}$$

where $v(12, U)$ is defined by

$$v(12, U) = \Sigma(12, U) + \delta(1-2)\, \varepsilon_{m_1} \tag{3.14}$$

and $\lambda(12)$ is assumed to be independent of the configuration of atoms.

By means of (3.13) the function

$$\mathcal{G}_\lambda(11', U) = \langle G(11', U)\rangle_\lambda\,, \tag{3.15}$$

where

$$\langle\ldots\rangle_\lambda = \left\langle \ldots \frac{\mathrm{e}^\Lambda}{\langle \mathrm{e}^\Lambda\rangle}\right\rangle \tag{3.16}$$

is introduced, which in the limit $U \to 0$, $\lambda \to 0$ reduces to the averaged one-particle Green's function $\mathcal{G}(11')$ from (2.3).

The Green's function (3.15) obeys the Dyson equation

$$\int \mathrm{d}2\, [G_{00}^{-1}(12) - U^{(2)}(12)]\, \mathcal{G}_\lambda(21', U) = i\hbar\delta(1-1') + \int \mathrm{d}2\, \vartheta_\lambda(12, U)\, \mathcal{G}_\lambda(21', U)\,, \tag{3.17}$$

where

$$G_{00}^{-1}(12) = \delta(t_1 - t_2)\left[i\hbar\delta_{m_1m_2}\frac{\partial}{\partial t_1} - V_{m_1m_2}\right], \tag{3.18}$$

and the self-energy ϑ_λ is defined by

$$\int \mathrm{d}2\, \vartheta_\lambda(12, U)\, \mathcal{G}_\lambda(21', U) = \int \mathrm{d}2\, \langle v(12, U)\, G(21', U)\rangle_\lambda\,. \tag{3.19}$$

A functional differential equation for ϑ_λ may be derived along the lines given in [3, 8, 9]. Using

$$\frac{\delta\mathcal{G}_\lambda(11', U)}{\delta\lambda(34)} = \frac{1}{i\hbar}\int \mathrm{d}2\, \mathrm{d}2'\, \mathcal{G}_\lambda(12, U)\, \frac{\delta\vartheta_\lambda(22', U)}{\delta\lambda(34)}\, \mathcal{G}_\lambda(2'1', U)\,, \tag{3.20}$$

one finds

$$\vartheta_\lambda(11', U) = \langle v(11', U)\rangle_\lambda + \frac{1}{i\hbar}\int \mathrm{d}2\, \mathrm{d}3\, \mathcal{G}_\lambda(23, U)\, \frac{\delta\vartheta_\lambda(31', U)}{\delta\lambda(12)}. \tag{3.21}$$

Further a function

$$\mathcal{L}(11'22', U) = \langle L(11'22', U)\rangle \tag{3.22}$$

is introduced, for which holds

$$\mathcal{L}(11'22', U)|_{t_{2'}=t_2^+} = -i\hbar\, \frac{\delta\mathcal{G}(11', U)}{\delta U^{(2)}_{m_{2'}m_2}(t_2)} \tag{3.23}$$

with $\mathcal{G}(11', U) \equiv \langle G(11', U)\rangle$. $\mathcal{L}(11', 22'\, U)$ reduces in the limit $U \to 0$ to the averaged two-particle Green's function $\mathcal{L}(11'22')$ from (2.3).

An equation of type (3.11) may be written down for the function (3.22) only if the dependence of the self-energy Σ on $U^{(2)}$ is ignored, i.e. if the approximation $L(11'22', U) \approx L_0(11'22', U)$ is used. In this approximation the self-energy ϑ connected with $\mathcal{G}(U)$ (which results from ϑ_λ in the limit $\lambda \to 0$) depends on

$U^{(2)}$ only as $\vartheta = \vartheta(\mathcal{G}(U^{(2)}))$ and it holds

$$\mathcal{L}(11'22',U) = \mathcal{L}^0(11'22', U) + \\ + \int d3\, d4\, d5\, d6\, \mathcal{L}^0(11'43, U)\, \Lambda(3456, U)\, \mathcal{L}(5622', U) \text{ for } t_{2'} = t_2^+, \quad (3.24)$$

where the following identifications have been made:

$$\mathcal{L}^0(11'22', U) = -\mathcal{G}(12', U)\, \mathcal{G}(21', U), \\ \Lambda(11'22', U) = \frac{i}{\hbar} \frac{\delta\vartheta(11', U)}{\delta\mathcal{G}(22', U)}. \quad (3.25)$$

4. A Single-Site Approximation for the Self-Energy

As above stated, Green's functions of physical interest are $\mathcal{G}(11')$ and $\mathcal{L}(11'22')$. These functions obey equations like (3.17) and (3.24), respectively, if in (3.17) the limit $U \to 0$ and $\lambda \to 0$, and in (3.24) the limit $U \to 0$ is taken.

The self-energy $\vartheta(11')$ and the vertex-part $\Lambda(11'22')$, occurring in these equations, result from (3.21) and (3.25) for $U \to 0$, $\lambda \to 0$ and $U \to 0$, respectively.

A calculation of the Green's functions $\mathcal{G}(11')$ and $\mathcal{L}(11'22')$ requires in general more or less drastic simplifications and approximations, respectively. In the following an incorporation of electron–phonon interaction into the customary single-site approximation concept is discussed. For this end in equation (3.21) only terms of $v(11', U)$ being diagonal in site indices are accounted for. Further each diagonal term is assumed to depend only on the type of atom belonging to its site index.

Thus $v(11', U)$ in (3.21) is thought to be

$$v(11', U) = \delta_{m_1 m_{1'}} \sum_i \eta^i_{m_1} v^i(t_1 t_{1'}, U), \quad (4.1)$$

$$\text{where } \eta^i_{m_1} = \begin{cases} 1 & \text{if atom of type } i \text{ is placed on site } m_1 \ (i = \text{A, B}), \\ 0 & \text{else}. \end{cases} \quad (4.2)$$

The quantity v^i occurring in (4.1) is discussed in more detail in the next section.

The following derivation for the single-site approximation for the self-energy runs along the lines used in [10] to derive an analogous expression for a static structurally disordered system.

Equation (3.21) may be rewritten as

$$\vartheta(11', U) = \langle v(11', U)\rangle + \frac{1}{i\hbar} \int d2\, d3\, \{\langle v(12, U)\, G(23, U)\, v(31', U)\rangle - \\ - \vartheta(12, U)\, \mathcal{G}(23, U)\, \vartheta(31', U)\} \quad (4.3)$$

by using (3.19) and taking the limit $\lambda \to 0$. $\vartheta(11', U)$ is the self-energy of the Green's function

$$\mathcal{G}(11', U) = \langle G(11', U)\rangle. \quad (4.4)$$

Substituting (4.1) into the second term on the right-hand side of (4.3) and using the conditional averaged Green's functions defined by

$$\langle \eta^i_m\rangle\, \mathcal{G}^i(m, 11', U) = \langle \eta^i_m G(11', U)\rangle, \quad (4.5)$$

$$\langle g^{ij}(mm')\rangle\, \mathcal{G}^{ij}(mm', 11', U) = \langle g^{ij}(mm')\, G(11', U)\rangle, \quad (4.6)$$

where

$$g^{ij}(mm') = \eta^i_m[\eta^j_{m'} - \delta_{ij}\delta_{mm'}] \tag{4.7}$$

one obtains

$$\vartheta(11', U) = \langle v(11', U)\rangle + \frac{1}{i\hbar}\int d2\,d3\,\Big\{\delta_{m_1m_2}\sum_i v^i(t_1t_2, U)\times .$$
$$\times\,[c^i\mathscr{G}^i(m_1, 23, U)\,\vartheta^i(m_1, 31', U) + c^i\mathscr{G}^i(m_1, 23, U)\,v^i(t_3t_{1'}, U)\times$$
$$\times\,\delta_{m_3m_{1'}}\delta_{m_3m_1}] - \vartheta(12, U)\,\mathscr{G}(23, U)\,\vartheta(31', U)\Big\}. \tag{4.8}$$

In (4.5) $\langle\eta^i_m\rangle$ equals c^i, the concentration of atoms of type i (i = A, B), and in (4.6) $\langle g^{ij}(mm')\rangle$ is the two-particle correlation function. $\vartheta^i(m, 11', U)$ in (4.8) represents the self-energy in the Dyson equation of the Green's function $\mathscr{G}^i(m, 11', U)$.

Starting from the equation of motion (3.4) and accounting for (4.5) and (4.6) the self-energies occurring in (4.8) may be expressed as

$$\vartheta(11', U) = \sum_i c^i\sigma^i(11', U)\,, \tag{4.9}$$

$$\vartheta^i(m, 11', U) = \frac{1}{c^i}\sum_j \langle g^{ij}(mm_1)\rangle\,\sigma^{ij}(m, 11', U)\,, \tag{4.10}$$

where

$$\sigma^i(11', U) = \int d2\,d\bar{2}\;\delta_{m_1m_2}v^i(t_1t_2, U)\;\mathscr{G}^i(m_1, 2\bar{2}, U)\;\mathscr{G}^{-1}(\bar{2}1', U)\,, \tag{4.11}$$

$$\sigma^{ij}(m, 11', U) = \int d2\,d\bar{2}\;\delta_{m_1m_2}v^j(t_1t_2, U)\;\mathscr{G}^{ij}(mm_1, 2\bar{2}, U)\;(\mathscr{G}^i)^{-1}(m, \bar{2}1', U)\,. \tag{4.12}$$

From (4.8) with (4.9) to (4.12) and the single-site approximation

$$\vartheta^i(m, 11', U) \approx \frac{1}{c^i}\sum_j \langle g^{ij}(mm_1)\rangle\,\sigma^j(11', U) \tag{4.13}$$

one obtains for total disorder, where $\langle g^{ij}(mm_1)\rangle = (1 - \delta_{mm_1})\,c^ic^j$ holds,

$$\sigma^i(11', U) = \delta_{m_1m_{1'}}v^i(t_1t_{1'}, U) +$$
$$+\frac{1}{i\hbar}\int d2\,d3\,\sigma^i(12, U)\;\mathscr{G}(23, U)\,\Big\{\delta_{m_3m_{1'}}\delta_{m_3m_1}v^i(t_3t_{1'}, U) - \sum_j \delta_{m_1m_3}c^j\sigma^j(31', U)\Big\}. \tag{4.14}$$

It is readily seen that $\sigma(11', U)$ and therefore also $\vartheta(11', U)$ determined from (4.14) are diagonal in site indices.

In a simple manner one obtains from (4.14) the following equation which determines the self-energy $\vartheta(11')$ in single-site approximation:

$$1 = c^A\left[1 - \frac{\mathscr{G}(z_\nu)}{i\hbar}\,(v^A(z_\nu) - \vartheta(z_\nu))\right]^{-1} + c^B\left[1 - \frac{\mathscr{G}(z_\nu)}{i\hbar}\,(v^B(z_\nu) - \vartheta(z_\nu))\right]^{-1}, \tag{4.15}$$

and from (4.14) and (3.25) the following expression for the vertex part $\Lambda(11'22')$ in single-site approximation:

$$\Lambda(11'22') = \frac{i}{\hbar}\left(\frac{1}{-i\hbar\beta}\right)^2 \delta_{m_1m_{1'}}\delta_{m_1m_2}\delta_{m_1m_{2'}}\sum_{z_\nu z_{\nu'}} e^{-iz_\nu(t_1-t_2)-iz_{\nu'}(t_{2'}-t_{1'})}\times$$
$$\times\,\frac{-\,v^A(z_\nu)\,v^B(z_{\nu'}) + v^A(z_\nu)\,\vartheta(z_{\nu'}) + \vartheta(z_\nu)\,v^B(z_{\nu'}) - \vartheta(z_\nu)\,\vartheta(z_{\nu'})}{[i/(-i\hbar\beta)^2] - v^A(z_\nu)\,\mathscr{G}(z_\nu) - \mathscr{G}(z_{\nu'})\,v^B(z_{\nu'}) + \mathscr{G}(z_{\nu'})\,\vartheta(z_{\nu'}) + \vartheta(z_\nu)\,\mathscr{G}(z_\nu)}. \tag{4.16}$$

In (4.15) and (4.16) $v^{A}(z_\nu)$, $v^{B}(z_\nu)$, $\mathcal{G}(z_\nu) \equiv \mathcal{G}(m_1m_1, z_\nu)$, $\vartheta(z_\nu) \equiv \vartheta(m_1m_1, z_\nu)$ are the coefficients of Fourier expansions like

$$\mathcal{G}(m_1t_1, m_1t_{1'}) = \mathcal{G}(m_1m_{1'}, t_1 - t_{1'}) = \frac{1}{-i\hbar\beta} \sum_\nu \mathcal{G}(m_1m_1, z_\nu)\, \mathrm{e}^{-iz_\nu(t_1 - t_{1'})}, \quad (4.17)$$

where $z_\nu = \dfrac{\pi\nu}{-i\hbar\beta} + \dfrac{\mu}{\hbar}$, $\nu = \pm 1, \pm 3, \pm \cdots$.

Ignoring electron–phonon interaction, i.e. putting $v^{A} = \varepsilon^{A}$ and $v^{B} = \varepsilon^{B}$ (cf. (3.14) and (4.1)), equation (4.15) reduces to the well-known CPA-equation [1]. Electron–phonon interaction obviously leads, if the assumption (4.1) is made, to a modification of this equation consisting in a z_ν-dependence of the "atomic potentials" v^{A} and v^{B}.

5. The Expression Used for $v(11', U)$ in Equation (4.1)

According to (4.15) and (4.16) it is sufficient to discuss $v(11', U)$ for $U \to 0$, which is called $v(11')$ in the following. According to (3.14) $v(11')$ is given by

$$v(11') = \delta(t_1 - t_{1'})\, \delta_{m_1m_{1'}}\varepsilon_{m_{1'}} + \Sigma(11'). \quad (5.1)$$

As ε_{m_1} may be written

$$\varepsilon_{m_1} = \sum_i \eta^{i}_{m_1}\varepsilon^{i} \quad (5.2)$$

the first term on the right-hand side of (5.1) has the form required in (4.1). The second term on the right-hand side of (5.1), $\Sigma(11')$, can be determined, at least in principle, from (3.8) by iteration and putting $U \to 0$. Accounting for the lowest order terms with respect to electron–phonon interaction iteration of (3.8) yields the following expression for the Fourier coefficients of $\Sigma(11')$:

$$\begin{aligned}\Sigma(m_1m_{1'}, z_\nu) = &-\delta_{m_1m_{1'}} \sum_{sm} \{\gamma_s(m_1)\,\gamma_s^*(m) + \gamma_s^*(m_1)\,\gamma_s(m)\}\, \frac{1}{\hbar\omega_s}\, G_0^<(mm, 0) + \\ &+ \left(\frac{1}{-i\hbar\beta}\right) \sum_{\mu s} G_0(m_1m_{1'}, z_\mu)\, \{D_0(ss, z_\nu - z_\mu)\,\gamma_s(m_1)\,\gamma_s^*(m_{1'}) + \\ &+ D_0(ss, z_\mu - z_\nu)\,\gamma_s^*(m_1)\,\gamma_s(m_{1'})\}, \end{aligned} \quad (5.3)$$

where $G_0(m_1m_{1'}, z_\mu)$ obeys the equation

$$\sum_{m_2} [(\hbar z_\mu - \varepsilon_{m_1})\, \delta_{m_1m_2} - v_{m_1m_2}]\, G_0(m_1m_{1'}, z_\mu) = \delta_{m_1m_{1'}}. \quad (5.4)$$

$D_0(ss, z_\nu - z_\mu)$ is the Fourier transform of a phonon Green's function and is given by

$$D_0(ss, z_\nu - z_\mu) = i/(z_\nu - z_\mu - \omega_s) \quad (5.5)$$

and

$$G_0^<(mm, 0) = \frac{\mathrm{Sp}\,\{\mathrm{e}^{-\beta(H_e - \mu N)}\, a_m^+ a_m\}}{\mathrm{Sp}\,\{\mathrm{e}^{-\beta(H_e - \mu N)}\}} = \int_{-\infty}^{\infty} \frac{\mathrm{d}\omega}{2\pi}\, \frac{A_0(mm, \omega)}{\mathrm{e}^{\beta(\hbar\omega - \mu)} + 1}, \quad (5.6)$$

$$A_0(mm, \omega) = \lim_{\varepsilon \to 0} \{G_0(mm, \omega + i\varepsilon) - G_0(mm, \omega - i\varepsilon)\}, \quad (5.7)$$

where the quantities on the right of (5.7) are the analytical continuation of $G_0(mm, z_\nu)$ into the complex z-plane taken at $z = \omega \pm i\varepsilon$.

In order to obtain from (5.3) an expression like (4.1) it is necessary to ignore on the right of (5.3) in the sum over m all terms with $m \neq m_1$ and to ignore all $G_0(m_1 m_{1'}, z_\mu)$ with $m_1 \neq m_{1'}$. The resulting expression is

$$\Sigma(m_1 m_{1'}, z_\nu) = \delta_{m_1 m_{1'}} \sum_s |\gamma_s(m_1)|^2 \Bigg\{ -\frac{2}{\hbar\omega_s} G_0^<(m_1 m_1, 0) + + \left(\frac{1}{-i\hbar\beta}\right) \sum_\mu G_0(m_1 m_{1'}, z_\mu) \left[D_0(ss, z_\nu - z_\mu) + D_0(ss, z_\mu - z_\nu)\right]\Bigg\}. \tag{5.8}$$

This expression can readily be written like (4.1), provided that the two configurationally dependent quantities, $|\gamma_s(m_1)|^2$ and $G_0(m_1 m_1, z_\nu)$, occurring in (5.8) depend only on the type of atom placed on the site m_1. However, $G_0(m_1 m_1, z_\nu)$ does not satisfy this condition.

The self-energy $\Delta_{m_1}(z_\nu)$ defined by

$$G_0(m_1 m_1, z_\nu) = [\hbar z_\nu - \varepsilon_{m_1} - \Delta_{m_1}(z_\nu)]^{-1} \tag{5.9}$$

depends on the atomic configuration of all sites but the site m_1 as seen e.g. from the representation of $\Delta_{m_1}(z_\nu)$ by a renormalized perturbation series, well known from localization investigations [11]. In order to get from (5.8) an expression like (4.1) it would be necessary to replace $\Delta_{m_1}(z_\nu)$ in (5.9) by a corresponding averaged quantity. But it is hard to see how to obtain explicitely such a quantity.

At this stage it seems to be reasonable to ignore in (5.8) completely the configuration dependence of G_0, i.e. to substitute in (5.8) G_0 by G_{00} defined in (3.18). If for the state density of the ordered system the semicircle model [1] is used, e.g., an analytical expression for $G_{00}(m_1 m_1, z_\mu)$ can be obtained and the calculation of Σ from (5.8) becomes a tractable task.

6. Concluding Remarks

In the theory presented here the equation (3.21) for the self-energy plays a decisive role. For a static system the quantitiy $v(11', U)$ entering into this equation equals $\delta(1 - 1')\, \varepsilon_{m_1}$, where ε_{m_1} takes either ε^A or ε^B depending on the type of atom on the site with index m_1. Taking into account electron–phonon interaction, $v(11', U)$ contains according to (3.14) another term, $\Sigma(11', U)$, the self-energy of the unaveraged one-electron Green's function (3.1). The latter term is very complicated, in contrast to the former one.

No doubt, from (3.8) and (3.7) $\Sigma(11', U)$ can be formally determined, at least in principle, by iteration. But an actual calculation of $\Sigma(11', U)$ for a definite disordered system is practically impossible. Thus, for instance, for the lowest order expression for $\Sigma(11')$ with respect to electron–phonon interaction the knowledge of $G_0(11')$, the unaveraged one-electron Green's function for a static disordered system, is required (cf. Section 5). But $G_0(11')$ cannot really be obtained from any calculation. In order to get any tractable expression for $\Sigma(11', U)$ or $\Sigma(11')$, respectively, more or less drastic simplifications are to be made, including possibly a certain averaging procedure, as discussed in Section 5.

Using such a concept it has been shown in this paper that contributions of the electron–phonon interaction can be incorporated into the customary single site approximation scheme.

Acknowledgements

The author gratefully acknowledges the warm hospitality of the Department of Theoretical Physics at the A. F. Ioffe Physico-Technical Institute, Academy of Sciences of the USSR, and of Prof. Yu. Firsov in particular.

References

[1] B. Velický, S. Kirckpatrick, and H. Ehrenreich, Phys. Rev. **175**, 747 (1968).
[2] L. Schwartz, F. Brouers, A. V. Vedyaev, and H. Ehrenreich, Phys. Rev. B **4**, 3383 (1971).
[3] L. Schwartz and E. Siggia, Phys. Rev. B **5**, 383 (1972).
[4] M. Tsukada, J. Phys. Soc. Japan **32**, 1475 (1972).
[5] A. Chen, G. Weisz, and A. Sher, Phys. Rev. B **5**, 2897 (1972).
[6] I. Sadakata, J. Phys. Soc. Japan **35**, 372 (1973).
[7] G. Baym and L. P. Kadanoff, Phys. Rev. **124**, 287 (1961).
[8] F. Klvaňa, Czech. J. Phys. B **17**, 400 (1967).
[9] H. Puff, phys. stat. sol. (b) **52**, 111 (1972).
[10] H. J. Fischbeck, phys. stat. sol. (b) **53**, 527 (1972).
[11] E. N. Economou and M. H. Cohen, Phys. Rev. B **5**, 2931 (1972).

(Received May 23, 1974)

phys. stat. sol. (b) **65**, 153 (1974)

Subject classification: 9; 22

Solid State and Materials Laboratory, Princeton University, New Jersey (a) and Brookhaven National Laboratory[1]), *Upton, New York (b)*

A Theoretical Analysis of Cation Diffusion in Potassium Azide[2])

By

Å. G. DANEMAR[3]) (a), B. S. H. ROYCE (a), and D. O. WELCH (b)

A theoretical analysis has been made of the temperature dependence and anisotropy of the self-diffusion of tracer potassium ions in KN_3. A vacancy mechanism is assumed, and the effects of correlation on diffusion are examined. An approximate analytic expression is obtained for the correlation factor, and its validity is determined by comparison with more accurate numerical calculations. The temperature dependence of the elements of the diffusion tensor including correlation is evaluated, making use of theoretical values of the migration energy of cation vacancies, and the role of correlation in determining the crystallographic anisotropy of cation diffusion is discussed. Comparison is made to available experimental data.

Es wird die theoretische Analyse der Temperaturabhängigkeit und Anisotropie der Selbstdiffusion von radioaktiv markierten Kaliumionen in KN_3 durchgeführt. Es wird ein Leerstellenmechanismus angenommen und die Einflüsse der Korrelation auf die Diffusion untersucht. Ein geeigneter analytischer Ausdruck für den Korrelationsfaktor wird erhalten und seine Gültigkeit durch Vergleich mit genaueren numerischen Rechnungen bestimmt. Die Temperaturabhängigkeit der Elemente des Diffusionstensors wird unter Einschluß der Korrelation berechnet, wobei theoretische Werte der Wanderungsenergie der Kationenleerstellen benutzt werden und die Rolle der Korrelation bei der Bestimmung der Kristallanisotropie der Kationendiffusion diskutiert wird. Die Ergebnisse werden mit erhältlichen experimentellen Werten verglichen.

1. Introduction

Potassium azide (KN_3) is an ionic crystal with a body centered tetragonal structure described by the space group I4/mcm. It is an ionic conductor at room temperature and at elevated temperatures decomposes into nitrogen gas and metallic potassium. Ionic conductivity data [1 to 3] indicates that transport within the crystal takes place via a vacancy mechanism on the potassium sublattice. No clear evidence for azide ion transport exists although ionic conductivity measurements close to the thermal decomposition temperature [4] may show a contribution from the azide sublattice. Tracer diffusion measurements [5] have also been made for the potassium sublattice in a direction parallel to the *c*-axis. These measurements show a temperature dependence of the diffusion coefficient that is consistent with the extrinsic ionic conductivity data

[1]) Supported by the United States Atomic Energy Commission.

[2]) Work supported in part by AROD Grants DAHC04-69-C-0049 and DAHC04-73-C-0019.

[3]) Present address: Capacitor Division, AB Liljeholmens Kabelfabrik, Box 42108, 126 12 Stockholm, Sweden.

for the same crystal orientation and also suggest that cation vacancies are the principal mobile species in the lattice below 250 °C. The single value obtained for the diffusion coefficient perpendicular to the crystalline c-axis is approximately 20% smaller then the corresponding value parallel to the c-axis at the same temperature.

As was first pointed out by Bardeen and Herring, a vacancy mechanism of ionic transport implies that tracer diffusion is non-random and that correlation effects must be taken into account in evaluating diffusion processes. Correlation effects have been shown to cause deviations from the Nernst-Einstein relationship [6], to reduce the isotope effect [7] and to result in a deviation from the Arrhenius temperature dependence of the diffusion coefficient [8].

Theoretical calculations [9] of the energies associated with potassium vacancy motion in KN_3 indicate that the energy of motion within the basal planes is higher than that parallel to the crystalline c-axis. These calculated energies are sufficiently divergent that, in the absence of correlation effects, a substantial difference between diffusion parallel and perpendicular to the c-axis is to be expected. Since the experimental results mentioned above do not show such anisotropy, it is important to examine the role of correlation in the diffusion coefficient of cations.

With regard to the potassium sublattice alone, the KN_3 structure may be considered as being a simple tetragonal lattice with the a and c lattice spacing having $1/\sqrt{2}$ and $1/2$ of the values of a_0 and c_0 for the crystallographic unit cell. Mullen [10] has treated the effects of correlation on vacancy diffusion in anisotropic crystals, including the simple tetragonal structure. However, the results of Mullen's numerical calculations are presented as a function of the ratio of the jump frequency for elementary jumps perpendicular and parallel to the tetragonal axis. In general, this ratio varies with temperature, owing to the difference in the activation energy for motion for the two jumps. The use of Mullen's results in discussing the temperature dependence of the diffusion coefficient is therefore rather unwieldy. Approximate analytical expressions for the correlation factors have been derived, with the method developed by Mullen and others [11], for use in the discussion of the temperature dependence. More exact numerical calculations of the correlation factor were also made and were used to estimate the errors resulting from the analytical expressions. Even though the approximate approach is less accurate, its use is justified by the additional insight and flexibility provided by the analytical representation of the correlation factors.

2. The Correlation Factor

The correlation factor is most easily discussed in the context of the generalized Einstein equation [11]

$$D_{ij} = \frac{\langle X_i X_j \rangle}{2\tau}, \tag{1}$$

where the indices $i, j(= x, y, z)$ label the cartesian coordinates, D_{ij} is one element of the diffusion coefficient tensor, $\boldsymbol{X}$ is the displacement of the tracer atom after a time τ, and the average is taken over many different jump sequences. If the coordinate axes are chosen to coincide with the crystal axes, the cation diffusion tensor for KN_3 is

$$\mathbf{D} = \begin{pmatrix} D_{xx} & 0 & 0 \\ 0 & D_{xx} & 0 \\ 0 & 0 & D_{zz} \end{pmatrix}, \tag{2}$$

where z is the tetragonal axis.

The correlation factor is obtained by first expanding equation (1) in terms of sequences of elementary jumps

$$D_{ij} = \frac{1}{2\tau}\left[\sum_{p=1}^{n} \langle x_{pi}x_{pj}\rangle + 2\sum_{p=1}^{n-1}\sum_{q>p}^{n} \langle x_{pi}x_{qj}\rangle\right], \tag{3}$$

where x_{pi} is the displacement along the i-axis of the p-th jump, and n is the number of jumps occurring in the interval τ. If there is no correlation between jumps, the double summation of equations (3) is zero, and the diffusion coefficient is the same as that in a purely random walk, $D_{ij}^{\text{random}} = n\overline{(x_i x_j)}/2\tau$, where the bar denotes an average over the different types of elementary jump permitted by the particular crystal. The correlation factor, f_{ij}, is defined as

$$f_{ij} = \frac{D_{ij}}{D_{ij}^{\text{random}}}. \tag{4}$$

There will be two elements of the correlation factor, f_{xx} and f_{zz}, for KN_3.

For crystals of sufficiently high symmetry, equation (3) can be summed to obtain a function of the correlation between successive jumps [11] and for tracer diffusion by the vacancy mechanism in the simple tetragonal structure [10]

$$f_{xx} = \frac{1+\xi}{1-\xi}, \qquad f_{zz} = \frac{1+\omega}{1-\omega}, \tag{5}$$

where

$$\xi = \frac{\langle x_0 x_1\rangle}{x^2}$$

and

$$\omega = \frac{\langle z_0 z_1\rangle}{z^2} \tag{6}$$

and the subscripts zero and one denote, respectively, the initial and succeeding jump of the tracer. The evaluation of these averages is carried out by considering the diffusion of the vacancy following the initial exchange with the tracer.

3. Evaluation of the Correlation Factor

The sites of the cation sublattice make up a simple tetragonal lattice, a portion of which is shown in Fig. 1. Suppose that a vacancy–tracer exchange has just taken place, and the vacancy is located on site "1". The neighboring sites to the vacancy may be formed into equivalent groups, several of which are shown in Fig. 1. (The fact that one of these sites is occupied by a tracer atom and is, thus, special is irrelevant in the evaluation of the correlation factor for this structure [11].) As the vacancy migrates the statistical population of each of

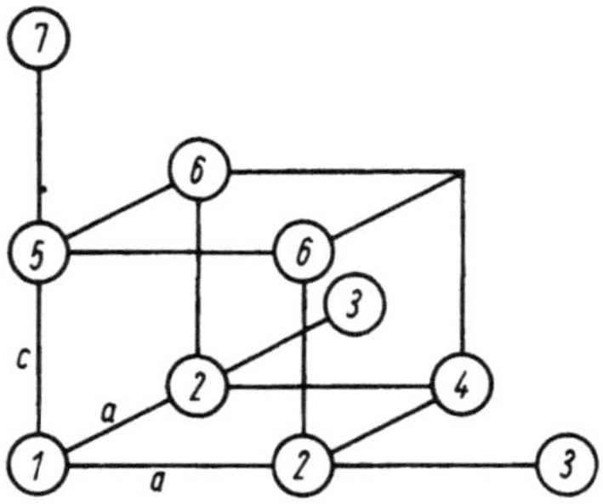

Fig. 1. Model lattice used in the evaluation of the correlation factors. $a = 4.309$ Å, $c = 3.528$ Å

the various groups of sites evolves in a well characterized way, and from these populations the probable direction of the next tracer jump can be found.

Let $p_g(k)$ denote the population of group g, defined as the sum over all sites in the group of the probability that the vacancy is on a particular site after the k-th jump of the vacancy. The various p_g act as the components of a vector which evolves according to

$$\boldsymbol{p}(k) = T \cdot \boldsymbol{p}(k-1) = T^k \cdot \boldsymbol{p}(0)\,, \tag{7}$$

where the T_{kg} is the probability that a vacancy on one of the sites in group g will land on any of the sites of group k after a jump. For the seven groups of sites shown in Fig. 1, the transition matrix is

$$T = \begin{pmatrix} 0 & A & 0 & 0 & C & 0 & 0 \\ 4A & 0 & A & 2A & 0 & C & 0 \\ 0 & A & 0 & 0 & 0 & 0 & 0 \\ 0 & 2A & 0 & 0 & 0 & 0 & 0 \\ 2C & 0 & 0 & 0 & 0 & A & C \\ 0 & 2C & 0 & 0 & 4A & 0 & 0 \\ 0 & 0 & 0 & 0 & C & 0 & 0 \end{pmatrix}. \tag{8}$$

The factors A and C are, respectively, the probabilities that a vacancy jump will be in a particular a- or c-direction and thus are related to the vacancy jump frequencies ν_a and ν_c by

$$A = \frac{\nu_a}{4\nu_a + 2\nu_c}, \qquad C = \frac{\nu_c}{4\nu_a + 2\nu_c}. \tag{9}$$

It is easily seen that $4A + 2C = 1$. Since the jump frequencies depend upon temperature, the T-matrix is temperature dependent. In principle the T-matrix is infinite in dimension, but its truncation is effected, as in equation (8), by assuming that if a vacancy escapes from the sites under explicit consideration, then it returns to the tracer atom from a random direction, thus contributing nothing to the correlation factor.

The direction cosines ξ and ω (6) are obtained from the population vector $\boldsymbol{p}(k)$. Suppose that the initial tracer–vacancy exchange took the tracer from site "1" to one of the sites "2" of Fig. 1. The value of x_0 is $+a$. The value of x_1 will be either $\pm a$ or 0 depending upon which site the vacancy occupies just prior to its next exchange with the tracer. The probability that a vacancy occupies a particular site after a given vacancy jump is obtained from the elements of the $\boldsymbol{p}$-vector after the jump, and a weighted average of the possible values of x_1 is constructed. This weighted average is summed over all possible numbers of vacancy jumps between tracer jumps. In this manner, it is readily shown that

$$\xi = A \sum_{k=0}^{\infty} \left(\tfrac{1}{2}\, p_3(k) - p_1(k)\right) . \tag{10}$$

A similar expression for ω is obtained if the initial tracer jump is from site "1", one of the sites "5", in which case

$$\omega = C \sum_{k=0}^{\infty} \left(\tfrac{1}{2}\, p_7(k) - p_1(k)\right) . \tag{11}$$

Mullen [10] was the first to show that the required sum over k of the $\boldsymbol{p}$-vector can be obtained by a matrix inversion

$$\sum_{k=0}^{\infty} \boldsymbol{p}(k) = \sum_{k=0}^{\infty} T^k \cdot \boldsymbol{p}(0) = (\mathbf{1} - T)^{-1} \cdot \boldsymbol{p}(0) . \tag{12}$$

Thus determining $(\mathbf{1} - T)_{11}^{-1}$, $(\mathbf{1} - T)_{31}^{-1}$, and $(\mathbf{1} - T)_{71}^{-1}$ yields ξ and ω from equations (10) and (11) and the correlation factors f_{xx} and f_{zz} from equation (5). The required inversion can be obtained explicitly for the seven-group matrix T of equation (8) yielding for the correlation factors

$$f_{xx} = \frac{1 + 29A - 2A^2 - 744A^3 + 1472A^4 - 576A^5}{1 + 35A + 62A^2 - 888A^3 + 832A^4 + 576A^5} \tag{13}$$

and

$$f_{zz} = \frac{A(10 + 46A - 252A^2 - 248A^3 + 864A^4)}{1 + 22A - 16A^2 - 564A^3 + 1400A^4 - 864A^5} . \tag{14}$$

Examination of the general form of the T-matrix (e.g. (8) and its generalization), its relationship to the correlation factors, and recognizing that $4A + 2C = 1$, shows that each of the correlation factors will be of the form

$$f_{ii} = \frac{\sum\limits_{n=0} a_n^{ii} A^n}{\sum\limits_{n\;0} b_n^{ii} A^n} ; \qquad i = x, z , \tag{15}$$

where the series are infinite for an infinite dimension T-matrix, but finite for the truncated versions. The value of the coefficients depend upon the size of the truncated T-matrix. In the limit as A goes to zero, diffusion in the c-direction becomes effectively one-dimensional, for which case the correlation factor for the vacancy mechanism goes to zero. This, in turn, implies that $a_0^{zz} = 0$, $b_0^{zz} \neq 0$, regardless of the size of the truncated T-matrix, a fact of importance in the discussion of the temperature dependence and anisotropy of the diffusion coefficients.

The accuracy of the analytical expressions for the correlation factors, equations (13) and (14), was tested by comparison with correlation factors obtained by numerical inversion of a T-matrix for 34 groups of sites, identical to that used by Mullen [10]. The results are shown in Fig. 2. The values of the f_{xx} obtained by the two methods agree within 2% over the entire range of A. Similar agreement is obtained for f_{zz} in the range $0.06 < A < 0.25$. As A goes to zero, discrepancies of as much as 20 to 30% result, but the qualitative behavior is the same.

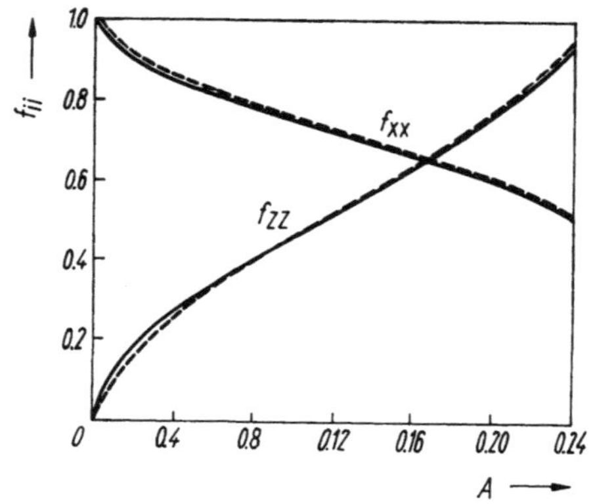

Fig. 2. The correlation factors for cation diffusion by the vacancy mechanism (as is defined in (9)). ---- 7 groups, —— 34 groups

4. The Temperature Dependence of the Diffusion Coefficient

When the number of tracer jumps in the time interval τ is written in terms of the vacancy jump frequencies, ν_a and ν_c, and the fraction of cation sites that are vacant, c_v, equations (3) and (4) yield the atomistic expressions for the elements of the diffusion coefficient

$$D_{xx} = a^2 f_{xx} \nu_a c_v \,, \qquad D_{zz} = c^2 f_{zz} \nu_c c_v \,. \tag{16}$$

The activation energy of cation diffusion in KN_3 by the vacancy mechanism, Q, defined as $- \partial \ln D/\partial(1/kT)$, is then

$$\left.\begin{aligned} Q_{xx} &= E_m^a - \frac{\partial \ln f_{xx}}{d(1/kT)} - \frac{d \ln c_v}{d(1/kT)} \,, \\ Q_{zz} &= E_m^c - \frac{\partial \ln f_{zz}}{d(1/kT)} - \frac{d \ln c_v}{d(1/kT)} \,, \end{aligned}\right\} \tag{17}$$

where the usual Arrhenius temperature dependence for the vacancy jump frequencies has been assumed and E_m is the migration energy for the vacancy. In the temperature range and for the crystal purity employed by Sharma and Laskar [5] in their diffusion measurements, the vacancy concentration is controlled by the crystal purity and may thus be considered essentially temperature independent (extrinsic conditions). For this case, the activation energy for diffusion is determined by the vacancy migration energy and the temperature dependence of the correlation factor. If E_m^c and E_m^a are sufficiently different, the latter factor can play a significant role in determining Q, as will be seen in the following discussion.

Theoretical calculations of the migration energy for cation vacancies in KN_3 have been published previously [9], and the results are summarized in Table 1. These calculations were made for three different descriptions of the repulsive "shape" of azide ions, two ellipsoidal and one di-spherical, denoted as GWI, GWII, and DS, respectively. The ellipsoidal potentials are modifications of a form originally proposed by Gray and Waddington [12]. In GWI, the ionic radii for the azide ellipsoid were obtained by fitting lattice energy and volume, while retaining the stiffness parameter appropriate to the potassium halides. This potential was found to give a poor account of azide elastic properties and a new stiffness parameter was obtained by fitting to the bulk modulus of KN_3.

Table 1

Theoretical values of the cation vacancy migration energy in KN_3, calculated with a polarizable, point-ion Born model and different repulsive potentials [9, 13]

potential	E_m^a (eV)	E_m^c (eV)	ε (18)
ellipsoidal, GWI	0.18	0.21	−0.14
ellipsoidal, GWII	0.72	0.51	+0.41
dispherical, DS	1.44	0.93	+0.55

This potential is denoted as GWII. The parameters of the di-spherical potential, DS, were obtained by fitting the equilibrium volume, bulk modulus, and lattice energy of KN_3. Potentials GWII and DS are both superior to GWI in accounting for several cohesive properties of the alkali azides [9, 13]. As shown in Table 1, the three potentials yield different values of the anisotropy in the vacancy migration energy, characterized by

$$\varepsilon = \frac{E_m^a - E_m^c}{E_m^c} . \tag{18}$$

The ellipsoidal potential GWI yields results which suggest that vacancy migration is somewhat easier perpendicular to the c-axis than parallel to it, while the ellipsoidal potential GWII and the di-spherical potential DS both lead to the conclusion that vacancy motion is markedly easier parallel to the c-axis. The consequences of these results for the interpretation of the diffusion activation energy will now be examined.

If the temperature is described by the dimensionless variable

$$\theta \equiv \frac{E_m^c}{kT} \tag{19}$$

then equations (9) and (17) to (19) can be combined to obtain an expression for the temperature dependence of the deviation of the activation energy for diffusion Q, from the activation energy for vacancy migration E_m:

$$\frac{\Delta Q_{\alpha\alpha}}{E_m^a - E_m^c} = A(1 - 4A) \frac{d(\ln f_{\alpha\alpha})}{dA} ; \qquad \alpha = x \text{ or } z , \tag{20}$$

where $\Delta Q_{\alpha\alpha} \equiv Q_{\alpha\alpha} - E_m^\alpha$ and A (defined by (9)) varies with θ as

$$A = \frac{1}{4} \left(\frac{2 \exp(-\varepsilon\theta)}{2 \exp(-\varepsilon\theta) + (\nu_c^0/\nu_a^0)} \right), \tag{21}$$

where ν_i^0 is the pre-exponential factor in the vacancy jump frequency for the i-th direction. Equation (20) is readily evaluated from the analytical expressions for the correlation factors, equations (13) and (14). This has been done for the case where $\nu_a^0 = \nu_c^0$ (isotropy of the pre-exponential factors), and the results are shown in Fig. 3. More exact expressions for the correlation factors than equations (13) and (14) will yield qualitatively the same behavior as is shown in Fig. 3.

Consideration of the results shown in Fig. 3 indicates that the effect of correlation on the temperature dependence of the diffusion coefficient has the

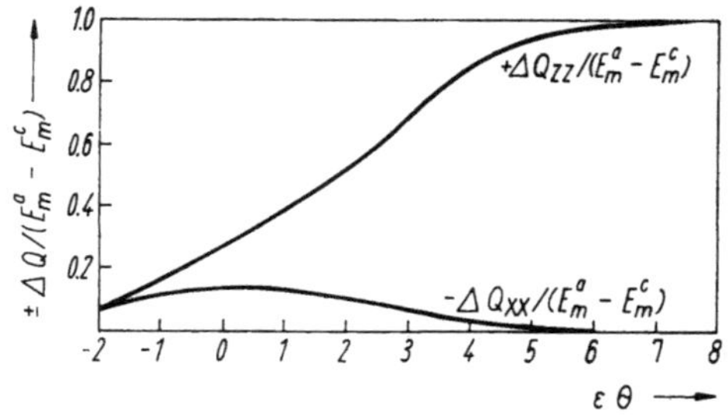

Fig. 3. The temperature dependence of the discrepancy between the activation energies for diffusion and vacancy migration

following consequences. The effect on diffusion perpendicular to the c-axis is small, the maximum deviation of Q_{xx} from E^a_m being about 15% of the difference between E^a_m and E^c_m. According to Table 1, this deviation would be, at most, 5% of E^a_m. For diffusion parallel to the c-axis, correlation has a much larger effect. For large positive $(\varepsilon\theta)$, the diffusion activation energy takes on the value of E^a_m, while at small positive and all negative values of $(\varepsilon\theta)$, it is very nearly equal to E^c_m. This can result in as much as a 50% change in the diffusion activation energy, provided that a sufficiently large range of temperature is employed. (This is unlikely to be the case for KN_3 since thermal decomposition commences at temperatures near 300 °C.) The diffusion activation energy is always greater than E^c_m for positive ε.

The experimental results of Sharma and Laskar [5], for diffusion parallel to the c-axis between 358 and 527 K, which yield $Q_{zz} = 0.8$ eV are not consistent with the predictions of any of the three potentials which imply $Q_{zz} \approx 0.2$, $0.5 \leqq Q_{zz} \leqq 0.7$; and $0.9 \leqq Q_{zz} \leqq 1.4$ eV for the GWI, GWII, and DS potentials respectively. A potential somewhere "between" GWII and DS appears to be necessary. Note that both GWII and DS predict similar values of the energy anisotropy factor ε.

While the absolute magnitude of the theoretical calculations are not in agreement with the experimental result of Laskar and Sharma, which lies between the prediction of GWII and DS, consider the consequences of the energy anisotropy factor ε. Approximate values of θ will be obtained by using the experimental Q_{zz}, rather than E^c_m, in equation (19). The range of $\varepsilon\theta$ in the Sharma-Laskar experiment is then found to be -2.6 to -2.3 for GWI, 7.2 to 10.2 for GWII, and 9.6 to 13.1 for DS. From Fig. 3, it can be seen that the range of $\varepsilon\theta$ is such that for all three potentials there will be no appreciable curvature of the Arrhenius plot (i.e. no appreciable change in Q_{zz}) as is the case in the Sharma-Laskar experiments. Furthermore, the GWI value of ε implies that $Q_{zz} \approx E^c_m \approx 0.8$ eV, $E^a_m = 0.7$ eV, while both GWII and DS imply that $Q_{zz} \approx E^a_m \approx 0.8$ eV and that $E^c_m \approx 0.5$–0.6 eV. "Reiterating" the θ values using the inferred values of E^c_m, rather than Q_{zz}, in equation (19) has no significant effect on these conclusions.

In view of the fact that the magnitude of the experimental Q_{zz} is significantly greater than that predicted by GWI while it lies between the predictions of GWII and DS (both of which are superior to GWI in describing perfect lattice properties) and that the value of ε for GWII and DS is similar, it is tempting to conclude from the above discussion of the temperature dependence that $E^c_m = 0.55$ eV and that $E^a_m = 0.8$ eV. However, there exists further experimental evidence on the anisotropy of the diffusion coefficient, which will be discussed below.

5. The Anisotropy of the Diffusion Coefficient

The anisotropy of the diffusion coefficient is markedly affected by correlation effects. From equation (16) it is seen that an appropriate measure of the diffusion anisotropy is

$$\frac{c^2 D_{xx}}{a^2 D_{zz}} = \frac{f_{xx} \nu_a}{f_{zz} \nu_c} = \frac{f_{xx}}{f_{zz}} \frac{\nu_a^0}{\nu_c^0} \exp\left(-\varepsilon\theta\right) . \tag{22}$$

In the absence of correlation effects, the ratio depends exponentially upon $\varepsilon\theta$ but including correlation effects causes the ratio to approach a constant value at large $\varepsilon\theta$. This is illustrated in Fig. 4, where it has been assumed, for simplicity, that $\nu_a^0/\nu_c^0 = 1$. The origin of the limiting value of the ratio can be seen by expanding equation (22) using equation (15) and putting $\nu_a^0/\nu_c^0 = 1$:

$$\frac{c^2 D_{xx}}{a^2 D_{zz}} = \exp\left(-\varepsilon\theta\right) \frac{\left(\sum_{n=0}^{\infty} a_n^{xx} A^n\right)\left(\sum_{n=0}^{\infty} b_n^{zz} A^n\right)}{\left(\sum_{n=0}^{\infty} a_n^{zz} A^n\right)\left(\sum_{n=0}^{\infty} b_n^{xx} A^n\right)} , \tag{23}$$

where the $\varepsilon\theta$ dependence of A is given by equation (21). In the discussion of Section 3, it was pointed out that a_0^{xx}, b_0^{xx}, $b_0^{zz} \neq 0$ while $a_0^{zz} = 0$. Thus in the large $\varepsilon\theta$ limit ($A \to 0$):

$$\lim_{A \to 0} \frac{c^2 D_{xx}}{a^2 D_{zz}} = \frac{2 a_0^{xx} b_0^{zz}}{a_1^{zz} b_0^{xx}} . \tag{24}$$

The magnitude of this limit depends upon the size of the truncated T-matrix of Section 3, as can be seen in Fig. 4. It does not, however, approach zero as does the uncorrelated ratio. For the 7-group T-matrix, the limit is 0.2, while for the 34-group T-matrix, the limit is approximately 0.12.

From the discussion of the temperature dependence in Section 4, the theoretical values of ε (at least for the GWII and DS potentials) imply that the temperature range of the Sharma-Laskar experiment falls in the large $\varepsilon\theta$ limit and hence the anisotropy ratio should take its limiting value of about 0.1. Sharma and Laskar report a single value of D_{xx} at 509 K and the resulting ratio is 0.53, too big a value for the large $\varepsilon\theta$ limit. However, measurements of the absolute value of diffusion coefficients are subject to larger errors than a determination of the temperature dependence. Thus firm conclusions on this matter

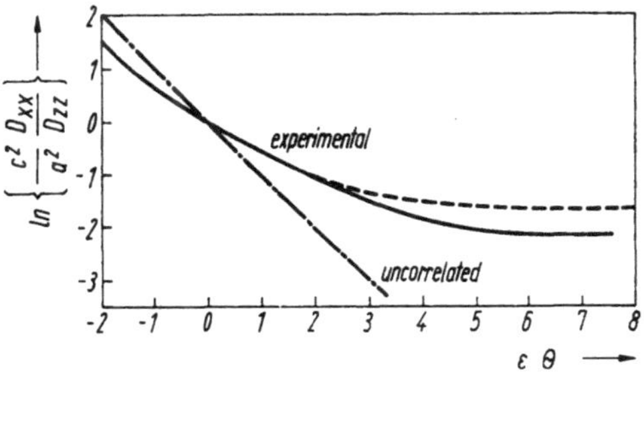

Fig. 4. The temperature dependence of the diffusion anisotropy. - - - - 7 groups, ——— 34 groups

should await further measurements of the diffusion perpendicular to the c-axis. Taking the results at face value yields $\varepsilon = 0.053$, which leads to only about a 5% difference between E_m^a and E_m^c. Taking $\varepsilon\theta = 1$, $Q_{zz} = 0.8$ eV and using Fig. 3, results in $E_m^a = 0.82$ eV and $E_m^c = 0.78$ eV.

6. Discussion

Previous theoretical calculations [9] have indicated that there is a significant anisotropy in the cation vacancy migration energy in KN_3. In this case correlation plays a significant role in the temperature dependence and anisotropy of the cation diffusion coefficient. Qualitatively, this can be seen easily, since if vacancy migration is much more difficult perpendicular than parallel to the c-axis, the cation sites, considered as parallel c-axis chains, behave as a "leaky" one-dimensional chain. Since the correlation factor for vacancy diffusion on a one-dimensional chain is zero [11], the "leaks" play the dominant role in determiningthe diffusion rate for tracer atoms parallel to the c-axis. The consequence of this is that the diffusion activation energy for this direction (in extrinsic crystals is) the vacancy migration energy *perpendicular* to the c-axis. The ratio of the diffusion coefficients parallel and perpendicular to the c-axis approaches a constant value at low temperatures, in contrast to the uncorrelated result.

Analysis of the experimental data of Sharma and Laskar [5] for the temperature dependence of tracer diffusion parallel to the c-axis, using theoretical values [9] for the anisotropy of vacancy migration energies, leads to values of the cation vacancy migration energy of 0.55 and 0.8 eV parallel and perpendicular to the c-axis respectively. However, analysis of the single datum for the anisotropy of the diffusion coefficient leads to migration energies for perpendicular and parallel vacancy motion of 0.82 and 0.78 eV respectively, i.e. essentially isotropic. Further experimental data for diffusion perpendicular to the c-axis would help to clarify the situation.

Rubidium and cesium azide have the same crystal structure as potassium azide and consequently the above considerations will be applicable to cation migration in these materials.

References

[1] J. N. MAYCOCK, V. R. PAI VERNEKER, and C. S. GORZYNSKI, JR., phys. stat. sol. **37**, 857 (1970).
[2] P. W. M. JACOBS and F. C. TOMPKINS, J. chem. Phys. **32**, 1445 (1955).
[3] Å. G. DANEMAR and B. S. H. ROYCE, Bull. Amer. Phys. Soc. **16**, 378 (1971).
[4] A. DE PANAFIEU, private communication.
[5] J. SHARMA and A. LASKAR, J. Phys. Chem. Solids **34**, 909 (1973).
[6] C. W. MCCOMBIE and A. B. LIDIARD, Phys. Rev. **101**, 1210 (1956).
[7] K. THARMALINGAM and A. B. LIDIARD, Phil. Mag. **44**, 899 (1955).
[8] J. R. MANNING, Phys. Rev. Letters **1**, 365 (1958).
[9] Å. G. DANEMAR, D. O. WELCH, and B. S. H. ROYCE, phys. stat. sol. (b) **55**, 201 (1973).
[10] J. G. MULLEN, Phys. Rev. **124**, 1723 (1961).
[11] J. R. MANNING, Diffusion Kinetics for Atoms in Crystals, Van Nostrand Co., Princeton 1968.
[12] P. GRAY and T. G. C. WADDINGTON, Proc. Roy. Soc. **235A**, 481 (1956).
[13] Å. G. DANEMAR, B. S. H. ROYCE, and D. O. WELCH, to be published.

(Received April 22, 1974)

phys. stat. sol. (b) **65**, 163 (1974)

Subject classification: 10

Institut für Physik am Max-Planck-Institut für Metallforschung, Stuttgart, and Institut für Theoretische und Angewandte Physik der Universität Stuttgart

Unpinning of Dislocations within the Framework of the Vibrating-String Model

By

W. FRANK and R. FRYDMAN

Within the framework of the vibrating-string model the frequency of thermally activated unpinning of dislocations from point obstacles is calculated with the aid of a dynamical theory. It is found that the unpinning frequency Γ has the form of an Arrhenius equation, $\Gamma = \Gamma_0 \, e^{-\Delta G/kT}$, where ΔG is the portion of the free enthalpy of activation which may be obtained by means of elasticity theory using isothermal elastic constants. The pre-exponential factor Γ_0 is shown to decrease with increasing length of the pinned dislocation segments, contrary to the result obtained on the basis of the absolute rate theory.

Im Rahmen des Saitenmodells wird die Frequenz der thermisch aktivierten Entankerung von Versetzungen von Punktfehlern mit Hilfe einer dynamischen Theorie berechnet. Es zeigt sich, daß die Entankerungsfrequenz Γ die Gestalt einer Arrhenius-Gleichung hat, $\Gamma = \Gamma_0 \, e^{-\Delta G/kT}$, wobei ΔG den Anteil der freien Aktivierungsenthalpie darstellt, welcher mit Hilfe der Elastizitätstheorie unter Verwendung isothermer elastischer Konstanten berechnet werden kann. Der Vorfaktor Γ_0 nimmt mit zunehmender Länge der verankerten Versetzungssegmente ab. Dies steht im Gegensatz zu dem Resultat, welches sich aus der absoluten Ratentheorie ergibt.

1. Introduction

In a preceding paper [1] a dynamical theory of the thermally activated unpinning of dislocations from point obstacles has been developed. It was found that the unpinning frequency Γ is given by an Arrhenius equation,

$$\Gamma = \Gamma_0 e^{-\Delta G/kT} , \tag{1.1}$$

with the Gibbs free energy of activation

$$\Delta G = \frac{1}{\sum\limits_{r=1}^{M/2} 1/\Delta G_r(q_c)} \tag{1.2}$$

and with the pre-exponential factor

$$\Gamma_0 = \frac{1}{2\pi} \left(\frac{\sum\limits_{r=1}^{M/2} 1/M_r}{\sum\limits_{r=1}^{M/2} 1/(M_r \omega_r^2)} \right)^{1/2} \tag{1.3}$$

(cf. [1], equations (3.9) to (3.11)). Here ω_r is the angular eigenfrequency of the

r-th normal mode of a pinned dislocation and

$$\Delta G_r(a_r) = \frac{1}{2} M_r \omega_r^2 a_r^2 \tag{1.4}$$

(cf. [1], equation (3.2)) the Gibbs free energy in the r-th mode if it is excited to the amplitude a_r measured at the location of the pinner. The quantity q_c entering (1.2) is the critical displacement of the dislocation at the obstacle, by which the dislocation must be separated from its equilibrium position E_1 in the pinned state before unpinning occurs. The summations in (1.2) and (1.3) run over all odd modes of the dislocation, which are denoted as $r = 1, 2, 3, \ldots, M/2$. The quantities M_r have been formally introduced through equation (3.2) of [1].[1]) They will be determined in this paper within the framework of the vibrating-string model [2, 3] of a dislocation. Since for a vibrating string also the eigenfrequencies may be easily obtained, the expressions (1.2) and (1.3) for ΔG and Γ_0 can be evaluated.

2. Evaluation of the Expressions for ΔG and Γ_0

Consider a vibrating string (Fig. 1) representing a dislocation loop, which is infinitely strongly pinned at I and III, whereas the pinning strength at II is finite and may thus be overcome with the aid of thermal fluctuations. The eigenfrequencies of the segment between I and III (length $2l$) are given by

$$\nu_r = \frac{c}{2l} r\alpha_r , \tag{2.1}$$

where c is the velocity of transverse sound waves [4] and α_r depends on the pinning strength at II, ranging from $\alpha_r = 1$ for an infinitely strong pinner to $\alpha_r = (2r - 1)/2r$ for a negligibly weak pinner. The total "energy" contained in mode r is

$$\Delta G_r = 2 \frac{\pi^2 c^2 \varrho}{4l} (rA_r\alpha_r)^2 , \tag{2.2}$$

which may easily be verified by realizing that $(\pi^2 c^2 \varrho/4l)(rA_r)^2$ is the energy of the r-th mode of a string of length l infinitely strongly anchored at both ends (A_r maximum amplitude, $\varrho = T/c^2$ dislocation mass per unit length, T line tension). The maximum amplitude A_r as well as the maximum force K_r exerted on pinner II by the dislocation during a fluctuation period in the r-th mode are

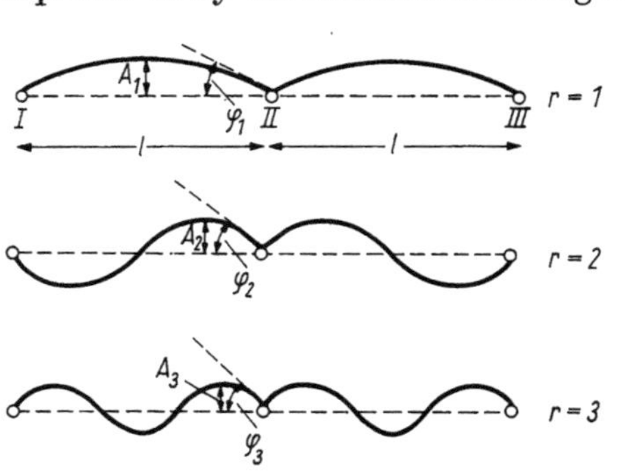

Fig. 1. First ($r = 1$), third ($r = 2$), and fifth ($r = 3$) mode of a pinned dislocation loop in the moment of maximum displacement (schematic diagram)

[1]) Further symbols used in the present paper have the same meaning as in [1].

related to the angle φ_r defined in Fig. 1 through the relations

$$\tan \varphi_r = \frac{A_r}{l/2r} \tag{2.3}$$

and

$$\sin \varphi_r = \frac{K_r}{2T}. \tag{2.4}$$

Since φ_r and A_r describe fluctuation-induced *deviations* from the equilibrium position E_1 (which is a function of the applied stress), the angles φ_r are sufficiently small in order to replace $\tan \varphi_r$ and $\sin \varphi_r$ in the above equations by their argument φ_r. Then (2.3) and (2.4) lead to

$$rA_r = \frac{K_r l}{4T}. \tag{2.5}$$

With the aid of (2.5) equation (2.2) may be rewritten as

$$\Delta G_r = \frac{\pi^2 c^2 \varrho l}{32 T^2} (\alpha_r K_r)^2 . \tag{2.6}$$

Now let attention be focused on the tiny kinked element of the dislocation that directly contacts pinner II. Its motion is governed by the differential equation[2])

$$m\ddot{q} + 2m\delta\dot{q} + m\omega_0^2 q = \sum_{r=1}^{M/2} K_r \sin(\omega_r t + \gamma_r') , \tag{2.7}$$

where m is the effective mass of this element describing its inertia against acceleration and straightening and where the term containing the damping coefficient δ allows for a frictional force acting on the element. The interaction between the pinner and the element is approximated by a parabolic interaction-energy profile characterized by a force constant f, which enters the eigenfrequency

$$\omega_0 = \sqrt{\frac{f}{m}} \tag{2.8}$$

of the element. The right-hand side of (2.7) is the sum of the periodic forces exerted on the dislocation element at the pinner by the rest of the dislocation loop. The well-known steady-state solution of (2.7) is [5]

$$mq = \sum_{r=1}^{M/2} F_r(\omega_r)\, K_r \sin(\omega_r t + \gamma_r') \tag{2.9}$$

with

$$F_r(\omega_r) = [(\omega_0^2 - \omega_r^2)^2 + 4\delta\omega_r^2]^{-1/2} . \tag{2.10}$$

A comparison between (2.9) and (3.1) of [1] yields a relationship between a_r and K_r,

$$ma_r = F_r(\omega_r)\, K_r , \tag{2.11}$$

2) It should be noted that the phases γ_r' differ from those entering the expression $q(t) = \sum_{r=1}^{M/2} a_r \sin(\omega_r t + \gamma_r)$ for the displacement $q(t)$ of the dislocation at the pinning point (cf. [1], equation (3.1)).

which may be used in order to replace K_r in (2.6) by a_r and then to determine M_r by comparison with (1.4) and (2.1). In this way one obtains

$$M_r = \frac{\varrho m^2 l^3}{16 T^2 r^2 F_r^2}. \tag{2.12}$$

Using (1.4), (2.1), (2.12) the expressions for ΔG (equation (1.2)) and Γ_0 (equation (1.3)) may be brought into the form

$$\Delta G = \frac{l(\pi m q_c)^2}{32 T \sum\limits_{r=1}^{M/2} (F_r/\alpha_r)^2} \tag{2.13}$$

and

$$\Gamma_0 = \frac{c}{2l} \sqrt{\frac{\sum\limits_{r=1}^{M/2} (r F_r)^2}{\sum\limits_{r=1}^{M/2} (F_r/\alpha_r)^2}}, \tag{2.14}$$

respectively.

3. Discussion

A general discussion of the results (2.13) and (2.14) is not possible, since neither the effective mass m nor the damping coefficient δ of the dislocation element at the pinner can be determined from first principles.[3] Fortunately, as shown in Appendix B, an average value $\langle F_r \rangle_r$ larger than ω_0^{-2} would lead to unreasonably small values for ΔG, so that $\langle F_r \rangle_r$ must lie below this upper limit. An inspection of F_r–ω_r curves for different values of δ (see, e.g., [5]) shows that the above condition is fulfilled either for sufficiently large ω_r-values and arbitrary values of δ or in the whole range of possible ω_r-values ($\omega_r \geqq 0$) for sufficiently large values of δ. The physically realistic regime is thus bordered at ω_r-values which are large compared to ω_0 by the F_r–ω_r curve for $\delta = 0$; for arbitrary ω_r-values the F_r–ω_r curves for which $\delta \gg |\omega_0^2 - \omega_r^2|/2\omega_r$ ($r = 1, 2, 3, \ldots, M/2$) is fulfilled certainly satisfy the condition $\langle F_r \rangle_r < \omega_0^{-2}$. These limiting cases will be discussed in the following.

For $\delta \gg |\omega_0^2 - \omega_r^2|/2\omega_r$ equation (2.10) reduces to $F_r \approx 1/(2\delta\omega_r)$, which leads to

$$\Delta G = \frac{3(\alpha_1^2 \pi c \delta m q_c)^2}{4 l T} \tag{3.1}$$

and

$$\Gamma_0 = \frac{2\sqrt{6}\alpha_1}{\pi} \sqrt{\frac{l}{b}}\, \nu_1^0 . \tag{3.2}$$

For $\omega_r \gg \omega_0$ and $\delta = 0$ we have $F_r \approx \omega_r^{-2}$ and hence

$$\Delta G = \frac{45(\alpha_1^3 c^2 \pi m q_c)^2}{16 l^3 T} \tag{3.3}$$

[3]) m and δ enter both ΔG and Γ_0 through F_r.

and

$$\Gamma_0 = \frac{2\sqrt{15}\alpha_1}{\pi} \nu_1^0 \tag{3.4}$$

($\nu_1^0 = c/4l = b\,\nu_D/2l$ fundamental mode of the unpinned dislocation segment of length $2l$). In deriving the expressions (3.1) through (3.4) we made the approximation $\alpha_r = \alpha_1$ for $r = 2, 3, 4, \ldots, M/2$.

It is absolutely unimportant that the expressions (3.1) and (3.3) for ΔG depend on the unknown quantities δ and m, since nobody would ever attempt to calculate the activation "energy" ΔG in the Arrhenius equation (1.1) with the aid of these relations. The only important result with respect to ΔG has already been stated in [1]: ΔG is the minimum of $\sum_{r=1}^{M/2} \Delta G_r(a_r)$ under the constraint $\sum_{r=1}^{M/2} a_r \geqq q_c$. Since $\sum_{r=1}^{M/2} \Delta G_r(a_r)$ may be directly obtained by means of elasticity theory or from atomistic calculations (cf. the Appendix of [1]), this holds also for the activation "energy" ΔG. The practical importance of this result is evident.

With regard to Γ_0 it is of great interest that the limits of the possible range of Γ_0 given by (3.2) and (3.4) do *not* depend on the undeterminable parameters m and δ. This is important, since Γ_0 can only be determined in the way described in [1] and the present paper, contrary to what is true for ΔG.

Let us now compare our results about Γ_0 with those obtained by Granato et al. [6]. These authors arrived at Γ_0-values between ν_D (infinitely strong pinning) and $1.6\,(b/l)\,\nu_D$ (extremely weak pinning). For intermediate pinning strengths they found $\Gamma_0 \approx \nu_D/100$. It should be noted that our results indicate a rather weak dependence of Γ_0 on the pinning strength ($\frac{1}{2} \leqq \alpha_1 \leqq 1$), contrary to the result of Granato et al. Furthermore, according to (3.2) and (3.4) Γ_0 varies with the loop length proportional to $l^{-\varkappa}$ ($\frac{1}{2} \leqq \varkappa \leqq 1$), whereas the rate theory [6] predicts Γ_0 to be fairly independent of l, except for unrealistically weak pinners.

For the limiting case $\varkappa = 1$, i.e. for $\Gamma_0 \sim \nu_1^0 = b\nu_D/2l$ two interesting consequences of practical importance arise:

1. For obstacles of uniform pinning strength the microscopic slip rate $\dot{a}$ may be put into the form

$$\dot{a} = \dot{a}_0 e^{-\Delta G/kT} \tag{3.5}$$

with a pre-exponential factor

$$\dot{a}_0 = \eta n b^2 \nu_D \tag{3.6}$$

(η numerical factor close to unity, n dislocation density) which is independent of l [7].

2. In the presence of pinners of different pinning strengths an Arrhenius equation (3.5) remains formally valid, though a_0 and ΔG have to be re-interpreted [8].

Acknowledgements

The authors are very grateful to Prof. Dr. G. Schöck and Prof. Dr. A. Seeger for fruitful discussions and advices. One of us (R.F.) gratefully acknowledges a fellowship granted by the Deutsche Akademische Austauschdienst.

Appendix A

Vibrational entropy versus configurational entropy

In the absolute rate theory the unpinning frequency of a dislocation may be written as [6, 9]

$$\Gamma = \frac{\prod_{r=1}^{M/2} \nu_r}{\prod_{r=2}^{M/2} \nu_r'} \, e^{-\Delta G/kT}, \tag{A.1}$$

where ν_r and ν_r' are the eigenfrequencies of the odd dislocation modes around the equilibrium position E_1 and saddle-point S, respectively, whereas ΔG is the difference between the Gibbs free energy at S and E_1 *without* entropy changes inherent to the dislocation line itself. (Entropy changes due to anharmonicities of the *lattice* modes are included in ΔG.) Conveniently, the pre-exponential factor

$$\Gamma_0 = \frac{\prod_{r=1}^{M/2} \nu_r}{\prod_{r=2}^{M/2} \nu_r'} \tag{A.2}$$

is brought into the form

$$\Gamma_0 = \nu_1 e^{\Delta S_{\mathrm{vib}}/k}, \tag{A.3}$$

where the attack frequency is identified with the frequency of the fundamental mode at E_1, ν_1, and

$$\Delta S_{\mathrm{vib}} = k \ln \frac{\prod_{r=2}^{M/2} \nu_r}{\prod_{r=2}^{M/2} \nu_r'} \tag{A.4}$$

is interpreted as the change of the vibrational entropy of the dislocation between E_1 and S due to anharmonicities of the dislocation modes. According to Section 2 in [1] it is rather improbable that thermodynamic equilibrium can be established at S, so that the definition of a vibrational entropy at S and hence of ΔS_{vib} is meaningless.

In the *non*-equilibrium transition-state theory introduced in [1] the activation "energy" ΔG entering the Arrhenius equation (1.1) has the same meaning as in the rate theory. However, the pre-exponential factor is

$$\Gamma_0 = \frac{c}{2l} \left[\frac{\sum_{r=1}^{M/2} (rF_r)^2}{\sum_{r=1}^{M/2} (F_r/\alpha_r)^2} \right]^{1/2}. \tag{A.5}$$

A physical interpretation of (A.5) becomes evident, if the notation "attack frequency" is literally taken as the number of fluctuations per unit time from the pinned state E_1 towards the hyper-surface B between E_1 and the unpinned state E_2. Since unpinning may be induced if any odd dislocation mode is excited to a sufficiently large amplitude, the attack frequency is

$$\nu_{\text{attack}} = \sum_{r=1}^{M/2} \nu_r \,. \tag{A.6}$$

This consideration implies to rewrite (A.5) as

$$\Gamma_0 = \nu_{\text{attack}}\, e^{\Delta S_{\text{conf}}/k} \tag{A.7}$$

with

$$\Delta S_{\text{conf}} = k \ln \left[\frac{\dfrac{c}{2l} \sqrt{\sum\limits_{r=1}^{M/2} (rF_r)^2}}{\left(\sum\limits_{r=1}^{M/2} \nu_r\right) \sqrt{\sum\limits_{r=1}^{M/2} (F_r/\alpha_r)^2}} \right]. \tag{A.8}$$

The entropy change defined by (A.8) may be interpreted as follows: In principle, all fluctuations which can make the representative point cross the hyper-surface B in the $M/2$-dimensional configuration space are able to induce unpinning. The result of [1] that the activation "energy" ΔG for unpinning is the difference between the Gibbs free energy at S and E_1 (*without* entropy changes inherent to the dislocation itself) indicates that practically all unpinning transitions go through the immediate vicinity of S. Hence, a successful "attack" of the hyper-surface by the representative point must hit the saddle-point area. This area comprises a fairly small number of dislocation configurations, whilst the mixture of dislocation modes around E_1 is completely random. This corresponds to a decrease of the configurational entropy when the dislocation moves from E_1 to S. In fact, a comparison of (A.7) with (3.2) or (3.4) shows that the change of the configurational entropy, ΔS_{conf}, defined by (A.8) is negative. Hence, though inconvenient,

$$\Gamma = \nu_{\text{attack}} e^{-\Delta G_{\text{total}}/kT} \tag{A.9}$$

appears to be the most appropriate representation of the unpinning rate. Here ν_{attack} is defined by (A.6), and

$$\Delta G_{\text{total}} = \Delta G - T\,\Delta S_{\text{conf}}\,, \tag{A.10}$$

where ΔG is the difference between the Gibbs free energy at S and E_1 (cf. (1.2)) *without* the change of configurational entropy of the dislocation, ΔS_{conf}, given by (A.8).

Appendix B

Exclusion of physically unrealistic values of $F_r(\omega_r)$

According to (2.13) the activation "energy" ΔG decreases with increasing values of F_r. It will now be demonstrated that for $F_r \equiv \omega_0^{-2}$ ($r = 1, 2, 3, \ldots, \ldots, M/2$) ΔG becomes unreasonably small. Since this would be even more true if any of the F_r's would be larger than ω_0^{-2}, the subsequent estimate of ΔG for $F_r = \omega_0^{-2}$ restricts the physically reasonable range of m and δ to values for which $\langle F_r \rangle_r = \left[\sum\limits_{r=1}^{M/2} F_r(\omega_r)\right] \Big/ (M/2)$ is smaller than ω_0^{-2}.

Putting $F_r = \omega_0^{-2}$ and $\alpha_r = 1$ for $r = 1, 2, 3, \ldots, M/2$ the sum $\sum\limits_{r=1}^{M/2} (F_r/\alpha_r)^2$ becomes $(M/2)\,\omega_0^{-4} = l/(b\omega_0^4)$, so that (2.13) reads

$$\Delta G = \frac{b(\pi m q_c \omega_0^2)^2}{32T}. \tag{B.1}$$

Inserting of $\omega_0 = \sqrt{f/m}$ makes the unknown effective mass m cancel out, whilst the force constant f enters the expression for ΔG:

$$\Delta G = \frac{b(\pi q_c f)^2}{32T}. \tag{B.2}$$

Since for a harmonic interaction energy profile the binding energy E_B of a dislocation to a pinner is related to q_c according to

$$E_B = \frac{1}{2} f q_c^2 , \tag{B.3}$$

(B.2) may be brought into the form

$$\Delta G = \frac{b(2\pi E_B/q_c)^2}{32T}. \tag{B.4}$$

For vanishing externally applied stress q_c approaches the obstacle diameter. Therefore, we put $q_c = b$ obtaining

$$\Delta G = \frac{\pi^2}{8} \frac{E_B^2}{Tb}. \tag{B.5}$$

With $T = \mu b^2/2$ (μ shear modulus) we finally arrive at

$$\Delta G = \frac{\pi^2}{8} \frac{E_B^2}{\mu b^3/2}. \tag{B.6}$$

Typical values for E_B and $\mu b^3/2$ are 0.5 eV and 5 eV, respectively, which leads to

$$\Delta G \approx 0.06 \text{ eV} . \tag{B.7}$$

This ΔG-value is about an order of magnitude smaller than the activation "energies" found experimentally for the unpinning of dislocations from point obstacles at room temperature [10 to 12].

References

[1] W. Frank and R. Frydman, phys. stat. sol. (b) **64**, 539 (1974).
[2] J. S. Koehler, in: Imperfections in Nearly Perfect Crystals, Ed. W. Shockley, J. H. Holloman, R. Maurer, and F. Seitz, John Wiley, New York 1952 (p. 197).
[3] A. V. Granato and K. Lücke, J. appl. Phys. **27**, 583 (1956).
[4] G. Leibfried, in: Dislocations and Mechanical Properties of Crystals, Ed. J. C. Fisher, W. G. Johnston, R. Thomson, and T. Vreeland, Jr., Wiley, New York 1957 (p. 495).
[5] I. Szabó, Einführung in die Technische Mechanik, Springer-Verlag, Berlin/Göttingen/Heidelberg 1959.
[6] A. V. Granato, K. Lücke, J. Schlipf, and L. J. Teutonico, J. appl. Phys. **35**, 2732 (1964).
[7] W. Frank, Z. Naturf. **22a**, 365 (1967).
[8] W. Frank, phys. stat. sol. **26**, 197 (1968).
[9] G. H. Vineyard, J. Phys. Chem. Solids **3**, 121 (1957).
[10] W. Frank, phys. stat. sol. **18**, 459 (1966).
[11] W. Frank, phys. stat. sol. **19**, 239 (1967).
[12] W. Frank, phys. stat. sol. **29**, 391 (1968).

(Received May 27, 1974)

phys. stat. sol. (b) **65**, 171 (1974)

Subject classification: 6 and 8; 4; 21.4

Hahn-Meitner-Institut für Kernforschung Berlin, GmbH (a), Neutron Physics, S.C.K./C.E.N., Mol (b), and Applied Mathematics, S.C.K./C.E.N., Mol (c)

Inelastic Neutron Scattering by Cerium Hydrides

II. Lattice Dynamical Approach[1])

By

P. VORDERWISCH (a), S. HAUTECLER (b), and H. DECKERS (c)

A central force lattice dynamical model, previously applied by Slaggie to the CaF_2 structure of ZrH_2, is extended to the BiF_3 structure. It is used for the analysis of inelastic neutron scattering data on cerium hydrides $CeH_{1.98}$ and $CeH_{2.72}$. For both samples the force constants are derived. The values of the corresponding force constants in both hydrides are different. The additional hydrogen in the CeH_2 structure is found to be weakly bound.

Ein Zentralkraft-Modell, von Slaggie zur Untersuchung der Gitterdynamik für die CaF_2-Struktur des ZrH_2-Gitters aufgestellt, wurde auf die BiF_3-Struktur erweitert und zur Analyse experimenteller Daten aus der inelastischen Neutronenstreuung an Cerhydrid $CeH_{1,98}$ und $CeH_{2,72}$ angewandt. Für beide Proben wurden Kraftkonstanten abgeleitet. Die in beiden Hydriden einander entsprechenden Kraftkonstanten haben unterschiedliche numerische Werte. Das zusätzlich in die CeH_2-Struktur eingefügte Wasserstoffatom ist schwach gebunden.

1. Introduction

Since for most of transition metal hydrides the phonon dispersion relations are not available because of difficulties for single crystal preparations, the models used in a lattice dynamical analysis should contain as few parameters as possible in order that the parameter set fitting the inelastic incoherent scattering cross-section should be unique. One way to reduce the number of parameters is to assume central forces between atom pairs. Central force models have been built by Karimov et al. [1] and by Slaggie [2] for the f.c.c. CaF_2-type structure, and by Lehman et al. [3] for the f.c.t. ZrH_2 structure.

In spite of some differences in the presentation, the three models are essentially equivalent due to the same physical description of the forces. Karimov et al. and Lehman et al. write the interaction potential $\varphi(|\boldsymbol{R}|)$ between two atoms as a function of the distance $|\boldsymbol{R}|$. The expansion in a Taylor's series in power of the displacement vectors contains a bond-stretching term as well as a bond-bending term (which both are familar in the description of molecular vibrations). From this expansion (limited to the second order terms), the following force constants are derived:

$$\Phi_{\alpha\beta}(r_0) = -\frac{r_0^\alpha r_0^\beta}{r_0^2}\left|\varphi''(r_0) - \frac{\varphi'(r_0)}{r_0}\right| - \frac{\varphi'(r_0)}{r_0}\,\delta_{\alpha\beta}\,, \qquad (1)$$

[1]) Part I see phys. stat. sol. (b) **64**, 495 (1974).

where α, β indicate cartesian components and the primes on $\varphi(r_0)$ denote differentiation calculated for the equilibrium distance r_0. On the other hand, Slaggie has written the interaction potential $\Psi(R^2)$ as depending on the square of the distance, from which the force constants

$$\Phi_{\alpha\beta}(r_0^2) = -2\Psi'(r_0^2)\,\delta_{\alpha\beta} - 4r_0^\alpha r_0^\beta \Psi''(r_0^2) \tag{2}$$

are derived.

Taking into account the relationships between φ', φ'', Ψ', and Ψ'',

$$\begin{aligned} \varphi' &= 2r_0\Psi' \\ \varphi'' &= 2\Psi' + 4r_0^2\,\Psi''\,, \end{aligned} \tag{3}$$

the two sets of force constants are seen to be equivalent.

For the dynamical matrix elements Slaggie and Karimov et al. have given expressions assuming interactions between metal–metal and metal–hydrogen nearest neighbours, hydrogen–hydrogen nearest, and next-nearest neighbours; the main difference is, apart in the formulation already mentioned, that Slaggie assumes the bending constants equal to zero keeping thus four parameters instead of eight in the Karimov treatment. (Lehman et al. did not give the matrix elements.)

The various authors have used their model in order to try to explain the observed width of the optical peaks in the scattered neutron energy distributions. However, whilst Slaggie has solved the general problem deriving the eigenvalues and eigenvectors of the dynamical matrix for a set of evenly spaced wave vectors $\boldsymbol{q}$, Karimov et al. limited themselves to derive, in an approximate manner, the dispersion relation for wave vectors along the $\langle 100 \rangle$ direction while Lehmann et al. only tried to get the optical frequencies at $\boldsymbol{q} = 0$.

The application of a central force lattice dynamical model to our experimental results on cerium dihydride and trihydride (described in the preceding paper, referred to as I) will be presented. In this treatment we assume, in order to reduce the number of parameters, the ideal CaF_2 and BiF_3 structures for CeH_2 and CeH_3, respectively.

2. Scattering Cross-Section with a Central Force Model for CaF_2- and BiF_3-Type Structures

We label by $k = 1$ to 4 the atoms in the primitive cell, respectively: the Ce atom at (000), the two H atoms on tetrahedral sites at $(a/4, a/4, a/4)$ and $(3a/4, 3a/4, 3a/4)$, and the H atom on octahedral site at $(a/2, a/2, a/2)$, a being the lattice constant.

In the harmonic approximation, the dynamical behaviour is determined by a dynamical matrix $D(\boldsymbol{q})$, whose elements are functions of the wave vector $\boldsymbol{q}$ and contain the force constants Φ. If the interaction potential $\Psi(R^2)$ between two atoms depends only on their distance R, then, assuming $\Psi' = 0$ throughout and $\Psi'' = 0$ for $r_0 > \sqrt{2}\,a/2$, the force constants in the system are defined in terms of the following seven parameters:

$$\mu = a^2\Psi''_{\text{Ce-H}_t}(r_0^2), \qquad \text{where} \qquad r_0 = \sqrt{3}\,\frac{a}{4}\,,$$

$$\gamma = a^2\Psi''_{\text{Ce-Ce}}(r_0^2), \qquad \text{where} \qquad r_0 = \sqrt{2}\,\frac{a}{2}\,.$$

$$\nu = a^2 \Psi''_{\mathrm{H_t-H_t}}(r_0^2), \qquad \text{where} \qquad r_0 = \frac{a}{2},$$

$$\delta = a^2 \Psi''_{\mathrm{H_t-H_t}}(r_0^2), \qquad \text{where} \qquad r_0 = \sqrt{2}\,\frac{a}{2},$$

$$\chi = a^2 \Psi''_{\mathrm{Ce-H_0}}(r_0^2); \qquad \text{where} \qquad r_0 = \frac{a}{2},$$

$$\xi = a^2 \Psi''_{\mathrm{H_t-H_0}}(r_0^2), \qquad \text{where} \qquad r_0 = \sqrt{3}\,\frac{a}{4},$$

$$\eta = a^2 \Psi''_{\mathrm{H_0-H_0}}(r_0^2), \qquad \text{where} \qquad r_0 = \sqrt{2}\,\frac{a}{2}.$$

For CeH_3 the dynamical matrix has the form

$$D(\boldsymbol{q}) = \begin{pmatrix} D_{11} & D_{12} & D_{13} & D_{14} \\ D_{21} & D_{22} & D_{23} & D_{24} \\ D_{31} & D_{32} & D_{33} & D_{34} \\ D_{41} & D_{42} & D_{43} & D_{44} \end{pmatrix},$$

where the $D_{kk'}$ are three-dimensional submatrices with elements $D_{\alpha\beta}\begin{pmatrix}\boldsymbol{q}\\ kk'\end{pmatrix}$; $\alpha, \beta = 1, 2, 3$. $D(\boldsymbol{q})$ is hermitian: $D_{\alpha\beta}\begin{pmatrix}\boldsymbol{q}\\ kk'\end{pmatrix} = D^*_{\beta\alpha}\begin{pmatrix}\boldsymbol{q}\\ k'k\end{pmatrix}$.

M_k being the mass of the k-th atom ($M_1 = M$, $M_2 = M_3 = M_4 = m$), one obtains the following elements:

$$\left.\begin{aligned}
D_{\alpha\alpha}\begin{pmatrix}\boldsymbol{q}\\ kk\end{pmatrix} &= M^{-1}(A_k - 4\tau_k C_\alpha)\,, \\
D_{\alpha\beta}\begin{pmatrix}\boldsymbol{q}\\ kk\end{pmatrix} &= M_k^{-1} 4\tau_k \sin \pi a q_\alpha \sin \pi a q_\beta \quad \text{for} \quad \alpha \neq \beta\,, \\
D_{\alpha\alpha}\begin{pmatrix}\boldsymbol{q}\\ 12\end{pmatrix} &= -(Mm)^{-1/2}\,\frac{\mu}{4}\,[1 + E_3 + E_2 + E_1]\,, \\
D_{12}\begin{pmatrix}\boldsymbol{q}\\ 12\end{pmatrix} = D_{21}\begin{pmatrix}\boldsymbol{q}\\ 12\end{pmatrix} &= -(Mm)^{-1/2}\,\frac{\mu}{4}\,[1 + E_3 - E_2 - E_1]\,, \\
D_{13}\begin{pmatrix}\boldsymbol{q}\\ 12\end{pmatrix} = D_{31}\begin{pmatrix}\boldsymbol{q}\\ 12\end{pmatrix} &= -(Mm)^{-1/2}\,\frac{\mu}{4}\,[1 - E_3 + E_2 - E_1]\,, \\
D_{23}\begin{pmatrix}\boldsymbol{q}\\ 12\end{pmatrix} = D_{32}\begin{pmatrix}\boldsymbol{q}\\ 12\end{pmatrix} &= -(Mm)^{-1/2}\,\frac{\mu}{4}\,[1 - E_3 - E_2 + E_1]\,, \\
D_{\alpha\beta}\begin{pmatrix}\boldsymbol{q}\\ 13\end{pmatrix} &= \exp\left[-i2\pi a\,(q_1 + q_2 + q_3)\right] D^*_{\alpha\beta}\begin{pmatrix}\boldsymbol{q}\\ 12\end{pmatrix}, \\
D_{\alpha\alpha}\begin{pmatrix}\boldsymbol{q}\\ 23\end{pmatrix} &= -m^{-1} 2\nu \cos \pi a q_\alpha \exp\left[-i\pi a\,(q_1 + q_2 + q_3)\right], \\
D_{\alpha\beta}\begin{pmatrix}\boldsymbol{q}\\ 23\end{pmatrix} &= 0 \quad \text{for} \quad \alpha \neq \beta\,,
\end{aligned}\right\} \qquad (4)$$

$$D_{\alpha\beta}\begin{pmatrix} \boldsymbol{q} \\ 14 \end{pmatrix} = \left(\frac{m}{M}\right)^{1/2} \frac{\chi}{\nu} D_{\alpha\beta}\begin{pmatrix} \boldsymbol{q} \\ 23 \end{pmatrix},$$

$$D_{\alpha\beta}\begin{pmatrix} \boldsymbol{q} \\ 24 \end{pmatrix} = \left(\frac{M}{m}\right)^{1/2} \frac{\xi}{\mu} D_{\alpha\beta}\begin{pmatrix} \boldsymbol{q} \\ 12 \end{pmatrix},$$

$$D_{\alpha\beta}\begin{pmatrix} \boldsymbol{q} \\ 34 \end{pmatrix} = D^*_{\alpha\beta}\begin{pmatrix} \boldsymbol{q} \\ 24 \end{pmatrix}$$

with

$$A_1 = 2\mu + 2\chi + 8\gamma\,, \quad A_2 = A_3 = \mu + \xi + 8\delta\,, \quad A_4 = 2\xi + 2\chi + 8\eta\,,$$

$$\tau_1 = \gamma\,, \quad \tau_2 = \tau_3 = \delta\,, \quad \tau_4 = \eta\,,$$

$$C_\alpha = \cos \pi a q_\alpha (\cos \pi a q_\beta + \cos \pi a q_\gamma)\,, \qquad E_\alpha = \exp\left[-i\pi a (q_\beta + q_\gamma)\right].$$

For CeH_2 the dynamical matrix reduces to

$$D(\boldsymbol{q}) = \begin{pmatrix} D_{11} & D_{12} & D_{13} \\ D_{21} & D_{22} & D_{23} \\ D_{31} & D_{32} & D_{33} \end{pmatrix}.$$

The required matrix elements $D_{\alpha\beta}\begin{pmatrix} \boldsymbol{q} \\ kk' \end{pmatrix}$ $(k, k' = 1, 2, 3)$ are obtained by setting $\chi = \xi = \eta = 0$ in the above expressions.

For each wave vector $\boldsymbol{q}$, the frequencies $\omega_j^2(\boldsymbol{q})$ and the polarization vectors $\boldsymbol{C}\left(k \middle| \begin{matrix} \boldsymbol{q} \\ j \end{matrix}\right)$ of the associated $3r$ lattice modes ($j = 1, 2, \ldots, 3r$; r being the number of atoms in the primitive cell) are given by the eigenvalues and the eigenvectors of the dynamical matrix.

For the computation of the specific heat and the neutron scattering cross-section, the frequency distribution $f(\omega)$ and the weighted frequency distributions $g_k(\omega)$ are required respectively. The eigenvalues and the eigenvectors of $D(\boldsymbol{q})$ have to be computed for a large number of $\boldsymbol{q}$-vectors uniformly distributed in the Brillouin zone (BZ). Due to the high symmetry of the f.c.c. lattice the labour can be limited to $\boldsymbol{q}$ contained in a reduced zone (RZ) whose volume is 1/48th the volume of the BZ. For a mesh of $\boldsymbol{q}$ restricted by the RZ [2] the eigenvalues of the complex matrices $D(\boldsymbol{q})$ were obtained using the subroutine DAMAT. The frequency distribution is derived by taking the square roots of these eigenvalues, and counting the number of frequencies within the interval $\omega, \omega + \Delta\omega$; a weight factor [2] takes into account the facts that q-vectors ending on a face of the BZ belong to several BZ, and that wave vectors in the faces of the RZ belong to several RZ. The weighted frequency distributions are obtained by the same procedure; they include normalized polarization vectors $\boldsymbol{C}\left(k \middle| \begin{matrix} \boldsymbol{q} \\ j \end{matrix}\right)$ $(k = 1, 2, \ldots, r)$ which are the eigenvectors of the dynamical matrix also furnished by the subroutine DAMAT.

With these weighted frequency distributions the scattering cross-section (I.1) is computed; the time integration is performed using a subroutine enclosed in the computer program GASKET [4]. The obtained cross-sections are converted to a time-of-flight scale and convoluted with the spectrometer resolution function.

3. Computation of the Specific Heats

The lattice contribution to the specific heat C_v of CeH_2 and CeH_3 can be calculated from their corresponding frequency distributions $f(\omega)$. For CeH_2 it appears that the contribution of the optical modes to C_v is negligible below ≈ 100 °K; for these temperatures the lattice contribution to the specific heat is thus only due to the acoustical modes, where the vibration of the rare-earth ion is dominating. However for a comparison with experimental specific heat data [5, 6] in the low-temperature region a difficulty arises due to a Schottky-type contribution.

In the cerium hydrides the degeneracy of the ground state $^2F_{5/2}$ of the Ce^{3+} ion is partially removed. In a cubic crystal field the $^2F_{5/2}$ level splits into a doublet Γ_7 and a quartet Γ_8. The sign of the coefficient A of the cubic crystal field (or the sign of the electrical charge on the hydrogen atoms surrounding the Ce ion) determines which level is the lowest. For $A > 0$ the doublet is the ground state, for $A < 0$ Γ_8 is the lowest level. The thermal population of the resulting two energy levels gives rise to the Schottky specific heat anomalies [7]:

$$C_{\text{Sch}} = \frac{Rg(a/T)^2 \exp(a/T)}{[g + \exp(a/T)]^2}, \tag{5}$$

where R is the gas constant, g the ratio of the degeneracies of the upper to the lower level and a the level splitting expressed in °K.

By comparing the experimental specific heats of CeH_2 and LaH_2 (where crystal field effects are absent and the lattice contribution to $C_p(T)$ in the low-temperature range should be similar to that of CeH_2) it was concluded [8, 9] that the lowest level was the quartet (corresponding to $A < 0$). A value $a \approx 156$ °K for the splitting parameter was derived from the temperature corresponding to the maximum of the Schottky contribution to the specific heat.

For a comparison of calculated and experimental specific heat data over the whole temperature range, the conduction electron contribution

$$C_{\text{E}}(T) = \gamma_{\text{E}} T \tag{6}$$

cannot be neglected in the case of CeH_2 which has a metallic character. As numerical value of γ_{E} is not known for cerium dihydride, the value $\gamma_{\text{E}} = 2.0 \times 10^{-3}$ cal mol^{-1} deg^{-2} for $LaH_{2.03}$ [10] was taken.

4. Fits to Experimental Data

In order to start the fit procedure of computed to experimental data, a first set of values for the coupling parameters can be derived from the eigenvalues of the dynamical matrix for $\boldsymbol{q} = 0$. Under the assumption $M = \infty$, one obtains from (4) the following four triply-degenerated eigenvalues for CeH_3:

$$\left.\begin{aligned} \omega_1^2(0) &= 0\,, \\ \omega_2^2(0) &= \frac{A + (A^2 - 8B)^{1/2}}{2m}\,, \\ \omega_3^2(0) &= \frac{\mu + 4\nu + \xi}{m}\,, \\ \omega_4^2(0) &= \frac{A - (A^2 - 8B)^{1/2}}{2m}\,, \end{aligned}\right\} \tag{7}$$

where

$$A = \mu + 3\xi + 2\chi\,,$$
$$B = \mu\xi + \mu\chi + \xi\chi\,.$$

These frequencies do not depend neither on γ, nor on δ and η. For CeH_2 the frequencies $\omega_i(0)$ are obtained setting $\chi = \xi = 0$ in the above expressions:

$$\omega_1^2(0) = 0\,, \quad \omega_2^2(0) = \frac{\mu}{m}, \quad \omega_3^2(0) = \frac{\mu + 4\nu}{m}, \quad \omega_4^2(0) = 0\,. \tag{8}$$

Thus ω_1 corresponds to the acoustical modes, ω_4 to the optical vibrations of the hydrogens at the octahedral sites, and ω_2, ω_3 to the optical vibrations of the hydrogens at the tetrahedral sites.

4.1 Cerium dihydride

According to (8) a first set of values for μ and ν was derived from the position and the width of the optical tetrahedral peak. Choosing δ smaller than ν, several computations of the frequency distributions and of the lattice contribution to the specific heat were performed varying γ between 0.5 and 1.0 × × 10^4 gs^{-2}. We found that

1. the optical peak in the frequency distribution is not sensitive to γ;
2. the optical modes do not contribute to the specific heat below 100 °K.

We could thus determine the value of γ by fitting below 100 °K the computed specific heat to the experimental one, taking into account the Schottky contribution in the way described in Section 3. The values of the other three force constants were determined by fitting the computed to the experimental neutron time-of-flight distributions. After several computer runs the best fit was obtained for the set of force constants given in Table 1.

Table 1

Values of the coupling parameters describing the lattice dynamics of $CeH_{1.98}$ and $CeH_{2.72}$ in a central force model

coupling parameter	coupling parameter values (10^4 g s^{-2})	
	$CeH_{1.98}$	$CeH_{2.72}$
γ	0.8	0.4
μ	3.0	2.6
χ		0.47
ξ		−0.15
ν	0.4	0.85
δ	0.1	0.2
η		0.15

Fig. 1 shows the frequency distributions for CeH_2 computed with the force constants of Table 1 and using 520 q-points in the reduced zone. Although the force constant δ is very small it plays an important role, bringing a smoothing

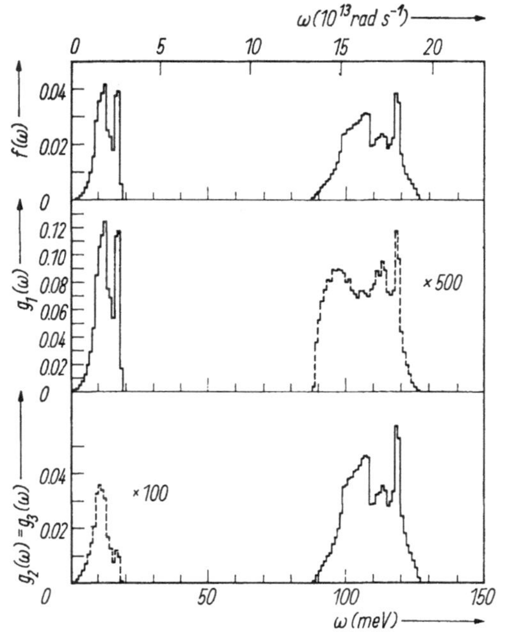

Fig. 1. Frequency distributions for CeH_2. a) Usual frequency distribution $f(\omega)$. b) Polarization vector weighted frequency distribution $g_k(\omega)$ ($k = 1 = Ce$; $k = 2,3 = H_t$)

at the upper and lower cut-off of the optical frequency distribution. Fig. 2 shows the computed specific heat $C_v(T)$; in order to compare with the experimental data $C_p(T)$, the Schottky and electronic contributions were added to the lattice contribution computed with the frequency distribution of Fig. 1a. A very good agreement is found over the whole temperature range. Fig. 3 shows the neutron time-of-flight distribution computed along the way described in Section 2 using the weighted frequency distribution of Fig. 1b; it is seen that, taking into account the resolution of the spectrometer, the structure shown in the computed $g_2(\omega)$ is smeared out.

4.2 Cerium trihydride

The sample used in our neutron measurement had the composition $CeH_{2.72}$; experimental data for the specific heat of cerium trihydride are only available for the composition $CeH_{2.86}$ [6]. To take this into account the frequency spectra were computed for an ideal CeH_3 lattice, but for a comparison with experimental

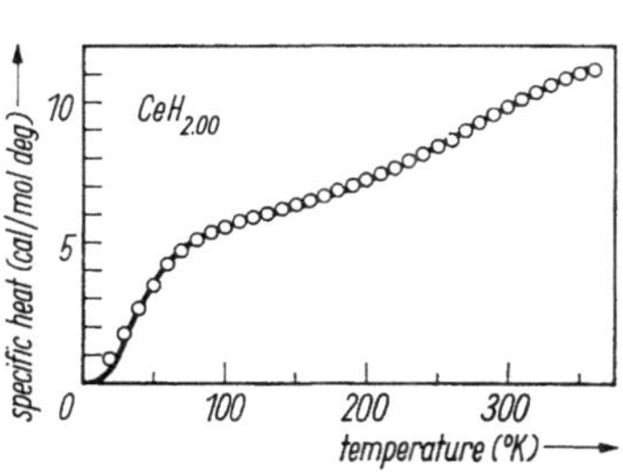

Fig. 2. Experimental and computed specific heats for CeH_2

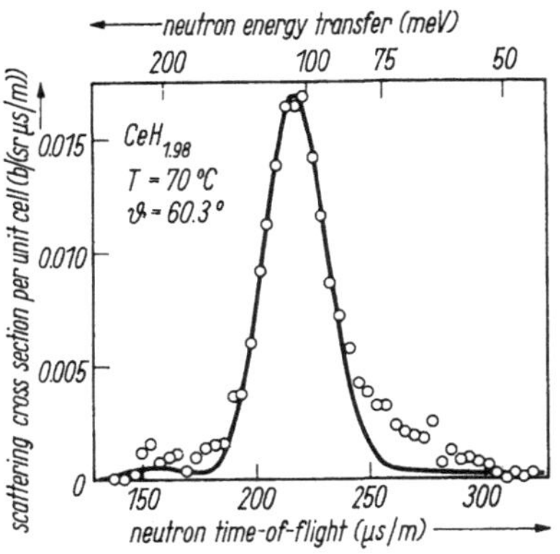

Fig. 3. Experimental and computed neutron time-of-flight distributions for $CeH_{1.98}$ at 70 °C under a scattering angle of 60.3°

data the octahedral part was weighted by the factors 0.78 and 0.92, respectively, the tetrahedral part being weighted by a factor 0.97; these values follow from accurate neutron diffraction measurements [12] which have shown that for $CeD_{2.75}$ 3% of the tetrahedral sites are unoccupied.

With the values for γ, μ, ν, and δ obtained from the CeH_2 analysis and varying only χ, ξ, and η we were not able to fit, even approximately, the experimental data. After several computer runs two important conclusions could be drawn:
1. Due to the low energy octahedral peak, the contribution of the optical modes to the specific heat starts at about 60 °K. However, in any case we could not fit the specific heat below 60 °K using the γ-value of CeH_2. Additional runs changing only γ showed that for $CeH_{2.72}$ a value as low as 0.4×10^4 gs^{-2} has to be attributed to the Ce–Ce coupling parameter.
2. Keeping μ, ν, and δ as in CeH_2, we could reproduce the experimentally observed shift in the frequency distribution of the tetrahedral hydrogens in $CeH_{2.72}$ with respect to the one in CeH_2, but never the broadening of the distribution as required by the neutron scattering experiments (in fact in all cases we got a narrowing). This means that we have to change the values of these parameters with respect to their values in $CeH_{1.98}$. Our best set of force constants resulting from many computer runs is given in the third column of Table 1.

Fig. 4 shows the frequency distribution for CeH_3 computed with the force constants of Table 1 and using 520 q-points in the reduced zone. Fig. 5 represents the neutron time-of-flight distribution for $CeH_{2.72}$ computed using the weighted frequency distribution of Fig. 4b. The overall agreement with the experimental data is satisfactory, except that the predicted intensity between the two maxima is too low. Fig. 6 shows the specific heat for $CeH_{2.86}$ computed by adding a Schottky contribution to the lattice contribution obtained from the frequency distribution of Fig. 4a. Above 60 °K the agreement with the experimental data is very poor. However, the anomaly observed at 250 °K in the experimental specific heat, which is associated with a metal-to-semiconductor transition, makes the comparison between computed and experimental curves

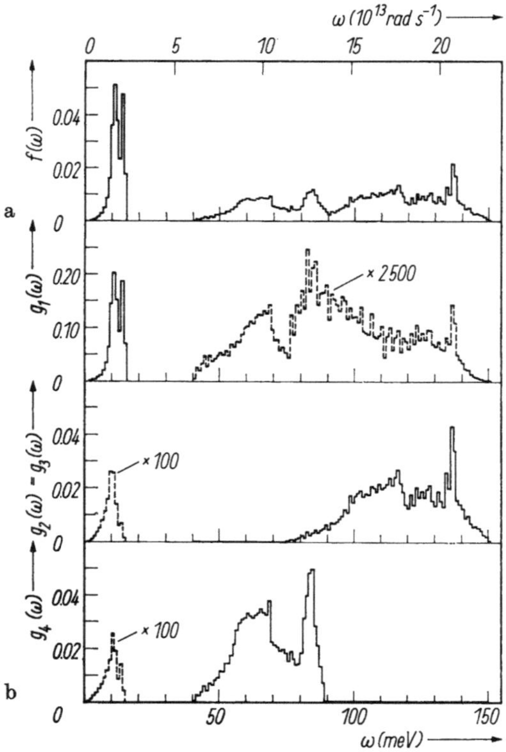

Fig. 4. Frequency distributions for CeH_3. a) Usual frequency distribution $f(\omega)$. b) Polarization vector weighted frequency distribution $g_k(\omega)$ ($k = 1 =$ Ce; $k = 2$, $3 = H_t$; $k = 4 = H_0$)

Fig. 5. Experimental and computed neutron time-of-flight distributions for $CeH_{2.72}$ at 70 °C under a scattering angle of 60.3°

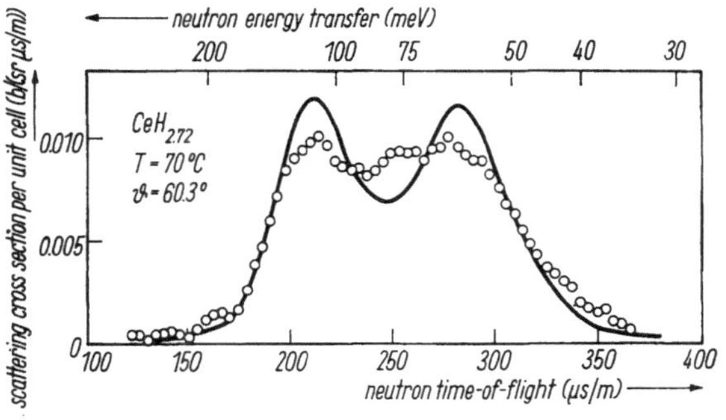

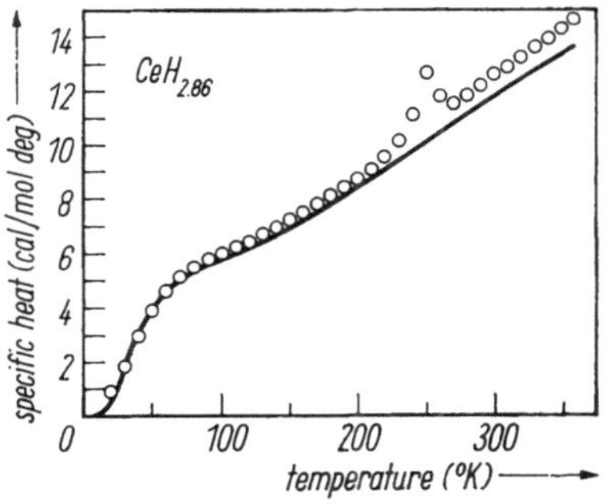

Fig. 6. Experimental and computed specific heats for $CeH_{2.86}$

very difficult; in particular we do not know how far extends this anomaly. Moreover, this could be an indication that also in this case we have to take into account an electronic contribution to the specific heat.

5. Discussion

5.1 Cerium dihydride

In view of the quality of the fits to specific heat and time-of-flight spectra, it appears that the original Slaggie central force model is satisfactory; neutron investigations performed under higher resolution should show if this statement holds. The excess of scattered neutrons which is observed in Fig. 3 at the low-energy side of the tetrahedral peak with respect to the predicted intensity is attributed by us as due to a localized vibration of hydrogen atoms already sitting at the octahedral sites for the CeH_2 composition; this point will be analysed in a separate paper.

5.2 Cerium trihydride

In this case the quality of the fits is somewhat less good than for $CeH_{1.98}$. This could be attributed to several reasons:

1. Deviation of the structure with respect to the assumed BiF_3 structure in the lattice dynamical analysis.
2. Limited number of computer runs.
3. The correction for vacant octahedral places is too simply done.
4. The interactions cannot be described in a satisfactory manner by a central force model.

Comparing now both sets of force constants given in Table 1 one remarks:
1. The values for the parameters involving Ce atoms (γ, μ) are decreased while the values for the parameters involving only the tetrahedral hydrogens (ν, δ) are increased going from $CeH_{1.98}$ to $CeH_{2.72}$.
2. The octahedral hydrogens are by far less strongly bonded than the tetrahedral hydrogens, since, assuming a pairwise interaction potential between the hydrogens, one would expect $\xi > \nu$ due to the respective distances. This is in agreement with the fact that in lanthanum hydrides for compositions beyond LaH_2 the activation energies for self-diffusion of hydrogen become smaller due to the weaker bonding of hydrogen in the octahedral sites [11].

Acknowledgements

The authors wish to thank Dr. M. Nève de Mévergnies for many helpful discussions and to express their gratitude to Mrs. C. Mongini-Tamagnini from the Euratom C.C.R., Ispra to place the subroutines GASKET and DAMAT to their disposal. Technical assistance by Mr. J. Brouwers is greatly appreciated. One of us (P.V.) is grateful to Euratom for financial support.

References

[1] I. KARIMOV, E. G. BROVMAN, M. G. ZEMLYANOV, V. A. SOMENKOV, and N. A. CHERNOPLEKOV, Soviet Phys. — Solid State **9**, 1566 (1968).
[2] E. L. SLAGGIE, General Atomic Report GA-8132 (1967); J. Phys. Chem. Solids **29**, 923 (1968).
[3] G. W. LEHMAN, T. WOLFRAM, and R. E. DE WAMES, Phys. Rev. **128**, 1593 (1962).
[4] J. U. KOPPEL, J. R. TRIPLETT, and Y. D. NALIBOFF, General Atomic Report GA-7417 (Rev.) (1967).
[5] B. STALIŃSKI and Z. BIEGAŃSKI, Bull. Acad. Polon. Sci., Sér. sci. chim. **12**, 331 (1964).
[6] Z. BIEGAŃSKI, W. FESENKO, and B. STALIŃSKI, Bull. Acad. Polon. Sci., Sér. sci. chim. **13**, 227 (1965).
[7] P. H. KEESOM and N. PEARLMAN, Hdb. Phys., Vol. 14, Springer-Verlag, 1956 (p. 282).
[8] Z. BIEGAŃSKI and B. STALIŃSKI, phys. stat. sol. (a) **2**, K161 (1970).
[9] Z. BIEGAŃSKI, Bull. Acad. Polon. Sci., Sér. sci. chim. **19**, 581 (1971).
[10] Z. BIEGAŃSKI, D. GONZALES ALVAREZ, and F. W. KLAAIJSEN, Physica (Utrecht) **37**, 153 (1967).
[11] D. S. SCHREIBER and R. M. COTTS, Phys. Rev. **131**, 1118 (1963).
[12] A. K. CHEETHAM and B. E. F. FENDER, J. Phys. C **5**, L35 (1972).

(Received March 11, 1974)

phys. stat. sol. (b) **65**, 181 (1974)

Subject classification: 9 and 19; 10.1; 21

Max-Planck-Institut für Metallforschung, Institut für Physik, Stuttgart and Institut für theoretische und angewandte Physik der Universität Stuttgart

The Contribution of Magnetic Dipolar Interactions to Nuclear Spin Relaxation by Vacancy-Type Self-Diffusion Mechanisms in Metals[1])

By

E. CAVELIUS[2])

The spectral densities determining the dipolar contribution to the nuclear spin relaxation times are calculated for a combined monovacancy–divacancy mechanism of self-diffusion in solids. The treatment is based on Wolf's encounter model for monovacancies and isotopically pure crystals. Numerical results are given for isotopically pure f.c.c. and b.c.c. crystals. The extension to more than one spin species is indicated.

Die für den dipolaren Beitrag zu den Kernspinrelaxationszeiten maßgebenden Spektraldichten werden für einen kombinierten Einfach–Doppelleerstellen-Mechanismus der Selbstdiffusion in Festkörpern berechnet. Die Behandlung basiert auf dem Wolfschen Begegnungsmodell für Einfachleerstellen in isotopenreinen Kristallen. Numerische Ergebnisse werden für isotopenreine k.f.z.- und k.r.z.-Kristalle angegeben. Es wird gezeigt, wie man das Vorhandensein von mehr als einer Spinsorte berücksichtigen kann.

1. Introduction

Since the fundamental papers of Bloembergen et al. (BPP) [1] and Pound [2] it is well known that both the time dependent nuclear magnetic spin–spin-coupling and the time dependent coupling of the nuclear electric quadrupole moment to electric field gradients are important mechanisms for bringing about spin–lattice and spin–spin relaxation in liquids and solids. In monoatomic solids this time dependence is due to atomic diffusion, which may be characterized by the mean time of stay of a nucleus at a given lattice site, τ, or by the self-diffusion coefficient D^{SD}, which is related to τ by the Einstein-Smoluchowski relationship

$$D^{SD} = \frac{b^2}{6\tau}. \tag{1.1}$$

Here b is the jump distance of the nuclei.

By an analysis of the temperature dependence of the spin–lattice relaxation time T_1, the spin–lattice relaxation time in the rotating frame $T_{1\rho}$, or the spin–spin relaxation time T_2 information on diffusional motions, i.e., on τ, may be obtained.

In the present paper we confine our analysis to magnetic dipolar interactions of the spins. Quadrupolar interactions are treated in a subsequent paper [3].

---

[1]) This work is based on the thesis of E. Cavelius, University of Stuttgart, 1973.

[2]) New address: Institut für theoretische Festkörperphysik der Universität des Saarlandes, Saarbrücken.

If relaxation takes place in fields $\boldsymbol{H}_0$ and $\boldsymbol{H}_1$ that are large compared to the internal local fields T_1, T_2 and $T_{1\varrho}$ may be calculated by means of perturbation theory [4, 5]. The results for magnetic dipolar interaction of N like spins are [4, 5]

$$\frac{1}{T_1} = \frac{3}{2} \gamma^4 \hbar^2 I\,(I+1)\,[J^{(1)}(\omega_L) + J^{(2)}(2\omega_L)]\,, \tag{1.2}$$

$$\frac{1}{T_2} = \frac{3}{8} \gamma^4 \hbar^2 I\,(I+1)\,[J^{(0)}(0) + 10 J^{(1)}(\omega_L) + J^{(2)}(2\omega_L)]\,, \tag{1.3}$$

$$\frac{1}{T_{1\varrho}} = \frac{3}{8} \gamma^4 \hbar^2 I\,(I+1)\,[J^{(0)}(2\omega_1) + 10 J^{(1)}(\omega_L) + J^{(2)}(2\omega_L)] \tag{1.4}$$

with

$$J^{(q)}(\omega) = \int_{-\infty}^{\infty} e^{i\omega t} k^{(q)}(t)\, dt\,, \tag{1.5}$$

$$k^{(q)}(t) = \sum_{j \neq i} \langle F_{ij}^{(q)}(t')\, F_{ij}^{(-q)}(t'+t)\rangle_{t'}\,. \tag{1.6}$$

$F_{ij}^{(q)}(t)$ denote functions depending on the relative distance and orientation of the spins i and j [4]. The summation over j in (1.6) has to be taken over all j except $j = i$. q may assume the values 0, 1, and 2. Through (1.2 to 1.4), (1.5) and (1.6) the precise shape of the correlation function $k^{(q)}(t)$ enters into the formulas for the relaxation times. This shape depends on the mechanism responsible for the time variation of the dipolar interaction, in our case on the mechanism of self-diffusion.

A random-walk model of self-diffusion, as treated by Torrey [6], is not realistic, since in most crystals diffusion takes place via the diffusion of point defects, e.g., vacancies. As pointed out first by Bardeen and Herring [7], the migration of nuclei caused by random jumps of vacancies is not completely random but the directions and time intervals of successive nuclear jumps are correlated. Eisenstadt and Redfield tried to incorporate these correlation effects into the NMR-theories in a qualitative way [8]. Wolf gave a more exact calculation of the correlation functions taking into account the correlations of successive jumps of a nucleus caused by a randomly migrating point defect [9]. In Section 2 we briefly recall Wolf's model and show how to extend it to the case of a simultaneously operating monovacancy–divacancy(1V/2V-)mechanism. Section 3 gives a short survey of the models of motion of mono- and divacancies in f.c.c. and b.c.c. metals. Section 4 presents numerical methods and results. In Section 5 the method is extended to the case of more than one sort of nuclear spins in the crystal.

2. Wolf's Encounter Model and Its Extension to Combined Mono-Divacancy-Mechanisms

The model of Wolf [9] is a generalization of Torrey's model of lattice diffusion in isotropic crystals [6]. While in Torrey's model the single jumps of an atom are considered as statistically independent ("random walk"), in Wolf's model it is the "encounters" of an atom with different vacancies that are treated as independent. An encounter is defined as the sequence of jumps of a nucleus produced by a given vacancy [8, 9]. The mean time between two such encounters, τ_{NMR}, is the significant time constant in Wolf's "random walk of encounters" [9].

Under conditions where perturbation theory is applicable, many encounters are necessary to produce appreciable relaxation. Therefore it is allowed to neglect the relaxation produced within a single encounter [9]. Encounters resulting in no total displacement of an atom at all are therefore not effective for the relaxation process and are omitted. Pair and higher correlations are also neglected, i.e., one neglects possible correlations of the jumps of adjacent spins by the same vacancy and considers only the correlation effects of a given atom ("tracer") with the vacancy [9]. In the interest of mathematical and numerical tractability, Torrey's isotropic model is used, which means that single jumps of atoms may occur in all directions with constant jump distance l [9].

Wolf's results for the spectral densities are

$$J^{(2)}(\omega) = 4J^{(1)}(\omega) \; ; \qquad J^{(0)}(\omega) = 6J^{(1)}(\omega) \, , \tag{2.1}$$

$$J^{(1)}(\omega) = \frac{8\pi n \tau_{\mathrm{NMR}}}{15a^3} \int_0^\infty J^2_{3/2}(a\varrho) \frac{1 - A(\varrho)}{(1-A(\varrho))^2 + (\omega\tau_{\mathrm{NMR}}/2)^2} \frac{\mathrm{d}\varrho}{\varrho} , \tag{2.2}$$

where n is the isotropic spin density, a is a normalization parameter [6], and $J_{3/2}(a\varrho)$ denotes the Bessel function of order 3/2 [6]. τ_{NMR} and $A(\varrho)$ are defined by [9]

$$\frac{1}{\tau_{\mathrm{NMR}}} = \frac{1}{\tau} \frac{1}{Z(0)} \left(1 - P(0)\right) , \tag{2.3}$$

$$A(\varrho) = \sum_{l_i > 0} P(l_i) \frac{\sin(\varrho l_i)}{\varrho l_i} . \tag{2.4}$$

Here, $P(l)$ denotes the probability that one encounter results in the total displacement l, $Z(0)$ is the mean number of jumps of an atom during its encounter with a vacancy. The summation in (2.4) has to be extended over all displacements $l_i > 0$ which can be reached by an encounter. The encounters which result in a total displacement 0 are considered in (2.3). The normalization parameter a in (2.2) is given by [6]

$$\left.\begin{aligned} k &= \frac{a}{l} = 0.76293 \text{ for a b.c.c. lattice } (l = a_0\sqrt{3}) , \\ k &= \frac{a}{l} = 0.7428 \text{ for a f.c.c. lattice } (l = a_0\sqrt{2}) , \end{aligned}\right\} \tag{2.5}$$

where a_0 denotes half the cube edge of the elementary cell. The diffusion mechanism enters into the spectral densities (2.1) and (2.2) via the values of $Z(0)$ and $P(l)$ which have to be calculated numerically for each mechanism. For practical purposes we write in the following for the f.c.c. lattice

$$J^{(1)}(\omega) = \frac{\sqrt{2}\,\pi c}{15k^3 a_0^6 \omega} \, \mathrm{INT}\,(F, \omega\tau); \qquad k = 0.7428; \; l = a_0\sqrt{2} \, , \tag{2.6}$$

and for the b.c.c. lattice

$$J^{(1)}(\omega) = \frac{2\pi c}{45\sqrt{3}\, k^3 a_0^6 \omega} \, \mathrm{INT}\,(F, \omega\tau); \qquad k = 0.76293; \; l = a_0\sqrt{3} \, , \tag{2.7}$$

where c denotes the ratio of the number of occupied lattice sites to the total number of lattice sites in the crystal. c may be replaced by 1 to a very good approximation since the equilibrium vacancy concentrations in metals are less than 10^{-3}. Furthermore we have

$$F(x, \omega\tau) = \frac{J_{3/2}^2(x)}{x} \frac{\left[1 - \sum_{l_i > 0} P(l_i) \frac{\sin\left(\frac{x l_i}{kl}\right)}{\frac{x l_i}{kl}}\right] \frac{Z(0)}{1 - P(0)} \omega\tau}{\left[1 - \sum_{l_i > 0} P(l_i) \frac{\sin\left(\frac{x l_i}{kl}\right)}{\frac{x l_i}{kl}}\right]^2 + \left[\frac{Z(0)}{1 - P(0)} \frac{\omega\tau}{2}\right]^2}, \tag{2.8}$$

$$\text{INT}\,(F, a, b, c, \ldots) = \int_0^\infty F(x, a, b, c, \ldots)\, \mathrm{d}x\,. \tag{2.9}$$

In the case of a combined 1V/2V-mechanism additional parameters enter into the function F and therefore into the spectral densities. Divacancies are known to contribute appreciably to the diffusion in a number of f.c.c. metals at high temperatures [10]. If two species of vacancies are present two types of encounters with different values of $Z(0)$ and $P(l)$ may occur. Since many encounters are necessary for relaxation we may weigh the different encounters to give a "mean encounter". We are thus led to introduce P_{2V}, the probability that a given encounter occurs with a divacancy and not with a monovacancy. We have then

$$Z(0) = (1 - P_{2V})\, Z^{1V}(0) + P_{2V} Z^{2V}(0)\,, \tag{2.10}$$

$$P(l) = (1 - P_{2V})\, P^{1V}(l) + P_{2V} P^{2V}(l)\,, \tag{2.11}$$

where the upper indices specify the mono- and divacancy values, respectively.

The parameter P_{2V} which depends on temperature is sufficient to characterize a 1V/2V-mechanism as long as the mechanism of motion of the divacancy is determined entirely by the lattice geometry. If the divacancies may assume several configurations during the migration, further temperature dependent parameters α, β, γ, ... may occur in $Z^{2V}(0)$ and $P^{2V}(l)$ and in the spectral densities. An example for a mechanism not entirely determined by geometry will be given in Section 3.

3. Models of Motion of Mono- and Divacancies

The simplest models for the motion of monovacancies in b.c.c. and f.c.c. lattices assume jumps of the vacancy to nearest neighbour positions only [10]. The motion is random and may be characterized by the mean time of stay of a vacancy at a lattice site, τ_V. A divacancy in an f.c.c. lattice is usually assumed to consist of two vacancies on nearest neighbour sites [10]. Divacancies are assumed to migrate by nearest-neighbour jumps leading again to a nearest-neighbour configuration of the divacancies [10]. Then the parameter P_{2V} is sufficient for describing the 1V/2V-mechanism in the f.c.c. lattice since the motion of the divacancy is determined entirely by the lattice geometry. In the b.c.c. lattice much less information concerning the configurations and mechanisms of motion of divacancies is available. The mechanism analogous to that discussed for the f.c.c. lattice is impossible since the nearest-neighbour configuration of a vacancy pair is destroyed by a nearest-neighbour jump of one of the constituent vacancies. The divacancy must therefore change its configuration during the migration. An extensive discussion of the possible models of motion is given in [11].

The most general mechanism used in [11] is a mechanism where jumps from the nearest-neighbour (1N)-configuration into a second nearest-neighbour (2N)-configuration occur. Starting from the 2N-configuration two types of jumps are allowed with different probabilities: a jump back to the 1N-configuration or a jump into a fourth nearest-neighbour (4N)-configuration. In every case the following jump leads to a 2N-configuration, and so on. Therefore we must introduce a second (temperature dependent) probability α that counts how many jumps starting from a 2N-configuration end in a 1N-configuration and not in a 4N-configuration. This mechanism is clearly not determined by the lattice geometry only but involves additional physical parameters. Using Mehrers notations [11] for the jump frequencies from one configuration to the other we may write

$$\alpha = \frac{\nu_{21}}{\nu_{21} + \nu_{24}} = \frac{R}{1 + R}, \tag{3.1}$$

where $R = \nu_{21}/\nu_{24}$ is the ratio of the jump frequency from the 2N- to the 1N-configuration to that from the 2N- to the 4N-configuration. For all mechanisms of motion considered the following equation may be derived for the f.c.c. and b.c.c. lattice [12]:

$$P_{2V}(T) = \frac{D_{2V}^{SD}}{D_{1V}^{SD} + D_{2V}^{SD}} = \frac{\dfrac{f_{1V}}{f_{2V}} D_{21}\, e^{-Q_{21}/kT}}{1 + \dfrac{f_{1V}}{f_{2V}} D_{21}\, e^{-Q_{21}/kT}}. \tag{3.2}$$

Here D_{1V}^{SD}, D_{2V}^{SD} are the contributions of the 1V or 2V-mechanism to the self-diffusion coefficient D^{SD}. $Q_{21} = Q_2 - Q_1$ is the difference of the activation energies Q_2 and Q_1 for the 2V- and the 1V-mechanism, and D_{21} is the ratio D_{20}/D_{10} of the preexponential factors of the tracer diffusion coefficients D_{2V}^{T} and D_{1V}^{T} [10]. The spatial correlation factors f_{1V} and f_{2V} are known: For the f.c.c. lattice we have $f_{1V} = 0.78145$, $f_{2V} = 0.468$ [10], and for the b.c.c. lattce $f_{1V} = 0.72149$, $0.345 \leqq f_{2V} \leqq 0.485$ [10, 11]. In the b.c.c. lattice f_{2V} is in geneial temperature dependent and equals the limits 0.345 for $\nu_{21} = 0$ and 0.485 for $\nu_{24} = 0$. Only in these two extreme cases (pure 1N–2N- or pure 2N–4N-mechanism of motion) f_{2V} is independent of temperature. If R or α can be estimated, f_{2V} may be deduced from Mehrers calculations [11].

The importance of the divacancy contribution for the nuclear spin relaxation is demonstrated in Fig. 1 which shows P_{2V} versus the absolute temperature T for aluminium metal. We used the values $D_{21} = 50$, $Q_{21} = 0.31$ eV given in [13]. It can be seen that at temperatures above 600 K more than 10% of all encounters occur with divacancies. At the melting point T_m more than 60% of all encounters are caused by divacancies.

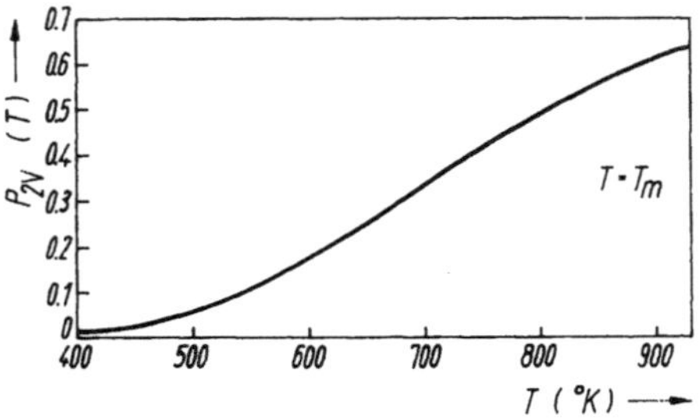

Fig. 1. P_{2V} vs. T for aluminium

4. Numerical Methods and Results

For the calculation of $Z(0)$ and $P(l)$ for a monovacancy mechanism Wolf used a modified matrix method well known in the theory of tracer diffusion [14, 15]. The application of this method to divacancies failed because of numerical problems (too big matrices, bad convergence). For our calculations we used the "Monte-Carlo-method", a computer simulation of the vacancy migration in the lattice similar to that used by Mehrer [16]. From the eight or twelve possible jump vectors to the nearest-neighbour positions in the crystal the computer selects one in a random manner by using a random-number generator. One simulates for example 1000 encounters of an atom with a vacancy and gets $Z(0)$ and $P(l)$ as mean values of the number of jumps and the displacements in one encounter. The precision of the results is limited by the simulation parameter NLS, the number of vacancy jumps per simulated encounter. The criterion for sufficient numerical convergence is given by the rate of change of $Z(0)$ and $P(l)$ with increasing NLS. Typical endvalues NLS were chosen between 710 (1V) and 1410 (2V). All numerical values of $Z(0)$ and $P(l)$ given in the following are exact within 4%. A further increase of precision by increasing NLS is possible but the necessary computer time becomes rather long. After computation of $Z(0)$ and $P(l)$ the spectral densities can be computed by numerical integration, using (2.6) and (2.7). Since the function $F(x)$ in (2.8) oscillates strongly, we employ the Gaussian quadrature of the 6th order. The integration is then correct up to the 11th derivative of $F(x)$ [17, 18]. The following intervals were used for the numerical quadrature:

125 intervals of equal length from $x = 0$ to $x = 10$,

125 intervals of equal length from $x = 10$ to $x = 30$,

125 intervals of equal length from $x = 30$ to $x = 300$.

Table 1

$Z(0)$ and $P(l)$ for the 1V-mechanisms (NLS = 710)

f.c.c. lattice	b.c.c. lattice
$Z(0) = 1.308$	$Z(0) = 1.37$
$P(0) = 0.08$	$P(0) = 0.14$
$P(a_0\sqrt{2}) = 0.86$	$P(a_0\sqrt{3}) = 0.775$
$P(2a_0) = 0.021$	$P(2a_0) = 0.042$
$P(a_0\sqrt{6}) = 0.031$	$P(2a_0\sqrt{2}) = 0.03$
$P(2a_0\sqrt{2}) = 0.006$	$P(a_0\sqrt{11}) = 0.007$
$P(a_0\sqrt{10}) = 0.002$	$P(2a_0\sqrt{3}) = 0.006$

Table 2

$Z(0)$ and $P(l)$ for the 2V-mechanism in the f.c.c. lattice (NLS = 1210)

$Z(0) = 2.426$
$P(0) = 0.106$
$P(a_0\sqrt{2}) = 0.752$
$P(2a_0) = 0.052$
$P(a_0\sqrt{6}) = 0.067$
$P(2a_0\sqrt{2}) = 0.013$
$P(a_0\sqrt{10}) = 0.004$
$P(a_0\sqrt{14}) = 0.005$
$P(a_0\sqrt{22}) = 0.001$

Table 3

$Z(0)$ and $P(l)$ for the 2V-mechanism in the b.c.c. lattice for various values of α (NLS = 1410)

α	0	0.25	0.5	0.75	1
$Z(0)$	2.448	2.368375	2.37575	2.450875	2.4965
$P(0)$	0.27	0.2395	0.204	0.164625	0.1515
$P(a_0\sqrt{3})$	0.585	0.586	0.59	0.595375	0.578
$P(2a_0)$	0.0805	0.100875	0.1235	0.13975	0.161
$P(2a_0\sqrt{2})$	0.039	0.036875	0.043	0.04775	0.059
$P(a_0\sqrt{11})$	0.013	0.020875	0.02475	0.031625	0.032
$P(2a_0\sqrt{3})$	0.01	0.00875	0.00825	0.013	0.0095
$P(4a_0)$	0	0.000625	0.0015	0.001125	0.003
$P(a_0\sqrt{19})$	0.002	0.00425	0.002	0.004125	0.0015
$P(a_0\sqrt{20})$	0	0.001625	0.0015	0.001375	0.002
$P(a_0\sqrt{24})$	0.0005	0	0.001	0	0.0005
$P(3a_0\sqrt{3})$	0	0.00025	0	0.0005	0.0015
$P(a_0\sqrt{32})$	0	0	0	0.000375	0
$P(a_0\sqrt{35})$	0	0	0.00025	0	0.0005
$P(6a_0)$	0	0.000375	0.00025	0.000375	0
R	0	1/3	1	3	∞

The contributions to the integral from $x > 300$ may be neglected in a very good approximation. The results of the integration (Table 4 to 5) are estimated to be correct up to the 4th position after the decimal point.

Tables 1, 2, and 3 show the values of $Z(0)$ and $P(l)$ for the various mechanisms and lattices. Table 4 and 5 show position and height of the maximum of INT $(F, \omega\tau, P_{2V})$ for various values of P_{2v} for the mechanisms and lattices treated in this section. In Fig. 2 a plot of INT $(F, \omega\tau, P_{2V})$ versus $\omega\tau$ is shown for $P_{2V} = 0, 0.5$, and 1 for the f.c.c. lattice.

Table 4

Position and height of the maximum of INT(F, $\omega\tau$, P_{2V}) for various values of P_{2V} for the f.c.c. lattice

P_{2V}	$(\omega\tau)_{max}$	INT(F, $(\omega\tau)_{max}$, P_{2V})
0.0	1.23	0.31127
0.1	1.20	0.31128
0.2	1.0 to 1.1	≈0.31181
0.3	1.0	0.31247
0.4	0.9	0.31291
0.5	0.8 to 0.9	≈0.31305
0.6	0.8	0.31373
0.7	0.7 to 0.8	≈0.31385
0.8	0.72	0.31451
0.9	0.69	0.31489
1.0	0.66	0.31525

Table 5

Position and height of the maximum of INT(F, $\omega\tau$, P_{2V}, α) for various values of P_{2V} and α

$\alpha = 0$

P_{2V}	$(\omega\tau)_{max}$	INT(F, $(\omega\tau)_{max}$, P_{2V}, α)
0.0	1.09	0.31386
0.1	1.0	0.31469
0.2	0.9	0.31544
0.3	≈0.85	≈0.3158
0.4	0.8	0.31699
0.5	≈0.75	≈0.3175
0.6	0.7	0.31839
0.7	0.64	0.31915
0.8	0.61	0.31978
0.9	0.56	0.32044
1.0	0.55	0.32091

$\alpha = 0.5$

P_{2V}	$(\omega\tau)_{max}$	INT(F, $(\omega\tau)_{max}$, P_{2V}, α)
0.0	1.09	0.31386
0.1	1.01	0.31398
0.2	≈0.95	≈0.3148
0.3	0.9	0.31543
0.4	≈0.85	≈0.3159
0.5	0.8	0.31641
0.6	≈0.75	≈0.3170
0.7	0.7	0.31750
0.8	0.65	0.31796
0.9	0.63	0.31845
1.0	0.6	0.31890

Table 5 (continued)

$\alpha = 1.0$

P_{2V}	$(\omega\tau)_{max}$	$\mathrm{INT}(F, (\omega\tau)_{max}, P_{2V}, \alpha)$
0.0	1.09	0.31386
0.1	1.01	0.31423
0.2	0.9 to 1.0	≈0.3144
0.3	0.9	0.31489
0.4	0.9 to 0.8	≈0.3152
0.5	0.8	0.31557
0.6	0.8 to 0.7	≈0.3157
0.7	0.7	0.31635
0.8	0.65	0.31662
0.9	0.63	0.31700
1.0	0.61	0.31731

5. The Case of More than One Spin Species

Most metals contain more than one isotope with spin different from zero. Since these isotopes have different masses they have different mean times of stay at a lattice site [10, 14]. This means that in principle we have as many values of τ as we have different isotopes. In the following we consider the case of two different isotopes I, S. Then we have N_I atoms of species I with γ_I, spin I, abundance f_I, Larmor frequency ω_I, mass m_I and N_S atoms of species S with γ_S, spin S, abundance f_S, Larmor frequency ω_S, mass $m_S \cdot N_I + N_S = N$, the total number of atoms in the crystal, $f_I + f_S = 1$, $n = (N_I + N_S)/V$ (V = crystal volume).

We assume that the NMR-experiments are performed at the resonance frequency ω_I of the I-spins, which are called "resonant spins". The mean times of stay at a lattice site of the I- and S-spins are called τ^I and τ^S, respectively. Then the following spectral densities enter into the equations (5.7 to 5.8) for the relaxation times:

$$\overline{J_{II}^{(1)}(\omega)} = f_I \frac{8\pi n \tau_{NMR}^{I}}{15 a^3} \int\limits_0^\infty J_{3/2}^2(a\varrho) \frac{1 - A(\varrho)}{(1 - A(\varrho))^2 + \left(\frac{\omega \tau_{NMR}^{I}}{2}\right)^2} \frac{d\varrho}{\varrho} = f_I J_{II}^{(1)}(\omega)\,, \tag{5.1}$$

$$\overline{J_{IS}^{(1)}(\omega)} = f_S \frac{8\pi n \tau_{NMR}^{IS}}{15 a^3} \int\limits_0^\infty J_{3/2}^2(a\varrho) \frac{1 - A(\varrho)}{(1 - A(\varrho))^2 + \left(\frac{\omega \tau_{NMR}^{IS}}{2}\right)} \frac{d\varrho}{\varrho} = f_S J_{IS}^{(1)}(\omega) \tag{5.2}$$

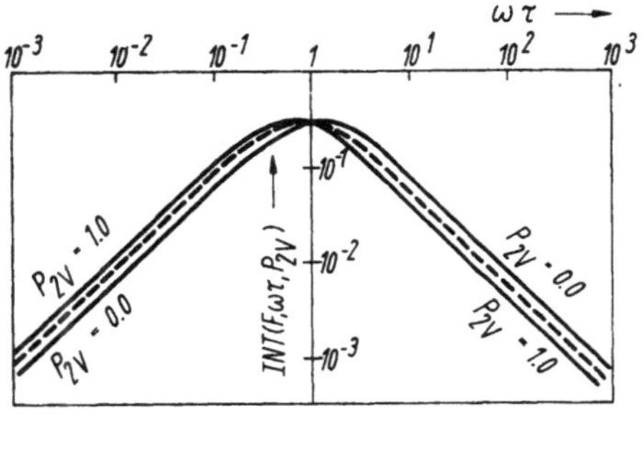

Fig. 2. INT(F, $\omega\tau$, P_{2V}) vs. $\omega\tau$ for P_{2V} = 0, 0.5, and 1 for the f.c.c. lattice. The dashed line is the curve for $P_{2V} = 0.5$

with

$$\left.\begin{aligned} \tau_{\mathrm{NMR}}^{\mathrm{I}} &= \frac{\tau^{\mathrm{I}} Z(0)}{1 - P(0)}, \\ \tau_{\mathrm{NMR}}^{\mathrm{S}} &= \frac{\tau^{\mathrm{S}} Z(0)}{1 - P(0)}, \\ \tau_{\mathrm{NMR}}^{\mathrm{IS}} &= \frac{2\tau_{\mathrm{NMR}}^{\mathrm{I}}}{1 + \dfrac{\tau_{\mathrm{NMR}}^{\mathrm{I}}}{\tau_{\mathrm{NMR}}^{\mathrm{S}}}}. \end{aligned}\right\} \tag{5.3}$$

Since the spectral densities are not very sensitive to small changes in $Z(0)$ and $P(l)$ we have neglected the mass dependence of $Z(0)$ and $P(l)$. In the theory of tracer diffusion one often assumes the relation [10, 14]

$$\frac{\tau_{\mathrm{NMR}}^{\mathrm{I}}}{\tau_{\mathrm{NMR}}^{\mathrm{S}}} = \frac{\tau^{\mathrm{I}}}{\tau^{\mathrm{S}}} = \left(\frac{m_{\mathrm{I}}}{m_{\mathrm{S}}}\right)^{1/2}. \tag{5.4}$$

(5.4) is valid if the jumps of the vacancies can be described as individual jumps of atoms [10, 14]. If it is necessary to consider the mechanical coupling of the jumping atom to its surroundings one introduces "ΔK-factors" [10, 14]. In ordinary cases, these ΔK-factors are nearly equal to 1 so that the model of individual atomic jumps is still realistic. From the equations in [10] we can derive easily

$$\begin{aligned} \frac{\tau^{\mathrm{I}}}{\tau^{\mathrm{S}}} = {} & (1 - P_{2\mathrm{V}}^{\mathrm{I}}) \frac{f_{1\mathrm{V}}^{\mathrm{I}}}{f_{1\mathrm{V}}^{\mathrm{S}}} \left(1 + f_{1\mathrm{V}}^{\mathrm{S}} \, \Delta K_{1\mathrm{V}} \left(\left(\frac{m_{\mathrm{I}}}{m_{\mathrm{S}}}\right)^{1/2} - 1\right)\right) + \\ & + P_{2\mathrm{V}}^{\mathrm{I}} \frac{f_{2\mathrm{V}}^{\mathrm{I}}}{f_{2\mathrm{V}}^{\mathrm{S}}} \left(1 + f_{2\mathrm{V}}^{\mathrm{S}} \, \Delta K_{2\mathrm{V}} \left(\left(\frac{m_{\mathrm{I}}}{m_{\mathrm{S}}}\right)^{1/2} - 1\right)\right). \end{aligned} \tag{5.5}$$

$f_{n\mathrm{V}}$ $(n = 1, 2)$ and $P_{2\mathrm{V}}$ have the same meaning as before, but now they are mass dependent [10] which is symbolized by the upper indices. The difference between (5.5) and (5.4) is very small in ordinary cases [10] ($\lesssim 1\%$), so that we can use the much simpler relation (5.4) for all practical purposes.

Then the spectral densities in (5.1) and (5.2) depend on τ^{I}, τ^{S}, the ratio $m_{\mathrm{I}}/m_{\mathrm{S}}$, $\omega\tau$, and $P_{2\mathrm{V}}$ (and α for the 1N–2N–4N-mechanism). If the resonances of the two spin species do not overlap, that means

$$|\omega_{\mathrm{I}} - \omega_{\mathrm{S}}| \gg \delta_{\mathrm{S}}, \tag{5.6}$$

where δ_{S} is the line width of the S-spins, one can use the concept of a single relaxation time if the coupling of the nonresonant spins to the lattice is stronger than their dipolar coupling to the I-spins. In metals this condition is always fulfilled because of the powerful conduction electron relaxation mechanism. Under the above conditions the following relationships for the relaxation times of the I-spins hold [4, 5, 8, 19]:

$$\begin{aligned} \frac{1}{T_1^{\mathrm{I}}} = {} & \frac{3}{2} \gamma_{\mathrm{I}}^4 \hbar^2 I\,(I + 1)\, f_{\mathrm{I}} \left(J_{\mathrm{II}}^{(1)}(\omega_{\mathrm{I}}) + J_{\mathrm{II}}^{(2)}(2\omega_{\mathrm{I}})\right) + \\ & + \gamma_{\mathrm{I}}^2 \gamma_{\mathrm{S}}^2 \hbar^2 S\,(S + 1)\, f_{\mathrm{S}} \left(\frac{1}{12} J_{\mathrm{IS}}^{(0)}(\omega_{\mathrm{I}} - \omega_{\mathrm{S}}) + \frac{3}{2} J_{\mathrm{IS}}^{(1)}(\omega_{\mathrm{I}}) + \frac{3}{4} J_{\mathrm{IS}}^{(2)}(\omega_{\mathrm{I}} + \omega_{\mathrm{S}})\right), \end{aligned} \tag{5.7}$$

$$\frac{1}{T_2^{\mathrm{I}}} = \gamma_{\mathrm{I}}^4 \hbar^2 I\,(I+1)\, f_{\mathrm{I}} \left(\frac{3}{8}\, J_{\mathrm{II}}^{(0)}(0) + \frac{15}{4}\, J_{\mathrm{II}}^{(1)}(\omega_{\mathrm{I}}) + \frac{3}{8}\, J_{\mathrm{II}}^{(2)}(2\omega_{\mathrm{I}})\right) +$$

$$+ \gamma_{\mathrm{I}}^2 \gamma_{\mathrm{S}}^2 \hbar^2 S\,(S+1)\, f_{\mathrm{S}} \left(\frac{1}{6}\, J_{\mathrm{IS}}^{(0)}(0) + \frac{1}{24}\, J_{\mathrm{IS}}^{(0)}\,(\omega_{\mathrm{I}} - \omega_{\mathrm{S}}) + \frac{3}{4}\, J_{\mathrm{IS}}^{(1)}(\omega_{\mathrm{I}}) +\right.$$

$$\left. + \frac{3}{2}\, J_{\mathrm{IS}}^{(1)}(\omega_{\mathrm{S}}) + \frac{3}{8}\, J_{\mathrm{IS}}^{(2)}\,(\omega_{\mathrm{I}} + \omega_{\mathrm{S}})\right). \qquad (5.8)$$

The expression for $1/T_{1\varrho}^{\mathrm{I}}$ can be obtained by simply replacing the frequency 0 in $J^{(0)}(0)$ in (5.8) by $2\omega_1^{\mathrm{I}}$ in the first bracket and by ω_1^{I} in the second one where $\omega_1^{\mathrm{I}} = \gamma_{\mathrm{I}} H_1$ [19].

Our numerical evaluation of (5.7) and (5.8) for some metals [12] showed that using (5.4) gives results which differ usually by less than $1^0/_{00}$ from the results obtained by setting $\tau^{\mathrm{I}} = \tau^{\mathrm{S}}$. For enriched lithium metal (50% Li^6, 50% Li^7) however, the difference was in the region of 1%.

Acknowledgements

The author is greatly indebted to Prof. A. Seeger for the initiation of this work and many helpful discussions and suggestions during its growth. I thank Dr. H. Mehrer, Dr. F. Noack, and Dr. D. Wolf for many stimulating discussions.

References

[1] N. Bloembergen, E. M. Purcell, and R. V. Pound, Phys. Rev. **73**, 679 (1948).
[2] R. V. Pound, Phys. Rev. **79**, 685 (1950).
[3] E. Cavelius, to be published.
[4] A. Abragam, The Principles of Nuclear Magnetism, Clarendon Press, Oxford 1961.
[5] D. C. Look and I. J. Lowe, J. chem. Phys. **44**, 2995 (1966).
[6] H. C. Torrey, Phys. Rev. **92**, 962 (1953).
[7] J. Bardeen and C. Herring, in: Imperfections in Nearly Perfect Crystals, Ed. W. Shockley, J. H. Holloman, R. Maurer, and F. Seitz, John Wiley & Sons, New York 1952 (p. 261).
[8] M. Eisenstadt and A. G. Redfield, Phys. Rev. **132**, 635 (1963).
[9] D. Wolf, Z. Naturf. **26a**, 1816 (1971).
[10] A. Seeger and H. Mehrer, in: Vacancies and Interstitials in Metals, North-Holland Publ. Co., Amsterdam 1970 (p. 1).
[11] H. Mehrer, J. Phys. F **3**, 543 (1973).
[12] E. Cavelius, Thesis, University of Stuttgart, 1973.
[13] A. Seeger, D. Wolf, and H. Mehrer, phys. stat. sol. (b) **48**, 481 (1971).
[14] A. D. Le Claire, Physical Chemistry, Vol. 10, Academic Press Inc., London/New York 1970.
[15] R. E. Howard, Phys. Rev. **144**, 650 (1966).
[16] H. Mehrer, Z. Naturf. **24a**, 358 (1969).
[17] A. Ralston, A First Course in Numerical Analysis, McGraw-Hill Publ. Co., New York 1965.
[18] V. I. Krylov, Approximate Calculations of Integrals, MacMillan Co., New York 1962.
[19] D. E. O'Reilly, E. M. Peterson, and T. Tsang, Phys. Rev. **160**, 333 (1967).

(Received June 10, 1974)

phys. stat. sol. (b) 65, 193 (1974)

Subject classification: 6 and 20.1; 22.4.4

Institute of Experimental Physics, University of Warsaw

Application of the Random-Element Isodisplacement Model to Long-Wavelength Optical Phonons in $CdSe_xTe_{1-x}$ Mixed Crystals

By

M. Górska[1]) and W. Nazarewicz

Infrared reflection spectra from mixed crystals $CdSe_xTe_{1-x}$ ($0 \leqq x \leqq 1$) have been investigated in the spectral region from 90 to 300 cm^{-1} at room temperature. The spectra contain two distinct reststrahlen bands whose strengths and frequencies depend on the composition, x. Using Kramers-Kronig analysis the frequencies of long-wavelength optical phonons were obtained. It is established, that the dependence of the phonon frequencies on the composition of crystals does not show any discontinuity when changing the structure of the mixture from zincblende to wurtzite. The calculations, based on the random-element isodisplacement (REI) model, of the dependence of long-wavelength optical phonon frequencies on the composition x are in good agreement with the experimental results.

Nous avons étudié les spectres de réflection infrarouge des cristaux mixtes de $CdSe_xTe_{1-x}$ ($0 \leqq x \leqq 1$) dans la région spectrale de 90 à 300 cm^{-1} à température ambiante. Les spectres contient deux bandes à rayons restants dont l'intensité et la fréquence depend de la composition x. Les fréquences des phonons optiques de grandes longueurs d'ondes ont été obtenues par inversion de Kramers-Kronig. Il a été etablie que les fréquences en fonction de la composition x ne démontrent aucune discontinuité dans la région où la structure change de la blende au wurtzite. Le calcul, basé sur le modèle de l'isodeplacement d'impurétés distribués aléatoirement (modèle REI), des fréquences des phonons optiques de grandes longueurs d'ondes en fonction de la composition x est en bon accord avec résultats expérimentaux.

1. Introduction

The properties of optical phonons in disordered systems of the types A_xB_{1-x} and AB_xC_{1-x} were recently the subject of many papers [1]. In these papers a lot of place is taken by the investigation of the relation between the frequency of the long-wavelength optical phonons and the composition of crystals, $\omega(x)$. This relation may be determined both theoretically and experimentally. The experimental results are obtained using far infrared spectroscopy or Raman spectroscopy. In the theoretical calculation one assumes such phenomenological models as virtual crystal model [2], linear chain model [3], cluster model [4], and random-element isodisplacement (REI) model [5] with its modifications [6, 7]. The models use different assumptions about dynamics and structure of the solid solutions. The comparison of experimental and theoretical curves allows to draw conclusions about the validity of the assumptions of the model.

Model calculations are known, among others, for two mixed crystals of II–VI compounds: $Zn_xCd_{1-x}Te$ [7] and CdS_xSe_{1-x} [8]. In the first case the REI model gave a satisfactory explanation of the experimental data. In the second

[1]) Present address: Institute of Physics, Polish Academy of Sciences, Warsaw.

case the fine structure of reststrahlen bands was interpreted in terms of the cluster model.

Up to now the dependences $\omega(x)$ were usually investigated in mixed crystals which have the same crystallographic structure in the whole composition range.

In the present paper we report results for $CdSe_xTe_{1-x}$. This system has a phase transition from zincblende to wurtzite structure for some x-value. In connection with this it was interesting to find out whether the transition influences the shape of $\omega(x)$ curves.

Preliminary experimental results on the long-wavelength optical phonons in $CdSe_xTe_{1-x}$ mixed crystals were recently published by Vinogradov et al. [9] and Górska and Nazarewicz [10]. The authors of both papers found that this mixed system has two-mode properties and from the reststrahlen spectra they obtained similar $\omega(x)$ dependences for optical phonons.

In this paper we present more detailed experimental results on the $\omega(x)$ dependence, and pay special attention to calculations of this dependence based on the REI model. The applied model allowed us to obtain good agreement of the calculated $\omega(x)$ curves with experimental data.

2. Samples and Apparatus

The $CdSe_xTe_{1-x}$ mixed crystals with nine different compositions x were prepared using the Bridgman method in the Semiconductor Technological Laboratory of the Institute of Physics of the Polish Academy of Sciences. They had the following x values: 0 (CdTe), 0.01, 0.15, 0.25, 0.35, 0.50, 0.70, 0.85, 1 (CdSe).

The Debye-Scherrer diffractograms were made for all samples. It was found that crystals with $x \leqq 0.35$ have the zincblende structure and those with $x \geqq 0.5$ have the wurtzite structure. These results are in agreement with earlier investigations by Litwin [11] and Stuckes and Farrel [12]. According to these authors the phase transition takes place in the range $0.4 \leqq x \leqq 0.6$ depending upon conditions of the growth of crystals.

The width at half maximum heights of the Bragg peaks was about 20′ for mixed crystals and it was a little wider than the width at half maximum heights for pure CdSe and CdTe crystals. The relatively small width of the Bragg peaks shows a good homogeneity of composition x in different samples.

The composition of mixed crystals was found from the known [11, 12] relation between lattice constants and the composition x.

The fundamental absorption edge was determined at room temperature. It follows from these measurements that the energy gap in mixed crystals changes from 1.48 eV for CdTe to 1.72 eV for CdSe, and has a minimum of 1.34 eV for $x = 0.35$. The resistivity of all samples at room temperature was higher than $1000\,\Omega$cm. It shows that the concentration of electrically active impurities is smaller than 10^{16} cm^{-3}. Thus in the following consideration one may neglect the effects connected with free carriers.

The reflection measurements were performed with flat parallel plates which had surfaces from 0.3 to 2.0 cm^2 and thickness about 1 mm. The flat surfaces of the samples were polished mechanically using aluminium oxide abrasive powder and diamond paste. It was established that further chemical etch polishing had no noticeable effect on the reflectivity spectra.

The measurements of the reflection coefficient were done by means of a vacuum grating spectrometer operating in a single-beam mode. As radiation detector the Golay cell was used. In the investigated spectral range (90 to 300 cm^{-1}) the spectral width of the monochromator slits was from 1 to 2 cm^{-1}. The total error in the measurements of the reflection coefficient was from 1 to 5% depending upon the surface of the samples. The reflection coefficient was found by comparison of reflection from samples with the reflection from a reference aluminium mirror. The angle of incidence of the light was 7°. Non-polarised light was used. In the zincblende crystals optical properties are isotropic and polarisation effects do not exist. In the wurtzite crystals the reflection spectra show relatively small polarisation effects, which in the present study may be neglected.

3. Experimental Results and Discussion

Fig. 1 contains the reflection spectra of $CdSe_xTe_{1-x}$ mixed crystals with different composition x ($x = 0.15, 0.25, 0.35, 0.50, 0.70, 0.85$). Also included are spectra of pure CdTe, CdSe, and CdTe crystals with small ($x = 0.01$) concentration of Se.

The reflection spectra of pure CdSe and CdTe crystals show one reststrahlen band. The position and intensity of this band agrees with the data of other authors [13].

The reflection spectra of mixed crystals contain two reststrahlen bands. The intensities of these bands clearly depend upon the composition x. In the crystals with wurtzite structure ($x = 0.85, 1.00$) one may notice a small splitting of one of bands. It is probably due to the fact that unpolarised light was used in the experiment. Therefore in the optical transitions optical phonons with polarisa-

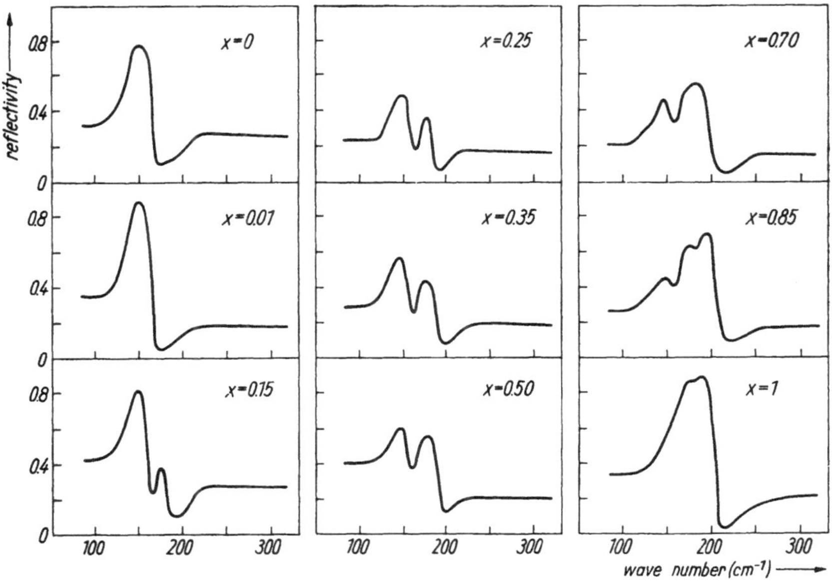

Fig. 1. The reflection spectra of $CdSe_xTe_{1-x}$ mixed crystals in the region 90 to 300 cm^{-1} at 300 K

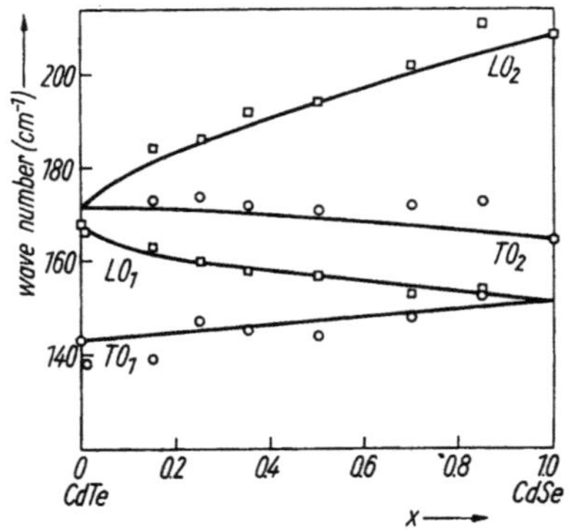

Fig. 2. The frequencies of the LO and TO phonons in $CdSe_xTe_{1-x}$ mixed crystals as a function of the composition x: points experimental data, solid lines REI model calculations

tion perpendicular and parallel to the c-axis of the crystal may take part. Those phonons have slightly different frequencies.

A Kramers-Kronig dispersion analysis was applied to the reflection spectra to obtain the following functions: the real part of the dielectric constant $\mathrm{Re}\,\varepsilon(\omega)$, the imaginary part of the dielectric constant $\mathrm{Im}\,\varepsilon(\omega)$, and the imaginary part of the reciprocal of the dielectric constant $-\,\mathrm{Im}\,(1/\varepsilon(\omega))$. The form of these functions allows in the known way to find the frequencies of transverse optical phonons (TO) and those of longitudinal optical phonons (LO). The frequencies of the TO phonons were obtained from the positions of the maxima of the $\mathrm{Im}\,\varepsilon(\omega)$ curve and the frequencies of the LO phonons from the positions of maxima of $-\,\mathrm{Im}\,(1/\varepsilon(\omega))$ or from the points where the $\mathrm{Re}\,\varepsilon(\omega)$ curve crosses the ω-axis.

The Kramers-Kronig analysis yields for each x-value two frequencies of the transverse optical phonons (TO_1 and TO_2) and two frequencies of the longitudinal optical phonons (LO_1 and LO_2). The results are presented as dots on Fig. 2. The full curves are from theoretical calculations and will be discussed later. Table 1 contains the phonon frequencies for different compositions x.

Table 1

The long-wavelength optical phonon frequencies in $CdSe_xTe_{1-x}$ mixed crystals

x	$\omega_{TO,1}$ (cm^{-1})	$\omega_{LO,1}$ (cm^{-1})	$\omega_{TO,2}$ (cm^{-1})	$\omega_{LO,2}$(cm^{-1})
0	143	168	—	—
0.01	138	166	—	—
0.15	139	163	173	184
0.25	147	160	174	186
0.35	145	158	172	192
0.50	144	157	171	194
0.70	148	153	172	202
0.85	153	154	173	211
1	—	—	165	209

The dependence of the frequencies of the long-wavelength optical phonons upon the composition of the $CdSe_xTe_{1-x}$ mixed crystal is similar to the results obtained by Vinogradov et al. [9] from the reststrahlen spectra by means of a graphic method (without Kramers-Kronig analysis). The TO_2 and LO_2 branches approach the common value 172 cm^{-1} for $x \approx 0$ and the TO_1 and LO_1 branches tend to the common value 151 cm^{-1} for $x \approx 1$. The value 172 cm^{-1}, probably corresponds to the frequency of the local mode of selenium in CdTe, this value is also near the theoretical one, calculated by Gaur et al. [14] and the experimental value measured by Beserman and Balkanski [15]. There are not enough data to establish whether the frequency 151 cm^{-1} corresponds to the gap mode of tellurium in CdSe. The frequency spectrum of CdSe was not investigated and it is not known whether it has a gap between acoustical and optical branches.

In crystals with zincblende structure the selection rules allow only one optical mode having $F(\Gamma_5)$ symmetry to be active in absorption. In crystals with wurtzite structure two modes may take part in the absorption: one has $A_1(\Gamma_1)$ symmetry and the second $E_1(\Gamma_5)$ symmetry. In ionic crystals each of the oscillation splits into a transverse mode and a longitudinal mode: A_1(TO), A_1(LO) and E_1(TO), E_1(LO). In pure crystals having wurtzite structure the frequencies of transverse oscillations are very near to each other; the same is true for longitudinal oscillation frequencies. In pure CdSe crystals the frequencies A_1(TO) and E_1(TO) are 166.5 and 172 cm^{-1} respectively and the frequencies A_1(LO) and E_1(LO) have the values 210 and 212 cm^{-1} [16]. One may expect that in the $CdSe_xTe_{1-x}$ mixed crystals the splitting of the frequencies A_1 and E_1 is of the order of a few cm^{-1}, similarly as in CdSe. The frequencies measured by us are a certain average, which characterises simultaneously the vibrations A_1 and E_1. The experimental curves do not show any clear discontinuity when the zincblende structure is changed to wurtzite for some composition x. It is so probably because in the mixed crystals during the phase transition from the zincblende to the wurtzite structure the local order does not change. In both structures there are four nearest neighbours in the corners of tetrahedra and twelve second neighbours. The distance between the nearest neighbours in both structures calculated from lattice constants differs very little. For $x = 0.35$ (zincblende structure) it is 2.74 Å and for $x = 0.5$ (wurtzite structure) it is 2.71 Å for three atoms and 2.73 Å for the fourth atom. The distance between second neighbours in the structures is also similar (for zincblende it is 4.47 Å and for wurtzite it is 4.43 and 4.44 Å). A similar distribution and nearly equal distances in both structures and short-range interactions could explain the lack of discontinuity of the $\omega(x)$ curves during the phase transition.

It should be noticed that the reflection spectra do not show any fine structure characteristic of clusters. It could be a proof that the distribution of tellurium and selenium atoms has random character. A fine structure of the reflection spectra was observed in some mixed crystals of the II–VI compounds, among others in CdS_xSe_{1-x} [8, 17].

4. Theoretical Model

In this part of the paper a phenomenological model for the $CdSe_xTe_{1-x}$ mixed system is presented. The model contains the basic assumptions of the REI model as proposed by Chen et al. [5] and subsequently developed by Chang and Mitra [6] and also by Harada and Narita [7].

The assumptions may be stated in two points:

1. The crystal lattice of a mixed crystal comprises two sublattices, one of them is filled by A atoms only and the second is filled by B and C atoms, randomly distributed.

2. The atoms of the same kind displace from equilibrium in the same phase and with the same amplitude.

The REI model was already successfully applied to calculations of the composition dependence of the long-wavelength optical phonon frequencies for some mixed crystals, among others for CdS_xSe_{1-x} [6] and $Zn_xCd_{1-x}Te$ [7].

In some papers the reflection spectra calculated according to this model were compared with the spectra obtained experimentally [7].

Now let us use the REI model for the determination of the frequencies of the long-wavelength optical phonons in the $CdSe_xTe_{1-x}$ mixed system. To do this one has to formulate and to solve the equations of motion for the atoms Cd, Se, and Te; let us call them A, B, and C atoms.

If one takes into account the influence of the effective electrical field on the motion of ions, the REI model supplies the following three equations of motion for A, B, and C atoms:

$$\left.\begin{aligned} m_A\ddot{\boldsymbol{u}}_A &= -(1-x)\,F_{AB}(\boldsymbol{u}_A-\boldsymbol{u}_B) - xF_{AC}\,(\boldsymbol{u}_A-\boldsymbol{u}_C) + e_A\boldsymbol{E}_{\text{eff}}\,,\\ m_B\ddot{\boldsymbol{u}}_B &= -\,F_{AB}(\boldsymbol{u}_B-\boldsymbol{u}_A) - xF_{BC}\,(\boldsymbol{u}_B-\boldsymbol{u}_C) - e_B\boldsymbol{E}_{\text{eff}}\,,\\ m_C\ddot{\boldsymbol{u}}_C &= -\,F_{AC}(\boldsymbol{u}_C-\boldsymbol{u}_A) - (1-x)\,F_{BC}(\boldsymbol{u}_C-\boldsymbol{u}_B) - e_C\boldsymbol{E}_{\text{eff}}\,,\end{aligned}\right\} \tag{1}$$

where m_A, m_B and m_C are the masses of atoms, $\boldsymbol{u}_A$, $\boldsymbol{u}_B$, and $\boldsymbol{u}_C$ are displacements from equilibrium.

In equations (1) one takes into account force constants between the nearest neighbours F_{AB} and F_{AC} and also one force constant between the second neighbours F_{BC}. One assumes their linear dependence upon the composition:

$$\frac{F_{AB}}{F_{AB0}} = \frac{F_{AC}}{F_{AC0}} = \frac{F_{BC}}{F_{BC0}} = 1 - \theta x\,. \tag{2}$$

The condition of electrical neutrality of the crystal has the form

$$e_A - (1-x)\,e_B - xe_C = 0\,. \tag{3}$$

The effective electrical field in a crystal of cubic structure is given by Lorentz' formula:

$$\boldsymbol{E}_{\text{eff}} = \boldsymbol{E} + \frac{4\pi}{3}\boldsymbol{P}\,. \tag{4}$$

Verleur and Barker [8] has shown that in the crystals with wurtzite structure the Lorentz formula is accurate to within 5%. In this paper the correction to the Lorentz formula was neglected and (4) was used for the cubic as well as hexagonal system.

For long-wavelength optical phonons it is possible to make the approximation $E \approx 0$. The polarisation is given by the equation

$$\boldsymbol{P} = N[e_A\boldsymbol{u}_A - (1-x)\,e_B\boldsymbol{u}_B - xe_C\boldsymbol{u}_C] + N\alpha\boldsymbol{E}_{\text{eff}}\,, \tag{5}$$

where N is the ion pair density, α is the electronic polarisability which is related to the high-frequency dielectric constant by the Clausius-Mossotti formula

$$\frac{4\pi}{3}N\alpha = \frac{\varepsilon_\infty - 1}{\varepsilon_\infty + 2}\,. \tag{6}$$

Inserting (3), (4), (5), and (6) into (1) one obtains the equations of motion with the new force constants

$$\left.\begin{aligned} m_A\ddot{\boldsymbol{u}}_A &= -(1-x)\,F'_{AB}(\boldsymbol{u}_A-\boldsymbol{u}_B)-xF'_{AC}\,(\boldsymbol{u}_A-\boldsymbol{u}_C)\,,\\ m_B\ddot{\boldsymbol{u}}_B &= -F'_{AB}(\boldsymbol{u}_B-\boldsymbol{u}_A)-xF'_{BC}\,(\boldsymbol{u}_B-\boldsymbol{u}_C)\,,\\ m_C\ddot{\boldsymbol{u}}_C &= -F'_{AC}(\boldsymbol{u}_C-\boldsymbol{u}_A)-(1-x)\,F'_{BC}(\boldsymbol{u}_C-\boldsymbol{u}_B)\,,\end{aligned}\right\} \tag{7}$$

where

$$\left.\begin{aligned} F'_{AB} &= F_{AB}-\frac{4\pi}{3}Ne_Ae_B\left(\frac{\varepsilon_\infty+2}{3}\right),\\ F'_{AC} &= F_{AC}-\frac{4\pi}{3}Ne_Ae_C\left(\frac{\varepsilon_\infty+2}{3}\right),\\ F'_{BC} &= F_{BC}+\frac{4\pi}{3}Ne_Be_C\left(\frac{\varepsilon_\infty+2}{3}\right).\end{aligned}\right\} \tag{8}$$

Assuming that the dependence of the displacements upon time is harmonic and following the procedure of Chang and Mitra [6] one obtains a biquadratic form for the normal modes frequencies of the $AB_{1-x}C_x$ system:

$$\omega^4-(K_1+K_2)\,\omega^2+(K_1K_2-K_{12}K_{21})=0\,, \tag{9}$$

where

$$\left.\begin{aligned} K_1 &= (1-x)\frac{F'_{AB}}{m_A}+\frac{F'_{AB}}{m_B}+x\frac{F'_{BC}}{m_B},\\ K_2 &= x\frac{F'_{AC}}{m_A}+\frac{F'_{AC}}{m_C}+(1-x)\frac{F'_{BC}}{m_C},\\ K_{12} &= x\frac{F'_{AC}}{m_A}-x\frac{F'_{BC}}{m_B}\\ K_{21} &= (1-x)\frac{F'_{AB}}{m_A}-(1-x)\frac{F'_{BC}}{m_C}.\end{aligned}\right\} \tag{10}$$

The procedure described above was good for long-wavelength transverse optical phonons. The frequencies of the longitudinal optical phonons are obtained from the condition $\varepsilon = 0$ which leads to the conclusions

$$\boldsymbol{E} = -4\pi\boldsymbol{P}\,, \tag{11}$$

$$\boldsymbol{E}_{\text{eff}} = -\frac{8\pi}{3}\boldsymbol{P}\,. \tag{12}$$

Inserting (3), (5), (6), and (12) into (1) one gets the equations of motion similar to (7) with the following force constants:

$$\left.\begin{aligned} F''_{AB} &= F_{AB}+\frac{8\pi}{3}Ne_Ae_B\left(\frac{\varepsilon_\infty+2}{3\varepsilon_\infty}\right),\\ F''_{AC} &= F_{AC}+\frac{8\pi}{3}Ne_Ae_C\left(\frac{\varepsilon_\infty+2}{3\varepsilon_\infty}\right),\\ F''_{BC} &= F_{BC}-\frac{8\pi}{3}Ne_Be_C\left(\frac{\varepsilon_\infty+2}{3\varepsilon_\infty}\right).\end{aligned}\right\} \tag{13}$$

The frequencies of longitudinal optical phonons are obtained from (9) and (10) where the force constants F'_{AB}, F'_{AC}, F'_{BC} are replaced by the force constants F''_{AB}, F''_{AC}, F''_{BC}.

One assumes a linear dependence of N and ε_∞ upon composition x:

$$N = (1 - x) N_{AB} + xN_{AC} , \tag{14}$$

$$\varepsilon_\infty = (1 - x) \varepsilon_{\infty, AB} + x\varepsilon_{\infty, AC} . \tag{15}$$

The further assumption is that the charges e_B and e_C are Szigeti effective charges:

$$e_B = e^*_{AB} , \qquad e_C = e^*_{AC} . \tag{16}$$

It follows from (3) that

$$e_A = (1 - x) e^*_{AB} + xe^*_{AC} . \tag{17}$$

In order to find the parameters F_{AB0}, F_{AC0}, F_{BC0}, and θ one has to solve equation (9) for $x = 0$ and $x = 1$. One gets the following initial conditions:

$$\left.\begin{aligned}
&x = 0 ,\\
\omega^2_{TO, AB} &= \frac{F_{AB0}}{\mu_{AB}} - \frac{4\pi}{3} \frac{N_{AB} e^{*2}_{AB}}{\mu_{AB}} \left(\frac{\varepsilon_{\infty, AB} + 2}{3}\right),\\
\omega^2_{loc, AB} &= \frac{F_{AC0} + F_{BC0}}{m_C} ;\\
&x = 1 ,\\
\omega^2_{TO, AC} &= \frac{F_{AC0}(1 - \theta)}{\mu_{AC}} - \frac{4\pi}{3} \frac{N_{AC} e^{*2}_{AC}}{\mu_{AC}} \left(\frac{\varepsilon_{\infty, AC} + 2}{3}\right),\\
\omega^2_{gap, AC} &= \frac{(F_{AB0} + F_{BC0}) (1 - \theta)}{m_B} .
\end{aligned}\right\} \tag{18}$$

μ_{AB} and μ_{AC} are the reduced masses of crystals AB and AC, defined by the relations

$$\frac{1}{\mu_{AB}} = \frac{1}{m_A} + \frac{1}{m_B} ; \qquad \frac{1}{\mu_{AC}} = \frac{1}{m_A} + \frac{1}{m_C} . \tag{19}$$

$\omega_{loc, AB}$ denotes the local frequency of impurity C in AB crystal, $\omega_{gap, AC}$ is the gap mode of impurity B in AC crystal. If these frequencies are known, then from (18) one may find four unknown parameters.

Similar initial conditions are obtained for longitudinal optical phonons:

$$\left.\begin{aligned}
&x = 0 ,\\
\omega^2_{LO, AB} &= \frac{F_{AB0}}{\mu_{AB}} + \frac{8\pi}{3} \frac{N_{AB} e^{*2}_{AB}}{\mu_{AB}} \left(\frac{\varepsilon_{\infty, AB} + 2}{3\varepsilon_{\infty, AB}}\right),\\
\omega^2_{loc, AB} &= \frac{F_{AC0} + F_{BC0}}{m_C} ;\\
&x = 1 ,\\
\omega^2_{LO, AC} &= \frac{F_{AC0}(1 - \theta)}{\mu_{AC}} + \frac{8\pi}{3} \frac{N_{AC} e^{*2}_{AC}}{\mu_{AC}} \left(\frac{\varepsilon_{\infty, AC} + 2}{3\varepsilon_{\infty, AC}}\right),\\
\omega^2_{gap, AC} &= \frac{(F_{AB0} + F_{BC0}) (1 - \theta)}{m_B} .
\end{aligned}\right\} \tag{20}$$

In the $CdSe_xTe_{1-x}$ mixed crystals only three initial conditions of the type (18) and (20) are available, because in the literature there is no data about the frequency of the gap mode of Te in CdSe and one does not know, whether such mode exists at all. The dependence of the frequency of transverse and longitudinal phonons upon composition was found taking the constant θ as a parameter to fit the experimental data. The frequency of the local mode of Se in CdTe was taken from the theoretical paper by Gaur et al. [14]. The dielectric constants in CdTe and CdSe were determined from reflection measurements in the region around 600 cm^{-1}. The number of ion pairs N in CdTe and in CdSe was calculated from the densities of these compounds, which according to Cline and Stephens [18] are $\varrho_{CdTe} = \varrho_{CdSe} = 5.854\ gcm^{-3}$.

Using (18) and (20), the Szigeti effective charges in CdTe and CdSe were also determined:

$$e^*_{CdTe} = 0.70\ ; \qquad e^*_{CdSe} = 0.85\ .$$

These values agree with the results of Manabe et al. [13]. The parameters used in the calculations are compiled in Table 2.

Table 2

The values of parameters used in the calculations

	CdTe	CdSe
ω_{TO} (cm^{-1})	143	165
ω_{LO} (cm^{-1})	168	209
ω_{loc} (cm^{-1})	171.6	—
ε_∞	7.2	5.8
N (cm^{-3})	147×10^{20}	184×10^{20}
μ	59.8	46.4

The best fit of the model to experimental data is obtained for $\theta = -0.217$. The frequencies of LO and TO phonons calculated as described above are presented as full lines in Fig. 2. The agreement between the calculated data and the experiment is satisfactory.

5. Conclusions

The $CdSe_xTe_{1-x}$ mixed system exhibits two-mode behaviour. A fine structure in the reflection bands, which is due to the formation of ion clusters was not observed.

The experimentally determined dependence of the frequency of long-wavelength TO and LO phonons upon the composition x agrees well with the expectations of the REI model. This leads to the conclusion that the distribution of Se and Te ions in the sublattice of anions has random character.

No discontinuities of the $\omega(x)$ curves were observed in the transition from zincblende to wurtzite structure. This may be connected with the following facts:

1. the phase transition does not change the local order,
2. in the dynamics of the mixed system the interactions between the nearest neighbours play the decisive role.

Acknowledgements

We are very much indebted to Doc. Dr. W. Giriat and his co-workers who made this investigation possible by providing the samples of $CdSe_xTe_{1-x}$ mixed crystals.

References

[1] I. F. Chang and S. S. Mitra, Adv. Phys. **20**, 359 (1971).
[2] J. S. Langer, J. Math. Phys. **2**, 584 (1961).
[3] F. Matossi, J. chem. Phys. **19**, 161 (1951).
[4] H. W. Verleur and A. S. Barker, Phys. Rev. **149**, 715 (1966).
[5] Y. S. Chen, W. Shockley, and G. L. Pearson, Phys. Rev. **151**, 648 (1966).
[6] I. F. Chang and S. S. Mitra, Phys. Rev. **172**, 924 (1968).
[7] H. Harada and S. Narita, J. Phys. Soc. Japan **30**, 1628 (1971).
[8] H. W. Verleur and A. S. Barker, Phys. Rev. **155**, 750 (1967).
[9] E. A. Vinogradov, L. K. Vodopyanov, and G. S. Oleinik, Soviet Phys. — Solid State **15**, 322 (1973).
[10] M. Górska and W. Nazarewicz, phys. stat. sol. (b) **57**, K65 (1973).
[11] J. Litwin, phys. stat. sol. **5**, 551 (1964).
[12] A. D. Stuckes and G. Farrel, J. Phys. Chem. Solids **25**, 477 (1964).
[13] A. Manabe, A. Mitsuishi, and H. Yoshinaga, Japan. J. appl. Phys. **6**, 593 (1967).
[14] S. P. Gaur, J. F. Vetelino, and S. S. Mitra, J. Phys. Chem. Solids **32**, 2737 (1971).
[15] R. Beserman and M. Balkanski, J. Phys. Chem. Solids **31**, 355 (1970).
[16] R. Geick, C. H. Perry, and S. S. Mitra, J. appl. Phys. **37**, 1994 (1966).
[17] M. Balkanski, R. Beserman, and J. M. Besson, Solid State Commun. **4**, 20 (1966).
[18] C. F. Cline and D. R. Stephens, J. appl. Phys. **36**, 2869 (1965).

(Received March 22, 1974)

phys. stat. sol. (b) 65, 203 (1974)

Subject classification: 20.1; 22.5.2

Istituto di Fisica dell' Università, Pisa,
and Gruppo Nazionale di Struttura della Materia del CNR, Pisa (a),
and Istituto di Fisica dell' Università, Messina,
and Gruppo Nazionale di Struttura della Materia del CNR, Messina and Pisa (b)

A Note on the Polarization Dependence of Two-Photon Absorption Coefficient in Crystals

By

E. Doni (a), R. Girlanda (b), and G. Pastori Parravicini (a)

The expression of the two-photon absorption coefficient in crystals is discussed using explicitly the Wigner-Eckart theorem for finite symmetry groups. It is discussed how to separate the dynamical part, which contains appropriate reduced matrix elements of the momentum operator, from the geometrical part, which depends on the polarization vectors of both the radiation beams. For the geometrical part the standard results of the literature are recovered; detailed new information for the two-photon absorption rate is supplied by the dynamical part. To illustrate the procedure, the case of TlCl is considered as a specific example.

Nous traitons le facteur d'absorption optique pour transitions à deux photons dans les cristaux en appliquant explicitèment le théorème de Wigner-Eckart au groupe fini des operateurs de symmétrie ponctuelle. On étudie comme on peut séparer la partie dynamique de la partie géométrique, la première étant déterminée par les éléments de matrice réduits de l'operateur d'impulsion, la deuxième contenant les vecteurs de polarization des deux faisceaux incidents. Pour le facteur géométrique, nous retrouvons les résultats bien connus; le facteur dynamique fournit des détaillées nouvelles conaissances à propos du facteur d'absorption à deux photons. L'application de notre procedure pour le cristal de TlCl est enfin donnée comme un exemple.

1. Introduction

The two-photon absorption spectroscopy is a very powerful tool to investigate the band structure in solids (see, for instance, the review articles [1, 2] and references quoted therein). It is well known that a wealth of information is obtained by varying the polarization vectors of the radiation beams with respect to one another or with respect to the crystallographic axes (see, for instance, [3 to 5]). The study of the polarization dependence of two-photon absorption to a given final state was initiated by Inoue and Toyozawa [6], and extended by Bader and Gold [7]; finally, Denisov and Makarov [8, 2] included also the effect of time-reversal symmetry.

In this paper (Section 2) we show that the symmetry of the crystal determines not only the polarization dependence of a given final state, but also supplies very useful information on the strength of contribution with different symmetry. For this purpose, we use explicitly the Wigner-Eckart theorem applied to finite symmetry groups [9] and we discuss, besides geometrical factors, also dynamical factors. In Section 3, we illustrate the usefulness of the procedure discussing some experimental features of the two-photon polarization dependence in TlCl. Section 4 contains the conclusions.

2. Geometrical and Dynamical Factors in Two-Photon Absorption

2.1 General considerations

From second order perturbation theory [10], the absorption coefficient for photons of energy $\hbar\omega_1$ and polarization $\boldsymbol{e}_1$ in the presence of N_2 photons per unit volume of energy $\hbar\omega_2$ and polarization $\boldsymbol{e}_2$ is given by

$$\alpha(\omega_1) = C \sum_{f_m^\mu} \left| (1 + \mathcal{P}_{12}) \sum_{\varphi_n^\nu} \frac{\boldsymbol{e}_1 \cdot \langle f_m^\mu | \boldsymbol{P} | \varphi_n^\nu \rangle \langle \varphi_n^\nu | \boldsymbol{P} | 0 \rangle \cdot \boldsymbol{e}_2}{E_{\varphi_n^\nu} - E_0 - \hbar\omega_2} \right|^2 \times \\ \times \delta(E_{f^\mu} - E_0 - \hbar\omega_1 - \hbar\omega_2) , \tag{1}$$

where $C = (8\pi^3\hbar e^2 N_2)/(cm^4 n_1 n_2 \omega_1 \omega_2)$ and $|0\rangle$ is the ground state of the crystal (of energy E_0). In (1) $|\varphi_n^\nu\rangle$ indicates an intermediate state of the crystal (with energy E_{φ^ν}) belonging to the n-th row of the irreducible representation D^ν of the little group at $\boldsymbol{k} = 0$, and the sum over φ_n^ν runs over all intermediate states with any symmetry (ν, n). Similarly, $|f_m^\mu\rangle$ indicates a final state of the crystal (with energy E_{f^μ}), which transforms according to the m-th row of the irreducible representation D^μ at $\boldsymbol{k} = 0$, and the sum over f_m^μ includes all final states with any symmetry (μ, m). The total momentum operator $\boldsymbol{P}$ is given by $\boldsymbol{P} = \sum_i \boldsymbol{p}_i$, where $\boldsymbol{p}_i$ is the momentum operator for the i-th electron. $\mathcal{P}_{12}$ interchanges $\boldsymbol{e}_1$ with $\boldsymbol{e}_2$ and ω_1 with ω_2. We notice explicitly that the only representations D^ν, D^μ of interest are those of the single group at $\boldsymbol{k} = 0$, which coincide with the representations of the single point group of the crystal.

The standard way [6 to 8] to handle with expression (1) is to define the quantities

$$\Lambda(\omega_i) = \sum_{\varphi_n^\nu} \frac{|\varphi_n^\nu\rangle \langle \varphi_n^\nu|}{E_{\varphi^\nu} - E_0 - \hbar\omega_i}, \qquad (i = 1, 2) \tag{2a}$$

$$\Lambda_\pm = \Lambda(\omega_2) \pm \Lambda(\omega_1) , \tag{2b}$$

containing contributions of all intermediate states, and to determine their symmetry properties. In this way it is, however, difficult to evaluate the strengths with which the various angular dependence functions appear in the absorption coefficient. In fact, in a number of actual situations only a few intermediate states do contribute significantly to the absorption rate. For instance, in alkali halides a two-band model [11] is sufficient to explain two-photon spectroscopy near the absorption edge; as another example, in CuCl it is enough to consider a very limited number of intermediate states [12, 13]. For this reason we handle expressions (2) with a different procedure which is well suited to include explicitly the intermediate states of interest; of course, if all intermediate states are included we recover the standard results of the literature [2, 6 to 8].

2.2 Case of cubic crystals (points group $\mathrm{T}, \mathrm{T_h}, \mathrm{T_d}, \mathrm{O}, \mathrm{O_h}$*)*

For simplicity, we consider first the case of cubic crystals. We use equation (A4) of the Appendix, and write

$$\boldsymbol{e}_1 \cdot \boldsymbol{P} = \sum_l e_{1l}^{\lambda *} P_l^\lambda , \tag{3}$$

where D^λ is the irreducible representation (at $\boldsymbol{k} = 0$) to which the vectorial operators $\boldsymbol{P}$ and $\boldsymbol{e}_1$ belong. Inserting (3) and the similar equation for $\boldsymbol{e}_2 \cdot \boldsymbol{P}$ into

(1), we obtain

$$\alpha(\omega_1) = C \sum_{f^\mu} \left[\sum_m \left| (1 + \mathcal{P}_{12}) \sum_{ll'\varphi_n^\nu} e_{2l'}^{\lambda *} \langle f_m^\mu | P_{l'}^\lambda | \varphi_n^\nu \rangle \times \right.\right.$$

$$\left.\left. \times \langle \varphi_n^\nu | P_l^\lambda | 0 \rangle e_{1l}^{\lambda *} \Lambda_{\varphi^\nu}(\omega_1) \right|^2 \right] \delta(E_{f^\mu} - E_0 - \hbar\omega_1 - \hbar\omega_2), \quad (4)$$

where

$$\Lambda_{\varphi^\nu}(\omega_1) = \frac{1}{E_{\varphi^\nu} - E_0 - \hbar\omega_1}. \quad (5)$$

In order to simplify expression (4), we use the Wigner-Eckart theorem, in a form suitable for finite symmetry groups (see, for instance, [9, 14]). The cubic groups T_d, O, O_h are simply reducible, i.e. [15]:

(i) the product of two irreducible representations of a group contains each irreducible representation no more than once.

(ii) No additional degeneracy is introduced by time-reversal symmetry.

For these groups, we immediately have

$$\langle \varphi_n^\nu | P_l^\lambda | 0 \rangle = \delta_{\nu\lambda} \delta_{nl} \langle \varphi^\nu || P^\lambda || 0 \rangle, \quad (6a)$$

$$\langle f_m^\mu | P_{l'}^\lambda | \varphi_n^\nu \rangle = (\mu m | \lambda l', \nu n) \langle f^\mu || P^\lambda || \varphi^\nu \rangle, \quad (6b)$$

where standard notations have been used to indicate the reduced matrix elements and Clebsch-Gordan coefficients [15]. The discussion of the other cubic groups T and T_h will be given at the end of this section. Since in all our treatment the only irreducible representations of interest are those at $\boldsymbol{k} = 0$, the Clebsch-Gordan coefficients of interest coincide with those of the crystal point groups, which are all tabulated in [14]. Notice that only single groups are involved. Using (6), we obtain for (4)

$$\alpha(\omega_1) = C \sum_{f^\mu} \left[\sum_m \left| \sum_{\varphi^\lambda} \langle f^\mu || P^\lambda || \varphi^\lambda \rangle \langle \varphi^\lambda || P^\lambda || 0 \rangle \times \right.\right.$$

$$\left.\left. \times \sum_{ll'} (1 + \mathcal{P}_{12}) \Lambda_{\varphi^\lambda}(\omega_1) e_{2l'}^{\lambda *} e_{1l}^{\lambda *} (\mu m | \lambda l', \lambda l) \right|^2 \right] \times$$

$$\times \delta(E_{f^\mu} - E_0 - \hbar\omega_1 - \hbar\omega_2). \quad (7)$$

We take advantage of the following properties of Clebsch-Gordan coefficients [15]:

$$(\mu m | \lambda l', \lambda l) = \pm (\mu m | \lambda l, \lambda l'), \quad (8)$$

where the sign + holds in the case D^μ appears in the decomposition of the symmetric product representation $[D^\lambda \times D^\lambda]$ and the sign − holds in the case D^μ belongs to the antisymmetric product representation $\{D^\lambda \times D^\lambda\}$. Using (8) and the definitions

$$\Lambda_{\varphi^\lambda}^{\pm} = \Lambda_{\varphi^\lambda}(\omega_1) \pm \Lambda_{\varphi^\lambda}(\omega_2), \quad (9)$$

we obtain the final result

$$\alpha(\omega_1) = C \sum_{f^\mu} \left[G_\mu(\boldsymbol{e}_1, \boldsymbol{e}_2) \left| \sum_{\varphi\lambda} \Lambda^{\pm}_{\varphi\lambda} \langle f^\mu || P^\lambda || \varphi^\lambda \rangle \langle \varphi^\lambda || P^\lambda || 0 \rangle \right|^2 \right] \times \\ \times \delta(E_{f^\mu} - E_0 - \hbar\omega_1 - \hbar\omega_2) , \tag{10}$$

where

$$G_\mu(\boldsymbol{e}_1, \boldsymbol{e}_2) = \sum_m \left| \sum_{ll'} e^{\lambda *}_{2l} e^{\lambda *}_{1l'} \, (\mu m | \lambda l, \lambda l') \right|^2 . \tag{11}$$

In (10) the sign + (−) holds if D^μ belongs to the symmetric (antisymmetric) part of $D^\lambda \times D^\lambda$.

We consider now the case of the cubic group T_h which is not simply reducible because of violation of both conditions (i) and (ii). The vector $\boldsymbol{P}$ belongs to the irreducible representation Γ_4^- and the product $\Gamma_4^- \times \Gamma_4^-$ contains twice the representation Γ_4^+, thus violating condition (i) [14]. We notice however that the symmetrized and antisymmetrized products $[\Gamma_4^- \times \Gamma_4^-]$ and $\{\Gamma_4^- \times \Gamma_4^-\}$ both contain Γ_4^+ only once. Moreover the relation (8) for Clebsch-Gordan coefficients still holds [14]. The same treatment as before leads to (10) and (11), where now one must bear in mind that Γ_4^+ belongs both to $[\Gamma_4^- \times \Gamma_4^-]$ and to $\{\Gamma_4^- \times \Gamma_4^-\}$ which must be summed up to obtain the total contribution to a final Γ_4^+ state. Finally, we notice that the violation of condition (ii) is trivially overcome by noting that (8) still holds and by summing up (10) with respect to the irreducible representations degenerate by time-reversal symmetry. Similar considerations apply to group T.

Equations (10) and (11) show that in the case of cubic crystals it is always possible to separate the dynamical part, which is expressed in terms of reduced matrix elements of the momentum operator, from the geometrical part, which depends on the polarization vectors of the two radiation beams. Equation (11) determines the polarization dependence for a given final state, while (10) supplies very useful information on the strength of contributions with different symmetry. Notice that these strengths are determined not only by intrinsic crystal properties (i.e. reduced matrix elements involving the actual free electron-hole or exciton states), but also by the factors $\Lambda^+_{\varphi\lambda}$ and $\Lambda^-_{\varphi\lambda}$. From (9) we see that it is possible in most practical situations to choose the frequency of the laser in such a way to make $\Lambda^-_{\varphi\lambda}$ negligible with respect to $\Lambda^+_{\varphi\lambda}$ thus obtaining additional information to identify symmetry properties. This fact was pointed out in the discussion of two-photon absorption to Landau levels in papers [16].

2.3 Case of non-cubic crystals

Our treatment of Section 2.2 for cubic crystals can be easily extended to any type of symmetry combining the procedure of Section 2.2 with the results of the Appendix. In the case of non-cubic crystals and for arbitrary polarization vectors the final expression of the two-photon absorption coefficient is a little more complicated than (10), because the more general expression (A3) must be used instead of (A4). Moreover, also in this case one can obtain very useful information on the strengths of contributions with different symmetry. A few more comments are briefly outlined in the following.

2.3.1 Point groups C_1, C_i, C_2, C_s, C_{2h}, D_2, C_{2v}, D_{2h}

These groups are simply reducible groups. The only generalization needed with respect to the treatment of Section 2.2 is to replace (A4) with (A3).

2.3.2 Point groups D_4, C_{4v}, D_{2d}, D_{4h}, D_3, D_{3v}, D_{3d}, D_6, C_{6v}, D_{3h}, D_{6h}

These groups are also simply reducible groups. The only generalization needed with respect to the treatment of Section 2.2 is to replace (A4) with (A3). An interesting particular case is obtained when the polarization vectors $\boldsymbol{e}_1$ and $\boldsymbol{e}_2$ are both perpendicular to the anisotropy axis. In this case the simpler expression (A4) holds instead of (A3) and results similar to those of Section 2.2 are recovered.

2.3.3 Point groups C_4, S_4, C_{4h}, C_3, C_{3i}, C_6, C_{3h}

These groups are not simply reducible groups, since condition (ii) is violated. To deal with these groups, besides substitution of (A4) with (A3), we must keep in mind that in the decomposition (A2) there are two irreducible representations which are degenerate by time-reversal symmetry. We finally consider these two irreducible representations as forming a unique representation, whose characters are the sum of the characters of the two component representations. It is now possible to consider symmetrized and antisymmetrized products of this new representation and apply the same considerations as in the case of groups T and T_h.

3. An Example: Angular Dependence of Two-Photon Absorption in TlCl

To illustrate the results of the previous section, we consider the case of TlCl. The point group of TlCl is O_h, and we use the notations of [14] to label the irreducible representations. The momentum operator $\boldsymbol{P}$ belongs to the irreducible representation Γ_4^-. The direct product $\Gamma_4^- \times \Gamma_4^-$ gives

$$\Gamma_4^- \times \Gamma_4^- = \Gamma_1^+ + \Gamma_3^+ + \Gamma_4^+ + \Gamma_5^+ \,, \tag{12a}$$

while the symmetrized and antisymmetrized products are:

$$[\Gamma_4^- \times \Gamma_4^-] = \Gamma_1^+ + \Gamma_3^+ + \Gamma_5^+ \,, \tag{12b}$$

$$\{\Gamma_4^- \times \Gamma_4^-\} = \Gamma_4^+ \,. \tag{12c}$$

For convenience the Clebsch-Gordan coefficients concerning the decomposition (12a) are reported in Table 1.

Let us now calculate the geometrical factors $G_\mu(\boldsymbol{e}_1, \boldsymbol{e}_2)$ defined in equation (11). Using Table 1, we obtain

$$G_{\Gamma_1^+}(\boldsymbol{e}_1, \boldsymbol{e}_2) = \left[\tfrac{1}{\sqrt{3}}\, e_{1x}e_{2x} + \tfrac{1}{\sqrt{3}}\, e_{1y}e_{2y} + \tfrac{1}{\sqrt{3}}\, e_{1z}e_{2z}\right]^2 = \tfrac{1}{3}\, |\boldsymbol{e}_1 \cdot \boldsymbol{e}_2|^2 \,, \tag{13a}$$

$$G_{\Gamma_3^+}(\boldsymbol{e}_1, \boldsymbol{e}_2) = \tfrac{2}{3}\,(e_{1x}^2e_{2x}^2 + e_{1y}^2e_{2y}^2 + e_{1z}^2e_{2z}^2) - \tfrac{2}{3}\,(e_{1x}e_{1y}e_{2x}e_{2y} + e_{1y}e_{1z}e_{2y}e_{2z} + e_{1z}e_{1x}e_{2z}e_{2x}) \,, \tag{13b}$$

$$G_{\Gamma_4^+}(\boldsymbol{e}_1, \boldsymbol{e}_2) = \tfrac{1}{2}\, |\boldsymbol{e}_1 \times \boldsymbol{e}_2|^2 \,, \tag{13c}$$

$$G_{\Gamma_5^+}(\boldsymbol{e}_1, \boldsymbol{e}_2) = \tfrac{1}{2}\,[(e_{1x}e_{2y} + e_{1y}e_{2x})^2 + (e_{1y}e_{2z} + e_{1z}e_{2y})^2 + (e_{1z}e_{2x} + e_{1x}e_{2z})^2] \,. \tag{13d}$$

Using the notations of Bader and Gold [7] we can express:

$$G_{\Gamma_1^+}(\boldsymbol{e}_1, \boldsymbol{e}_2) = \tfrac{1}{3}\, A_1(\boldsymbol{e}_1, \boldsymbol{e}_2) \,; \quad G_{\Gamma_3^+}(\boldsymbol{e}_1, \boldsymbol{e}_2) = \tfrac{2}{3}\, E(\boldsymbol{e}_1, \boldsymbol{e}_2) \,,$$

$$G_{\Gamma_4^+}(\boldsymbol{e}_1, \boldsymbol{e}_2) = \tfrac{1}{2}\, T_1(\boldsymbol{e}_1, \boldsymbol{e}_2) \,; \quad G_{\Gamma^+}(\boldsymbol{e}_1, \boldsymbol{e}_2) = \tfrac{1}{2}\, T_2(\boldsymbol{e}_1, \boldsymbol{e}_2) \,.$$

Table 1

Clebsch-Gordan coefficients for the decomposition of the product representation $\Gamma_4^- \times \Gamma_4^-$ of the group O_h (after reference [14]). u_i $(i = x, y, z)$ and v_i $(i = x, y, z)$ denote two independent sets of functions transforming as the i-th row of Γ_4^-

	$u_x v_x$	$u_x v_y$	$u_x v_z$	$u_y v_x$	$u_y v_y$	$u_y v_z$	$u_z v_x$	$u_z v_y$	$u_z v_z$
Γ_1^+	$1/\sqrt{3}$	0	0	0	$1/\sqrt{3}$	0	0	0	$1/\sqrt{3}$
Γ_3^+	$-1/\sqrt{6}$	0	0	0	$-1/\sqrt{6}$	0	0	0	$2/\sqrt{6}$
	$1/\sqrt{2}$	0	0	0	$-1/\sqrt{2}$	0	0	0	0
Γ_4^+	0	0	0	0	0	$1/\sqrt{2}$	0	$-1/\sqrt{2}$	0
	0	0	$-1/\sqrt{2}$	0	0	0	$1/\sqrt{2}$	0	0
	0	$1/\sqrt{2}$	0	$-1/\sqrt{2}$	0	0	0	0	0
Γ_5^+	0	0	0	0	0	$1/\sqrt{2}$	0	$1/\sqrt{2}$	0
	0	0	$1/\sqrt{2}$	0	0	0	$1/\sqrt{2}$	0	0
	0	$1/\sqrt{2}$	0	$1/\sqrt{2}$	0	0	0	0	0

We now discuss briefly the band structure of TlCl in order to get information of the reduced matrix elements appearing in (10). The band structure calculations [17] show that the maxima of the valence band occur at the points $\boldsymbol{k}_X = 2\pi/a\,(1, 0, 0)$, $\boldsymbol{k}_Y = 2\pi/a\,(0, 1, 0)$, $\boldsymbol{k}_Z = 2\pi/a\,(0, 0, 1)$ with symmetry X_1^+ for the small group D_{4h} of $\boldsymbol{k}_X$ (s-like symmetry). The minima of the conduction bands occur at the same points with symmetry X_2^- for the small group of $\boldsymbol{k}_X$ (x-like symmetry). Since band-to-band transitions at the fundamental edge are allowed in the dipole approximation and since the inversion is a symmetry operation of the small group of $\boldsymbol{k}_X$ (or $\boldsymbol{k}_Y$ or $\boldsymbol{k}_Z$), two-photon transitions near the absorption edge are possible through the two-band model considered by Mahan [11]. We are thus left to consider transitions to discrete or continuum exciton states with p-like envelope functions. The p-like excitons involving valence and conduction states at the extrema $\boldsymbol{k}_X$, $\boldsymbol{k}_Y$, $\boldsymbol{k}_Z$ constitute a basis for a representation at $\boldsymbol{k} = 0$, which can be decomposed into $\Gamma_1^+, \Gamma_3^+, \Gamma_4^+, \Gamma_5^+$ (while the s-like excitons involving the same extrema belong to Γ_4^-). This can be easily proved either using the projection operators technique (see, for instance, the paper [18] for the treatment of a similar situation) or applying the even simpler procedure of [19]. Furthermore, the reduced matrix elements involving s-like excitons, the momentum operator $\boldsymbol{P}$ and the final p-like excitons are the same regardless of the nature of the final state, as can be easily verified by carrying out the decomposition of the final states and using Table 1. The polarization dependence of the two-photon absorption coefficient becomes for our case

$$\alpha(\omega_1) \propto |\Lambda^+|^2\,[G_{\Gamma_1^+}(\boldsymbol{e}_1, \boldsymbol{e}_2) + G_{\Gamma_3^+}(\boldsymbol{e}_1, \boldsymbol{e}_2) + G_{\Gamma_5^+}(\boldsymbol{e}_1, \boldsymbol{e}_2)] + |\Lambda^-|^2\,G_{\Gamma_4^+}(\boldsymbol{e}_1, \boldsymbol{e}_2)\,. \quad (14)$$

Notice the detailed information contained in (14) with respect to the less stringent expression

$$\alpha(\omega_1) \propto a_1 A_1(\boldsymbol{e}_1, \boldsymbol{e}_2) + a_2 E(\boldsymbol{e}_1, \boldsymbol{e}_2) + a_3 T_1(\boldsymbol{e}_1, \boldsymbol{e}_2) + a_4 T_2(\boldsymbol{e}_1, \boldsymbol{e}_2)\,, \quad (15)$$

with a_i ($i = 1, 2, 3, 4$) arbitrary constants used by Bader and Gold [18].

By means of (13), we now discuss the experimental results of Matsuoka and Yajima [4, 20] on the angular dependence of two-photon absorption in TlCl. Two particularly convenient geometrical configurations of light beams were chosen [4]. In the parallel configuration we have

$$\boldsymbol{e}_1 \equiv (\cos\theta, \sin\theta, 0)\ ; \qquad \boldsymbol{e}_2 \equiv (1, 0, 0)\ .$$

In the $\pi/4$ configuration we have

$$\boldsymbol{e}_1 \equiv (\cos\theta, \sin\theta, 0)\ ; \qquad \boldsymbol{e}_2 \equiv \left(\tfrac{1}{\sqrt{2}}, \tfrac{1}{\sqrt{2}}, 0\right)\ .$$

For the parallel configuration, equations (13) give

$$G_{\Gamma_1^+}(\boldsymbol{e}_1, \boldsymbol{e}_2) = \tfrac{1}{3}\cos^2\theta\ ; \qquad G_{\Gamma_3^+}(\boldsymbol{e}_1, \boldsymbol{e}_2) = \tfrac{2}{3}\cos^2\theta\ ;$$

$$G_{\Gamma_4^+}(\boldsymbol{e}_1, \boldsymbol{e}_2) = \tfrac{1}{2}\sin^2\theta\ ; \qquad G_{\Gamma_5^+}(\boldsymbol{e}_1, \boldsymbol{e}_2) = \tfrac{1}{2}\sin^2\theta\ .$$

For the $\pi/4$ configuration equations (13) give:

$$G_{\Gamma_1^+}(\boldsymbol{e}_1, \boldsymbol{e}_2) = \tfrac{1}{6}(1 + \sin 2\theta)\ ; \qquad G_{\Gamma_3^+}(\boldsymbol{e}_1, \boldsymbol{e}_2) = \tfrac{1}{6}(2 - \sin 2\theta)\ ;$$

$$G_{\Gamma_4^+}(\boldsymbol{e}_1, \boldsymbol{e}_2) = \tfrac{1}{2}(1 - \sin 2\theta)\ ; \qquad G_{\Gamma_5^+}(\boldsymbol{e}_1, \boldsymbol{e}_2) = \tfrac{1}{2}(1 + \sin 2\theta)\ .$$

Expression (14) for parallel and for $\pi/4$ configuration becomes, respectively,

$$\alpha_{||}(\omega_1) \propto \tfrac{1}{2}|\Lambda^+|^2(1 + \cos^2\theta) + \tfrac{1}{2}|\Lambda^-|^2\sin^2\theta\ , \tag{16}$$

$$\alpha_{\pi/4}(\omega_1) \propto \tfrac{1}{4}|\Lambda^+|^2(3 + \sin 2\theta) + \tfrac{1}{4}|\Lambda^-|^2(1 - \sin 2\theta)\ . \tag{17}$$

In the experimental situation of the paper [4], we have $|\Lambda^+|^2 \approx 10|\Lambda^-|^2$. From Fig. 3 of paper [4] we see that the experimental angular dependence of $\alpha_{||}$ and $\alpha_{\pi/4}$ (at least up to ≈ 3.80 eV) is reasonably well described by (16) and (17) as far as positions of maxima and minima are concerned. From (16) and (17) we see that the maximum and the minimum values of $\alpha_{||}$ and $\alpha_{\pi/4}$ are the same. This "consistency relation" which is more stringent than the similar consistency relation discussed theoretically in [18] and [5] and experimentally in [5], is clearly violated by Matsuoka's experiments.

Detailed results and symmetry assignments of two-photon transitions in TlCl have been performed by Fröhlich et al. [5]. Measurements with linearly and circularly polarized light were used to decompose the two-photon absorption coefficient into its different components. The coefficients a_i of equation (15) were thus determined. With our more stringent equation (14) we have

$$a_{\Gamma_1^+} \propto \tfrac{1}{3}|\Lambda^+|^2\ , \quad a_{\Gamma_3^+} \propto \tfrac{2}{3}|\Lambda^+|^2\ , \quad a_{\Gamma_4^+} \propto \tfrac{1}{2}|\Lambda^-|^2\ , \quad a_{\Gamma_5^+} \propto \tfrac{1}{2}|\Lambda^+|^2\ , \tag{18}$$

with $|\Lambda^+|^2 \approx 10\,|\Lambda^-|^2$ in the experimental situation of paper [5]. Qualitatively, our results (18) are in good agreement with experimental results (Fig. 3 of paper [5]), which show definitely that the strength of Γ_4^+ symmetry is much smaller than the ones of Γ_1^+, Γ_3^+, Γ_5^+ symmetry. From a quantitative point of view the experimental strength of the contributions of Γ_1^+ and Γ_4^+ symmetry seems to be too high compared with what expected from the two-band model (i.e. equation (18)). At this stage, however, it is not clear if experiments of paper [5] are accurate enough to indicate contributions besides the Mahan model [11].

4. Conclusions

The main results of this paper can be summarized as follows:

1. A new very simple procedure is given to study the polarization dependence in two-photon spectroscopy. As a result of this procedure one obtains, besides the standard geometrical factors, very useful information on the strength of the various contributions of transitions to final states with different symmetry.

2. The application of the procedure requires the knowledge of Clebsch-Gordan coefficients for the single point groups only. Our treatment clearly shows that the possible angular dependence functions of two-photon absorption are uniquely determined by the point symmetry group of the crystal (and not, for instance, by the position of band extrema in the Brillouin zone). The detailed band structure determines the relative importance of the different components.

3. The case of TlCl has been discussed in detail and some experimental features of two-photon absorption have been interpreted.

Acknowledgement

It is a pleasure to thank Prof. F. Bassani for many stimulating discussions.

Appendix

The result of this Appendix allows us to write the inner product $\boldsymbol{e} \cdot \boldsymbol{P}$ in a form which is very suitable for application in the main text.

Let P_x, P_y, P_z be the components of the vectorial operator $\boldsymbol{P}$. Under the operations of the point symmetry group $G^0(R)$ (of order g), P_x, P_y, P_z transform as rows of an appropriate representation $D(R)$ (in general reducible) defined by

$$RP_iR^{-1} = \sum_j D_{ji}(R)\, P_j \qquad (i, j = x, y, z)\,. \tag{A1}$$

Let $D^\lambda_{ll'}(R)$ $(l, l' = 1, \ldots, d_\lambda)$ be the irreducible representations, of order d_λ, contained in $D(R)$ (the same irreducible representation can appear more than once). We indicate by $M_{\lambda l, i}$ $(i = x, y, z)$ a unitary matrix which transform $D(R)$ into block form and by P^λ_l the transformed operators

$$P^\lambda_l = \sum_i M_{\lambda l, i} P_i\,, \tag{A2}$$

belonging to the l-th row of the irreducible representation $D^\lambda(R)$. The same considerations apply also to the vectorial operator $\boldsymbol{e}$.

We now use the invariance property of the inner product $\boldsymbol{e} \cdot \boldsymbol{P}$ under the operations $R \in G^0(R)$, and write

$$\begin{aligned} \boldsymbol{e} \cdot \boldsymbol{P} &= \sum_{i=x,y,z} e_i^* P_i \\ &= \frac{1}{g} \sum_R \sum_{i=x,y,z} R e_i^*\, R^{-1} R P_i R^{-1}\,. \end{aligned}$$

Using (A1) we have

$$\begin{aligned} \boldsymbol{e} \cdot \boldsymbol{P} &= \frac{1}{g} \sum_R \sum_{i,l,j=x,y,z} D_{ji}(R)^*\, D_{li}(R)\, e_j^* P_l \\ &= \frac{1}{g} \sum_R \sum_{\lambda,\mu,l,l',m,m'} D^\lambda_{ll'}(R)^*\, D^\mu_{mm'}(R)\, e^{\lambda *}_l P^\mu_m \delta_{l'm'}\,, \end{aligned}$$

where the last step implies use of (A2) and $\delta_{l'm'}$ appears because of unitarity of $M_{\lambda l, i}$. Using the orthogonality relation

$$\sum_R D^{\lambda}_{ll'}(R)^* D^{\mu}_{mm'}(R) = \frac{g}{d_\lambda} \delta_{\lambda\mu}\delta_{lm}\delta_{l'm'}$$

we finally obtain

$$\boldsymbol{e} \cdot \boldsymbol{P} = \sum_{\lambda l} e^{\lambda *}_l P^{\lambda}_l . \tag{A3}$$

We notice explicitely that, in the case of cubic crystals, the operators P_x, P_y, P_z span an irreducible space. In this case (A3) takes the simpler form

$$\boldsymbol{e} \cdot \boldsymbol{P} = \sum_{l=1,2,3} e^{\lambda *}_l P^{\lambda}_l , \tag{A4}$$

which has been used in the main text.

References

[1] J. M. Worlock, Laser Handbook, Vol. 2, Ed. F. T. Arecchi and E. O. Schulz-Dubois, North-Holland Publ. Co., Amsterdam 1972 (p. 1323).
F. Bassani, International School of Physics "E. Fermi", course LII, Academic Press, Inc., London/New York 1972.
[2] M. M. Denisov and V. P. Makarov, phys. stat. sol. (b) **56**, 9 (1973).
[3] D. Fröhlich, B. Staginnus, and E. Schönherr, Phys. Rev. Letters **19**, 1032 (1967).
D. Fröhlich and B. Staginnus, Phys. Rev. Letters **19**, 496 (1967).
D. Fröhlich, B. Staginnus, and T. Caps, Solid State Commun. **6**, 173 (1968).
K. Park and R. G. Stafford, Phys. Rev. Letters **22**, 1426 (1969).
H. S. Waff and K. Park, Phys. Letters **A32**, 109 (1970).
F. Pradere and A. Mysyrowicz, Proc. X. Internat. Conf. Phys. Semicond., Cambridge (Mass.) 1970, USAEC, Oak Ridge 1970 (p. 101); Proc. XI. Internat. Conf. Phys. Semicond., Warsaw 1972, Elsevier, Amsterdam 1972 (p. 1363).
[4] M. Matsuoka, J. Phys. Soc. Japan **23**, 1028 (1967).
[5] D. Fröhlich, B. Staginnus, and S. Thurm, phys. stat. sol. **40**, 287 (1970).
[6] M. Inoue and Y. Toyozawa, J. Phys. Soc. Japan **20**, 363 (1965).
[7] T. R. Bader and A. Gold, Phys. Rev. **171**, 997 (1968).
[8] M. M. Denisov and V. P. Makarov, J. Phys. C **5**, 2651 (1972).
[9] G. F. Koster, Phys. Rev. **109**, 227 (1958).
[10] M. Göppert-Mayer, Ann. Phys. (Germany) **9**, 273 (1931).
[11] G. D. Mahan, Phys. Rev. Letters **20**, 332 (1968); Phys. Rev. **170**, 825 (1968).
[12] A. Bivas, C. Marange, J. B. Grun, and C. Schwab, Opt. Commun. **6**, 142 (1973).
[13] E. Doni, R. Girlanda, and G. Pastori Parravicini, Solid State Commun. **14**, 873 (1974).
[14] G. F. Koster, J. O. Dimmock, R. J. Wheeler, and H. Statz, Properties of the Thirty-Two Point Groups, MIT Press, Cambridge (Mass.) 1963.
[15] M. Hamermesh, Group Theory and Its Application to Physical Problems, Addison-Wesley, Reading (Mass.) 1962.
[16] F. Bassani and R. Girlanda, Opt. Commun. **1**, 359 (1970).
R. Girlanda, Nuovo Cimento B **6**, 53 (1971).
[17] H. Overhof and J. Treusch, Solid State Commun. **9**, 53 (1971).
M. Inoue and M. Okazaki, J. Phys. Soc. Japan **31**, 1313 (1971).
[18] T. R. Bader and A. Gold, Phys. Rev. **181**, 1369 (1969).
[19] E. Doni and G. Pastori Parravicini, J. Phys. C **6**, 2859 (1973); **7**, 1786 (1974).
[20] M. Matsuoka and T. Yajima, Phys. Letters (Netherlands) **23**, 54 (1966).

(Received May 6, 1974)

phys. stat. sol. (b) **65**, 213 (1974)

Subject classification: 14.3.4 and 20.1; 3; 13.1; 22.2.4

Physics Department, University of Ottawa

Electroreflectance Spectra of $Al_xIn_{1-x}Sb$ Alloys

By

S. Isomura[1]), F. G. D. Prat, and J. C. Woolley

The preparation of homogeneous samples of $Al_xIn_{1-x}Sb$ alloys from a stoichiometric melt is considered. A slow horizontal Bridgman method with ingots encapsulated in liquid B_2O_3 has been used to provide samples suitable for optical reflectance measurements and details of the variation of composition with position along the ingot are given. Electroreflectance measurements, using the electrolyte technique, have been made on samples cut from cross sectional slices, and the variation of the transition energy with composition has been determined for the E_0, $E_0 + \Delta_0$, E_1 and $E_1 + \Delta_1$ peaks. The experimental values of E_0, Δ_0 and Δ_1 as a function of x are compared with the previously proposed theoretical equations based on the concept of interband mixing due to disorder of the lattice, and the significance of some of the energy parameters in these equations is discussed.

Es wird die Herstellung von homogenen $Al_xIn_{1-x}Sb$-Legierungsproben aus der stöchiometrischen Schmelze betrachtet. Eine langsame horizontale Bridgmanmethode, bei der der Barren in flüssigem B_2O_3 eingeschlossen ist, wird zur Herstellung von für optische Reflexionsmessungen geeigneten Proben benutzt und Einzelheiten der Änderung der Zusammensetzung mit dem Ort längs des Barrens werden angegeben. Mit der Elektrolyttechnik werden Elektroreflexionsmessungen an Proben aus Querschnittsscheiben durchgeführt. Die Änderung der Energie der Übergänge mit der Zusammensetzung wurde für die E_0-, $E_0 + \Delta_0$-, E_1- und $E_1 + \Delta_1$-Maxima bestimmt. Die experimentellen Werte für E_0, Δ_0 und Δ_1 in Abhängigkeit von x werden mit kürzlich vorgeschlagenen theoretischen Gleichungen, die auf dem Konzept der Interbandmischung infolge der Gitterfehlordnung basieren, verglichen und es wird die Bedeutung einiger Energieparameter in diesen Gleichungen diskutiert.

1. Introduction

Little work has been carried out so far on alloys of the system $Al_xIn_{1-x}Sb$. The existence of a large range of solid solution was shown by various workers [1, 2] and complete solid solution over the whole composition range was confirmed by Baranov and Goryunova [3], who prepared homogeneous alloy samples by annealing material which had initially been rapidly quenched from the melt. Such samples were used by Agaev and Nasledov [4] for measurements of some electrical properties. Later Agaev and Bekmedova [5] made measurements of optical absorption and reflectance to determine energy gap values for a number of alloys of this system, the method of preparation in this case not being indicated.

As has been shown in various work [6], it is difficult to obtain good homogeneity in III–V alloys involving arsenic or antimony by means of an annealing method and various directional freezing techniques are to be preferred. In the present work, preparation of $Al_xIn_{1-x}Sb$ samples by a step-freeze (horizontal Bridgman) method has been investigated. Suitable specimens cut from the

[1]) Permanent address: National Research Institute for Metals, Tokyo 153, Japan.

resulting ingots have then been used for measurements of electroreflectance spectra.

2. Preparation and Metallurgical Data

Two main problems in the preparation of these alloys by a stoichiometric melt method are firstly the reaction of AlSb with quartz etc. at the temperatures to be used and secondly the loss of aluminium when the alloy is held for long periods in the liquid state. To overcome these difficulties, here the alloy was contained in a high purity graphite boat and was effectively encapsulated by covering with a layer of liquid B_2O_3.

The ingots produced were about 28 g in weight, 10 cm long and had a roughly oval cross-section with maximum and minimum dimensions of about 1 and 0.5 cm respectively. From a consideration of the phase diagram [6], the mean composition of each ingot was chosen to be $x = 0.2$, as this would give a liquidus temperature below 1000 °C but would still yield a suitable composition range in the finished ingot. For each ingot, the initial charge was made up from appropriate weights of the three elements, all of at least 5 9's purity. This charge was placed in a graphite boat without B_2O_3 and sealed in a quartz ampoule back filled with argon. The charge was heated to 1000 °C long enough for complete melting to occur and was then cooled. This initial step was necessary to produce in the boat a continuous ingot over which the B_2O_3 could be placed. If the B_2O_3 were added to the initial charge, on melting the charge tended to break up into short ingots separated by layers of B_2O_3 and was unsuitable for the directional freezing process. The single ingot obtained from the initial melt was covered with pieces of degassed transparent B_2O_3 and resealed in a quartz ampoule. This ampoule was then placed in the high temperature region of the step furnace and held above the liquidus temperature of the alloy for a day or two to homogenize. At the same time, a smooth encapsulating layer of liquid B_2O_3 was formed.

The step furnace consisted of two sections, a hot zone and a cool zone, separated by a water cooled jacket to permit a steep temperature gradient to be established. This system has been described in detail previously [7]. Here, the hot zone was held at 980 °C and the cool zone at 470 °C. The ampoule was slowy drawn from the hot to the cool zone, solidification occurring in the region of the temperature gradient.

Initially, the ingot was moved through the furnace at a rate of approximately 14 mm/d. When such an ingot was investigated by routine X-ray powder photograph methods, the samples were found to be inhomogeneous, the photographs in some cases showing multiphase conditions and broad lines. Thus the rate was reduced to 3.3 mm/d so that the complete freezing process required about 75 days. As indicated below, this cooling rate produced ingots which were found to be satisfactorily homogeneous. The ingot so produced was soaked in alcohol to remove the B_2O_3 coating and then cross sectional slices of about 0.5 mm thickness were cut from it, each slice being through both boat and sample. These slices were then cut up to provide specimens suitable for Hall coefficient and electroreflectance measurements, in addition to small samples for X-ray powder photographs. Because of highly hygroscopic nature of AlSb and its alloys, all samples were kept under vacuum when not in use.

Fig. 1. Variation of composition x with distance along ingot, for the ingot used to provide the samples for electroreflectance measurements. ○ Single-phase sample; × two-phase sample

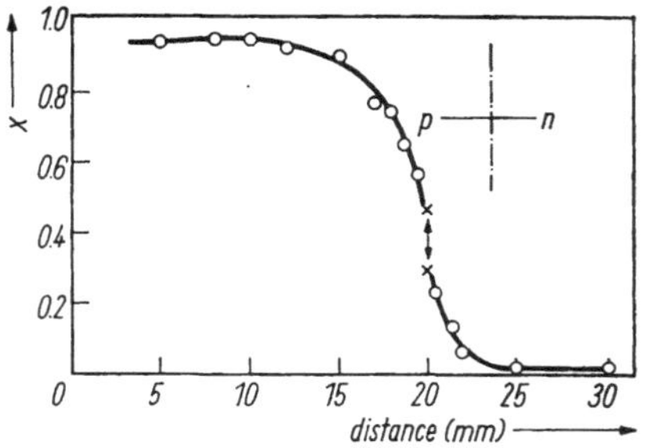

For the slowly cooled ingot, the X-ray photographs showed that almost all the samples were reasonably homogeneous, although at points near the centre of the composition range, the lines were broadened to a small extent. Only at one point in the ingot was a two phase condition obtained, and in this case two clearly separate phases with sharp X-ray lines were observed. No accurate data are available on the variation of lattice parameter with composition for this system. However, in the majority of cases for these mixed III–V alloys, a Vegard law is a very good approximation [6] and hence a linear variation of a_0 with x has been assumed in this case. The resulting variation of composition with distance along the ingot is shown in Fig. 1. It is seen that the main variation in composition occurs in a range of a little less than 1 cm, a not-unexpected result in view of the form of the phase diagram. Thus in this case cross-sectional slices of 0.5 mm thickness will have some composition gradient in them, and this could account for the broadening of the X-ray lines mentioned above.

While samples cut from such slices are not satisfactory for some physical measurements, in the case of modulated reflectance work a very thin layer of sample is involved in the measurements and the composition of this layer can be reasonably determined from Fig. 1. Also, in the case of the two-phase sample, this normally would be useless for physical measurements, but in the modulated reflectance work, two separate spectra are obtained, each characteristic of one of the phases present. Thus in this case, a non-equilibrium two phase sample can provide more data than a single phase one.

3. Electrical and Optical Results

Standard measurements of Hall coefficient and electrical conductivity were made by the van der Pauw method on specimens taken from cross-sectional slices of the ingot. The samples were found to be p-type for the first 2.3 cm of the ingot (see Fig. 1) which included all compositions with x greater than 0.04. The values of mobility determined for these samples were low compared with these of Agaev et al. [4, 5] and also with the accepted value for p-type AlSb. This was taken to indicate that the samples showed a considerable degree of compensation, and hence no further electrical measurements were made.

For the electroreflectance measurements, rectangular specimens were cut from the cross-sectional slices, the specimen dimensions being in the range 4×3 mm to 2×2 mm. Each specimen was mechanically polished with oil-suspended powders down to 1 μm alumina and was then etched in a methanol–5% bromine solution at room temperature. The electroreflectance spectra were determined using the electrolytic technique described previously [8]. In this case

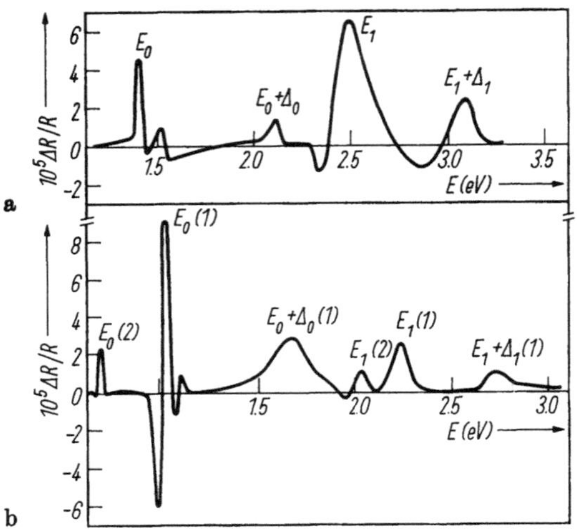

Fig. 2. Typical electroreflectance spectra. a) Single-phase sample: $x = 0.65$; b) two-phase sample: $x = 0.47$ (1), $x = 0.30$ (2)

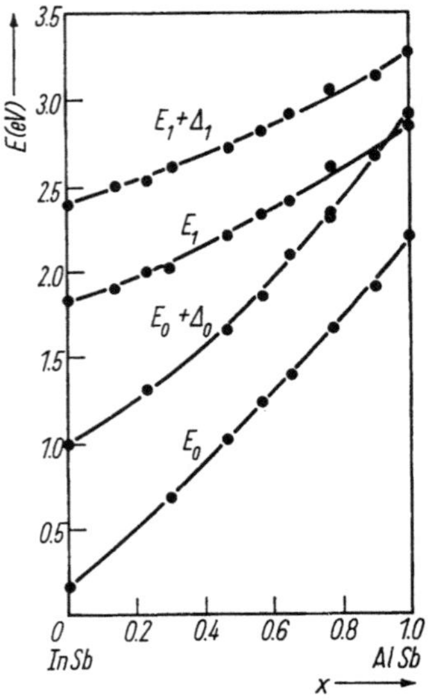

Fig. 3. Variation of transition energy with composition x for the E_0, $E_0 + \Delta_0$, E_1, and $E_1 + \Delta_1$ peaks

careful choice of the electrolyte was necessary since all aqueous electrolytes and also other solutions of inorganic salts would attack the specimen surfaces and hence could not be used. It was found that acetonitrile was satisfactory from this point of view, and also, because of its absorption coefficient, measurements could be made at wavelengths out to 1.8 μm. With this electrolyte, the suitable voltage conditions were found to be zero dc bias and ac voltages up to 20 V peak-to-peak.

Typical examples of the spectra obtained are shown in Fig. 2 where the spectrum of a single phase specimen and also that for the two phase material are shown. The transition energy associated with each peak was determined by the method of Aspnes and Rowe [9]. The resulting curves of transition energy against composition for the E_0, $E_0 + \Delta_0$, E_1, and $E_1 + \Delta_1$ peaks are shown in Fig. 3.

4. Discussion

In previous work [10, 11], expressions have been given for the variation of E_0, Δ_0, and Δ_1 as a function of alloy composition. Thus, E_0 is given as

$$E_0 = a + bx + cx^2 \,, \tag{1}$$

where a and b are constants determined by the E_0-value for the compounds concerned, the bowing parameter $c = c_i + c_e$ with $c_e = C_{FG}^2/A$, C_{FG} being the Phillips electronegativity difference [12] between the mixed elements, A a constant bandwidth parameter found by comparison with experiment to be 1.0 eV for all alloy systems and c_i being the bowing parameter obtained from a virtual crystal analysis (VCA). Van Vechten and Bergstresser [13] have determined values of c_i for a number of alloy systems by using the dielectric model of band structure analysis.

In the case of Δ_1, the variation with x is given as [11]

$$\Delta_1 = \Delta_{1v} + x(1-x)\,K\,\frac{C_{FG}^2}{\Delta_{1v}}, \tag{2}$$

where Δ_{1v} is the VCA splitting at the L-point, taken to be linear with x, and K a constant, determined by comparison with experiment to be 0.14.

The variation of Δ_0 with x is given as

$$\Delta_0 = \Delta_{0v}\left\{1 - \frac{x(1-x)\,C_{FG}^2}{A\tilde{E}_{0v}}\right\}, \tag{3}$$

where Δ_{0v} is the VCA splitting at the Γ-point again assumed to be linear with x, and $\tilde{E}_{0v}$ is a weighted mean VCA band gap, assumed to be given by

$$\frac{3}{\tilde{E}_{0v}} = \frac{2}{E_{0v}} + \frac{1}{E_{0v} + \Delta_{0v}}. \tag{4}$$

If the experimental values of E_0 versus x are fitted to equation (1), the value obtained for c is 0.43 eV. As indicated above, this is to be compared with the theoretical value of $c_i + c_e$. No value of C_{FG} has been published for this system but a value can be calculated from the equations of van Vechten [13, 14] and has been found here to be 0.65 eV, giving a value for c_e of 0.42 eV. Van Vechten and Bergstresser [13] give $c_i = 0.57$ eV for this system, and hence the predicted value for c is 0.99 eV. The disagreement between this and the experimental value requires further investigation of the values of c_i and C_{FG}. While, as is seen from (3), Δ_0 depends upon both C_{FG} and c_i (since E_{0v} is a function of c_i) Δ_1 depends on C_{FG} only (equation (2)). Using the value of $C_{FG} = 0.65$ eV, the variation of Δ_1 with x has been calculated from equation (2) and the predicted values are shown in Fig. 4. Because of the scatter in the data, best lines have been drawn through the experimental values of E_1 and $E_1 + \Delta_1$ versus x in Fig. 3 and the difference between these two lines taken as the experimental variation of Δ_1 with x. This curve also is shown in Fig. 4 and it is seen to be in good agreement with the predicted values. This would indicate that the values of $C_{FG} = 0.65$ eV is a reasonable one and that c_i must be the parameter incorrectly predicted. Since the value of $c_e = 0.42$ eV should be reasonable also, from the experimental value of $c = 0.43$ eV, c_i should have a value of 0.01 eV, i.e. can be taken as zero.

Using this zero value for c_i, the variation of Δ_0 with x can be predicted from (3). One question which arises is the form $\tilde{E}_{0v}$. This parameter appears in (3) as a measure of the transfer of s-like states from the conduction to valence bands as a consequence of the effect of disorder in the lattice [10]. The value of Δ_0

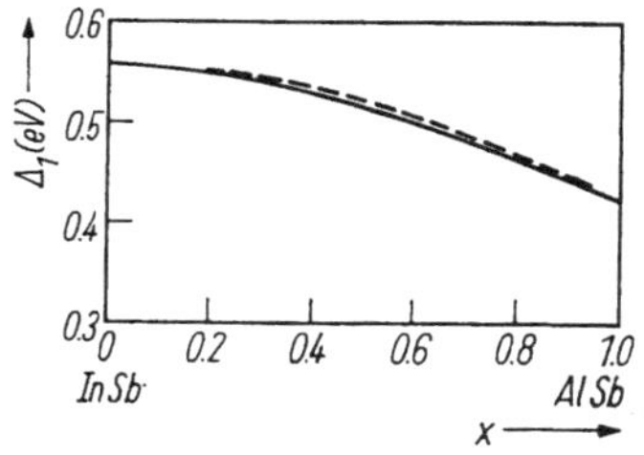

Fig. 4. Variation of spin-orbit splitting Δ_1 with composition x. ——— Best curve through experimental data; ---- theoretical curve from (2)

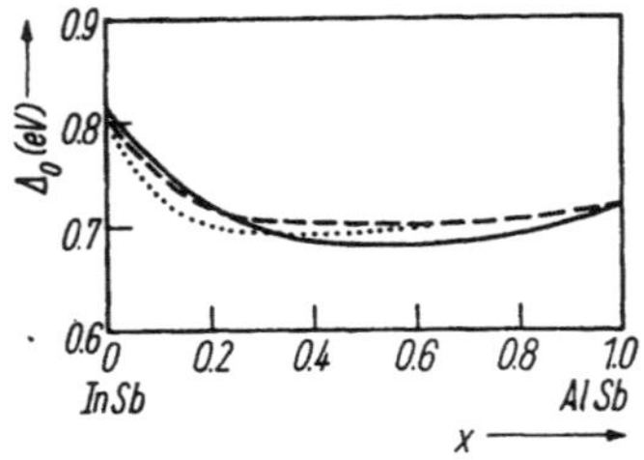

Fig. 5. Variation of spin–orbit splitting Δ_0 with composition x. ——— Best curve through experimental data; ---- theoretical curve from (3) with arithmetic mean $\tilde{E}_{0v}$; ······ theoretical curve from (3) with geometric mean $\tilde{E}_{0v}$

is reduced because these s-like states will not contribute to the valence band spin–orbit splitting. Although in previous work $\tilde{E}_{0v}$ has been taken to be a weighted geometric mean of E_{0v} and $E_{0v} + \Delta_{0v}$, it may be argued that if in the analysis the state mixing is considered before spin–orbit splitting, an arithmetic mean would be more appropriate, i.e.

$$\tilde{E}_{0v} = \frac{1}{3}(2E_{0v} + E_{0v} + \Delta_{0v}) = E_{0v} + \frac{\Delta_{0v}}{3}. \tag{5}$$

This question is not important for most III–V alloys, since the differences in the values of Δ_0 predicted by using equation (4) or (5) are within the limits of experimental error. However, in the case of InSb, because of the very large difference between E_0 and $E_0 + \Delta_0$, the arithmetic is approximately twice the geometric mean. Thus for the $Al_xIn_{1-x}Sb$ alloys, the predicted curves of Δ_0 versus x corresponding to the use of the two means are quite different at the low x-values, although almost identical for $x > 0.5$. Both curves are plotted in Fig. 5 and again the experimental curve is the difference between the best curves drawn through the experimental values of E_0 and $E_0 + \Delta_0$ in Fig. 3. It is seen that for $x < 0.5$, the values of Δ_0 predicted using the arithmetic mean E_{0v} are in good agreement with the experimental values, while those from the geometric mean fall well below. For $x > 0.5$, where the two predicted curves are very similar, both lie above the experimental values. Similar discrepancies have been seen for other III–V systems [11]. However, bearing in mind the approximations made in (3), i.e. (a) that the VCA value of Δ_{0v} is linear with x and (b) that the effects of intraband and inter-valence band mixing are small compared with those of conduction–valence band mixing, the agreement between the experimental and predicted values is satisfactory.

The electroreflectance measurements give only direct gap values, E_0 being the direct gap at Γ. However, since for AlSb the conduction band minimum is near the X-point, the main band gap will change from direct to indirect at some value of x. This value may be estimated approximately by assuming a linear variation with x of the energy difference E_{in} between the X-conduction band minima and the valence band maximum. Taking values of E_{in} of 1.6 eV for AlSb [15] and 2.2 eV for InSb [16], the values for the transition point are predicted to be $x = 0.80$ and $E_0 = 1.72$ eV. Because of the uncertainty in the value of E_2' for InSb, there is no point in attempting to allow for any bowing of the E_2' versus x curve.

Acknowledgement

The authors wish to thank Dr. O. Berolo for cooperation during the work and for useful discussion.

References

[1] W. Koster and W. Ulrich, Z. Metallk. **49**, 147 (1958).
[2] J. C. Woolley and B. A. Smith, Proc. Phys. Soc. **72**, 214 (1958).
[3] B. V. Baranov and N. A. Goryunova, Soviet Phys. — Solid State **2**, 262 (1960).
[4] Ya. Agaev and D. N. Nasledov, Soviet Phys. — Solid State **2**, 758 (1960).
[5] Ya. Agaev and N. G. Bekmedova, Soviet Phys. — Semicond. **5**, 1330 (1972).
[6] J. C. Woolley, Compound Semiconductors, Chap. I, Ed. R. K. Willardson and H. L. Goering, Reinhold Publ. Co., New York 1962.
[7] G. Dionne and J. C. Woolley, J. Electrochem. Soc. **119**, 784 (1972).
[8] O. Berolo and J. C. Woolley, Canad. J. Phys. **49**, 1335 (1971).
[9] D. E. Aspnes and J. E. Rowe, Phys. Rev. Letters **27**, 188 (1971).
[10] J. A. van Vechten, O. Berolo, and J. C. Woolley, Phys. Rev. Letters **29**, 1400 (1972).
[11] O. Berolo, J. C. Woolley, and J. A. van Vechten, Phys. Rev. B **8**, 3794 (1973).
[12] J. C. Phillips, Rev. mod. Phys. **42**, 317 (1970).
[13] J. A. van Vechten and T. K. Bergstresser, Phys. Rev. **187**, 1007 (1969).
[14] J. A. van Vechten, Phys. Rev. **182**, 891 (1969).
[15] O. Madelung, Physics of III–V Compounds, John Wiley & Sons, Inc., New York 1964 (p. 44).
[16] C. M. de Alvarez, J. P. Walter, R. W. Boyd, and M. L. Cohen, J. Phys. Chem. Solids **34**, 337 (1973).

(Received May 31, 1974)

phys. stat. sol. (b) 65, 221 (1974)

Subject classification: 13.3; 13.5.2

Sektion Physik der Technischen Universität Dresden, Arbeitsgruppe Theoretische Physik

The Dynamic Contribution to the Surface Energy

By

G. PAASCH

The dynamic or surface-plasmon contribution to the surface energy of the electron gas derived by Craig is evaluated taking into account correctly the electron–hole excitations, dispersion of the plasmon and surface-plasmon modes, and the damping of these modes by electron–hole excitations. The resulting surface energy vs. the density parameter is close to the earlier approximation which was criticized since all these influences were neglected there. The success of the simple approximation is explained using the simplified theory of electron correlations by Overhauser.

Der dynamische oder Oberflächenplasmonen-Beitrag zur Oberflächenenergie des Elektronengases, der von Craig abgeleitet wurde, wird unter Berücksichtigung von Paaranregungen sowie von Dispersion und Dämpfung der Volumen- und Oberflächenplasmonen ausgewertet. Die so berechneten Oberflächenenergien stimmen in ihrer Abhängigkeit vom Dichteparameter nahezu mit den früheren Näherungswerten überein, die unter Vernachlässigung der genannten Einflüsse gewonnen wurden und wegen dieser Vernachlässigung als nicht aussagekräftig angesehen werden. Die Ursache dafür, daß die früheren Näherungswerte trotz der groben Annahmen sehr gut sind, wird unter Verwendung eines einfachen Modells für die Elektronenkorrelation erläutert.

1. Introduction

Recent investigations of the electronic structure of the surface of solids gave insight into the static redistribution of the electron density and of the ionic positions and into the corresponding contributions to the surface energy [1 to 4]. Peuckert [5], Schmit and Lucas [6] and Craig [7] focused the attention on the change in the zero-point energy of the electron gas due to the occurrence of new collective modes when a surface is created. Surface plasmons have a sufficiently high energy to give an important contribution to the surface energy. Neglecting the electron–hole excitations and assuming that the plasmons and the surface plasmons have the same cut-off wave number k_c, the authors [5 to 7] found

$$\sigma = \text{const}\,(\sqrt{2} - 1)\,\omega_p k_c^2\,,$$

$$\omega_p = \left(\frac{ne^2}{\varepsilon_0 m}\right)^{1/2} = \text{plasma frequency}\,.$$

The correct value of the constant is given in equation (12) of this paper. With $k_c = \alpha\omega_p/v_f$ a surprising fit of the surface energies of the simple metals is possible with the same value α for all metals. A further success of this simple model is the explanation of the energies of voids and void lattices [8]. But on the other hand Jonson and Srinivasan [9], Feibelman [10], and Kohn [11] objected against this theory:

(i) Due to the different damping of the volume and surface plasmons one has to use different cut-off wave numbers.

(ii) k_c itself was only fitted.

(iii) The electron–hole excitations must be taken into account.

(iv) It is difficult to understand that this dynamic contribution alone gives nearly the total surface energy — this would imply that all other contributions should cancel out. As a conclusion Jonson and Srinivasan believe that the apparent success of this plasmon theory of the surface energy is a spurious one, arising from the arbitrary choice of the cut-off wave number.

In this paper, in Section 2 the general expression for the dynamic contribution to the surface energy given by Craig [7] is evaluated using the frequency- and wave number-dependent dielectric function of the electron gas and avoiding the approximations (i) to (iii). In Section 3 the reason for the success of the original simple approximation [5 to 7] is explained using the simplified theory of electron correlations by Overhauser [12]. Conclusions are given in Section 4.

2. Evaluation of the Dynamic Contribution

We start with the expression for the dynamic contribution to the surface energy of the electron gas derived by Craig [7]:

$$\sigma = \frac{1}{8}\int_0^{e^2}\frac{\mathrm{d}g}{g}\int\frac{\mathrm{d}^2k}{(2\pi)^2}\int\frac{\mathrm{d}\omega}{2\pi}\,\mathrm{Im}\,\frac{(1-\varepsilon(k,\omega))^2}{\varepsilon(k,\omega)\,\big(1+\varepsilon(k,\omega)\big)}\,. \tag{1}$$

The case of a metal–vacuum surface is considered here. The metal is characterized by the frequency- and wave number-dependent dielectric function (DF) $\varepsilon(k,\omega)$ of the homogeneous electron gas. The DF contains the density and the coupling constant g as parameters. In the calculation we use the expression [13, 18]

$$\varepsilon(k,\omega) = 1 - \frac{\Lambda(k,\omega)}{1 + G(k)\,\Lambda(k,\omega)}\,, \tag{2}$$

$$\Lambda(k,\omega) = A(k,\omega) + i\,\Sigma(k,\omega)\,. \tag{3}$$

$A(k,\omega)$ is the RPA polarizability given by

$$A(k,\omega) = -\frac{\xi}{x^3}\left[x + \frac{1}{2}\left\{1 - \frac{1}{4}\left(\frac{y}{x} - x\right)^2\right\}\ln\left|\frac{y - x(x+2)}{y - x(x-2)}\right| + \right.$$
$$\left. + \frac{1}{2}\left\{1 - \frac{1}{4}\left(\frac{y}{x} + x\right)^2\right\}\ln\left|\frac{y + x(x+2)}{y + x(x-2)}\right|\right], \tag{4}$$

$$\Sigma(k,\omega)\begin{cases} = -\,0 & y > x(x+2) \text{ or } x > 2 \text{ and } y < x(x-2)\,,\\ = -\dfrac{\pi}{2}\dfrac{\xi}{x^3}\left[1 - \dfrac{1}{4}\left(\dfrac{y}{x} - x\right)^2\right] & \begin{cases} x > 2 \text{ and } x(x-2) < y < x(x+2)\\ x < 2 \text{ and } -x(x-2) < y < x(x+2)\,,\end{cases}\\ = -\dfrac{\pi}{2}\dfrac{\xi\, y}{x^3} & x < 2 \text{ and } 0 < y < -x(x-2)\,.\end{cases} \tag{5}$$

Reduced units $x = k/k_\mathrm{f}$ and $y = \hbar\omega/E_\mathrm{f}$ are used. For $\omega < 0$ one has $\Sigma(k,-\omega) = \Sigma(k,\omega)$. The constant $\xi = e^2k_\mathrm{f}/(\pi E_\mathrm{f}) = 0.33\,r_\mathrm{s}$ contains the usual density parameter r_s and the coupling constant e^2. The function $G(k)$ describes the influence of exchange and correlation beyond the RPA. There exist several expressions for this XC correction $G(k)$ to the DF [14 to 16]. In our calculations we arbitrarily used the modified Hubbard expression [14]

$$G(k) = \frac{1}{2} \frac{k^2}{k^2 + k_f^2 + k_T^2/2}; \quad k_T^2 = \frac{4me^2k_F}{\pi\hbar^2} \tag{6}$$

and that one of Overhauser [12]. However, the influence of $G(k)$ on the surface energy (1) is neglegible as will be shown later. This is in contrast to the correlation energy of the bulk electron gas and also to calculations of the surface energy with the assumption of a slowly varying electron density [17].

With (2) and (3) one gets for the integrand of (1)

$$\begin{aligned} \operatorname{Im} \frac{(1-\varepsilon)^2}{\varepsilon(1+\varepsilon)} &= \operatorname{Im} \frac{1}{\varepsilon} - 4 \operatorname{Im} \frac{1}{\varepsilon + 1} \\ &= \frac{\Sigma}{(1 - A')^2 + \Sigma'^2} - 4 \frac{\Sigma}{(2 - A'')^2 + \Sigma''^2}, \end{aligned} \tag{7}$$

$$A' = (1 - G) A, \qquad \Sigma' = (1 - G) \Sigma, \tag{8}$$

$$A'' = (1 - 2G) A, \qquad \Sigma'' = (1 - 2G) \Sigma. \tag{9}$$

At this point one immediately can get the contribution of the collective modes to (1). As shown in Fig. 1, for small wave numbers in the volume and surface plasmon range there is $\Sigma = -0$. Then the two terms of (7) are for $k \to 0$ (and for simplicity for $G = 0$)

$$\lim_{\Sigma \to -0} \frac{\Sigma}{(1 - A)^2 + \Sigma^2} = -\pi\,\delta(1 - A) = -\frac{\pi}{2} \omega_p \{\delta(\omega + \omega_p) + \delta(\omega - \omega_p)\}, \tag{10}$$

$$\lim_{\Sigma \to -0} \frac{-4\Sigma}{(2 - A)^2 + \Sigma^2} = 4\pi\,\delta(2 - A) = \frac{\pi}{2} \frac{\omega_p}{\sqrt{2}} \left\{\delta\left(\omega + \frac{\omega_p}{\sqrt{2}}\right) + \delta\left(\omega - \frac{\omega_p}{\sqrt{2}}\right)\right\}. \tag{11}$$

ω_p and $\omega_p/\sqrt{2}$ are the plasma and surface plasma frequencies, respectively. Then with the same cut-off wave number k_c for both the volume and surface plasmons, from (1), (7) (10), and (11) one gets

$$\sigma = \frac{k_c^2 \omega_p}{32\pi} (\sqrt{2} - 1) \to \frac{\omega_p^3}{32\pi v_f^2} (\sqrt{2} - 1)\, \alpha^2 . \tag{12}$$

The result of fitting k_c by $k_c = \alpha\omega_p/v_f$ with $\alpha^2 = 0.731$ is shown in Fig. 2. This is the result of Craig [7] (in Craig's work there is an additional factor 1/3 in (12) in error and hence we used $\alpha^2 = \alpha_{\text{Craig}}^2/3$). It should be mentioned that the same approximation — i.e. account of only the collective mode for $k < k_c$ — yields a positive correlation energy for the bulk electron gas whereas this quantity in fact is a negative one. Therefore the electron–hole excitations are very important for the correlation energy and one has to study their influence on the surface energy.

As in Hubbard's work on the correlation energy [13] the integration over the coupling constant in (1) can be done analytically using (7) to (9) and 4), (5).

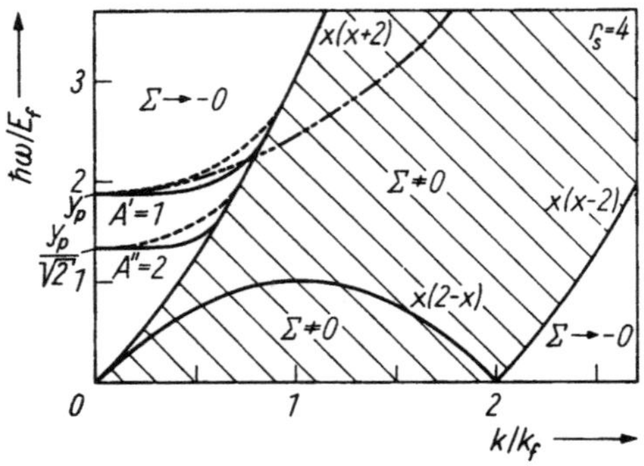

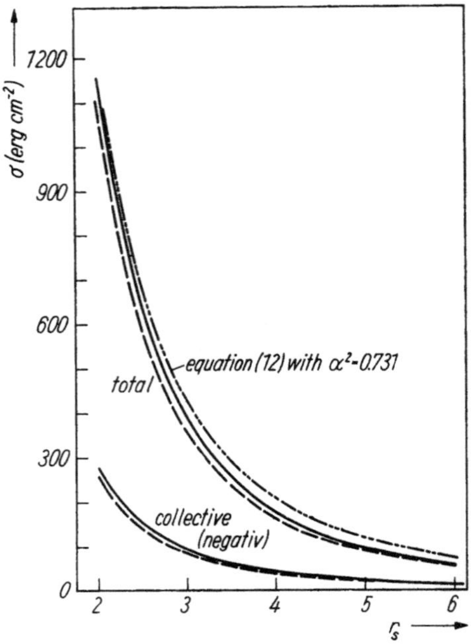

Fig. 1. Spectrum of excitation energies vs. wave number transfer for the interacting electron gas: Electron–hole excitations in the shadowed region where $\Sigma \neq 0$ and the volume and surface plasmon excitations ($A' = 1$ and $A'' = 2$, respectively according to equations (4), (8), and (9)). The latter are calculated for $r_s=4$ for the XC correction $G = 0$ (– – – –) and for G from (6) (———). In the simplified model of electron correlations described in Section 3 the whole spectrum is replaced by a single mode $\hbar\omega_k$ indicated schematically by the line (— · · · —)

Fig. 2. Calculated values for the dynamic contribution to the surface energy. The total value according to equation (15) is given for $G = 0$ (– – –) and for G from (6) (———). The meaning of the collective part is described in the text. The approximate expression (12) is given with $\alpha^2 = 0.731$ (— · · · —)

One gets

$$\int_0^{e^2} \frac{dg}{g} \frac{(1-\varepsilon)^2}{\varepsilon(1+\varepsilon)} = \frac{\Sigma}{\Sigma'} \arctan \frac{\Sigma'}{1-A'} - 2\frac{\Sigma}{\Sigma''} \arctan \frac{\Sigma''}{2-A''}, \tag{13}$$

where the branch $0 < \arctan \cdots < \pi$ has to be used. Since for $x \to \infty$ (13) is given by

$$\frac{\Sigma}{\Sigma'} \Sigma'(1-A') - 2\frac{\Sigma}{\Sigma''} \frac{\Sigma''}{2}\left(1-\frac{A''}{2}\right) = \Sigma(A' - A''/2) \sim \frac{1}{x^6} \tag{14}$$

the d^2k-integration in (1) converges very well. With (13) for (1) one gets the final formula for the surface energy:

$$\sigma = \frac{E_f k_f^2}{16\pi^2} \int_0^\infty dx\, x \int_0^\infty dy \left\{ -\frac{\Sigma}{\Sigma'} \arctan \frac{-\Sigma'}{1-A'} + 2\frac{\Sigma}{\Sigma''} \arctan \frac{-\Sigma''}{2-A''} \right\}. \tag{15}$$

The range of this two-dimensional integration can be divided naturally into two parts. The first one is that of the electron–hole excitations which is shadowed in Fig. 1. There the arctan . . . in (15) have values between 0 and π. The integration is done numerically. In the remaining region of the xy plane $\Sigma = -0$ and hence the arctan ... have the values 0 or π depending on the sign of the denominator of the argument. In the first term of (15) the arctan ... is π between the plasmon line $A' = 1$ (Fig. 1) and the shadowed region and zero elsewhere. For the second term one has the value π between the surface plasmon line $A'' = 2$ and the shadowed region and the value zero elsewhere. In this way one gets a division of the surface energy (1) or (15) into one part arising from individual electron–hole excitations and a remaining one which can be attributed to the collective modes. Fig. 2 shows the total value of the surface energy (15) vs. r_s and in addition the collective part as just defined — this part has negative sign.

Fig. 2 shows that the correct evaluation of the total dynamic contribution (1) to the surface energy yields nearly the same values as the rough approximation (12) with arbitrary choice of the cut-off wave number k_c as described in connection with (12). In addition it is interesting that the XC correction $G(k)$ results in only small changes in the surface energy: according to (15), σ is the difference of two terms which are influenced by the function $G(k)$ in a similar manner (compare with (8), (9)).

3. Application of a Simplified Theory of Electron Correlations

In the preceding section an evaluation of the dynamic contribution to the surface energy (1) is given using the full DF (2) to (6) and hence including not only the plasmon contribution as in [4 to 6] but also the electron–hole excitations. To understand the reason for the success of the rough approximation (12) an evaluation of (1) is carried out with the simplified theory of electron correlations. Overhauser [12] replaced in his theory the complete excitation spectrum of Fig. 1 by a single "plasmon" branch $\hbar\omega_k$. For small k it approaches the plasmon frequency $\hbar\omega_p$, and for large k it approaches $\hbar^2k^2/2m$ as scetched in Fig. 1. This plasmon mode is given by

$$\omega_k^2 = \omega_p^2 \frac{\varepsilon_k}{\varepsilon_k - 1}; \quad \varepsilon_k = \varepsilon(k, \omega = 0) . \tag{16}$$

Here the static limit of the DF (2) to (6) occurs. All oscillator strengths including those of the electron–hole excitations are concentrated in the single mode (16). Overhauser did'nt derive a frequency dependent DF for his model. But for our purpose it is useful to do so. We start from the exact linear response expression for the DF [18]:

$$\frac{1}{\varepsilon(k, \omega)} = 1 + \frac{4\pi e^2}{\Omega k^2} \sum_k |(\varrho_k)_{n0}|^2 \left\{\frac{1}{\omega - \omega_{n0} + i\eta} - \frac{1}{\omega + \omega_{n0} - i\eta}\right\}; \quad \eta \to +0. \tag{17}$$

ω_{n0} is the energy for the excitation of the interacting electron system from the ground state "0" to the excited state "n", $(\varrho_k)_{n0}$ is the corresponding matrix element of the density fluctuation and Ω the normalization volume. As shown in Appendix A from (17) one gets for the Overhauser model of electron correlations the following expression for the DF:

$$\frac{1}{\varepsilon(k, \omega)} = 1 + \frac{\omega_p^2}{2\omega_k} \left\{\frac{1}{\omega - \omega_k + i\eta} - \frac{1}{\omega + \omega_k - i\eta}\right\} \tag{18}$$

with ω_k from (16). Due to the neglect of the true spectral width of the excitation spectrum such an expression is most useful in integrals as e.g. (1) or the correlation energy.

The ω-integration in (1) can be done using (18) and

$$\operatorname{Im} \frac{(1-\varepsilon)^2}{\varepsilon(1+\varepsilon)} = \operatorname{Im} \frac{1}{\varepsilon} + 4 \operatorname{Im} \frac{1}{\varepsilon^{-1}+1}. \tag{19}$$

The result is

$$\sigma = \frac{1}{16} \int_0^{e^2} \frac{\mathrm{d}g}{g} \int \frac{\mathrm{d}^2k}{(2\pi)^2} \left\{ \frac{\omega_\mathrm{p}^2}{\sqrt{\omega_k^2 - \omega_\mathrm{p}^2/2}} - \frac{\omega_\mathrm{p}^2}{\omega_k} \right\}. \tag{20}$$

In this expression ω_p and ω_k from (16) depend on the coupling constant. The integration over the coupling constant is elementary and as the final result one gets the one-dimensional integral

$$\sigma = \frac{\omega_\mathrm{p}}{8} \int_0^\infty \frac{\mathrm{d}k\, k}{2\pi} \left\{ \frac{\sqrt{(\varepsilon_k+1)/2} - \sqrt{1+G(k)\,(\varepsilon_k-1)}}{(0.5-G(k))\,\sqrt{\varepsilon_k-1}} - \frac{\sqrt{\varepsilon_k} - \sqrt{1+G(k)\,(\varepsilon_k-1)}}{(1-G(k))\,\sqrt{\varepsilon_k-1}} \right\}. \tag{21}$$

In this result the XC correction function $G(k)$ arises only from the static ε_k in (16) but is not included directly in the frequency dependent DF $\varepsilon(k, \omega)$ (18) beyond the dependence through ω_k. The modification of (18) to (20) due to a proper use of the XC correction in the frequency dependent DF is discussed in Appendix B. With this modification instead of (21) one obtains

$$\sigma = \frac{\omega_\mathrm{p}}{8} \int_0^\infty \frac{\mathrm{d}k\, k}{2\pi} \left\{ 2\, \frac{\sqrt{1+(\varepsilon_k-1)\,(0.5+G(k))} - \sqrt{1+G(k)\,(\varepsilon_k-1)}}{\sqrt{\varepsilon_k-1}} - \frac{\sqrt{\varepsilon_k} - \sqrt{1+G(k)\,(\varepsilon_k-1)}}{(1-G(k))\,\sqrt{\varepsilon_k-1}} \right\}. \tag{22}$$

The integration is done numerically. The results are shown in Fig. 3 in comparison with the result of the calculation of Section 2. It is seen that the XC correction has a somewhat larger influence on the result as in the exact calculation. But including the XC correction one obtains from the approximation (22) within 10% the same results as from the full expression (15) from Section 2. This fact allows one to explain the apparent success of the original approximation (12) of Craig [7] and Schmit and Lucas [6] as follows. In the simplified model of electron correlations all oscillator strengths are concentrated in the single "plasmon" mode (16). Our results show that this total oscillator strength for each momentum transfer k is most important to obtain the correct value of σ whereas the true width of the excitation spectrum influences the result only to within 10%. Further our calculation showed that both in (15) and (22) the main contribution to σ arises from momentum transfers $k \lessapprox (1.2 \text{ to } 1.5)\, k_\mathrm{f}$. The model underlying the approximation of Craig (12) describes a similar situation: there one has also a single plasmon mode (although without dispersion) for

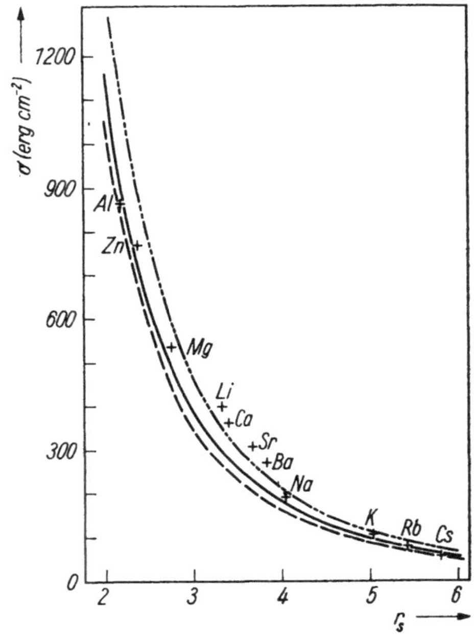

Fig. 3. Dynamic contribution to the surface energy: Values of (15) with G from (6) (———) and the values calculated with the simplified model of electron correlations according to (22) with G from (6) (- - -) and with $G = 0$ (— · · · —). In addition experimental values for the surface energy are given (from Table I of [7])

momentum transfers $k < k_c$. Therefore it is actually not surprising that a suitable choice of the cut-off wave number as described above yields values for σ near to the correct ones.

4. Conclusions

In this paper an evaluation of the general expression (1) of Craig [7] for the dynamic contribution or surface plasmon contribution to the surface energy is given. By using the frequency- and wave number-dependent DF of the homogeneous electron gas the influence of electron–hole excitations and of dispersion and damping of the collective modes are taken into account. The resulting surface energy vs. density parameter agrees surprisingly well with the approximate expression (12) [5 to 7] which was criticized by several authors [9 to 11] since all the above-mentioned influences were neglected and an arbitrary cut-off for the volume and surface plasmon mode was used. Using the simplified theory of electron correlations [12] we showed the reason for the success of the approximation (12): Only small wave numbers, that means long-range interactions, contribute effectively to the expression (1). In this region it is sufficient to take into account correctly the sum of all oscillator strengths of excitations for a given wave number whereas the width of the excitation spectrum has only a small influence on the surface energy (1).

Let us now comment the situation that the dynamic contribution to the surface energy alone agrees well with the experimental values of the surface energy (Fig. 3). First this resembles the rough agreement of the correlation energy of the homogeneous electron gas with the binding energy per electron being about one eV for simple metals. Secondly, in his derivation of expression (1) Craig [7] assumed the electron density to be constant within the metal and cut-off at the surface. For such a model all other contributions to the surface energy (see Lang and Kohn [1]) are identically zero except the so-called classical cleavage energy. The latter is the energy to split a crystal without any change of charge densities at the cleavage face. Addition of this contribution to the dynamic contribution (1) would destroy the agreement with the experimental values seen in Fig. 3. This is what one would expect since the electron density cut-off at the surface is highly unstable. For a realistic electron density showing a decrease in a region of about $2a_0$ [1 to 3] and with ions present instead of the homogeneous background a refined formulation of the dynamic contribution to the surface energy is needed. Finally it should be mentioned that the change of the zero-point energy of the collective modes due to the existence of a surface is included at least to some extent in all other calculations of the surface energy when any approximation for the correlation energy is included (e.g. a local approximation or a gradient expansion for the Hohenberg-Kohn energy functional [19]): The correlation energy contains the collective modes and hence their change when a surface with any decreasing electron density is created.

Acknowledgements

The author acknowledges the suggestion of Dr. W. John to investigate this problem, and helpful discussions with him and with Prof. Dr. P. Rennert, Prof. Dr. P. Ziesche, and Dr. H. Eschrig.

Appendix A

Derivation of the DF (18)

In the simplified model of electron correlations of Overhauser [12] there exists only one "plasmon" mode $\hbar\omega_k$ (16). The corresponding density fluctuation operator is given by the plasmon creation and annihilation operators a_k^+ and a_k as

$$\varrho_k = \Omega\lambda_k\,(a_{-k} + a_k^+) \tag{A1}$$

with a parameter λ_k which is to be determined. Since there is only one excitation for each wave number k, the matrix element of the density fluctuation for excitations of the interacting system from the ground state "0" to the excited state "n" is

$$|(\varrho_k)_{n0}|^2 = \Omega^2\lambda_k^2\,. \tag{A2}$$

Using (A2) and the f-sum rule

$$\sum_n \hbar\omega_{n0}\,|(\varrho_k)_{n0}|^2 = \frac{N\hbar^2k^2}{2m}; \quad \omega_{n0} = \omega_k \tag{A3}$$

one gets for the parameter λ_k

$$\Omega^2\lambda_k^2 = \frac{N\hbar^2k^2}{2m\hbar\omega_k}. \tag{A4}$$

Then with (A2) and (A4) the expression (17) for the DF takes the form (18).

Appendix B

XC correction in the simplified theory of electron correlations

Without the XC correction $G(k)$ in the DF (2) one has

$$\varepsilon_{\text{RPA}}(k, \omega) = 1 - \Lambda(k, \omega) \equiv 1 - \frac{4\pi e^2}{k^2}\,Q(k, \omega)\,. \tag{B1}$$

The approximation (18) is an RPA result but due to the singular imaginary part it is not given in the form (B1). But with the XC corrected DF (2) one gets immediately

$$\operatorname{Im}\frac{1}{\varepsilon(k,\omega)}=\frac{1}{1-G(k)}\operatorname{Im}\frac{1}{\varepsilon'(k,\omega)}, \tag{B2}$$

where in $\varepsilon'(k, \omega)$ the Coulomb interaction is reduced due to the exchange and correlation interaction in comparison with the RPA-DF (B1):

$$\frac{4\pi e^2}{k^2}\to\frac{4\pi e^2}{k^2}\bigl(1-G(k)\bigr). \tag{B3}$$

This result can be used in the DF (18) where according to the derivation in Appendix A the second term of the right-hand side has to be multiplied with $\bigl(1-G(k)\bigr)$ (λ_k from (A4) is not changed since the sum rule (A3) is not affected by the XC correction). Similar for the second term of (19) one obtains

$$4\operatorname{Im}\frac{1}{\varepsilon^{-1}+1}=\frac{1}{1-2G}\,4\operatorname{Im}\frac{1}{(\varepsilon'')^{-1}+1}; \tag{B4}$$

in $(\varepsilon'')^{-1}$:

$$\frac{4\pi e^2}{k^2}\to\frac{4\pi e^2}{k^2}(1-2G). \tag{B5}$$

With (B2) to (B4) and (18) to (19) one gets from (1) instead of (20)

$$\sigma=\frac{1}{16}\int\limits_0^{e^2}\frac{\mathrm{d}g}{g}\int\frac{\mathrm{d}^2k}{(2\pi)^2}\left\{\frac{1}{1-2G}\,\frac{\omega_p^2(1-2G)}{\sqrt{\omega_k^2-\omega_p^2(1-2G)/2}}-\frac{1}{1-G}\,\frac{\omega_p^2(1-G)}{\omega_k}\right\}. \tag{B6}$$

Then the integration over the coupling constant gives (22).

References

[1] N. D. Lang and W. Kohn, Phys. Rev. B **1**, 4555 (1970).
[2] N. D. Lang and W. Kohn, Phys. Rev. B **3**, 1215 (1971).
[3] G. Paasch, H. Eschrig, and W. John, phys. stat. sol. (b) **51**, 283 (1972).
[4] G. Paasch, Exper. Tech. Phys., to be published.
[5] V. Peuckert, Z. Phys. **241**, 191 (1971).
[6] J. Schmit and A. A. Lucas, Solid State Commun. **11**, 415 (1972).
[7] R. A. Craig, Phys. Rev. B **6**, 1134 (1972).
[8] A. A. Lucas, Phys. Rev. B **7**, 3527 (1973).
[9] M. Jonson and G. Srinivasan, Phys. Letters **43A**, 427 (1973).
[10] P. J. Feibelman, Solid State Commun. **13**, 319 (1973).
[11] W. Kohn, Solid State Commun. **13**, 323 (1973).
[12] A. W. Overhauser, Phys. Rev. B **3**, 1888 (1971).
[13] J. Hubbard, Proc. Roy. Soc. **A243**, 336 (1957).
[14] V. Heine and I. V. Abarenkov, Phil. Mag. **9**, 451 (1964).
[15] P. Vashista and K. S. Singwi, Phys. Rev. B **6**, 875 (1972).
[16] F. Toigo and T. O. Woodruff, Phys. Rev. B **2**, 3958 (1970); B **4**, 371 (1971).
[17] G. Paasch and M. Hietschold, to be published.
[18] D. Pines, Elementary Excitations in Solids, W. A. Benjamin, Inc., New York 1963.
[19] P. Hohenberg and W. Kohn, Phys. Rev. **136**, B864 (1964).

(Received May 24, 1974)

phys. stat. sol. (b) **65**, 231 (1974)

Subject classification: 13.1 and 18.4

Sektion Physik der Bergakademie Freiberg

On the Chemical Potential of the Hubbard Model

By

J. Monecke

On the basis of a cluster approach to the Hubbard model it is proved that the chemical potential μ as a function of the number of electrons per atom n has a jump at $n = I$ on the insulating site and that n as a function of μ has a jump at $\mu = I/2$ on the metallic site of the metal–insulator transition. In terms of an one-electron picture this corresponds to an infinite density of states at the Fermi level in the metallic phase.

Im Rahmen einer Clusternäherung für das Hubbardmodell wird gezeigt, daß das chemische Potential μ als Funktion der Zahl der Elektronen pro Atom n auf der Isolatorseite des Metall–Isolator-Überganges einen Sprung bei $n = I$, und daß n als Funktion von μ auf der Metallseite einen Sprung bei $\mu = I/2$ besitzen. In einem Einelektronenbild entspricht dies einer unendlichen Zustandsdichte am Ferminiveau in der metallischen Phase.

1. Introduction

The Hubbard Hamiltonian [1]

$$H = \sum_{\sigma;i,j=1}^{N} t_{i,j} c_{i,\sigma}^{+} c_{j,\sigma}^{-} + I \sum_{i=1}^{N} n_{i\uparrow} n_{i\downarrow} \tag{1}$$

is widely used as a basis in order to understand the metal–insulator transitions in some transition metal compounds [2]. Within the model described by this Hamiltonian all interactions between electrons in a s-band, described by the hopping probabilities $t_{i,j}$ between the Wannier sites i and j, except the interaction between two electrons at the same Wannier site, described by the interaction energy I, are neglected. In two limiting cases the solution of the Hubbard Hamiltonian is trivial: if $I = 0$ the eigenvalues are sums of the eigenvalues $\varepsilon(\boldsymbol{K})$ of one-electron Bloch eigenfunctions

$$\varepsilon(\boldsymbol{K}) = \frac{1}{N} \sum_{i,j=1}^{N} t_{i,j} \, \mathrm{e}^{-i\boldsymbol{K}(\boldsymbol{R}_i - \boldsymbol{R}_j)} \tag{2}$$

and if the $t_{i,j}$ are zero the eigenvalues are sums of the two values 0 and I which are the eigenvalues of zero, one and two electron Slater determinants built of from Wannier functions. Due to the different nature of the wave functions in the two limiting cases it is not easy to interpolate in a simple way between the results for $I = 0$ and $t_{i,j} = 0$. One can, of course, determine one-electron bands for all I and $t_{i,j}$ from the poles of the corresponding Green function [1], but the band energies depend, in general, on higher correlation functions [3], and even within the simplest decoupling schemes the shapes and weights of the bands depend not only on I and on the $t_{i,j}$ [4] but on the number of electrons per atom, too. In order to obtain the same information as in conventional band theory one has to determine, therefore, the infinite set of bands corresponding to the different numbers of electrons per atom.

In this paper the chemical potential μ as a function of the number of electrons per atom n is discussed within some different approximations. The function $\mu(n)$ contains essentially the same information about the system as the infinite set of bands for different n. Furthermore, $\mu(n)$ is a well defined function whereas the definition of band energies itself depends on special decoupling (or equivalent) procedures.

In Section 2 the results of Gutzwiller's variational approach [5] are discussed in terms of a one-electron band picture and the behaviour of the chemical potential is deduced qualitatively. It is found that μ has a jump at $n = I$ on the insulating site and that n has a jump at $\mu = I/2$ on the metallic site of the metal–insulator transition. The result is compared with that of Hubbard [1].

In Section 3 $n(\mu)$ is calculated on the basis of the cluster approach of Caron and Pratt [6 to 8] with qualitatively the same result as according to the approach of Gutzwiller. It is proposed that the jump of n on the metallic site, which corresponds to an infinite density of states at the Fermi level, is an inherent feature of the Hubbard model and not only the result of the approximation used.

2. On the Chemical Potential in the Variational Approach of Gutzwiller

In order to obtain the ground-state wave function of the Hubbard Hamiltonian Gutzwiller [5] performed an approximate variational calculation starting with a trial wave function constructed from conventional Bloch states but reducing the amplitudes of all components in which ν atoms are doubly occupied with a factor η^ν, $0 \leqq \eta \leqq 1$. The matrix elements have been calculated neglecting the kinetic energy of opposite spin electrons. Then the energy was minimized with respect to η. The result for the occupation number $n_{\boldsymbol{K}}$ in the case $n = I$ is represented in the first column of Fig. 1.

In the second column a two band model which would just give the same occupation numbers is constructed in a qualitative way. The last two columns give the corresponding density of states and the chemical potential. The construction of the band energies $E_{\boldsymbol{K}}$ is not without ambiguity. For $I \ll I_0$, e.g., the flat band could be replaced by a band with a finite width lying entirely below the Fermi level. For $I \lesseqgtr I_0$, however, the flatness of the upper band for $|\boldsymbol{K}| < K_{\mathrm{F}}$ and of the lower band for $|\boldsymbol{K}| > K_{\mathrm{F}}$ is required in order to obtain one single Fermi surface without very artificial assumptions concerning the weights of the two bands. According to our construction we have to assume only that the weight of the flat band for $I < I_0$ vanishes with $I \to 0$.

It is interesting to compare Fig. 1 with the results of Hubbard's first solution [1] according to which the band energies are given by

$$E_{\boldsymbol{K}}^{1/2} = \frac{1}{2}\left[\varepsilon(\boldsymbol{K}) + I\right] \pm \frac{1}{2}\sqrt{[\varepsilon(\boldsymbol{K}) - I]^2 + 2I\varepsilon(\boldsymbol{K})n} \tag{3}$$

with the weights

$$A_{\boldsymbol{K}}^{1/2} = \frac{1}{2}\left[1 \pm \frac{[\varepsilon(\boldsymbol{K}) - I] + In}{\sqrt{[\varepsilon(\boldsymbol{K}) - I]^2 + 2I\varepsilon(\boldsymbol{K})\,n}}\right] = \frac{\mathrm{d}E_{\boldsymbol{K}}^{1/2}}{\mathrm{d}\varepsilon(\boldsymbol{K})}. \tag{4}$$

These two bands can be interpreted as the result of the hybridization of a level at I and a band $\varepsilon(\boldsymbol{K})$ with a hybridization parameter $4\Delta\Delta^* = 2I\varepsilon(\boldsymbol{K})\,n$ which vanishes for $I \to 0$. The band structure for $I \to 0$ qualitatively is the

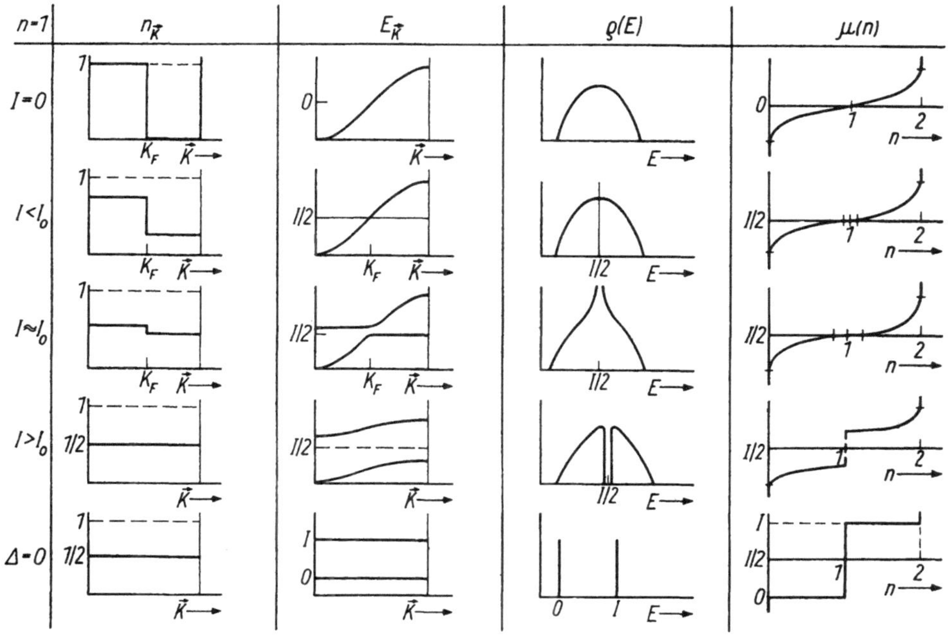

Fig. 1. Occupation number $n_{\boldsymbol{K}}$, band structure $E_{\boldsymbol{K}}$, density of states $\varrho(E)$, and chemical potential $\mu(n)$ within Gutzwiller's approach to the Hubbard model

same as in Gutzwiller's approach for $I \to I_0$: the upper band becomes flat for $|\boldsymbol{K}| < K_F$ and the lower band becomes flat for $|\boldsymbol{K}| > K_F$ with K_F lying just in the middle of the Brillouin zone in the case of $n = I$ and symmetric $\varepsilon(\boldsymbol{K})$. The weights of the flat parts vanish with $I \to 0$. Hubbard's first solution would be equivalent to the result of Gutzwiller if one improvement could be performed: the hybridization parameter should vanish at some $I \to I_0 \neq 0$ and the weight of the flat parts for $I < I_0$ should remain nonzero for nonzero I. We are not aware of any decoupling procedure bearing out this suggestion. In the next section, however, it is shown that the simple cluster approach of Caron and Pratt [6 to 8] essentially gives this result.

In the context of these considerations Hubbard's improved solution [9], which yields a finite I_0, does not represent a real improvement: it does not yield a sharp Fermi surface in the metallic phase [10] and the density of states goes to zero and not to infinity at the critical value I_0. It should be pointed out that both the specific heat and susceptibility density of states in the metallic phase of V_2O_3 are experimentally found to be quite enhanced [11].

Inspection of Fig. 1 shows that the knowledge of the function $\mu(n)$ gives the same information as the knowledge of the bands or of the density of states: a low density of states is connected with a rapid increase and a high density of states with a slow increase of $\mu(n)$. A gap corresponds to a jump in $\mu(n)$ and an infinite density of states to a jump in $n(\mu)$. Whereas, however, in the presence of electron–electron interactions both $\varrho(E)$ and $E_{\boldsymbol{K}}$ depend on higher correlation functions and are defined as functions of n only on the basis of a given decoupling (or equivalent) procedure, $\mu(n)$ is a physically well defined function. Knowing the chemical potential on can interpret all results in a simple way in terms of a one-electron picture.

3. On the Chemical Potential in the Cluster Approach to the Hubbard Model

According to the cluster approach of Caron and Pratt [6 to 8] the Hubbard Hamiltonian (1) with nearest neighbour hopping only is replaced by $H' = \sum_i H_i$ with

$$H_i = \frac{zt}{2} \sum_\sigma \left[c_{i,\sigma}^+ \langle c_{i,\bar\sigma}^- \rangle + \langle c_{i,\bar\sigma}^+ \rangle\, c_{i,\sigma}^-\right] + I n_{i\uparrow} n_{i\downarrow} - \mu \sum_\sigma n_{i,\sigma}\,, \tag{5}$$

where z is the number of nearest neighbours. It can be proved [8] that H' is the best single-site approximation to the Hubbard Hamiltonian in the sense of $\delta\langle (H - H')^2\rangle = 0$ if the anticommutative properties of electrons at different sites are neglected.

The complete basis of H_i is given by the states

$$\psi_1 = |0\rangle\,, \quad \psi_{2/3} = \frac{1}{\sqrt{2}}(|\uparrow\rangle \pm |\downarrow\rangle)\,, \quad \psi_4 = |\downarrow\uparrow\rangle\,.$$

The corresponding 4×4-matrix can be easily diagonalized. The eigenvalues are

$$\left.\begin{aligned} E_{1/2} &= -\frac{\mu}{2} \pm \sqrt{\frac{z^2t^2}{2}|\alpha|^2 + \frac{\mu^2}{4}}\,,\\ E_{3/4} &= -\frac{3\mu - I}{2} \pm \sqrt{\frac{z^2t^2}{2}|\alpha|^2 + \frac{1}{4}(\mu - I)^2}\,. \end{aligned}\right\} \tag{6}$$

The corresponding eigenstates are

$$\left.\begin{aligned} \varphi_{1/2} &= \psi_1 + \frac{\sqrt{2}}{zt\alpha}\left[-\frac{\mu}{2} \pm \sqrt{\frac{z^2t^2}{2}|\alpha|^2 + \frac{1}{4}\mu^2}\,\right]\psi_2\,,\\ \varphi_{3/4} &= \psi_3 + \frac{\sqrt{2}}{zt\alpha}\left[-\frac{\mu - I}{2} \pm \sqrt{\frac{z^2t^2}{2}|\alpha|^2 + \frac{1}{4}(\mu - I)^2}\,\right]\psi_4\,, \end{aligned}\right\} \tag{7}$$

where we have used the abbreviation $\alpha = \langle c_{i,\uparrow}^- \rangle = \langle c_{i,\downarrow}^- \rangle$.

In the sense of any cluster or decoupling approximation α has to be determined self-consistently from the derivation of the partition function $Z = \mathrm{Sp}\, e^{-\beta H_i}$:

$$\langle c_{i,\sigma}^\pm \rangle = -\frac{2}{\beta z t}\, \frac{\partial \ln Z}{\partial \langle c_{i,\sigma}^\mp \rangle}. \tag{8}$$

With the help of the eigenvalues (6) the derivation is easily carried out with the result

$$\langle c_{i,\sigma}^\pm \rangle = -\frac{zt}{2} \langle c_{i,\sigma}^\pm \rangle \left[\frac{e^{\beta\mu/2}\dfrac{1}{a}\sinh \beta a + e^{\beta(3\mu - I)/2}\dfrac{1}{b}\sinh \beta b}{e^{\beta\mu/2}\cosh \beta a + e^{\beta(3\mu - I)/2}\cosh \beta b}\right], \tag{9}$$

where a and b are the roots

$$a = \sqrt{\frac{z^2t^2}{2}|\alpha|^2 + \frac{1}{4}\mu^2}$$

$$b = \sqrt{\frac{z^2t^2}{2}|\alpha|^2 + \frac{1}{4}(\mu - I)^2}.$$

For $kT = 1/\beta = 0$ we get from this

$$|\alpha|^2 = \begin{cases} \frac{1}{8z^2t^2}[z^2t^2 - 4\mu^2] & \text{if } I \geqq 2\mu \text{ and } z^2t^2 \geqq 4\mu^2, \\ 0 & \text{if } I \geqq 2\mu \text{ and } z^2t^2 \leqq 4\mu^2, \\ \frac{1}{8z^2t^2}[z^2t^2 - 4(\mu - I)^2] & \text{if } I \leqq 2\mu \text{ and } z^2t^2 \geqq 4(\mu - I)^2, \\ 0 & \text{if } I \leqq 2\mu \text{ and } z^2t^2 \leqq 4(\mu - I)^2. \end{cases} \tag{10}$$

It has been proposed by Caron and Pratt [6] that $\alpha \neq 0$ and $\alpha = 0$ describe the metallic and insulating phases, respectively. The phase diagram for $n = I$ in the $kT - I/\Delta$-plane, where $\Delta = 2z|t|$ is the bandwidth, has been calculated [7] and it has been proved that the optical mass

$$m_{xx}^{-1} = \frac{1}{N}\sum_{K,\sigma}\frac{\partial^2\varepsilon(\boldsymbol{K})}{\partial \boldsymbol{K}_x^2}\langle n_{\boldsymbol{K},\sigma}\rangle \tag{11}$$

is connected with α by

$$m_{xx}^{-1} = 4a^2|t|\,|\alpha|^2, \tag{12}$$

where a is the lattice constant of a cubic lattice. The optical mass on the other hand is connected with the frequency dependent conductivity $\sigma_{xx}(\omega)$ by

$$\frac{1}{\pi}\int_{-\infty}^{+\infty} \operatorname{Re}\sigma_{xx}(\omega)\, d\omega = e^2 m_{xx}^{-1}. \tag{13}$$

The regions in which $\alpha \neq 0$ or $\alpha = 0$, therefore, indeed can be identified with the metallic or insulating phase, respectively.

In the metallic phase with $\alpha \neq 0$ the ground state of H_i is φ_2 or φ_4 for $I \geqq 2\mu$ or $I \leqq 2\mu$. The mean number of particles $n = \langle 0|\, n_{i\uparrow} + n_{i\downarrow}|0\rangle$ as a function of μ in the ground state easily can be calculated and is given by

$$\begin{aligned} n &= 0 && \text{if } \mu \leqq -\frac{z|t|}{2} \\ n &= \frac{1}{2} + \frac{\mu}{z|t|} && \text{if } -\frac{z|t|}{2} \leqq \mu < \frac{I}{2}, \\ n &= \frac{3}{2} + \frac{\mu - I}{z|t|} && \text{if } \frac{I}{2} < \mu \leqq I + \frac{z|t|}{2}, \\ n &= 2 && \text{if } I + \frac{z|t|}{2} \leqq \mu, \end{aligned} \tag{14}$$

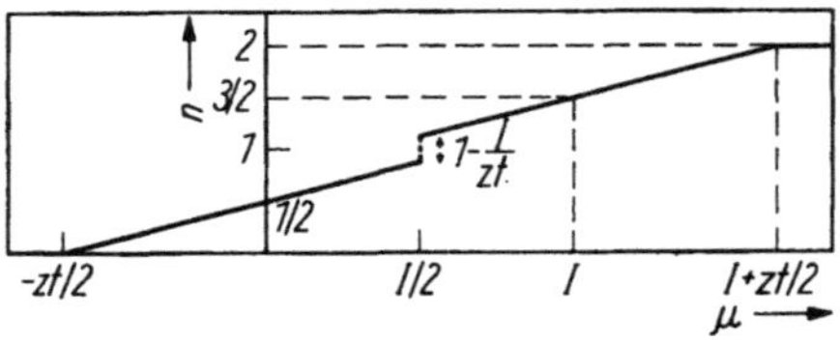

Fig. 2. $n(\mu)$ in the metallic phase within the cluster approximation to the Hubbard model

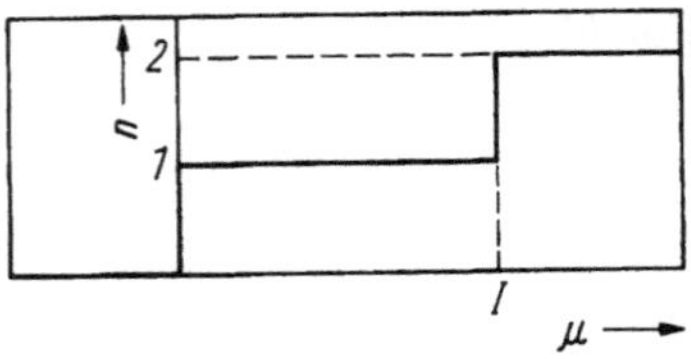

Fig. 3. $n(\mu)$ in the insulating phase within the cluster approximation to the Hubbard model

(see Fig. 2). There is a jump in n of the magnitude $1 - I/z|t|$ at $\mu = I/2$ corresponding to an infinite density of states at the Fermi level in a one-electron picture. This jump vanishes at the metal–insulator transition which takes place at $I = z|t|$.

In the insulating phase with $\alpha = 0$ we get in the same way

$$\left.\begin{aligned} n &= 0 \quad \text{if} \quad \mu < 0\,, \\ n &= 1 \quad \text{if} \quad 0 < \mu < I\,, \\ n &= 2 \quad \text{if} \quad I < \mu \end{aligned}\right\} \tag{15}$$

(see Fig. 3). In this phase $H' = \sum_i H_i$ is identical with the atomic limit of the Hubbard Hamiltonian and we get, therefore, the same function $n(\mu)$ as if we would have only the two levels $E = 0$ and $E = I$.

The general behaviour of $n(\mu)$ according to this simple cluster approach is in full accordance with that deduced from Gutzwiller's result, the jump of n at $\mu = I/2$ on the metallic site hardly can be supposed to be a consequence of the approximation only.

4. Conclusions

On the basis of the cluster approach of Caron and Pratt to the Hubbard model it has been proved that $n(\mu)$ has a jump at $\mu = I/2$ on the metallic site of the metal–insulator transition. This jump corresponds to an infinite density of states at the Fermi level. Comparing this result with those of Gutzwiller and Hubbard we propose this to be an inherent feature of the Hubbard model and not a consequence of the approximation used. The density of states deduced from experimental results both of the susceptibility and the specific heat in the metallic phase of V_2O_3 is unusually large [11], and can be explained in this way. An infinite density of states at the Fermi level is connected with an infinite renormalized mass [8, 11]. In difference to a Wilson band overlap transition the metal–insulator transition described by the Hubbard model is characterized by a finite number of charge carriers of infinite mass and not by a vanishing number of charge carriers of finite mass.

Acknowledgements

The author wants to express his gratidude to the members of the research group "Theory of Magnetism" of the Technical University of Dresden and especially to Mr. K.-H. Heinig for the opportunity of many fruitful discussions.

References

[1] J. Hubbard, Proc. Roy. Soc. **A276**, 238 (1963).
[2] J. A. Wilson, Adv. Phys. **21**, 143 (1972).
[3] W. Nolting, Z. Phys. **255**, 25 (1972).
[4] A. B. Harris and R. V. Lange, Phys. Rev. **157**, 295 (1967).
[5] M. C. Gutzwiller, Phys. Rev. **137**, A1726 (1965).
[6] L. G. Caron and G. W. Pratt, Rev. mod. Phys. **40**, 802 (1968).
[7] J. Monecke, phys. stat. sol. (b) **51**, K81 (1972).
[8] J. Monecke, phys. stat. sol. (b) **57**, K21 (1973).
[9] J. Hubbard, Proc. Roy. Soc. **A281**, 401 (1964).
[10] D. M. Edwards and A. C. Hewson, Rev. mod. Phys. **40**, 810 (1968).
[11] W. F. Brinkman and T. M. Rice, Phys. Rev. B **2**, 4302 (1970).

(Received May 29, 1974)

phys. stat. sol. (b) 65, 239 (1974)

Subject classification: 13; 14.4; 22.9

Institut für Angewandte Physik der Universität Hamburg

Electron Energy Loss Measurements on Fluorene ($C_{13}H_{10}$) Vapour and Single Crystals[1])

By

H. VENGHAUS[2]) and H.-J. HINZ

Electron energy loss spectra from 0 to 40 eV were measured with 55 keV electrons on fluorene single crystals. The momentum transferred to the crystal by the electrons scattered inelastically was taken to be parallel to the main axes of the dielectric tensor $\boldsymbol{\varepsilon}(E)$. The energy loss functions Im $(-1/\varepsilon_{ii})$ $(i = x, y, z)$ were derived from the spectra and the tensor elements $\varepsilon_{ii}(E)$ calculated from Im $(-1/\varepsilon_{ii})$ by Kramers-Kronig analysis. The results obtained are discussed in terms of excited molecular states determined by electron energy loss measurements on fluorene vapour in the present investigation. The results are also related to optical absorption measurements performed by other authors on fluorene in solution.

Elektronen-Energieverlustspektren wurden mit 55 keV-Elektronen an Fluoren-Einkristallen im Bereich von 0 bis 40 eV gemessen. Der von den unelastisch gestreuten Elektronen übertragene Impuls wurde parallel zu den Hauptachsen des dielektrischen Tensors $\boldsymbol{\varepsilon}(E)$ gewählt. Aus den Spektren wurden die Energieverlustfunktionen Im $(-1/\varepsilon_{ii})$ $(i = x, y, z)$ bestimmt und daraus durch Kramers-Kronig-Transformation die Hauptachsenelemente $\varepsilon_{ii}(E)$ des dielektrischen Tensors berechnet. Die angeregten Festkörperzustände werden diskutiert ausgehend von der Kenntnis angeregter Molekülzustände, die im Rahmen der vorliegenden Arbeit durch Energieverlustmessungen an Fluorendampf bestimmt wurden. Darüber hinaus werden die Resultate dieser Arbeit in Beziehung gesetzt zu optischen Absorptionsmessungen anderer Autoren an Fluoren in Lösung.

1. Introduction

Fast electrons passing through thin films of a solid or the vapour of a given substance may be scattered inelastically and excite electronic states of the specimen. By measuring the energy loss of the primary electrons information can be obtained about the excited electronic states of the substance under investigation, in particular, the dielectric function $\boldsymbol{\varepsilon}$, which reflects the electronic properties of a solid, can be determined. The electron energy loss function is an isotropic function for cubic crystals for small wave vectors. For crystals with lower symmetry the loss function has anisotropic character, demonstrated e.g. by Zeppenfeld in the case of graphite (cf. [1]), where the dielectric function $\boldsymbol{\varepsilon}$ is a tensor with two different main axis elements.

Energy loss experiments were also done on several biaxial, molecular crystals, and the dielectric tensor of anthracene [2], phenanthrene and chrysene [3], and pyrene [4] was derived from the energy loss functions determined experimentally. It turned out in those investigations that many of the low lying π–π^* excitations ($E \lesssim 8$ eV) of molecular crystals can be explained in a straight-

[1]) Paper partly presented on the VI. Molecular Crystal Symposium, Schloß Elmau (FRG) 1973.

[2]) Present address: Max-Planck-Institut für Festkörperforschung, D 7000 Stuttgart 1, Postfach 1099.

forward manner, relating the crystal excitations to excited states of the corresponding molecule. For this reason electron energy loss measurements were performed not only on fluorene single crystals, but also on the corresponding vapour in order to determine the excited states of the free molecule, and the results of the present investigation demonstrate the close relation between excited molecular and crystal states.

The electronic properties of a substance in the vapour phase can be characterized by a dielectric function $\varepsilon_{\text{vap}}(E)$. The propability p for the excitation of an excited state with energy E and wave vector $\boldsymbol{q}$ by electron impact (in the non-relativistic limit and without occurrence of surface excitations and energy losses due to radiation) is [1]

$$p \sim q^{-2}\,\text{Im}\left[-\frac{1}{\varepsilon_{\text{vap}}(E)}\right] \approx q^{-2}\,\text{Im}\,[\varepsilon_{\text{vap}}(E)]\,, \tag{1}$$

where the approximation is due to the fact that

$$\text{Re}\,(\varepsilon_{\text{vap}}) \approx 1 \qquad \text{and} \qquad \text{Im}\,(\varepsilon_{\text{vap}}) \ll 1\,.$$ [3])

If the same substance is dissolved in a liquid, the electronic properties of the substance can be described by a dielectric function $\varepsilon_{\text{sol}}(E)$ or by the extinction coefficient $\mu(E)$, which can be determined by optical measurements. $\mu(E)$ and $\varepsilon_{\text{sol}}(E)$ are related by

$$\mu = \frac{2E}{\hbar c}\,k = \frac{2E}{\hbar c}\left\{\frac{[\text{Re}^2\,(\varepsilon_{\text{sol}}) + \text{Im}^2\,(\varepsilon_{\text{sol}})]^{1/2} - \text{Re}\,(\varepsilon_{\text{sol}})}{2}\right\}^{1/2} \approx \frac{2E}{\hbar c}\,\frac{\text{Im}\,(\varepsilon_{\text{sol}})}{2}\,; \tag{2}$$

$$\varepsilon = (n + ik)^2\,, \qquad \text{Re}\,(\varepsilon_{\text{sol}}) \approx 1\,, \quad \text{Im}\,(\varepsilon_{\text{sol}}) \ll 1\,.$$

In a good approximation ε_{sol} and ε_{vap} are related by [5]

$$\text{Im}\,[\varepsilon_{\text{vap}}(E)] = \text{const}\;\text{Im}\,\{\varepsilon_{\text{sol}}[E + \Delta_{\text{s}}(E)]\}\,. \tag{3}$$

$\text{Im}(\varepsilon_{\text{vap}})$ and $\text{Im}(\varepsilon_{\text{sol}})$ (or $\mu(E)/E$) exhibit the same shape, but corresponding structure occurs with an energy displacement Δ_{s}, where Δ_{s} represents the so-called "solution shift". In the case of aromatic hydrocarbon molecules, Δ_{s} is in general negative with absolute values up to some 0.1 eV for π–π^* excitations. The actual value of Δ_{s} depends on the oscillator strength (i.e. the intensity) of the different transitions, and to a smaller extent on the solvent.

As mentioned at the beginning, excited states of molecular crystals are closely related to excited states of the corresponding isolated molecule. If the electronic properties of a crystal are expressed in terms of the dielectric tensor $\boldsymbol{\varepsilon}$, this relation can be written in a first approximation [6, 7]

$$\text{Im}\,[\varepsilon_{ii}(E)] = \text{const}\;\text{Im}\,\{\varepsilon^{i}_{\text{vap}}[E + \Delta_{\text{c}}(E) + \Delta^{i}_{\text{D}}(E)]\}\,; \qquad i = x, y, z\,, \tag{4}$$

where the superscript "i" of $\varepsilon^{i}_{\text{vap}}$ indicates that from all molecular transitions contributing to ε_{vap} only those have to be taken into account which give rise to crystal states polarized along the i-axis of the dielectric tensor. If one molecular state gives rise to more than one crystal state, the total oscillator strength of the molecular transition is distributed over these states. Their relative inten-

[3]) Real and imaginary part of a complex quantity z are denoted by Re (z) and Im (z), respectively.

sities correspond to the projection of the molecular transition moments on the dielectric tensor main axes. Δ_c represents the "crystal shift" and has in general negative values of some 0.1 eV for π–π^* excitations of aromatic hydrocarbon crystals. Δ_D^i represents the "Davydov splitting". The actual value of Δ_D^i depends mainly on the oscillator strength of the transition and can be of the order of several 0.1 eV in the case of strong excitations.

Im (ε_{ii}) may exhibit additional structure, which is not expressed by (4), if symmetry-forbidden molecular transitions become allowed on account of lower symmetry of the crystal, and due to "stealing" of oscillator strength such crystal excitations may be observed with considerable intensity.

Summarizing the foregoing section yields: Optical absorption measurements on solution and electron energy loss measurements on vapour yield spectra of similar shape; the maxima of the solution spectra occur in general at slightly different (lower) energies than in vapour measurements. In the case of molecular crystals the imaginary parts of the dielectric tensor elements Im (ε_{ii}) exhibit maxima approximately at the energies of maxima observed in solution spectra. Additional maxima of the functions Im(ε_{ii}) may occur due to the removal of symmetry restrictions by lower crystal symmetry.

2. Experimental

Energy loss spectra on fluorene vapour were taken using an electron spectrometer with monochromator described in detail in [8] and [9]. A description of the filter lens serving as monochromator is given in [10].

The substance to be investigated was put into a copper chamber of $(3\ \text{cm})^3$ size with apertures of 200 µm diameter on the top and the bottom. Heating the chamber produced a vapour of approximately 1 Torr pressure within the chamber, whereas the vacuum within the recipient of the spectrometer simultaneously changed from $\lessapprox 10^{-6}$ to 10^{-5} Torr.

The spectra were taken with the following experimental data:

50 keV energy of the primary electrons;

75 meV half-width of the energy distribution of the primary electron beam;

2.5×10^{-4} rad half-width of the angular distribution of the primary electrons;

$\approx 5\ \mu\text{A/cm}^2$ electron current density within the chamber.

All spectra were recorded with zero scattering angle, i.e. in the forward direction.

The experiments on the single crystals were performed using the same electron spectrometer as in the foregoing experiments, with the same experimental data [4].

Single crystals were obtained from a solution of fluorene in toluene or benzene (for details see [3] and [11]). In order to prevent the fluorene crystals from evaporation during the experiments they were cooled to approximately -100 °C.

3. Results

3.1 Energy loss functions, dielectric tensor elements, and vapour spectrum

The energy loss functions Im $(-1/\varepsilon_{ii})$ are shown in Fig. 1. Fluorene crystallizes in the orthorhombic system and forms thin plates with the crystallographic a- and b-axis (x- and y-axis of the dielectric tensor) within the plane of these plates [12]. In order to get the momentum $\hbar \boldsymbol{q}$ — transferred by the scattered electrons — parallel to the tensor's z-axis when Im $(-1/\varepsilon_{zz})$ is to be determined,

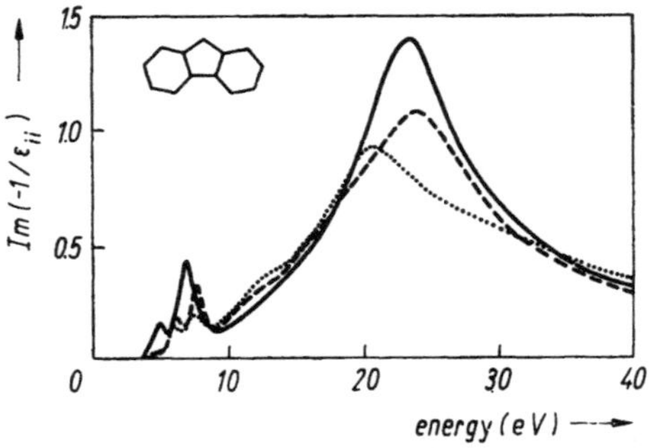

Fig. 1. Energy loss functions of fluorene. Dotted line: Im $(-1/\varepsilon_{xx})$; dashed line: Im $(-1/\varepsilon_{yy})$; full line: Im $(-1/\varepsilon_{zz})$

the angle α between the crystal's ab plane and the direction of the incoming electrons should be very small (since the direction of $\hbar \boldsymbol{q}$ is nearly perpendicular to the direction of the incoming electrons) [1, 2]. In the experiments α was approximately 25°. The angle between $\hbar \boldsymbol{q}$ and the tensor z-axis depends on the energy loss E and decreases from 25° for $E = 0$ eV to $\approx 10°$ for $E = 25$ eV. Accordingly the energy loss spectra measured under those conditions, where $\hbar \boldsymbol{q}$ and the z-axis are not parallel, are determined mainly by ε_{zz}, but a dependence on ε_{xx}, ε_{yy}, or ε_{xx} and ε_{yy} — depending on the actual orientation of the crystal — is given as well. For this reason ε_{xx} and ε_{yy} were determined first, and then from the corresponding energy loss spectra Im $(-1/\varepsilon_{zz})$ was derived including appropriate corrections.

Table 1

Energy values of maxima of Im $(-1/\varepsilon_{ii})$

Im $(-1/\varepsilon_{xx})$	Im $(-1/\varepsilon_{yy})$	Im $(-1/\varepsilon_{zz})$
		5.0 ± 0.1 eV
5.95 ± 0.1 eV	6.0 ± 0.1 eV	
		6.95 ± 0.1 eV
7.5 ± 0.1 eV	7.6 ± 0.1 eV	
≈13 eV (shoulder)		
20.8 ± 0.4 eV	23.9 ± 0.3 eV	23.4 ± 0.3 eV

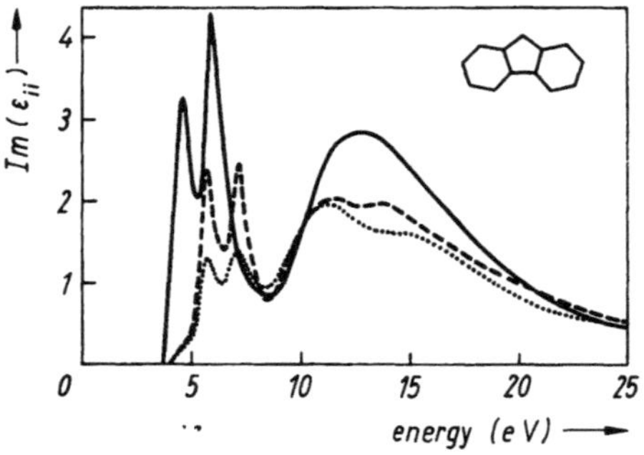

Fig. 2. Imaginary parts of dielectric tensor elements of fluorene. Dotted line: Im (ε_{xx}); dashed line: Im (ε_{yy}); full line: Im (ε_{zz})

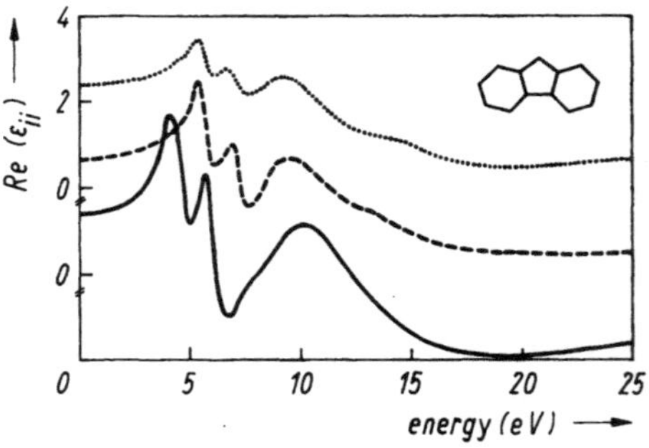

Fig. 3. Real parts of dielectric tensor elements of fluorene. Dotted line: Re (ε_{xx}); dashed line: Re (ε_{yy}); full line: Re (ε_{zz})

Fig. 4. Energy loss functions of fluorene vapour

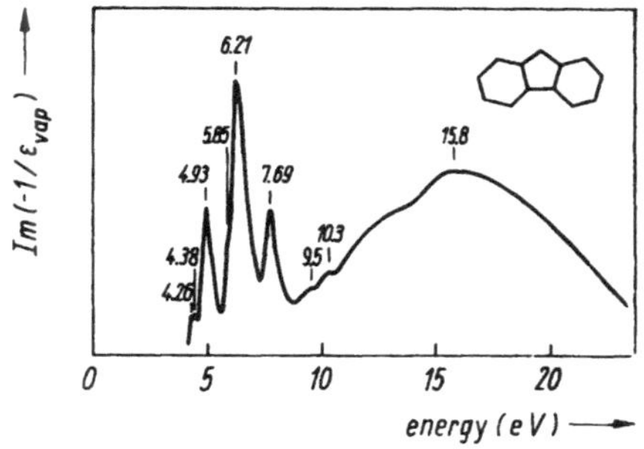

From Im $(-1/\varepsilon_{ii})$ the tensor elements ε_{ii} (see Fig. 2 and 3) were calculated by means of a Kramers-Kronig analysis. The absolute values of Im $(-1/\varepsilon_{ii})$ were determined using the equation

$$\mathrm{Re}\left[\frac{1}{\varepsilon_{ii}(E)}\right] = \frac{1}{n_i^2(E)}, \tag{5}$$

with the following values for the refractive indices n_i corresponding to light vibrating along the i-axis of the dielectric tensor:

$$n_x = 1.578\,, \quad n_y = 1.663\,, \quad n_z = 1.919\,,$$

for $E = 2.27\ \mathrm{eV} \equiv \lambda = 5460\ \text{Å}$ [12].

For more details how to deduce the energy loss functions Im $(-1/\varepsilon_{ii})$ from energy loss spectra and how to calculate the tensor elements $\varepsilon_{ii}(E)$ from Im $[-1/\varepsilon_{ii}(E)]$, see [1 to 3].

The energy loss function Im $(-1/\varepsilon_{\mathrm{vap}}) \approx$ Im $(\varepsilon_{\mathrm{vap}})$ of fluorene vapour is shown in Fig. 4 in arbitrary units.

3.2 Discussion of low-energy excitations

3.2.1 General considerations

The space group of the orthorhombic fluorene crystal is D_{2h}^{16} or Pnam, respectively [20]. There are 4 molecules within the unit cell with general coordinates: (x, y, z), $(x + \frac{1}{2}, \frac{1}{2} - y, z)$, $(\bar{x}, \bar{y}, \bar{z})$, and $(\frac{1}{2} - x, \frac{1}{2} + y, \bar{z})$. The group consisting of all symmetry operations applicable to the fluorene molecule is C_{2v}. Dipole transitions are allowed with three polarization directions (i.e. orientations of the dipole transition moment) only: longitudinal (i.e. in the direction of the longest extension of the molecule; irreducible representation of the wave function: B_1), transverse (A_1), and perpendicular to the plane of the molecule (B_2).

If a molecular crystal has N unit cells and contains n molecules per unit cell, the crystal wave functions Ψ are linear combinations of nN wave functions χ_{ik} ($i = 1, \ldots, n$; $k = 1, \ldots, N$; χ_{ik} is the wave function of a state where the i-th molecule in the k-th cell is excited, while all other molecules are in the ground state):

$$\Psi = \sum_{i,k} c_{ik}\chi_{ik}\,.$$

In order to have the requirement

$$S\Psi = \pm\Psi$$

fulfilled, where S is a symmetry operation transforming the crystal into itself, the following restrictions apply to the coefficients c_{ik}:

1. All c_{ik} with fixed i and $k = 1, \ldots, N$ must be equal.
2. $c_{ik} = \pm c_{jk} (i \neq j)$.
3. If c_{ik} with different sign occur, the number of c_{ik} with either sign must be equal.

Due to these conditions the number of crystal states corresponding to one excited molecular state and allowed by symmetry is n, i.e. equal to the number of molecules per unit cell. Whether a particular of these n states is actually allowed or not depends on the related transition moment. A state is regarded as allowed only if the transition moment is $\neq 0$. Dipole transition moments of crystal states are obtained combining the corresponding molecular transition moments in the same way as the molecular wave functions were combined in order to get the crystal wave function.

In the case of fluorene excited molecular states with symmetry A_1 or B_2 give rise to two allowed crystal states each, characterized by the following wave functions:

$$\Psi(B_{2u}) = \varphi_1(M) - \varphi_2(M) - \varphi_3(M) + \varphi_4(M) \qquad [\|\boldsymbol{b}, \varepsilon_{yy}] , \tag{6}$$

$$\Psi(B_{3u}) = \varphi_1(M) + \varphi_2(M) - \varphi_3(M) - \varphi_4(M) \qquad [\|\boldsymbol{a}, \varepsilon_{xx}] ; \tag{7}$$

$$M = A_1, B_2 .$$

Here the symmetry of wave functions is added in parentheses, the polarization direction and the related tensor element of the crystal state are added in brackets,

$$\varphi_i = \frac{1}{\sqrt{4N}} \sum_{k=1}^{N} \chi_{ik} .$$

Excited molecular states with symmetry B_1 give rise to one crystal state each, where

$$\Psi(B_{1u}) = \varphi_1(B_1) + \varphi_2(B_1) - \varphi_3(B_1) - \varphi_4(B_1) \quad [\|\boldsymbol{c}, \varepsilon_{zz}] , \tag{8}$$

is the related wave function.

3.2.2 Discussion

The lowest excitations of the fluorene molecule are observed in the vapour spectrum at 4.26 and 4.38 eV respectively and at slightly lower energies in solution spectra (see Table 2), but corresponding crystal states could not be observed. Assuming longitudinal polarization for the molecular excitation [13], maxima of $\mathrm{Im}(\varepsilon_{zz})$ at 4.1 to 4.2 eV are expected, but apparently they are too weak to be observed in the present investigation.

$\mathrm{Im}(\varepsilon_{zz})$ exhibits the first maximum at 4.7 eV, the corresponding crystal state has symmetry B_{1u} and is related to a longitudinal polarized molecular excitation. This is in accordance with earlier investigations. A molecular excitation is observed in absorption measurements on fluorene in solution at 4.7 to 4.75 eV and assumed to be longitudinal polarized. In the vapour measurements the corresponding maximum occurs at 4.93 eV yielding ≈ -0.2 eV for the crystal shift and the solution shift of this excitation.

The first maxima of $\mathrm{Im}(\varepsilon_{xx})$ and $\mathrm{Im}(\varepsilon_{yy})$ are observed at 5.75 eV. They are related to crystal states with symmetry B_{3u} and B_{2u} respectively. The polari-

Table 2

excitation energies					symmetry of molecular state
of crystal states			of molecular states		
Im (ε_{xx})	Im (ε_{yy})	Im (ε_{zz})	in solution	vapour	
			4.12 eV	4.26 ± 0.03 eV	
			4.28 eV	4.38 ± 0.03 eV	
		4.7 ± 0.1 eV	4.70 to 4.75 eV	4.93 ± 0.03 eV	B_1
5.75 ± 0.1 eV	5.75 ± 0.1 eV		5.46 to 5.6 eV	5.85 eV (shoulder)	A_1
		6.0 ± 0.1 eV	5.95 to 6.1 eV	6.21 ± 0.03 eV	B_1
7.1 ± 0.1 eV	7.25 ± 0.1 eV		7.5 eV	7.69 ± 0.03 eV	A_1
				9.5 ± 0.1 eV	
				10.3 ± 0.2 eV	
11.2 ± 0.3 eV	11.6 ± 0.5 eV			12.7 eV (shoulder)	
		13.0 ± 0.5 eV			
	13.8 ± 0.5 eV				
14.8 ± 0.5 eV			[13 to 19]		
				15.8 ± 0.5 eV	

zation of the underlying molecular transition cannot be deduced unambiguously from the measurements on the crystal (cf. equations (6) and (7)). The higher intensity of the Im(ε_{yy}) maximum compared to the maximum of Im(ε_{xx}) indicates transverse polarization for the molecular transition. This conclusion, however, is rigorously valid only as long as second-order effects are negligible. The assignment "transverse" is in accordance with absorption measurements with polarized light on fluorene single crystals [13]. The electron energy loss experiments on fluorene vapour exhibit only a weak shoulder in the energy region of the excitation under consideration (cf. Fig. 4).

The second maximum of Im (ε_{zz}) is observed at 6.0 eV and, similar to the 4.7 eV maximum, corresponds to a crystal state with symmetry B_{1u}; it is related to a longitudinal polarized molecular excitation, observed at 6.21 eV in the vapour spectrum.

Two more maxima occur at 7.1 eV in Im (ε_{xx}) and at 7.25 eV in Im (ε_{yy}), and the most probable interpretation is the same as in the case of the 5.75 eV maxima. But two additional facts should be mentioned: The maxima are observed with an energy separation of 0.15 eV, which is a rather large Davydov splitting, and secondly the excitation energy is strongly shifted to lower energies compared to the corresponding molecular excitation. The latter is observed at 7.69 eV in the vapour and at 7.5 eV in solution, thus showing a solution shift of ≈ -0.2 eV as in the case of the other excitations. An explanation of the strong crystal shift of ≈ -0.5 eV cannot be given in a straightforward manner, but requires the consideration of interactions between different crystal excitations.

3.3 High-energy excitations

3.3.1 Im (ε_{ii})

Beyond the π–π^* excitation region ($E \lessapprox 8$ eV) additional structures occur in the vapour spectra and in the functions Im $(-1/\varepsilon_{ii})$ as well.

Photoelectron spectra of aromatic hydrocarbons like anthracene, phenanthrene, or pyrene [21, 22], covering the range 7 eV $\lessapprox E \lessapprox$ 16 eV, exhibit two different regions: For 7 eV $\lessapprox E \lessapprox$ 10 to 11 eV a certain number of very sharp maxima is observed in contrast to broad maxima at higher energies (10 to 11 eV $\lessapprox E \lessapprox$ 16 eV). In accordance with theoretical calculations the narrow maxima are assumed to be due to the ionization of π-states, whereas the broad structures are attributed to σ-states.

Taking into account these results, the 9.5 and 10.3 eV maxima of fluorene vapour may be regarded as due to π-state ionizations and the remaining structures related to σ-states, the latter being the most probable explanation for the broad maxima of Im $(-1/\varepsilon_{ii})$ in the 10 eV $\lessapprox E \lessapprox$ 17 eV region as well.

A more detailed and quantitative discussion taking also into consideration the possible occurrence of maxima below the ionization energy of a given state due to Rydberg series, will not be attempted in this paper.

3.3.2 Im $(-1/\varepsilon_{ii})$

In addition to the structure corresponding to the excitations discussed above, the energy loss functions Im$(-1/\varepsilon_{ii})$ exhibit a strong maximum in the 20 to 24 eV region.

Im (ε_{ii}) decreases monotonously at these energies and Re (ε_{ii}) has rather small values. For this reason the broad maxima of the energy loss functions are attributed to collective excitations. Calculating a plasma frequency according to the free electron gas formula

$$\omega_{\text{p}}^2 = \frac{4\pi n e^2}{m}, \tag{9}$$

where e is the electron charge, m the free electron mass, and n the electron density, (taking into account 4 electrons per carbon atom and 1 electron per hydrogen atom), yields $\omega_{\text{p}} \approx$ 20 to 21 eV. This is approximately the energy of the Im $(-1/\varepsilon_{xx})$ maximum, whereas the maxima of Im $(-1/\varepsilon_{yy})$ and Im $(-1/\varepsilon_{zz})$ occur at higher energies. The same holds for other hydrocarbons, e.g., anthracene [2], phenanthrene, chrysene [3], pyrene [4], and hence the similarity between Im $(-1/\varepsilon_{yy})$ and Im $(-1/\varepsilon_{zz})$ in contrast to Im $(-1/\varepsilon_{xx})$ in the high-energy region should be emphasized, since in the low-energy region Im$(-1/\varepsilon_{xx})$ and Im$(-1/\varepsilon_{yy})$ are closely related to one another differing significantly from Im$(-1/\varepsilon_{zz})$. More accurate results compared to those obtained from (9) can be expected from a calculation taking into account an effective mass m_i and an effective electron density n_i.

4. Conclusion

The low-energy excitations of fluorene single crystals can be related to excitations observed in the fluorene vapour spectrum. The crystal excitations can be interpreted satisfactorily in terms of excited molecular states, yielding simultaneously the polarization of the latter. The crystal transitions occur generally some 0.1 eV below the transition energies of the free molecule. The "crystal shift" is in general equal to or greater than the "solution shift".

Maxima in the vapour spectra in the 10 eV region can be attributed to the ionization of π-states, broad maxima of the vapour spectra and of the functions Im (ε_{ii}) in the 10 to 16 eV region are related to σ-excitations.

A strong maximum of the energy loss function $\mathrm{Im}\,(-1/\varepsilon_{xx})$ is observed at ≈ 20 eV. $\mathrm{Im}\,(-1/\varepsilon_{yy})$ and $\mathrm{Im}\,(-1/\varepsilon_{zz})$ exhibit corresponding maxima at ≈ 24 eV, attributed to collective excitations.

Acknowledgements

The authors are indebted to Prof. H. Raether for discussions. Support by the Deutsche Forschungsgemeinschaft (to H. V.) is gratefully acknowledged.

References

[1] J. Daniels, C. von Festenberg, H. Raether, and K. Zeppenfeld, Springer Tracts mod. Phys. **54**, 77 (1970).
[2] H. Venghaus, Z. Phys. **239**, 289 (1970).
[3] H. Venghaus, J. chem. Phys. **57**, 3478 (1972).
[4] H. Venghaus, phys. stat. sol. (b) **54**, 671 (1972).
[5] A. E. Gillam and E. S. Stern, Introduction to Electronic Absorption Spectroscopy in Organic Chemistry, F. Arnold, London 1958.
[6] D. P. Craig and S. H. Walmsley, in: Physics and Chemistry of the Organic Solid State, Vol. I, Ed. D. Fox, M. M. Labes, and A. Weissberger, Wiley, New York/London 1963 (p. 585).
[7] A. S. Davydov, Theory of Molecular Excitons, Plenum Press, New York/London 1971.
[8] H. Meyer-Wachsmuth, Diploma Work, Universität Hamburg, 1968.
[9] U. Killat, J. Phys. C, in the press.
[10] W. Hartl, Z. Phys. **191**, 487 (1967).
[11] J. Jäger, Ann. Phys. (Germany) (7), **22**, 147 (1969).
[12] N. Winchell, The Optical Properties of Organic Compounds, 2nd edition, Academic Press, 1954.
[13] A. Bree and R. Zwarich, J. chem. Phys. **51**, 903 (1969).
[14] V. Henri, in: International Critical Tables of Numerical Data — Physics, Chemistry, and Technology, Vol. V, Ed. E. W. Washburn, McGraw-Hill Publ. Co., New York 1929 (p. 359).
[15] G. J. Hoijtink and P. J. Zandstra, Mol. Phys. **3**, 371 (1960).
[16] E. M. Layton, J. mol. Spectroscopy **5**, 181 (1960).
[17] R. L. Hummel and K. Ruedenberg, J. chem. Phys. **66**, 2334 (1962).
[18] H. H. Perkampus, I. Sandemann, and C. J. Timmons (Ed.), UV Atlas of Organic Compounds, Butterworths, London 1966 (spectrum E 7/8).
[19] V. I. Danilova, O. K. Bazyl, and A. N. Novikov, Izv. vuzov Fiz. **11**, 121 (1968).
[20] R. W. G. Wykoff, Crystal Structures, Vol. V, Interscience Publishers, New York 1960
[21] R. Boschi and W. Schmidt, Tetrahedron Letters **25**, 2577 (1972).
[22] R. Boschi, J. N. Murrell, and W. Schmidt, Faraday Disc. Chem. Soc. **54**, 116 (1972).

(Received June 14, 1974)

phys. stat. sol. (b) **65**, 249 (1974)

Subject classification: 6 and 14.3; 13.1; 20.3; 22.3

Laboratoire de Physique Appliquée, Ecole Polytechnique Fédérale, Lausanne

Electron–Lattice Interaction in Gallium Selenide

By

PH. SCHMID and J. P. VOITCHOVSKY

In the layered semiconductor GaSe, the interaction of holes with non-polar, fully symmetric, optical phonons governs both the Hall mobility of the carriers and the temperature shift of the direct energy gap. The results are in agreement with theory and indicate that the interaction of charge carriers with acoustic phonons is rather weak.

Dans le sélénïure de gallium, un semiconducteur à structure en couches, l'interaction des trous avec des phonons optiques non polaires et totalement symétriques détermine la mobilité des porteurs d'une part, et le coefficient de température du gap direct d'autre part. Les résultats sont en accord avec la théorie. Ils indiquent que la diffusion des porteurs de charge par les phonons acoustiques est faible.

1. Introduction

Gallium selenide crystallizes in a layer structure. Each layer consists essentially of two layers of Ga atoms sandwiched between two layers of Se atoms. Inside the layers, the bonding is largely covalent, between them it is much weaker [1]. This high dissymmetry between intra- and inter-layer bond strengths gives rise to a strong scattering of the charge carriers by optical phonons polarized perpendicular to the layers [2, 3]. The present paper shows how this specific scattering mechanism provides a consistent interpretation of the Hall mobility of holes and the temperature coefficient of the direct energy gap of GaSe.

It is usually surmised that in layered crystals, the electronic properties reflect the high anisotropy of the structure. For GaSe, however, measurements of the magneto-optical properties of the direct exciton [4] show that the reduced mass tensor of the exciton is nearly isotropic. Furthermore, measurements of the hole mobility both parallel and perpendicular to the layer normal have shown that

$$m_{\parallel}^* \approx 0.25 m_{\perp}^* , \tag{1}$$

i.e., the holes are actually lighter when they move across the layers than when they move along the layers [4]. Moreover, measurements of the Hall constant of p-type material [5] indicate that the density-of-states effective mass of the valence band is close to

$$m_{\mathrm{v}}^* \approx 0.5 m_0 , \tag{2}$$

where m_0 is the free electron mass. All this experimental evidence is in agreement with recent band structure calculations [6], which show that the electronic density of the top valence band extends considerably into the inter-layer space. The electronic properties of GaSe in the vicinity of the Fermi level are therefore clearly three-dimensional. Accordingly, the experiments presented in this paper will be discussed on the basis of a model which takes into account the three-dimensional character of the bands [3].

2. Experiment

2.1 Hall effect

The Hall mobility of holes moving in an electric field parallel to the layer plane has been measured on a series of samples of p-GaSe. The samples were cleaved perpendicularly to the *c*-axis of monocrystalline ingots grown by the Bridgman technique. Typical sample dimensions were 10 mm in diameter and 100 μm in thickness. The measurements were carried out in the van der Pauw [7] geometry in a magnetic field of 8 kG. Since no magnetoresistance effect could be detected when the magnetic field was parallel to the layers, measurements in this geometry were equivalent to measurements with zero magnetic field. The Hall effect was measured by rotating the samples through an angle of $\pm 90^\circ$ with respect to the magnetic field. Constant currents of 10 nA to 10 mA were sent through the samples, and the resulting voltages were measured by a digital voltmeter operated by an automatic data acquisition system. The system controlled the functions of the contacts in the various arrangements of the measurement procedure, as well as the orientation of the sample with respect to the magnetic field.

2.2 Luminescence

The temperature dependence of the direct energy gap can be accurately determined from the shifts of the direct exciton lines. Radiative recombinations of free excitons are easily detected in the photoluminescence spectra of our purest crystals in the whole temperature range extending from liquid helium to room temperature. This rather unusual efficiency of the free exciton luminescence over such a wide temperature range has tentatively been attributed to a resonance between the excitonic levels and the indirect valleys of the conduction band [11].

All luminescence measurements were carried out in a reflection arrangement, most of them with the exciting beam normal to the cleaved surface of the samples, i.e. parallel to the *c*-axis of the crystals. The polarization of the luminessence was analyzed when the emitted light was collected at an angle of 45° with respect to the *c*-axis. This information was of considerable help in identifying the $n = 1$ singlet line, which is very strongly polarized with $\boldsymbol{E} \parallel \boldsymbol{c}$. For excitation we used the green line at 5320 Å of a frequency doubled, pulsed YAG:Nd laser. The maximum peak power of the pulses was 60 kW/cm², pulse duration was 200 ns at a repetition rate of 2 kHz. The emitted light was dispersed by a Spex 1402 double spectrometer and the signal to noise ratio was improved by the use of a boxcar integrator. The luminescence results for the $n = 1, 2$ direct exciton lines have been checked against optical absorption data at 4.2 and 77 °K. The halfwidth of the emission lines is close to 3 meV at low temperatures. This width probably originates from structural imperfections. At high temperatures, the width of the lines is of the order of kT.

3. Hall Mobility of Holes

In Fig. 1, the mobility of a typical p-type sample is plotted as a function of temperature between 80 and 450 °K. At temperatures higher than 300 °K, the present mobility values correspond to the highest values observed by Fivaz

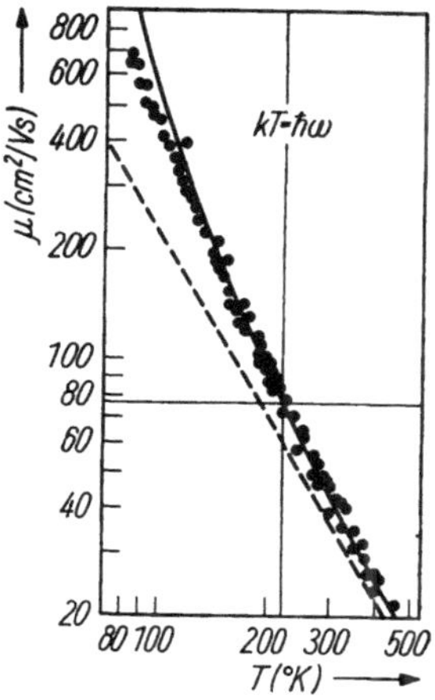

Fig. 1. Hall mobility of holes parallel to the layers in p-GaSe as a function of temperature. The dashed line is the high temperature asymptote of the solid, theoretical curve

and Mooser [1]. The temperature dependence can be expressed approximately as

$$\mu \sim T^{-\gamma} \tag{3}$$

with

$$\gamma = 1.9 \pm 0.1\,. \tag{4}$$

This positive value of γ is indicative of electron-lattice scattering. In contrast with Fivaz and Mooser's findings, γ remains nearly constant down to temperatures as low as 80 °K. The discrepancy is most likely due to impurity scattering. The total ionized impurity concentration deduced from the temperature dependence of the Hall constant is of the order of 3×10^{17} cm^{-3} [5].

Band structure calculations [6] indicate that the valence band of GaSe is non-degenerate and has its maximum at Γ. The closest valence band lies about 0.5 eV below. Since γ is appreciably larger than 3/2, we conclude that the mobility of holes is limited by intraband interactions with long-wavelength optical phonons. In layered structures with sufficient ionicity, scattering by polar phonons can be significant. This is not the case in GaSe: infrared measurements [8] have established that the energy of the LO polar modes is close to 30 meV at the Γ-point. This value is too high to fit the temperature dependence of μ. Furthermore, evaluating the Fröhlich coupling constant α [9] with the dielectric constants given by Leung et al. [8], one finds $\alpha \approx 0.3$, which is much too small to explain the observed values of the mobility. The hole mobility must therefore be governed by short-range interaction with homopolar optical phonons. The relevant phonons are the fully symmetric A_1' modes which are polarized normally to the layers [3]. Their energies, deduced from Raman experiments [10], are 16.7 meV ($A_1'^{(1)}$) and 38.1 meV ($A_1'^{(2)}$). In the temperature range of our experiments, only the energy of the former is low enough to affect significantly the mobility of the charge carriers. The coupling constant for the intraband interactions of three-dimensional carriers with homopolar optical phonon branch is [3]

$$g^2 = \frac{\varepsilon^2 m^{*3/2}}{2\sqrt{2}\,\pi\hbar M N (\hbar\omega)^{3/2}}\,, \tag{5}$$

where M is a reduced ionic mass, N the number of cells per unit volume, $\hbar\omega$ the energy of the optical phonons, m^* the density-of-states mass, and ε the deformation potential for the valence band with respect to the normal coordinate of the $A_1'^{(1)}$ phonon. Fitting the theoretical mobility curve of [3] with the data of Fig. 1 taking $m_\perp = 0.8\, m_0$ [4], one deduces a value of

$$g^2 = 0.25\,, \tag{6}$$

which corresponds to a deformation potential $\varepsilon = 6.6$ eV/Å of the normal phonon coordinate. In this fit, the ratio of Hall to drift mobility was taken equal to 1. This is correct at low temperature, the limiting value at high temperature being $3\pi/8$. Schlüter [6] has evaluated some deformation potentials by varying the structure factors in his band structure calculations. His results indicate that the top of the valence band is most sensitive to a stretching of the Ga–Ga bond. If one computes the polarization vectors of the $A_1'^{(1)}$ phonon at Γ according to the force constant model proposed by Wieting and Verble [10], the fit of the mobility data implies a deformation potential $\varepsilon_g \approx 8$ eV/Å for the Ga–Ga bond, which compares reasonably well with the value $\varepsilon_g \approx 5.8$ eV/Å given by Schlüter.

In the next section we will see that the self-energy of the hole-optical phonon interaction, computed with the same value of the coupling constant, explains the temperature shift of the direct energy gap in a satisfactory manner.

4. Temperature Coefficient of the Gap

The temperature dependence of the energy gap has a two-fold origin:

the thermal expansion of the lattice due to anharmonic terms in the vibrational potential and

the influence of thermal phonons on the energies of the electron states.

With a knowledge of the compressibility and of the coefficient of thermal expansion, the part of the temperature coefficient of the gap due to lattice expansion can be related by a scalar to the pressure coefficient of the gap in cubic AB semiconducting compounds. Such a relation is much more difficult to establish in a layer compound like GaSe because, due to the asymmetry of the chemical bonding, it is of tensorial nature. It has been observed in several materials that the lattice expansion accounts for some 20% only of the temperature coefficient of the gap [12, 13]. This strongly suggests that to a large degree the temperature coefficient of the gap is due to electron–phonon interaction in the crystal. Part of the effect of the phonons can be accounted for by introducing Debye-Waller factors in the band structure calculations. A second contribution to the temperature coefficient arises from the self-energy of the electron–lattice interaction, as was first pointed out by Fan [14]. In view of the critical discussion by Guenzer and Bienenstock [13], it does not seem possible to assign a priori a dominant role to either of these two contributions.

Since the mobility of holes in GaSe can be explained completely by the interaction of the carriers with non-polar, fully symmetric, optical phonons, it is reasonable to evaluate first of all the temperature coefficient of the gap due to the self-energy of this specific interaction. Taking into account the three-dimensional character of the density of states, and neglecting interband transitions, the self-energy contribution is [3]

$$\Delta E_g = -\frac{4 \ln 2}{\pi}\left(\frac{g_v^2}{\sqrt{2m_v^*}} + \frac{g_c^2}{\sqrt{2m_c^*}}\right) 2n\hbar Q \sqrt{\hbar\omega}\,, \tag{7}$$

where the subscripts v and c refer to the valence and conduction band, respectively, n is the number of $A_1'^{(1)}$ phonons in one particular mode and Q is the radius of the Brillouin zone. The energy $\hbar\omega$ of the $A_1'^{(1)}$ modes is taken to be constant over the whole Brillouin zone. The deformation potentials for the conduction band are not known experimentally. Because the electron density corresponding to the conduction band is more diffuse than that corresponding to the valence band [6], the deformation potentials of the conduction band are expected to

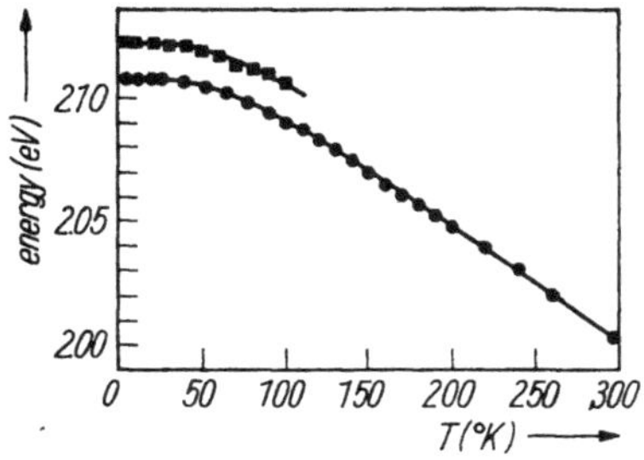

Fig. 2. Energy of the $n = 1, 2$ free exciton lines of GaSe as a function of temperature. ■ FE, $n = 2$; ● FE, $n = 1$; — — ■ theoretical curve

be smaller than those of the valence band. This is confirmed by the deformation potentials deduced from band calculations [15]. Furthermore, the value of $m_c^* \approx 0.2m_0$ deduced from the magneto optical properties of the exciton [4], is significantly smaller than the valence band mass $m_v^* \approx 0.5m_0$. One expects, then, that the coupling constant for the conduction band is significantly smaller than that of the valence band. Thus, nearly all of the self-energy contribution to the temperature coefficient is due to the valence band. The self-energy can now be computed without any adjustable parameter. The result is shown as solid curves in Fig. 2, which fit the experimental data in a very satisfactory manner: Firstly, the temperature dependence of the gap follows very closely that of the occupation number of the $A_1'^{(1)}$ phonon. Secondly, the good fit of the slopes at high temperatures tends to confirm the prevailing importance of self-energy effects on the temperature coefficient of the gap. It should be noted that the polar coupling constant $\alpha = 0.3$ is once again much too small to account for the shift of the gap.

5. Summary

In conclusion we may say that, in GaSe, the interaction of holes with fully symmetric $A_1'^{(1)}$ homopolar optical phonons governs both the mobility of the holes and the temperature dependence of the energy gap. This interpretation is in agreement with the theory of electron–lattice interactions in layered structures. The agreement between theory and experiment justifies a posteriori the assumption that the contribution of acoustic scattering to the self-energy of the hole–lattice interaction can be neglected. It also confirms the value $m_v^* = 0.5m_0$ of the density-of-states effective mass of the valence band deduced from Hall measurements. It is not very common that transport parameters can be quantitatively compared to optical parameters such as the temperature shift of the gap. We believe that this is rendered possible by the coincidence in GaSe of a series of favourable circumstances:

GaSe has a layered structure, giving rise to a strong coupling of charge carriers to one branch of fully symmetrical optical phonons;

the valence band maximum at the centre of the Brillouin zone is well separated from deeper bands;

the minimum of the conduction band at the zone centre has a low effective mass;

the valence band is weakly anisotropic and allows three-dimensional band description of electronic transport;

GaSe has a low ionicity.

Acknowledgement

Part of this work was carried out with the financial support of the Swiss National Research Fund.

References

[1] R. FIVAZ and E. MOOSER, Phys. Rev. **163**, 743 (1967).
[2] R. FIVAZ, Nuovo Cimento X, **63B**, 10 (1969).
[3] PH. SCHMID, Nuovo Cimento **21B**, 258 (1971).
[4] G. OTTAVIANI, C. CANALI, F. NAVA, PH. SCHMID, E. MOOSER, R. MINDER, and I. ZSCHOKKE, Solid State Commun. **14**, 933 (1974).
[5] PH. SCHMID, J. P. VOITCHOVSKY, and A. MERCIER, phys. stat. sol. (a) **21**, **443** (1974).
[6] M. SCHLÜTER, Nuovo Cimento **13B**, 313 (1973).
[7] L. J. VAN DER PAUW, Philips Res. Rep. **13**, 1 (1958).
[8] P. C. LEUNG, G. ANDERMANN, and W. G. SPITZER, J. Phys. Chem. Solids **27**, 849 (1966).
[9] H. FRÖHLICH, in: Polarons and Excitons, Ed. C. G. KUPER and G. D. WHITFIELD, Oliver and Boyd, London 1963. (Chap. I).
[10] T. J. WIETING and J. L. VERBLE, Phys. Rev. B **5**, 1473 (1972).
[11] J. P. VOITCHOVSKY and A. MERCIER, Nuovo Cimento B, to be published.
[12] J. R. DIXON and J. M. ELLIS, Phys. Rev. **123**, 1560 (1961).
[13] C. S. GUENZER and A. BIENENSTOCK, Phys. Rev. B **8**, 4655 (1973).
[14] H. Y. FAN, Phys. Rev. **82**, 900 (1951).
[15] M. SCHLÜTER, private communication.

(Received June 18, 1974)

phys. stat. sol. (b) 65, 255 (1974)
Subject classification: 16 and 20.1; 2; 13.1; 22.1.3

Universités Paris VI et XI, Laboratoire de Génie Electrique de Paris, Fontenay-aux-Roses

Optical Absorption and Photoconductivity in Liquid Selenium[1])

By
J. RABIT and J. C. PERRON

The optical absorption and photoconductivity is measured in liquid selenium. Urbach's spectral rule is satisfied and explained with the Dow and Redfield model. The optical gap is close to 1.8 eV at 230 °C and its temperature dependence is 8×10^{-4} eV/K. Optical absorption and quantum efficiency curves are not energy shifted in contrast to amorphous selenium. The low energy photogeneration is due to a "dynamically"-assisted separation of electron–hole pairs. Two lifetimes are found and the shorter is related to the structural relaxation. The values of photoconduction and conduction gaps are about 1.8 eV in agreement with the optical result but the Fermi level must be shifted towards the valence band.

Es wird die optische Absorption und Photoleitfähigkeit von flüssigem Selen gemessen. Die spektrale Urbach-Regel ist erfüllt und wird mit dem Modell von Dow und Redfield erklärt. Die optische Bandlücke beträgt etwa 1,8 eV bei 230 °C mit einer Temperaturabhängigkeit von 8×10^{-4} eV/K. Die Kurven der optischen Absorption und Quantenausbeute sind nicht energieverschoben wie im amorphen Selen. Die Trägererzeugung bei niedrigen Photoenergien wird durch eine „dynamisch" assistierte Trennung von Elektron–Loch-Paaren hervorgerufen. Es werden zwei Lebensdauern gefunden, und die kürzere ist mit der Strukturrelaxation verbunden. Die Werte für die Photoleitungs- und Leitfähigkeits-Bandlücken von etwa 1,8 eV befinden sich in Übereinstimmung mit dem optischen Ergebnis, jedoch muß das Ferminiveau zum Valenzband verschoben werden.

1. Introduction

In the last decade the electronic properties of various liquid semiconductors have been investigated to some extent. The optical properties were less studied mainly because of the experimental difficulties involved in such measurements.

In liquid selenium the electrical conductivity and thermopower results show a true semiconducting behaviour as has been shown by Henkels and Maczuk [1], Baker [2] and Gobrecht et al. [3]. The first investigations on optical absorption near the fundamental edge were made at low temperatures and weak absorption coefficients (400 °C, 20 cm^{-1}) by Saker [4] and Caldwell and Fan [5]. Siemsen and Fenton [6] have studied amorphous selenium between −200 to 200 °C and they have extended previous data up to 10^3 cm^{-1} in the liquid (300 to 400 °C). The absorption shows a 0.4 to 0.8 eV wide exponential tail. Siemsen and Fenton have related the tail observed in the amorphous phase to a non-photoconductive component arising from excitons localized by strong lattice interactions. We report here complementary results on the optical absorption up to about 10^4 cm^{-1} in the temperature range 230 to 600 °C.

Recently Vengris et al. [7] have studied the drift mobility and the photogeneration in liquid selenium near the melting point. Independently we have

[1]) Paper has been presented at the 5th International Conference on Amorphous and Liquid Semiconductors, held at Garmisch-Partenkirchen, September 3 to 8, 1973.

measured the transverse photoconductivity as a function of temperature and photon energy. All the results display important differences between the amorphous and liquid phases. The role of the liquid structure and its dynamical properties must be emphasized.

2. Experimental Techniques

Along this work we have used commercial selenium of 99.999% purity without further purification.

The optical absorption was obtained by comparison of the transmission of the empty and filled cell. A Perkin Elmer spectrometer (model 112) was used in the wavelength range 0.7 to 2.5 μm, and some modifications were made for high temperature runs. The optical paths of the cells were varied from 10 μm to a few millimetres. The reflectivities of the various interfaces were taken into account. The sample temperature was checked carefully and sufficiently accurate (±1 °C) so that data from different path cells overlap fairly well. Measurements of the transverse photoconductivity were made in a glass cell with two graphite electrodes. The light was focused onto the liquid but the electrodes are masked to avoid photoemission. In steady-state conditions, the light of a tungsten filament lamp was mechanically chopped at a rate of 10 to 500 Hz. The photosignal was synchronously detected and chart recorded. In transient conditions we have used a xenon flash. After low-noise amplification, the photosignal was observed on an oscilloscope and photographically recorded. The spectral dependence was studied between 0.4 to 0.9 μm by using a set of interference filters. The low photovoltages (≈0.1 μV) obtained with the first method are coherent with those obtained in pulse technique but the weak emission of the tungsten lamp causes a strong dispersion in the data so we reported here (see Section 3) only pulse measurements. The incident photon flux density was raised from about 10^{15} photons/cm^{2}s in steady-state conditions to 10^{18} in transient measurements. Since all the photons were absorbed in the sample (see Section 3) we had to make sure that the observed photocurrent was not simply due to the heating of the liquid. Below 500 °C the electrical conductivity of liquid selenium is thermally activated with an energy of about 0.9 eV [7, 8] so that a thermal effect can occur. To check this we have coated the glass window with a completely opaque carbon layer. No measurable voltage was detected with the maximum light intensity used in this work. Furthermore the voltage, measured with a constant light intensity, decreases when the temperature increases. Since the thermal impedance of the system does not change appreciably, it is unlikely that we have measured thermal effects. However, a small heating contribution cannot be completely excluded.

3. Results

Fig. 1 shows a semilog plot of the absorption coefficient K of liquid selenium versus photon energy in the temperature range 230 to 600 °C. The values agree well with previous determinations and the spectral Urbach rule is satisfied. The results can be fitted by the relationship

$$K(h\nu, T) = K_0 \{\exp a(T) [h\nu - E_K(T)]\} ,$$

where K_0 is a constant, $a(T)$ the temperature dependent slope, and $E_K(T)$ an

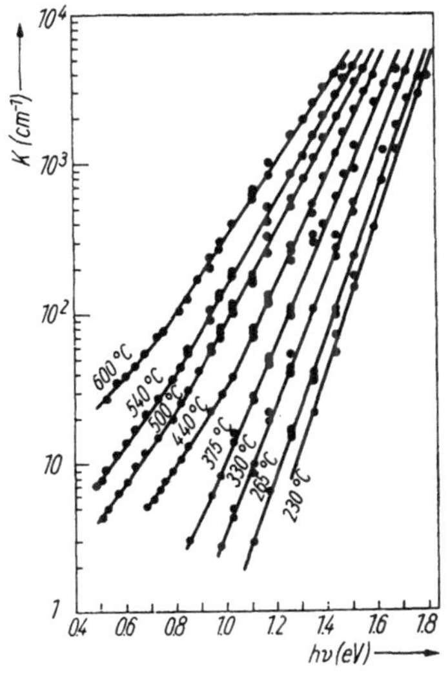

Fig. 1. Optical absorption in liquid selenium

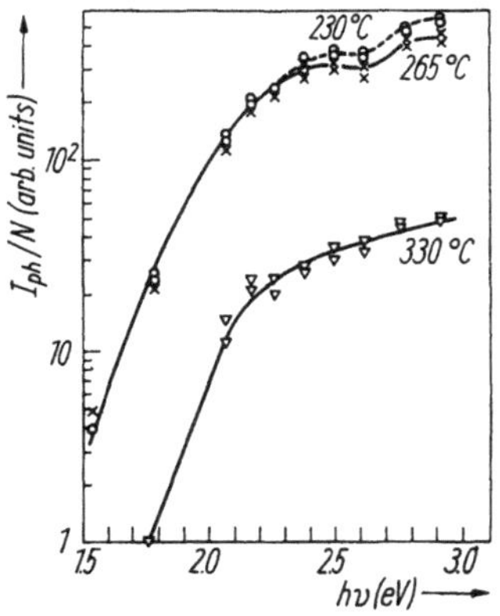

Fig. 2. Photocurrent per absorbed photon in arbitrary units

energy parameter. In Fig. 2 we have the semilog plot, in arbitrary units, of the normalized photocurrent peak I_{ph}/N versus photon energy. In the experimental range 1.6 to 3 eV all the incident photons are absorbed in the sample ($Kd \gg 1$ with $K > 10^3$ cm^{-1} and $d \approx 0.1$ cm) so we can easily normalize the photocurrent I_{ph} by absorbed photons N. Therefore the results are proportional to the quantum efficiency. Photoconductivity is given for three temperatures (230, 265, and 330 °C) corresponding to the absorption data. Fig. 3 displays the temperature dependence of the photocurrent measured at 4490 Å. Between the melting point and about 260 °C the signal is practically constant, then it decreases at higher temperature.

4. Discussion

The results will be discussed by comparisons between the amorphous and liquid phases. Amorphous selenium is a mixture of atomic chains and Se_8 rings.

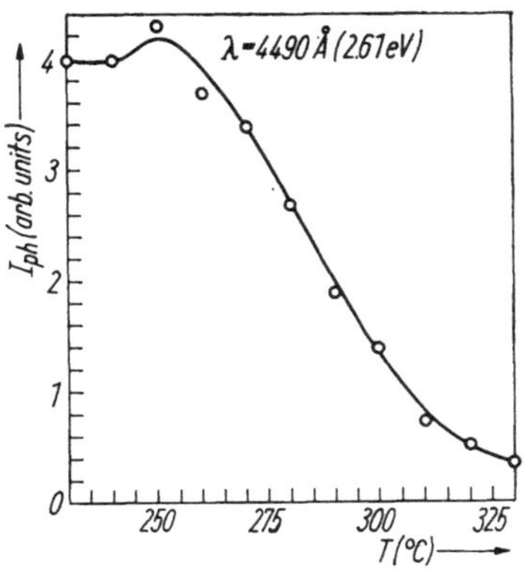

Fig. 3. Photocurrent (arb. units) versus temperature for $\lambda = 4490$ Å

The chains are probably distributed at random. The X-ray and neutron diffraction data by Kaplow et al. [9], Henninger et al. [10], and Richter [11] cannot give the accurate topology of rings and chains. The liquid structure is even more complex. By selective CS_2 dissolution of vitreous selenium quenched from the melt at various temperatures an estimation of the weight fraction of rings gives 40% at 220 °C and 15% at 600 °C [12]. However, this assumes that the quenching time is sufficiently short to leave the ring-chain equilibrium unperturbed, but this is not proved. Several estimations of the average chain length give 10^4 to 10^5 atoms at 220 °C and 10^3 atoms at 600 °C [13 to 16]. From X-ray diffraction studies, Richter [11] postulates the existence of plane zig-zag chains becoming unbounded above 400 °C. In some respect, this model agrees with the recent determination of the structure factor by neutron diffraction by Tourand [17] at three temperatures (258, 358, and 460 °C). Low temperature data suggest that the chain arrangement is hampered by Se_8 rings. At higher temperature the structure seems to be more similar to that of the trigonal form. The smaller ring concentration and the shortened chains allow a better structural arrangement. These considerations are consistent with the views of Cherin and Unger [18] on the transformation of α-monoclinic selenium into the hexagonal form. By linking together closely approaching atoms of different Se_8 molecules, a chain-like arrangement emerges.

Now we can discuss the optical absorption results and the existence of spectral Urbach's rule. Recently Dow and Redfield [19] give a unified theory of Urbach's rule based on the dissociation of excitons by the internal electric microfields distributed at random. The disordered structure of liquid selenium might produce such microfields with fluctuations not only in space but in time because of the liquid dynamics. The absorption tail may be due to the relative motion of electron and hole, and its slope is inversely proportional to the r.m.s. value of the microfield. Furthermore Chopra and Bahl [20] think that the temperature dependence of the slope is related to the extent of long-range order in the material. Fig. 4 shows a schematic representation of the density of states in liquid selenium. In the insert we have plotted a^{-1}, reciprocal of the tail slope, versus kT. Although an approximate linear dependence is found at low temperature, a careful study on the whole temperature range gives the relationship

$$a(T) = -1.57 + \frac{6.81\times 10^3}{T}$$

with $a(T)$ in eV^{-1}. We have $a^{-1} \approx 85$ meV at 230 °C and 170 meV at 600 °C. In the Dow and Redfield model this corresponds to a two-fold increase in the microfield strength due to a greater disorder in the structure. However, difficulties arise if the exciton formation is related to the Se_8 rings as the electroreflectance measurements seem to indicate [21]. In the liquid, at high temperature, the ring population decreases strongly so that the exciton existence is unlikely. Another way to calculate the optical absorption is to use the density of state model given in Fig. 4 and previously discussed by Perron [22]. We assume that the absorption is mainly due to the optical transitions between a parabolic valence band and an exponential tail in the conduction band. Assuming constant matrix elements we find the formula given in Section 3 where a^{-1} represents the extension of the conduction band tail and $E_K = E_0 - E_A$. E_0 is the junction point energy of exponential tail and conduction band. W.

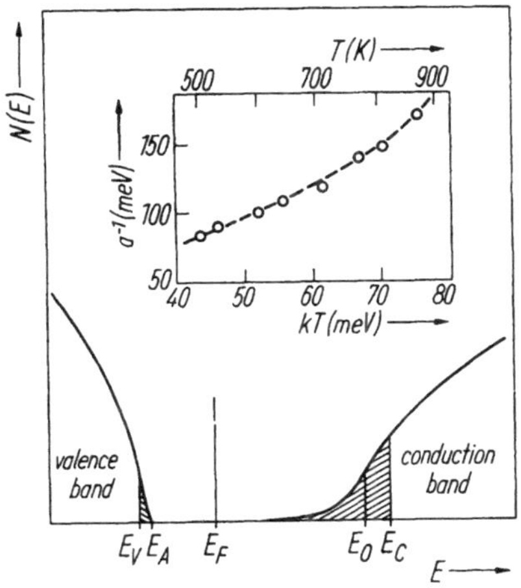

Fig. 4

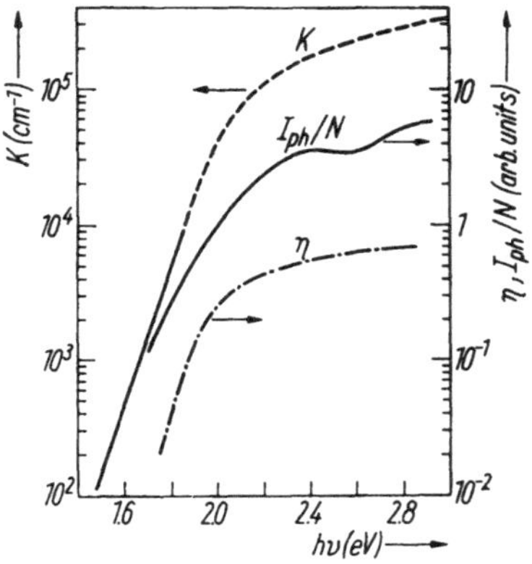

Fig. 5

Fig. 4. Schematic density of states in liquid selenium. The insert shows the a^{-1} variation versus kT (see text)

Fig. 5. Optical absorption K (cm^{-1}, left scale) and normalized photocurrent I_{ph}/N (arb. units, right scale) at 230 °C. The dashed part is extrapolated by assuming an amorphous-like behaviour. The quantum efficiency η (right scale) at 222 °C is taken from Vengris et al. [7]. Note the negligible energy shift between the three curves (see text)

write the temperature dependence $E_K = E_K^0 - \beta T$ and isoabsorption curves give $E_K^0 = 2.2$ eV and $\beta = 8 \times 10^{-4}$ eV/K ($K = 10^4$ cm^{-1}). We can take the E_K value as an estimation of the optical gap and we have 1.8 eV at 230 °C. In amorphous selenium above the exponential tail Davis and Mott [23] have found the relationship

$$h\nu\, K(h\nu, T) \sim h\nu - E_G(T)$$

at room temperature $E_G \approx 2.1$ eV corresponding to an absorption value close to 10^4 cm^{-1}. Fig. 5 shows the optical absorption of liquid selenium (230 °C) measured up to 10^4 cm^{-1} (experimental limit). Above, the extrapolated curve is estimated by assuming an amorphous-like behaviour. The 10^4 cm^{-1} value gives an optical gap of 1.83 eV in agreement with the E_K value given above.

In Fig. 5 we have also plotted two other curves: the normalized photocurrent I_{ph}/N (230 °C), which is proportional to the quantum efficiency η, and this latter measured directly by Vengris et al. [7]. They have used the drift mobility method at 222 °C. We observe that the energy shift between the three curves is very small compared to the 0.5 eV shift found in amorphous selenium by Hartke and Regensburger [24]. This latter phenomenon has received a simple explanation by Davis and Mott [23]: the generation of free carriers via an absorbed photon requires the spatial separation of the initially created electron–hole pair out of their mutual Coulomb potential the excess energy being supplied by the incident photon. This idea and the Dow and Redfield model give a good basis to explain the liquid results. In the liquid phase distorted chains and rings give rise to internal microfields stronger than in the amorphous phase (slope of the absorption curves). These fields are time fluctuating because the liquid motions and the photogeneration occurs because the electron–hole separation is "dynamically"-aided. The hole contribution seems to dominate the charge transport. The thermopower according to Henkels and Maczuk [1] is highly positive and observation of electron transit in drift mobility measurements is impossible [7]. Therefore, after photon absorption, the hole is tunneling away from a localized electron towards a delocalized valence state. By assuming monomolecular recombination characterized by a lifetime τ our results give $\eta\mu\tau \approx 5\times10^{-10}$ cm²/V and with $\eta \approx 0.5$ we obtain $\mu\tau \approx 10^{-9}$ cm²/V. Now we can assume that the final state of the hole is on the left of E_V (Fig. 4) with a mobility around 0.1 to 1 cm²/Vs (diffusive motion regime) and we get $\tau \approx 10^{-8}$ to 10^{-9} s. A similar lifetime was found for the rapid component of photoconductivity in some amorphous semiconductors, e.g. for GeTe by Howard and Tsu [25] and for As_2Te_3 by Weiser and Brodsky [26]. In liquid selenium the excess holes captured by localized states will have a different lower mobility and probably larger lifetime. For these carriers, a rough estimate, from the decay of the observed photovoltages, gives about 1 ms. The hole drift mobility was measured by Vengris et al. [7]. Its value is 10^{-2} cm²/Vs at 220 °C and the activation energy is strong (0.4 eV), so the motion of trapped carriers seems involved. The short hole lifetime (10^{-8} to 10^{-9} s) must be related to the structural relaxation. The correlation time for highly viscous motion is about 10^{-7} s at 250 °C (Brown et al. [27]) and decreases to 5×10^{-9} s at 400 °C (Gitis et al. [28]). On the other hand, ESR signals (Koningsberger [16]) are produced by dangling bonds at the chain ends. These bonds give strong local microfields and their lifetime is close to 10^{-9} s at 300 °C. The strong decrease of photocurrents with temperature (Fig. 3) is probably due to the shortened relaxation time of the atomic structure.

Finally we can remark the difficulty to specify a photoconductive gap. If we take the onset of the photosignal we get about 1.8 eV (230 °C) in agreement with the absorption gap. The activation energy of the electrical conductivity is close to 0.9 eV, if we assume intrinsic behaviour we obtain a conductivity gap of 1.8 eV. However, at low temperature the hole mobility is thermally activated (0.4 eV) so $E_F - E_A$ (Fig. 4) is about 0.5 eV and the agreement between the various gaps is obtained only if the Fermi level E_F is shifted towards the valence band.

References

[1] H. W. Henkels and J. Maczuk, J. appl. Phys. **24**, 1056 (1953).
[2] E. H. Baker, J. Chem. Soc. 1089 (1968).
[3] H. Gobrecht, D. Gawlik, and F. Mahdjuri, Phys. kondens. Materie **13**, 156 (1971).
[4] E. W. Saker, Proc. Phys. Soc. **B65**, 785 (1952).
[5] R. S. Caldwell and H. Y. Fan, Phys. Rev. **114**, 664 (1959).
[6] K. J. Siemsen and E. W. Fenton, Phys. Rev. **161**, 632 (1967).
[7] S. A. Vengris, Yu. K. Vishchakas, A. P. Sakalas, and G. B. Yushka, Soviet. Phys. — Semicond. **6**, 903 (1972).
[8] J. C. Perron, to be published.
[9] R. Kaplow, T. A. Rowe, and B. L. Averbach, Phys. Rev. **168**, 1068 (1968).
[10] E. H. Henninger, R. C. Buschert, and L. Heaton, J. chem. Phys. **46**, 586 (1967).
[11] H. Richter, J. non-crystall. Solids **8/10**, 388 (1972).
[12] G. Briegleb, Z. phys. Chem. **A144**, 321 (1929).
[13] A. Eisenberg and A. V. Tobolsky, J. Polymer Sci. **46**, 19 (1960).
[14] C. H. Massen, A. G. L. Weijts, and J. A. Poulis, Trans. Faraday Soc. **60**, 317 (1964).

[15] R. C. Keezer and M. W. Bailey, Mater. Res. Bull. **2**, 185 (1967).
[16] D. C. Koningsberger, Thesis, University of Technology, Eindhoven 1971.
[17] G. Tourand, private communication and J. Physique, to be published.
[18] P. Cherin and P. Unger, Acta Cryst. **B28**, 313 (1972).
[19] J. D. Dow and D. Redfield, Phys. Rev. B **5**, 594 (1972).
[20] K. L. Chopra and S. K. Bahl, Thin Solid Films **11**, 377 (1972).
[21] G. Weiser and J. Stuke, phys. stat. sol. **35**, 747 (1969).
[22] J. C. Perron, J. non-crystall. Solids **8/10**, 272 (1972).
[23] E. A. Davis and N. F. Mott, Phil. Mag. **22**, 903 (1970).
[24] J. L. Hartke and P. J. Regensburger, Phys. Rev. **139**, A970 (1965).
[25] W. E. Howard and R. Tsu, Phys. Rev. B **1**, 4709 (1970).
[26] K. Weiser and M. H. Brodsky, Phys. Rev. B **1**, 791 (1970).
[27] D. Brown, D. S. Moore, and E. F. W. Seymour, J. non-crystall. Solids **8/10**, 256 (1972).
[28] M. B. Gitis, I. G. Mikhailov, and S. Niyazov, Soviet. Phys. — Acoustics **15**, 259 (1969).

(Received June 24, 1974)

phys. stat. sol. (b) 65, 263 (1974)

Subject classification: 10.2 and 13.5.3; 22.5.1; 22.5.2

Institute of Semiconductors, Academy of Sciences of the Ukrainian SSR, Kiev

F-Centre Theory in Multivalley Cubic Crystals

By

L. S. KHAZAN, S. I. PEKAR, and V. I. SHEKA

The interaction between an electron bound to a Coulomb potential and the LO lattice vibrations in a cubic multivalley ionic crystal is considered for not too weak coupling. The vibrational degrees of freedom in the partition function of the system are eliminated by Feynman's ordered-operator technique. The complicated operator thus obtained depends only on the electron variables and is approximated by a suitable operator containing some available parameters. The latter provides to calculate the partition function of the F-centre by means of a variational principle at arbitrary temperatures and thus to obtain the ground state energy of the F-centre as a function of coupling parameters and of the anisotropy of electron band masses.

В работе рассмотрено взаимодействие электрона, связанного с кулоновским центром, с продольными оптическими колебаниями решетки в многодолинном ионном кубическом кристалле при не слишком слабой связи. Сложный оператор, получающийся после исключения методом упорядоченных операторов Фейнмана колебательных степеней свободы в статистической сумме системы, аппроксимируется некоторым другим оператором, что позволяет при помощи вариационного принципа определить статистическую сумму F-центра при произвольных температурах, а из нее и энергию основного состояния как функцию параметров связи и анизотропии зонных масс электрона.

1. Introduction

In [1] we have considered a free polaron at arbitrary coupling in a multivalley cubic crystal. In the present work we modify the method of [1] for the case of an F-centre (or polaron bound to a Coulomb potential). One may show like in [1] that in passing to the effective mass method in the F-centre Hamiltonian containing the periodic lattice potential (in this work we consider F-centres of large radius for which the effective mass method is valid), the matrix elements of the Hamiltonian between wave functions of different valleys are small, i.e. the F-centre wave function is formed from Bloch functions of only one valley.

Taking this into account, the effective mass Hamiltonian of the system may be written as

$$\left.\begin{aligned} H &= -\sum_{\alpha=1}^{3} \frac{\hbar^2}{2\mu_\alpha}\frac{\partial^2}{\partial r_\alpha^2} + \sum_{\varkappa} \hbar\omega_\varkappa a_\varkappa^+ a_\varkappa + \sum_{\varkappa} (V_\varkappa a_\varkappa \,\mathrm{e}^{i\varkappa\cdot\boldsymbol{r}} + V_\varkappa^* a_\varkappa^+ \,\mathrm{e}^{-i\varkappa\cdot\boldsymbol{r}}) - \frac{e^2}{\varepsilon r}, \\ V_\varkappa &= \frac{e}{\varkappa}\sqrt{\frac{2\pi\hbar\omega c}{V}}, \qquad c = \frac{1}{\varepsilon_\infty} - \frac{1}{\varepsilon}, \end{aligned}\right\} \tag{1}$$

where μ_α are the electron effective masses and e its charge; $a_\varkappa^+$, $a_\varkappa$ are creation and annihilation operators of LO phonons with wave vector $\varkappa$ and frequency $\omega_\varkappa$ (henceforth we shall neglect the LO phonon dispersion supposing $\omega_\varkappa = \omega$, where ω is a cut-off frequency of longitudinal optical vibrations); ε, ε_∞ are the static and high-frequency dielectric constants, respectively; V is the normalization volume.

As this Hamiltonian is translationally noninvariant (due to the Coulomb term) we choose the trial Hamiltonian in contrast to [1] in the form $H' = \sum_{\alpha=1}^{3} H'_\alpha$,

$$H'_\alpha = -\frac{\hbar^2}{2\mu_\alpha}\frac{\partial^2}{\partial r_\alpha^2} + \sum_{i=1}^{s}\left[-\frac{\hbar^2}{2\mu_{\alpha i}}\frac{\partial^2}{\partial r_{\alpha i}^2} + \frac{c_{\alpha i}}{2}(r_\alpha - r_{\alpha i})^2\right] + \frac{c_\alpha}{2} r_\alpha^2\,, \qquad c_{\alpha i} = \mu_{\alpha i}\omega_{\alpha i}^2\,. \tag{2}$$

The trial Hamiltonian (2) corresponds to a model where the particle with coordinates r_α and effective masses μ_α ($\alpha = 1, 2, 3$) represents an electron, while the fictitious particles with coordinates $r_{\alpha i}$ and masses $\mu_{\alpha i}$ ($i = 1, 2, \dots, s$) harmonically bound to the electron represent a "polarizational potential well" for the electron, capable to oscillate around the electron and to vary its shape. Here the masses of fictitious particles $\mu_{\alpha i}$ and the constants $c_{\alpha i}$ of their spring interactions with the electron as well as the numbers c_α are variational parameters. The Hamiltonian (2) is translationally noninvariant like (1), due to its last term.

2. Partition Function of an F-Centre and Its Ground State Energy

Let us consider the partition function

$$Z = \operatorname{tr} e^{-\lambda H} \tag{3}$$

of the Hamiltonian (1) and that of the Hamiltonian (2),

$$Z' = \operatorname{tr} e^{-\lambda H'}\,. \tag{4}$$

Using the method of [1, 2] one can order the operators standing under the trace in (3) and (4), disentangle the phonon variables from Z and the coordinates of fictitious particles from Z', and take the trace over them. As a result one obtains

$$\left.\begin{aligned}
Z &= X \operatorname{tr} \exp\left\{-\int_0^\lambda H_{r_{\lambda_1}}\,d\lambda_1 + U\right\}, \qquad U = \int_0^\lambda \frac{e^2}{\varepsilon r_{\lambda_1}}\,d\lambda_1 + \sum_\varkappa |V_\varkappa|^2\, U_\varkappa\,,\\
U_\varkappa &= \int_0^\lambda d\lambda_1 \int_0^{\lambda_1} d\lambda_2\, e^{i\varkappa\cdot(r_{\lambda_1} - r_{\lambda_2}) - \hbar\omega_\varkappa(\lambda_1-\lambda_2)} + \bar n_\varkappa \int_0^\lambda d\lambda_1 \int_0^\lambda d\lambda_2\, e^{i\varkappa\cdot(r_{\lambda_1} - r_{\lambda_2}) - \hbar\omega_\varkappa(\lambda_1-\lambda_2)}\,,\\
H_r &= -\sum_{\alpha=1}^{3}\frac{\hbar^2}{2\mu_\alpha}\frac{\partial^2}{\partial r_\alpha^2}\,, \qquad \bar n_\varkappa = \frac{1}{e^{\lambda\hbar\omega_\varkappa} - 1}\,, \qquad X = \prod_\varkappa(\bar n_\varkappa + 1)\,,
\end{aligned}\right\} \tag{5}$$

$$\left.\begin{aligned}
Z' &= \operatorname{tr}\exp\left\{-\int_0^\lambda H_{r_{\lambda_1}}\,d\lambda_1 + U'\right\},\\
U' &= -\sum_{\alpha, i}\ln\left(2\sinh\frac{\lambda}{2}\hbar\omega_{\alpha i}\right) + \sum_\alpha\left(-\frac{c_\alpha}{2}\int_0^\lambda r_{\alpha_{\lambda_1}}^2\,d\lambda_1 + \sum_{i=1}^{s} c_{\alpha i}\,U'_{\alpha i}\right),\\
U'_{\alpha i} &= -\frac{1}{2}\int_0^\lambda r_{\alpha_{\lambda_1}}^2\,d\lambda_1 + \frac{\hbar\omega_{\alpha i}}{2}\int_0^\lambda\int_0^{\lambda_1} r_{\alpha_{\lambda_1}} r_{\alpha_{\lambda_2}}\, e^{\hbar\omega_{\alpha i}(\lambda_2-\lambda_1)}\,d\lambda_1\,d\lambda_2 +\\
&\quad + \frac{\hbar\omega_{\alpha i}}{2(e^{\lambda\hbar\omega_{\alpha i}} - 1)}\int_0^\lambda\int_0^\lambda r_{\alpha_{\lambda_1}} r_{\alpha_{\lambda_2}}\, e^{\hbar\omega_{\alpha i}(\lambda_2-\lambda_1)}\,d\lambda_1\,d\lambda_2\,.
\end{aligned}\right\} \tag{6}$$

The operators standing in the exponentials in (5) and (6) depend on the electron coordinates only, and we may write down the following variational principle for Z [1, 2]:

$$Z = X \operatorname{tr} \exp\left\{-\int_0^\lambda H_{r_{\lambda_1}}\, d\lambda_1 + U\right\} = X \operatorname{tr} \exp\left\{-\int_0^\lambda H_{r_{\lambda_1}}\, d\lambda_1 + U' + (U - U')\right\} \geqq$$

$$\geqq X \operatorname{tr} \exp\left\{-\int_0^\lambda H_{r_{\lambda_1}}\, d\lambda_1 + U'\right\} \exp\,(\bar{U} - \bar{U}') = XZ'\, e^{(\bar{U}-\bar{U}')} \equiv Z_1\,, \quad (7)$$

where

$$\left.\begin{aligned} \bar{U} &= \frac{1}{Z'} \operatorname{tr}\left\{U \exp\left(-\int_0^\lambda H_{r_{\lambda_1}}\, d\lambda_1 + U'\right)\right\}, \\ \bar{U}' &= \frac{1}{Z'} \operatorname{tr}\left\{U' \exp\left(-\int_0^\lambda H_{r_{\lambda_1}}\, d\lambda_1 + U'\right)\right\}. \end{aligned}\right\} \quad (8)$$

As the trial partition function Z_1 is always smaller than the true one, Z_1 has to be maximized with respect to the variational parameters $\mu_{\alpha i}$, $c_{\alpha i}$, c_α of H'. After such a procedure, Z_1 will represent a lower bound for the F-centre partition function. Therefore, our task reduces to calculate Z', $\bar{U}'$, and $\bar{U}$.

Z' given by (4) may be easily calculated after the Hamiltonian H' (which represents a system of $s + 1$ harmonically bound particles) is diagonalized, i.e. after introducing the Bose operators $b_{\alpha i}^+$ and $b_{\alpha i}$ of normal modes and the normal frequencies $\Omega_{\alpha i}$ and reducing H' to the form

$$H' = \sum_\alpha \sum_{i=1}^{s+1} \hbar \Omega_{\alpha i}\, (b_{\alpha i}^+ b_{\alpha i} + \tfrac{1}{2})\,, \qquad r_\alpha = \sum_{i=1}^{s+1} \beta_{\alpha i} (b_{\alpha i}^+ + b_{\alpha i})\,, \quad (9)$$

where $\Omega_{\alpha i}$ are functions of parameters $c_{\alpha i}$ and $\mu_{\alpha i}$ and the subscript $s + 1$ corresponds to the electron ($\mu_\alpha \equiv \mu_{\alpha, s+1}$, $c_\alpha \equiv c_{\alpha, s+1}$). As a result,

$$Z' = \prod_\alpha \prod_{i=1}^{s+1} \frac{1}{2 \sinh \dfrac{\lambda \hbar \Omega_{\alpha i}}{2}}\,. \quad (10)$$

U' given in (6) is the sum of the constant term $-\sum_\alpha \sum_{i=1}^{s} \ln\left(2 \sinh \frac{\lambda}{2} \hbar \omega_{\alpha i}\right)$ and the terms linear in $c_{\alpha i}$ ($i = 1, \ldots, s + 1$). Therefore $\bar{U}'$ defined by (8) may be calculated as

$$\bar{U}' = -\sum_\alpha \sum_{i=1}^{s} \ln\left(2 \sinh \frac{\lambda}{2} \hbar \omega_{\alpha i}\right) + \frac{1}{Z'} \sum_{\alpha=1}^{3} \sum_{i=1}^{s+1} c_{\alpha i} \left(\frac{\partial Z'}{\partial c_{\alpha i}}\right)_{\omega_{\alpha i} = \text{const}}\,. \quad (11)$$

Similarly one obtains $\bar{U}$ (see (5) and (8)) as

$$\bar{U} = \frac{1}{Z'}\left[\gamma \left(\frac{\partial Z''}{\partial \gamma}\right)_{\gamma \to 0} + \sum_\varkappa |V_\varkappa|^2 \left(\frac{\partial Z''}{\partial |V_\varkappa|^2}\right)_{|V_\varkappa|^2 \to 0}\right], \quad (12)$$

where

$$\left.\begin{aligned} Z'' &= \operatorname{tr} \mathrm{e}^{-\lambda(\tilde{H}_0+\tilde{H}_1)} \cdot 2 \sinh \frac{\lambda \hbar \omega_\varkappa}{2}, \\ \tilde{H}_0 &= H' + \hbar\omega_\varkappa \left(a_\varkappa^+ a_\varkappa + \tfrac{1}{2}\right), \qquad \tilde{H}_1 = V_\varkappa a_\varkappa \mathrm{e}^{i\varkappa\cdot r} + V_\varkappa^* a_\varkappa^+ \mathrm{e}^{-i\varkappa\cdot r} - \frac{\gamma}{r}, \\ \gamma &= \frac{e^2}{\varepsilon}. \end{aligned}\right\} \tag{13}$$

Like in [1] it is sufficient to restrict ourselves in (13) by the Schwinger expansion [3] up to the first order in γ and $|V_\varkappa|^2$ (as a consequence of (12)). After tedious calculations we obtain

$$\begin{aligned} \overline{U} = {} & \frac{\lambda e^2}{2\pi^2 \varepsilon} \int \frac{\mathrm{d}\varkappa}{\varkappa^2} \exp\left\{-\frac{1}{2} \sum_\alpha \sum_{j=1}^{s+1} \varkappa_\alpha^2 \beta_{\alpha j}^2 \coth \frac{\lambda}{2} \hbar\Omega_{\alpha j}\right\} + \\ & + \frac{\lambda^2}{2} \sum_\varkappa |V_\varkappa|^2 \int_0^1 \mathrm{d}v \frac{\cosh \frac{\lambda}{2} \hbar\omega_\varkappa v}{\sinh \frac{\lambda}{2} \hbar\omega_\varkappa} \exp\left\{-\sum_\alpha \sum_{j=1}^{s+1} \varkappa_\alpha^2 \beta_{\alpha j}^2 \times \right. \\ & \left. \times \left[\coth \frac{\lambda}{2} \hbar\Omega_{\alpha j} - \frac{\cosh \frac{\lambda}{2} \hbar\Omega_{\alpha j} v}{\sinh \frac{\lambda}{2} \hbar\Omega_{\alpha j}}\right]\right\} \end{aligned} \tag{14}$$

and the partition function at finite temperatures becomes

$$\begin{aligned} Z_1 = X & \left(\prod_{\alpha=1}^{3} \frac{1}{2 \sinh \frac{\lambda}{2} \hbar\Omega_{\alpha,s+1}} \prod_{i=1}^{s} \frac{\sinh \frac{\lambda}{2} \hbar\omega_{\alpha i}}{\sinh \frac{\lambda}{2} \hbar\Omega_{\alpha i}} \right) \times \\ & \times \exp\left\{\frac{\lambda}{2} \sum_{\alpha=1}^{3} \sum_{i,j=1}^{s+1} \coth\left(\frac{\lambda}{2} \hbar\Omega_{\alpha i}\right) c_{\alpha j} \left(\frac{\partial \hbar\Omega_{\alpha i}}{\partial c_{\alpha j}}\right)_{\omega_{\alpha i}=\text{const}} + \overline{U}\right\}. \end{aligned} \tag{15}$$

For the case of an F-centre, after inserting $V_\varkappa$ from (1) into (14) and replacing the sum over $\varkappa$ by an integral, we finally obtain

$$\begin{aligned} \overline{U} = {} & \frac{e^2 \hbar\omega c \lambda^2 \sqrt{\pi}}{16\pi^2} \int_0^1 \mathrm{d}v \frac{\cosh \frac{\lambda\hbar\omega v}{2}}{\sinh \frac{\lambda\hbar\omega}{2}} \int \mathrm{d}\Omega \left\{\sum_{\alpha,j} s_\alpha^2 \beta_{\alpha j}^2 \left[\coth \frac{\lambda\hbar\Omega_{\alpha j}}{2} - \right.\right. \\ & \left.\left. - \frac{\cosh \frac{\lambda}{2} \hbar\Omega_{\alpha j} v}{\sinh \frac{\lambda}{2} \hbar\Omega_{\alpha j}}\right]\right\}^{-1/2} + \frac{\lambda e^2}{(2\pi)^{3/2}\, \varepsilon} \int \mathrm{d}\Omega \left\{\sum_{\alpha,j} s_\alpha^2 \beta_{\alpha j}^2 \coth \frac{\lambda\hbar\Omega_{\alpha j}}{2}\right\}^{-1/2}. \end{aligned} \tag{16}$$

Here $\mathrm{d}\Omega$ is an element of spatial angle; s_α are the components of a unit vector $\mathbf{s}$ lying in $\mathrm{d}\Omega$.

In order to determine the ground state energy of the F-centre (which is our main concern), we shall consider the limit of low temperatures, i.e. large λ. Expanding $\ln Z_1$ in a power series with respect to λ^{-1}, we obtain

$$\ln Z_1 = \ln X - \lambda E_0 + \cdots, \tag{17}$$

where

$$-\frac{E_0}{\hbar\omega} = \frac{1}{2}\Biggl\{\frac{1}{\omega}\sum_{\alpha=1}^{3}\left[\sum_{i=1}^{s}\omega_{\alpha i} + \sum_{i=1}^{s+1}\left(-\Omega_{\alpha i} + \sum_{j=1}^{s+1} c_{\alpha i}\left(\frac{\partial \Omega_{\alpha j}}{\partial c_{\alpha i}}\right)_{\omega_{\alpha i}}\right)\right] +$$

$$+\frac{e^2 c}{4\hbar\omega\pi^{3/2}}\int_0^\infty \frac{e^{-x}\,dx\,d\Omega}{\left[\sum_\alpha \sum_{j=1}^{s+1} s_\alpha^2\beta_{\alpha j}^2\,(1 - e^{-x\Omega_{\alpha j}/\omega})\right]^{1/2}} + \frac{e^2}{\sqrt{2}\,\pi^{3/2}\,\varepsilon\hbar\omega}\int \frac{d\Omega}{\left[\sum_\alpha \sum_{j=1}^{s+1} s_\alpha^2\beta_{\alpha j}^2\right]^{1/2}}\Biggr\},$$

$$x = (1 - v)\,\frac{\lambda\hbar\omega}{2}. \tag{18}$$

In the case of one fictitious particle ($s = 1$), $\Omega_{\alpha i}$ imply the secular equation

$$\Omega_\alpha^4 - \Omega_\alpha^2\left[\omega_{\alpha 1}^2 + \frac{1}{\mu_{\alpha 2}}(c_{\alpha 1} + c_{\alpha 2})\right] + \frac{c_{\alpha 2}\omega_{\alpha 1}^2}{\mu_{\alpha 2}} = 0, \tag{19}$$

and

$$\beta_{\alpha i}^2 = \frac{\hbar}{2\mu_{\alpha 2}}\left|\frac{1}{\Omega_{\alpha 1}^2 - \Omega_{\alpha 2}^2}\left(\Omega_{\alpha i} - \frac{\omega_{\alpha 1}^2}{\Omega_{\alpha 1}}\right)\right|. \tag{19a}$$

In this case

$$-\frac{E_0}{\hbar\omega} =$$

$$= \frac{1}{2\omega}\sum_\alpha f_{0\alpha} + \frac{e^2 c}{8\pi^{3/2}\hbar\omega}\int_0^\infty\!\!\int \frac{e^{-x}\,dx\,d\Omega}{\left\{\sum_\alpha s_\alpha^2\,[\beta_{\alpha 1}^2\,(1 - e^{-x\Omega_{\alpha 1}/\omega}) + \beta_{\alpha 2}^2\,(1 - e^{-x\Omega_{\alpha 2}/\omega})]\right\}^{1/2}} +$$

$$+\frac{e^2}{(2\pi)^{3/2}\,\varepsilon\hbar\omega}\int \frac{d\Omega}{\left\{\sum_\alpha s_\alpha^2\,(\beta_{\alpha 1}^2 + \beta_{\alpha 2}^2)\right\}^{1/2}}. \tag{20}$$

The maximum of the trial partition function Z_1 with respect to the parameters $\mu_{\alpha i}$ and $c_{\alpha i}$ was found with the help of a computer for the cases when the surface of constant energy of the band electron is an ellipsoid of revolution. The F-centre ground state energy was obtained as a function of the dimensionless coupling parameters $\alpha = \frac{ce^2}{2\hbar}\sqrt{\frac{2\mu_\perp}{\hbar\omega}}$ and εc, and also of the anisotropy parameter $\chi = \mu_{\|}/\mu_\perp$ of the electron effective masses, where $\mu_\perp = \mu_{12} = \mu_{22}$, $\mu_{\|} = \mu_{32}$; and the approximation was kept for one fictitious particle.

In this case, after a transformation to the dimensionless parameters

$$\omega'_{\alpha 1} = \frac{\omega_{\alpha 1}}{\omega}, \quad \mu'_{\alpha 1} = \frac{\mu_{\alpha 1}}{\mu_{\alpha 2}}, \quad c'_{\alpha i} = \frac{c_{\alpha i}}{\mu_{\alpha 2}\omega^2}, \quad \Omega'_{\alpha i} = \frac{\Omega_{\alpha i}}{\omega}, \qquad (i = 1, 2)$$

we finally obtain

$$-\frac{E_0}{\hbar\omega} = \frac{1}{2} f_{03} + f_{01} + 2\sqrt{\frac{2}{\pi}}\,\frac{\alpha}{\varepsilon c}\sqrt{\chi}\,\frac{1}{\sqrt{z_3}}\,f\left(\chi\,\frac{z_1}{z_3} - 1\right) + \\ + \frac{\alpha}{\sqrt{\pi}}\sqrt{\chi}\int\limits_0^\infty \frac{e^{-x}}{\sqrt{z_3(x)}}\,f\left(\chi\,\frac{z_1(x)}{z_3(x)} - 1\right) dx\,, \tag{21}$$

where

$$f_{0\alpha} = \omega'_{\alpha 1} - \Omega'_{\alpha 1} - \Omega'_{\alpha 2} + \frac{1}{2(\Omega'_{\alpha 1} + \Omega'_{\alpha 2})}\,(c'_{\alpha 1} + c'_{\alpha 2} + \omega'_{\alpha 1}\sqrt{c'_{\alpha 2}})\,,$$

$$z_\alpha = z_{\alpha 1} + z_{\alpha 2}\,, \quad z_\alpha(x) = z_{\alpha 1}\left(1 - e^{-\Omega'_{\alpha 1}x}\right) + z_{\alpha 2}\left(1 - e^{-\Omega'_{\alpha 2}x}\right),$$

$$z_{\alpha i} = \left|\frac{1}{\Omega'^2_{\alpha 1} - \Omega'^2_{\alpha 2}}\left(\Omega'_{\alpha i} - \frac{\omega'^2_{\alpha 1}}{\Omega'_{\alpha i}}\right)\right|, \qquad f(y) = \begin{cases} \dfrac{1}{\sqrt{y}}\,\operatorname{arc\,cot}\sqrt{y}, & y \geqq 0 \\ \dfrac{1}{2\sqrt{-y}}\ln\dfrac{1+\sqrt{-y}}{1-\sqrt{-y}}, & y < 0 \end{cases}$$

$$\alpha = 1, 3; \qquad i = 1, 2$$

and the equation for $\Omega'_{\alpha i}$ is

$$\Omega'^4_{\alpha i} - \Omega'^2_{\alpha i}(\omega'^2_{\alpha 1} + c'_{\alpha 1} + c'_{\alpha 2}) + c'_{\alpha 2}\omega'^2_{\alpha 1} = 0\,.$$

3. Results and Discussion

The results for some values of α and εc are presented in Fig. 1. It should be noted that an application of this "translationally-non-invariant" approximation (2) to calculate the ground state energy of the free polaron (Hamiltonian (1) without the Coulomb term) gives values that are only on parts of 1%

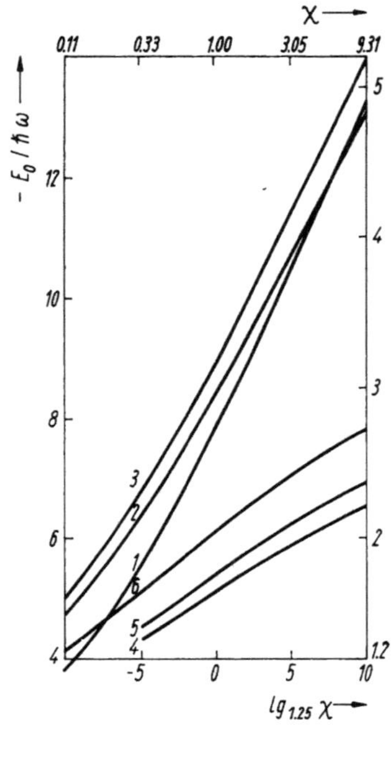

Fig. 1. Dependence of the F-centre ground state energy on the anisotropy of band electron masses χ for some values of α and εc: (1) $\alpha = 4.35$, $\varepsilon c = 3.5$ (LiF); (2) $\alpha = 1.6$, $\varepsilon c = 1.49$ (AgBr); (3) $\alpha = 1.7$, $\varepsilon c = 1.54$ (AgCl). The lines 4 and 5 are plotted to show the energy dependence on the magnitude of polaron binding to the Coulomb centre (for the lines 4 and 5 the Coulomb binding is smaller than for the lines 2 and 3). (4) $\alpha = 1.6$. $\varepsilon c = 11.5$; (5) $\alpha = 1.7$, $\varepsilon c = 10.3$. The line 6 represents the polaron ground state energy with $\alpha = 2$. (The left-hand scale is for line 1, the right-hand scale for all the others)

larger than those obtained in [1]. The difference in energies lowers as the coupling increases, in accordance with the localization of a "free" polaron at extremely strong coupling.

It should be noted that our method provides a rough estimation of excited F-centre state energies and of oscillator strengths for the transitions into them from the ground state. Namely, if we imagine the $\mu_{\alpha i}$ and $c_{\alpha i}$ values to be those that yield the maximum of Z_1 for given α, εc, and χ, the distance between the first excited state energies $E_{\alpha i}$ and the ground state energy will be

$$E_{\alpha i} - E_0 \approx \hbar \Omega_{\alpha i} \, ,$$

and the oscillator strengths

$$t_{\alpha i} \approx \Omega'_{\alpha i} z_{\alpha i} \, .$$

The values of excited state energies and of oscillator strengths calculated in such a way give rather satisfactory coincidence when compared with those obtained by some another procedure (for instance, by the adiabatic method). Particularly, the difference in the oscillator strengths (our and adiabatic) is about 15% (for not too weak coupling). It indicates that our approximation is rather good.

Finally we would like to note that in the case of a polaron [1], the approximation was almost of the same accuracy independent of electron–phonon coupling strength. In the case of the F-centre, our method is formally relevant for all values of both electron–phonon and Coulomb couplings, but in contrast to the polaron case its accuracy increases with the growth of electron–phonon coupling (the Coulomb binding being fixed). This is connected with the following fact. Setting in (1) $V_{\varkappa} = 0$, (1) will represent a hydrogen-like Hamiltonian. The approximation by the oscillator-like Hamiltonian (2) gives a ground state energy 14% larger than the true value. When $V_{\varkappa} \neq 0$ this deviation decreases as $V_{\varkappa}$ increases, because the growing contribution of the polarizational well corresponds to the more oscillator-like character of the resulting potential. This means our results are true in the whole range of electron–phonon coupling, except for the weak coupling limit with the large inaccuracy $\lessapprox 14\%$. Mor precisely, our results are good when the first term in (21) is smaller than the second.

References

[1] S. I. Pekar, L. S. Khazan, and V. I. Sheka, Soviet Phys. — J. exper. theor. Phys. **65**, 1999 (1973).
[2] M. A. Krivoglaz and S. I. Pekar, Izv. Akad. Nauk SSSR, Ser. fiz. **21**, 16 (1957).
[3] J. Schwinger, Phys. Rev. **82**, 664 (1951).

(Received May 13, 1974)

phys. stat. sol. (b) 65, 271 (1974)

Subject classification: 13.5.2 and 19; 18.1

Institute of Semiconductors, Academy of Sciences of the Ukrainian SSR, Kiev

Spin–Plasmon Relaxation in Solids

By

M. F. Deigen and L. A. Suslin

The paramagnetic relaxation of impurity centres in solids due to the interaction between these centres and Langmuir's oscillations of the current carrier plasma is investigated. The transition probabilities of direct and combined relaxation processes as well as of the process accompanied by the resonance fluorescence of plasmons are calculated. The dependences of the transition probabilities on transition frequency, carrier concentration, and temperature are obtained. Comparison with experimental data shows agreement between theory and the observed pattern of the paramagnetic relaxation in semiconductors.

Рассмотрена парамагнитная релаксация примесных центров в твердых телах, обязанная взаимодействию этих центров с ленгмюровскими колебаниями плазмы носителей тока. Рассчитаны вероятности прямых, комбинационных процессов и процессов резонансной флюоресценции плазмонов. Получены зависимости вероятностей релаксационных переходов от частоты перехода, концентрации носителей тока и температуры. Сопоставление с экспериментом показывает согласие теории с наблюдаемой картиной парамагнитной релаксации в полупроводниках.

1. Introduction

In this paper the paramagnetic relaxation process connected with the interaction between the paramagnetic centre and oscillations of the current carrier plasma in crystals is suggested. Plasma oscillations cause alternating electric fields influencing the orbital movement of the electron of the paramagnetic centre (electric field effect — EFE — in the paramagnetic resonance). This influence leads to relaxation transitions in the energy spectrum of the paramagnetic centre due to the spin–orbit coupling. Transitions between magnetic energy levels of the impurity will be accompanied either by the absorption or irradiation of plasmons. In this case the energy of the paramagnetic centre is transferred, at first, to current carriers (precisely, to the system of collective oscillations of current carriers–plasmons).

It must be noted that similar phenomena were already considered as the possible recombination mechanism of current carriers in semiconductors [1].

Indeed, of the variety of plasma oscillations only those accompanied by electric field fluctuations are essential to the suggested relaxation processes. Among them there are longitudinal Langmuir oscillations.

It should be noted that in the examined process the one-plasmon transitions are also possible. The conformity between the transition frequency and that of the oscillation is easily accessible because of the wide range of the carrier concentration changes. This in particular is the case in semiconductors.

For the magitudes of external magnetic fields used in the X-band radiospectroscopy the semiconductor plasma may be assumed isotropic and nonmagnetized. Relaxation processes connected with magnetized plasma will be considered elsewhere. We shall also confine ourselves to the consideration of the one-component plasma. The results may be easily expanded to the case of the multi-component plasma.

Nonscattered plasma which is realized (even) in the intermediate doped semiconductors is considered below.

Section 2 is devoted to obtaining the operator of the interaction between the paramagnetic centre and the current carrier plasma oscillations. In Section 3 the transition probability of the direct (one-plasmon) process is calculated and its dependences on the magnitude of the external magnetic field, on carrier concentration, and on temperature are analysed. The calculation of the probabilities of relaxation transitions accompanied by the combined scattering of plasmons is given in Section 4.

Section 5 contains the consideration of the resonance fluorescence of plasmons. In Section 6 the electronic paramagnetic relaxation of iron-transition-group ions in silicon is discussed from the point of view of the suggested theory.

2. The Operator of the Spin–Plasmon Interaction

The interaction between the paramagnetic centre and the alternating electric field can be described by the following spin Hamiltonian:

$$\hat{\mathscr{H}}_1(\boldsymbol{R}, t) = \sum_{ijk} \alpha_{ijk} E_i(\boldsymbol{R}, t) \{\hat{S}_j \hat{S}_k\} + \sum_{ijk} \beta_{ijk} H_i E_j(\boldsymbol{R}, t) \hat{S}_k . \tag{1}$$

Here $\boldsymbol{R}$ is the coordinate of the paramagnetic centre; $E_j(\boldsymbol{R}, t)$ is the component of the electric field strength; α_{ijk}, β_{ijk} are the components of the EFE tensors; H_i is the component of the external magnetic field strength,

$$i, j, k = x, y, z; \quad \{\hat{S}_j \hat{S}_k\} = \hat{S}_j \hat{S}_k + \hat{S}_k \hat{S}_j .$$

It is implied that the position of the paramagnetic impurity in the lattice is not the inversion symmetry centre.

Let us consider that there is a concentration n of carriers (electrons, holes) in the crystal. Then the fluctuation charge density arising with the displacement of those carriers equals

$$\varrho(\boldsymbol{r}, t) = -ne\Delta(\boldsymbol{r}) . \tag{2}$$

Here $\Delta(\boldsymbol{r})$ is the "volume widening operator" at small oscillations and e is the electron charge.

With the accuracy to the terms containing the squared derivatives from the difference of displacements $\boldsymbol{u}$ by the coordinates, the operator $\Delta(\boldsymbol{r})$ can be written in the form

$$\Delta(\boldsymbol{r}) = \operatorname{div} \boldsymbol{u} + \frac{1}{2} (\operatorname{div} \boldsymbol{u})^2 + \sum_{i \neq k} \left(\frac{\partial u_i}{\partial x_k}\right)^2 . \tag{3}$$

The "longitudinal part" of $\Delta(\boldsymbol{r})$ may be rewritten as

$$\Delta(\boldsymbol{r}) = \sum_{\boldsymbol{k}} ik \left(\frac{\hbar}{2Nm^*\omega_{\boldsymbol{k}}}\right)^{1/2} (\hat{a}_{\boldsymbol{k}} \, e^{i\boldsymbol{kr}} - \hat{a}_{\boldsymbol{k}}^+ \, e^{-i\boldsymbol{kr}}) -$$
$$- \frac{\hbar}{4Nm^*} \sum_{\boldsymbol{k}} \sum_{\boldsymbol{k}'} \frac{kk'}{(\omega_{\boldsymbol{k}}\omega_{\boldsymbol{k}'})^{1/2}} [\hat{a}_{\boldsymbol{k}} \hat{a}_{\boldsymbol{k}'} \, e^{i(\boldsymbol{k}+\boldsymbol{k}')\boldsymbol{r}} +$$
$$+ \hat{a}_{\boldsymbol{k}}^+ \hat{a}_{\boldsymbol{k}'}^+ \, e^{-i(\boldsymbol{k}+\boldsymbol{k}')\boldsymbol{r}} - \hat{a}_{\boldsymbol{k}}^+ \hat{a}_{\boldsymbol{k}'} \, e^{i(\boldsymbol{k}'-\boldsymbol{k})\boldsymbol{r}} - \hat{a}_{\boldsymbol{k}} \hat{a}_{\boldsymbol{k}'}^+ \, e^{i(\boldsymbol{k}'-\boldsymbol{k})\boldsymbol{r}}] . \tag{4}$$

Here Nm^* is the effective mass of the oscillating current carriers; $\hat{a}_{\boldsymbol{k}}^+(\hat{a}_{\boldsymbol{k}})$ is the creation (annihilation) operator of plasmon with the frequency $\omega_{\boldsymbol{k}}$ and the momentum $\hbar\boldsymbol{k}$ at the time t.

The potential created by charge density (2) in the point of observation is apparently equal to

$$V(\boldsymbol{R}, t) = \frac{1}{\varepsilon} \int \frac{\varrho(\boldsymbol{r}, t)}{|\boldsymbol{r} - \boldsymbol{R}|} \, \mathrm{d}^3 r \,, \tag{5}$$

where ε is the dielectric constant of a crystal. Substituting (2) into (5) and using (4), the following expression for $V(\boldsymbol{R}, t)$ is obtained:

$$\begin{aligned} V(\boldsymbol{R}, t) = {} & \frac{4\pi nei}{\varepsilon} \left(\frac{\hbar}{2Nm^*}\right)^{1/2} \sum_{\boldsymbol{k}} k^{-1} \omega_{\boldsymbol{k}}^{-1/2} \, [\hat{a}_{\boldsymbol{k}}^+ \, \mathrm{e}^{-i\boldsymbol{k}\boldsymbol{R}} - \hat{a}_{\boldsymbol{k}} \, \mathrm{e}^{i\boldsymbol{k}\boldsymbol{R}}] - \\ & - \frac{\pi\hbar en}{\varepsilon Nm^*} \sum_{\boldsymbol{k}, \boldsymbol{k}'} \frac{kk'}{(\omega_{\boldsymbol{k}} \omega_{\boldsymbol{k}'})^{1/2}} \, q^{-2} [\hat{a}_{\boldsymbol{k}}^+ \hat{a}_{\boldsymbol{k}'} \, \mathrm{e}^{i\boldsymbol{q}\boldsymbol{R}} + \hat{a}_{\boldsymbol{k}} \hat{a}_{\boldsymbol{k}'}^+ \, \mathrm{e}^{-i\boldsymbol{q}\boldsymbol{R}}] + \\ & + \frac{\pi\hbar en}{\varepsilon Nm^*} \sum_{\boldsymbol{k}\boldsymbol{k}'} \frac{kk'}{(\omega_{\boldsymbol{k}} \omega_{\boldsymbol{k}'})^{1/2}} \, p^{-2} \, [\hat{a}_{\boldsymbol{k}} \hat{a}_{\boldsymbol{k}'} \, \mathrm{e}^{i\boldsymbol{p}\boldsymbol{R}} + \hat{a}_{\boldsymbol{k}}^+ \hat{a}_{\boldsymbol{k}'}^+ \, \mathrm{e}^{-i\boldsymbol{p}\boldsymbol{R}}] ; \end{aligned} \tag{6}$$

$$\boldsymbol{q} = \boldsymbol{k}' - \boldsymbol{k} \,, \quad \boldsymbol{p} = \boldsymbol{k}' + \boldsymbol{k} \,.$$

Calculating the component of the electric field strength $E_i(\boldsymbol{R}, t) = [-\nabla V(\boldsymbol{R}, t)]_i$ and substituting the result into (1) we obtain the expression for the operator of spin–plasmon interaction:

$$\begin{aligned} \hat{\mathscr{H}}_1(\boldsymbol{R}, t) = {} & \frac{4\pi ne}{\varepsilon} \left(\frac{\hbar}{2Nm^*}\right)^{1/2} \sum_{\boldsymbol{k}} k^{-1} \omega_{\boldsymbol{k}}^{-1/2} \, (\hat{a}_{\boldsymbol{k}} \, \mathrm{e}^{i\boldsymbol{k}\boldsymbol{R}} - \hat{a}_{\boldsymbol{k}}^+ \, \mathrm{e}^{-i\boldsymbol{k}\boldsymbol{R}}) \times \\ & \times \left(\sum_i k_i \hat{Q}_i + \sum_j k_j \hat{P}_j\right) + \frac{\pi i n e \hbar}{\varepsilon Nm^*} \sum_{\boldsymbol{k}, \boldsymbol{k}'} \frac{kk'}{(\omega_{\boldsymbol{k}} \omega_{\boldsymbol{k}'})^{1/2}} \, q^{-2} [\hat{a}_{\boldsymbol{k}}^+ \hat{a}_{\boldsymbol{k}'} \, \mathrm{e}^{i\boldsymbol{q}\boldsymbol{R}} - \\ & - \hat{a}_{\boldsymbol{k}} \hat{a}_{\boldsymbol{k}'}^+ \, \mathrm{e}^{-i\boldsymbol{q}\boldsymbol{R}}] \left(\sum_i q_i \hat{Q}_i + \sum_j q_j \hat{P}_j\right) + \\ & + \frac{\pi i n e \hbar}{\varepsilon Nm^*} \sum_{\boldsymbol{k}\boldsymbol{k}'} \frac{k'k}{(\omega_{\boldsymbol{k}} \omega_{\boldsymbol{k}'})^{1/2}} \, p^{-2} [\hat{a}_{\boldsymbol{k}}^+ \hat{a}_{\boldsymbol{k}'}^+ \, \mathrm{e}^{-i\boldsymbol{p}\boldsymbol{R}} - \hat{a}_{\boldsymbol{k}} \hat{a}_{\boldsymbol{k}'} \, \mathrm{e}^{i\boldsymbol{p}\boldsymbol{R}}] \times \\ & \times \left(\sum_i p_i \hat{Q}_i + \sum_j p_j \hat{P}_j\right) ; \end{aligned} \tag{7}$$

$$\hat{P}_j = \sum_{ik} \beta_{ijk} H_i \hat{S}_k \,,$$

$$\hat{Q}_i = \sum_{jk} \alpha_{ijk} \{\hat{S}_j \hat{S}_k\} \,.$$

The first term in (7) is the interaction operator leading to one-plasmon transitions. The second one describes the combined scattering of plasmons and the third one — the simultaneous absorption or irradiation of two plasmons.

3. Direct (One-Plasmon) Processes

The probability of a direct relaxation transition between the magnetic energy levels $a \to b$ of the impurity under the action of (7) is equal to (the transition is accompanied by the creation of one plasmon)

$$W(a \to b) = \\ = \frac{16\pi^3 n^2 e^2}{N m^* \hbar \varepsilon^2} \sum_{\boldsymbol{k}} \frac{k^{-2}\omega_{\boldsymbol{k}}^{-1} \exp(\hbar\omega_{\boldsymbol{k}}/k_0 T)}{\exp(\hbar\omega_{\boldsymbol{k}}/k_0 T) - 1} \left|\sum_i k_i Q_i^{ab} + \sum_j k_j P_j^{ab}\right|^2 \delta(\omega_{\boldsymbol{k}} - \omega_{ab}) \,. \tag{8}$$

Here statistical averaging in the plasmon system was made; ω_{ab} is the transition frequency.

Let us consider for example a crystal of T_d symmetry. Then $\beta_{ijk} = \beta_0 \neq 0$; $\alpha_{ijk} = \alpha_0 \neq 0$ and (8) acquires the form

$$W(a \to b) = \frac{16\pi^3 n^2 e^2}{N m^* \hbar \varepsilon^2} \sum_{\boldsymbol{k}} \frac{\omega_{\boldsymbol{k}}^{-1} \exp(\hbar\omega_{\boldsymbol{k}}/k_0 T)}{\exp(\hbar\omega_{\boldsymbol{k}}/k_0 T) - 1} \sum_i n_i^2 \, |D_{ab}|^2 \, \delta(\omega_{\boldsymbol{k}} - \omega_{ab}); \tag{9}$$

$$n_i = \frac{k_i}{k} \,.$$

In (9)

$$D_{ab} = \alpha_0 \sum_{j \neq k} \{\hat{S}_j \hat{S}_k\}_{ab} + \beta_0 \sum_{i \neq k} H_i (\hat{S}_k)_{ab} \,. \tag{10}$$

Summation over $\boldsymbol{k}$ in (9) carried out with the help of integration,

$$\sum_{\boldsymbol{k}} \to (2\pi)^{-3} \, V \int \mathrm{d}^3 k \,,$$

gives for the Langmuir oscillations of a current carrier plasma ($\omega_{\boldsymbol{k}}^2 = \Omega^2 + \frac{3}{2} k^2 s^2$, $\Omega^2 = 4\pi n e^2/\varepsilon m^*$, $s^2 = 2 k_0 T/m^*$, $0 \leqq k \leqq a^{-1}$, $a = s/\Omega\sqrt{2}$ Debye radius)

$$W(a \to b) = 2 \left(\frac{2}{3}\right)^{3/2} \frac{\Omega^2}{\hbar \varepsilon s^3} |D_{ab}|^2 \frac{\exp(\hbar\omega_{ab}/k_0 T)}{\exp(\hbar\omega_{ab}/k_0 T) - 1} (\omega_{ab}^2 - \Omega^2)^{1/2} \,; \tag{11a}$$

$$\Omega < \omega_{ab} < 2\Omega \,.$$

Usually $\hbar\omega_{ab} \ll k_0 T$ and (11a) takes the form

$$W(a \to b) = 2 \left(\frac{2}{3}\right)^{3/2} |D_{ab}|^2 \frac{k_0 T \Omega^2}{\hbar^2 \varepsilon s^3} \left(1 - \frac{\Omega^2}{\omega_{ab}^2}\right)^{1/2} . \tag{11b}$$

It is interesting to note in (11) that the transition probability of a direct process is inversely proportional to the cubed velocity of the oscillator expansion unlike the dependence v^{-5} (v is the sound velocity) for the corresponding spin–phonon process. This fact reflects the different nature of dissipative systems (plasmons and phonons).

As follows from (10) in the example considered there are possible relaxation transitions between the magnetic energy levels of a paramagnetic centre with the selection rules: $|\Delta M| = (|M_a - M_b|) = 1, 2$; ($M_a$, M_b are the corresponding eigenvalues of the $\hat{S}_z$ operator.) Therefore $\omega_{ab} = g\beta H \, \Delta M/\hbar$, β Bohr's magneton.

Then (11) can be presented in the form (we consider for simplicity that $\boldsymbol{H}$ is parallel to the Z-axis of the crystallographic reference frame)

$$W(a \to b) = \begin{cases} (C_1 H^2 + C_2 H + C_0)\left(1 - \dfrac{H_0^2}{H^2}\right)^{1/2}; & |\Delta M| = 1 \,. \\ C_0 \left(1 - \dfrac{H_0^2}{4H^2}\right)^{1/2}; & |\Delta M| = 2 \,. \end{cases} \tag{12}$$

Here $H_0 = \hbar \Omega / g\beta$; C_0, C_1, C_2 are dimension constants. The region of changing for H is determined by the bounds of application of (11): $\Omega < \omega_{ab} < 2\Omega$. It is clear that with the transition frequency and that of the plasma coinciding (H_0, $\frac{1}{2}H_0$ for $|\Delta M| = 1, 2$) the corresponding relaxation transition probability would be turned to zero. The latter is a consequence of equality of the plasma oscillation density to zero if the oscillation frequency is equal to Ω. The quality difference of the dependences (12) on the corresponding ones in the spin–phonon relaxation processes must be noted.

Substituting the evident expression of the plasma frequency Ω into (11b) we obtain

$$W(a \to b) = An \left(1 - \frac{n}{n_0}\right)^{1/2}, \tag{13}$$

where $n_0 = \omega_{ab}^2 m^* \varepsilon / 4\pi e^2$, A is a dimension constant. One can see that $W(a \to b)$ has a maximum at $n = \frac{2}{3} n_0$ and equals zero if $n = n_0$ (see Fig. 1).

The temperature dependence of $W(a \to b)$ is closely connected with the concentration one. Indeed (11b) contains in addition to typical proportionality to T (for the one-quantum relaxation processes) the nonevident dependence on temperature through the functions $s = s(T)$ and $n = n(T)$. The latter may be quite different in semiconductors. Considering, for example, that the current carriers arise in the conduction band due to the thermal ionization of donors we obtain

$$W(a \to b) \sim T^{1/4} \exp\left(-\frac{\mathscr{E}_\mathrm{D}}{2k_0 T}\right), \tag{14}$$

where $\mathscr{E}_\mathrm{D}$ is the ionization energy of donors counted from the bottom of the conduction band.

It must be noted that through n the transition probability may depend on the compensation of the sample.

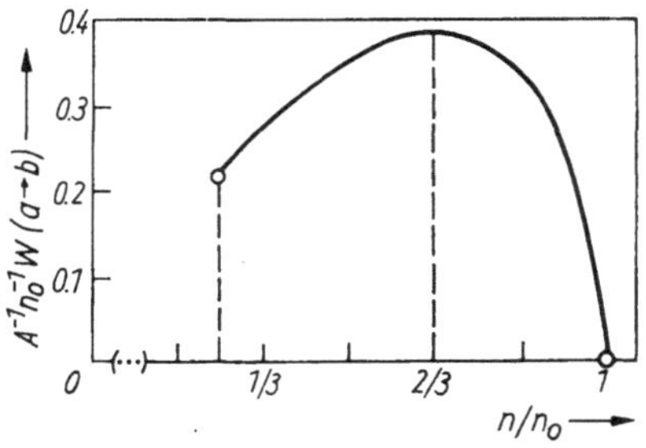

Fig. 1. The concentration dependence of the probability of the direct relaxation process

4. Combined Scattering of Plasmons

The probability of the combined scattering of plasmons is equal to the sum of probabilities of two processes: 1. plasmon ($\boldsymbol{k}, \omega_{\boldsymbol{k}}$) is irradiated and plasmon ($\boldsymbol{k}', \omega_{\boldsymbol{k}'}$) is absorbed or 2. vice versa.

$$W_{\text{cs}}(a \to b) =$$
$$= 4\pi^3 \left(\frac{ne}{\varepsilon N m^*}\right)^2 \sum_{\boldsymbol{k}, \boldsymbol{k}'} \frac{(k'k)^2}{\omega_{\boldsymbol{k}} \omega_{\boldsymbol{k}'}} \frac{\bar{\bar{n}}_{\boldsymbol{k}}(\bar{\bar{n}}_{\boldsymbol{k}'} + 1)}{q^4} \left|\sum_i q_i Q_i^{ab} + \sum_j q_j P_j^{ab}\right|^2 \delta(\omega_{ab} - \omega_{\boldsymbol{k}} + \omega_{\boldsymbol{k}'}), \tag{15}$$

where $\bar{\bar{n}}_{\boldsymbol{k}}$ means the average number of plasmons in the state with the wave vector $\boldsymbol{k}$.

Proceeding from the summation over $\boldsymbol{k}$ and $\boldsymbol{k}'$ in (15) to integrals and carrying out the tedious integration over the spherical angles $d\Omega_{\boldsymbol{k}}$ and $d\Omega_{\boldsymbol{k}'}$ and, finally, using the δ-function we obtain

$$W_{\text{cs}}(a \to b) = \frac{8}{243} s^{-8} \left(\frac{e}{m^* \varepsilon}\right)^2 \left(\frac{k_0 T}{\hbar}\right)^5 \gamma^5 J(\gamma, \delta) \sum_i |P_i^{ab} + Q_i^{ab}|^2 . \tag{16}$$

Here

$$J(\gamma, \delta) = \int_1^2 \frac{[(x+\delta)^2 - 1]\,[x^2 - 1]\,e^{\gamma x}}{[e^{\gamma x} - 1]^2} \ln \left| \frac{\sqrt{(x+\delta)^2 - 1} + \sqrt{x^2 - 1}}{\sqrt{(x+\delta)^2 - 1} - \sqrt{x^2 - 1}} \right| dx; \tag{17}$$

$$\gamma = \frac{\hbar \Omega}{k_0 T}, \qquad \delta = \frac{\omega_{ab}}{\Omega}.$$

The investigation shows that for the actual values $0.03 \leqq \delta \leqq 0.3$ the integral (17) depends on δ weakly and it may be well approximated by the following expression:

$$J(\gamma) = \int_1^2 \frac{x^5 e^{\gamma x}}{[e^{\gamma x} - 1]^2} dx \,.^{1)} \tag{18}$$

It can be seen that in the interval mentioned the transition probability of the combined scattering of plasmons does not depend on the transition frequency. It must be noted that for $\delta < 0.03$ $W_{\text{cs}}(a \to b)$ may become the essential function of ω_{ab}.

The concentration dependence of the transition probability is determined by the product $\gamma^5 J(\gamma)$.

Fig. 2 presents the dependence $W_{\text{cs}}(a \to b)$ on γ for $0.1 \leqq \gamma \leqq 5.0$. One can see that for some values of carrier concentration the transition probability has a maximum as well as for the direct processes.

---

1) The expression (18) may be presented by the combination of integrals

$$I_n = \int_0^\lambda \frac{x^n e^{\gamma x}}{[e^{\gamma x} - 1]^2} dx$$

which are tabulated in the theory of spin-lattice relaxation (see, for example, [2]).

Fig. 2. The concentration dependence of the probability of the combined relaxation process. $\gamma = \hbar\Omega/k_0T$

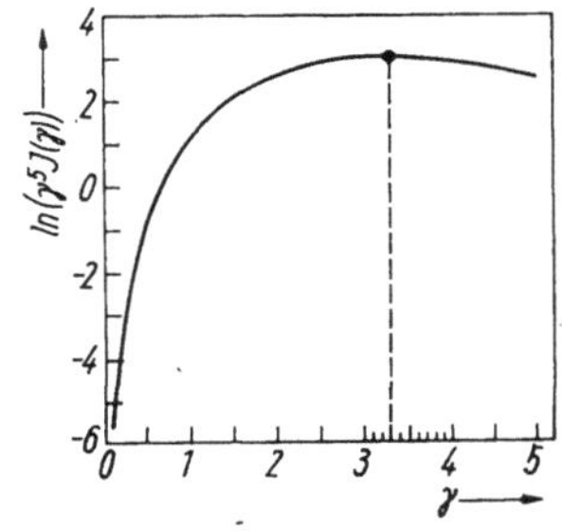

The temperature dependence of $W_{cs}(a \to b)$ is determined by the product $T\gamma^5 J(\gamma)$. If the carrier concentration does not depend on temperature, $W_{cs}(a \to b) \sim T^{-4}J(T^{-1})$. Usually the carrier concentration (and consequently γ) is essentially dependent on temperature. In this case the evident form $W_{cs}(T)$ may be ascertained as soon as the concrete dependence $n = n(T)$ is presented.

5. Resonance Fluorescence of Plasmons

Let us consider that the plasmon energy exceeds the interval Δ_l between the lowest magnetic level of the paramagnetic ion and the level l corresponding to the excited state of the centre. Then it is well known that the relaxation process $m' \to m$ can be carried out in the two stages $m' \to l \to m$ and is the alternation of the one-quantum transitions.

Following the traditional calculation scheme (see, for example, [2]) we obtain for the transition probability $(m' \to m)$:

$$W_l(m' \to m) = 2\left(\frac{2}{3}\right)^3 \frac{\Omega^4}{\pi\varepsilon^2 s^6} |D_{lm}|^2 \, |D_{m'l}|^2 \, \{J_2 - \Omega^2 J_0\}\,; \tag{19}$$

$$(\hbar\Omega < \Delta_l < 2\hbar\Omega)\,.$$

$$J_n = \int\limits_{\Omega}^{2\Omega} \frac{\omega^n \exp(\hbar\omega/k_0T)}{[\exp(\hbar\omega/k_0T) - 1]^2} \, \frac{\Delta_l^2 + (\hbar\omega)^2}{[\Delta_l^2 - (\hbar\omega)^2]^2} \, d\omega\,. \tag{20}$$

Substituting into (20):

$$\hbar\omega - \Delta_l \to \left|\hbar\omega - \Delta_l + \frac{i}{2}\hbar\Gamma\right|;$$

$$\Gamma = (2\pi)^{-1}\,[W(l \to m) + W(l \to m')]\,,$$

$W(l \to m)$ and $W(l \to m')$ are the one-plasmon transition probabilities.

Then, considering $[(\hbar\omega - \Delta_l)^2 + \frac{1}{4}\hbar^2\Gamma^2]^{-1} \to (2\pi/\Gamma)\,\delta(\hbar\omega - \Delta_l)$, we have

$$\{J_2 - \Omega^2 J_0\} = \frac{4\pi^2\Delta_l^2 \exp(-\Delta_l/k_0T)}{\hbar^4[W(l \to m) + W(l \to m')]} \left\{1 - \left(\frac{\hbar\Omega}{\Delta_l}\right)^2\right\} \tag{21}$$

and consequently the "resonance part" of transition probability (19) takes the following form:

$$W_l^{\text{res}}(m' \to m) = 2\left(\frac{2}{3}\right)^3 \frac{4\pi^2 \Delta_l^2 \Omega^4 \exp\left(-\Delta_l/k_0T\right) |D_{lm}|^2 |D_{m'l}|^2}{\hbar^4 \varepsilon^2 s^6 \left[W(l \to m) + W(l \to m')\right]} \left\{1 - \left(\frac{\hbar\Omega}{\Delta_l}\right)^2\right\}. \quad (22)$$

Substituting (11a) into (22) and using the approximation $\Delta_l \gg k_0T$ we obtain

$$W_l^{\text{res}}(m' \to m) =$$
$$= \left(\frac{2}{3}\right)^{3/2} \frac{4\pi\Delta_l \Omega^2 \exp\left(-\Delta_l k_0 T\right)}{\hbar^2 \varepsilon s^3} \left\{1 - \left(\frac{\hbar\Omega}{\Delta_l}\right)^2\right\}^{1/2} \left[|D_{lm}|^{-2} + |D_{m'l}|^{-2}\right]^{-1}. \quad (23)$$

Here only one (the nearest) excited level l was considered. If there are several excited levels in the plasma spectra bounds the summation over the corresponding probabilities must be carried out.

The concentration dependence of $W_l^{\text{res}}(m' \to m)$ is determined by the function $\Omega = \Omega(n)$ and is analytically similar to (13), where $\tilde{n}_0 = (\Delta_l/\hbar)^2\, m^* \varepsilon/4\pi e^2$. The transition probability has a maximum if $n = \frac{2}{3}\tilde{n}_0$ and becomes zero if $n = \tilde{n}_0$.

The temperature dependence of $W_l^{\text{res}}(m' \to m)$ is due to $s(T)$, $\exp\left(-\Delta_l/k_0T\right)$, and also to $n = n(T)$.

Assume for example that current carriers arise in the conduction band due to the thermal ionization of donors. Then

$$W_l^{\text{res}}(m' \to m) \sim T^{-3/4} \exp\left(-\frac{\Delta_l + \frac{1}{2}\mathscr{E}_\text{D}}{k_0T}\right). \quad (24)$$

If the carrier concentration does not depend on temperature (the region of impurity exhaustion or injection of carriers into the crystal) we obtain

$$W_l^{\text{res}}(m' \to m) \sim T^{-3/2} \exp\left(-\frac{\Delta_l}{k_0T}\right). \quad (25)$$

Finally, it must be noted that for the paramagnetic centres with $\beta_{ijk} \approx 0$, $W_l^{\text{res}}(m' \to m)$ does not depend on the transition frequency. In other cases $W_l^{\text{res}}(m' \to m) \sim \omega_{m'm}^2$.

6. Paramagnetic Relaxation of Deep Centres in Semiconductors

As an example of application of the above theory let us consider the paramagnetic relaxation of iron-transition-group ions in silicon. The choice of these objects is stimulated by well known EPR spectra, the known relaxation rates, and also by the investigated EFE in EPR (see the detailed bibliography in [3, 4]).

The iron-transition-group ions form deep energy levels in the forbidden energy gap of silicon. Various charge states of the centres are due to the changing of a shallow donor (acceptor) concentration in the initial silicon. Thus the latter may be effective current carrier suppliers to the conduction (or valence) band of the crystal even at low temperatures. EPR spectra of ions $Cr^+(3d^5)$ and $Fe^0(3d^8)$ are well observed at 20.4 °K and sometimes at 78 °K. It is for these ions that $\beta_0 \approx 0$ and α_0 values are small [5]. EPR spectra of ions $Cr^0(3d^6)$, $Mn^+(3d^6)$, and $Fe^+(3d^6)$ at $T = 10$ °K consist of broadened lines corresponding to the relaxation rates of the order of and less than one second. The comparison between the observed relaxation rates and those calculated from (11b) shows good agreement. For reasonable values of parameters:

$$\tau(Cr^0\!:Si) \approx 1.7\ s\,,$$
$$\tau(Mn^+\!:Si) \approx 0.2\ s\,,$$
$$\tau(Fe^+\!:Si) \approx 2.3\ s\,.$$

It must be noted that the order of magnitudes of the relaxation rates is also explained by the mechanisms considered in [6, 7], which are characterized by other dependences on the transition frequency and on temperature with respect to (12) and (14). The absence of the corresponding experimental data makes the separation of concrete mechanisms difficult. However, it is known that the observation of EPR spectra of ions Cr^0, Mn^+, and Fe^+ becomes impossible when the temperature increases over 10 °K. It testifies the existence of the paramagnetic relaxation mechanism whose effectiveness is essentially dependent on temperature. Just this temperature dependence was obtained for the relaxation rate in the present paper (see (14)).

It is clear that direct processes and combined scattering are corresponding to different value of the current carrier concentration.

The estimation of the relaxation rates at the concentrations where the combined scattering is actual leads to the following values:

$$\tau(Cr^0\!:Si) \approx 1.1\times10^{-4}\ s\,,$$
$$\tau(Mn^+\!:Si) \approx 1.3\times10^{-5}\ s\,,$$
$$\tau(Fe^+\!:Si) \approx 1.5\times10^{-4}\ s\,.$$

One can see that the combined scattering leads to considerably shorter paramagnetic relaxation rates.

The above considerations lead to the conclusion that Langmuir oscillations of the current carrier plasma may play an important role in the paramagnetic relaxation processes explaining, in particular, the relaxation of the deep centres in semiconductors.

It must be noted that the dependence of the relaxation transition probability on the plasma frequency Ω makes it possible in principle to determine it from the relaxation rate measurements.

Thus it is possible to use the impurity centres as the "paramagnetic probe" to obtain the solid state plasma characteristics. If the current carrier concentration in the sample is known (e.g. from independent experiments) it is possible to consider the effective masses of carriers as the values to be determined.

7. Conclusion

A new relaxation mechanism of paramagnetic centres in solids is suggested originating due to the interaction between paramagnetic centres and current carrier plasma oscillations.

The realization of the spin–plasmon relaxation mechanism is possible unless the position of the paramagnetic impurity in the crystal is the centre of inversion symmetry.

The Hamiltonian of the interaction between paramagnetic centres and current carrier plasma oscillations is obtained, which makes it possible to calculate the probabilities of the one- and two-plasmon transitions between magnetic energy levels of the impurity.

Relaxation, rates of direct processes, combined scattering processes, and resonance fluorescence of plasmons are calculated.

Relaxation rate dependences on temperature, transition frequency, and current carrier concentration are obtained.

Comparison of the theory and the experiment was made for deep centres in silicon.

The suggested mechanism proved to give the correct order of magnitudes of relaxation rates. The correlation between the measured relaxation rates of different impurity ions and the corresponding magnitudes of electric field effect tensor components acquires natural explanation.

References

[1] V. L. Bonch-Bruevich, Izv. Akad. Nauk SSSR, Ser. fiz. **21**, 87 (1957).
[2] S. A. Altshuler and B. M. Kozirev, Electronic Paramagnetic Resonance, Izd. Nauka, Moscow 1972 (in Russian).
[3] G. W. Ludwig and G. Woodbury, Electronic Spin Resonance in Semiconductors, Izd. Mir, Moscow 1964 (in Russian).
[4] A. B. Roitsin, Uspekhi fiz. Nauk **105** (4), 677 (1971).
[5] G. W. Ludwig and F. S. Ham, Phys. Rev. Letters **8**, 210 (1962).
[6] M. F. Deigen, M. D. Glinchuk, and L. A. Suslin, Fiz. tverd. Tela **12**, 2371 (1970).
[7] M. D. Glinchuk, M. F. Deigen, and L. A. Suslin, phys. stat. sol. (b) **46**, 501 (1971).

(Received March 8, 1974)

phys. stat. sol. (b) 65, 281 (1974)

Subject classification: 13.5.2; 13.3

Polytechnical Institute, Leningrad

Theory of Plasmon Excitation in the Case of Electron Diffuse Reflection at the Surface

By

B. N. LIBENSON and V. V. RUMYANTSEV

A theory is given of the excitation of plasmons by fast electrons in solids with a spatially inhomogeneous surface. The properties of the surface are described similarly as in the theory of the anomalous skin effect. The probability of the excitation of plasmons is expressed through the retarded Green's function of the electric field. The Dyson equation for this Green's function is solved by the Fock method. It is shown that the surface roughness influences the probability of the excitation of the bulk as well as of the surface plasmons. The latter exhibit additional damping due to the surface roughness.

Построена теория возбуждения плазмонов быстрыми электронами в твердом теле с пространственно неоднородной границей. Свойства поверхности описываются способом, аналогичным тому, который используется в теории аномального скин-эффекта. Вероятность возбуждения плазмонов выражается через запаздывающую функцию Грина электрического поля твёрдого тела. Уравнение Дайсона, определяющее эту функцию Грина, решено методом Винера-Фока. Показано, что диффузность влияет на вероятность возбуждения, как объёмных, так и поверхностных плазмонов. В последнем случае диффузность вызывает дополнительное затухание поверхностных плазмонов.

1. Introduction

The probability of the excitation of surface plasmons by fast electrons in solids of finite dimensions was originally calculated by Ritchie [1] using a semiclassical method and ignoring spatial dispersion. The theory was developed further considering such factors as the finiteness of the damping of plasmons, the anisotropy of solids, quantum corrections, electrodynamic retardation, etc. In addition to those questions we are faced with the problem of a more realistic description of the influence of the surface on the properties and the probability of excitation of plasmons and, primarily, of surface plasmons. Two aspects should be noted here. First of all one should consider that at the boundary solid–vacuum there is not a rectangular potential step (normal to the surface), by which this boundary is usually approximated in microscopic treatments. This problem has been discussed in papers of Romanov [2] and Bennett [3]. Secondly, one should note that there is no spatial homogeneity parallel to the surface. Classically speaking this would mean that the electrons suffer diffuse reflection at the surface. The consideration of this latter factor in quantum theory is a complicated problem. There are a number of works in which some aspects of this problem are discussed within the framework of the classical kinetic theory. Mekhtiev [4] considered the diffuse reflection of bulk plasmons at the surface. Romanov [5] solving the kinetic equation together with the Poisson equation found the dispersion and the damping of surface plasmons. The probability of the excitation of plasmons was calcualted directly in the microscopic theory [2, 6] assuming that the electrons suffer only specular reflection at the surface.

The present work is devoted to the calculation of the probability of excitation of bulk and surface plasmons by fast electrons taking into account the diffuse character of the reflection of electrons.

According to Romanov [2], the probability of a transition of a fast electron from a state with momentum $\boldsymbol{p}$ to a state with momentum $\boldsymbol{p}'$ (initial state of the solid N, final state N') can be written in the following way:

$$W_{\boldsymbol{p},\boldsymbol{p}'} = \sum_{N'} |\int H_{\boldsymbol{p},N,\boldsymbol{p}',N'}(t)\,\mathrm{d}t|^2 = \int \mathrm{d}^4x\,\mathrm{d}^4x'\,\langle \boldsymbol{p}|\,\hat{\varrho}(x)\,|\boldsymbol{p}'\rangle\,\langle \boldsymbol{p}'|\,\hat{\varrho}(x')\,|\boldsymbol{p}\rangle\,D(x, x')\,, \tag{1.1}$$

where

$$D(x, x') = \langle \tilde{\varphi}(x)\,\tilde{\varphi}(x')\rangle\,.$$

$\hat{H}$ is the operator of the interaction of the fast electron with the electron of the solid $\hat{H} = \int \mathrm{d}^3x\,\hat{\varrho}(x)\,\hat{\varphi}(x)$. The operator of the charge density $\hat{\varrho}(x)$ of the fast incident electron in second quantization has the form $\hat{\varrho}(x) = \psi^+(x)\,\psi(x)$. $\hat{\varphi}(x)$ is the operator of the electrostatic potential of the solid. The tilded operators are to be taken in the Heisenberg representation. D is the correlation function of the scalar potentials of the electrons of the solid. It is assumed that $e = m = \hbar = 1$.

The fast electron exchanges with the surface a momentum of the order of $q|U_0|(pp')^{-1}$, where q is the momentum lost by the electron inside the target, and U_0 is the depth of the model potential box. This quantity can be neglected in comparison with q and the wave function of the fast electron can be written in the form of a plane wave. In this case the transition probability can be written in the following way:

$$W_{\boldsymbol{p},\boldsymbol{p}-\boldsymbol{q}} = 2\,\mathrm{Im}\,D^{\mathrm{R}}(\boldsymbol{q}, -\boldsymbol{q}, \omega)\,. \tag{1.2}$$

This formula is correct both for normal and inclined incidence of the fast electron. Thus, we must find the double Fourier transform of the Green function D. The latter is defined by the equation

$$\Delta D(x, x') + 4\pi \int \Pi(x, x_1)\,D(x_1, x')\,\mathrm{d}^4x_1 = -4\pi\,\delta(x - x') \tag{1.3}$$

or equivalent Dyson equation. To write the equation in our particular case, considering the spatial inhomogeneity of the boundary, we shall proceed in the following way: we first establish the relationship between the functions D and the classical electrostatic potential φ. The latter is the average of the quantum operator $\hat{\varphi}$. $\varphi = \langle \hat{\tilde{\varphi}}(\boldsymbol{r}, t)\rangle$.

Following [7], we introduce external charges with the density $\hat{\varrho}_{\mathrm{cm}}(\boldsymbol{r}, t)$. Then

$$\varphi(\boldsymbol{r}, t) = \langle S_{\mathrm{cm}}^{-1}(t)\,\hat{\varphi}(\boldsymbol{r}, t)\,S_{\mathrm{cm}}(t)\rangle\,, \tag{1.4}$$

where $\hat{\varphi}(\boldsymbol{r}, t)$ is the operator of the potential φ in the interaction representation, assuming that the energy of interaction is equal to $\hat{H}_i = \int \varrho_{\mathrm{cm}}\varphi\,\mathrm{d}\boldsymbol{r}$.

$$S_{\mathrm{cm}} = T\exp\left\{-i\int_{-\infty}^{t}\mathrm{d}t'\int\mathrm{d}\boldsymbol{r}'\,\varrho_{\mathrm{cm}}(\boldsymbol{r}', t')\,\hat{\varphi}(\boldsymbol{r}', t')\right\}. \tag{1.5}$$

Expanding (1.5) with respect to H_i we find that only the zeroth and first terms in the series differ from zero. The latter term gives

$$\varphi_{\mathrm{cm}}(\boldsymbol{r}, t) = i\int_{-\infty}^{+\infty}\mathrm{d}t'\,\mathrm{d}\boldsymbol{r}'\,\varrho_{\mathrm{cm}}(\boldsymbol{r}', t')\,D^{\mathrm{R}}(\boldsymbol{r}, \boldsymbol{r}', t - t')\,. \tag{1.6}$$

Since ϱ_{cm} has the form of δ-function we have

$$\varphi_{\mathrm{cm}}(\boldsymbol{r}, t) = D^{\mathrm{R}}(\boldsymbol{r}, t) . \tag{1.7}$$

So, using the relationship (1.7) we can write the equation (1.3) also for $\varphi(\boldsymbol{r}, t)$. The kernel of the integral equation which includes the polarization of the substance can be found in the classical approximation by solving the classical kinetic equation with appropriate boundary conditions taking into account diffuse reflection and expressing ϱ through φ [8]. Then (1.3) will be the equation for the Green function $D(\boldsymbol{r}, \boldsymbol{r}', \omega)$ in the classical limit allowing for spatial inhomogeneity of the target. Solving this equation for $D(\boldsymbol{r}, \boldsymbol{r}', \omega)$ we find the transition probability from (1.2).

In correspondence with the developed scheme we consider in Section 2 the kinetic equation for the distribution function of electrons, derive the bulk and surface charge-densities, and obtain (1.3) in explicit form. In Section 3 this equation is solved with the help of the Wiener-Fock method [9]. In Section 4 the probability of excitation is found. Section 5 contains the discussion of the results.

2. The Distribution Function of Electrons and the Electric Field in the Metal

The equation for the distribution function f has the usual form

$$\frac{\partial f}{\partial t} + \frac{\partial f}{\partial \boldsymbol{r}} \cdot \boldsymbol{v} + \frac{\partial f}{\partial \boldsymbol{v}} \frac{e}{m} \operatorname{grad} \varphi = (f_0 - f)\, \nu , \tag{2.1}$$

where ν is the scattering frequency and $\boldsymbol{v}$ is a velocity. The potential φ can be considered as a harmonic function of time. Thus the linearized kinetic equation can be written in the form

$$(i\omega + \nu)\, f_1 + \boldsymbol{v} \cdot \frac{\partial f_1}{\partial \boldsymbol{r}} = -\frac{e}{m} \frac{\partial f_0}{\partial \boldsymbol{v}} \cdot \operatorname{grad} \varphi , \tag{2.2}$$

where f_1 is the change of the equilibrium function f_0 (Fermi function) linear in the field. We choose a Cartesian system of coordinates with the axis z normal to the surface and pointing into the metal and the origin on the interface. It is convenient to pass from $f_1(x, y, z, \boldsymbol{v}, \omega)$ to $f_1(\boldsymbol{q}, z, \boldsymbol{v}, \omega)$ where $\boldsymbol{q}$ is the component of the wave vector parallel to the interface. The solution of (2.2) can be written in the following form:

$$f_1(\boldsymbol{q}, z, \boldsymbol{v}, \omega) = \\ = \exp(-Az) \left\{ C \int_0^z \exp(A\tau) \frac{\mathrm{d}\varphi}{\mathrm{d}\tau} \mathrm{d}\tau + B \int_0^z \exp(A\tau)\, \varphi(\tau, \boldsymbol{q}, \omega)\, \mathrm{d}\tau + K \right\}, \tag{2.3}$$

where

$$\left.\begin{aligned} A &= \frac{\nu + i[\omega + v \sin\theta\, (q_x \cos\alpha + q_y \sin\alpha)]}{v \cos\theta}, \\ B &= -i \frac{e}{mv} \operatorname{tg}\theta\, (q_x \cos\alpha + q_y \sin\alpha) \frac{\partial f_0}{\partial v}, \\ C &= -\frac{e}{mv} \frac{\partial f_0}{\partial v}, \end{aligned}\right\} \tag{2.4}$$

and where K is a constant which must be obtained from the boundary conditions, $\theta = \sphericalangle (\boldsymbol{v}, \boldsymbol{n})$, $\boldsymbol{n}$ is a unit vector normal to the surface, α is an azimuth angle.

To consider the diffuse reflection of electrons from the boundary it is convenient to break the function $f_1(\boldsymbol{q}, z, \boldsymbol{v}, \omega)$ into two parts $f_1^{(1)}$ and $f_1^{(2)}$ in analogy to the theory of the anomalous skin effect [8]: $f_1^{(1)}$ corresponds to the electrons incident on the boundary ($\cos \theta < 0$) and $f_1^{(2)}$ corresponds to the electrons leaving the boundary ($\cos \theta > 0$). The corresponding constants $K^{(1)}$ and $K^{(2)}$ are determined from the following conditions:

$$f_1^{(1)}(z = +\infty) = 0 , \quad f_1^{(2)}(x, y, z = 0, v_{||}, +v_z) = 0 . \tag{2.5}$$

The second condition means that the anisotropic part of the distribution function, $f_1^{(2)}$, vanishes on the boundary. From (2.5) one obtains

$$K^{(1)} = C\varphi(z = 0, \boldsymbol{q}, \omega) + (AC - B) \int_0^\infty d\tau\, \varphi(\tau) \exp(A\tau) , \qquad K^{(2)} = 0 . \tag{2.6}$$

The bulk electronic charge density is

$$\varrho_{\text{bulk}}(z, \boldsymbol{q}, \omega) = -2e \left(\frac{m}{h}\right)^3 \int_{(\boldsymbol{v})} d\boldsymbol{v}\, f_1(\boldsymbol{v}, \boldsymbol{q}, z, \omega) . \tag{2.7}$$

With the help of (2.3) and (2.6) ϱ is calculated.

$$\varrho_{\text{bulk}}(z, \boldsymbol{q}, \omega) = -\frac{3\omega_p^2}{4\pi v_F^2} \left\{ \varphi(z, \boldsymbol{q}, \omega) - \frac{\varphi(z = 0, \boldsymbol{q}, \omega)}{2} \times \right.$$

$$\times \int_0^1 d\gamma \exp\left[-\frac{z(\nu + i\omega)}{\gamma v_F} \sqrt{1 - \left(\frac{q\, v_F \gamma}{\omega - i\nu}\right)^2} \right] - \frac{(\nu + i\omega)}{2v_F} \int_0^\infty d\tau\, \varphi(\tau, \boldsymbol{q}, \omega) \times$$

$$\left. \times \int_1^\infty du \exp\left[-|z - \tau| \left(\frac{\nu + i\omega}{v_F}\right) \sqrt{u^2 - \left(\frac{q v_F}{\omega - i\nu}\right)^2} \right] \left[u^2 - \left(\frac{q v_F}{\omega - i\nu}\right)^2 \right]^{-1/2} \right\}, \tag{2.8}$$

where ω_p is the frequency of the bulk plasma-oscillations, v_F is the velocity of electrons at the Fermi surface, $\varphi(z = 0, \boldsymbol{q}, \omega)$ is the scalar potential at the surface. Formula (2.7) gives only the bulk density of charges. The density of the surface charge is

$$\sigma(t, \omega, \boldsymbol{q}) - \sigma(t = 0, \omega, \boldsymbol{q}) = -\int_0^t dt'\, j_n(z = 0, \boldsymbol{q}, \omega)\, e^{i\omega t'} , \tag{2.9}$$

where

$$j_n(z = 0, \boldsymbol{q}, \omega) = -2e \left(\frac{m}{h}\right)^3 \int d\boldsymbol{v}\, v_z f_1(\boldsymbol{v}, \boldsymbol{q}, \omega, z = 0) . \tag{2.10}$$

Having in mind (2.3) the density of the surface charge can be written

$$\varrho_{\text{surf.}}(z, \boldsymbol{q}, \omega) = \lim_{s \to +0} \delta(z - s) \left\{ \frac{3\omega_p^2 i\varphi(z = 0, \boldsymbol{q}, \omega)}{16\pi v_F \omega} + \right.$$

$$+\frac{3\omega_p^2}{8\pi v_F^2}\left(1-i\frac{\nu}{\omega}\right)\int_0^\infty d\tau\,\varphi(\tau,\boldsymbol{q},\omega)\int_0^1 d\gamma\,\exp\left[-\frac{\tau(\nu+i\omega)}{\gamma v_F}\sqrt{1-\left(\frac{qv_F\gamma}{\omega-i\nu}\right)^2}\right]\Bigg\}. \tag{2.11}$$

The potential $\varphi(\boldsymbol{r},\omega)$ is determined by the Poisson equation

$$\Delta\varphi(\boldsymbol{r},\omega) = -4\pi\left[\varrho_{\text{bulk}}(\boldsymbol{r},\omega)+\varrho_{\text{surf.}}(\boldsymbol{r},\omega)\right]. \tag{2.12}$$

Substituting here (2.8) and (2.11) we obtain the following equation for the potential $\varphi(z,\boldsymbol{q},\omega)$ of the electric field in the metal

$$\frac{d^2\varphi(z,\boldsymbol{q},\omega)}{dz^2} - q^2\varphi(z,\boldsymbol{q},\omega) + \int_0^\infty d\tau\,\varphi(\tau,\boldsymbol{q},\omega)\left\{\frac{3\omega_p^2(\nu+i\omega)}{2v_F^3}\int_1^\infty du\,\times\right.$$

$$\left.\times\frac{\exp\left[-|z-\tau|\left(\frac{\nu+i\omega}{v_F}\right)\sqrt{u^2-(qv_F)^2(\omega-i\nu)^{-2}}\right]}{[u^2-(qv_F)^2(\omega-i\nu)^{-2}]^{1/2}} - \frac{3\omega_p^2}{v_F^2}\delta(z-\tau)\right\} +$$

$$+\frac{3\omega_p^2}{2v_F^2}\varphi(z=0,\boldsymbol{q},\omega)\int_0^1 d\gamma\,\exp\left[-\frac{z(\nu+i\omega)}{\gamma v_F}\sqrt{1-\left(\frac{qv_F\gamma}{\omega-i\nu}\right)^2}\right] +$$

$$+\frac{3i\omega_p^2}{4v_F\omega}\varphi(z=0,\boldsymbol{q},\omega)\,\delta(z-s) - \frac{3i\omega_p^2}{2\omega v_F^2}\delta(z-s)\,\times$$

$$\times\int_0^\infty d\tau\,\varphi(\tau,\boldsymbol{q},\omega)\int_0^1 d\gamma\,\exp\left[-\frac{\tau(\nu+i\omega)}{\gamma v_F}\sqrt{1-\left(\frac{qv_F\gamma}{\omega-i\nu}\right)^2}\right] = 0\,, \tag{2.13}$$

here $s\to+0$.

3. Dyson Equation for the Green Function of Plasmons

According to the formulae (1.3), (1.7), and (2.13) the Dyson equation for the Green function D of the electric field can be written

$$\frac{d^2D(z,z',\boldsymbol{q},\omega)}{dz^2} - q^2D(z,z',\boldsymbol{q},\omega) + 4\pi\int_0^\infty dz_1\,D(z_1,z',\boldsymbol{q},\omega)\,\Pi_0(|z-z_1|,\boldsymbol{q},\omega) +$$

$$+\frac{3i\omega_p^2}{4\omega v_F}D(+0,z',\boldsymbol{q},\omega)\,\delta(z-s) +$$

$$+\frac{3\omega_p^2}{2v_F^2}D(+0,z',\boldsymbol{q},\omega)\int_0^1 d\gamma\,\exp\left[-\frac{z(\nu+i\omega)}{\gamma v_F}\sqrt{1-\left(\frac{qv_F\gamma}{\omega-i\nu}\right)^2}\right] +$$

$$+\frac{3\omega_p^2\left(1-i\frac{\nu}{\omega}\right)}{2v_F^2}a(z',\boldsymbol{q},\omega)\,\delta(z-s) = -4\pi\delta(z-z') \tag{3.1}$$

with $s\to+0$; $4\pi\Pi_0(|z-z'|,\boldsymbol{q},\omega)$ stands for the expression in the braces in

(2.13), the function $a(z', \boldsymbol{q}, \omega)$ is equal to

$$\int_0^\infty \mathrm{d}y\, D(y, z', \boldsymbol{q}, \omega) \int_0^1 \mathrm{d}\gamma \exp\left[-\frac{y(\nu + i\omega)}{\gamma v_\mathrm{F}} \sqrt{1 - \left(\frac{q v_\mathrm{F} \gamma}{\omega - i\nu}\right)^2}\right]$$

and does not depend on z. Equation (3.1) defines the function D for $z > 0$. For $z < 0$ the equation for the function D is trivial:

$$\frac{\mathrm{d}^2 D(z, z', \boldsymbol{q}, \omega)}{\mathrm{d}z^2} - q^2 D(z, z', \boldsymbol{q}, \omega) = 0\,. \tag{3.2}$$

The integral equation (3.1) is solved by the Wiener-Fock approach [9]. Let us introduce the following Fourier transform of the function $D(z, z', \boldsymbol{q}, \omega)$:

$$D(\varkappa, z', \boldsymbol{q}, \omega) = \int_{-\infty}^{+\infty} D(z, z', \boldsymbol{q}, \omega)\, \mathrm{e}^{-i\varkappa z}\, \mathrm{d}z\,. \tag{3.3}$$

The function $D(z)$ in the integrand of (3.3) has been defined only for $z > 0$. It will be continued into the region $z < 0$ in the zeroth way. In this case the function $D(\varkappa, z', \boldsymbol{q}, \omega)$ is analytical in the region $\operatorname{Im} \varkappa < 0$. Consider first the case $z' > 0$. We perform a Fourier transformation of (3.1) taking into account the zeroth character of the continuation of $D(z, z')$ and then represent $D(\varkappa, z')$ and all Fourier transforms of the inhomogeneous terms as Cauchy integrals in the region $\operatorname{Im} \varkappa < 0$. Having in mind lemma I of Fock's work [9] we obtain

$$\begin{aligned}
&- D(\varkappa, z', \boldsymbol{q}, \omega)\,[\varkappa^2 + q^2 - 4\pi \Pi_0(\varkappa, \boldsymbol{q}, \omega)] - i\varkappa D(0, z', \boldsymbol{q}, \omega) - D'(0, z', \boldsymbol{q}, \omega) + \\
&+ 4\pi\, \mathrm{e}^{-i\varkappa z'} + \frac{3\omega_\mathrm{p}^2}{2v_\mathrm{F}^2} D(0, z', \boldsymbol{q}, \omega)\, F(\varkappa, \boldsymbol{q}, \omega) + \frac{3i\omega_\mathrm{p}^2}{4\omega v_\mathrm{F}} \times \\
&\times D(0, z', \boldsymbol{q}, \omega)\, \mathrm{e}^{-i\varkappa s} + \frac{3\omega_\mathrm{p}^2}{2v_\mathrm{F}^2}\left(1 - i\frac{\nu}{\omega}\right) \mathrm{e}^{-i\varkappa s}\, a(z', \boldsymbol{q}, \omega) = M^+(\varkappa, \boldsymbol{q}, \omega, z')\,.
\end{aligned} \tag{3.4}$$

Here

$$F(\varkappa, \boldsymbol{q}, \omega) = -i \int_0^1 \frac{\mathrm{d}\gamma}{\varkappa + \dfrac{\omega - i\nu}{\gamma v_\mathrm{F}} \sqrt{1 - \left(\dfrac{q v_\mathrm{F} \gamma}{\omega - i\nu}\right)^2}} \tag{3.5}$$

and $M^+(\varkappa, \boldsymbol{q}, \omega, z')$ is a certain function analytical in the region $\operatorname{Im} \varkappa > 0$.

Here the relation $\varkappa^2 + q^2 - 4\pi \Pi_0(\varkappa, \boldsymbol{q}, \omega) = (\varkappa^2 + q^2)\, \varepsilon(\varkappa, \boldsymbol{q}, \omega)$ holds, where

$$\varepsilon(\varkappa, \boldsymbol{q}, \omega) = 1 - \frac{3\omega_\mathrm{p}^2}{v_\mathrm{F}^2(\varkappa^2 + q^2)} \left[\int_1^\infty \frac{\mathrm{d}u}{u^2 - (\varkappa^2 + q^2)\, v_\mathrm{F}^2 (\omega - i\nu)^{-2}} - 1\right]. \tag{3.6}$$

Following the Wiener-Fock method, we obtain, by some simple formal transformations of (3.4)

$$\begin{aligned}
D(\varkappa, z', \boldsymbol{q}, \omega) = &-D(0, z', \boldsymbol{q}, \omega) \frac{Q(\varkappa, q, \omega) + \dfrac{3\omega_\mathrm{p}^2}{8\pi\omega v_\mathrm{F}} P(\varkappa, q, \omega, s)}{(\varkappa - iq)\, \varepsilon_-(\varkappa, q, \omega)} - \\
&- \frac{2P(\varkappa, q, \omega, z')}{i(\varkappa - iq)\, \varepsilon_-(\varkappa, q, \omega)} - \frac{3\omega_\mathrm{p}^2 \left(1 - i\dfrac{\nu}{\omega}\right) a(z', q, \omega)\, P(\varkappa, q, \omega, s)}{4\pi i v_\mathrm{F}^2 (\varkappa - iq)\, \varepsilon_-(\varkappa, q, \omega)},
\end{aligned} \tag{3.7}$$

where the factorization $\varepsilon(\varkappa, q, \omega) = \varepsilon_-(\varkappa, q, \omega)/\varepsilon_+(\varkappa, q, \omega)$ is characteristic of the Wiener-Fock method. The functions P and Q are defined by the integrals

$$\left.\begin{aligned} P(\varkappa, q, \omega, z') &= \int\limits_{-\infty}^{+\infty} \frac{du\, e^{-iuz'}\, \varepsilon_+(u, q, \omega)}{(u + iq)(u - \varkappa)}; \\ Q(\varkappa, q, \omega) &= i + \frac{3\omega_p^2}{4\pi i v_F^2} \int\limits_{-\infty+i\alpha}^{+\infty+i\alpha} \frac{du\, F(u, q, \omega)\, \varepsilon_+(u, q, \omega)}{(u + iq)(u - \varkappa)}. \end{aligned}\right\} \tag{3.8}$$

We now perform the reverse Fourier transformation from $D(\varkappa, z')$ to the function $D(z, z')$. From this function and the solution of (3.2) we determined the constant $D(0, z', q, \omega)$. In analogy we can obtain the solution in the case $z' < 0$. The final full D function in coordinate representation can be written in the following way:

$$\begin{aligned} D(z, z', \boldsymbol{q}, \omega) = 4\pi &\left[\theta(-z)\, e^{qz} - \frac{\theta(z)}{2\pi} \int\limits_{-\infty}^{+\infty} \frac{d\varkappa\, e^{i\varkappa z}\, Q(\varkappa)}{(\varkappa - iq)\, \varepsilon_-(\varkappa)}\right] \times \\ \times &\left[\theta(-z')\, e^{qz'} - \frac{\theta(z')}{2\pi} \int\limits_{-\infty}^{+\infty} \frac{d\varkappa\, e^{i\varkappa z'}\, Q(\varkappa)}{(\varkappa - iq)\, \varepsilon_-(\varkappa)}\right] \times \left\{q + \frac{i}{2\pi} \int\limits_{-\infty}^{+\infty} \frac{d\varkappa\, \varkappa Q(\varkappa)\, e^{i\varkappa s}}{(\varkappa - iq)\, \varepsilon_-(\varkappa)} - \right. \\ &\left. - \frac{3\omega_p^2 i}{4\omega v_F}\left[1 + \frac{i\omega\left(1 - i\dfrac{\nu}{\omega}\right)}{\pi v_F} \int\limits_{-\infty}^{+\infty} \frac{d\varkappa\, e^{i\varkappa s}\, Q(\varkappa)\, F(-\varkappa)}{(\varkappa - iq)\, \varepsilon_-(\varkappa)}\right]\right\}^{-1} + 2\theta(-z)\, \theta(-z') \times \\ &\times \int\limits_{-\infty}^{+\infty} d\varkappa\, \frac{e^{i\varkappa z}\, (e^{-i\varkappa z'} - e^{qz'})}{\varkappa^2 + q^2} - \frac{\theta(z)\, \theta(z')}{\pi i} \int\limits_{-\infty}^{+\infty} \frac{d\varkappa\, e^{i\varkappa z}\, P(\varkappa, z')}{(\varkappa - iq)\, \varepsilon_-(\varkappa)}. \end{aligned} \tag{3.9}$$

For all functions under the integrals in (3.9) only those arguments are given over which the integration is performed. A calculation of the integrals in (3.9) without using approximations for the functions $\varepsilon_+(\varkappa, q, \omega)$ and $\varepsilon_-(\varkappa, q, \omega)$ is impossible. It is convenient to approximate (3.6) by the following function:

$$\varepsilon(\varkappa, q, \omega) = \frac{\varepsilon(\omega, \nu) - b^2 \dfrac{(\varkappa^2 + q^2)\, v_F^2}{(\omega - i\nu)^2}}{1 - b^2 \dfrac{(\varkappa^2 + q^2)\, v_F^2}{(\omega - i\nu)^2}}. \tag{3.10}$$

Here $\varepsilon(\omega, \nu)$ is the dielectric function with $\varkappa = q = 0$. In the extreme cases of great and small $\varkappa$ and q the function (3.10) tends to the correct asymptotic values and is close to (3.6) in the region of the intermediate values of $\varkappa$ and q for an appropriate choice of the coefficient b. Due to the same reason the function $F(\varkappa, q, \omega)$ must be also presented in a simplified form. This approximation of F must be in agreement with (3.10). The necessity of such an agreement is caused by the fact that the logarithmic singularities in the exact expressions for ε and F are the same. Therefore and because of the pole singularities both the function (3.10) and the approximation function F have to coincide, too. It is convenient to approximate the function F in the following way:

$$F(\varkappa, q, \omega) = -\frac{3i}{4}\left[\varkappa + \sqrt{\frac{9(\omega - i\nu)^2}{4v_F^2} - q^2}\right]^{-1}. \tag{3.11}$$

The function (3.10) will be adjusted in the above sense if $b^2 = 4/9$. According to the method of factorization used in the Wiener-Fock theory ε_+ and ε_- has to be written

$$\varepsilon_-(\varkappa, q, \omega) = \frac{\varkappa + \sqrt{\frac{9(\omega^2 - \omega_p^2)}{4v_F^2} - q^2 - i\frac{\nu}{v_F}}}{\varkappa + \sqrt{\frac{9\omega^2}{4v_F^2} - q^2 - i\frac{\nu}{v_F}}}, \tag{3.12}$$

$$\varepsilon_+(\varkappa, q, \omega) = \frac{\varkappa - \sqrt{\frac{9\omega^2}{4v_F^2} - q^2 + i\frac{\nu}{v_F}}}{\varkappa - \sqrt{\frac{9(\omega^2 - \omega_p^2)}{4v_F^2} - q^2 + i\frac{\nu}{v_F}}}. \tag{3.13}$$

With the help of (3.10), (3.11), (3.12), and (3.13) the Green function (3.9) can be calculated in explicit form.

4. The Probability of Excitation

According to (1.2) the probability of a transition with $\omega > 0$ is proportional to the imaginary part of the Green function $D(+\varkappa, -\varkappa, \boldsymbol{q}, \omega)$. Using the Green function in the z, $\boldsymbol{q}$ representation calculated as discussed in the previous paragraph, we obtain after a double Fourier transformation,

$$D(\varkappa, -\varkappa, \boldsymbol{q}, \omega) = D_B(\varkappa, -\varkappa, \boldsymbol{q}, \omega) + D_S(\varkappa, -\varkappa, \boldsymbol{q}, \omega) + D_v(\varkappa, -\varkappa, \boldsymbol{q}, \omega), \tag{4.1}$$

where

$$D_B(\varkappa, -\varkappa, \boldsymbol{q}, \omega) = \frac{4\pi L}{(\varkappa^2 + q^2)\,\varepsilon(\varkappa, q, \omega)} - \frac{2\pi\,(3q^2 - \varkappa^2)}{q(\varkappa^2 + q^2)^2\,\varepsilon(\omega, \nu)} - \frac{2\pi i\,[\varepsilon(\omega, \nu) - 1]\,(\varkappa^2 + 3r^2)}{r\varepsilon(\omega, \nu)\,(\varkappa^2 - r^2)^2} + \frac{4\pi i(r - l)\,(l + iq)}{(\varkappa^2 + q^2)\,(r - iq)\,(\varkappa^2 - r^2)}; \tag{4.2}$$

$$D_S(\varkappa, -\varkappa, \boldsymbol{q}, \omega) = \frac{4\pi q}{(\varkappa^2 + q^2)^2}\left[2 + \frac{(r - l)\,(\varkappa - iq)}{(l - iq)\,(\varkappa - r)}\right]\times\left[2 + \frac{(r - l)\,(\varkappa + iq)}{(l - iq)\,(\varkappa + r)}\right]\times$$
$$\times\left\{1 + \varepsilon(\omega, \nu) - \frac{iqv_F\,[1 - \varepsilon(\omega, \nu)]}{3\omega\left(1 + \frac{2v_F l}{3\omega}\right)} + \frac{4iqv_F^2\,(r - iq)\,(r - l)}{9\omega\,(\omega - i\nu)\,(l - iq)} - \right.$$
$$\left. - i\frac{\nu}{\omega}\left[\varepsilon(\omega, \nu) - \frac{r - iq}{l - iq}\right]\right\}^{-1}; \tag{4.3}$$

$$D_v(\varkappa, -\varkappa, \boldsymbol{q}, \omega) = \frac{4\pi L}{\varkappa^2 + q^2} - \frac{2\pi\,(3q^2 - \varkappa^2)}{q(\varkappa^2 + q^2)^2}. \tag{4.4}$$

Here the notation

$$r^2 = \frac{9(\omega - i\nu)^2\,\varepsilon(\omega, \nu)}{4v_F^2} - q^2; \qquad l^2 = \frac{9(\omega - i\nu)^2}{4v_F^2} - q^2$$

has been introduced.

In (4.1) there are some terms whose denominators may become zero. These zeroes correspond to resonances in the transition probability. Im D_S describes the excitation of surface plasmons. The resonance corresponding to the excitation of bulk plasmons is described by Im $(D_B + D_S)$. The function D_v has no imaginary part and does not contribute to the transition probability. In particular it is evident from (4.3) that the dispersion of the frequency of surface plasmons has the form

$$\omega_S = \frac{\omega_p}{\sqrt{2}}\left(1 + \frac{qv_F}{3\sqrt{2}\,\omega_p} + \dots\right). \tag{4.5}$$

The coefficient in (4.5) linear in q is equal to 1/6. For specular reflection of the electrons at the surface this coefficient has twice this value (under the present approximations). From (4.3) it follows that

$$\text{Im}\,\omega_S = + \frac{qv_F}{4} \tag{4.6}$$

in the limit of small q. In this form the damping which is described by (4.6) reminds of the Romanov damping [10] which takes place in the case when the electrons suffer only specular scattering from the surface. However, with $\varepsilon(\varkappa, q, \omega)$ and $F(\varkappa, q, \omega)$ approximated by the formulae (3.10) and (3.11), it is impossible to obtain Romanov damping. To yield this type of damping ε and F must have imaginary parts of a certain form. It is easy to complete the imaginary parts in such a way that Romanov damping occurs, in the case of spatially uniform surfaces. But in the case of diffuse scattering of electrons and for Fermi statistics it is rather difficult to factorize the function $\varepsilon(\varkappa, q, \omega)$ completed in such a way. According to these considerations the damping connected with (4.6) should be exclusively due to the diffuse character of the scattering of electrons from the boundary. To conclude this paragraph we give a formula for $W_{p,p-q}$ for frequencies near the surface resonance and for $\nu < qv_F$:

$$W_{p,p-q} = \frac{32\sqrt{2}\,\pi q^2 v_F}{\left[q^2 + \frac{\omega_p^2}{2p^2}\right]^2 \omega_p \left\{\left[1 + \varepsilon(\omega) + \frac{2\sqrt{2}\,qv_F}{3\omega_p}\right]^2 + \frac{2q^2v_F^2}{\omega_p^2}\right\}} \tag{4.7}$$

which does not include the Romanov damping.

5. Discussion

According to (4.6) the diffuse reflection of electrons at the surface causes a additional damping of surface plasmons. This damping is linear in q as Romanov damping, has the same order of magnitude, and is also connected with spatial dispersion. (However, Romanov damping also occurs in the case of specular reflection.) Of course, in real solids one should also consider the presence of lattice imperfections and impurities [11]. The additional diffuse damping has a simple physical origin. The electrons arriving at the surface and "feeling" the field of the surface plasmon come from a region of depth q^{-1}. Their velocity is approximately v_F, the distance q^{-1} they pass during the time $(qv_F)^{-1}$. At the surface the electrons either suffer diffuse scattering or are "trapped", getting out of the process in this way. Thus, the dimensionless parameter which characterizes the diffuse damping must be of the order qv_F/ω in agreement with (4.6).

The surface roughness should also affect the probability of excitation of bulk plasmons. As evident from (4.2) one has to add some terms to the corresponding expression whose structure is close that of the well-known Ritchie formula [1]. The existence of these new terms is caused by two reasons: surface roughness and spatial dispersion. To find the part resulting from surface roughness it is convenient to subtract from (4.2) the expression obtained for pure specular scattering. This difference has the form

$$\Delta \operatorname{Im} D = -\frac{4\pi}{(\varkappa^2 + q^2)} \operatorname{Re} \frac{(r - l)}{(\varkappa^2 - r^2)} - \frac{8\pi q}{(\varkappa^2 + q^2)} \operatorname{Im} \frac{1}{(\varkappa^2 - r^2)}. \tag{5.1}$$

For $\varepsilon(\omega) \approx 0$ the first term is dominant in (5.1). It is of the order

$$-\frac{8\pi v_F}{3\omega_p (\varkappa^2 + q^2)} \operatorname{Re} \frac{1}{\varepsilon(\omega)}. \tag{5.2}$$

The ratio of (5.2) to the term describing the size effect is proportional q. The physical meaning of the term (5.2) lies in the fact that it describes either the creation or the damping of a bulk plasmon at the rough surface. The influence of a rough surface on the behaviour of a bulk plasmon was considered by Mekhtiev [4]. His approach to a lesser extent corresponds to the experimental situation. Due to this reason it is difficult to compare his results with ours. Recently the effect of spatial dispersion on the damping of surface plasmons seems to have been observed experimentally (for example in the work of Bagchi et al. [12]). According to the present results surface roughness should be taken into account in analyzing experimental data on surface plasmons.

Appendix

By carrying out a Fourier transformation of the function $D(\varkappa, z', \boldsymbol{q}, \omega)$ given by (3.7) one can obtain the function $D(z, z', \boldsymbol{q}, \omega)$. The limiting value of this function for $z \to \pm 0$ must coincide with the constant in our solution, $D(0, z', \boldsymbol{q}, \omega)$: $\lim D(z, z', \boldsymbol{q}, \omega) = D(0, z', \boldsymbol{q}, \omega)$. Two integral rules which are connected with different contours around the singularities of the functions ε and F are obtained from this equation:

$$\int_{-\infty}^{+\infty} \frac{d\varkappa\, e^{i\varkappa\beta} P(\varkappa, q, \omega, z')}{(\varkappa - iq)\, \varepsilon_-(\varkappa, q, \omega)} = 0\,, \qquad \text{with } \beta \to 0 \quad \text{and} \quad z' > \beta\,, \tag{A.1}$$

$$\int_{-\infty}^{+\infty} \frac{d\varkappa\, e^{i\varkappa\beta} Q(\varkappa, q, \omega)}{(\varkappa - iq)\, \varepsilon_-(\varkappa, q, \omega)} = -2\pi\,, \quad \text{with } \beta \to 0\,. \tag{A.2}$$

These integral rules are fulfilled also by our approximate functions ε and F. It is this fact that confirms the adequacy of the approximations chosen for ε and F which are given by (3.10) and (3.11). With the help of these approximations two other integral rules can be obtained:

$$\int_{-\infty}^{+\infty} \frac{d\varkappa\, e^{i\varkappa\beta}\, P(\varkappa, q, \omega, s)\, F(-\varkappa, q, \omega)}{(\varkappa - iq)\, \varepsilon_-(\varkappa, q, \omega)} = 0\,, \qquad \text{with} \qquad s \to \beta \to 0\,, \tag{A.3}$$

$$\int_{-\infty}^{+\infty} \frac{d\varkappa\, e^{i\varkappa\beta}\, \varkappa P(\varkappa, q, \omega, s)}{(\varkappa - iq)\, \varepsilon_-(\varkappa, q, \omega)} = -4\pi^2\,, \ \text{with} \qquad s \to \beta \to 0\,. \tag{A.4}$$

By means of the integral rules (A.1) to (A.4) calculations are considerably facilitated.

In conclusion we present the calculation of the resonance denominator in (4.3) using the approximations for ε and F. The resonance denominator is

$$I = q + \frac{i}{2\pi} \int_{-\infty}^{+\infty} \frac{d\varkappa\, e^{+i\varkappa\beta}\, \varkappa Q(\varkappa, q, \omega)}{(\varkappa - iq)\, \varepsilon_-(\varkappa, q, \omega)} - \frac{3\omega_p^2 i}{4\omega v_F} \Bigg[1 + \frac{i(\omega - i\nu)}{\pi v_F} \times \\ \times \int_{-\infty}^{+\infty} \frac{d\varkappa\, e^{+i\varkappa\beta}\, Q(\varkappa, \xi, \omega)\, F(-\varkappa, q, \omega)}{(\varkappa - iq)\, \varepsilon_-(\varkappa, q, \omega)} \Bigg] = I_1 + I_2\,.$$

At first let us consider

$$I_1 = q + \frac{i}{2\pi} \int_{-\infty}^{+\infty} \frac{d\varkappa\, e^{+i\varkappa\beta}\, \varkappa Q(\varkappa, q\, \omega)}{(\varkappa - iq)\, \varepsilon_-(\varkappa, q\, \omega)} = 2q - \frac{1}{2\pi i} \int_{-\infty}^{+\infty} \frac{d\varkappa\, e^{+i\varkappa\beta}\, Q(\varkappa, q, \omega)}{\varepsilon_-(\varkappa, q, \omega)}\,,$$

here it has been used formula (A.2)

$$I_1 = 2q - \frac{1}{2\pi i} \int_{-\infty}^{+\infty} \frac{d\varkappa\, e^{+i\varkappa\beta}\, (\varkappa + l)}{(\varkappa + r)} \times \\ \times \left\{ i + \frac{9\omega_p^2}{8 i v_F^2} \frac{1}{2\pi i} \int_{-\infty + i\alpha}^{+\infty + i\alpha} \frac{du (u - l)}{(u - \varkappa)\, (u + iq)\, (u + l)\, (u - r)} \right\} = 2q + q\, \frac{(r - l)}{(l - iq)}\,.$$

The integration path for the integral in the braces is closed in the lower half-plane of the complex variable $\varkappa$.

The integral I_2 is calculated in a similar way:

$$I_2 = -\frac{3\omega_p^2 i}{4\omega v_F} \left[1 + \frac{i\omega\left(1 - i\dfrac{\nu}{\omega}\right)}{\pi v_F} \int_{-\infty}^{+\infty} \frac{d\varkappa\, e^{+i\varkappa\beta}\, Q(\varkappa, q, \omega)\, F(-\varkappa, q, \omega)}{(\varkappa - iq)\, \varepsilon_-(\varkappa, q, \omega)} \right] =$$

$$= -\frac{3\omega_p^2 i}{4\omega v_F} \left[1 - \frac{3(\omega - i\nu)}{2 v_F (l - iq)} - \frac{3(\omega - i\nu)}{4\pi v_F (l - iq)} \int_{-\infty}^{+\infty} \frac{d\varkappa\, e^{+i\varkappa\beta}\, Q(\varkappa, q, \omega)}{\varepsilon_-(\varkappa, q, \omega)\, (\varkappa - l)} \right] =$$

$$= -\frac{3\omega_p^2 i}{4\omega v_F}\left[1 - \frac{3(\omega - i\nu)}{2v_F(l - iq)} - \frac{3(\omega - i\nu)}{4\pi v_F(l - iq)} \int\limits_{-\infty}^{+\infty} \frac{d\varkappa\, e^{+i\varkappa\beta}(\varkappa + l)}{(\varkappa + r)(\varkappa - l)} \left\{ i - \frac{9\omega_p^2}{8iv_F^2} \times \right.\right.$$

$$\times \left[\frac{(\varkappa - l)}{(\varkappa + iq)(\varkappa + l)(\varkappa - r)} - \frac{(l + iq)}{(l - iq)(r + iq)(\varkappa + iq)} - \right.$$

$$\left.\left.\left. - \frac{(r - l)}{(r + iq)(r + l)(\varkappa - r)}\right]\right\}\right] =$$

$$= -\frac{3\omega_p^2 i}{4\omega v_F}\left[1 - \frac{3(\omega - i\nu)}{2v_F(l - iq)} + \frac{27iq(\omega - i\nu)^3(l^2 - r^2)}{8v_F^2(l^2 + q^2)(l - iq)^2(r + l)^2}\right] =$$

$$= -\frac{q}{2}\left(1 - i\frac{\nu}{\omega}\right)\left\{1 + \varepsilon(\omega, \nu) - \frac{(r - l)^2}{(l - iq)^2} + \frac{2iqv_F[1 - \varepsilon(\omega, \nu)]}{3(\omega - i\nu)\left(1 + \dfrac{2v_F l}{3\omega}\right)}\right\}.$$

Then

$$I = I_1 + I_2 = q\left\{1 + \varepsilon(\omega, \nu) - \frac{iqv_F[1 - \varepsilon(\omega, \nu)]}{3\omega\left(1 + \dfrac{2v_F l}{3\omega}\right)} + \right.$$

$$\left. + \frac{4iqv_F^2(r - iq)(r - l)}{9\omega(\omega - i\nu)(l - iq)} - i\frac{\nu}{\omega}\left[\varepsilon(\omega, \nu) - \frac{r - iq}{l - iq}\right]\right\}.$$

References

[1] R. H. RITCHIE, Phys. Rev. **106**, 874 (1957).
[2] YU. A. ROMANOV, Zh. eksper. teor. Fiz. **47**, 2119 (1964).
[3] A. J. BENNETT, Phys. Rev. B **1**, 203 (1970).
[4] M. A. MEKHTIEV, Fiz. tverd. Tela **9**, 2098 (1967).
[5] YU. A. ROMANOV, Tezisy doklada, Vsesoyuznoe soveshchanie po teorii poluprovodnikov, Tartu 1966.
[6] V. V. RUMYANTSEV, Trudy politekh. Inst., Leningrad, **328**, 100 (1973).
[7] A. A. ABRIKOSOV, L. P. GORKOV, and I. E. DZYALOSHINSKII, Metody kvantovoi teorii polya v statisticheskoi fizike, GIFML 1962.
[8] G. REUTER and E. SONDHEIMER, Proc. Roy. Soc. **A195**, 338 (1948).
[9] V. A. FOCK, Matematicheskii Sbornik **14**, 3 (1944).
[10] YU. A. ROMANOV, Izv. vuzov, radiofizika **7**, 242 (1964).
[11] V. V. RUMYANTSEV, Fiz. tverd. Tela **13**, 2038 (1971).
[12] A. BAGCHI, C. B. DUKE, P. J. FEIBELMANN, and J. O. PORTEUS, Phys. Rev. Letters **27**, 998 (1971).

(Received June 11, 1974)

phys. stat. sol. (b) **65**, 293 (1974)

Subject classification: 14.3; 16; 22.1.3

Sektion Physik der Humboldt-Universität zu Berlin, Bereich Experimentelle Halbleiterphysik

Longitudinal Photon Drag in p-Type Tellurium

By

J. Auth, D. Genzow, K. H. Herrmann, and M. Wendt

The longitudinal photon drag has been studied for free holes in tellurium using a Q-switched CO_2 laser. A great effect is expected due to the great interband matrix element for allowed direct hole transitions between the sub-valence bands H_4 and H_5. Due to the low crystal symmetry interesting anisotropy phenomena take place. The measurements have been performed for a set of samples, oriented either parallel or perpendicular to the c-axis. The drag effect connected with sub-band transitions is dominating as predicted by the theory (tensor coefficient $\sigma_{1133} = 5 \times 10^{-12}$ cm/V at $T = 77$ K). The coefficient σ_{3311} connected with the absorption of radiation polarized $\boldsymbol{F}^\omega \perp \boldsymbol{c}$ was found to be $-1 \times$ $\times\ 10^{-12}$cm/V at 77 K.

Der longitudinale photon-drag-Effekt an quasifreien Löchern in Tellur wird mit einem gütegeschalteten CO_2-Laser untersucht. Wegen des großen Interbandmatrixelements für direkte erlaubte Löcherübergänge zwischen den Subvalenzbändern H_4 und H_5 ist ein großer Effekt zu erwarten. Wegen der geringen Gittersymmetrie treten interessante Anisotropieeffekte auf. Die Messungen an einem Satz Proben, die entweder parallel oder senkrecht zur c-Achse des Tellurs orientiert waren, zeigen, daß erwartungsgemäß der mit den Subbandübergängen verknüpfte Effekt dominiert (Tensorkoeffizient $\sigma_{1133} = 5 \times 10^{-12}$ cm/V bei $T = 77$ K). Der mit der Absorption von Strahlung $\boldsymbol{F}^\omega \perp \boldsymbol{c}$ verknüpfte Koeffizient σ_{3311} wurde zu -1×10^{-12} cm/V bei 77 K bestimmt.

1. Introduction

In thermal equilibrium the momentum distribution of quasi-free carriers in a semiconductor is symmetric with respect to the $\boldsymbol{k}_0$-point. The electric current $\boldsymbol{j} \sim \int f(\boldsymbol{k})\, \boldsymbol{k}\, d^3\boldsymbol{k}$ turns out to be zero. If electromagnetic radiation is being absorbed energy as well as momentum of the photons are transferred to the charge carriers. Since the momentum of a photon is quite small at optical wavelengths, usually it is neglected; in this case the momentum distribution of charge carriers remains symmetric.

If the photon momentum is taken into account, the momentum distribution of charge carriers becomes asymmetric if photons are absorbed; from that a non-zero electric current results (light pressure [1], photon drag). Under open-circuit conditions a voltage is being measured which corresponds to the electric field built up within the sample to compensate the photon-drag current. According to Moss [2] diffusion must be taken into account if charge carriers of different polarity are present. For the first time this effect was observed in p-type germanium (experiment [3], theory [4]), for later investigations see [17 to 29].

Up to now the photon-drag effect was found with sub-band excitation [3], intraband excitation [17, 20], excitation from impurities [23], and multi-quantum excitation [6]. All photoeffects of free carriers are favoured by a high absorption cross-section of the carriers. The absorption cross-section S_p of holes in tellurium for CO_2 laser radiation ($\lambda = 10.6\,\mu m$) is the highest of any semiconductor known, since

(i) the quantum energy agrees very well with the distance of sub-valence bands H_4 and H_5 ($h\nu = 117$ meV; $\delta E = 122$ meV at 77 K) [7];

(ii) the interband matrix element is high due to direct allowed transitions for $\boldsymbol{F}^\omega \parallel \boldsymbol{c}$ (electric vector of light parallel to the c-axis).

It holds $S_{p\parallel}(10.6) = (1.5 \pm 0.5) \times 10^{-15}$ cm² for $\boldsymbol{F}^\omega \parallel \boldsymbol{c}$, $T = 77$ K, however, $S_{p\perp}(10.6) = 3 \times 10^{-17}$ cm² for $\boldsymbol{F}^\omega \perp \boldsymbol{c}$; the latter value is governed by Drude absorption [7] and possibly by inter-subband absorption $H_4 \rightarrow H_6$ [9].

Up to now the photon-drag effect was observed in germanium, in some $A^{III}B^{V}$ alloys, and in tellurium. As to tellurium, up to now only a short note [10] was published.

2. Theory

In a phenomenologic attempt the photon-drag effect can be described by the tensor relation

$$j_i = I \sigma_{iklm} e_l e_m \varkappa_k \tag{1}$$

with $\boldsymbol{j}$ photon-drag current density, $\hbar\boldsymbol{\varkappa}$ photon momentum, $\boldsymbol{e}$ unit vector in direction of the electric vector of the wave, I light intensity.

The tensor σ_{iklm} has the symmetry properties following from point group symmetry D_3. Furthermore it is symmetric with respect to the last two indices, because $e_l e_m$ and $e_m e_l$ are indistinguishable. In cubic semiconductors like Ge and Si the photon-drag tensor contains three independent coefficients. If light is propagated in [111] direction, e.g., a smaller transverse photon drag is found in addition to the longitudinal effect [26]. This is specific for light pressure onto crystal electrons in a semiconductor with anisotropic band structure. Due to the low symmetry of tellurium especially interesting anisotropic phenomena in this material are expected. The tensor σ_{iklm} has ten independent coefficients (see Appendix). A Cartesian coordinate system was chosen as follows: The z-axis parallel to the optical c-axis and x-axis parallel to one of the electric axes.

In this paper the study is limited to longitudinal photon drag in samples which have their long axis parallel or perpendicular to the c-axis. In this case only the coefficients A, B, C, E are retained. For determining these four coefficients measurements have to be carried out with polarized radiation, applying different sample configurations (type I and II samples, see Table 1). The laser beam is propagated normally with respect to the front face. Type III samples can be used to check the results obtained from type I and II, type IV samples will be discussed below. Sample configuration, polarization of the light wave as well as the corresponding expressions for the current density of longitudinal photon drag in tellurium are given in Table 1. The σ_{iklm} in (1) are proportional to the absorption coefficient. As a consequence of $K_\parallel > K_\perp$ it is expected that the coefficient A connected with absorption of radiation $\boldsymbol{F}^\omega \parallel \boldsymbol{c}$ is high as compared with B and C which are proportional to $K_\perp$. The expressions for the current density given in Table 1 take into account only the photon-drag current. On the other hand, also in the case of inter-subband absorption, concentration gradients of non-equilibrium heavy and light holes $\Delta p^{(1)}$, $\Delta p^{(2)} = -\Delta p^{(1)}$, respectively, result from optical excitation. These gradients cause diffusion currents.

Table 1

sample type	sample configuration	orientation	polarization	current density equation	components of the tensor
I		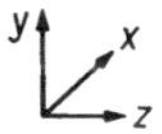	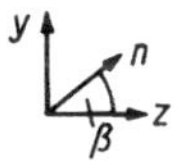	$j_x = (Be_y^2 + Ae_z^2 + 2Ee_y e_z)\, I\varkappa_x$ $= (B \sin^2\beta + A \cos^2\beta +$ $+ 2E \sin\beta \cos\beta)\, I\varkappa_x$	$A = \sigma_{1133}$ $B = \sigma_{1122}$ $E = \sigma_{1132}$
II	as type I	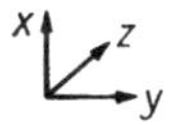	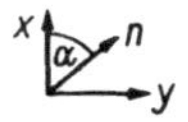	$j_z = C(e_x^2 + e_y^2)\, I\varkappa_z$ $= CI\varkappa_z$	$C = \sigma_{3311}$
III	as type I	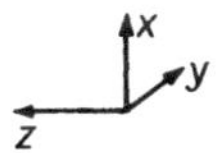	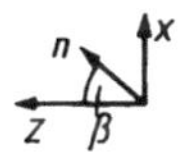	$j_y = (Ae_z^2 + Be_x^2)\, I\varkappa_y$ $= (A \cos^2\beta + B \sin^2\beta)\, I\varkappa_y$	$A = \sigma_{1133}$ $B = \sigma_{1122}$
IV	as type I	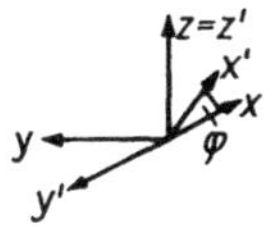		$j_x = [Ae_3'^2 + Be_2'^2 + 2 \cos\varphi\, Ee_2' e_3' \times$ $\times (\cos^2\varphi - 3 \sin^2\varphi)]\, I\varkappa_{x'}$ $= [A \cos^2\beta + B \sin^2\beta + 2E \cos\varphi \cos\beta \sin\beta \times$ $\times (\cos^2\varphi - 3 \sin^2\varphi)]\, I\varkappa_{x'}$	

The total current is given by

$$j_i = \sum_\nu j_i^{(\nu)} = \sum_\nu \left(\sigma_{iklm}^{(\nu)} e_l e_m \varkappa_k I - \frac{kT}{e} \frac{\sigma_{ik}^{(\nu)}}{p^{(\nu)}} \frac{\partial p^{(\nu)}}{\partial x_k} + \sigma_{ik}^{(\nu)} E_k \right). \tag{2}$$

Usually in experiments the open-circuit voltage is measured, therefore (2) has to be resolved with respect to E_j

$$E_j = \varrho_{ji} s_i$$

with

$$s_i = \sum_{\nu=1,2} \left(j_i^{(\nu)} + \frac{kT}{e} \frac{\sigma_{ik}^{(\nu)}}{p^{(\nu)}} \frac{\partial p^{(\nu)}}{\partial x_k} - I \sigma_{iklm}^{(\nu)} e_l e_m \varkappa_k \right). \tag{3}$$

The ϱ_{ji} are the components of the tensor

$$\hat{\boldsymbol{\rho}} = \left(\sum_\nu \hat{\boldsymbol{\sigma}}^{(\nu)}\right)^{-1}.$$

The photon-drag field

$$E_j = -\varrho_{ji} I \left(\sum_\nu \sigma_{iklm}^{(\nu)} e_l e_m \varkappa_k\right)$$

is always accompanied by the pseudo-Dember effect

$$E_j = \frac{kT}{e} \frac{(\mu_\perp^{(2)} - \mu_\perp^{(1)})}{p^{(1)}\mu_\perp^{(1)} + p^{(2)}\mu_\perp^{(2)}} \frac{\partial \Delta p^{(2)}}{\partial x_j}.$$

(This is not the usual Dember effect caused by excitation of electron–hole pairs but the corresponding effect resulting from light-induced hole transitions between sub-valence bands. Since this excitation is equivalent to hole heating by radiation, an alternative description would be thermoelectric power of optically heated carriers.)

We want to point out that Dember [1] discussed the momentum transfer from photons to crystal electrons as a possibility for generating a photo-e.m.f. in the direction of light propagation as early as in 1931.

3. Experiment

Use was made of a Q-switched CO_2 laser (peak pulse power 2.5 kW, half-width 200 ns, degree of polarization $>99.9\%$). The peak pulse power has been determined by measuring average power, pulse shape, and repetition rate; the measuring error was not greater than $\pm 20\%$. A set of teflon sheets was used as a beam attenuator. The investigations were carried out in the temperature range between 77 and 300 K. The samples have been prepared from single crystals grown by the Czochralski method. The acceptor concentration was within the limits 1.5×10^{14} and 3.5×10^{14} cm^{-3}. A comparison was made between unannealed samples (Hall mobility $\mu_H = 500$ to 2700 cm^2/Vs) and samples annealed by a special treatment to get high mobility and low concentration simultaneously [11] ($\mu_H = 9000$ cm^2/Vs at 77 K).

According to the methodical hints by Danishevskii et al. [3] initially samples have been prepared of the complicated shape shown in Table 1. In a later stage of the work simple rod-like samples have been used. The sample was mounted onto a holder free to rotate around the vertical axis of a liquid N_2 cryostat. The

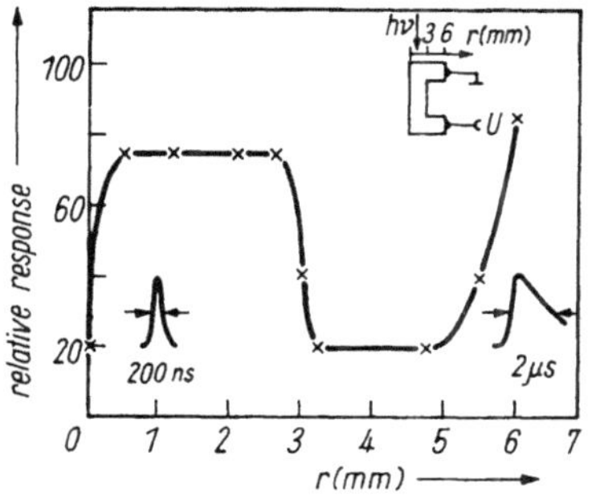

Fig. 1. Response and pulse shape profile. Scanning spot diameter 0.4 mm

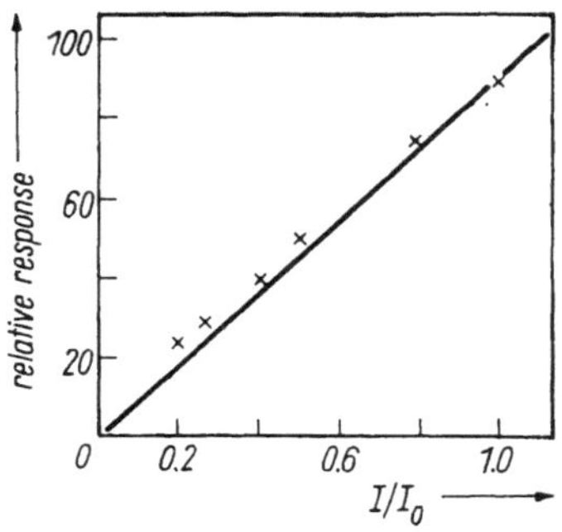

Fig. 2. Variation of photon-drag e.m.f. with laser power

laser beam was incident onto the front face of the sample from below, parallel to the axis. The signals were amplified by a pre-amplifier (input resistance 60 Ω) and detected by a real-time pulse oscilloscope.

4. Results

To avoid the competition processes observed earlier [12], the laser beam was focused only weakly ($I < 100$ kW/cm^2). Signals were identified as photon drag

— exhibiting the same pulse shape as the laser pulse (as monitored by a detector on the basis of photon drag in germanium);

— inversing its polarity, if the laser beam was propagated in the opposite direction;

— showing constant pulse height, if the position of the laser beam was changed within the central region of the sample front surface (Fig. 1);

— exhibiting linear dependence on laser power (Fig. 2).

The pseudo-Dember effect mentioned above exhibits all these properties, too; for distinction see Section 5.

4.1 Photon drag with polarization $\boldsymbol{F}^\omega \parallel \boldsymbol{c}$

With $\boldsymbol{F}^\omega \parallel \boldsymbol{c}$ the highest signals were observed in accordance with the estimate given above (type I and type II samples, respectively). The temperature dependence of these signals (Fig. 3) shows no change of polarity. The sign of coefficient A is always positive (see Table 2).

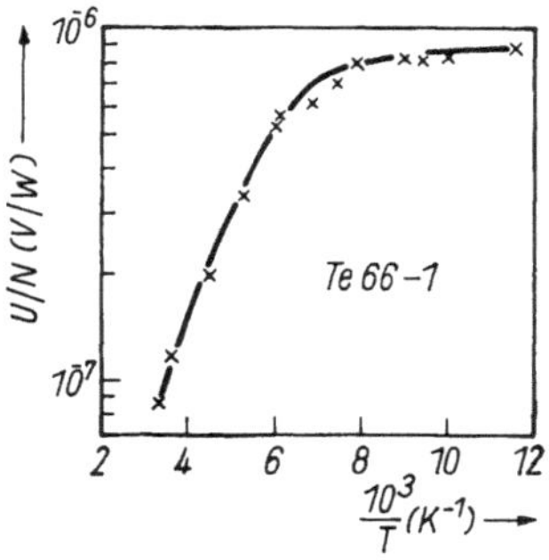

Fig. 3. Variation of photon-drag e.m.f. with temperature. Type I sample

Table 2

Results obtained for type I (III) samples

sample No. and sample type	length l (cm)	mobility $\mu_\perp \left(\frac{cm^2}{Vs}\right)$	hole concentration p_0 (cm^{-3})	$\left(\frac{U}{N_0}\right)^{*)} \left(\frac{mV}{kW}\right)$ $T = 77$ K	$\left(\frac{U}{N_0}\right)^{*)} \left(\frac{mV}{kW}\right)$ $T = 300$ K	$A \left(\frac{cm}{V}\right)$ $T = 77$ K	$A_0 \left(\frac{cm^2}{V}\right)$ $T = 77$ K exper.	$A_0 \left(\frac{cm^2}{V}\right)$ $T = 77$ K theor.
45–2 III	1.14	500	2.3×10^{14}	3	0.14	0.25×10^{-11}	0.712×10^{-11}	2.8×10^{-12}
45–3 I	0.85	500	2.3×10^{14}	3	0.18	0.31×10^{-11}	0.875×10^{-11}	2.8×10^{-12}
66–1 I	0.73	2700	1.6×10^{14}	1.03	0.10	0.52×10^{-11}	2.18×10^{-11}	1.5×10^{-11}
66–4 III	0.68	2700	1.6×10^{14}	1	0.10	0.54×10^{-11}	2.26×10^{-11}	1.5×10^{-11}
66–2 I	0.67	9000	3.3×10^{14}	0.55	0.06	1.93×10^{-11}	3.87×10^{-11}	5.1×10^{-11}

*) Reduced to open-circuit conditions.

4.2 Photon drag with polarization $F^\omega \perp c$

a) Type II samples: For $\boldsymbol{F}^\omega \perp \boldsymbol{c}$, $\boldsymbol{\varkappa} \parallel \boldsymbol{c}$ a photon-drag effect was found in the temperature range between 77 and 180 K. The effect was much smaller in this configuration than the effect discussed in Section 4.1 (see Table 3). The sign

Table 3

Results obtained for type II samples

sample No.	length l (cm)	mobility $\mu_{\parallel}\left(\frac{\text{cm}^2}{\text{Vs}}\right)$	hole concentration p_0 (cm^{-3})	$\left(\frac{U}{N_0}\right)^{*)}\left(\frac{\text{mV}}{\text{kW}}\right)$ $T = 77$ K	$C\left(\frac{\text{cm}}{\text{V}}\right)$ $T = 77$ K	$C_0\left(\frac{\text{cm}^2}{\text{V}}\right)$ $T = 77$ K
37–1 II	0.79	3500	3.8×10^{14}	0.1	-1.07×10^{-12}	-9.7×10^{-11}
66–3 II	0.57		1.6×10^{14}	0.18		

*) Reduced to open-circuit conditions.

was always negative, below 140 K a strong temperature dependence was observed (Fig. 4). The signal is not changed if the electric vector is being rotated in the xy plane.

b) With $\boldsymbol{F}^\omega \perp \boldsymbol{c}$ and type I (or type III) samples we were not able to detect any longitudinal signal, which could be identified as photon drag. Whether no signals greater than noise were present or the effect looked for was masked by other ones not changing polarity or exhibiting longer pulses as compared with the laser pulse shape could not be stated. From that behaviour we conclude that the coefficient B is smaller than 1.7×10^{-14} cm²/V (see Table 1).

c) Identical dependences of photon drag on the angle β between the vector of polarization and the c-axis are observed for type I and type III samples,

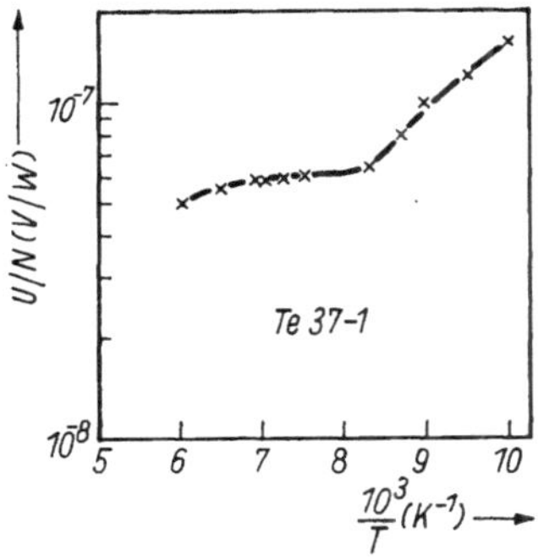

Fig. 4. As for Fig. 3, type II sample

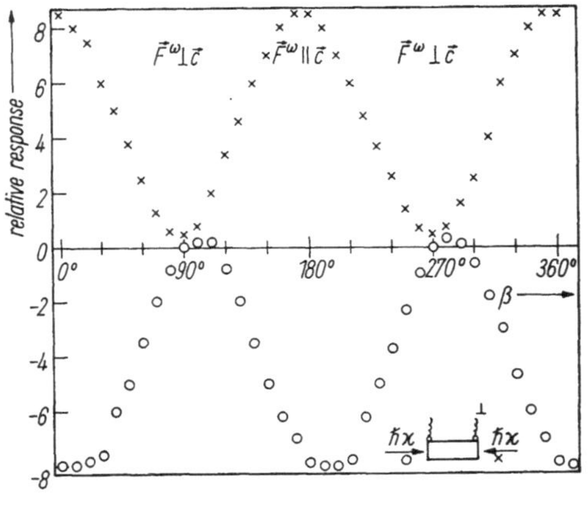

Fig. 5. Dependence of the longitudinal photon-drag e.m.f. on the angle β between the electric vector of the light wave and c-axis (sample type I or III)

respectively. For type I samples a dependence on angle β was measured corresponding very well to the function $\cos^2\beta$ (Fig. 5). Comparing this result with the theoretical equations given in Table 1 we conclude that the coefficient E is not greater than the coefficient B. This is confirmed by measurements of the transverse photon drag [30].

5. Discussion

As mentioned above, it is difficult to decide between photon-drag and pseudo-Dember effect: To estimate the magnitude of the latter effect use was made of the result concerning the photomagnetoelectric effect in tellurium by CO_2 laser [13]. An analysis shows that in the case of weak absorption realized in our experiments the pseudo-Dember effect is negligible as compared with the photon-drag effect.

In order to get explicit formulae containing the tensor coefficients A and C, we have to evaluate (3) in accordance with the equations given in Table 1. All terms containing tensor coefficients B and E are neglected as suggested by our experimental results. Under open-circuit conditions we have from $i = \int j_{\parallel}\, dS_{\perp} = 0$ for type I samples

$$\boldsymbol{E}(x) = -\frac{1}{\sigma_{\perp}}\frac{1}{S} N(x)\, A \cos^2\beta\, \boldsymbol{\varkappa}$$

(S sample cross-section, S' beam cross-section, $N(x) = S' I(x)$ effective laser power, measured in W). Taking into consideration multiple reflections we have

$$N(x) = \frac{N_0(1 - R_{\parallel})}{1 - R_{\parallel}^2\, e^{-2K_{\parallel}l}} (e^{-K_{\parallel}x} - R_{\parallel}\, e^{-2K_{\parallel}l}\, e^{K_{\parallel}x})$$

($K_{\parallel}$, $R_{\parallel}$ absorption and reflection coefficient, respectively, for light with $\boldsymbol{F}^{\omega} \parallel \boldsymbol{c}$, l sample length). With $A = A_0 K_{\parallel}$ we have

$$\boldsymbol{E}(x) = -\frac{A_0}{S}\frac{K_{\parallel}N(x)}{ep_0\mu_{\perp}^{(1)}} \cos^2\beta\, \boldsymbol{\varkappa} \tag{4}$$

and

$$U = -\int_0^l E(x)\, dx = \frac{N_0}{S}\frac{(1 - R_{\parallel})(1 - e^{-K_{\parallel}l})}{(1 + R_{\parallel}\, e^{-K_{\parallel}l})}\frac{A_0 \cos^2\beta\, \varkappa}{ep_0\mu_{\perp}^{(1)}}. \tag{5}$$

In this equation we wrote $\sigma_{\perp} = e(p^{(1)}\mu_{\perp}^{(1)} + p^{(2)}\mu_{\perp}^{(2)}) \approx ep_0\mu_{\perp}^{(1)}$ for weak excitation.

For type II samples we have

$$\boldsymbol{E}(z) = -\frac{1}{\sigma_{\parallel}}\frac{1}{S} N(z)\, C\boldsymbol{\varkappa}$$

(with $N(z)$ as above, but containing $K_{\perp}$ and $R_{\perp}$ instead of $K_{\parallel}$ and $R_{\parallel}$, respectively) and with $\sigma_{\parallel} = ep_0\mu_{\parallel}^{(1)}$ we finally have

$$U = -\int_0^l E(z)\, dz = \frac{N_0}{S}\frac{(1 - R_{\perp})(1 - e^{-K_{\perp}l})}{(1 + R_{\perp}\, e^{-K_{\perp}l})ep_0\mu_{\parallel}^{(1)}} C_0\varkappa\,. \tag{6}$$

Taking into account the parameters given in Tables 2 and 3 and assuming $K_{||} = S_{p||}p_0$, $K_{\perp} = S_{p\perp}p_0$ we finally have the magnitudes of A, A_0, C, C_0 as given in Tables 2 and 3.

Any microscopic theory of photon drag in tellurium has not been given so far. The model developed by Grinberg et al. [26] and Imamov [27] for the case of p-type germanium is to be modified in the following sense:

— There is a non-zero energy gap between sub-valence bands H_4 and H_5 in the H-point of the first BZ.

— Transitions $H_4 \to H_5$ are allowed for radiation $\boldsymbol{F}^{\omega} \parallel \boldsymbol{c}$.

— The anisotropy of isoenergetic surfaces is much more complicated in tellurium than in germanium.

As to the photon-drag tensor coefficient A_0, we explain our results in the following simple model:

$\boldsymbol{\varkappa}$ is oriented perpendicularly with respect to the c-axis. The first order $\boldsymbol{k}\cdot\boldsymbol{p}$ perturbation calculation yields for this case $m_{\perp}^{(1)} = m_{\perp}^{(2)} = m_{\perp}$. From fitting this theory to Shubnikov-de Haas experiments $m_{\perp} = 0.12 m_0$ was found [14]. (It is worthwile to note that different $m_{\perp}^{(i)}$ have been determined from magneto-absorption experiments by Hardy et al. [15] and Miura and Tanaka [16].) At $T = 77$ K it holds $\Delta E \approx \hbar\omega_{CO_2}$, if collision broadening of energy eigenstates is taken into account [7]. For this case of accidental coincidence between ΔE and $\hbar\omega$ we have from energy conservation

$$\frac{\hbar^2 k^{(1)2}}{2m_{\perp}} + (\hbar\omega - \Delta E) = \frac{\hbar^2 k^{(2)2}}{2m_{\perp}}$$

and momentum conservation

$$\hbar\boldsymbol{k}^{(1)} + \hbar\boldsymbol{\varkappa} = \hbar\boldsymbol{k}^{(2)}\,,$$

the wave vectors of holes undergoing transitions $H_4 \to H_5$:

$$\boldsymbol{k}^{(i)} = (-1)^i \frac{\boldsymbol{\varkappa}}{2}, \qquad \text{where} \qquad i = 1, 2\,.$$

For the antisymmetric generation terms we get

$$\left[\frac{\mathrm{d}f_{\mathrm{as}}^{(i)}(\boldsymbol{k})}{\mathrm{d}t}\right]_{\mathrm{gen}} = (-1)^i \frac{K_{||} I}{\hbar\omega}\, \delta(\boldsymbol{k} - \boldsymbol{k}^{(i)})\,.$$

Solving the antisymmetric part of the Boltzmann equation [4] in τ-approximation we have from

$$\boldsymbol{j} = \sum_i e\, \frac{2}{(2\pi)^3} \int \boldsymbol{v}_{\boldsymbol{k}}\, f_{\mathrm{as}}^{(i)}(\boldsymbol{k})\, \mathrm{d}^3\boldsymbol{k} \tag{7}$$

the photon-drag current

$$\boldsymbol{j}(x) = \frac{eK_{||} I(x)\, [\tau^{(1)}(\boldsymbol{k}^{(1)}) + \tau^{(2)}(\boldsymbol{k}^{(2)})]}{2m_{\perp}\omega}\, \boldsymbol{\varkappa} \tag{8}$$

and finally the tensor coefficient

$$A_0 = \frac{e[\tau^{(1)}(\boldsymbol{k}^{(1)}) + \tau^{(2)}(\boldsymbol{k}^{(2)})]}{2m_{\perp}\omega} \approx \frac{\mu_{\perp}^{(1)}}{\omega}. \tag{9}$$

The latter step has to be justified by a more detailed treatment taking into account the real valence band structure of tellurium and collision broadening.

The generation terms will be a superposition of contributions from transitions across different energy gaps [7]. For any $\Delta E < \hbar\omega$ the solution of energy and momentum conservation given above has to be modified. The analysis shows that for the dispersion law [14]

$$E^{(i)}(\boldsymbol{k}) = A_1 k_z^2 + A_2 k_\perp^2 + (-1)^i \sqrt{\left(\frac{\Delta E}{2}\right)^2 + C^2 k_z^2}$$

and in the case of $\boldsymbol{\varkappa} = (\varkappa, 0, 0)$ hole transitions take place for any k_y, k_z, and for $k_x^{(1)}$ given by

$$\sqrt{\Delta E^2 + 4C^2 k_z^2} - \hbar\omega = -2A_2 k_x^{(1)} \varkappa ,$$

where $C^2 = 6 \times 10^{-16}\ \text{eV}^2\ \text{cm}^2$, $A_2 = 0.326 \times 10^{-14}\ \text{eV cm}^2$ [14]. For that reason the integration in (7) gives some average momentum relaxation time $[\tau^{(i)}]$ instead of $\tau^{(i)}(k^{(i)})$ which might be not too different from the thermal average $\langle\tau^{(i)}\rangle = (m/e)\,\mu$. The coefficients A_0 calculated from the mobility measured are given in Table 2, too. The agreement with experimental values is surprisingly good. The temperature dependence of open-circuit voltage (see Fig. 3) is mainly governed by the increasing carrier concentration under intrinsic conditions. This is due to the high hole absorption cross-section in tellurium, because with $l \approx 1$ cm it holds $K_{||} l = p_0 S_{p||} l \gg 1$ in the intrinsic region; from (4) we have $U \sim (1 - e^{-K_{||} l})/p_0 \approx 1/p_0$. Only in very short samples with $l < 1$ mm we have $U \sim K_{||} l/p_0 = S_{p||} l$ which is approximately independent of temperature.

As to the coefficient C_0, a microscopic theory to be developed in the future, will have to consider forbidden transitions $H_4 \to H_5$ for $\boldsymbol{F}^\omega \perp \boldsymbol{c}$, Drude absorption within H_4 and allowed transitions $H_4 \to H_6$.

Acknowledgement

We are indebted to Dr. P. M. Valov from the A. F. Ioffe Physico-Technical Institute of the Academy of Sciences of the USSR for stimulating discussions.

Appendix

Non-zero coefficients of the photon-drag tensor of tellurium (Cartesian coordinate system as mentioned above):

$$A = \sigma_{1133} = \sigma_{2233} ,$$

$$B = \sigma_{1122} = \sigma_{2211} ,$$

$$C = \sigma_{3311} = \sigma_{3322} ,$$

$$D = -\sigma_{2322} = \sigma_{2311} = \sigma_{1312} = \sigma_{1321} ,$$

$$E = -\sigma_{2223} = \sigma_{1123} = \sigma_{1213} = \sigma_{2113} = -\sigma_{2232} = \sigma_{2131} = \sigma_{1132} = \sigma_{1231} ,$$

$$-\sigma_{3222} = \sigma_{3121} = \sigma_{3211} = \sigma_{3112} ,$$

$$\sigma_{1212} = \sigma_{2112} = \sigma_{1221} = \sigma_{2121} ,$$

$$\sigma_{3223} = \sigma_{3113} = \sigma_{3131} = \sigma_{3232} ,$$

$$\sigma_{2332} = \sigma_{1331} = \sigma_{1313} = \sigma_{2323} ,$$

$$\sigma_{3333} ,$$

$$\sigma_{1111} = \sigma_{2222} = 2\sigma_{1212} + \sigma_{1122} .$$

Only those coefficients denoted by capital latters A to E are relevant to the experiments presented in this paper. In the same coordinate system the conductivity tensor $\boldsymbol{\sigma}_{ik}$ looks as follows:

$$\sigma_{11} = \sigma_{22} = \sigma_{\perp} = ep\mu_{p_{\perp}}\,,$$

$$\sigma_{33} = \sigma_{||} = ep\mu_{p_{||}}\,.$$

References

[1] H. Dember, Phys. Z. **33**, 207 (1931).
[2] T. S. Moss, phys. stat. sol. (a) 8, 223 (1971).
[3] A. M. Danishevskii, A. A. Kastalskii, S. M. Ryvkin, and I. D. Yaroshetskii, Zh. eksper. teor. Fiz. **58**, 544 (1970).
[4] A. A. Grinberg, Zh. eksper. teor. Fiz. **58**, 989 (1970).
[5] P. M. Valov, B. S. Ryvkin, S. M. Ryvkin, and I. D. Yaroshetskii, phys. stat. sol. (b) **53**, 65 (1972).
[6] V. G. Agafonov, P. M. Valov, B. S. Ryvkin, and I. D. Yaroshetskii, Fiz. Tekh. Poluprov. **6**, 2219 (1972).
[7] D. Hardy and C. Rigaux, phys. stat. sol. **38**, 799 (1970).
[8] P. Grosse, Z. Phys. **193**, 318 (1966).
[9] E. Bangert, R. Fischer, and P. Grosse, phys. stat. sol. (b) **59**, 419 (1973).
[10] S. Panyakeow, J. Shirafuji, and Y. Inuishi, Appl. Phys. Letters **21**, 314 (1972).
[11] D. Kusnick and M. Spitzer, Dissertation, Humboldt-Univ. Berlin, 1973 (unpublished).
[12] K. H. Herrmann and R. Vogel, Proc. XI. Internat. Conf. Phys. Semicond., Warsaw 1972 (p. 870).
[13] D. Genzow, K. H. Herrmann, and B. Viehweger, phys. stat. sol. (a) **15**, K171 (1973).
[14] M. S. Bresler, V. G. Veselago, Yu. V. Kosichkin, G. E. Pikus, I. I. Farbshtein, and S. S. Shalyt, Zh. exper. teor. Fiz. **57**, 1479 (1969).
[15] D. Hardy, C. Rigaux, J. P. Vieren, and N. Hy Hau, phys. stat. sol. (b) **47**, 643 (1971).
[16] N. Miura and S. Tanaka, phys. stat. sol. **42**, 257 (1970).
[17] A. F. Gibson, M. F. Kimmit, and A. C. Walker, Appl. Phys. Letters, **17**, 75 (1970).
[18] P. M. Valov, A. A. Grinberg, A. M. Danishevskii, A. A. Kastalskii, S. M. Ryvkin, and I. D. Yaroshetskii, Proc. X. Internat. Conf. Phys. Semicond., Cambridge (Mass.) 1970 (p. 683).
[19] A. F. Gibson, M. F. Kimmit, and A. C. Walker, Proc. X. Internat. Conf. Phys. Semicond., Cambridge (Mass.) 1970 (p. 690).
[20] P. M. Valov, A. M. Danishevskii, A. A. Kastalskii, B. S. Ryvkin, S. M. Ryvkin, and I. D. Yaroshetskii, Zh. eksper. teor. Fiz. **59**, 1919 (1970).
[21] V. G. Agafonov, P. M. Valov, B. S. Ryvkin, and I. D. Yaroshetskii, Fiz. Tekh. Poluprov. **6**, 909 (1972).
[22] A. A. Grinberg and L. L. Marovskii, Fiz. Tekh. Poluprov. **4**, 1162 (1970).
[23] P. M. Valov, B. S. Ryvkin, S. M. Ryvkin, E. V. Titova, and I. D. Yaroshetskii, Fiz. Tekh. Poluprov. **5**, 1772 (1971).
[24] P. M. Valov, A. M. Danishevskii, and I. D. Yaroshetskii, Zh. eksper. teor. Fiz. **59**, 722 (1970).
[25] L. L. Makovskii, Fiz. Tekh. Poluprov. **4**, 1563 (1970).
[26] A. A. Grinberg, N. A. Brynskikh, and E. Z. Imanov, Fiz. Tekh. Poluprov. **5**, 148 (1971).
[27] E. Z. Imamov, Fiz. Tekh. Poluprov. **6**, 1693 (1972).
[28] P. M. Valov, B. S. Ryvkin, S. M. Ryvkin, E. V. Titova, and I. D. Yaroshetskii, Fiz. Tekh. Poluprov. **6**, 123 (1972).
[29] A. F. Gibson and A. C. Walker, J. Phys. C **4**, 2209 (1971).
[30] J. Auth et al., to be published.

(Received June 25, 1974)

phys. stat. sol. (b) 65, 305 (1974)

Subject classification: 13 and 18

Institute of Physics, Belgrade[1])

A New Exchange Interaction Operator for Magnetic Materials

By

M. POPOVIĆ-BOŽIĆ

Girardeau's second quantization representation for compound particles is used to derive, starting from first principles, a new approximate Hamiltonian for the exchange interaction of two atoms in their ground states. As distinct from the conventionally used Dirac-Heisenberg Hamiltonian, the new operator comprises the effects of exchange of one, two, three etc. pairs of electrons. Assuming that the exchange interaction in a many-atom system can be taken as a sum over the operators for interaction of pairs of atoms, a new approximate Hamiltonian for magnetic crystals is derived.

La représentation de la seconde quantification pour les particules composées, proposée par Girardeau, est utilisée pour obtenir, à partir des premiers principes, un nouveau approximative opérateur hamiltonien d'interaction d'échange de deux atoms, chaqun dans son état fondamental. A la différence d'opérateur hamiltonien de Dirac-Heisenberg, dans le nouveau opérateur les effects d'échange d'un, deux, trois etc. couples d'électrons sont inclus. Supposant que l'interaction d'échange dans un système contenant N particules composées peut être approchée par la somme d'opérateurs d'interaction des couples d'atoms, on a dérivé un nouveau opérateur pour les cristeaux magnetiques.

1. Introduction

Recently in a number of papers [1 to 3] a second quantization representation for treating systems of many atoms with any number of electrons has been developed, in which electrons and nuclear variables do not appear explicitly, but which takes the atomic internal structure properly into account. Even if atomic states are taken as variables, the exchange effects of any number of electrons between two atoms are involved as an effective interatomic exchange interaction which can be treated along with the interatomic Coulomb interaction.

We consider this representation to be a proper formalism which can be used to obtain, starting from first principles, a new approximate Hamiltonian to describe low-lying states of a system of fairly well separated magnetic atoms (atoms with spin different from zero). According to the Dirac-Heisenberg method [4 to 6] only the interchange of a single pair of electrons between neighbouring atoms is taken into account, whereas the proposed new exchange interaction Hamiltonian involves the exchange of one, two or more pairs of electrons between neighbouring atoms.

In Section 2 Girardeau's second quantization representation for compound particles is briefly exposed. In Section 3 the new representation is used to obtain the Hamiltonian of two fixed atoms, while in Section 4 an exchange interaction operator acting only on the spin part of the wave function of two fixed atoms is obtained as an approximation to the exact two-atom exchange

[1]) P. O. Box 57.

interaction operator given in the preceding section. Assuming that exchange interaction in a many-atom system can be taken as a sum over two-atom interactions, a new approximate Hamiltonian for magnetic crystals is derived in Section 5, while in Section 6 the new operator is compared with the conventionally used Heisenberg model.

2. The Second Quantization Representation for Compound Particles

In a number of papers [1 to 3] the second quantization representation for compound particles has been derived. In fact, the atomic annihilation and creation operators a_α and a_α^+ for atoms in orthonormal single-atom states $\varphi_\alpha(Xx_1 \cdots x_N)$ have been introduced. The functions $\varphi_\alpha(Xx_1 \cdots x_N)$ satisfy the Schrödinger equation

$$\left[T(X) + \sum_{j=1}^{N} T(x_j) + \sum_{\substack{j,k=1\\ j<k}}^{N} V(x_j x_k) + \sum_{j=1}^{N} V(X x_j)\right] \varphi_\alpha(Xx_1 \cdots x_N) = \\ = E_\alpha \varphi_\alpha(Xx_1 \cdots x_N) . \tag{2.1}$$

Here $x_j = (r_j \sigma_j)$ denotes both the position and the z-component of the spin of the electron, $X = (R\sigma)$ denotes the position of the nucleus of spin J and the z-component of the spin. Single-particle operators T represent the nuclear and electron kinetic energies, and two-particle operators V are the electron–electron and nucleus–electron Coulomb interactions. The functions $\varphi_\alpha(Xx_1 \cdots x_N)$ are antisymmetric with respect to the permutation of the electron coordinates x_i, and form an orthonormal complete set

$$\left.\begin{aligned} &\int \varphi_\alpha^*(Xx_1 \cdots x_N)\, \varphi_\beta(Xx_1 \cdots x_N)\, \mathrm{d}X\, \mathrm{d}x_1 \cdots \mathrm{d}x_N = \delta_{\alpha\beta} , \\ &\sum_\alpha \varphi_\alpha(Xx_1 \cdots x_N)\, \varphi_\alpha(X'x_1' \cdots x_N') = \\ &= (l!)^{-1}\, \delta(X - X') \sum_{P'} (-1)^{p(P')}\, P'[\delta(x_1 - x_1') \cdots \delta(x_N - x_N')] . \end{aligned}\right\} \tag{2.2}$$

The symbol $\int$ means integration over positions and summation over spins, $\sum_\alpha$ is a sum over discrete quantum numbers and an integral over continuous ones, $\sum_{P'}$ denotes a sum over all permutations P' of primed variables, and $p(P')$ is the parity of the permutation P'.

Atomic creation and annihilation operators a_α^+ and a_α satisfy the Bose or Fermi commutation relations:

$$a_\alpha a_\beta^+ - (-1)^{2J+N}\, a_\beta^+ a_\alpha = \delta_{\alpha\beta} , \qquad a_\alpha a_\beta - (-1)^{2J+N}\, a_\beta a_\alpha = 0 . \tag{2.3}$$

In the new representation the Hamiltonian has the form [3, 7]

$$\mathcal{H} = H + V^{\text{ex}} = H + \sum_{j=1}^{N} V^{(j)} , \tag{2.4}$$

where

$$H = H_0 + V' = \sum_\alpha E_\alpha a_\alpha^+ a_\alpha + \tfrac{1}{2} \sum_{\alpha\beta\gamma\delta} (\alpha\beta|\, V'|\gamma\delta)\, a_\alpha^+ a_\beta^+ a_\delta a_\gamma , \tag{2.5}$$

$$V^{(j)} = (-1)^j \binom{N}{j}^2 \sum_{\alpha\beta\gamma\delta} (\alpha\beta|\, V_j|\gamma\delta)\, a_\alpha^+ a_\beta^+ a_\delta a_\gamma . \tag{2.6}$$

Matrix elements in (2.5) and (2.6) are

$$(\alpha\beta|V'|\gamma\delta) = \int \varphi_\alpha^*(Xx_1 \cdots x_N)\, \varphi_\beta^*(X'x_1' \cdots x_N') \Big\{V(XX') + \sum_{j,k=1}^{N} V(x_j x_k') + \sum_{j=1}^{N} V(Xx_j') + \sum_{j=1}^{N} V(X'x_j)\Big\} \varphi_\gamma(Xx_1 \cdots x_N)\, \varphi_\delta(X'x_1' \cdots x_N') \times \mathrm{d}X\, \mathrm{d}x_1 \cdots \mathrm{d}x_N\, \mathrm{d}X'\, \mathrm{d}x_1' \cdots \mathrm{d}x_N' \,, \tag{2.7}$$

$$(\alpha\beta|V_j|\gamma\delta) = (E_\alpha + E_\beta)\, (\alpha\beta|I_j|\gamma\delta) + (\alpha\beta|V_j'|\gamma\delta) \,, \tag{2.8}$$

$$\left.\begin{aligned}(\alpha\beta|I_j|\gamma\delta) = \int \varphi_\alpha^*(Xx_1 \cdots x_N)\, \varphi_\beta^*(X'x_1' \cdots x_N')\, \varphi_\gamma(Xx_1' \cdots x_j'\, x_{j+1} \cdots x_N) \,,\\ \varphi_\delta(Xx_1 \cdots x_j x_{j+1}' \cdots x_N')\, \mathrm{d}X\, \mathrm{d}x_1 \cdots \mathrm{d}x_N\, \mathrm{d}X'\, \mathrm{d}x_1' \cdots \mathrm{d}x_N' \,,\end{aligned}\right\} \tag{2.9}$$

$$(\alpha\beta|V_j'|\gamma\delta) = \int \varphi_\alpha^*(Xx_1 \cdots x_N)\, \varphi_\beta^*(X'x_1' \cdots x_N') \times \Big\{V(XX') + \sum_{j,k=1}^{N} V(x_j x_k') + \sum_{j=1}^{N} [V(Xx_j') + V(X'x_j)]\Big\} \times \varphi_\gamma(Xx_1' \cdots x_j'\, x_{j+1} \cdots x_N)\, \varphi_\delta(X'x_1 \cdots x_j x_{j+1}' \cdots x_N') \times \mathrm{d}X\, \mathrm{d}x_1 \cdots \mathrm{d}x_N\, \mathrm{d}X'\, \mathrm{d}x_1' \cdots \mathrm{d}x_N' \,. \tag{2.10}$$

The operator (2.6) is exact for a two-atom system. In the case of a many-atom system, the Hamiltonian (2.6) is an approximation to the exact one, because the third-, fourth- and higher-order terms in the annihilation and creation operators are neglected.

The Hamiltonian (2.4) comprises both the effects of exchange of atoms as a whole and the effects of exchange of electrons between atoms. The former are involved in the operator H through the commutation relation for creation and annihilation operators, whereas the latter are contained in the operator V^{ex}. In fact, the operator H has the same structure as the Hamiltonian of elementary bosons or fermions. Examining matrix elements (2.8), (2.9), and (2.10) it can be noted that the operator $V^{(j)}$ is attributed to the exchange of j pairs of electrons between any two atoms.

3. The Hamiltonian of a Pair of Fixed Atoms in the Second Quantization Representation for Compound Particles

In solid state the effects of exchange of electrons between atoms are important only because atoms as a whole do not change their sites. That means that the coordinate of the atomic site is one of the atomic state labels. In fact, the atomic quantum number $\alpha = (\bar{\alpha}n)$ contains the coordinate n of the atomic site and the label $\bar{\alpha}$ which stands for the set of all internal quantum numbers necessary to specify the atomic wave function in the centre-of-mass system.

The atomic annihilation and creation operators $a_\alpha = a_{n\bar{\alpha}}^-$ and $a_\alpha^+ = a_{n\bar{\alpha}}^{+-}$ operate on the state vector of the atom at the site n. Because of the conservation of the number of particles in the operators of physical quantities, the annihilation and creation operators always appear as a normal product of one creation and one annihilation operator which operate at the same site. For example, the normal product $a_{n\bar{\alpha}_1}^{+-} a_{n\bar{\alpha}_2}^{-}$ operates as follows on the N_0-atom function:

$$a_{n\bar{\alpha}_1}^{+-} a_{n\bar{\alpha}_2}^{-} |m_1\bar{\beta}_1 \cdots n\bar{\gamma} \cdots m_{N_0}\bar{\beta}_{N_0}\rangle = \delta_{\bar{\gamma}\bar{\alpha}_2} |m\bar{\beta}_1 \cdots n\bar{\alpha}_1 \cdots m_{N_0}\bar{\beta}_{N_0}\rangle \,. \tag{3.1}$$

The Hamiltonian of the whole crystal is usually taken as a sum of two-atom Hamiltonians. Hence, in order to use Girardeau's formalism to study the exchange interaction in magnetic crystals, let us first find, in the new representation, the Hamiltonian for a pair of fixed atoms.

The eigenvalue problem for an atom fixed at point r (here the approximation of identification of the centre of mass with the position of the nucleus can be used) is of the form

$$\left[\sum_{j=1}^{N} T(y_j) + \sum_{j=1}^{N} V(Yy_j) + \sum_{j,k=1}^{N} V(y_j y_k)\right] \delta_{Rr}\overline{\varphi}_{r\overline{\alpha}}(\sigma y_1 \cdots y_N) =$$

$$= E_{\overline{\alpha}}\delta_{rR}\overline{\varphi}_{r\overline{\alpha}}(\sigma y_1 \cdots y_N)\,, \tag{3.2}$$

where $y_j = (r_j - r, \sigma_j) \equiv (q_j, \sigma_j)$, $Y = (R, \sigma) = (r, \sigma)$, and $\overline{\varphi}_{r\overline{\alpha}}(\sigma y_1 \cdots y_N)$ is the usual form of the internal wave function in which the electron positions are measured in a coordinate system whose origin is located in the nucleus.

In order to obtain the Hamiltonian for two atoms, one fixed at point n and the other at point m, we have to do the following:

1. insert in matrix elements (2.7), (2.9), and (2.10) the functions $\delta_{Rr}\overline{\varphi}_{r\overline{\alpha}}(\sigma y_1 \cdots y_N)$ instead of functions $\varphi_\alpha(Yy_1 \cdots y_N)$,
2. express the atomic quantum numbers α, β, γ, δ in (2.5) and (2.6) in terms of $(\overline{\alpha}r)$, $(\overline{\beta}r')$, $(\overline{\gamma}r'')$, $(\overline{\delta}r''')$,
3. substitute (n, m) and (m, n) for the pair (r, r').

In such a way we obtain the following expression for the Hamiltonian sought for:

$$\mathscr{H}_{nm} = H_{nm} + V_{nm}^{\text{ex}} = H_{nm} + \sum_{j=1}^{N} V_{nm}^{(j)}\,, \tag{3.3}$$

$$H_{nm} = \sum_{\overline{\alpha}} E_{\overline{\alpha}}(a^+_{n\overline{\alpha}}a_{n\overline{\alpha}} + a^+_{m\overline{\alpha}}a_{m\overline{\alpha}}) + \sum_{\overline{\alpha}\overline{\beta}\overline{\gamma}\overline{\delta}} (n\overline{\alpha}\, m\overline{\beta}\,|V'|\, n\overline{\gamma}m\overline{\delta})\, a^+_{n\overline{\alpha}}a_{n\overline{\gamma}}a^+_{m\overline{\beta}}a_{m\overline{\delta}}\,, \tag{3.4}$$

$$V_{nm}^{(j)} = (-1)^j \binom{N}{j}^2 \sum_{\overline{\alpha}\overline{\beta}\overline{\gamma}\overline{\delta}} (n\overline{\alpha}m\overline{\beta}\,|V_j|\, n\overline{\gamma}m\overline{\delta})\, a^+_{n\overline{\alpha}}a_{n\overline{\gamma}}a^+_{m\overline{\beta}}a_{m\overline{\delta}}\,. \tag{3.5}$$

The matrix elements in (3.4) and (3.5) are

$$(n\overline{\alpha}m\overline{\beta}\,|V'|\, n\overline{\gamma}m\overline{\delta}) = \sum_{\sigma\sigma'} \int \overline{\varphi}^*_{n\overline{\alpha}}(\sigma y_1 \cdots y_N)\, \overline{\varphi}^*_{m\overline{\beta}}(\sigma' y_1' \cdots y_N') \times$$

$$\times \left\{V(nm\sigma\sigma') + \sum_{j,k=1}^{N} V(y_j y_k') + \sum_{j=1}^{N} [V(y_j'\sigma n) + V(y_j\sigma' m)]\right\} \times$$

$$\times\, \overline{\varphi}_{n\overline{\gamma}}(\sigma y_1 \cdots y_N)\, \overline{\varphi}_{m\overline{\delta}}(\sigma' y_1' \cdots y_N')\, \mathrm{d}y_1 \cdots \mathrm{d}y_N\, \mathrm{d}y_1' \cdots \mathrm{d}y_N'\,, \tag{3.6}$$

$$(n\overline{\alpha}m\overline{\beta}|\, V_j|\, n\overline{\gamma}m\overline{\delta}) = \int \overline{\varphi}^*_{n\overline{\alpha}}(\sigma y_1 \cdots y_N)\, \overline{\varphi}^*_{m\overline{\beta}}(\sigma' y_1' \cdots y_N') \times$$

$$\times \left\{(E_{\overline{\alpha}} + E_{\overline{\beta}}) + V(nm\sigma\sigma') + \sum_{j,k=1}^{N} V(y_j y_k') + \sum_{j=1}^{N} V(y_j'\sigma n) + \sum_{j=1}^{N} V(y_j\sigma' m)\right\} \times$$

$$\times\, \overline{\varphi}_{n\overline{\gamma}}(\sigma y_1' \cdots y_j' y_{j+1} \cdots y_N)\, \overline{\varphi}_{m\overline{\delta}}(\sigma y_1 \cdots y_j y_{j+1}' \cdots y_N')\, \mathrm{d}y_1 \cdots \mathrm{d}y_N\, \mathrm{d}y_1' \cdots \mathrm{d}y_N' =$$

$$= (E_{\overline{\alpha}} + E_{\overline{\beta}})\, (\overline{\alpha}\overline{\beta}\,|I_j|\, \overline{\gamma}\overline{\delta}) + (\overline{\alpha}\overline{\beta}\,|V_j'|\, \overline{\gamma}\overline{\delta})\,. \tag{3.7}$$

4. Approximate Exchange Interaction Operator for a Pair of Fixed Atoms in the Case where the only Degeneracy of the Atomic Ground State is that Associated with Various Possible Orientations of the Atomic Spin

Using the methods presently available we cannot find the eigenvalues and eigenfunctions of the exact Hamiltonian (3.3). However, sometimes it is possible to separate out of $\mathcal{H}_{nm}$ those terms which can be shown to give the greatest contribution to the effect considered.

In this section we shall derive from $\mathcal{H}_{nm}$ an approximate Hamiltonian for studying the low-lying states of two fairly separated magnetic atoms.

The assumptions that we are going to make are applicable mainly in the case of 3d metals.

We shall neglect the electrons in filled shells. Consequently henceforth we shall denote the number of electrons in an unfilled shell by N. Furthermore, we shall suppose that the magnetic interaction can be neglected, i.e. that the set of atomic internal quantum numbers is $\bar{\alpha} = (\omega L M_L S M_S)$, where ω stands for the set of additional quantum numbers needed to specify the states fully. Assuming also that the energy difference between the atomic ground state and the first excited state is much larger than the two-atom exchange interaction energy in the sum over atomic quantum numbers, we shall retain only the sum over the quantum numbers of the atomic ground state. But since we also assume that the total atomic spin is not destroyed by the mutual interaction of two atoms and that the only degeneracy of the atomic ground state is that associated with various possible orientations of the atomic spin, the sum $\sum_{\bar{\alpha}\bar{\beta}\bar{\gamma}\bar{\delta}}$ in the operators (3.4) and (3.5) can be approximated by $\sum_{S_1S_2S_3S_4=-S^g}^{S^g}$ (in order to abbreviate the notation, we use in the sums the symbols S_1, S_2, S_3, S_4 instead of M_{S_1}, M_{S_2}, M_{S_3}, M_{S_4} and we omit the orbital ground state quantum numbers L^g and M_L^g). The effects related to the spin of the nucleus can be neglected, too. That means that we can set the nuclear spin to be equal to zero, and thus the sum over projections of the nuclear spin in matrix elements is to be omitted.

Finally, we have the following expression for the approximate two-atom Hamiltonian:

$$\mathcal{H}^{a}_{nm} = H^{a}_{nm} + H_{\text{ex}}(nm)\,, \tag{4.1}$$

$$H^{a}_{nm} = E_g \sum_{S_1=-S^g}^{S^g} (a^{+}_{nS_1}a_{nS_1} + a^{+}_{mS_1}a_{mS_1}) + \sum_{S_1S_2=-S^g}^{S^g} V'_{nmg}a^{+}_{nS_1}a_{nS_1}a^{+}_{mS_2}a_{mS_2}\,, \tag{4.2}$$

$$H_{\text{ex}}(nm) = \sum_{j=1}^{N} H^{(j)}_{nm}\,, \tag{4.3}$$

$$H^{(j)}_{nm} = (-1)^j \binom{N}{j}^2 \sum_{\substack{S_1S_2S_3S_4=-S^g \\ S_1+S_2=S_3+S_4}}^{S^g} A^{(j)}_{nm}(S_1S_2S_3S_4)\, a^{+}_{nS_1}a_{nS_3}a^{+}_{mS_2}a_{mS_4}\,, \tag{4.4}$$

where V'_{nmg} is the matrix element of the Coulomb interaction of two atoms in their ground states, while $A^{(j)}_{nm}(S_1S_2S_3S_4)$ can be written as follows:

$$\begin{aligned} A^{(j)}_{nm}(S_1S_2S_3S_4) &\equiv \\ &\equiv (nl^NL^gM^g_LS^gS_1,\, ml^NL^gM^g_LS^gS_2\,|V_j|\, nl^NL^gM^g_LS^gS_3,\, ml^NL^gM^g_LS^gS_4) = \\ &= 2E_g I^{(j)}_{nm}(S_1S_2S_3S_4) + J^{(j)}_{nm}(S_1S_2S_3S_4)\,, \end{aligned} \tag{4.5}$$

$$I^{(j)}_{nm}(S_1S_2S_3S_4) = \int \bar{\varphi}^*_{ngS_1}(y_1 \cdots y_N)\, \bar{\varphi}^*_{mgS_2}(y'_1 \cdots y'_N) \times$$
$$\times\, \bar{\varphi}_{ngS_3}(y'_1 + m - n, \cdots, y'_j + m - n, y_{j+1} \cdots y_N) \times$$
$$\times\, \bar{\varphi}_{ngS_4}(y_1 + n - m, \cdots, y_j + n - m, y'_{j+1} \cdots y'_N)\, \mathrm{d}y_1 \cdots \mathrm{d}y_N\, \mathrm{d}y'_1 \cdots \mathrm{d}y'_N\,, \tag{4.6}$$

$$J^{(j)}_{nm}(S_1S_2S_3S_4) = \int \bar{\varphi}^*_{ngS_1}(y_1 \cdots, y_N)\, \bar{\varphi}^*_{mgS_2}(y'_1 \cdots y'_N) \times$$
$$\times \left\{ V(nm\sigma\sigma') + \sum_{j,k=1}^{N} V(y_j y'_k) + \sum_{j=1}^{N} [V(y'_j \sigma n) + V(y_j \sigma' m)] \right\} \times$$
$$\times\, \bar{\varphi}_{ngS_3}(y'_1 + m - n, \cdots, y'_j + m - n, y_{j+1} \cdots y_N) \times$$
$$\times\, \bar{\varphi}_{mgS_4}(y_1 + n - m, \ldots, y_j + n - m, y'_{j+1} \cdots y'_N)\, \mathrm{d}y_1 \cdots \mathrm{d}y_N\, \mathrm{d}y'_1 \cdots \mathrm{d}y'_N\,. \tag{4.7}$$

In the above expressions $y_j + n - m$ is the symbolic expression for $(r_j - m, \sigma_j)$, while l is the orbital angular momentum of a single electron in an unfilled shell.

The operator H^{a}_{nm} commutes with the square of the operator of the total spin and with the operator of the z-component of the total spin. The eigenvalues of H^{a}_{nm} do not depend on the total spin of the two atoms, whereas the eigenvalues of $H_{ex}(nm)$ do. Therefore, in order to find the energy difference between the eigenvalues of our approximate Hamiltonian $\mathcal{H}^{a}_{nm}$, it suffices to study the operator $H_{ex}(nm)$ alone.

All operators $H^{(j)}_{nm}$ $j = 1, \ldots, N$ have the same structure and each commutes with the square of the operator of the total spin and with the operator of the z-component of the spin. Each eigenvalue $E^{(j)}_{S_{tot}}(nm)$ of the operator $H^{(j)}_{nm}$ $j = 1, \ldots, N$ is $(2S_{tot} + 1)$-fold degenerate. The situation is the same with the eigenvalues $E_{S_{tot}}(nm)$ of the operator $H_{ex}(nm)$, which are

$$E_{S_{tot}}(nm) = \sum_{j=1}^{N} E^{(j)}_{S_{tot}}(nm)\,. \tag{4.8}$$

But before starting to find the eigenvalues of the operators $H^{(j)}_{nm}$ we have to simplify the expressions for matrix elements $A^{(j)}_{nm}(S_1S_2S_3S_4)$.

Because of Hund's rule, in the case of atoms with $N \leqq 2l + 1$ the spin part of the atomic function is symmetric in σ_i, whereas the coordinate part is antisymmetric in q_i. That means that the atomic function $\bar{\varphi}_{ngS_k} \equiv [y_1 \cdots y_N | nl^N \omega L^g \times \times M^g_L S^g M_{S_k}]$ is simply a product of the function $\chi_{S_k}(\sigma_1 \cdots \sigma_N) \equiv [\sigma_1 \cdots \sigma_N | S^g M_{S_k}]$ of spin variables and of the function $u_{ng}(q_1 \cdots q_N) = [q_1 \cdots q_N | nL^g M^g_L]$ of orbital variables. In this case the expression for matrix element $A^{(j)}_{nm}(S_1S_2S_3S_4)$ can be simplified in the following way:

$$A^{(j)}_{nm}(S_1S_2S_3S_4) = Q^{(j)}_{nm} I_j(S_1S_2S_3S_4)\,, \tag{4.9}$$

where

$$Q^{(j)}_{nm} = 2E_g I^{(j)}_{nm} + J^{(j)}_{nm}\,, \tag{4.10}$$

$$I^{(j)}_{nm} = \int u^*_{ng}(q_1 \cdots q_N)\, u^*_{mg}(q'_1 \cdots q'_N)\, u_{ng}(q'_1 + m - n, \cdots, q'_j + m - n, q_{j+1} \cdots q_N) \times$$
$$\times\, u_{mg}(q_1 + n - m, \cdots, q_j + n - m, q'_{j+1} \cdots q'_N)\, \mathrm{d}q_1 \cdots \mathrm{d}q_N\, \mathrm{d}q'_1 \cdots \mathrm{d}q'_N, \tag{4.11}$$

$$J^{(j)}_{nm} = l^2 \int u^*_{ng}(q_1 \cdots q_N)\; u^*_{mg}(q'_1 \cdots q'_N) \left\{ \frac{N^2}{|n - m|} - \sum_{j=1}^{N} \frac{N}{|q_j + n - m|} - \right.$$
$$\left. - \sum_{j=1}^{N} \frac{N}{|q'_j + m - n|} + \sum_{j,k=1}^{N} \frac{1}{|q'_j - q_k + m - n|} \right\} \times$$
$$\times\, u_{ng}(q'_1 + m - n, \cdots, q'_j + m - n,\; q_{j+1} \cdots q_N) \times$$
$$\times\, u_{mg}(q_1 + n - m,\; \ldots, q_j + n - m, q'_{j+1} \cdots q'_N)\, \mathrm{d}q_1 \cdots \mathrm{d}q_N\; \mathrm{d}q'_1 \cdots \mathrm{d}q'_N\,, \tag{4.12}$$

$$I_j(S_1S_2S_3S_4) = \sum_{\substack{\sigma_1\cdots\sigma_N\\ \sigma_1'\cdots\sigma_N'}} \chi^*_{S_1}(\sigma_1\cdots\sigma_N)\,\chi^*_{S_2}(\sigma_1'\cdots\sigma_N')\,\times$$
$$\times\,\chi_{S_3}(\sigma_1'\cdots\sigma_j'\,\sigma_{j+1}\cdots\sigma_N)\,\chi_{S_4}(\sigma_1\cdots\sigma_j\,\sigma_{j+1}'\cdots\sigma_N')\,. \tag{4.13}$$

By means of (4.9) the expression for the exchange interaction operator $H_{\text{ex}}(nm)$ can be simplified as well:

$$H_{\text{ex}}(nm) = \sum_{j=1}^{N} H_{nm}^{(j)} =$$
$$= \sum_{j=1}^{N} (-1)^j \binom{N}{j}^2 Q_{nm}^{(j)} \sum_{\substack{S_1S_2S_3S_4=-S^g\\ S_1+S_2=S_3+S_4}}^{S^g} I_j(S_1S_2S_3S_4)\, a^+_{nS_1}a_{nS_3}a^+_{mS_2}a_{mS_4}\,,$$
$$N \leqq 2l+1\,. \tag{4.14}$$

Each term in (4.14) can be illustrated by the following diagram:

The diagram in Fig. 1 illustrates the transition due to the exchange of j pairs of electrons between atoms n and m whose spin z-components in the initial and final states are S_3 and S_4, and S_1 and S_2, respectively, whereas the z-components of the atomic angular momenta are not changed. To each such transition an energy $(-1)^j\binom{N}{j}^2 I_j(S_1S_2S_3S_4)\,Q_{nm}^{(j)}$ is assigned.

It is not easy to calculate the coordinate part $Q_{nm}^{(j)}$ of the exchange interaction matrix element. Hence it is necessary to take the coefficients $Q_{nm}^{(j)}$ as parameters. On the other hand, using atomic spin functions the spin factors $I_j(S_1S_2S_3S_4)$ can easily be calculated for any combination of the numbers S_1, S_2, S_3, S_4. Since the eigenstates of the exchange interaction operator are also the eigenstates of $\hat{S}^2_{\text{tot}}$ and $\hat{S}^z_{\text{tot}}$, if matrix elements $I_j(S_1S_2S_3S_4)$ are known the problem of calculating the eigenvalues of the operator (4.14) as functions of parameters $Q_{nm}^{(j)}$ is solved in principle.

In the case of atoms with more than half-filled shell ($N > 2l+1$) the exchange interaction operator has a somewhat more complex structure, since the atomic function is then not the simple product of the coordinate and spin parts.

In fact, each operator $H_{nm}^{(j)}$ $j = 1, \ldots, N$ has to be simplified separately. In the present paper we shall simplify only the operator $H_{nm}^{(1)}$. As regards the other operators, $H_{nm}^{(j)}$ $j > 1$, it is more convenient to simplify them in each special case of atomic configuration, as will be done in the papers [11] to be published.

In order to simplify the expression $A_{nm}^{(1)}(S_1S_2S_3S_4)$ it is convenient to write the atomic wave function using the coefficients of fractional parentage (cfp)

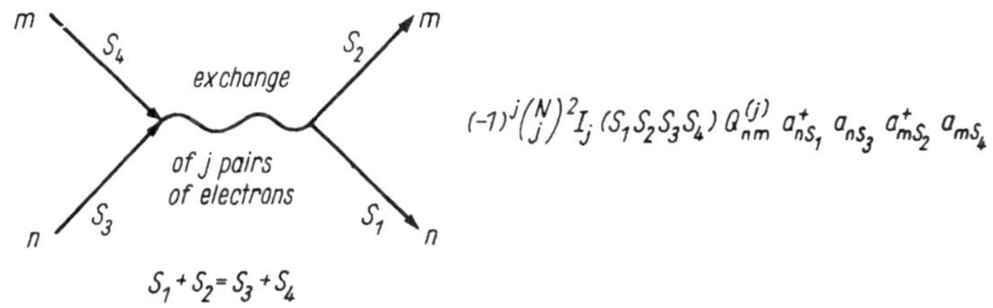

Fig. 1. The diagram for the transition caused by exchange of j pairs of electrons between two atoms at sites n and m

which describe how the state of N equivalent electrons $|l^N\omega LSM_LM_S]$ (where l is the orbital angular momentum of a single electron) may be constructed out of its possible parent states obtained by removing one electron:

$$|l^N\omega LSM_LM_S] = \sum_{\bar\omega\bar L\bar S} |l^{N-1}(\bar\omega\bar L\bar S)\, lLSM_LM_S]\, (l^{N-1}(\bar\omega\bar L\bar S)\, l||l^N\omega LS)\,. \quad (4.15)$$

Here $(l^{N-1}(\bar\omega\bar L\bar S)l||l^N\omega LS)$ is by definition the coefficient of fractional parentage. It represents the amplitude of the probability that the quantum numbers of the parent state are $(\bar\omega\bar L\bar S)$. For our purposes it is more convenient to rewrite (4.15) in the form

$$|l^N\omega L^g S^g M_L^g M_S] = \sum_{\bar S} \sqrt{\sum_{\bar L} (l^{N-1}(\bar L\bar S)l||\,l^N L^g S^g)^2}\, |(\bar S)_{N-1} S^g M_S] \times$$

$$\times \sum_{\bar L} \frac{(l^{N-1}(\bar L\bar S)\, l\,||\, l^N L^g S^g)}{\sqrt{\sum_{\bar L} (l^{N-1}(\bar L\bar S)\, l\,||\, l^N L^g S^g)^2}}\, |l^{N-1}(\bar L)\, lL^g M_L^g] =$$

$$= b^2\,|\,(S^g - \tfrac{1}{2})_{N-1}\, S^g M_S]\, |l^{N-1}(\bar L)\, lL^g M_L^g] +$$

$$+ d^2\,|\,(S^g + \tfrac{1}{2})_{N-1}\, S^g M_S]\,|\, l^{N-1}(\bar L)\, lL^g M_L^g]\,,$$

$$b^2 + d^2 = 1\,. \quad (4.16)$$

The coefficients $b^2 = \sum_{\bar L} (l^{N-1}(\bar L, S^g - \tfrac{1}{2})\, l\,||\, l^N L^g S^g)^2$

and

$$d^2 = \sum_{\bar L} (l^{N-1}(\bar L, S^g + \tfrac{1}{2})\, l\,||\, l^N L^g S^g)^2$$

are equal to the probability that the spin of the parent state equals $S^g - \frac{1}{2}$ and $S^g + \frac{1}{2}$ respectively. By means of the symbols

$$\left.\begin{aligned}
\chi^b_{S_k}\big(\sigma_1(\sigma_2 \cdots \sigma_N)\big) &= [\sigma_1(\sigma_2 \cdots \sigma_N)\,|(S^g - \tfrac{1}{2})_{N-1}\, S^g S_k]\,,\\
\chi^d_{S_k}\big(\sigma_1(\sigma_2 \cdots \sigma_N)\big) &= [\sigma_1(\sigma_2 \cdots \sigma_N)\,|(S^g + \tfrac{1}{2})_{N-1}\, S^g S_k]\,,\\
u^b_{ng}\big(q_1(q_2 \cdots q_N)\big) &= \frac{1}{b^2} \sum_{\bar L} (l^{N-1}(\bar L, S^g - \tfrac{1}{2})\, l\,||\, l^N L^g S^g)\, u^{\bar L}_{ng}\big(q_1(q_2 \cdots q_N)\big)\,,\\
u^d_{ng}\big(q_1(q_2 \cdots q_N)\big) &= \frac{1}{d^2} \sum_{\bar L} (l^{N-1}(\bar L, S^g + \tfrac{1}{2})\, l\,||\, l^N L^g S^g)\, u^{\bar L}_{ng}\big(q_1(q_2 \cdots q_N)\big)
\end{aligned}\right\} \quad (4.17)$$

and the expression (4.16) the coefficient $A^{(1)}_{nm}(S_1S_2S_3S_4)$ takes the form

$$A^{(1)}_{nm}(S_1S_2S_3S_4) = b^4\, {}^{bb}Q^{(1)}_{nm}\, {}^{bb}I_1(S_1S_2S_3S_4) +$$

$$+ 2b^2d^2\, {}^{bd}Q^{(1)}_{nm}\, {}^{bd}I_1(S_1S_2S_3S_4) + d^4\, {}^{dd}Q^{(1)}_{nm}\, {}^{dd}I_1(S_1S_2S_3S_4)\,. \quad (4.18)$$

The expression for the coefficient ${}^{bb}Q^{(1)}_{nm}$ is obtained by inserting in (4.11) and (4.12) $u^b_{ng}(q_1 \cdots q_N)$ instead of $u_{ng}(q_1 \cdots q_N)$. At the same time one obtains ${}^{bb}I_1(S_1S_2S_3S_4)$ by inserting in (4.13) $\chi^b_{S_k}\big(\sigma_1(\sigma_2 \cdots \sigma_N)\big)$ instead of $\chi_{S_k}(\sigma_1 \cdots \sigma_N)$.

Substituting (4.18) into (4.4) we obtain

$$H^{(1)}_{nm} = -N^2 \sum_{\substack{S_1S_2S_3S_4=-S^g \\ S_1+S_2=S_3+S_4}}^{S^g} \{b^4\, {}^{bb}Q^{(1)}_{nm}\, {}^{bb}I_1(S_1S_2S_3S_4) +$$

$$+ d^4\, {}^{dd}Q^{(1)}_{nm}\, {}^{dd}I_1(S_1S_2S_3S_4) + 2b^2d^2\, {}^{bd}Q^{(1)}_{nm}\, {}^{bd}I_1(S_1S_2S_3S_4)\}\, a^+_{nS_1}a_{nS_3}a^+_{mS_2}a_{mS_4}\,,$$
$$N > 2l + 1\,. \tag{4.19}$$

5. A New Approximate Exchange Interaction Hamiltonian for Magnetic Materials

As the exchange interaction of atoms with spin different from zero is especially interesting in magnetic materials, we shall use our two-atom exchange interaction Hamiltonian to derive an improved Hamiltonian for a crystal.

We have first to examine whether the assumptions which we have made in deriving the operator for exchange interaction are appropriate for a crystal.

The crystalline field diminishes the orbital degeneracy of the atomic ground state, so that, when spin–orbit coupling is negligible, the z-component of the atomic ground state orbital angular momentum in the crystal is zero [12]. Therefore, if we take as the atomic ground state wave function $\bar{\varphi}_{ng}(y_1 \cdots y_N)$, the function of the atomic ground state in the crystal, then the assumption that the only degeneracy of the atomic ground state is that associated with various possible orientations of the atomic spin is appropriate. At low temperatures the oscillations of atoms about nodes can be neglected. Furthermore, magnetic materials belong to the group of materials called "high spin complexes" [13], which means that the influence of the crystalline field is not sufficiently strong to destroy the Hund intra-atomic interaction which is responsible for the value of the atomic spin.

However, before applying our exchange interaction operator in a crystal, we have to consider the following problem: can the exchange interaction operator for the crystal be approximated by the sum over two-atom interactions? In the paper [3] it has been shown that in the case of a many-atom system the Hamiltonian of the fourth order in creation and annihilation operators is the first approximation to the exact one in the sense that the matrix element of interaction, when electrons exchange between more than two atoms at the same time, is assumed to be small. Disregarding the fact that we cannot check this assumption we shall write the exchange interaction of the whole crystal as a sum of two-atom Hamiltonians:

$$H_{\text{ex}} = \tfrac{1}{2} \sum_{nm}' H_{\text{ex}}(nm) = \tfrac{1}{2} \sum_{nm}' \sum_{j=1}^{N} H^{(j)}_{nm} = \sum_{j=1}^{N} H^{(j)}\,, \tag{5.1}$$

where

$$H^{(j)} = \frac{(-1)^j}{2} \binom{N}{j}^2 \sum_{nm}' \sum_{S_1S_2S_3S_4=-S^g}^{S^g} A^{(j)}_{nm}(S_1S_2S_3S_4)\, a^+_{nS_1}a_{nS_3}a^+_{mS_2}a_{mS_4}\,. \tag{5.2}$$

The product $a^+_{nS_1}a_{nS_3}$ operates on the spin state of the atom at the node n as in (3.1). Matrix elements $A^{(j)}_{nm}$ are defined in (4.5). $\sum\limits_{nm}'$ denotes the sum over all nodes of the crystal, except when $n = m$. This sum can be approximated by the sum over nearest neighbours.

H_{ex} commutes with the square of the operator of the total spin

$$\hat{S}^2_{\text{tot}} = \Big(\sum_n \hat{S}_n\Big)^2 \tag{5.3}$$

and with the operator of the z-component of the total spin

$$\hat{S}^z_{\text{tot}} = \sum_n \hat{S}^z_n \,. \tag{5.4}$$

As in the case of a two-atom system, the eigenvalues of the Hamiltonian (5.1) can be found using its eigenstates, which are known as they are the eigenstates of $\hat{S}^2_{\text{tot}}$ and $\hat{S}^z_{\text{tot}}$.

6. Comparison of the New Operator with the Heisenberg Model

The Dirac-Heisenberg two-atom exchange interaction operator is obtained by summing the corresponding operators, in the spin representation, of all pairs of electrons on atomic orbitals ψ_n, ω_m [4 to 6]:

$$H_{\text{DH}}(nm) = - \sum_{\psi_n \omega_m} J_{\psi_n\omega_m}(1 + \hat{s}_{\psi_n}\hat{s}_{\omega_m}) \,. \tag{6.1}$$

Here $\hat{s}_{\psi_n}$ and $\hat{s}_{\omega_m}$ are the spin operators of the electrons occupying the single-electron orbitals ψ_n, ω_m. The coefficient $J_{\psi_n\omega_m}$ is defined by

$$J_{\psi_n\omega_m} = e^2 \int \psi_n(q)\, \omega_m(q') \left[\frac{1}{|n-m|} - \frac{1}{|q+n-m|} - \frac{1}{|q'+m-n|} + \right.$$
$$\left. + \frac{1}{|q'-q+m-n|}\right] \psi_n(q')\, \omega_m(q)\, \mathrm{d}q\, \mathrm{d}q' \,. \tag{6.2}$$

Assuming that the coefficients $J_{\psi_n\omega_m}$ have the same value J_{nm} for all pairs of orbitals ψ_n, ω_m, the summation in (6.1) can be carried out and we have, within the two-atom subspace for which each atom is in its ground state,

$$H_{\text{DH}}(nm) = - J_{nm}\big(\text{const}\,(N) + 2\hat{S}_n\hat{S}_m\big) \,, \tag{6.3}$$

where $\hat{S}_n = \sum \hat{s}_{\psi_n}$, $\hat{S}_m = \sum \hat{s}_{\omega_m}$ are the total spin vectors of the atoms at points n and m.

Many-atom cases have been treated with similar assumptions and one gets the so-called Heisenberg Hamiltonian

$$H_{\text{H}} = \tfrac{1}{2} \sum_{nm}{}' H_{\text{DH}}(nm) = \text{const} - \sum_{nm}{}' J_{nm} S_n S_m \,, \tag{6.4}$$

where the summation is over all pairs of atoms n and m with J_{nm} only appreciable of course, when n and m are reasonably close to each other.

Since both of the operators H_{H} and H_{ex} are the sum of corresponding two-atom Hamiltonians, the comparison of the Heisenberg model with the new exchange interaction operator means in fact the comparison of the operators (6.3) and (4.3).

If we analyze the set of assumptions used in deriving the Dirac-Heisenberg model we conclude that in addition to the assumptions cited in Section 4, where the new exchange interaction operator is obtained as an approximate form of the exact one, it contains three more assumptions. These are:

1. The effects associated with the simultaneous exchange of two, three etc. pairs of electrons can be neglected.

2. The interaction of a pair of the considered electrons with other electrons from the unfilled shell which can be exchanged, too, as well as the symmetry properties of the atomic function can be neglected (this is assumed implicitly when $J_{\psi_n\omega_m}$ is written in the form (6.2)).

3. The coefficients $J_{\psi_n \omega_m}$ have the same value J_{nm} for all pairs of orbitals ψ_n, ω_m.

In fact, the above three assumptions are plausible, especially in the case of magnetic materials.

The symmetry properties of the atomic wave function are very important for the determination of the ground state of magnetic atoms. It is unjustified to suppose that these properties play no role in the interaction of two magnetic atoms.

The third assumption is unrealistic as well. Namely, in a number of papers [8 to 10] it has been shown that the coefficients $J_{\psi_n \omega_m}$ have not the same value for all pairs of orbitals ψ_n, ω_m. Their values calculated from (6.2) differ considerably from the measured ones [8 to 10].

But the most interesting question is: are many-electron exchange effects negligible?

Examining the operator (4.3) we conclude that in order to answer the above question it is necessary to know at least the values of the matrix elements $A^{(1)}_{nm}(S_1S_2S_3S_4)$ and $A^{(2)}_{nm}(S_1S_2S_3S_4)$. However, unfortunately it is possible to calculate only the spin part of these matrix elements, whereas it is very difficult to calculate their coordinate parts. One intuitively supposes that the absolute value of the matrix element $A^{(2)}_{nm}(S_1S_2S_3S_4)$ is less than that of the matrix element $A^{(1)}_{nm}(S_1S_2S_3S_4)$. But, on the other hand, if $l \gg 1$ the ratio $\binom{l}{2}^2 \Big/ \binom{l}{1}^2 = (l-1)^2/4$ is a large number, i.e. two-electron exchange gives the major contribution. In fact, in magnetic materials the number of electrons in the unfilled shell is always very large. In the 3d, 4f, (5f, 6d) groups the elements which have small number of electrons are paramagnetic. With increasing number of exchange electrons, in the periodic system of elements first antiferromagnetic elements appear. Ferromagnetics are always situated near the end of the group.

The important consequence of neglect of many-electron exchange effects is the conclusion that in the ground state of two atoms each of which having spin S^g the total spin in the ground state of the two atoms equals zero or $2S^g$. Hence for a magnetic material which has N_0 atoms each of which having spin S^g the only possible value different from zero of the macroscopic spin appears to be equal to $S_{tot} = N_0S^g$.

In real magnetic materials, however, the total macroscopic spin is very often less than N_0S^g.

In all theories which try to explain this as a strange effect it is usually assumed that an ideal ferromagnetic with N_0 atoms should have the value S^gN_0 for the total spin in the ground state, i.e. it is implicitly assumed that many-electron exchange effects are negligibly small. The fact that this spin is less than N_0S^g is attributed to the interaction of d–s or f–s electrons or alternatively it is assumed that d and f electrons are delocalized.

With the new exchange interaction operator which comprises many-electron exchange effects it is possible to study the influence of these effects on the ground state value of the total spin in magnetic materials.

In the works [11] the two-atom operator $H_{ex}(nm)$ is studied for all elements in the 3d group. Since the operator $H_{ex}(nm)$ can be approximated by the operator $H^{(1)}_{ex}$ when many-electron exchanges are neglected, we have first examined whether the operators $H^{(1)}_{nm}$ and H_{DH} are differents representations of the same operator. The eigenstates of the operators $H_{nm}^{(1)}$ and H_{DH} are identical because they are the eigenstates of the operator of the total spin and of its z-component. In the above mentioned papers [11] it has been shown that the eigenvalues of $H_{nm}^{(1)}$ and H_{DH} are the same functions of their coefficients $\bar{Q}_{nm}^{(1)}$ and J_{nm} respectively. (If $N \leqq 2l + 1$, then $\bar{Q}_{nm}^{(1)}$ is equal to the coefficient $Q_{nm}^{(1)}$ defined by (4.10). If $N > 2l + 1$, then $\bar{Q}_{nm}^{(1)}$ is a certain linear combination of coefficients ${}^{bb}Q_{nm}^{(1)}$, ${}^{bd}Q_{nm}^{(1)}$, and ${}^{dd}Q_{nm}^{(1)}$.) That means that the operator $H_{nm}^{(1)}$ is the second quantization representation of the Dirac-Heisenberg two-atom Hamiltonian. Hence in the ground state of the operator $H_{nm}^{(1)}$ the total spin can have only the values 0 and $2S^{\mathrm{g}}$. Furthermore, in the operator $H_{nm}^{(1)}$ the matrix elements of the exchange interaction are defined taking into account the Coulomb interaction between all electrons which take part in the exchange, as well as the symmetry properties of the atomic ground state wave function.

Corrections to the exchange energy due to the operator $H_{nm}^{(2)}$ have also been calculated in the above mentioned papers. It has been concluded that if $Q_{nm}^{(2)}$ is not negligibly small, the ground state two-atom total spin may have one of the values $2S^{\mathrm{g}} - 1, \ldots, 1$. (It is interesting to note that the contribution due to the operator $H_{nm}^{(2)}$ is important, especially if N is large.)

As a matter of fact, in the papers [11] it has been shown that the total spin in the ferromagnetic ground state may have a value $0 < S_{\mathrm{tot}} < N_0 S^{\mathrm{g}}$ if certain relations between matrix elements $\bar{Q}_{nm}^{(1)}$ and $\bar{Q}_{nm}^{(2)}$ are satisfied. These relations have been established for all elements in the 3d group.

Acknowledgement

The author wishes to thank Prof. Z. Marić for many valuable discussions by which he encouraged and supported the present work.

References

[1] M. D. Girardeau, J. Math. Phys. **4**, 1096 (1963).
[2] M. D. Girardeau, J. Math. Phys. **11**, 681 (1970).
[3] M. D. Girardeau, J. Math. Phys. **12**, 1799 (1971).
[4] W. Heisenberg, Z. Phys. **49**, 619 (1928).
[5] P. A. M. Dirac, Principles of Quantum Mechanics, 4th edition, Oxford University Press, London/New York 1958.
[6] P. A. M. Dirac, Proc. Roy. Soc. **A 123**, 714 (1929).
[7] M. Popović-Božić, Fizika, **6**, No. 2 (1974).
[8] A. J. Freeman and R. E. Watson, Phys. Rev. **124**, 1439 (1961).
[9] A. J. Freeman, R. K. Nesbet, and R. E. Watson, Phys. Rev. **125**, 6 (1962); **125**, 1978 (1962).
[10] R. Stuart and W. Marshall, Phys. Rev. **120**, 353 (1970).
[11] M. Popović-Božić, to be published.
[12] R. M. White, Quantum Theory of Magnetism, McGraw Hill Publ. Co., New York 1970.
[13] J. N. Murrell, S. F. A. Kettle, and J. M. Tedder, Valence Theory, London 1965.

(Received May 6, 1974)

phys. stat. sol. (b) 65, 317 (1974)

Subject classification: 18.2; 22

Institute for Low Temperature and Structure Research, Polish Academy of Sciences, Wrocław (a), and Argonne National Laboratory, Argonne, Illinois (b)

Magnetic Susceptibility of UP and UAs

By

R. Troć (a) and D. J. Lam (b)[1])

The magnetic susceptibility of UP and UAs have been measured as a function of temperature from 4.2 to 1000 K and the results are compared with published results in the literature. From the temperature dependence of the magnetic susceptibility, the most likely $5f^3$ electronic configuration of the uranium ion in these two compounds are deduced using a nonperturbative (full J-mixing) crystal-field theory.

Es wurde der Temperaturverlauf der magnetischen Suszeptibilität von UP und UAs im Bereich von 4,2 bis 1000 K gemessen. Die erhaltenen Resultate werden mit den Ergebnissen anderer Autoren verglichen. Die Analyse der Temperaturabhängigkeit der magnetischen Suszeptibilität mit Kristallfeld-Theorie ohne Störung (volle J-Mischung) führt zu der Schlußfolgerung, daß die Elektronenkonfiguration des Uranions in diesen beiden Verbindungen $5f^3$ ist.

1. Introduction

Although the antiferromagnetic transition temperatures of NaCl-type compounds of uranium with group V elements have been established in the early sixties [1 to 3], the complex magnetic properties below the Néel temperature of UP and UAs have not been investigated until recently. UP orders antiferromagnetically at about 125 K with a type 1 magnetic structure and the ordered moment changes abruptly at 23 K with no change of the magnetic structure [4 to 6]. In addition to the change of the ordered moment, a drop in NMR frequency [7], a sharp peak in the specific heat versus temperature curve [8], a change of the thermal expansion coefficient from positive to negative [9], and a steep increase in the electrical resistivity also occur at near 23 K [10]. UAs orders antiferromagnetically at 127 K with type 1 structure. Neutron diffraction results indicated that the magnetic structure changes from type 1 to type 1a at 64 K with a concomitant change of the ordered moment [11, 12].

Several attempts have been made to account for the complex magnetic behavior of UP and UAs below the Néel temperature. Long and Wang [13] invoked an electrostatic multipole interaction model to account for the moment jump observed in UP, whereas Robinson and Erdös [14] proposed that the moment changing observed in both UP and UAs is caused by a partial delocalization of the 5f electrons in the uranium ions.

The difficulty in assigning the valence state to the uranium ions is, in part, due to the lack of high temperature magnetic susceptibility results and the unsatisfactory description of the paramagnetic moments by crystal-field perturbation of the Hund's rule ground state. Recently, Allbutt et al. [15] published susceptibility results for UP in the temperature range from 100 to 1000 K, but

[1]) Work performed under the auspices of the U.S. Atomic Energy Commission.

no high-temperature susceptibility results have been reported for other monopnictides.

In this paper, we present magnetic susceptibility results for UP and UAs measured in the temperature range 4.2 to 1000 K. From the temperature dependence of the magnetic susceptibility, we derived the most likely 5f electronic configuration for the uranium ion in these two compounds using a non-perturbative (full J-mixing) crystal-field theory [16].

2. Experimental Procedure

The uranium monophosphide and monoarsenide were prepared by a method similar to that previously described [1, 3]. The X-ray examination showed the samples to be single phase with the following lattice parameters:

$$(5.589 \pm 0.001)\ \text{Å for UP and } (5.789 \pm 0.001)\ \text{Å for UAs}\,.$$

No contamination from UO_2 has been seen on the X-ray diagrams.

Magnetic measurements by the Faraday method were carried out using an RH Cahn-Electrobalance in the temperature range between 4.2 and 300 K and in applied fields up to 7.2 kOe. A gold–iron versus chromel thermocouple was used for measuring temperature with accuracy ±0.5 K. Mercury cobaltithiocyanate was used as a standard calibrant. The results of measurements were automatically recorded on an X–Y recorder having sensitivity of 10 μg/mm. High-temperature susceptibility measurements were performed on samples enclosed in quartz ampoules using a simple balance and Pt–PtRh thermocouple, as described previously [17]. No field dependence of the susceptibility was observed.

3. Experimental Results

3.1 UP

The magnetic susceptibility χ of UP as a function of temperature from 4.2 to 300 K is shown in Fig. 1. Precipitous drops in χ occur at 23 K and at the Néel temperature T_N. The low temperature anomaly corresponds to the abrupt change in the ordered moment from 1.9 μ_B to 1.72 μ_B observed by neutron diffraction experiments and to the anomalies by different experiments mentioned previously. Our low-temperature susceptibility results agree fairly well

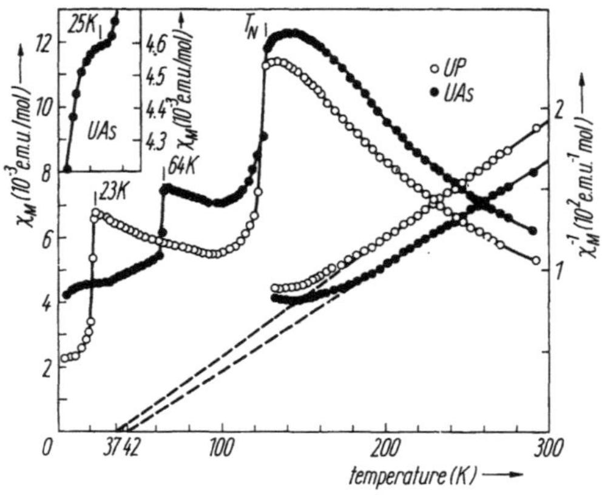

Fig. 1. Susceptibility and reciprocal susceptibility of UP and UAs as a function of temperature. The insert shows on enlarged scale a temperature variation of the susceptibility of UAs in the lowest temperature region

with that obtained by Gardner and Penfold [18] and by Adachi et al. [19] but higher than that reported by Gulick and Moulton [20]. Table 1 summarizes all the published results on UP.

Table 1
Magnetic measurement results for UP

No.	$10^6 \chi_g$ (e.m.u./mol) at temp. (K)				θ	μ_{eff}	T_N	temperature range investigated	Ref.
	4.2	23	T_N	290	(K)	(μ_B)	(K)	(K)	
1	—	—	44.4	20.4	3	3.56	123	80 to 300	[1]
2	8.2	24.8	42.5	19.7	37	3.26	125	4.2 to 1000	this paper
3	—	—	43.0	19.8	36	3.31	130	77 to 300	[2]
4	—	—	41.0	17.5	30	3.11	125	77 to 950	[15]
5	—	—	51.0	20.2	49	3.24	129	77 to 600	[27]
6	6.0	17.5	31.5	15.0	0	3.15	121	1.3 to 300	[20]
7	6.7	22.5	40.5	—	—	—	125	12 to 200	[18]
8	—	—	—	—	−15	3.48	116	77 to 300	[28]
9	8.0*)	22.3	41.5	19.0	6	3.45	121	4.2 to 300	[19]

*) Extrapolated value.

The χ^{-1} vs. T curve (Fig. 1) shows marked deviations from the Curie-Weiss behaviour below about 200 K. The magnetic constants θ and μ_{eff} listed in Table 1 are derived by a usual method, e.g., from the slope of the approximately straight line part of the χ^{-1} vs. T dependence.

Fig. 2 shows the results of high-temperature (up to 1000 K) susceptibility measurements. From fitting the susceptibility data to the equation

$$\chi = \chi_0 + \frac{C}{T - \theta}$$

the temperature-independent term χ_0 has been estimated to be about $40 \times \times 10^{-6}$ e.m.u./mol. This value is in good agreement with that obtained by Allbutt et al. [15]. Although the absolute values of the high-temperature susceptibility obtained in this investigation are higher than that reported by Allbut et al., the temperature dependences are very similar.

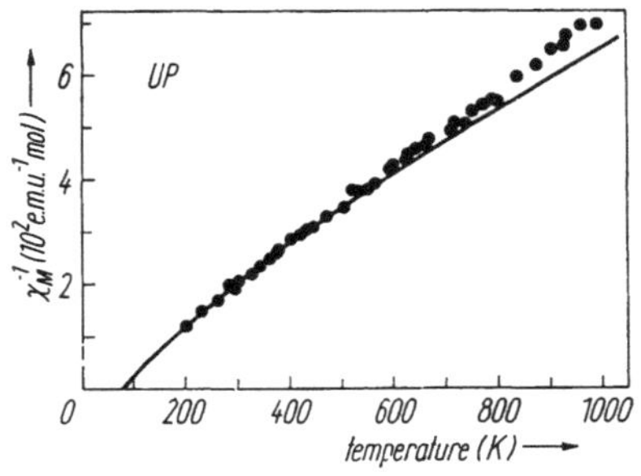

Fig. 2. Experimental (solid circles) and calculated (solid line) inverse molar susceptibility of UP above 200 K

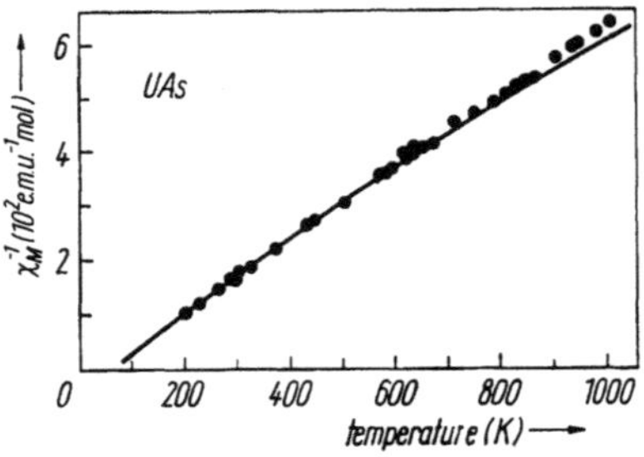

Fig. 3. Experimental (solid circles) and calculated (solid line) inverse molar susceptibility of UAs above 200 K

3.2 *UAs*

The temperature-dependent susceptibility of UAs below 80 K and above 300 K has not been reported previously.

The magnetic susceptibility results obtained in this work on UAs are shown in Fig. 1 and Fig. 3. Precipitous drops in χ occur at 64 K and at the Néel temperature. In addition, a small broad maximum occurs in the χ vs. T curve of UAs at about 25 K, the reason of which cannot be given at present (see insert in Fig. 1). The change in χ at 64 K corresponds to the type 1 to type 1a magnetic transition with the concomitant change of ordered moments from 2.24 μ_B to 1.92 μ_B [11, 12]. Although the anomalous behaviour of the χ vs. T curve of both UP and UAs are very similar, the moment-jump in UP takes place within 0.2 K [7] whereas the ordered moment in UAs changes rather gradually and shows hysteresis with temperature [12]. Recent electrical resistivity results on UAs [10] revealed a "hump-back" peak in the ϱ vs. T curve typical for transitions involving a change in magnetic ordering. The magnetic constants, similarly derived as in the case of UP are: $\theta = 42$ K, $\mu_{eff} = 3.46\ \mu_B$, and $\chi_0 \approx \approx 50 \times 10^{-6}$ emu/mol.

4. Numerical Calculation of the Susceptibility and the Spin Projections of $5f^n$ Configurations in Cubic Crystal Field

To evaluate the temperature-dependent magnetic susceptibility of the uranium ion in UP and UAs in terms of the crystal-field model, we have to know the ionicity of the magnetic ion in the compound, the effective charge of the ligands, the crystal-field parameters, and the molecular-field constants. Since the ionicity of the uranium ion and the effective charge of the ligands are not known in conducting compounds, a large amount of numerical computation is required to obtain a configuration that uniquely yields the experimental results. Inasmuch as the method used to calculate the magnetic susceptibility using accurate low-lying crystal-field states has been described elsewhere [16] only a brief outline will be presented. The free-ion eigenvalues and eigenvectors of the $5f^n$ configuration were calculated with the Coulomb and spin–orbit parameters for uranium ions derived from spectroscopic data. These free-ion eigenvectors are then used as basis functions to construct the crystal-field matrix, thereby obtaining, after diagonalization, the crystal-field states of the uranium ion. The magnetic susceptibility as a function of temperature is expressed as

$$\chi_f = \frac{\mu_B}{Z} \left\{ \frac{1}{kT} \sum_{\Gamma} e^{-E_\Gamma/kT} \sum_{\Gamma'}' |\langle\Gamma| \mu |\Gamma'\rangle|^2 + \right.$$
$$\left. + \sum_{\Gamma} \sum_{\Gamma'}'' |\langle\Gamma| \mu |\Gamma'\rangle|^2 \frac{e^{-E_\Gamma/kT} - e^{E_{\Gamma'}/kT}}{E_{\Gamma'} - E_\Gamma} \right\}, \qquad (1)$$

where μ_B is the Bohr magneton, μ is the magnetic moment operator. $\sum'$ sums over states $|\Gamma'\rangle$ that are identical to or degenerate with $|\Gamma\rangle$, and $\sum''$ sums over states $|\Gamma'\rangle$ that are distinct from and nondegenerate with $|\Gamma\rangle$. Z is the partition function. The exchange effect is included in terms of the molecular-field approximation $\chi_0(T) = \chi_f(1 - \lambda\chi_f)^{-1}$, where λ is the molecular-field constant.

The Knight shift is given in the spin-coupled model by

$$K = K_0 + H_{\text{eff}} \frac{\langle S_z \rangle}{H}, \tag{2}$$

where H_{eff} is an effective field at the nucleus, and K_0 represents the shift (approximately temperature-indepenent) in the absence of a localized magnetic moment. If we define $K_f \equiv K - K_0$, then

$$\frac{\mathrm{d}K_f}{\mathrm{d}\chi_f} = H_{\text{eff}} \frac{\mathrm{d}\langle S_z \rangle / H}{\mathrm{d}\chi_f}. \tag{3}$$

The expression for the spin projection of the localized f^n configuration is [16]

$$\frac{\langle S_z \rangle}{H} = \frac{\mu_B}{Z} \left\{ \frac{1}{kT} \sum_{\Gamma} e^{-E_\Gamma/kT} \sum_{\Gamma'}' \langle \Gamma | \mu_z | \Gamma' \rangle \langle \Gamma' | S_z | \Gamma \rangle + \right.$$
$$\left. + \sum_{\Gamma} \sum_{\Gamma'}'' \langle \Gamma | \mu_z | \Gamma' \rangle \langle \Gamma' | S_z | \Gamma \rangle \frac{e^{-E_\Gamma/kT} - e^{E_{\Gamma'}/kT}}{E_{\Gamma'} - E_\Gamma} \right\}, \tag{4}$$

where $\sum'$ and $\sum''$ have the same meanings as in (1).

We have calculated the temperature-dependent susceptibility and the spin projection of the localized f electrons for $5f^n$ configurations, where $2 \leqq n \leqq 6$, and with crystal-field strength ranges from 100 to 4000 cm^{-1} for the fourth-order crystal-field potential $A_4\langle r^4 \rangle$ and from zero to 1000 cm^{-1} for the sixth-order potential $A_6\langle r^6 \rangle$. Since we do not know the effective charge of the ligands, we must try different signs for the crystal-field parameters.

Reasonable fits between theoretical and experimental data can only be obtained based on the $5f^3$ configuration for the uranium ions in both compounds. Fig. 2 shows the experimental data (solid circles) for UP and the calculated temperature-dependent susceptibility of the $5f^3$ configuration with $A_4\langle r^4 \rangle = 2500\ cm^{-1}$ and $A_6\langle r^6 \rangle = 50\ cm^{-1}$ (solid line). The high-temperature susceptibility results of Allbutt et al. [15] can be fitted with the same crystal-field potentials but a smaller molecular-field constant. Fig. 3 shows the results for UAs. The calculated susceptibility curve (solid line) is based on the $5f^3$ configuration with $A_4\langle r^4 \rangle = 2100\ cm^{-1}$ and $A_6\langle r^6 \rangle = 0\ cm^{-1}$. The positive signs for both the fourth- and the sixth-order potentials are consistent with that obtained for the isostructural monophosphides of neptunium and plutonium [21]. The Coulomb, spin–orbit, and crystal-field parameters for the $5f^3$ configuration used in the present calculation are listed in Table 2. The calculated results of the reciprocal thermal average of the spin projection as a function of temperature for UP and UAs are shown in Fig. 4.

Although the calculated values of both the reciprocal susceptibility, χ^{-1}, and spin projection $H/\langle S_z \rangle$, are not linear with temperature, the plots of $\langle S_z \rangle / H$ versus χ are linear (see Fig. 5).

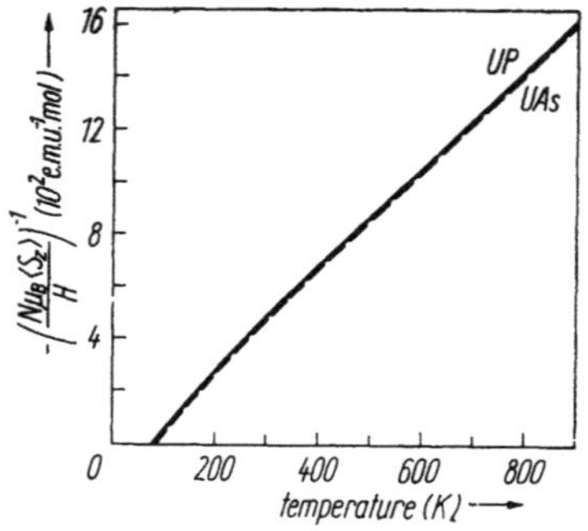

Fig. 4. Calculated temperature dependence of the reciprocal thermal average of the spin projection for the U^{3+} ions in UP and UAs

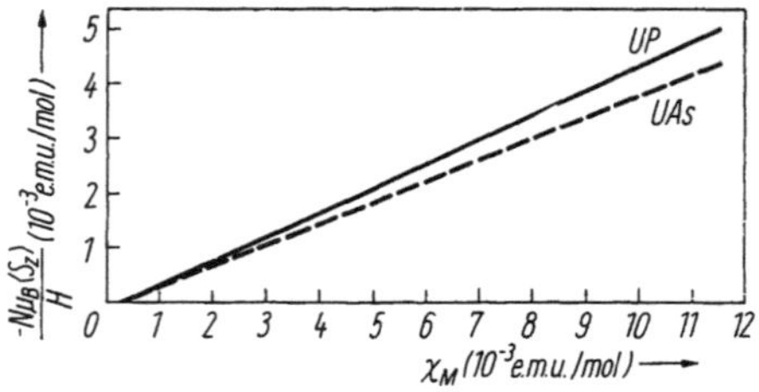

Fig. 5. Calculated thermal average of the spin operator as a function of the magnetic susceptibility for the $5f^3$ configuration of UP and UAs

Table 2

Coulomb, spin–orbit, and crystal-field parameters

compound	N_{5f}	E^1 (cm^{-1})	E^2 (cm^{-1})	E^3 (cm^{-1})	ξ_{5f} (cm^{-1})	$A_4\langle r^4\rangle$ (cm^{-1})	$A_6\langle r^6\rangle$ (cm^{-1})
UP	3	2941.8	14.94	286.7	1666	2500	50
UAs	3	2941.8	14.94	286.7	1666	2100	0

5. Discussion and Conclusion

The measured susceptibility $\chi(T)$ includes, besides $\chi_0(T)$, the diamagnetic contribution χ_D from both the ion core, χ_i, and from the conduction electrons, χ_L. For conducting compounds like UP and UAs, $\chi(T)$ includes also the paramagnetism of the conduction electrons, χ_p. So, the total measured susceptibility is

$$\chi(T) = \chi_0(T) + \chi_D + \chi_p \, .$$

For UP and UAs, assuming the uranium ions to be in the +3 state, the value of χ_i is about -150×10^{-6} emu/mol with small variation depending on the ionicity of the anions. The magnitude of χ_L, approximately -20×10^{-6} emu/mol, can be estimated by assuming that the density of states of the s-electron in UP and UAs is about the same as in NpP and PuP [21]. The total diamagnetism is, therefore, in the order of -170×10^{-6} emu/mol. The magnitude and the temperature dependence of the conduction electron paramagnetism will be difficult to estimate. If we assume that the conduction electrons in these compounds are completely free and that the exchange interaction effects can be neglected, the theoretical χ_p may be calculated at 0 K using the $N(E)$ value derived from the experimental electronic specific heat coefficient, γ. The magnitude of χ_p estimated from the measured γ for UP [8] is approximately 130×10^{-6} emu/mol at 0 K (about the same magnitude as the total diamagnetism).

The Knight shift K of the non-magnetic nucleus in rare-earth and actinide intermetallic compounds has been found to be proportional to the thermal expectation of the spin projection of the localized magnetic moment $\langle S_z\rangle$, and

Table 3
Calculated values for UP and UAs

compound	$E_{\Gamma_8-\Gamma_6}$ (cm^{-1})	μ_s (μ_B)	$\frac{d\langle S_z\rangle/H}{d\chi_f}$
UP	638	1.7	−0.45
UAs	48	1.9	−0.39

the effective indirect coupling between the rare-earth or actinide ion and the nucleus of the nonmagnetic ion is via the conduction electrons [21, 22]. χ_p in UP and UAs should, therefore, have a temperature dependence very similar to that shown in Fig. 4. In any case, we can conclude that the measured temperature-dependent susceptibility of UP and UAs is almost entirely derived from the paramagnetism of the uranium ions. Our results indicated that only the $5f^3$ configuration U^{3+} ion gives a resonable fit to the measured temperature-dependent susceptibility of UP and UAs, within the uncertainty of small corrections which arise from the resulting effect of the χ_D and χ_p terms. However, we cannot ascertain the crystal-field parameters because of these corrections.

The crystal-field level scheme for the uranium ions in UP derived from the high-temperature susceptibility results indicates that the Γ_6 first-excited state and the $\Gamma_8^{(2)}$ second-excited state are 638 and 2500 cm^{-2}, respectively, above the $\Gamma_8^{(1)}$ ground state. For the uranium ions in UAs, the Γ_6 and $\Gamma_8^{(2)}$ are 48 and 1750 cm^{-1} above $\Gamma_8^{(1)}$, respectively. The calculated values for the energy separation ΔE between the ground and the first excited crystal-field states, the saturation moment, and the derivative of the spin polarization with respect to the susceptibility are listed in Table 3. The value of ΔE derived from the high-temperature susceptibility results for UP is much larger than the value obtained by Long and Wang from the moment jump in the magnetic ordered state using an electrostatic multipole interaction model. The +4 valence state assumed by Grunzweig-Genossar et al. [23] for uranium ion in all the uranium monopnictides is inconsistant with the high-temperature susceptibility results. Since no reasonable fit can be obtained with the $5f^2$ configuration regardless of what sign and magnitude of the crystal-field parameters we used. The $5f^2$ stable configuration at $T = 0$ K assumed by Robinson and Erdös for the uranium ion in UP is also incorrect because the ground state of the uranium ion will be a nonmagnetic singlet instead of a magnetic triplet as they indicated. The calculated order moment for the U^{3+} ion decreases with increasing crystal-field interaction due to larger admixture of higher J-manifolds into the ground state wavefunction and is consistent with the experimental results of UP and UAs in their magnetic order state. The reported values of $dK/d\chi$ for UP ranges from 3.9 to 4.7 mol/emu [24 to 26]. An average value of 3.9 mol/emu for $dK/d\chi$ can be obtained if the reported three sets of ^{31}P Knight shift values were replotted with our susceptibility results. Using the average experimental $dK/d\chi$ value for UP and the calculated $(d\langle S_z\rangle/H)/(d\chi_f)$ value listed in Table 3, we obtained a H_{eff} value of −49 kOe.

Acknowledgement

We are grateful to Prof. W. Trzebiatowski for his continuing interest in this work.

References

[1] W. Trzebiatowski and R. Troć, Bull. Acad. Polon. Sci., sér. sci. chim. **11**, 661 (1963).
[2] M. Allbutt, A. R. Junkison, and R. M. Dell, Compounds of Interest in Nuclear Reactor Technology in Nuclear Metallurgy, Vol. 10, A.I.M.E., New York 1964 (p. 65).
[3] W. Trzebiatowski, A. Sępichowska, and A. Zygmunt, Bull. Acad. Polon. Sci., sér. sci. chim. **12**, 687 (1964).
W. Trzebiatowski and A. Zygmunt, Bull. Acad. Polon. Sci., sér. sci. chim. **14**, 495 (1966).
[4] N. A. Curry, Proc. Phys. Soc. **89**, 427 (1966).
[5] L. Heaton, M. H. Mueller, K. D. Anderson, and D. D. Zauberis, J. Phys. Chem. Solids **30**, 453 (1969).
[6] L. Heaton, M. H. Mueller, M. F. Adams, and R. L. Hitterman, J. appl. Cryst. **3**, 289 (1970).
[7] S. L. Carr, C. Long, W. C. Moulton, and M. Kuznietz, Phys. Rev. Letters **23**, 786 (1969).
[8] J. F. Counsell, R. M. Duell, A. R. Junkison, and J. F. Martin, Trans. Faraday. Soc. **63**, 72 (1967).
[9] J. A. C. Marples, J. Phys. Chem. Solids **31**, 2431 (1970).
[10] R. Troć and Z. Kletowski, Bull. Acad. Polon. Sci., in the press.
[11] R. Troć, A. Murasik, A. Zygmunt, and J. Leciejewicz, phys. stat. sol. **23**, K123 (1967).
J. Leciejewicz, A. Murasik, and R. Troć, phys. stat. sol. **30**, 157 (1968).
J. Leciejewicz, A. Murasik, T. Palewski, and R. Troć, phys. stat. sol. **38**, K89 (1970).
[12] G. H. Lander, M. H. Mueller, and J. F. Reddy, Phys. Rev. B **6**, 1880 (1972).
[13] C. Long and Yung-Li Wang, Phys. Rev. B **3**, 1656 (1971).
[14] J. M. Robinson and P. Erdös, AIP Conf. Proc. **10**, 1070 (1973); Phys. Rev. B **8**, 4233 (1973).
[15] M. Allbutt, R. M. Dell, A. R. Junkison, and J. A. Marples, J. inorg. nuclear Chem. **32**, 2159 (1970).
[16] S.-K. Chan and D. J. Lam, in: Plutonium 1970 and Other Actinides, Vol. 1, Ed. W. N. Miner, The Metallurgical Society of A.I.M.E., New York 1970 (p. 219); and in: Electronic Structure and Related Properties of the Actinides, Ed. A. J. Freeman and J. B. Darby, Academic Press, New York, to be published.
[17] W. Trzebiatowski and J. Mulak, Bull. Acad. Polon. Sci., sér. sci. chim. **13**, 759 (1965).
[18] W. E. Gardner and J. Penfold, in: Thermal Analysis, Vol. 2, Ed. H. G. Wiedemann, 1972 (p. 151).
[19] H. Adachi, S. Imoto, and T. Kuki, Phys. Letters A **44**, 491 (1973).
[20] J. M. Gulick and G. Moulton, Phys. Letters A **35**, 429 (1971).
[21] D. J. Lam and F. Y. Fradin, Phys. Rev. B **9**, 15 (1974).
[22] D. D. Jones, Phys. Rev. **180**, 445 (1969).
[23] J. Grunzweig-Genossar, M. Kuznietz, and F. Friedman, Phys. Rev. **173**, 562 (1968).
[24] B. A. Scott, K. A. Gingerich, and R. A. Bernheim, Phys. Rev. **178**, 387 (1967).
[25] M. Kuznietz, G. A. Matzkanin, and Y. Baskin, Phys. Letters A **28**, 122 (1968).
[26] F. Friedman and J. Grunzweig-Genossar, Phys. Rev. B **4**, 180 (1971).
[27] V. I. Chechernikov, T. M. Shavishvili, A. A. Pletyshkin, and V. K. Slovyanskikh, Zh. eksper. teor. Fiz. **55**, 152 (1968).
[28] H. Adachi and S. Imoto, J. nuclear Sci. Tech. **6**, 532 (1970).

(Received June 25, 1974)

phys. stat. sol. (b) 65, 325 (1974)

Subject classification: 10.2; 13.4; 22.5.2

*Department of the Mathematical Institute,
Academy of Sciences of the USSR, Leningrad*

F-Centre in KCl with the Extended Polarizable Ion Lattice Model

By

I. V. Abarenkov and I. M. Antonova

The F-centre in KCl has been calculated within the same approximations as those used in calculations of the bottom of the conduction band in KCl. It is found that if the electronic polarization of the crystal is neglected, the calculated values of the optical excitation and optical ionization energies are 1.5 and 2.5 times the corresponding experimental values. The difference between the calculated and the experimental values is reduced considerably when the polarization potential is incorporated to account for the electronic polarization.

Es werden die F-Zentren in KCl mit den gleichen Näherungen berechnet, wie sie bei den Berechnungen der Leitungsbandkante in KCl benutzt wurden. Für den Fall, daß die elektronische Polarisation des Kristalls vernachlässigt wird, wird gefunden, daß die berechneten Werte der optischen Anregungs- und optischen Ionisierungsenergien 1,5 bis 2,5mal größer als die entsprechenden experimentellen Werte sind. Die Differenz zwischen den berechneten und experimentellen Werten läßt sich beträchtlich verringern, wenn das Polarisationspotential für die elektronische Polarisation eingeführt wird.

1. Introduction

The F-centre in alkali halides has been calculated many times (see [1] for references) since Tibbs' first calculations [2], but still papers concerning the problem continue to appear. One of the possible reasons is the fact that despite the simplicity of the physical model (de Boer model) the F-centre is not at all a simple object. Even the one-electron approximation must be used here with caution and the corrections for the polarization effects must be taken into account in the case of the diffuse excited state in particular.

Different models were used more or less successfully to account for the one-electron part of the electron–ionic core interaction and for polarization effects. In general, these methods are adapted to the case of the F-centre. The method equally suitable for the local and band states has been developed by Tolpygo (see, for example, [3]). Being based on the particular representation of the wave function, this method seems to be sensitive to the choice of the trial function.

The purpose of the present paper is to calculate the F-centre with the same approximation as the position of the bottom of the conduction band has been recently calculated [5] and to compare different effects in the compact, diffuse, and band states.

One essential approximation was used in the present paper which makes the comparison with the experiment less instructive. We assumed that the nuclei of ions are fixed at the corresponding ideal lattice sites. The experimental values relate to the distorted lattice, the distortion being correspond to the ground state of the F-centre (absorption), to the excited state of the F-centre (emission) or to the single anion vacancy (recombination). Therefore the calculated values will coincide with none of the experimental ones. But the distortion of the lattice around the vacancy is known to be small in the ground state of the F-centre. Hence the calculated values will relate to the experimental values taken from the absorption data.

2. Method of Computations

In calculating the electronic structure of the F-centre we used basically the model potential method [5] with an appropriate change in the polarization potential made in accordance with [6]. Below we will write down the final equations only referring the reader to the papers [5, 6] for details.

Let us denote by $\boldsymbol{R}_j$ the radius vector of the ideal lattice site corresponding to the ion j and by e_j, α_j^*, r_{0j} the charge, the effective polarizability, and the radius of the ion j. We assign the index $j = 0$ to the vacancy site and therefore in every sum over j in what follows the term corresponding to $j = 0$ will be excluded. We devided all ions of the crystal into two groups: C and D. The group C comprises ions situated on M successive spherical shells nearest to the vacancy. All other ions constitute the group D. In the present calculation M was taken to be equal to 1 (the first spherical shell), 3 (the first cubic shell) and 7, the latter being used only in the case of the ground state.

The Hamiltonian of the extra electron in ionic crystal can be expressed as a sum of the three terms, of the main part $\hat{H}_0$ and of two corrections $\hat{H}_{\text{ov}}$ and $\hat{H}_{\text{pol}}$. In the case of the F-centre these terms are as follows:

$$\hat{H}_0 = \hat{T} + \sum_j \hat{V}_j^{(\text{M})}(\boldsymbol{r} - \boldsymbol{R}_j, E + u_j^{(p_i)}) . \tag{1}$$

Here T is the kinetic energy operator and $V_j^{(\text{M})}(\boldsymbol{r}, \varepsilon)$ is the operator of the model potential of the ion j [7, 8]. The model potentials of ions constituting the group C were treated exactly. The approximate local model potentials were used for ions of the group D, the local potential being chosen so that to coincide with the s-part of the corresponding non-local model potential.

The correction for the overlapping of the core orbitals is

$$\hat{H}_{\text{ov}} = \sum_{\substack{i,j \\ i \neq j}} \hat{P}_i \left(\hat{T} - \frac{e_i}{|\boldsymbol{r} - \boldsymbol{R}_i|} - \frac{e_j}{|\boldsymbol{r} - \boldsymbol{R}_j|} - E \right) \hat{P}_j . \tag{2}$$

Here $\hat{P}_j$ is the projector onto the subspace of the orbitals of the ion j. The orbitals of the free ion calculated by Clementi [9] were used in the present calculations. The summation in (2) is over the ions of the group C only. The overlapping of the nearest and the next nearest neighbouring ions was taken into account. Only outer s and p orbitals were retained in (2).

The correction for the polarization of the crystal is

$$H_{\text{pol}} = V_{\text{e}}(r) + V_{\text{v}} + V_{\text{ve}}(r) , \tag{3}$$

$$V_{\text{e}}(r) = -\frac{1}{2} \sum_j \frac{\alpha_j^*}{\left((r - R_j)^2 + r_{0j}^2\right) \left((r + R_j)^2 + r_{0j}^2\right)} , \tag{3a}$$

$$V_{\text{v}} = -\frac{1}{2} \sum_j \frac{\alpha_j^*}{R_j^4} , \tag{3b}$$

$$V_{\text{ve}}(r) = \sum_j \frac{\alpha_j^* \xi_j}{2r R_j^4} \left\{1 + \frac{R_j^2 - r^2 - r_{0j}^2}{2r_{0j}\xi_j} \operatorname{arctg} \frac{2r_{0j}\xi_j}{r_{0j}^2 + r^2 + R_j^2 - 2\xi_j^2}\right\}, \tag{3c}$$

$$\xi_j = \begin{cases} r & \text{if} \quad r < R_j \\ R_j & \text{if} \quad r \geqq R_j . \end{cases} \tag{3d}$$

Here V_e and V_v correspond to the polarization of the lattice by the single electron and by the single vacancy. V_{ve} is an additional term due to the simultaneous presence of both the electron and the vacancy.

Three cases were considered:

$$\hat{H}_0 \varphi = E\varphi , \tag{4a}$$

$$(\hat{H}_0 + \hat{H}_{\text{ov}})\, \varphi = E\varphi , \tag{4b}$$

$$(H_0 + \hat{H}_{\text{ov}} + \hat{H}_{\text{pol}})\, \varphi = E\varphi . \tag{4c}$$

To obtain the solution, the integro-differential equation (4) has been converted into the matrix problem by expanding φ in terms of Slater type atomic orbitals (STO)

$$\varphi(\boldsymbol{r}) = \frac{1}{r} \sum_l \left\{\sum_{n=1}^{N_l} c_{nl} \sqrt{\frac{(2\alpha_l)^{2(n+l)+1}}{(2(n+l))!}}\, r^{l+n}\, \mathrm{e}^{-\alpha_l r}\right\} \mathrm{K}_l(\vartheta, \varphi) . \tag{5}$$

Here $\mathrm{K}_l(\vartheta, \varphi)$ is the normalized to 1 cubic harmonic (CH) of an appropriate irreducible representation [10]. One CH K_0 and two CH K_0 and K_4 were used for the ground state of the F-centre and one CH K_1 was used for the first excited state.

The process of obtaining the solutions was as follows. Some value of α_l was chosen. The matrix problem for a given α_l was solved. As $\hat{H}_0$ and $\hat{H}_{\text{ov}}$ depend on the energy E the problem was solved self-consistently, the direct iterations being used. Then another value of α_l was chosen and the process was repeated until the least possible value of the energy E was obtained.

Thus we can calculate the pseudo-one-electron function $\varphi(\boldsymbol{r})$. To obtain the one-electron function $\psi(\boldsymbol{r})$ we must orthogonalize $\psi(\boldsymbol{r})$ to the orbitals of ions. The latter being nonorthogonal, the symmetrized orbitals $\psi_j(\boldsymbol{r})$ were constructed from the orbitals of the equivalent ions of the group C and the equation

$$\psi(\boldsymbol{r}) = N \left\{\varphi(\boldsymbol{r}) - \sum_{i,k} \psi_i(\boldsymbol{r})\, G_{ik}^{-1} S_k\right\} \tag{6}$$

was used for $\psi(\boldsymbol{r})$. Here G is the matrix with elements

$$G_{ik} = \int \psi_i^*(\boldsymbol{r})\, \psi_k(\boldsymbol{r})\, \mathrm{d}r , \tag{7}$$

$$S_k = \int \psi_k^*(\boldsymbol{r})\, \varphi(\boldsymbol{r})\, \mathrm{d}\boldsymbol{r} , \tag{8}$$

and

$$|N|^2 = 1 - \sum_{i,k} S_i^* G_{ik}^{-1} S_k \tag{9}$$

if $\varphi(\boldsymbol{r})$ is normalized to 1. The overlapping between the nearest and the next nearest neighbouring ions was taken into account in (7).

As an additional characteristic the mean value of r

$$\bar{r} = \int \varphi^*(\boldsymbol{r})\, r\varphi(\boldsymbol{r})\, \mathrm{d}\boldsymbol{r} \tag{10}$$

was calculated with the pseudo-one-electron function $\varphi(\boldsymbol{r})$.

Table 1

Ground state, 7 shells

		$N_0 = 1$	$N_0 = 2$	$N_0 = 3$	$N_0 = 4$	$N_0 = 5$	$N_0 = 6$	$N_0 = 6$	$N_4 = 2$
α	(a)	0.413	0.71	0.97	1.20	0.65	0.80	0.7	1.2
	(b)	0.404	0.70	0.96	1.20	0.60	0.80	0.7	1.2
	(c)	0.358	0.65	0.89	0.40	0.60	0.75	0.7	1.2
E	(a)	−0.112	−0.132	−0.136	−0.137	−0.137	−0.138	−0.139	
	(b)	−0.116	−0.134	−0.138	−0.139	−0.140	−0.140	−0.141	
	(c)	−0.138	−0.150	−0.152	−0.154	−0.156	−0.156	−0.157	
$\bar{r}$	(a)	3.6	3.3	3.2	3.2	3.5	3.3	3.3	
	(b)	3.7	3.3	3.3	3.2	3.4	3.3	3.4	
	(c)	4.2	3.6	3.4	4.2	3.9	3.8	3.9	

Table 2

The value $|\psi|^2 \times 10^4$ in at. units, ground state, 7 shells

site	case	$N_0 = 1$	$N_0 = 2$	$N_0 = 3$	$N_0 = 4$	$N_0 = 5$	$N_0 = 6$	$N_0 = 6$ $N_4 = 2$	experiment [11]
(001) K^+	(a)	726	563	518	498	496	501	399	
	(b)	745	595	535	509	537	525	448	993
	(c)	905	719	657	692	665	665	566	
(011) Cl^-	(a)	185	78	51	39	49	50	59	
	(b)	195	86	55	40	58	59	64	158
	(c)	284	124	82	137	125	123	134	
(111) K^+	(a)	23	3	1	0	2	3	5	
	(b)	25	4	1	0	4	5	6	15
	(c)	45	8	3	19	21	21	25	
(002) Cl^-	(a)	16	3	1	1	5	6	4	
	(b)	17	3	1	1	8	8	7	24
	(c)	32	5	2	26	28	28	27	
(012) K^+	(a)	6	2	1	1	5	4	5	
	(b)	7	2	1	1	7	5	7	6
	(c)	12	3	2	18	17	15	18	
(112) Cl^-	(a)	3	0	0	0	3	2	3	
	(b)	3	0	0	0	4	3	4	2
	(c)	6	1	0	16	12	9	13	
(022) Cl^-	(a)	1	0	0	0	2	1	1	
	(b)	1	0	0	0	2	1	2	1
	(c)	2	0	0	11	5	3	5	

3. Numerical Results

The calculations were performed for the KCl crystal. In the present calculations we used the same input data as in [5]: the same type of the model potential, the same values of a_0, α_j^*, r_{0j} ($\gamma = 1$), the same orbitals of ions. A rather flexible function was used to eliminate the effect of the crudeness of the trial function. Therefore all the states considered, both bound and band, correspond to the same model of the crystal. The value 5.89 at. units taken for the a_0 corresponds to 0 °K.

The results of the calculation of the ground state are shown in Tables 1 and 2. First of all the s-type ($l = 0$) function was used and the number of STO was increased until the results were considered to be convergent. The dependence of E on α weakens gradually as the number of STO is increasing. It is worth noting that the second minimum appears on the curve $E(\alpha)$ in the region of small α when $N_0 = 3$ (case c) or $N_0 = 4$ (cases a, b) and this minimum becomes the principal one when N_0 is increasing. This results in a sudden change of α for a smaller value as it can be seen from Table 1.

Next, some combination of the s-type function ($l = 0$) and of the g-type function ($l = 4$) was used as a trial function. The contribution of the g-part, essential in the beginning, reduces considerably with the increased number of STO in the s-part. The results of the calculation with 6 STO in the s-part and 2 STO in the g-part are given in Tables 1 and 2. In this case the energy E is almost insensitive to the variation of α_0 and α_4 in the vicinity of the minimum.

The first excited state was calculated with the use of the p-type ($l = 1$) function only. The results of the calculation are shown in Table 3. It was found

Table 3

Excited state, 3 shells

		$N_1 = 1$	$N_1 = 2$	$N_1 = 3$	$N_1 = 4$
α	(a)	0.52	0.83	1.0	1.2
	(b)	0.50	0.80	1.0	1.2
	(c)	0.10	0.1	0.3	0.4
E	(a)	0.020	0.005	0.003	0.002
	(b)	0.008	−0.005	−0.006	−0.007
	(c)	−0.045	−0.046	−0.047	−0.048
$\bar{r}$	(a)	4.8	4.3	4.2	4.1
	(b)	5.0	4.4	4.3	4.2
	(c)	25.0	20.6	14.5	13.0

that the wave function and the energy of the p-type excited state are rather sensitive to the potential of far ions. In our calculation of the excited state only three spherical shells of ions were treated exactly within the adopted model. Therefore we consider the results of calculation in cases a and b as consistent ($\bar{r} < a_0$) and those in the case c as unreliable ($\bar{r} > a_0 \sqrt{4}$).

In Table 4 the calculated energies of the ground state (E_g), of the excited state (E_e) and of the bottom of the conduction band (E_b) are shown as well as the calculated and the experimental values of the optical transition (ΔE_F) and optical ionisation (ΔE_I) energies of the F-centre. The energy of the bottom of the

Table 4
Energy values (in eV)

	(a)	(b)	(c)	experiment
E_g	−3.7	−3.8	−4.3	
E_e	0.0	−0.2	−1.3	
E_b	2.8	2.1	−2.1	
ΔE_F	3.7	3.6	—	2.3 [12]
ΔE_I	6.5	5.9	2.2	2.7 [13]

conduction band was taken from the paper [5] and the energy of the polarization of the lattice by the vacancy (V_v) was added in the case c. It can be seen from Table 4 that in the case c the calculated energy level of the excited state lies well inside the conduction band. This is another evidence that the results of the calculation in this particular case are unreliable. For the estimation of the experimental value of the optical ionization energy the transition energy in the maximum of the K band was made use of.

4. Discussion

The results of the present calculations enable us to make some conclusions about the models and approximations usually used in the theory of electron excess colour centres.

Several approximations have been used by different authors to calculate the energy of the ground state of the F-centre in KCl.

The point ion lattice approximation (PILA) [14] yielded −6 eV for the E_g. It should be noted that this value was obtained with the use of K_0 CH only. The value of E_g is lowering to −7 eV if K_4 CH is included and approximately to −10 eV if all the CH of the identical irreducible representation are incorporated. The latter value is $\frac{\alpha_{Mad} - 1}{a_0} - 0.5$ in atomic units and it was obtained with the help of the simple LCAO function constructed from 1 s hydrogen orbitals centred on positive charges.

The ion-size effect (ISE) has been taken into account following the pseudo-potential method. Different approximations of the pseudo-potential yielded the values of E_g within 0.6 eV: −3.5 eV [4], −3.7 eV [15] and the present work, −4.0 eV [16, 17], −4.1 eV [18]. The value $E_g = -4$ eV in this approximation can be considered reasonable and hence the order of magnitude of the ISE in the ground state of the F-centre can be estimated as 5 eV.

It was found that the value of E_g was lowered by about 0.5 eV (0.5 eV [4], 0.6 eV the present work, 0.8 eV [17]) due to the lattice polarization effects (LPE). It means that in the ground state of the F-centre the LPE is one order of magnitude smaller than the ISE. Moreover, the whole LPE is of the same order of magnitude as the deviations of the ISE due to different approximations of the pseudo-potential. At present the LPE could be taken into account only making use of some plausible expression with physical but nevertheless adjustable parameters. Therefore it is necessary to employ the precise enough approximation of the pseudo-potential and only then to incorporate the LPE. Otherwise the expression intended to account for the LPE will serve to improve the inaccuracy of the pseudo-potential used.

The details of the pseudo-potential become more important in the case of excited or conduction states of the F electron. In the case of the conduction band in the perfect crystal the PILA yields $E_b = -6$ eV, this estimation being made with the same LCAO model as that in the ground state. The ISE makes the E_b go up. The values -0.5, 0.9, 2.1 eV were obtained for E_b in [18, 16] and in the present work respectively. Consequently the order of magnitude of the ISE is 5 eV as previously, but the deviations due to the approximations of the pseudo-potential are now almost one order of magnitude higher (2 to 3 eV).

The LPE lowers the E_b by 1 to 3 eV. Hence again the LPE and the deviations of ISE are of the same order of magnitude. Therefore the above comment on the use of the LPE seems to be valid not only for the ground but for every state of the F-centre.

In the present paper different types of the model potential, the "step" potential (equation (29) in [7]) and two "exponential" potentials (equations (2) and (3) in [8]), were tried. (The results shown in Tables 1 to 4 correspond to the "exponential" potential (3) in [8].) The values of E_g obtained with the help of a flexible trial function differ by hundredths of eV. This test is an additional argument in favour of the accuracy of the model potential used in the present paper. It should be noted that the difference can be as large as 0.5 eV if a simple trial function is used. This is a consequence resulting from the fact that different types of the model potential correspond to different linear combinations of the valence and the core orbitals, that is to different pseudo-one-electron functions. Therefore a simple trial function suitable for one model potential can fail to be such for another.

The LPE in the present work was taken into account by means of the polarization potential, whose parameters were adjusted in the case of the conduction band electron, where the influence of the LPE was most pronounced. Therefore the error involved can be considered as small.

The results of the present calculations (Table 4) show that with the LPE omitted the calculated values of ΔE_F and especially of ΔE_I do not agree with the experimental values completely. On the other hand the polarization potential reduces the difference between the calculated and the experimental values of ΔE_I considerably. The residual difference is due to the rigid lattice approximation used in the present paper.

The energy test is essential but not sufficient in this complex situation with many effects of different types and comparable order of magnitude. As an additional test the ENDOR data are often used (see [19] for example). Unfortunately the experiment yields the values of the electron spin density and the computation yields the value of the square of the one-electron wave function of the F-electron. These two quantities will coincide only if the exchange polarization effects (EPE) are negligible. In the case of a single K atom the Hartree-Fock calculations give $|\psi|^2 = 0.7$ at. units [20] compared to 1.1 at. units in the experiment. The difference between these values is the measure of the EPE in the free atom. Note that this difference (0.4 at. units) is greater than the value of $|\psi|^2$ at the nucleus nearest to the vacancy potassium ion (0.1 at. units). Of course in the absolute scale the EPE influence in the crystal will be smaller than in the free atom because only a small amount of the F electron density is concentrated in the vicinity of a given ion, but still the disagreement between the calculated and measured values must not be unexpected.

With this in mind we nethertheless calculated the values of $|\psi|^2$ (Table 2) because they provide us with a good check of convergency of computations, a much better check than any of E or $\bar{r}$ values could provide. It occurred that though there is a difference between the calculated and the experimental values the overall behaviour is similar including a small rise at the (002) chlorine ion. Here again one can see that the LPE reduces the difference between the calculated and the experimental values.

The above discussion shows that even in the case of the simplest colour centre rather precise approximations should be employed to eliminate the accidental coincidence of the calculated and the experimental values. It follows from the present calculations that a reasonably accurate description of the electron excess colour centres can be provided by the model potential method with the polarization potential incorporated. To obtain better agreement with the experiment it is necessary to take into account displacements of crystal ions.

References

[1] W. B. FOWLER, in: Physics of Color Centers, Ed. W. B. FOWLER, Academic Press, New York 1968.
[2] S. R. TIBBS, Trans. Faraday Soc. **35**, 1471 (1939).
[3] Z. YA. EVSEEV and K. B. TOLPYGO, Fiz. tverd. Tela **10**, 1193 (1968).
[4] C. H. WOO and S. WANG, Phys. Rev. B **7**, 2810 (1973).
[5] I. V. ABARENKOV, phys. stat. sol. (b) **61**, 757 (1974).
[6] I. V. ABARENKOV and I. M. ANTONOVA, phys. stat. sol. (b) **64**, 747 (1974).
[7] I. V. ABARENKOV and I. M. ANTONOVA, phys. stat. sol. **20**, 643 (1967).
[8] I. M. ANTONOVA, phys. stat. sol. (b) **62**, K21 (1974).
[9] E. CLEMENTI, Suppl. IBM J. Res. Developm. **9**, 2 (1965).
[10] S. L. ALTMAN and A. P. CRACKNELL, Rev. mod. Phys. **37**, 19 (1965).
[11] R. KERSTEN, phys. stat. sol. **29**, 575 (1968).
[12] H. RABIN and C. C. KLICK, Phys. Rev. **117**, 1005 (1960).
[13] F. LÜTY, Z. Phys. **160**, 1 (1960).
[14] B. S. GOURARY and F. J. ADRIAN, Solid State Phys. **10**, 127 (1960).
[15] R. A. EVARESTOV, Optika i Spektroskopiya **28**, 712 (1970).
[16] R. F. WOOD and U. ÖPIK, Phys. Rev. **179**, 783 (1969).
[17] K. THOMMEN, Z. Phys. **186**, 347 (1965).
[18] J. K. KÜBLER, and R. F. FRIAUF, Phys. Rev. **140**, A1742 (1965).
[19] R. F. WOOD, phys. stat. sol. **42**, 849 (1970).
[20] D. A. GOODINGS, Phys. Rev. **123**, 1706 (1961).

(Received April 15, 1974)

phys. stat. sol. (b) 65, 333 (1974)

Subject classification: 13.5.3 and 14.3.4

V. I. Lenin State University, Kishinev (a)
and I. Kreanga State Pedagogical Institute, Kishinev (b)

The Interaction of an Optical Polaron with Zero Vibrations of the Lattice in High Electric Fields

By

G. M. SHMELEV (a) and V. A. SIBIRSKII (b)[1])

The mobility of polarons interacting with zero acoustic vibrations of the lattice in high electric fields is evaluated. The polaron is described by the Feynmann-Kadanoff model. The distribution function of such carriers is obtained as a solution of the Boltzmann equation, which is found to be never normalized. The effect of the escape of polarons in high electric fields leads, in principle, to the appearance of regions with negative differential conductivity in the voltage–current curve.

Вычисляется подвижность поляронов, взаимодействующих с нулевыми акустическими колебаниями решетки в сильном электрическом поле. Полярон описывается по Фейнману-Каданову. Функция распределения таких носителей находится как решение уравнения Больцмана, которое оказывается ненормируемым. Эффект убегания поляронов в сильном поле приводит, в принципе, к появлению на вольтамперной характеристике участков с отрицательной дифференциальной проводимостью.

1. Introduction

The investigation of the behaviour of polarons in high electric field represents a principal interest, because the polaron specificity results in some new qualitative effects. It is shown in paper [1] that interaction of polarons with acoustic phonons, generally speaking, is not a retaining mechanism, as in the case of band electrons. Because of this circumstance the polaron distribution function in electric field $\boldsymbol{E}$ can never be normalized. The presence of a tail in the distribution function is bound up with effect of carriers' escape in the region of high energies, where different kinds of inelastic processes are playing the main role. In such a situation it is possible that a region with negative differential conductivity appears in the current–voltage curve.

From the experimental investigations of nonlinear phenomena in ionic crystals the work of Masumi [2] should be pointed out, in which the data of very careful measurements of mobility in AgCl and AgBr crystals at temperatures $T = 6.5$ to 77 K are presented. These data can be satisfactorily explained by the results of [1], in which the case where equilibrium thermal distribution of phonons is determined by the law of equipartition is considered. This law may not hold at low lattice temperatures, and in such a situation the interaction of the carriers with zero vibrations (0-vibrations) becomes more significant. It will be shown that the region of temperatures for which this kind of interaction is

[1]) Present address: Tolbukhin St. 20, Apt. 61, Kishinev 32, USSR, MSSR.

actual, extends when the polaron effect is taken into account, i.e. when interaction with optical phonons is considered.

Interaction of hot carriers (but not polarons) with zero acoustic vibrations of the lattice was studied by Stratton [3]. In the case when interelectronic collisions do not influence the carrier distribution, the distribution function may be written as follows:

$$f_0(\varepsilon) \sim \exp\left\{-\left(\frac{\varepsilon}{k_0T}\right)^{5/2}\right\}, \tag{1}$$

where ε is the energy of the electron; k_0 is the Boltzmann constant; T is the temperature of the lattice. Using function (1) Stratton has calculated the mobility and current

$$u \sim E^{0.8}\,; \qquad j \sim E^{0.2}\,.$$

A consequent description of a theory of electron interaction with 0-vibrations is presented also by Conwell in [4], where, like in [3], the problem is solved quasi-classicaly, i.e. with the use of the Boltzmann equation and without consideration of the influence of the field on the scattering. It should be noted that a single solution of the transport equation in this case but in a weak electric field ($\boldsymbol{E} \to 0$), is zero, and it is clear from the physical point of view: the electron interacting with 0-vibrations is able only to emit the energy and there is no equilibrium in the electron–phonon system. Thus a field not only warms up carriers, but becomes a necessary condition for establishing the equilibrium. The interaction with 0-vibrations is essential when [4]

$$\frac{\varepsilon}{k_0T} \gg \frac{k_0T}{8m^*v_s^2}, \tag{2}$$

m^* is the effective mass of the carrier; v_s is the velocity of longitudinal sound waves. Since the effective mass of the polaron is greater than of the band electron (for example for KCl crystals $m^* \approx 3.1\,\mu$), from inequality (2) it follows that the region of lattice temperature extends for the polarons interacting with 0-vibrations.

2. Relaxation Time and Polaron Distribution Function

The polaron will be described in the Feynmann-Kadanoff model according to which it can be represented as an electron with mass μ elastically connected with a fictitious particle with mass M [5]. At temperatures $k_0T \ll \hbar\omega_0$ (ω_0 is the limiting frequency of an optical vibration), the polaron is in its ground state and the interaction with 0-vibrations can only change the momentum and the energy of its translational motion. In paper [6] it is shown that the transition probability for the Feynmann polaron to change its translational momentum by $\hbar\boldsymbol{q}$ (without change of its inner state and with consideration of its interaction with acoustic vibrations in the deformation potential approximation) differs from the corresponding probability for the band electron by the factor

$$\exp\left\{-\frac{\hbar q^2 M}{2m^*v\mu}\right\}, \tag{3}$$

where v is the inner frequency of the polaron; $m^* = \mu + M$.

Using the standard method [4] one can obtain the relaxation time:

$$\frac{1}{\tau_a(\varepsilon)} = \frac{\sqrt{2}\,(k_0T)^4}{m^{*5/2} l_{ae} v_s^4 \varepsilon^{3/2}} \int\limits_0^{\chi_0} \exp\left[-\lambda^2 \left(\frac{2k_0T}{\hbar v_s}\right)^2 \chi^2\right] \chi^4 \operatorname{cth} \chi \, d\chi \,; \quad \left(\lambda^2 = \frac{\hbar M}{2m^* \mu v}\right), \tag{4}$$

where l_{ae} is the mean free path of band electrons interacting with acoustic phonons for the case of equipartition and is determined by

$$l_{ae} = \frac{\pi \hbar^4 \varrho v_s^2}{\mathscr{E}_1^2 m^{*2} k_0 T} \tag{5}$$

($\mathscr{E}_1$ is the deformation potential constant; ϱ is the density of the crystal). In the discussed case, when $\chi_0 = (2m^* v_s^2 \varepsilon)^{1/2}/k_0T \gg 1$, the oscillations of the lattice have only the energy of 0-vibrations $\hbar v_s q/2$. At small values of χ the function $F(\chi)$ under the integral is proportional to χ^3, and when $\chi \to \infty$ it tends to zero, attaining the maximum at $\chi_{max} = \frac{1}{\sqrt{2}} \frac{\hbar v_s}{\lambda k_0 T}$. For real parameters of crystals similar, for example, to alkali-halides ($\mu = 0.65\, m_e$; $m^* = 2m_e$; $v = 4\omega_0$; $\omega_0 \approx 10^{14}$ to 10^{13} s^{-1}; $v_s = 4.5 \times 10^5$ cm/s) and at temperatures $T < 20$ K, $\chi_{max} \gg 1$. Thus, if we want to take into account the interaction with 0-vibrations, we must, in that situation, substitute in formula (4) cth $\chi \approx 1$, and a simple integration gives the relaxation time for the scattering of polarons by 0-vibrations:

$$\tau_{a0}(\varepsilon) = \frac{2^6 k_0 T \lambda^5 m^{*5/2} l_{ae} \varepsilon^{3/2}}{\sqrt{2}\, \hbar^5 v_s \gamma\left(\frac{5}{2},\ \lambda^2 \frac{8m^*\varepsilon}{\hbar^2}\right)}, \tag{6}$$

where $\gamma(\nu, x)$ is the incomplete gamma function. In the absence of the polaron effect, when $M \to 0$ ($\lambda \to 0$), (6) transforms into the corresponding formula for the relaxation time of band electrons (but with mass m^*) interacting with 0-vibrations [3]

$$\tau_{a0}^{(0)} = \frac{5k_0 T l_{ae}}{8\varepsilon v_s}.$$

In the discussed case the phonon population is determined by

$$\bar{\nu}_q \sim \exp\left\{-\frac{2(\bar{\varepsilon} m^* v_s^2)^{1/2}}{k_0 T}\right\} \ll 1 \tag{7}$$

($\bar{\varepsilon}$ is the average energy of carriers) and putting $\bar{\nu}_q = 0$ in the collision integral [1] and taking into account the expression for the transition probability [6], one obtains

$$\left(\frac{\partial f_0(\varepsilon)}{\partial t}\right)_{col} = \frac{8m^* v_s^2 k_0 T}{l_{ae}(2m^*\varepsilon)^{1/2} \gamma_0^2} \frac{d}{d\varepsilon} \left\{ S\left(\gamma_0 \frac{\varepsilon}{k_0 T}\right) f_0(\varepsilon) + \right.$$
$$\left. + \left(\frac{2m^* v_s^2 k_0 T}{\gamma_0}\right)^{1/2} \gamma\left(\frac{5}{2};\ \gamma_0 \frac{\varepsilon}{k_0 T}\right) \frac{d f_0(\varepsilon)}{d\varepsilon} \right\}. \tag{8}$$

Here $S(x) = \gamma(2, x)$, and

$$\gamma_0 = \frac{4k_0 T}{\hbar v} \frac{M}{\mu}. \tag{9}$$

Applying further the method which was used in the case of equipartition one obtains the transport equation

$$\frac{\mathrm{d}}{\mathrm{d}\varepsilon}\left\{\frac{8m^*v_s^2k_0TS\left(\gamma_0\frac{\varepsilon}{k_0T}\right)}{\gamma_0^2\varepsilon^{1/2}}f_0(\varepsilon)+\left[\frac{8\sqrt{2}\,v_s^3(k_0T)^{3/2}\,m^{*3/2}}{\gamma_0^{5/2}}\frac{\gamma\left(\frac{5}{2},\gamma_0\frac{\varepsilon}{k_0T}\right)}{\varepsilon^{1/2}}+\right.\right.$$
$$\left.\left.+\frac{\sqrt{2}\,(\gamma_0\,\varepsilon)^{5/2}\,E^2(m^*v_s^2)^{1/2}}{(k_0T)^{1/2}\,E_c^2\gamma\left(\frac{5}{2},\,\gamma_0\frac{\varepsilon}{k_0T}\right)}\right]\frac{\mathrm{d}f_0(\varepsilon)}{\mathrm{d}\varepsilon}\right\}=0\,. \tag{10}$$

The characteristic field E_c is defined by

$$E_c=\frac{(6m^*v_s^2k_0T)^{1/2}}{e\,l_{ae}}, \tag{11}$$

e being the charge of the electron. Equation (10) has the following form:

$$\frac{1}{g(\varepsilon)}\frac{\mathrm{d}}{\mathrm{d}\varepsilon}[g(\varepsilon)\,J(\varepsilon)]=0\,, \tag{12}$$

where $g(\varepsilon)$ is the density of states. Equation (10) having been written in form (12) represents the equation of continuity

$$\mathrm{div}_\varepsilon\,\boldsymbol{J}=0$$

and $\boldsymbol{J}$ is the flow of polarons in ε-space. The flow through the sphere of radius p equal to $8\pi m^*\varepsilon J$ is directed towards decreasing energy (at $E=0$), so that when $\varepsilon=0$ it must be equal to zero. Taking into consideration this requirement one obtains a solution of equation (10), which has no singularities at zero:

$$f_0(x)=\exp\left\{-\,8m^*\left(\frac{v_sk_0T}{\gamma_0}\right)^2\int^x\left[\frac{8\sqrt{2}\,m^{*3/2}v_s^3\,(k_0T)^{3/2}}{\gamma_0^{5/2}}\gamma\left(\frac{5}{2},\,\gamma_0x\right)+\right.\right.$$
$$\left.\left.+\,\eta\frac{\sqrt{2}\,\gamma_0^{5/2}x^3\,(k_0T)^{5/2}\,(m^*v_s^2)^{1/2}}{\gamma(\frac{5}{2},\,\gamma_0x)}\right]^{-1}S(\gamma_0x)\,\mathrm{d}x\right\};\qquad\left(x=\frac{\varepsilon}{k_0T}\,;\;\;\eta=\left(\frac{E}{E_c}\right)^2\neq0\right). \tag{13}$$

3. Escape Effect. Mobility of Polarons

The supposition that $\gamma_0=0$ in (13) (extremely weak polaron coupling) gives for the distribution function the following expression:

$$f_0(x)=A_1\exp\left\{-\frac{8\sqrt{2}}{25}\left(\frac{m^*v_s^2}{k_0T}\right)^{1/2}\frac{1}{\eta}\,x^{5/2}\right\} \tag{14}$$

which coincides with (1). If $\gamma_0\neq0$ function (13) differs sufficiently from zero; namely, when $x\to\infty$ it does not tend to zero (like a Stratton function) but to some constant value, which is the greater and can the more rapidly be achieved the greater the magnitudes of γ_0 and η are. Indeed, at $x\gg x_0=\gamma_0^{-1}$ we have

$$f_0(x)=A_2\exp\left\{\frac{3\sqrt{\pi}}{2\eta\gamma_0^{9/2}}\left(\frac{2m^*v_s^2}{k_0T}\right)^{1/2}\frac{1}{x^2}\right\} \tag{15}$$

and if $\eta\neq0$, with increasing x, $f_0(x)$ tends to a nonzero limiting value. Such behaviour of the distribution function is the display of the so-called escape of carriers [7]. $x_0 = \gamma_0^{-1}$ is the characteristic relative energy of carriers. The mean free path of polarons interacting with 0-vibrations $l_{a0}(\gamma_0, x)$ defined by

$$l_{a0}(\gamma_0, x) = \frac{l_{ae}(2k_0T)^{1/2}\,\gamma_0^{5/2}x^2}{4m^{*1/2}\,v_s\gamma(\frac{5}{2}, \gamma_0 x)} \tag{16}$$

is proportional to $x^{-1/2}$, when $x \ll x_0$ and at $x \gg x_0$, $l_{a0}(\gamma_0, x) \sim x^2$. Thus a rapid growth (more rapid than x) of the function $l_{a0}(\gamma_0, x)$ when $x \to \infty$ results in the appearance of a tail in the distribution function.

The emission of optical phonons is, apparently, the energy limiting mechanism in most of the ionic crystals. This circumstance may be taking into account by putting the upper limit in all integrals at $x > z = \dfrac{\hbar\omega_0}{k_0T} \gg 1$ equal to $\dfrac{\hbar\omega_0}{k_0T}$. Practically there is no escape effect if $x_0 > z$. For the existence of that effect it is necessary to have $z > x_0$, i. e.

$$\frac{v}{4\omega_0}\frac{\mu}{M} < 1\,. \tag{17}$$

Inequality (17) does not hold for crystals with weak polaron coupling ($\alpha \lessapprox 1$, α being the polaron coupling constant), when $v \approx 3\omega_0$, $M \approx \alpha/6$, but when $\alpha \gtrapprox 3$ the requirement (17) is already satisfied. In crystals with $\alpha \gg 1$ $v\mu/4\omega_0M \approx 1.8/\alpha^2 \ll 1$ and $z - x_0 \approx (\hbar\omega_0)/(k_0T) \gg 1$. This effect can be more readily observed when x_0 is small, the difference $z - x_0$ is large and the electric field is high enough. The field, however, must not be too high, so that the interaction with acoustic phonons would predominate over the interaction with optical phonons, a requirement that is expressed by the inequality

$$\hbar\omega_0 \gg eEl_{a0}\,(\gamma_0, z)\,. \tag{18}$$

When $\gamma_0 z \ll 1$ requirement (18) is satisfied, for example, at $\hbar\omega_0 = 0.02$ eV, $T \lesssim 10$ °K and at $l_{a0}(\gamma_0, z) \approx 3\times10^{-5}$ cm in fields $E \ll 1000$ V/cm. In the contrary case, when $\gamma_0 z > 1$ (f.e. $\gamma_0 z \approx 5$), $l_{a0}(\gamma_0, z) \approx 10^{-4}$ cm, and, for example, for crystals with parameters similar to alkali-halides ($v \approx 4\omega_0$; $m^* \approx 4\mu$) at $T \lesssim 10$ K, $\hbar\omega_0 = 0.03$ eV and in fields $E \ll 1000$ V/cm the inequality (18) is realized well.

When $x \ll x_0$

$$f_0(x) = A_3 \exp\left\{-\frac{8\sqrt{2}}{25}\left(\frac{m^*v_s^2}{k_0T}\right)^{1/2}\frac{1}{\eta}\left(x^{5/2} + \frac{25}{49}\gamma_0 x^{7/2}\right)\right\}. \tag{19}$$

The polaron mobility can be found from

$$u = \frac{2}{3E}\,\frac{\int\limits_0^\infty f_1(x)\,x^{3/2}\,dx}{\int\limits_0^\infty f_0(x)\,x^{1/2}\,dx}, \tag{20}$$

where

$$f_1(x) = -\frac{eE\tau_{a0}\,(\gamma_0, x)}{m^*}\,\frac{df_0(x)}{dx}. \tag{21}$$

In the absence of the escape effect ($x > z \gg 1$) the upper limit in integrals (20) may be supposed to be infinity, because of the rapid decrease of the function

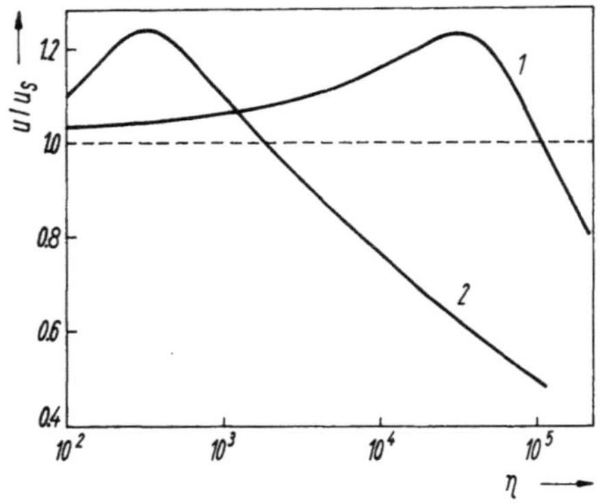

Fig. 1. The mobility of the polaron interacting with zero acoustic vibrations as a function of the electric field η. u_s is the mobility of band electrons obtained by Stratton in [3]. ($\mu = 0.65 m_e = 0.65 \times 9.1 \times \times 10^{-28}$ g; $m^* \approx 2.1 m_e$; $v_s = 4.5 \times 10^5$ cm/s) (1) $T = 1$ K, (2) $T = 2.5$ K

(19) at greater x. The mobility is given by

$$u = u_s \left(1 - 0.86\, \gamma_0 \left[\left(\frac{m^* v_s^2}{k_0 T}\right)^{1/2} \frac{1}{\eta}\right]^{-2/5}\right), \tag{22}$$

where u_s is the mobility calculated by Stratton with the distribution function (14)

$$u_s = \frac{5}{24} \frac{e l_{ae}}{m^* v_s} \left(\frac{8\sqrt{2}}{25}\right)^{2/5} \left(\frac{m^* v_s^2}{k_0 T}\right)^{1/5} \frac{\Gamma\left(\frac{1}{5}\right)}{\Gamma\left(\frac{3}{5}\right)} \left(\frac{1}{\eta}\right)^{2/5}. \tag{23}$$

At $\gamma_0 = 0$ (22) and (23) coincide. In the case of high electric field and strong polaron coupling, when the escape effect is possible ($x_0 < z$), the integration interval should be divided into two: $0 \leqq x \leqq x_0$ and $x_0 \leqq x \leqq z$. In the first interval $f_0(x)$ is determined by (19) and in the second one by (15). The mobility obtained with the use of (15), (19), (20) represents a very cumbrous expression and we shall discuss only some results which follow from it. The polaron mobility in the escape regime is

$$u = \sqrt{2}\, e\, l_{ae}\, [\hbar \omega_0 m^* \gamma_0^4 \eta^2]^{-1/2}. \tag{24}$$

The irregular dependence $u(E) \sim E^{-2}$ results in the appearance, on the voltage–current curve, of a region which corresponds to negative differential conductivity ($j \sim E^{-1}$).

For crystals with parameters similar to alkali-halides the mobility as a function of η for two lattice temperatures is presented in Fig. 1. It should be mentioned that with rising temperature the maximum u/u_s displaces towards weaker fields.

References

[1] G. M. SHMELEV, Fiz. tverd. Tela **13**, 3664 (1971).
[2] T. MASUMI, Phys. Rev. **159**, 761 (1967).
[3] R. STRATTON, Proc. Roy. Soc. **A242**, 355 (1957).
[4] E. M. CONWELL, High Field Transport in Semiconductors, Academic Press, New York/London 1967.
[5] L. KADANOFF, Phys. Rev. **130**, 1364 (1963).
[6] G. M. SHMELEV, Fiz. tverd. Tela **9**, 488 (1967).
[7] I. V. LEVINSON, Fiz. tverd. Tela **6**, 2113 (1964).

(Received May 3, 1974)

phys. stat. sol. (b) 65, 339 (1974)

Subject classification: 13.4; 13.1; 22.1.3

Department of Physics, State University of New York, Albany (a) and Physikalisches Institut der Universität Würzburg (b)

Semi-Empirical LCAO Calculations for Tellurium Crystals with Vacancies [1)] [2)]

By

C. Weigel (a), G. Landwehr (b), and J. W. Corbett (a)

The results of semi-empirical LCAO calculations based on the extended Hückel theory are presented for trigonal tellurium model crystals with vacancies. It is shown that an empty gap-level close to the valence band is associated with a vacancy. The migration of the vacancy takes place preferentially in the *c*-direction. The double vacancy is less mobile than the single vacancy, for which the migration energy is of the order of several hundred meV. The results are compared with experimental findings.

Ergebnisse von halbempirischen LCAO-Rechnungen auf der Grundlage der erweiterten Hückel-Theorie für trigonale Tellurmodellkristalle mit Leerstellen werden beschrieben. Es wird gezeigt, daß von einer Leerstelle ein unbesetztes Niveau in der Energielücke nahe am Valenzband erzeugt wird. Die Leerstelle wandert vorzugsweise in *c*-Richtung. Die Doppelleerstelle ist weniger beweglich als die einfache Leerstelle, für die die Wanderungsenergie einige hundert meV beträgt. Die Ergebnisse werden mit experimentellen Befunden verglichen.

1. Introduction

In an earlier paper [1], we reported that reasonable band structures for tellurium crystals could be obtained by semi-empirical LCAO (Linear Combination of Atomic Orbitals) calculations based on the Extended Hückel Theory (EHT) [2]. In particular, it was shown that a set of EHT parameters could reproduce many features of the band structure of trigonal Te as established by experiment and other more accurate theoretical approaches. Also, the elastic constants of trigonal Te were calculated in reasonable agreement with experiment. The parameters yielded, however, a smallest energy gap that was three times as large as the experimental one. But the predicted relative variations of this gap with the lattice parameters were in good agreement with cyclotron resonance measurements under uniaxial stress [3].

All other EHT band structures were calculated for group IV elements [4, 5]. For diamond crystals, bulk calculations have been carried out, which showed EHT to be very successful in the treatment of defects such as the vacancy [6], substitutional impurities [6], and self-interstitials [7]. Many of these defects could be treated by EHT for the first time, and for other defects better understanding was obtained.

---

1) Work supported in part by the office of Naval Research under Contract No. N 00014-70-C-0296.

2) This work is part of the doctoral thesis of C. Weigel which was submitted to the Universität Würzburg.

For trigonal Te there are a number of defect problems that await theoretical treatment, among them are the vacancy and the p-type character of all Te samples at low temperatures. We therefore applied the EHT formalism to Te model crystals with lattice defects. In this paper we report the results of calculations for the simplest lattice defect type, the vacancy. The results for interstitials and Frenkel pairs will be reported elsewhere.

Since EHT with its implications, successes, and limitations has been discussed in detail (e.g. [6, 7] and references therein), we give only a description of the basic concepts of EHT in Section 2. In Section 3, the model crystals are explained. In Section 4, we consider crystals without defects, and in Section 5 the results for vacancy calculations are presented. In these two sections, comparison is made to pertinent experiments, too.

2. Basic EHT Equations

EHT is an LCAO scheme, i.e. the molecular orbitals (MO), φ_i, are given by

$$\varphi_i = \sum_{\nu}^{n} c_{\nu i} \chi_\nu \,, \tag{1}$$

where the χ_ν's are the atomic orbitals (AO), and n is the number of AO's. The MO energies, E_i and the MO coefficients, $c_{\nu i}$, are the solutions of the secular equations

$$\sum^{n} (H_{\mu\nu} - E_i S_{\mu\nu})\, c_{\nu i} = 0 \qquad (\mu = 1, 2, \ldots, n)\,. \tag{2}$$

Thus $S_{\mu\nu}$'s are overlap integrals between AO's. The $H_{\mu\nu}$'s are matrix elements of an effective one-electron Hamiltonian with respect to the AO's, and are given by

$$H_{\mu\nu} = K(H_{\mu\mu} + H_{\nu\nu})\, S_{\mu\nu}/2 \qquad (\mu \neq \nu) \tag{3}$$

and

$$H_{\mu\mu} = -I_\mu \,, \tag{4}$$

where I_μ is the empirical ionization energy of an AO, μ, and K is an empirical constant between 1 and 2.

We use as AO's one 5s and three 5p Slater type orbitals; their orbital exponents are taken to be the same as those for which a reasonable Te band structure was obtained (2.75 for 5s; 2.35 for 5p) [1]. We employ also the other parameters from [1]: $K = 1.75$ for s–s and p–p interactions, $K = 1.4$ for s–p interactions, $I(5s) = 19.11$ eV, $I(5p) = 9.54$ eV.

The total energy of a system is measured by a quantity

$$E_{\text{EHT}} = \sum_{i}^{n} n_i E_i \,, \tag{5}$$

n_i being the occupation number of the MO, i; n_i can be 0, 1, or 2, since our MO's are spin-degenerate (the omission of spin effects was discussed in [1]). The n_i's are chosen such that all states in the valence bands are filled for the neutral perfect crystal [8]. We consider only neutral systems, since EHT works best for them [9, 10].

3. Model Crystals

The lattice of trigonal Te consists of parallel chains with three-fold screw axes. In the infinite lattice, each chain is hexagonally surrounded by other chains. The lattice basis vectors are

$$\boldsymbol{t}_1 = (a, 0, 0)\,, \qquad \boldsymbol{t}_2 = (-a/2, \sqrt{3}a/2, 0)\,, \qquad \boldsymbol{t}_3 = (0, 0, c)\,. \tag{6}$$

The $\boldsymbol{t}_3$ axis is also called c-axis. Considering right-handed screws, a unit cell contains atoms at

$$\boldsymbol{r}_1 = (ua, 0, 0)\,, \quad \boldsymbol{r}_2 = (-ua/2, \sqrt{3}ua/2, c/3)\,, \quad \boldsymbol{r}_3 = (-ua/2, -\sqrt{3}ua/2, 2c/3)\,. \tag{7}$$

The values of the lattice parameters are $a = 4.47$ Å, $c = 5.91$ Å, and $u = 0.269$ [11].

The space group of this lattice is D_3^4 (or $P3_121$). Its symmetry elements are the identity, translations by any linear combination of the basis vectors, three two-fold rotations perpendicular to the c-axis, and two three-fold rotations around the c-axis followed by non-primitive translations $\pm\boldsymbol{t}_3/3$.

For the simulation of defects, finite model crystals should be large enough to reproduce the essential features (notably the density of states) of the infinite crystal and keep the defects away from the crystal surfaces; on the other hand, the model crystals should not be too large so that they are compatible with the available computer memory and allow for practical computation times.

Our model crystals are chosen to have 7 or 10 chains, as displayed in Fig. 1, so that a defect can be assumed to be within one chain or between two chains and still be surrounded by other chains.

Fig. 1 shows the labels we will use to describe the individual chains, which are represented by hexagons. The labeling of specific atoms is shown in Fig. 2a and b. The radius of a sphere representing an atom is Slater's atomic radius for Te [12], which is also about the mean radius of self-consistent-field 5p orbitals for Te [13]. Here the spheres are drawn to give a better understanding of how the Te atoms fill the space in the lattice.

In our nomenclature, McNa is a model crystal that has M chains consisting of N atoms each. An atom, m/n, is the n-th atom in the chain with label m.

For the finite model crystals, the number of symmetry elements is less than for the infinite lattice. There are no translations and no three-fold rotations. The crystals with 7 chains have only one two-fold rotation and the identity; therefore their symmetry group is C_2 [14]. The crystals with 10 chains have

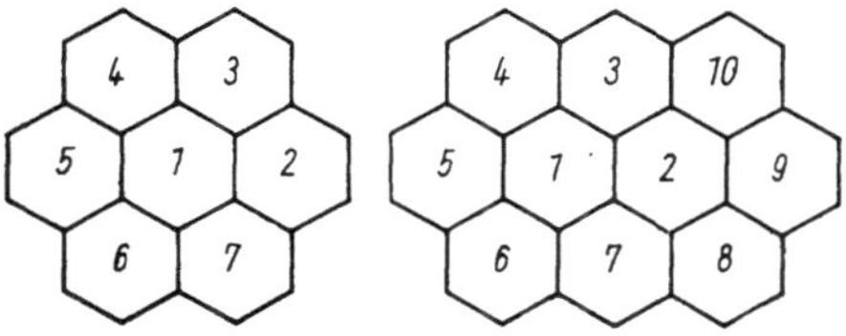

Fig. 1. Model crystals with 7 and 10 chains as seen in the z-direction

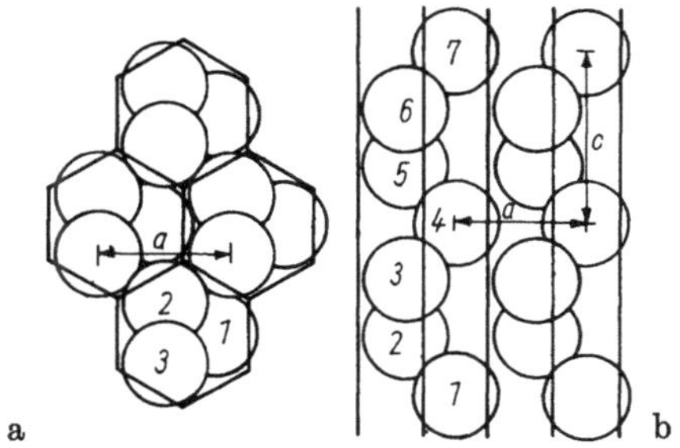

Fig. 2. Positions and labels of atoms in a chain as seen a) in the z-direction and b) in the y-direction

only the identity and therefore C_1 symmetry [14]. Both C_1 and C_2 have only one-dimensional representations. Hence all MO levels encountered in the following results will be non-degenerate (apart from the spin-degeneracy).

4. Model Crystals without Defects

4.1 Energy levels

Table 1
The energy gaps (eV) for various model crystals; "infinite" stands for the band structure gap [1]

crystal	7c4a	7c5a	7c6a	7c7a	10c4a	10c5a	infinite
gap	1.20	1.16	1.07	1.13	1.18	1.11	1.04

Table 1 lists the energy gaps of various model crystals and compares them to the band gap obtained in band structure calculations [1]. As pointed out, the EHT parameters employed in these calculations yield a gap about three times as large as the experimental value. But we can also see that this gap value, however its accuracy, is better approximated when more chains are added to a crystal. It also looks as if the gaps of crystals with even and the gaps of those with odd numbers of atoms per chain form two monotonically decreasing series converging at the band structure gap. Similar observations concerning the decrease of the gap with crystal size were made for EHT calculations on diamond [6].

We do not list the other MO energies, E_i, for the different crystals, but present them in form of energy histograms; we indicate how many levels fall into energy intervals of 0.2 eV width. In Fig. 3 we show the results for the two largest model crystals considered, 7c7a and 10c5a. The histograms are compared to the density of states (DOS) obtained from EHT band structure calculations [1] for evenly distributed points in the Brillouin zone; they are also compared to the DOS based on a pseudopotential calculation [15]. We further plot, for comparison with experiment, the energy distribution of photo-emitted electrons [16] and an X-ray photo-electron spectrum [17].

The overall picture looks similar in all cases. There is a low-lying band of states that are predominantly s-like, as can be verified by inspection of the respective MO coefficients, and there are the higher-lying, predominantly p-like bands, viz. valence and conductions bands. For the valence bands, the experimental curves confirm this overall picture, but results of this kind are not detailed and exact enough for closer comparison with a theoretical DOS. As far as the widths and locations of the p-like bands are concerned, we observe reasonable agreement between our results and the pseudopotential DOS.

A major point of interest are the highest peaks in the valence bands for the model crystals. They are located in the very intervals in which the energies of the atomic 5s and 5p states (indicated by thick lines in Fig. 3a and 3b) lie. The peaks in the DOS at these intervals are not so pronounced for the infinite lattice, where the orbitals on every atom interact with the AO's on neighboring atoms to form bonding and antibonding orbitals. The surface atoms in finite crystals clearly have less neighbors to interact with; therefore the corresponding AO

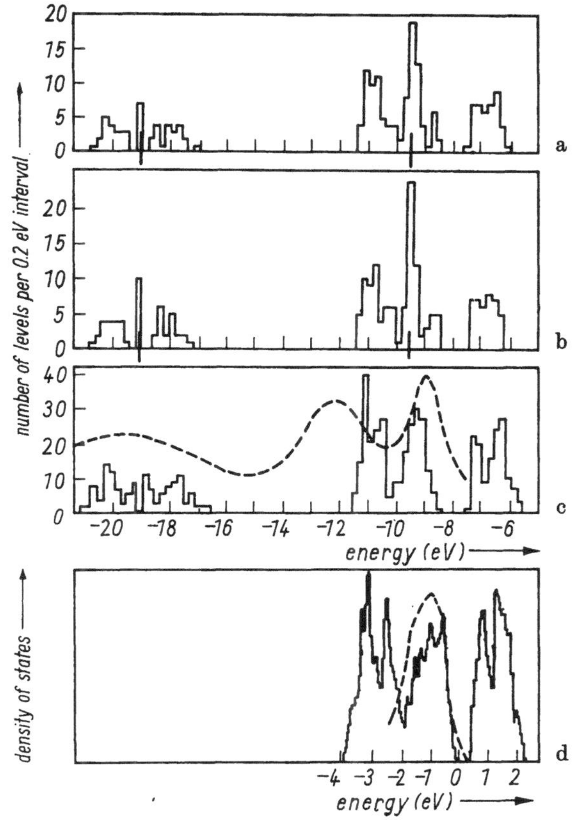

Fig. 3. Energy level distributions: a) 7c7a crystal; b) 10c5a crystal; c) density of states obtained by EHT band structure calculations [3], dashed curve: spectrum of X-ray photoelectrons [17]; d) density of states obtained by pseudopotential band structure calculations [15], dashed curve: energy distribution of photo-emitted electrons [16]. Reference [15] does not provide information on the lower-lying band. Note the heavy bars in a) and b) indicating the location of atomic levels. The energy scale of d) has been correlated to that of the other plots in such a way that the highest peaks in c) and d) are at about the same positions. In a) to c) the density of states is measured per 0.2 eV; in d) in arb. units

states do not spread and the MO energies stay at about the atomic values. The MO coefficients for the levels in the peak intervals show that these MO's are localized at chain end atoms. We also observe that the high peaks are higher for the 10c5a than the 7c7a because of its greater number of chain end atoms.

This result is reasonable, because in trigonal Te there are covalent bonds only between atoms within the same chain (for more details of the bonding situation in trigonal Te, see e.g. [18]); therefore the number of chain end atoms and the number of states that are not spread into bonding and antibonding states are proportional.

4.2 Charge distribution

We fill all MO levels in the valance bands, which is the best way to reproduce the elastic properties [8], and carry out a Mulliken population analysis [19]

Table 2

Charges on the individual atoms in various model crystals. The charges are given in units of the elementary charge; a neutral Te atom has charge 6

7c5a crystal

chain	1	2	3	4	5	6	7
atom							
5	7.00	6.92	6.93	7.09	7.09	7.08	6.89
4	5.94	6.13	6.13	6.04	5.84	5.88	5.94
3	5.98	5.97	5.83	5.97	6.03	6.21	6.03
2	5.94	6.04	6.13	6.13	5.94	5.88	5.84
1	7.00	7.09	6.93	6.92	6.89	7.08	7.09

10c5a crystal

chain	1	2	3	4	5	6	7	8	9	10
atom										
5	7.00	7.01	7.02	7.09	7.09	7.08	6.99	6.89	6.92	6.93
4	5.95	5.96	6.06	6.04	5.84	5.88	5.89	5.94	6.13	6.13
3	5.98	5.99	5.91	5.97	6.03	6.21	6.13	6.03	5.97	5.83
2	5.96	5.95	6.13	6.13	5.94	5.88	5.86	5.84	6.04	6.13
1	6.99	7.00	6.92	6.92	6.89	7.08	7.07	7.09	7.09	6.93

7c7a crystal

chain	1	2	3	4	5	6	7
atom							
7	7.00	7.10	7.10	7.08	6.89	6.91	6.92
6	5.93	6.03	5.84	5.88	5.92	6.10	6.09
5	5.98	5.99	6.05	6.20	6.01	5.96	5.82
4	6.00	6.27	6.08	5.98	5.80	5.98	6.08
3	5.98	5.99	5.82	5.96	6.01	6.20	6.05
2	5.93	6.03	6.09	6.10	5.92	5.88	5.84
1	7.00	7.10	6.92	6.91	6.89	7.08	7.10

which yields the charge distribution over the atoms of the crystal. For some typical crystals, Table 2 lists the results which are given in units of elementary charges; due to its 6 valence electrons, a neutral Te atom has charge 6.

The chain end atoms have roughly one electron more than the neutral value, because the broken bond states are in the valence band (cf. Section 4.1) and therefore filled [8]. This extra charge is found on all chain end atoms, independently on whether the chain is entirely within the cluster or not. This supports our results from Section 4.1, which indicate that covalent bonds occur only within chains.

The smaller deviations from the neutral value are due to the fact that some atoms protrude out of the model crystals to varying degrees. When the crystal is enlarged and a particular atom is no longer at the surface, its charge is closer to the neutral value. This can be seen for the atoms of chain 2 in the 7c5a and 10c5a crystals, and also for the atoms No. 5 in the 7c5a and 7c7a crystals.

This indicates that all these deviations are really only due to the crystal surfaces. An infinite lattice would indeed have a uniform charge distribution, thereby fulfilling an important requirement for the applicability of EHT to a system [9, 10].

We note there is experimental evidence apparently supporting our results concerning the pile-up of charges at chain end atoms: In surface quantum oscillation experiments [20] the surface state density was larger for sample surfaces perpendicular to the c-direction than for $(10\bar{1}0)$ surfaces which are parallel to the chains. A more quantitative comparison does not seem justified, however, because real surfaces have a considerably more complicated structure (due to structural defects and possible rearrangements) than the surfaces in our idealized model crystals.

Also, the decrease of the second component τ_2 of the positron lifetime in tellurium powder with decreasing grain size can be explained by an increased electron density at the chain ends [22].

In summary, the results for the perfect crystals presented in this Section 4 show that our 7c7a and 10c5a model crystals are large enough as to approximate reasonably some features of the infinite lattice, such as the MO level distribution. We can therefore hope to get some meaningful results when we introduce defects into these crystals.

5. Model Crystals with Vacancies

5.1 Vacancies without nearest neighbor relaxations

As pointed out by Messmer and Watkins [6], the lattice vacancy is the type of defect that is most appropriate to study by EHT. A vacancy is simulated by a crystal with one empty atomic site, preferentially close to the crystal center, such as the 1/4 position in the 7c7a crystal. First we leave all other atoms at their lattice positions.

The solution of equation (2) yields one unoccupied level slightly above the valence band edge for all model crystals considered. The valence band edge is the highest filled level when one takes into account that there are six electrons less than in the respective perfect crystal due to the missing atom. Table 3 lists

Table 3

The energy gaps and Δ (gap level—valence band edge) of various model crystals with a vacancy. No lattice relaxations included, all values in eV

crystal	7c4a	7c5a	7c6a	7c7a	10c4a	10c5a
gap	1.28	1.22	1.20	1.05	1.19	1.16
Δ	0.06	0.07	0.19	0.03	0.03	0.05

the energy gaps for various crystals and also Δ, which we define as the energy difference between the gap level and the valence band edge.

As in the case of the unperturbed crystals (Table 1), the gaps become smaller when more chains are added and/or the number of atoms per chain increases. The values of Δ also shows the tendency of eventually getting smaller with increasing crystal size.

For the 7c7a crystal with a vacancy, Table 4 shows the results of a Mulliken population analysis which reveals the localization of charge on the various vacancy neighbors for the individual MO levels in the vicinity of the gap. The charge in each MO level is normalized to 2, because each level can hold two elec-

Table 4

The MO levels near the gap and their localized charges, normalized to 2, on one of the 1st, 2nd, 3rd, and 4th nearest neighbors of the vacancy at 1/4 in the undistorted 7c7a crystal. The valence band edge is set at 0 eV

MO level (eV)	localized charge on one			
	1st n.n.	2nd n.n.	3rd n.n.	4th n.n.
1.096	0.01	0.08	0.04	0.05
1.056	0.00	0.00	0.00	0.05
0.031	0.20	0.00	0.17	0.02
0.000	0.18	0.00	0.21	0.02
−0.164	0.02	0.00	0.06	0.01
−0.174	0.01	0.00	0.09	0.01
−0.214	0.00	0.00	0.00	0.00
atoms at	1/3, 1/5	2/3, 2/5 3/3, 7/5	1/2, 1/6	2/4, 3/4, 4/4 5/4, 6/4, 7/4

trons. From this, it can be determined if a level has sizeable concentration on the vicinity of a defect or if it is spread out all over the crystal. As can be seen, only the 1st and 3rd nearest neighbors have any sizeable localized charges. These localized charges are found for the gap level and the valence band edge level only. These two levels can therefore justifiably be called defect levels.

We note that the vacancy has introduced localization of charges only for the atoms in the very chain that contains the vacancy. This observation that the vacancy affects predominantly its host chain can be made also for the charge distribution over all atoms of the crystal (Table 5).

Table 5

Charges on the individual atoms of the undistorted 7c7a crystal with a vacancy at 1/4

chain	1	2	3	4	5	6	7
atom							
7	6.95	7.09	7.09	7.00	6.78	6.91	6.92
6	5.74	6.01	5.77	5.78	5.91	6.10	6.00
5	6.68	6.06	6.03	6.20	6.02	5.95	5.83
4	—	6.25	6.05	5.98	5.80	5.98	6.05
3	6.68	6.06	5.83	5.95	6.02	6.20	6.03
2	5.74	6.01	6.00	6.10	5.91	5.78	5.77
1	6.95	7.09	6.92	6.91	6.78	7.00	7.09

For the 1st nearest neighbors (1/3 and 1/5), there is a pile-up of charges similar to that for chain end atoms. For the 3rd nearest neighbors (1/2 and 1/6), there is a charge deficit. For the rest of the crystal, the charge distribution displays many features of the one found for the perfect crystal (Table 2).

5.2 *The effects of distortion*

If we allow the vacancy neighbors to leave their original unrelaxed positions we can lower E_{EHT}, i.e. get a more stable system. Lacking other suggestions, the most plausible way of simulating distortions is to move the two 1st nearest neighbors of the vacancy by equal distances along directions that point from their original lattice positions to the vacancy. For the 7c7a crystal, an E_{EHT} minimum with respect to these displacements is found when the 1st nearest neigbors are moved 1/20 of the bond length away from their positions towards the vacancy. It is then found that, with respect to the unrelaxed case,

(i) E_{EHT} is 0.025 eV lower.

(ii) The vacancy level is lower (Δ is 29 meV as compared to 31 meV).

(iii) Gap level and valence band edge level have less localized charge on the 1st nearest neighbors (0.14 and 0.10), but more on the 3rd nearest neighbors (0.22 and 0.26).

(iv) The atomic charges on the 1st nearest neighbors increase to 6.77, but decrease for the 3rd nearest neighbors to 5.70, whereas there is little or no change for the rest of the crystal.

The same trend — summarized in the points (i) through (iv) — can be observed in the results for the other model crystals, too.

To take further lattice distortions into account did not seem worth while because the 1st nearest neighbor distortion gained already so little E_{EHT}. All these distortions, as well as those mentioned in Section 5.3 and 5.4, are the most plausible for us to assume. To our knowledge, there is nothing in the literature that elaborates on specific distortions around the Te vacancy. We think our distortion results ought to be seriously taken into consideration, because EHT results for distortions around defects in diamond have been shown to be in excellent agreement with experiment [6].

We can compare our Δ values to experimental results. The smallest values obtained so far are about 30 meV. Larger model crystals would probably lower this value somewhat. Before we make a direct comparison with experiment, we must remember that our calculated energy gaps are about three times too large. But we can still argue that the position of the gap level in the gap is proportionally correct. We carried out test calculations for the 7c5a crystal with a vacany and varied the lattice parameter c. The gap varied the same way as in band structure calculations [3], and Δ varied proportionally. Therefore, we predict on the basis of our calculations that a vacancy is associated with a single level about 10 meV above the valence band. This compares rather well with electron-irradiation experiments on Te [21], where a defect level is observed several meV above the valence band.

More generally, our calculations predict the vacancy in Te to act as an acceptor. This has been assumed for vacancies in covalent crystals [23]. For Te, in particular, this assumption has been used to interpret experimental results [21, 24 to 29].

5.3 Migration of single vacancies

For a migration of a vacancy in the trigonal Te crystal, there are two distinct possibilities: within a chain or between chains. To check the relative importance of these two cases, we carried out some preliminary calculations in the 10c5a crystal. This crystal is large enough to allow for all plausible split vacancy configurations[3]); the underlying assumption is that the migration barrier is biggest at this configuration. The 10c5a crystal is, however, not large enough

[3]) A split vacancy is a configuration where there is an atom half-way between two vacant lattice positions.

to allow for realistic simulations of distortions around the defects; in some cases the vacancy neighbors are chain end atoms. These preliminary tests showed a considerably lower barrier for the migration along a chain than for migration between chains; the same conclusion has been drawn on the basis of diffusion experiments [26, 29].

Therefore we investigate the vacancy migration in the chain direction in more detail. The calculations were carried out in the 7c7a and 7c6a crystals. In the 7c6a crystal it was also verified by detailed calculations that the migration barrier is highest for the split vacancy configuration.

The height of the migration barrier was calculated as the energy difference between the relaxed split vacancy and the relaxed vacancy. The relaxed vacancy has been described in Section 5.2. The relaxed split vacancy is simulated in an analogous way: In the 7c6a crystal for instance, the atom which actually migrates is kept fixed between the empty 1/3 and 1/4 sites, while the two other vacancy neighbors, 1/2 and 1/5, are moved by equal distances towards their respective vacancy until a minimum E_{EHT} is found. Farther neighbor displacements cannot be included in these crystals due to their finite sizes.

The migration barriers calculated in this way are 350 meV for the 7c7a and 167 meV for the 7c6a crystal. The question arises which result ought to be given more credit. The 7c7a is larger and the vacancy in it has the right symmetry; the split vacancy has the right symmetry in the 7c6a crystal (right symmetry means the same as in the infinite crystal). In any case, the order of magnitude of the results compares rather well with the activation energy of 170 meV that was obtained in the analysis of annealing stage II after electron-irradiation [21].

5.4 Double vacancies

The existence of several vacancies lying closely together has been conjectured. Especially at low temperatures, several successive lattice sites in a chain are thought to be vacant [24, 25]. We consider the simplest of such defects, viz. the double vacancy; it can be represented by the 7c7a crystal having vacant sites at 1/3 and 1/4. The calculations carried out for the defect were of the type described for the single vacancy (cf. Sections 5.1 to 5.3). The objective was to obtain results that can be compared to the single vacancy results.

Among the available model crystals, only the 7c6a and 7c7a can realistically simulate a double vacancy. Calculations for both model crystals yielded a single unoccupied defect level above the valance band edge, very much like for the single vacancy. The results were, however, inconclusive as far as the value Δ was concerned. For the undistorted crystals, Δ is 0.088 eV for the 7c6a and 0.127 eV for the 7c7a. That means that we get for the 7c6a a smaller Δ for the double vacancy, whereas we get a smaller Δ for the single vacancy in the 7c7a (cf. Table 3).

Nearest neighbor relaxations were taken into account in an analogous way as for the single vacancy. They did not change the relations of the Δ's of single and double vacancies just described; otherwise, they had similar effects as for the single vacancy, which are described in points (i) through (iv) in Section 5.2.

The migration of the double vacancy along a chain can be realistically simulated in the 7c7a crystal only. The highest E_{EHT} value was again found for a split vacancy configuration (an atom at 1/4, while 1/3 and 1/5 are empty). The E_{EHT} difference between the relaxed split double vacancy and the relaxed double vacancy is 1.157 eV. This is about three times more than the migration barrier for the single vacancy in the same crystal, which was 0.350 eV. We are not in a position to draw a firm conclusion from the results for just this one crystal, but we may note that this result is in agreement with Link [24] who assumes that multiple vacancies become mobile at higher temperatures than single vacancies.

The other main difference between single and double vacancy, as far as obtained from our calculations, is that per vacant site the double vacancy can cause only half as many holes in the valence band as the single vacancy, since both defects have only one acceptor level.

References

[1] C. WEIGEL, R. P. MESSMER, and J. W. CORBETT, phys. stat. sol. (b) 57, 455 (1973).
[2] R. HOFFMANN, J. chem. Phys. 39, 1397 (1963).
[3] C. WEIGEL and J. W. CORBETT, phys. stat. sol. (b) 59, K115 (1973).
[4] R. P. MESSMER, Chem. Phys. Letters 11, 589 (1971).
[5] R. P. MESSMER, B. MCCARROLL, and C. M. SINGAL, J. Vacuum Sci. Technol. 9, 891 (1972).
[6] R. P. MESSMER and G. D. WATKINS, Phys. Rev. B 7, 2568 (1973).
[7] C. WEIGEL, D. PEAK, J. W. CORBETT, G. D. WATKINS, and R. P. MESSMER, Phys. Rev. B 8, 2906 (1973).
[8] G. D. WATKINS and R. P. MESSMER, Phys. Rev. B 4, 2066 (1971).
[9] G. BLYHOLDER and C. A. COULSON, Theor. Chim. Acta (Berlin) 10, 316 (1968).
[10] L. C. ALLEN and J. D. RUSSEL, J. chem. Phys. 46, 1029 (1967).
[11] J. C. SLATER, Quantum Theory of Solids, Vol. 2, McGraw-Hill Publ. Co., New York 1965 (p. 49).
[12] J. C. SLATER, Quantum Theory of Solids, Vol. 2, McGraw-Hill Publ. Co., New York 1965 (p. 55).
[13] L. C. CUSACHS and J. H. CORRINGTON, in: Sigma Molecular Orbital Theory, Ed. O. SINANOGLU and K. B. WIBERG, Yale University Press, New Haven 1970 (p. 260).
[14] M. TINKHAM, Group Theory and Quantum Mechanics, McGraw-Hill Publ. Co., New York 1964 (p. 324).
[15] B. KRAMER, K. MASCHKE, and L. D. LAUDE, Phys. Rev. B 8, 5781 (1973).
[16] L. D. LAUDE, B. KRAMER, and K. MASCHKE, Phys. Rev. B 8, 5794 (1973).
[17] R. A. POLLACK, S. KOWALCZYK, L. LEY, and D. A. SHIRLEY, Phys. Rev. Letters 29, 274 (1972).
[18] P. GROSSE, Die Festkörpereigenschaften von Tellur, Springer-Verlag, Berlin 1969 (p. 22).
[19] R. S. MULLIKEN, J. chem. Phys. 23, 1833 (1955).
[20] R. SILBERMANN, G. LANDWEHR, J. C. THUILLIER, and J. BOUAT, Proc. Internat. Conf. Solid Surfaces, Kyoto 1974, Japan. J. appl. Phys. Suppl., to be published.
[21] R. SAFFERT, G. LANDWEHR, and E. GMELIN, Proc. Internat. Conf. Lattice Defects in Semiconductors, Freiburg 1974.
R. SAFFERT, Thesis, Universität Würzburg, 1973.
[22] K. DIETRICH, Thesis, Universität Würzburg, 1974.
K. DIETRICH and G. LANDWEHR, to be published.
[23] H. JAMES and K. LARK-HOROVITZ, Z. phys. Chem. 198, 107 (1951).
[24] R. LINK, Z. phys. Chem. 232, 418 (1966).
[25] K. H. HERRMANN, R. LINK, and W.-D. RENTSCH, phys. stat. sol. 8, 719 (1965).
[26] W. HÖRSTEL and G. KRETZSCHMAR, phys. stat. sol. 23, 639 (1967).
[27] V. A. J. VAN LINT and H. ROTH, J. appl. Phys. 30, 1235 (1959).
[28] V. N. LANGE and A. R. REGEL, Soviet. Phys.—Solid State 2, 2173 (1961).
[29] R. N. GHOSHTAGORE, Phys. Rev. 155, 598 (1967).

(Received June 7, 1974)

phys. stat. sol. (b) 65, 351 (1974)

Subject classification: 13

Tata Institute of Fundamental Research, Colaba, Bombay

On the Relativistic Electronic States of a Monatomic and a Diatomic One-Dimensional Lattice

By

N. D. SEN GUPTA

A new method has been developed to obtain directly wave functions and relations between the energy and momentum of an electron in a linear array of δ-potentials obeying the Dirac equation, both for a monatomic and a diatomic one-dimensional lattice.

Es wird eine neue Methode für die direkte Bestimmung von Wellenfunktionen entwickelt, und es werden Beziehungen zwischen der Energie und dem Impuls eines Elektrons in einer linearen Anordnung von δ-Potentialen, die der Dirac-Gleichung genügen, sowohl für ein einatomares als auch für ein zweiatomares eindimensionales Gitter erhalten.

1. Introduction

In order to investigate the relativistic effect on the electronic band structures of crystals composed of heavy atoms at high energy, Subramanian and Bhagwat [1] have obtained the relation between the energy and momentum of an electron in a linear array of δ-potentials for a monatomic lattice obeying the Dirac equation with the help of the transfer matrix. Fairbairn et al. [2] obtained the same with the help of Green's function. The usual methods of matching the solutions cannot be directly applied in this case, without adequate modifications, because of the discontinuity of the wave function at the lattice site (Stęślicka and Davison [3]. However, an interesting property of a linear array of δ-function potentials (equation (4)) may be used to obtain directly the exact solution of the Dirac equation and the energy band spectra. The method is easily applicable to monatomic and diatomic lattices and may be extended to more complicated systems.

The procedure followed in this paper is reminiscent of the classical method of Hill (Whittaker and Watson [4]) for the solution of differential equation with periodic coefficients. A simple property of a linear array of δ-function potentials reduces Hill's determinant to the sum of infinite series of elementary nature. Hence, the energy momentum dispersion relations and the wave functions may be obtained directly without recourse to the usual process of matching the wave functions and their derivatives at the points of discontinuities of the potential. As a preliminary application of the method, the non-relativistic Schrödinger equation both for the monatomic and diatomic lattice is examined very briefly in Section 2. The same problem with the Dirac equation is investigated in Section 3. Finally, the last section is devoted to the discussion of the results obtained.

2. The One-Dimensional Schrödinger Equation

The time-independent Schrödinger equation in one dimension may be written as

$$\left(\frac{d^2}{dx^2} + \varkappa^2\right) \Psi(x) = \frac{2m}{\hbar^2} V(x) \Psi(x) , \tag{1}$$

where

$$\varkappa^2 = \frac{2m}{\hbar^2} E . \tag{2}$$

Since $V(x)$ is periodic with period a, the lattice distance, from Floquet's theorem $\Psi(x)$ can be written as

$$\Psi(x) = e^{ikx} \sum_{-\infty}^{\infty} \psi_n \, e^{i2\pi nx/a} . \tag{3}$$

(I): monatomic lattice

$$V(x) = V_0 \sum_{-\infty}^{\infty} \delta(x - na) = \frac{V_0}{a} \sum_{-\infty}^{\infty} e^{i2\pi nx/a} . \tag{4}$$

The last step can be easily verified. If we substitute the expressions (3) and (4) in equation (1), we obtain

$$-\psi_n = \xi_0 \left\{\left(k + \frac{2\pi n}{a}\right)^2 - \varkappa^2\right\}^{-1} \Psi(0) , \tag{5}$$

where

$$\xi_0 = \frac{2mV_0}{a\hbar^2} \quad \text{and} \quad \Psi(0) = \sum_{-\infty}^{\infty} \psi_j . \tag{6}$$

Now, summing equation (5) for all values of n one gets (see Appendix (i)),

$$\left[1 + \xi_0 \sum_{-\infty}^{\infty} \left\{\left(k + \frac{2\pi n}{a}\right)^2 - \varkappa^2\right\}^{-1}\right] \Psi(0) = 0 . \tag{7}$$

Hence, for a non-trivial solution, this leads to the usual relation between k and E, for monatomic lattice as

$$F(\xi_0) = 0 \tag{8}$$

where

$$F(y) = \cos ak - \cos a\varkappa - \frac{ya}{2\varkappa} \sin a\varkappa . \tag{9}$$

(II): diatomic lattice

$$V(x) = \sum_{-\infty}^{\infty} \left\{V_1 \delta (x - na) + V_2 \delta\left(x + \frac{a}{2} - na\right)\right\}. \tag{10}$$

Proceeding as before

$$\psi_n = \left\{\varkappa^2 - \left(k + \frac{2\pi n}{a}\right)^2\right\}^{-1} \{\xi_1 \Psi(0) + \xi_2 (-1)^n \, e^{-ika/2} \, \Psi(a/2)\} , \tag{11}$$

where

$$\xi_i = \frac{2mV_i}{a\hbar^2}\,, \qquad (i = 1, 2) \quad \text{and} \quad \Psi(a/2) = e^{ika/2} \sum_{-\infty}^{\infty} (-1)^p \psi_p\,. \tag{12}$$

Now, summing equation (11) for all values of n, once directly and again after multiplying both sides by $(-1)^n$, one gets (See Appendix (ii)),

$$\left.\begin{aligned} F(\xi_1)\,\Psi(0) - \frac{\xi_2 a}{\varkappa} \sin\frac{a\varkappa}{2} \sin\frac{ak}{2}\,\Psi(a/2)\,e^{-ika/2} &= 0\,,\\ \xi_1 \frac{a}{\varkappa} \sin\frac{a\varkappa}{2} \sin\frac{ak}{2}\,\Psi(0) + F(\xi_2)\,\Psi(a/2)\,e^{-ika/2} &= 0\,. \end{aligned}\right\} \tag{13}$$

The condition for non-trivial solution leads to the desired relation between k, E (Saxon and Hunter [5], Stęślicka et al. [6]),

$$\cos^2\frac{ak}{2} = \left(\cos\frac{a\varkappa}{2} + \frac{\xi_1 a}{2\varkappa} \sin\frac{a\varkappa}{2}\right)\left(\cos\frac{a\varkappa}{2} + \frac{\xi_2 a}{2\varkappa} \sin\frac{a\varkappa}{2}\right). \tag{14}$$

3. The One-Dimensional Dirac Equation

The Dirac equation in one dimension corresponding to a stationary state with energy E may be written as

$$\left(\frac{E}{c} + i\hbar\alpha \frac{d}{dx} + \beta mc - \frac{V}{c}\right)\Psi(x) = 0\,. \tag{15}$$

Since only one 'α' matrix appears, the subscript is redundant. As before $\Psi(x)$ can be expressed as

$$\Psi(x) = e^{ikx} \sum_{-\infty}^{\infty} \psi_j\, e^{i2\pi jx/a}\,. \tag{16}$$

(I): monatomic lattice.

Substituting the expressions (16) and (4) in equation (15), one gets

$$\psi_p = \zeta_0 \frac{\varepsilon\alpha\, e^{-\beta\theta} + \alpha(ak + 2\pi p)}{\varepsilon^2 a^2 - (ak + 2\pi p)^2}\,\Psi(0)\,, \tag{17}$$

where

$$E = \varepsilon\hbar c \cosh\theta\,, \qquad mc = \varepsilon\hbar \sinh\theta\,, \qquad \varepsilon\hbar c = \sqrt{E^2 - m^2c^4} \tag{18}$$

and

$$\zeta_0 = \frac{V_0}{\hbar c}\,, \qquad \Psi(0) = \sum_{-\infty}^{\infty} \psi_j\,. \tag{19}$$

ζ_0 is a dimensionless parameter depending on V_0. From equation (17) one gets (See Appendix (iii) (a)),

$$M(\zeta_0)\,\Psi(0) = 0\,, \tag{20}$$

where

$$M(y) = \cos ak - \cos a\varepsilon - \frac{y}{2}\left(e^{-\beta\theta} \sin a\varepsilon + \alpha \sin ak\right). \tag{21}$$

Hence, k and E must satisfy the following condition

$$S(k, a, \zeta_0) = 0\,, \tag{22}$$

where

$$S_\pm(k, a, y) = \left(1 + \frac{y^2}{4}\right) \cos ak \mp \left(1 - \frac{y^2}{4}\right) \cos a\varepsilon \mp y \cosh \theta \sin a\varepsilon\,. \tag{23}$$

This is the same relativistic dispersion relation as obtained by Subramanian and Bhagwat [1] and Fairbairn et al. [2].

(II): diatomic lattice.

Proceeding as before with V, now given by the expression (10), one obtains

$$\psi_p = \frac{\varepsilon a\, \mathrm{e}^{-\beta\theta} + \alpha(ak + 2\pi p)}{\varepsilon^2 a^2 - (ak + 2\pi p)^2} \{\zeta_1 \Psi(0) + \zeta_2 (-1)^p\, \Psi(a/2)\, \mathrm{e}^{-ika/2}\}\,, \tag{24}$$

where

$$\zeta_i = \frac{V_i}{\hbar c} \quad (i = 1, 2)\,, \qquad \Psi(a/2) = \mathrm{e}^{ika/2} \sum_{-\infty}^{\infty} (-1)^j\, \psi_j\,. \tag{25}$$

Forming $\sum_{-\infty}^{\infty} \psi_p$ and $\sum_{-\infty}^{\infty} (-1)^p\, \psi_p$ from equation (24), we obtain

$$\left.\begin{aligned} M(\zeta_1)\, \Psi(0) + \zeta_2 N \Psi(a/2)\, \mathrm{e}^{-ika/2} = 0\,, \\ \zeta_1 N \Psi(0) + M(\zeta_2)\, \Psi(a/2)\, \mathrm{e}^{-ika/2} = 0\,, \end{aligned}\right\} \tag{26}$$

where

$$N = \mathrm{e}^{-\beta\theta} \sin \frac{a\varepsilon}{2} \cos \frac{ak}{2} + \alpha \cos \frac{a\varepsilon}{2} \sin \frac{ak}{2}\,. \tag{27}$$

See Appendix (iii) (a) and (b). Equation (26) may be written in a much simpler form

$$\left.\begin{aligned} W_+ \Psi(0) + (W_- - \zeta_2)\, \Psi(a/2)\, \mathrm{e}^{-ika/2} = 0\,, \\ (W_- - \zeta_1)\, \Psi(0) + W_+ \Psi(a/2)\, \mathrm{e}^{-ika/2} = 0\,, \end{aligned}\right\} \tag{28}$$

with

$$\begin{aligned} W_\pm = {} & \left(\mathrm{e}^{\beta\theta} \sin \frac{a\varepsilon}{2} - \alpha \sin \frac{ak}{2}\right)\left(\cos \frac{ak}{2} + \cos \frac{a\varepsilon}{2}\right)^{-1} \pm \\ & \pm \left(\mathrm{e}^{\beta\theta} \sin \frac{a\varepsilon}{2} + \alpha \sin \frac{ak}{2}\right)\left(\cos \frac{ak}{2} - \cos \frac{a\varepsilon}{2}\right)^{-1}. \end{aligned} \tag{29}$$

For the existence of non-trivial solutions of equation (28), it follows that

$$\| W_+^{-1}(W_- - \zeta_1)\, W_+^{-1}(W_- - \zeta_2) - 1 \| = 0\,. \tag{30}$$

This leads to the relation between k, and ε (i.e. E),

$$\begin{aligned} \left(1 + \frac{\zeta_1^2}{4}\right)\left(1 + \frac{\zeta_2^2}{4}\right) \cos ak = {} & (\zeta_1 + \zeta_2)\left(1 - \frac{\zeta_1 \zeta_2}{4}\right) \cosh \theta \sin a\varepsilon + \\ & + \zeta_1 \zeta_2 \left(\cosh 2\,\theta \sin^2 \frac{a\varepsilon}{2} - \cos^2 \frac{a\varepsilon}{2}\right) + \left(1 - \frac{\zeta_1^2}{4}\right)\left(1 - \frac{\zeta_2^2}{4}\right) \cos a\varepsilon\,. \end{aligned} \tag{31}$$

It goes over into the corresponding relations for the non-relativistic case only when terms in second and higher powers of ζ_1 and ζ_2, which depend on strengths of the potential, are neglected. The above relation may be written in the form

$$S_+\left(k, \frac{1}{2}(\zeta_1+\zeta_2), \frac{a}{2}\right) S_-\left(k, \frac{1}{2}(\zeta_1+\zeta_2), \frac{a}{2}\right) = \frac{1}{8}(\zeta_1-\zeta_2)^2 \left[\sin^2\frac{ak}{2} - \right.$$

$$- \cosh 2\theta \sin^2 \frac{a\varepsilon}{2} + \frac{1}{4}(\zeta_1+\zeta_2)\cosh\theta \sin a\varepsilon +$$

$$\left. + \frac{1}{64}(\zeta_1^2 + 6\zeta_1\zeta_2 + \zeta_2^2)(\cos ak + \cos a\varepsilon)\right]. \tag{31'}$$

Special Cases:

(i) It can be easily checked that in either case $\zeta_2 = 0$ and $\zeta_1 = \zeta$ or $\zeta_1 = 0$ and $\zeta_2 = \zeta$ the relation (31) leads to that of (22).

(ii) for $\zeta_1 = \zeta_2 = \zeta$, the problem effectively reduces to that of a monatomic lattice with a replaced by $a/2$; the relation (31′) leads to

$$S_+\left(k, \frac{1}{2}(\zeta_1+\zeta_2), \frac{a}{2}\right) S_-\left(k, \frac{1}{2}(\zeta_1+\zeta_2), \frac{a}{2}\right) = 0\,. \tag{32}$$

The situation is exactly similar to the non-relativistic case, relation (14). It is of interest to note that

$$S_-\left(k, \zeta, \frac{a}{2}\right) = -S_+\left(k + \frac{2\pi}{a}, \zeta, \frac{a}{2}\right). \tag{33}$$

In the expressions (3) and (16), k is defined but for modulo $2\pi n(a/2)^{-1}$ (n an integer), so that for $|k| < \pi(a/2)^{-1}$ there is no ambiguity. Another interesting special case is worth noting, namely,

(iii) $\zeta_1 = -\zeta_2$ i.e., the strength of the sources are of same magnitude but alternately changes sign. The expression (32) in this case reduces simply to

$$\left(1 + \frac{\zeta^4}{16}\right)(\cos ak - \cos a\varepsilon) + \zeta^2\left(\cosh 2\theta \sin^2\frac{a\varepsilon}{2} - \sin^2\frac{ak}{2}\right) = 0 \tag{34}$$

where $\zeta = |\zeta_1| = |\zeta_2|$. The dispersion relation depends only on ζ^2 as is expected from the symmetry of the problem. For small ζ the electronic states are almost those of a free electron except in the neighbourhood of a very small kinetic energy of the electron.

4. Discussion

It is clear that explicit expressions for discontinuities are not needed. At the Nth lattice point, they may be directly obtained from (4) and (15) as

$$\Psi(Na+0) = e^{-i2\alpha\tan^{-1}\zeta/2}\,\Psi(Na-0)\,. \tag{35}$$

However, physically significant quantities namely

$$\tilde{\Psi}(x)\,\Psi(x) \quad \text{and} \quad \tilde{\Psi}(x)\,\alpha\Psi(x)$$

are continuous at the lattice points.

Apart from these higher order terms, the relativistic treatment of the problem introduces another difference which is worth noting. From equation (18), ε may be taken both positive and negative and the usual non-relativistic limit is obtained only with the positive sign of ε. But this restriction is nowhere imposed on the derivation of the relation (22), hence, the Dirac equation leads to two distinct branches of dispersion relations.

In connection to the special case (ii), i.e., $\zeta_1 = \zeta_2 = \zeta$, it may be noted that both factors in equation (32) cannot be simultaneously zero. It may be noted directly from equation (26) either

$$\left.\begin{aligned} S_+(k, \zeta, a/2) &= 0 \quad \text{and} \quad \Psi(0) = \Psi(a/2)\, e^{-ika/2}\,, \\ S_-(k, \zeta, a/2) &= 0 \quad \text{and} \quad \Psi(0) = -\Psi(a/2)\, e^{-ika/2}\,. \end{aligned}\right\} \tag{36}$$

The method developed here may be extended to the case of surface states and it can also be generalized for other simple types of potentials.

I wish to express my sincere thanks to the referee for his critical remarks.

Appendix

All the infinite sums referred to in the text follow basically from the two well-known formulae, namely

$$\cot y = \frac{1}{y} + \sum_1^\infty \frac{2y}{y^2 - n^2\pi^2} = \sum_{-\infty}^\infty \frac{1}{y + n\pi}, \tag{A1}$$

and

$$\operatorname{cosec} y = \frac{1}{y} + \sum_1^\infty \frac{(-1)^n\, 2y}{y^2 - n^2\pi^2} = \sum_{-\infty}^\infty \frac{(-1)^n}{y + n\pi}\,. \tag{A2}$$

(i): $$\sum_{-\infty}^\infty \frac{1}{\left(k + \frac{2\pi n}{a}\right)^2 - \varkappa^2} = \frac{a}{4\varkappa} \sum_{-\infty}^\infty \left\{\frac{1}{\frac{a}{2}(k - \varkappa) + n\pi} - \frac{1}{\frac{a}{2}(k + \varkappa) + n\pi}\right\} =$$

$$= \frac{a}{4\varkappa}\left\{\cot\frac{a}{2}(k - \varkappa) - \cot\frac{a}{2}(k + \varkappa)\right\} =$$

$$= \frac{a}{2\varkappa}\,\frac{\sin a\varkappa}{\cos a\varkappa - \cos ak}\,. \tag{A3}$$

From which follow (8) and (9).

(ii): $$\sum_{-\infty}^\infty \frac{(-1)^n}{\left(k + \frac{2\pi n}{a}\right)^2 - \varkappa^2} = \frac{1}{2\varkappa} \sum_{-\infty}^\infty (-1)^n \left\{\frac{1}{k + \frac{2\pi n}{a} - \varkappa} - \frac{1}{k + \frac{2\pi n}{a} + \varkappa}\right\} =$$

$$= \frac{a}{4\varkappa}\left\{\operatorname{cosec}\frac{a}{2}(k - \varkappa) - \operatorname{cosec}\frac{a}{2}(k + \varkappa)\right\} =$$

$$= \frac{a}{\varkappa}\,\frac{\cos ak/2\,\sin a\varkappa/2}{\cos a\varkappa - \cos ak}\,. \tag{A4}$$

Equation (13) follows from equation (A3) and (A4).

$$\text{(iii) (a):}\quad \sum_{-\infty}^{\infty} \frac{ak + 2\pi n}{\varepsilon^2 a^2 - (ak + 2\pi n)^2} = \frac{1}{4} \sum_{-\infty}^{\infty} \left\{ \frac{1}{\frac{a}{2}(\varepsilon - k) + \pi n} - \frac{1}{\frac{a}{2}(\varepsilon + k) + \pi n} \right\} =$$

$$= \frac{1}{4} \left\{ \cot \frac{a}{2} (\varepsilon - k) - \cot \frac{a}{2} (\varepsilon + k) \right\} =$$

$$= \frac{1}{2} \frac{\sin ak}{\cos ak - \cos a\varepsilon} \,. \tag{A5}$$

$$\text{(b):}\quad \sum_{-\infty}^{\infty} \frac{(-1)^n (ak + 2\pi n)}{\varepsilon^2 a^2 - (ak + 2\pi n)^2} = \frac{1}{4} \sum (-1)^n \left\{ \frac{1}{\frac{a}{2}(\varepsilon - k) + \pi n} - \frac{1}{\frac{a}{2}(\varepsilon + k) + \pi n} \right\} =$$

$$= \frac{1}{4} \left\{ \operatorname{cosec} \frac{a}{2} (\varepsilon - k) - \operatorname{cosec} \frac{a}{2} (\varepsilon + k) \right\} =$$

$$= \frac{\sin ak/2 \cos a\varepsilon/2}{\cos ak - \cos a\varepsilon} \,. \tag{A6}$$

From equations (A5) and (A6) follow equations (26) and (27).

References

[1] R. Subramanian and K. V. Bhagwat, phys. stat. sol. (b) **48**, 399 (1971).
[2] W. M. Fairbairn, M. L. Glasser, and M. Stęślicka, Surface Sci. **36**, 462 (1973).
[3] M. Stęślicka and S. G. Davison, Phys. Rev. B **1**, 1858 (1970).
[4] E. T. Whittaker and G. N. Watson, A Course of Modern Analysis, Cambridge University Press, 1927 (Chap. XIX).
[5] D. S. Saxon and R. A. Hunter, Philips Res. Rep. **4**, 81 (1949).
[6] M. Stęślicka, S. G. Davison, and U. Srinivasan, J. Phys. Chem. Solids **32**, 1917 (1971).

(Received July 2, 1974)

phys. stat. sol. (b) **65**, 359 (1965)

Subject classification: 13.5.1; 20.1; 20.3; 22.1.2; 22.4.1

Institute of Applied Physics, Academy of Sciences of the Moldavian SSR, Kishinev

Effect of Exciton–Exciton Interaction on the Free Exciton Absorption and Luminescence Spectra of Si and CdS Crystals under High Intensity Excitation

By

A. V. Lelyakov

Lorentz shapes of the excitonic absorption and luminescence bands are obtained by Matsubara's Green's function technique in second order of the perturbation theory. The shapes of the excitonic absorption and luminescence bands are studied for the gaseous phase of free excitons in Si and CdS crystals under high intensity excitation. It is demonstrated that the elastic collisions of excitons lead to a broadening and shift of excitonic absorption and luminescence bands even when phonons are absent. The effect of phonons in the present case has not been taken into account. The possibility of smoothly varying the frequency of laser emission by changing the pumping level is pointed out.

Методом мацубаровских функций Грина во втором порядке теории возмущений получены лоренцовские формы экситонных полос поглощения и люминесценции. Формы полос исследуются для газообразной фазы экситонов в кристаллах Si и CdS при больших интенсивностях возбуждения. Показано, что упругие столкновения экситонов приводят к уширению и сдвигу экситонных полос поглощения и люминесценции даже в отсутствие фононов, влияние которых в данном случае во внимание не принимается. Указана возможность плавного изменения частоты лазерного излучения путем изменения уровня накачки.

1. Introduction

For the true absorption of light in pure crystals besides a photon and an exciton it is necessary to have a dissipative system to which part of the light energy is transmitted [1]. The excitons are converted into photons again with the same wave vectors and may leave the crystal if there is no dissipative system. The phonons are such a dissipative system in relatively pure crystals. A lot of papers describing the experimental process (see for example the review [2]) as well as theoretical ones [3 to 6] have been dedicated to the exciton–phonon interaction and the phonon effect on the shape of absorption and luminescence bands in semiconducting, organic, and other crystals. But the interaction processes between excitons become more and more important as the concentration of excitons is increased. The excitons themselves may act as a dissipative system owing to their interaction when the concentration of excitons is sufficiently high. The concentration of excitons N_{ex} is considered large if the mean spacing between the excitons becomes of the order of their radius r_{ex}, i.e. $N_{ex} r_{ex}^3 < 1$. The exciton–exciton interaction results in new phenomena in solid state physics, as for instance the excitonic molecule (biexciton) [7 to 10], the Bose-Einstein condensation of excitons and (or) biexcitons [7, 11 to 17], the formation of exciton dielectric drops [18], the screening and destruction of excitons [12], the metallization and Mott transition of excitons [19, 20], and the formation of metallic drops [14, 19 to 21].

The exciton interaction can lead to elastic collisions which are followed by a redistribution of their kinetic energy and by the establishment of a quasi-equilibrium. The exciton interaction can also lead to inelastic collisions in which one of the excitons which gains energy is scattered into a higher excited state or dissociates into an electron and hole while the other exciton which has lost part of its energy converts into a photon and then leaves the crystal [9, 22]. The luminescence intensity in such a process is proportional to N_{ex}^2. (N_{ex} is the concentration of excitons in the crystal.) In the case of indirect transition a phonon also takes part. A theoretical prediction of the inelastic collision of two excitons was made in [13]. An interpretation of the experiment from this point of view has been proposed for the first time in [22] for the explanation of a new luminescence band which was observed in CdS crystals excited by electron bombardment. In the some manner the authors of [23] interpret their experimental results on the observation of an extra luminescence band in CdS, CdSe, and ZnO crystals under intense laser illumination. A new luminescence band appearing in ZnO under electron excitation has been also explained on this basis [24]. Büttner [25] has calculated in the Born approximation a cross-section scattering which leads to an excited exciton and a polariton in CdS. In our paper we report the effect of elastic collisions on the shape of the exciton absorption and luminescence bands in Si and CdS crystals at $T > T_c$, i.e. when the excitons are in the gaseous phase. T_c is the temperature of the phase transition of excitons. This problem has been primely regarded in [26]. In general when a complex system is investigated which includes dipole active excitons, photons, and phonons the following processes should be regarded: 1. the intercollision of excitons, 2. the transformation of light into excitons and vice versa, 3. the scattering of the excitons by phonons. The exciton–exciton interaction dominates if

$$\nu_{ex-ex} < \nu_{ex-ph},\ \nu_{ex-lat}\,.$$

Here ν_{ex-ex} is the intercollision frequency of excitons. ν_{ex-ph} is the frequency of transformation of an exciton into a photon leading to the formation of a polariton when no other interactions are observed. ν_{ex-lat} is the scattering frequency of the excitons by phonons. ν_{ex-ex} and ν_{cx-lat} may be determined from the shift and attenuation of the exciton level due to the interaction between the excitons and phonons, respectively. ν_{ex-ph} is determined by the splitting δ between the frequencies of the longitudinal and transverse excitons in cubic crystals. The value δ may be determined from the study of the linear Zeeman effect. If the magnetic field is perpendicular to the direction of observation, the long-range Coulomb potential which determines the difference $\hbar\omega_{\parallel} - \hbar\omega_{\perp}$, distort the linear Zeeman effect. From the degree of distortion one can determine δ. In this manner it was demonstrated that $\delta < 10^{-4}$ eV for the main bands of the yellow exciton series of Cu_2O and the long-range Coulomb potential has no effect. The same is found for CdS crystals of hexagonal modification. In this case the crystal field splits the previously three-fold degenerate level with basis functions x, y, z into a two-fold degenerate level Γ_6 (basis functions x, y) and one non-degenerate level Γ_1 (basis function z). If the light wave vector is perpendicular to the axis of six-fold symmetry, the level Γ_6 splits into longitudinal and transverse components due to the long-range Coulomb potential.

The distance between them, as stated in [1], is 0.9 meV. Our calculations show that the shift of the exciton level due to the exciton–exciton interaction may amount up to 3.2 meV when the concentration of excitons is increasing from 10^{16} to 5×10^{16} cm^{-3} at $T = 60$ °K. This value exceeds the shift due to the Coulomb potential (0.9 meV) which shows that the polariton effects play a secondary part when the concentrations of excitons are large enough. Even though at some concentrations $\nu_{\text{ex-ex}} < \nu_{\text{ex-ph}}$ at low temperatures, this inequality may become reversed with increasing temperature due to the increasing kinetic energy of excitons and their more effective collisions. The same situation takes place when the concentration of excitons is insufficient and the scattering of excitons is due to phonons. In this case it is well known [27] that the polariton effects may appear only at low temperatures and therefore we neglect them.

Strictly speaking the shapes of exciton absorption and luminescence bands obtained by us may be applicable to direct band gap semiconductors like CdS where the exciton–phonon interaction is weak and may be neglected. In indirect gap semiconductors like Si it is practically impossible to eliminate the phonons from the considerations since the momentum conservation law is not fulfilled in the quantum transitions. This means that phonons will make a certain contribution to the shape of absorption and luminescence bands of indirect excitons. But in this case it can be also expected that the exciton–exciton interaction will behave in the same manner as in the case of direct excitons and will result in specific properties of exciton absorption and luminescence bands. This is proved by our results as well as by a comparison of theory with the experiment.

2. The Green's Functions for Excitons at $T \neq 0$

The Hamiltonian in the isotropic mass approximation for two energy bands (valence and conduction band) has the form

$$H = H_0 + H_i\,; \tag{1}$$

$$H_0 = \sum_{\boldsymbol{p}} (\Delta + T_{\boldsymbol{p}})\, a^+_{\boldsymbol{p}} a_{\boldsymbol{p}}\,, \tag{2}$$

$$H_i = \frac{1}{2V} \sum_{\boldsymbol{p}_1\boldsymbol{p}_2\boldsymbol{p}_3\boldsymbol{p}_4} \nu(\boldsymbol{p}_1 - \boldsymbol{p}_4)\, \delta_{kr}(\boldsymbol{p}_1 + \boldsymbol{p}_2 - \boldsymbol{p}_3 - \boldsymbol{p}_4)\, a^+_{\boldsymbol{p}_1} a^+_{\boldsymbol{p}_2} a_{\boldsymbol{p}_3} a_{\boldsymbol{p}_4}\,. \tag{3}$$

Here Δ is the exciton creation energy, $T_{\boldsymbol{p}}$ is the kinetic energy of the exciton, $a^+_{\boldsymbol{p}}$, $a_{\boldsymbol{p}}$ are the exciton creation and annihilation operators, respectively $\nu(\boldsymbol{p})$ is the Fourier transform of the exciton–exciton interaction energy, δ_{kr} is the Kronecker symbol.

The Green's function is according to [28] defined by

$$G_{\boldsymbol{p}}(\tau - \tau') = -\langle T a_{\boldsymbol{p}}(\tau)\, a^+_{\boldsymbol{p}}(\tau')\, U(\beta)\rangle_c\,. \tag{4}$$

Here

$$\tau = it\,,\quad a_{\boldsymbol{p}}(\tau) = e^{\tau H_0}\, a_{\boldsymbol{p}}\, e^{-\tau H_0}\,,\qquad U(\beta) = T\, e^{-\int_0^\beta H_i(\tau_1)\,d\tau_1}\,.$$

T is the chronological operator. The index c in equation (4) indicates that it is necessary to take into account only the connected diagrams in which all points are connected by some lines with external ends. $\beta = 1/k_0T$ where k_0 is the Boltzmann constant and T the absolute temperature.

In the second order of perturbation theory we obtain

$$U(\beta) = 1 - \int_0^\beta H_\mathrm{i}(\tau_1)\,\mathrm{d}\tau_1 + \tfrac{1}{2}\,T \iint_0^\beta H_\mathrm{i}(\tau_1)\,H_\mathrm{i}(\tau_2)\,\mathrm{d}\tau_1\,\mathrm{d}\tau_2\,. \tag{5}$$

Substituting (5) into (4) we obtain corrections of the first and second order to the Green's function:

$$G_{\boldsymbol p}(\tau - \tau') =$$
$$= -\left\langle T a_{\boldsymbol p}(\tau)\,a_{\boldsymbol p}^+(\tau')\left[1 - \int_0^\beta H_\mathrm{i}(\tau_1)\,\mathrm{d}\tau_1 + \tfrac{1}{2}\,T \iint_0^\beta H_\mathrm{i}(\tau_1)\,H_\mathrm{i}(\tau_2)\,\mathrm{d}\tau_1\,\mathrm{d}\tau_2\right]\right\rangle_\mathrm{c}\,. \tag{6}$$

The zero-order Green's function is

$$G_{\boldsymbol p}^0(\tau - \tau') = -\,\langle T a_{\boldsymbol p}(\tau)\,a_{\boldsymbol p}^+(\tau')\rangle\,. \tag{7}$$

The correction of first order to the Green's function has the form

$$G_{\boldsymbol p}^1(\tau - \tau') = \int_0^\beta \mathrm{d}\tau_1\,\langle T a_{\boldsymbol p}(\tau)\,a_{\boldsymbol p}^+(\tau')\,H_\mathrm{i}(\tau_1)\rangle_\mathrm{c}\,. \tag{8}$$

Using Wick's theorem we find the connected diagrams in the first order of perturbation theory. In order to single out the tops of interaction we introduce the δ-function: $\nu(\boldsymbol p_1 - \boldsymbol p_4)\,\delta(\tau_1 - \tau_1') = \nu(\boldsymbol p_1 - \boldsymbol p_4; \tau_1 - \tau_1')$ (see Fig. 1). In the first order of perturbation theory only two connected topologically inequivalent diagrams exist which do not follow from each other by the exchange of tops.

Using the rules according to which the analytical expressions corresponding to the diagrams of Fig. 2 are written we find

$$G_{\boldsymbol p}^{1\mathrm{a}}(\tau - \tau') = \frac{1}{V}\sum_{\boldsymbol p_2}\int_0^\beta G_{\boldsymbol p}^0(\tau - \tau_1)\,G_{\boldsymbol p_2}^0(\tau_1 - \tau_1')\,\nu(\boldsymbol p - \boldsymbol p_2; \tau_1 - \tau_1')\,G_{\boldsymbol p}^0(\tau_1' - \tau')\,\mathrm{d}\tau_1\,, \tag{9}$$

$$G_{\boldsymbol p}^{1\mathrm{b}}(\tau - \tau') = \frac{1}{V}\sum_{\boldsymbol p_2}\int_0^\beta G_{\boldsymbol p}^0(\tau - \tau_1)\,G_{\boldsymbol p}^0\,(\tau_1 - \tau')\,\nu(0; \tau_1 - \tau_1')\,G_{\boldsymbol p_2}^0(\tau_1 - \tau_1')\,\mathrm{d}\tau_1\,. \tag{10}$$

Let us expand the Green's function into a Fourier series in the fourth component. The argument $\tau - \tau'$ varies only in the interval $-\beta \leqq \tau - \tau' \leqq \beta$. For the expansion of the Green's function $G_{\boldsymbol p}(\tau - \tau')$ into a Fourier series it is necessary first of all to continue it up to $\pm\infty$ periodically. After the continuation the

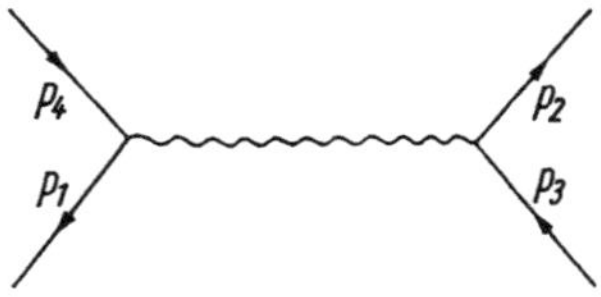

Fig. 1

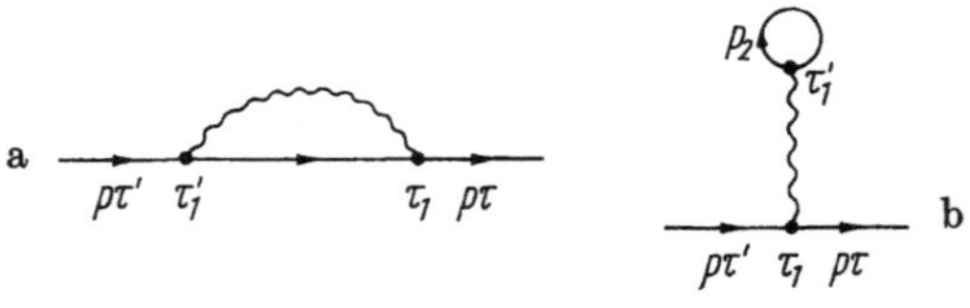

Fig. 2

expansion has the form

$$G_{\boldsymbol{p}}(\tau - \tau') = \frac{1}{\beta} \sum_{\omega_n} G_{\boldsymbol{p}}(\omega_n)\, e^{-i\omega_n(\tau-\tau')}\,; \qquad \omega_n = 2\pi n k_0 T\,, \tag{11}$$

$$\delta(\tau_1 - \tau_1') = \frac{1}{\beta} \sum_{\omega_n} e^{-i\omega_n(\tau_1-\tau_1')}\,. \tag{12}$$

Now formula (9) taking into account (11) and (12) looks as follows:

$$G_{\boldsymbol{p}}^{1a}(\tau - \tau') = \frac{1}{\beta^2 V} \sum_{\boldsymbol{p}_2\omega_1\omega_2} G_{\boldsymbol{p}}^0(\omega_1)\, G_{\boldsymbol{p}_2}^0(\omega_2)\, G_{\boldsymbol{p}}^0(\omega_1)\, \nu(\boldsymbol{p} - \boldsymbol{p}_2)\, e^{-i\omega_1(\tau-\tau')}\,.$$

Hence

$$G_{\boldsymbol{p}}^{1a}(\omega_n) = G_{\boldsymbol{p}}^0(\omega_n)\, M_{\boldsymbol{p}}^{1a}(\omega_n)\, G_{\boldsymbol{p}}^0(\omega_n) \tag{13}$$

where

$$M_{\boldsymbol{p}}^{1a}(\omega_n) = \frac{1}{\beta V} \sum_{\boldsymbol{p}_2\omega_2} \nu(\boldsymbol{p} - \boldsymbol{p}_2)\, G_{\boldsymbol{p}_2}^0(\omega_2)\,. \tag{14}$$

Transforming then formula (10) we find

$$G_{\boldsymbol{p}}^{1b}(\omega_n) = G_{\boldsymbol{p}}^0(\omega_n)\, M_{\boldsymbol{p}}^{1b}(\omega_n)\, G_{\boldsymbol{p}}^0(\omega_n) \tag{15}$$

where

$$M_{\boldsymbol{p}}^{1b}(\omega_n) = \frac{1}{\beta V} \sum_{\boldsymbol{p}_2\omega_2} \nu(0)\, G_{\boldsymbol{p}_2}^0(\omega_2)\,. \tag{16}$$

Summarizing all the diagrams in all orders of perturbation theory we obtain the complete Green's function:

$$G_{\boldsymbol{p}}(\omega_n) = \frac{G_{\boldsymbol{p}}^0(\omega_n)}{1 - M_{\boldsymbol{p}}(\omega_n)\, G_{\boldsymbol{p}}^0(\omega_n)} \tag{17}$$

where

$$M_{\boldsymbol{p}}(\omega_n) = M_{\boldsymbol{p}}^{1a}(\omega_n) + M_{\boldsymbol{p}}^{1b}(\omega_n) + M_{\boldsymbol{p}}^2(\omega_n) + \cdots \tag{18}$$

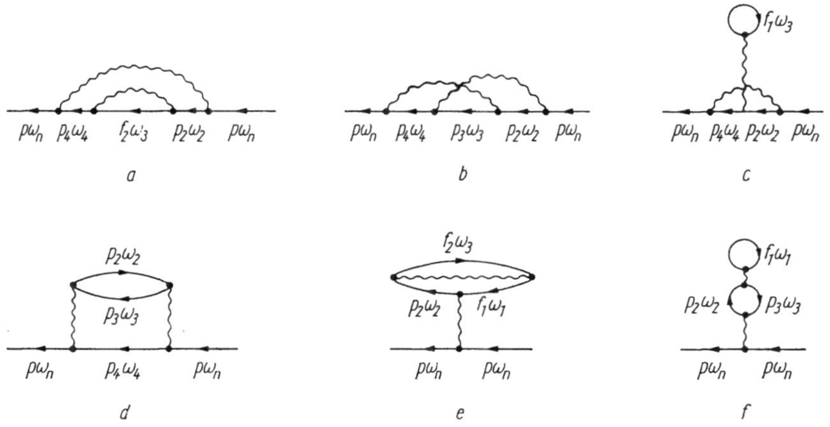

Fig. 3

is the mass operator which includes only the connected self-energy diagrams (without the external ends). In the second order of perturbation theory there exist six topologically inequivalent diagrams for the mass operator $M_{\boldsymbol{p}}(\omega_n)$ They are given in Fig. 3.

The contributions of the self-energy diagrams in Fig. 3 are as follows:

$$\text{a)}\quad \frac{1}{\beta^2 V^2} \sum_{\boldsymbol{f}_2 \boldsymbol{p}_4 \omega_3 \omega_4} \nu(\boldsymbol{p} - \boldsymbol{p}_4)\, \nu(\boldsymbol{p}_4 - \boldsymbol{f}_2)\, [G^0(\boldsymbol{p}_4 \omega_4)]^2\, G^0(\boldsymbol{f}_2 \omega_3)\, \mathrm{e}^{i\omega_3 0^+}\,, \tag{19}$$

$$\text{b)}\quad \frac{1}{\beta^2 V^2} \sum_{\boldsymbol{p}_3 \boldsymbol{p}_4 \omega_3 \omega_4} \nu(\boldsymbol{p}-\boldsymbol{p}_4)\, \nu(\boldsymbol{p}_4-\boldsymbol{p}_3)\, G^0(\boldsymbol{p}_4 \omega_4)\, G^0(\boldsymbol{p}_3 \omega_3)\, G^0(\boldsymbol{p}+\boldsymbol{p}_3-\boldsymbol{p}_4\ \omega_n + \omega_3 - \omega_4)\,, \tag{20}$$

$$\text{c)}\quad \frac{1}{\beta^2 V^2} \sum_{\boldsymbol{p}_2 \boldsymbol{f}_1 \omega_2 \omega_3} \nu(0)\, \nu(\boldsymbol{p} - \boldsymbol{p}_2)\, [G^0(\boldsymbol{p}_2 \omega_2)]^2\, G^0(\boldsymbol{f}_1 \omega_3)\, \mathrm{e}^{i\omega_3 0^+}\,, \tag{21}$$

$$\text{d)}\quad \frac{1}{\beta^2 V^2} \sum_{\boldsymbol{p}_2 \boldsymbol{p}_4 \omega_2 \omega_4} \nu^2(\boldsymbol{p} - \boldsymbol{p}_4)\, G^0(\boldsymbol{p}_4 \omega_4)\, G^0(\boldsymbol{p}_2 + \boldsymbol{p} - \boldsymbol{p}_4 \omega_2 + \omega_2 + \omega_n - \omega_4) G^0(\boldsymbol{p}_2 \omega_2)\,, \tag{22}$$

$$\text{e)}\quad \frac{1}{\beta^2 V^2} \sum_{\boldsymbol{p}_2 \boldsymbol{f}_2 \omega_2 \omega_3} \nu(0)\, \nu(\boldsymbol{f}_2 - \boldsymbol{p}_2)\, [G^0(\boldsymbol{p}_2 \omega_2)]^2\, G^0(\boldsymbol{f}_2 \omega_3)\, \mathrm{e}^{i\omega_3 0^+}\,, \tag{23}$$

$$\text{f)}\quad \frac{1}{\beta^2 V^2} \sum_{\boldsymbol{p}_2 \boldsymbol{f}_1 \omega_2 \omega_1} \nu^2(0)\, [G^0(\boldsymbol{p}_2 \omega_2)]^2\, G^0(\boldsymbol{f}_1 \omega_1)\, \mathrm{e}^{i\omega_1 0^+}\,. \tag{24}$$

Here

$$G^0(\boldsymbol{p}\omega_n) = \frac{1}{i\hbar\omega - \varepsilon(\boldsymbol{p})}\,. \tag{25}$$

For the summation over frequencies in (19 to 24) we use the formula

$$\frac{1}{\beta} \sum_{\omega_n} f(\omega_n) = \tag{26}$$

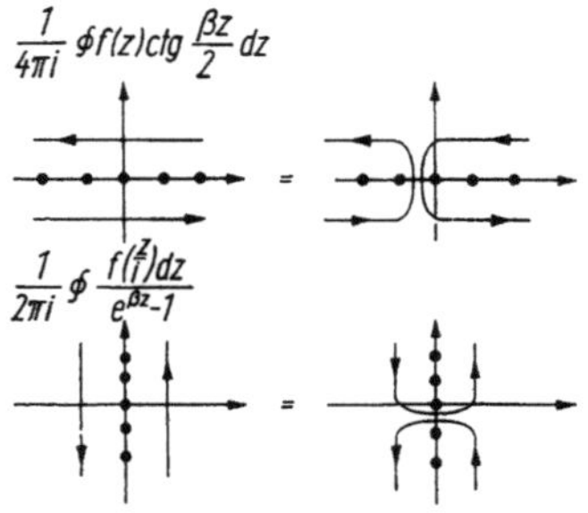

In formula (26) the contour integral is taken in the complex plane. The points on the real axis denote the poles of the function $\operatorname{ctg} \beta z/2$, and on the imaginary axis those of the function $(\exp \beta z - 1)^{-1}$. The contour is closed at infinity including these poles. It is assumed that the functions $f(z)$ and $f(z/i)$ do not have any peculiarities of these poles. By expanding around these poles it is not difficult to obtain the left part of formula (26). Then the formulae (14),

(16), and (19 to 24) take the form

$$M_{\boldsymbol{p}}^{1a}(\omega_n) = \frac{1}{V} \sum_{\boldsymbol{p}_2} \nu(\boldsymbol{p} - \boldsymbol{p}_2)\, n_{\boldsymbol{p}_2}\,, \tag{27}$$

$$M_{\boldsymbol{p}}^{1b}(\omega_n) = \frac{1}{V} \sum_{\boldsymbol{p}_2} \nu(0)\, n_{\boldsymbol{p}_2} = \frac{\nu(0)\, N}{V}, \tag{28}$$

$$M_{\boldsymbol{p}}^{2a}(\omega_n) = -\frac{\beta}{V^2} \sum_{\boldsymbol{f}_2 \boldsymbol{p}_4} \nu(\boldsymbol{p} - \boldsymbol{p}_4)\, \nu(\boldsymbol{p}_4 - \boldsymbol{f}_2)\, (n_{\boldsymbol{f}_2} n_{\boldsymbol{p}_4}^2 + n_{\boldsymbol{f}_2} n_{\boldsymbol{p}_4})\,, \tag{29}$$

$$M_{\boldsymbol{p}}^{2b}(\omega_n) = \frac{1}{V^2} \sum_{\boldsymbol{p}_3 \boldsymbol{p}_4} \nu(\boldsymbol{p} - \boldsymbol{p}_4)\, \nu(\boldsymbol{p}_4 - \boldsymbol{p}_3)\, \frac{n_{\boldsymbol{p}_4} n_{\boldsymbol{p}_3} - n_{\boldsymbol{p}+\boldsymbol{p}_3-\boldsymbol{p}_4}(n_{\boldsymbol{p}_4} - n_{\boldsymbol{p}_3}) + n_{\boldsymbol{p}_3}}{i\hbar\omega_n + \varepsilon(\boldsymbol{p}_3) - \varepsilon(\boldsymbol{p}_4) - \varepsilon(\boldsymbol{p} + \boldsymbol{p}_3 - \boldsymbol{p}_4)}, \tag{30}$$

$$M_{\boldsymbol{p}}^{2c}(\omega_n) = -\frac{\beta}{V^2} \sum_{\boldsymbol{f}_1 \boldsymbol{p}_2} \nu(0)\, \nu(\boldsymbol{p} - \boldsymbol{p}_2)\, (n_{\boldsymbol{f}_1} n_{\boldsymbol{p}_2}^2 + n_{\boldsymbol{f}_1} n_{\boldsymbol{p}_2})\,, \tag{31}$$

$$M_{\boldsymbol{p}}^{2d}(\omega_n) = \frac{1}{V^2} \sum_{\boldsymbol{p}_2 \boldsymbol{p}_4} [\nu(\boldsymbol{p} - \boldsymbol{p}_4)]^2\, \frac{n_{\boldsymbol{p}_4} n_{\boldsymbol{p}_2} - n_{\boldsymbol{p}+\boldsymbol{p}_2-\boldsymbol{p}_4}(n_{\boldsymbol{p}_4} - n_{\boldsymbol{p}_2}) + n_{\boldsymbol{p}_2}}{i\hbar\omega + \varepsilon(\boldsymbol{p}_2) - \varepsilon(\boldsymbol{p}_4) - \varepsilon(\boldsymbol{p} + \boldsymbol{p}_2 - \boldsymbol{p}_4)}, \tag{32}$$

$$M_{\boldsymbol{p}}^{2e}(\omega_n) = -\frac{\beta}{V^2} \sum_{\boldsymbol{p}_2 \boldsymbol{f}_2} \nu(0)\, \nu(\boldsymbol{f}_2 - \boldsymbol{p}_2)\, (n_{\boldsymbol{f}_2} n_{\boldsymbol{p}_2}^2 + n_{\boldsymbol{f}_2} n_{\boldsymbol{p}_2})\,, \tag{33}$$

$$M_{\boldsymbol{p}}^{2f}(\omega_n) = -\frac{\beta}{V^2} \sum_{\boldsymbol{f}_1 \boldsymbol{p}_2} \nu^2(0)\, (n_{\boldsymbol{f}_1} n_{\boldsymbol{p}_2}^2 + n_{\boldsymbol{f}_1} n_{\boldsymbol{p}_2})\,. \tag{34}$$

In the intermediate calculations we used the relation

$$\operatorname{cth} \frac{\beta \varepsilon(\boldsymbol{p})}{2} = 2 n_{\boldsymbol{p}} + 1 \tag{35}$$

where $n_{\boldsymbol{p}_i}$ and $n_{\boldsymbol{f}_i}$ are the occupancy numbers of excitons in the band. The mass operator in the second order of perturbation theory has the form

$$\begin{aligned} &M_{\boldsymbol{p}}^{1a}(\omega_n) + M_{\boldsymbol{p}}^{1b}(\omega_n) + M_{\boldsymbol{p}}^{2a}(\omega_n) + M_{\boldsymbol{p}}^{2b}(\omega_n) + M_{\boldsymbol{p}}^{2c}(\omega_n) + M_{\boldsymbol{p}}^{2d}(\omega_n) + \\ &+ M_{\boldsymbol{p}}^{2e}(\omega_n) + M_{\boldsymbol{p}}^{2f}(\omega_n) = \frac{1}{V} \sum_{\boldsymbol{p}_2} \nu(\boldsymbol{p}_2)\, n_{\boldsymbol{p}_2} + \frac{N\nu(0)}{V} - \frac{\beta}{V^2} \sum_{\boldsymbol{p}_2 \boldsymbol{f}_2} \times \\ &\times [\nu(0) + \nu(\boldsymbol{p} - \boldsymbol{p}_2)]\, [\nu(0) + \nu(\boldsymbol{p}_2 - \boldsymbol{f}_2)]\, [n_{\boldsymbol{f}_2} n_{\boldsymbol{p}_2}^2 + n_{\boldsymbol{f}_2} n_{\boldsymbol{p}_2}] + \\ &+ \frac{1}{V^2} \sum_{\boldsymbol{p}_2 \boldsymbol{p}_4} [\nu^2(\boldsymbol{p} - \boldsymbol{p}_4) + \nu(\boldsymbol{p}_4 - \boldsymbol{p}_2)\, \nu(\boldsymbol{p} - \boldsymbol{p}_4)]\, \frac{n_{\boldsymbol{p}_4} n_{\boldsymbol{p}_2} - n_{\boldsymbol{p}+\boldsymbol{p}_2-\boldsymbol{p}_4}(n_{\boldsymbol{p}_4} - n_{\boldsymbol{p}_2}) + n_{\boldsymbol{p}_2}}{i\hbar\omega_n + \varepsilon(\boldsymbol{p}_2) - \varepsilon(\boldsymbol{p}_4) - \varepsilon(\boldsymbol{p} + \boldsymbol{p}_2 - \boldsymbol{p}_4)}. \end{aligned} \tag{36}$$

Substituting the formulae (25) and (36) into (17) we find the temperature causal Green's function in the second order of perturbation theory:

$$G_{\boldsymbol{p}}^{e}(\omega_n) = \frac{1}{i\hbar\omega_n - \varepsilon(\boldsymbol{p}) - M_{\boldsymbol{p}}^{c}(\omega_n)}\,. \tag{37}$$

3. The Shapes of the Exciton Absorption Band Calculated for Si and CdS Crystals at High Excitations

The shape of the exciton absorption band is determined by the formula [6]

$$f_{\boldsymbol{p}}^{a}(\omega) = -\frac{\hbar}{\pi} \operatorname{Im} G_{\boldsymbol{p}}^{R}(\omega) . \tag{38}$$

Here $G_{\boldsymbol{p}}^{R}(\omega)$ is the Fourier transform of the retarded commutator Green's function:

$$G_{\boldsymbol{p}}^{R}(t-t') = -i\theta(t-t')\,\langle[\tilde{a}_{\boldsymbol{p}}(t), \tilde{a}_{\boldsymbol{p}}^{+}(t')]\rangle\,; \quad \tilde{a}_{\boldsymbol{p}}(t) = e^{iHt/\hbar}\, a_{\boldsymbol{p}}\, e^{-iHt/\hbar} . \tag{39}$$

The transition from the temperature causal Green's function (37) to the retarded commutator Green's function (39) is performed as follows. We consider (37) at $\omega_n > 0$ and analytically continue this function from the discrete points $z = i\omega_n$ to the full upper half-plane of the complex variable z. Further z has to approach the real axis from above, i.e. $z = \omega + i\delta$ with $\delta \to +0$.

$$G_{\boldsymbol{p}}^{R}(\omega) = G_{\boldsymbol{p}}^{c}(-i\omega + \delta) = \frac{1}{\hbar\omega - \varepsilon(\boldsymbol{p}) - M_{\boldsymbol{p}}^{R}(\omega)}, \tag{40}$$

where $M_{\boldsymbol{p}}^{R}(\omega) = M_{\boldsymbol{p}}(\omega) + i\gamma_{\boldsymbol{p}}(\omega)$.

For the real and imaginary parts of the mass operator we obtain

$$\begin{aligned} M_{\boldsymbol{p}}(\omega) = {} & \frac{\nu(0)N}{V} + \frac{1}{V}\sum_{\boldsymbol{p}_1} \nu(\boldsymbol{p}_1)\, n_{\boldsymbol{p}_1} - \frac{\beta}{V^2}\sum_{\boldsymbol{p}_1\boldsymbol{p}_2} [\nu(0) + \nu(\boldsymbol{p}-\boldsymbol{p}_1)]\,[\nu(0) + \nu(\boldsymbol{p}_1-\boldsymbol{p}_2)] \times \\ & \times [n_{\boldsymbol{p}_2} n_{\boldsymbol{p}_1}^2 + n_{\boldsymbol{p}_2} n_{\boldsymbol{p}_1}] + \frac{1}{V}\sum_{\boldsymbol{p}_1\boldsymbol{p}_2} [\nu^2(\boldsymbol{p}-\boldsymbol{p}_1) + \nu(\boldsymbol{p}-\boldsymbol{p}_1)\,\nu(\boldsymbol{p}_1-\boldsymbol{p}_2)] \times \\ & \times \frac{n_{\boldsymbol{p}_1} n_{\boldsymbol{p}_2} - n_{\boldsymbol{p}_2+\boldsymbol{p}-\boldsymbol{p}_1}(n_{\boldsymbol{p}_1} - n_{\boldsymbol{p}_2}) + n_{\boldsymbol{p}_2}}{\hbar\omega + \varepsilon(\boldsymbol{p}_2) - \varepsilon(\boldsymbol{p}_1) - \varepsilon(\boldsymbol{p}+\boldsymbol{p}_2-\boldsymbol{p}_1)}, \end{aligned} \tag{41}$$

$$\begin{aligned} \gamma_{\boldsymbol{p}}(\omega) = {} & -\frac{\pi}{V^2}\sum_{\boldsymbol{p}_2\boldsymbol{p}_1} [\nu^2(\boldsymbol{p}-\boldsymbol{p}_1) + \nu(\boldsymbol{p}-\boldsymbol{p}_1)\,\nu(\boldsymbol{p}_1-\boldsymbol{p}_2)] \times \\ & \times [n_{\boldsymbol{p}_1} n_{\boldsymbol{p}_2} - n_{\boldsymbol{p}+\boldsymbol{p}_2-\boldsymbol{p}_1}(n_{\boldsymbol{p}_1} - n_{\boldsymbol{p}_2}) + n_{\boldsymbol{p}_2}]\, \delta[\hbar\omega + \varepsilon(\boldsymbol{p}_2) - \varepsilon(\boldsymbol{p}_1) - \varepsilon(\boldsymbol{p}+\boldsymbol{p}_2-\boldsymbol{p}_1)] . \end{aligned} \tag{42}$$

Here $\varepsilon(\boldsymbol{p}_i) = \boldsymbol{p}_i^2/2m_{\text{ex}}$; $i = (1, 2)$; m_{ex} is the mass of the exciton; N/V is the number of excitons per unit volume.

Substituting the formulae (40 to 42) into (38) we obtain an expression for the unknown shape of the exciton absorption band:

$$f_{\boldsymbol{p}}^{a}(\omega) = -\frac{\hbar}{\pi}\,\frac{\gamma_{\boldsymbol{p}}(\omega)}{[\hbar\omega - \varepsilon(\boldsymbol{p}) - M_{\boldsymbol{p}}(\omega)]^2 + \gamma_{\boldsymbol{p}}^2(\omega)} . \tag{43}$$

Owing to the small momentum of the photon the $\boldsymbol{p}$ dependence in formulae (41 to 43) may be ignored.

Analysing the formula for the shape of the exciton absorption band we find that the band is shifted compared to the case when the exciton concentration is low. The halfwidth of the exciton absorption band, $\gamma_{\boldsymbol{p}}(\omega)$, and the shift of the maximum of the exciton absorption band $M_{\boldsymbol{p}}(\omega)$, are smooth functions of $\boldsymbol{p}$ and ω. In this case the shape of the exciton absorption band does not have the form ω/p^4 as in [6] when discussing the Bose-Einstein condensation of excitons at $T = 0$.

Let us assume that the excitons have a Maxwell distribution:

$$n_{\boldsymbol{p}_i} = A \exp\left(-\frac{p_i^2}{2m_{\text{ex}}}\right); \quad A = \frac{N}{V}\left(\frac{2\pi\hbar^2\beta}{m_{\text{ex}}}\right)^{3/2}$$

where $\boldsymbol{p}_i$ is the quasi-momentum of the exciton. In order to obtain convenient formulae for the calculation of the halfwidth $\gamma_0(\omega)$ and the shift of the maximum $M_0(\omega)$ of exciton absorption bands it is necessary to make specific assumptions about the interaction mechanism between excitons and the dependence of the Fourier transform of the interaction energy of two excitons on the magnitude of momentum transfer in the moment of collision. The last magnitude is unknown and may essentially change from crystal to crystal. Taking into account the experimental data for Si [29] and CdS [17] we may assume that in these crystals there exists attractive interaction between excitons. We approximate the function $\nu(\boldsymbol{p})$ by a step function:

$$\nu(\boldsymbol{p}) = \begin{cases} \nu(0) < 0 & \text{at} \quad \boldsymbol{p} \leqq \dfrac{\hbar}{r_{\text{ex}}}, \\ 0 & \text{at} \quad \boldsymbol{p} > \dfrac{\hbar}{r_{\text{ex}}}, \end{cases} \tag{44}$$

where r_{ex} is the exciton radius. This model corresponds to a contact attraction between excitons. In this case formulae (41) and (42) have the form

$$\left.\begin{aligned} \gamma_0(\omega) &= -\frac{A\nu^2(0)\, m_{\text{ex}}^3\, \omega}{4\pi^3\hbar^5\beta}\left[e^{\hbar\omega\beta} K_1(\hbar\omega\beta) + \sqrt{2}A\, e^{\hbar\omega\beta} K_1(\sqrt{2}\hbar\omega\beta) - AK_1(\hbar\omega\beta)\right] \\ &\qquad\qquad \text{at} \quad \omega \neq 0, \\ \gamma_0 &= -\frac{A\nu^2(0)\, m_{\text{ex}}^3}{4\pi^3\hbar^6\beta^2} \qquad \text{at} \quad \omega = 0. \end{aligned}\right\} \tag{45}$$

Here $K_n(z)$ is the Macdonald function with asymptotical expressions

$$K_n(z) = \begin{cases} \dfrac{2^{n-1}(n-1)!}{z^n} & \text{at} \quad z \ll 1, \\ \sqrt{\dfrac{\pi}{2}}\dfrac{e^{-z}}{\sqrt{z}} & \text{at} \quad z \gg 1; \end{cases}$$

$$M_0(\omega) = \frac{2\nu(0)N}{V} - \frac{A^2\nu^2(0)m_{\text{ex}}^3}{2\pi^3\hbar^6\beta^2}\left(\frac{A\sqrt{2}}{4} + 1\right) + \frac{A\nu^2(0)m_{\text{ex}}}{4\pi^4\hbar^6}\left\{\iint_0^\infty p_1\, dp_1 p_2\, dp_2 \times \right.$$

$$\times \exp\left(-\frac{\beta(p_1^2 + p_2^2)}{2m_{\text{ex}}}\right)\ln\left|\frac{m_{\text{ex}}\hbar\omega - p_2^2 + p_1p_2}{m_{\text{ex}}\hbar\omega - p_2^2 - p_1p_2}\right| -$$

$$- A\exp(-\hbar\omega\beta)\iint_0^\infty p_1\, dp_1 p_2\, dp_2\left[\exp\left(-\frac{\beta p_1^2}{2m_{\text{ex}}}\right) - \right.$$

$$- \exp\left(-\frac{\beta(2p_1^2 + p_2^2)}{2m_{ex}}\right)\Bigg]\Bigg[\mathrm{Ei}\left(\frac{[m_{ex}\hbar\omega - p_2^2 + p_1p_2]\beta}{m_{ex}}\right) -$$

$$- \mathrm{Ei}\left(\frac{[m_{ex}\hbar\omega - p_2^2 - p_1p_2]\beta}{m_{ex}}\right)\Bigg] +$$

$$+ \int\limits_0^{\hbar/r_{ex}} p_2\,\mathrm{d}p_2 \int\limits_0^{\infty} p_1\,\mathrm{d}p_1 \exp(-p_1^2\beta/2m_{ex}) \ln\left|\frac{m_{ex}\hbar\omega - p_2^2 + p_1p_2}{m_{ex}\hbar\omega - p_2^2 - p_1p_2}\right|\Bigg\}. \qquad (46)$$

Here $\mathrm{Ei}(z)$ is the integral exponential function. All integrals with the exception of one quickly converge. Therefore, only in the last term of (46) we limited the range of p_2 integration following the approximation (44).

For simplifying the expression (46) we use the fact that most the probable values of p_1 and p_2 coincide; therefore we substitute p_1 by p_2 in the logarithm so that neither the numerator nor the denominator converts into zero due to such an approximation. Then the formula (46) takes the form

$$M_0(\omega) = \frac{2\nu(0)N}{V} - \frac{A^2\nu^2(0)m_{ex}^3}{2\pi^3\hbar^6\beta^2}\left(\frac{A\sqrt{2}}{4} + 1\right) + \frac{A\nu^2(0)m_{ex}^3\omega}{(2\pi)^4\hbar^5\beta}\left[\frac{4A}{\hbar\omega\beta}\exp\left(-\frac{\hbar\omega\beta}{4}\right)\times\right.$$

$$\left.\times\mathrm{Ei}\left(\frac{\hbar\omega\beta}{4}\right) + \left(1 - \frac{2\hbar}{r_{ex}^2m_{ex}\omega}\right)\ln\left|1 - \frac{2\hbar}{r_{ex}^2m_{ex}\omega}\right| + \frac{2\hbar}{r_{ex}^2m_{ex}\omega}\right] \quad \text{at} \quad \omega \neq 0\,. \qquad (47)$$

The point $\omega = 0$ is calculated separately. In this case the previous approximations are not sufficient.

$$M_0 = \frac{2\nu(0)N}{V} - \frac{A^2\nu^2(0)m_{ex}^3}{2\pi^3\hbar^6\beta^2}\left(\frac{A\sqrt{2}}{4} + 1\right) - \frac{A\nu^2(0)m_{ex}^3}{4\pi^4\hbar^6\beta^2}\left(A\ln 2 + \frac{\hbar^2\beta}{2r_{ex}^2m_{ex}}\ln 2 +\right.$$

$$\left. + \ln\frac{\hbar^2\beta}{2r_{ex}^2m_{ex}} - c\right) \quad \text{at} \quad \omega = 0\,. \qquad (48)$$

Here c is a constant equal to 0.43.

The shapes of exciton absorption bands for Si and CdS crystals are calculated by formula (43) by taking into account (45), (47), and (48). For Si the calculation is carried out at $T = 4.2$ °K for the concentrations of free excitons $N/V = 10^{14}$ to 10^{16} cm^{-3} and at $T = 77$ °K for the concentrations $N/V = 10^{14}$ to 10^{17} cm^{-3} with the following parameters of free excitons: $m_{ex} \approx m_0$ (m_0 is the free electron mass), $r_{ex} = 75$ Å, $E_{ex} = 8$ meV (binding energy of the free exciton) [30].[1]) For the Fourier transform of the interaction energy of two excitons we took $\nu(0) = -10^{-32}$ erg cm^3; ω is measured from the bottom of the exciton band without taking into account the renormalization caused by the interaction of excitons. The shapes of exciton absorption bands (normalized to unity) at $T = 77$ °K and a concentration of free excitons of 10^{17} cm^{-3} are presented in Fig. 4.

It follows from calculations that at $T = 77$ °K the halfwidth and the shift of the maximum of exciton absorption bands to the long-wave side of the spectrum depend on the exciton concentration more sharply than at $T = 4.2$ °K. When the concentration of excitons is not changed the maximum of the band

[1]) A different value of E_{ex} is given in [31].

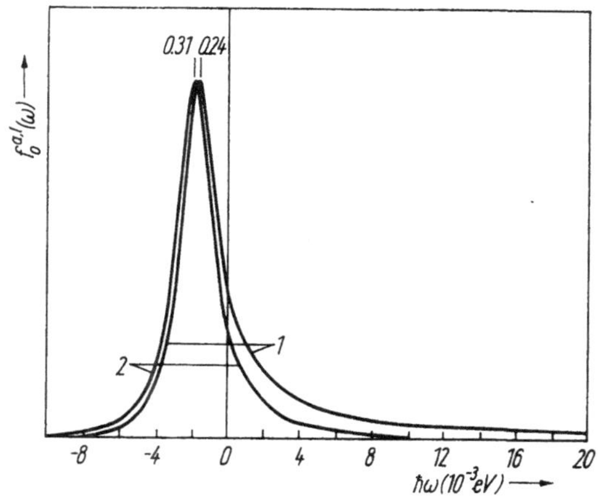

Fig. 4. Curve 1: the shape of the exciton absorption band of Si at $T = 77$ °K and $N/V = 10^{17}$ cm^{-3}. $f_0^{a,l}(\omega)$ is given in units of 10^{-3} eV. The figures at the peaks denote the peak heights

Curve 2: the shape of the exciton luminescence band of Si at $T = 77$ °K and $N/V = 10^{17}$ cm^{-3}. $f_0^{a,l}(\omega)$ is given in units of 10^{-3} eV. The figures at the peaks denote the peak heights

is practically not shifted in the temperature interval 4.2 to 77 °K, whereas the halfwidth of exciton absorption bands in this case is obviously increased. The halfwidth of exciton absorption bands is changed insignificantly when the concentration of excitons is changed by two orders at $T = 4.2$ °K and it is significantly increased when the concentration of excitons is increased by one order at $T = 77$ °K. The shift of the maximum of the exciton absorption band to the long-wave side of the spectrum with increasing concentration of excitons can be seen already at $T = 4.2$ °K and it becomes much larger at $T = 77$ °K. The exciton absorption band is symmetrical when $N/V = 10^{14}$ cm^{-3} and $T = 4.2$ °K, while the rest of the bands is asymmetrical. The asymmetry and the short-wave tail of exciton absorption bands increase at constant temperature with increasing exciton concentration. But when the concentration of excitons is constant the asymmetry and the short-wave tail of exciton absorption bands are increased when the temperature rises. The absorption coefficient is the highest possible one at the frequency of the maximum of the exciton absorption band for $T = 4.2$ °K and on exciton concentration $N/V = 10^{14}$ cm^{-3}. We have calculated the shapes of the exciton absorption band of CdS at $T = 10$ and 60 °K. The parameters of the free exciton have the following values: $N/V = 10^{16}$ cm^{-3}, $m_{ex} = 1.6m_0$ [22], $r_{ex} = 27$ Å, $E_{ex} = 28$ meV [32], and $\nu(0) =$

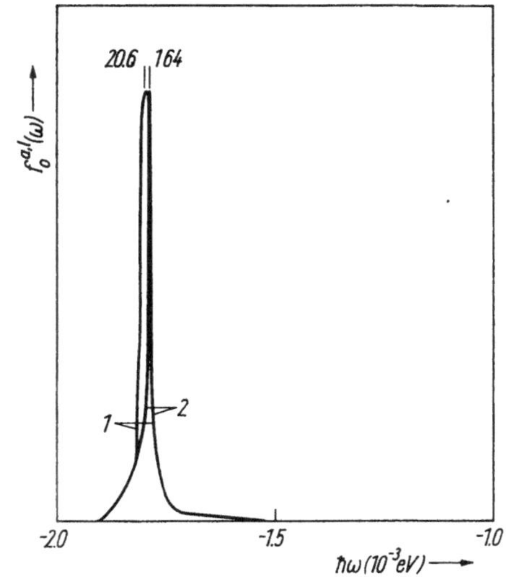

Fig. 5. Curve 1: the shape of the absorption band of CdS at $T = 10$ °K and $N/V = 10^{16}$ cm^{-3}.

Curve 2: the shape of the exciton luminescence band of CdS at $T = 10$ °K and $N/V = 10^{16}$ cm^{-3}

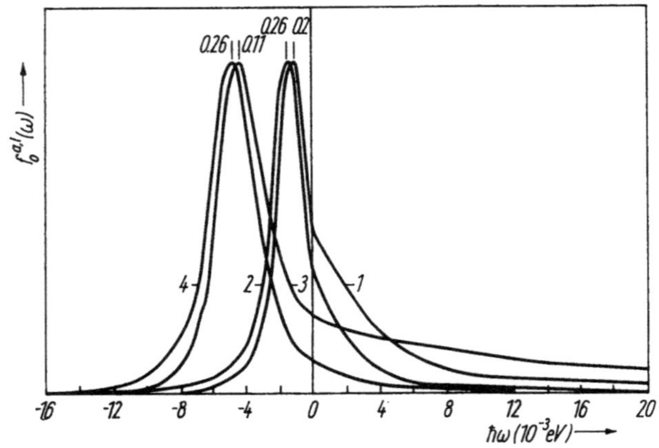

Fig. 6. Curve 1: the shape of the exciton absorption band of CdS at $T = 60$ °K and $N/V = 10^{16}$ cm^{-3}. Curve 2: the shape of the exciton luminescence band of CdS at $T = 60$ °K and $N/V = 10^{16}$ cm^{-3}. Curve 3: the shape of the exciton absorption band of CdS at $T = 60$ °K and $N/V = 5 \times 10^{16}$ cm^{-3}. Curve 4: the shape of the exciton luminescence band of CdS at $T = 60$ °K and $N/V = 5 \times 10^{16}$ cm^{-3}. $f_0^{a,l}(\omega)$ is given in units of 10^{-3} eV. The figures at the peaks denote the peak heights

$= -\frac{26\pi}{3} E_{ex}\, r_{ex}^3$ [33]. They are given in Fig. 5 and 6. Here as well as in the case of Si the shapes of exciton absorption bands are asymmetric and have short-wave tails. Similarly in Si the half-width of the exciton absorption band and the short-wave tail are increasing with the rise of temperature from 10 to 60 °K at constant concentration of excitons as well as with increasing exciton concentration at constant temperature (60 °K). But the maximum of the exciton absorption band in CdS, when the temperature changes from 10 to 60 °K at constant exciton concentration, is shifted by a larger amount than in Si at the same concentration when the temperature rises from 4.2 to 77 °K.

4. The Shapes of the Exciton Luminescence Band Calculated for Si and CdS Crystals at High Excitation Levels

The shape of the exciton luminescence band has been determined from the formula [10]

$$f_p^{l}(\omega) \sim f_p^{a}(\omega) \exp\left(-\hbar\omega\beta\right).$$

Here as well as in formula (43), when calculating the shapes of exciton luminescence bands, the dependence on the light wave vector was neglected. The shapes of the exciton luminescence band for Si at $T = 77$ °K are shown in Fig. 4 and for CdS at $T = 10$ and 60 °K in Fig. 5 and 6.

Everything said above about the exciton absorption bands in these crystals may be applied to the exciton luminescence bands with the exception of the case when the concentration of excitons is increased at $T = 77$ °K. Then the short-wave tail of exciton luminescence bands is decreased in comparison with that of exciton absorption bands, whereas the shapes of the exciton luminescence band become more symmetrical. When comparing the obtained shapes of exciton luminescence bands for Si with the experimental ones of [29] one can see that the bands agree in the following respects: The halfwidth $\gamma_0(\omega)$ of the exciton luminescence band increases with the rise of temperature at a fixed concentration of excitons; the halfwidth $\gamma_0(\omega)$ of the band increases with the growth of exciton concentration at constant temperature; the shape of exciton luminescence bands is asymmetric and it has a short-wave tail; the maximum of bands is shifted to the long-wave side of the spectrum with increasing exciton concentration at constant temperature; the maximum of bands is not practically shifted with the change of temperature at a fixed concentration of excitons; the shape of the luminescence band becomes more symmetrical with increasing exciton concentration at $T = 77$ °K.

The calculated exciton luminescence bands for CdS at 10 and 60 °K are compared with the luminescence bands for Si. At the same temperature (60 °K) and the same concentration of excitons (see Fig. 6) it is seen that the luminescence band has a less pronounced short-wave tail than the absorption band and the band itself becomes more symmetrical. When the concentration of excitons increases by a factor 5 at 60 °K, the maximum of the exciton luminescence band shifts to lower energies by 3.2 meV.

The shift of the exciton level with exciton concentration in the case of laser generation will lead to a change of emission frequency which depends upon the pumping intensity. A similar effect will also take place in the case of biexcitons, since there is a possibility to vary smoothly the frequency of laser generation by changing the level pumping.

Acknowledgements

The author would like to express his deep gratitude to Prof. V. A. Moskalenko and Prof. S. A. Moskalenko for fruitful discussions and essential advices. The author thanks Dr. S. L. Pyshkin and L. V. Dogaru for their help in translating this article in English.

References

[1] R. Knox, Theory of Excitons, Izd. Mir, Moscow 1966.
[2] E. F. Gross, S. A. Permogorov, and B. S. Razbirin, Uspekhi fiz. Nauk **103**, 431 (1971).
[3] Y. Toyozawa, Progr. theor. Phys. **20**, 53 (1958); **27**, 89 (1962); J. Phys. Chem. Solids **25**, 59 (1964).
[4] A. S. Davydov and E. N. Myasnikov, phys. stat. sol. **20**, 153 (1967).
[5] B. Segall and G. D. Mahan, Phys. Rev. **171**, 935 (1968).
[6] A. V. Lelyakov and S. A. Moskalenko, Fiz. tverd. Tela **11**, 3260 (1969).
[7] S. A. Moskalenko, Optika i Spektroskopiya **5**, 146 (1958); Fiz. tverd. Tela **4**, 276 (1962); Zh. eksper. teor. Fiz. **45**, 1160 (1963); Bose-Einsteinovskaya kondensatsiya eksitonov i bieksitonov, Redak. izdat. otdel Akad. Nauk MSSR, Kishinev 1970.
[8] M. A. Lampert, Phys. Rev. Letters **1**, 450 (1958).
[9] J. R. Heynes, Phys. Rev. Letters **17**, 860 (1966).
[10] J. D. Cuthbert, J. Luminescence **1/2**, 307 (1970).
[11] J. M. Blatt, K. W. Böer, and W. Brandt, Phys. Rev. **126**, 1691 (1962).
[12] R. C. Casella, J. Phys. Chem. Solids **24**, 19 (1963); J. appl. Phys. **34**, 1703 (1963); **36**, 2485 (1965).
[13] L. V. Keldysh and A. N. Kozlov, Zh. eksper. teor. Fiz., Pisma **5**, 238 (1967); Zh. eksper. teor. Fiz. **54**, 978 (1968).
[14] L. V. Keldysh, Proc. IX. Internat. Conf. Phys. Semicond., Moscow 1968, Vol. 2, Izd. Nauka, Leningrad 1969 (p. 1384).

[15] V. A. Gergel, R. F. Kazarinov, and R. A. Suris, Zh. eksper. teor. Fiz. **53**, 544 (1967); **54**, 298 (1968).
[16] I. Kh. Akopyan, E. F. Gross, and B. S. Razbirin, Zh. eksper. teor. Fiz., Pisma **12**, 366 (1970).
[17] H. Kuroda, S. Shionoya, H. Saito, and E. Hanamura, Techn. Rep. ISSP A562, 1972.
[18] D. B. Chesnut, J. chem. Phys. **40**, 405 (1964); **41**, 472 (1964).
[19] V. M. Asnin, A. A. Rogachev, and N. I. Sablina, Zh. eksper. teor. Fiz., Pisma **11**, 162 (1970).
V. M. Asnin, B. V. Zubkov, T. M. Murina, A. M. Prokhorov, A. A. Rogachev, and N. I. Sablina, Zh. eksper. teor. Fiz. **62**, 737 (1972).
[20] Ya. E. Pokrovskii and K. I. Svistunova, Zh. eksper. teor. Fiz., Pisma **9**, 435 (1969); **13**, 297 (1971).
[21] V. S. Bagaev, T. I. Galkina, O. V. Gogolin, and L. V. Keldysh, Zh. eksper. teor. Fiz., Pisma **10**, 309 (1969).
V. S. Vavilov, A. V. Zayats, and V. N. Murina, Zh. eksper. teor. Fiz., Pisma **10**, 304 (1969).
[22] C. Benoit à la Guillaume, J. M. Debever, and F. Salvan, Phys. Rev. **177**, 567 (1969).
[23] D. Magde and H. Mahr, Phys. Rev. Letters **24**, 890 (1970).
[24] J. M. Hvam, Solid State Commun. **12**, 95 (1973).
[25] H. Büttner, phys. stat. sol. **42**, 775 (1970).
[26] A. V. Lelyakov, Thesis, Inst. Appl. Phys. AN MSSR, Kishinev 1970; Optika i Spektroskopiya **30**, 1172 (1971); Fiz. tverd. Tela **15**, 309 (1973).
L. A. Ablova, A. I. Bobrysheva, A. V. Lelyakov, S. A. Moskalenko, and P. I. Khadzhi, J. Luminescence **8**, 82 (1973).
[27] J. J. Hopfield, Phys. Rev. **112**, 1555 (1958).
[28] A. A. Abrikosov, L. P. Gorkov, and I. E. Dzyaloshinskii, Metody teorii polya v statisticheskoi fizike, Fizmatgiz, Moscow 1962.
[29] B. M. Ashkinadze, I. P. Kretsu, S. M. Ryvkin, and I. D. Yaroshetskii, Zh. eksper. teor. Fiz. **58**, 507 (1970).
[30] A. S. Kaminskii, Ya. E. Pokrovskii, and N. V. Alkeev, Zh. eksper. teor. Fiz. **59**, 1936 (1970).
[31] K. L. Shaklee and R. E. Nahory, Phys. Rev. Letters **24**, 942 (1970).
[32] D. S. Reynolds, C. W. Litton, and T. C. Collins, phys. stat. sol. **12**, 3 (1965).
[33] A. I. Bobrysheva, M. F. Miglei, S. A. Moskalenko, and M. I. Shmiglyuk, Slozhnye poluprovodniki i ikh fizicheskie svoistva, Shtiintsa, Kishinev 1971 (p. 25).
E. Hanamura, Solid State Commun. **11**, 485 (1972).

(Received June 21, 1974)

phys. stat. sol. (b) 65, 373 (1974)

Subject classification: 13

Lumumba University, Moscow

The Local and Quasi-Local Impurity States in the Finite One-Dimensional Hubbard Model

By

I. D. MIKHAILOV and I. YU. SHELESTINA

An analytic expression for the Green's function of the ground state of the finite one-dimensional Hubbard model and the density of the electronic states on the sites are obtained. It is shown that the introduction of a local perturbation can cause the formation of quasilocalized levels which may transform into localized states when the pertubation reaches some critical value.

Es wird ein analytischer Ausdruck für die Greensche Funktion des Grundzustands des endlichen eindimensionalen Hubbard-Modells und die Dichte der elektronischen Zustände erhalten. Es wird gezeigt, daß die Einführung einer lokalen Störung die Bildung von quasilokalisierten Niveaus verursachen kann, die in lokalisierte Zustände übergehen können, wenn die Störung einen kritischen Wert erreicht.

1. Introduction

The one-dimensional Hubbard model [1 to 3] is interesting from the point of view of studying the spectrum and electric properties of polymeric semiconductors with conjugated bonds and also in case the organic complexes with charge transfer [4]. This model is the simplest of the one-dimensional systems which takes into consideration the electron–electron interaction. The impurity states in the semi-infinite one-dimensional Hubbard model were studied in [5 to 7], and it was shown that as soon as the impurity approaches the end of the chain the conditions for the formation of localized levels become more complicated. Different from [5 to 7] in this work the finite Hubbard model is considered for which an analytical expression for the Green's function is obtained. It gives the opportunity of obtaining not only the conditions of the formation of the localized levels and their energies but makes it possible to investigate the density of the electronic states inside the band when the impurity is present. The conditions for the formation of quasi-localized levels [8] is studied.

2. Green's Function for the Finite One-Dimensional Hubbard Model

In case the impurity is absent the Hubbard Hamiltonian has the form

$$H_0 = \beta \sum_{n,\sigma} (a^+_{n,\sigma} a_{n+1,\sigma} + a^+_{n,\sigma} a_{n-1,\sigma}) + \frac{\gamma}{2} \sum_{n,\sigma} a^+_{n,\sigma} a_{n,\sigma} a^+_{n,-\sigma} a_{n,-\sigma}\,, \qquad (1)$$

where $a^+_{n,\sigma}$ and $a_{n,\sigma}$ are the operators of electron creation and annihilation on site n with spin σ ($\sigma = \pm\frac{1}{2}$), γ and β are the Coulomb and resonance integrals, respectively.

Let us introduce the advanced Green's function [9]

$$G_{n,\sigma;m,\sigma'}(t) = \langle\langle a_{n,\sigma}(t)\, a^+_{m,\sigma'}(0)\rangle\rangle = -i\theta(t)\,\langle[a_{n,\sigma}(t), a^+_{m,\sigma'}(0)]\rangle\,, \qquad (2)$$

where $\theta(t)$ is the step function; $\theta(t) = 1$ if $t > 0$, $\theta(t) = 0$ otherwise. These

Green's functions for the Hamiltonian (1) can be shown to satisfy the equation of motion [9]

$$i\frac{\mathrm{d}G^{(0)}_{n,\sigma;m,\sigma'}(t)}{\mathrm{d}t} = \delta(t)\,\delta_{n,m}\delta_{\sigma,\sigma'} + \beta\,[G^{(0)}_{n+1,\sigma;m,\sigma'}(t) + G^{(0)}_{n-1,\sigma;m,\sigma'}(t)] + $$
$$+ \gamma\langle\langle a_{n,-\sigma}(t)\,a^+_{n,-\sigma}(t)\,a_{n,\sigma}(t);\,a^+_{m,\sigma'}(0)\rangle\rangle\,. \tag{3}$$

A simple approximate method for the solution of the system (3) is offered in [2] which being expressed in formal symbols used here leads to a decoupling of the system of equations for the Green's function with the help of the following correlations:

$$\langle\langle a^+_{n,-\sigma}(t)\,a_{n,-\sigma}(t)\,a_{n,\sigma}(t);\,a^+_{m,\sigma'}(0)\rangle\rangle \approx \langle a^+_{n,-\sigma}(t)a_{n,-\sigma}(t)\rangle\,\langle\langle a_{n,\sigma}(t)\,a^+_{m,\sigma'}(0)\rangle\rangle\,, \tag{4}$$

$$\langle a^+_{n,\sigma}a_{n,\sigma}\rangle = \tfrac{1}{2} + (-1)^n\,\delta\;;\;\langle a^+_{n,-\sigma}a_{n,-\sigma}\rangle = \tfrac{1}{2} - (-1)^n\,\delta;\;\langle a^+_{n,\sigma}a_{n,-\sigma}\rangle = 0\,. \tag{5}$$

The correlation (5) takes into consideration the fact that in case the electron–electron interaction is strong enough the ground state of the one-dimensional Hubbard model is antiferromagnetic [10].

Subsituting (4) and (5) in (3) and using the Fourier component for the Green's function

$$G^{(0)}_{n,\sigma;m,\sigma'}(t) = \frac{1}{2\pi}\int\limits_{-\infty}^{+\infty} G^{(0)}_{n,\sigma;m,\sigma'}(E)\,\mathrm{e}^{iEt}\,\mathrm{d}E\,, \tag{6}$$

we can write

$$\left.\begin{array}{l} A_nG^{(0)}_{n,m}(E) = \delta_{n,m} + \beta\,[G^{(0)}_{n+1,m}(E) + G^{(0)}_{n-1,m}(E)]\;;\quad n = 2, 3, \ldots, l-1\,,\\ A_1G^{(0)}_{1,m}(E) = \delta_{1,m} + \beta G^{(0)}_{2,m}(E)\,,\quad A_lG^{(0)}_{l,m}(E) = \delta_{l,m} + \beta G^{(0)}_{l-1,m}(E)\,, \end{array}\right\} \tag{7}$$

$$A_n = E - \tfrac{1}{2}\gamma - (-1)^n\,\gamma\delta\,, \tag{8}$$

where l is the length of the chain.

Let us assume that $\sigma = \sigma' = +\frac{1}{2}$ in (7) and further. The equation for the state with $\sigma = \sigma' = -\frac{1}{2}$ can be obtained by analogy.

One can always express the solution of the system of recurrence correlations (7) in the form of a linear combination of Chebyshev polynomials [11, 12] where the solution will be different for chains with an even and an odd number of electrons.

1. Chain with an even number of atoms ($l = 2N$):

$$G^{(0)}_{2k,2k}(E) = \frac{A_1U_{k-1}(x)}{\beta^2}\,\frac{U_{N-k}(x) + U_{N-k-1}(x)}{U_N(x) + U_{N-1}(x)};\quad k = 1, 2, 3, \ldots, N\,, \tag{9}$$

$$G^{(0)}_{2k+1,2k+1}(E) = \frac{A_2}{\beta^2}\,[U_k(x) + U_{k-1}(x)]\,\frac{U_{N-k-1}(x)}{U_N(x) + U_{N-1}(x)};\quad k = 1, 2, 3, \ldots, N\,. \tag{10}$$

2. Chain with an odd number of atoms ($l = 2N + 1$):

$$G^{(0)}_{2k,2k}(E) = \frac{A_2}{\beta^2}\,U_{k-1}(x)\,\frac{U_{N-k}(x)}{U_N(x)};\quad k = 1, 2, 3, \ldots, N\,, \tag{11}$$

$$G^{(0)}_{2k+1,2k+1}(E) = \frac{1}{A_1}[U_k(x) + U_{k-1}(x)]\frac{U_{N-k}(x) + U_{N-k-1}(x)}{U_N(x)};$$

$$k = 1, 2, 3, \ldots, N, \tag{12}$$

where

$$U_N(x) = \frac{\sin[(N+1)\arccos x]}{\sin(\arccos x)}; \quad x = \frac{A_1A_2 - 2\beta^2}{2\beta^2} = \frac{(E - \frac{1}{2}\gamma)^2 - \gamma^2\delta^2 - 2\beta^2}{2\beta^2}. \tag{13}$$

The parameter δ in (5) can be found from the self-consistent equation for the Green's functions (9) to (12) which in this case has the form [2, 9]:

$$\frac{1}{2} - (-1)^n\,\delta = \frac{2}{\pi l}\int\limits_{-\infty}^{\mu} \mathrm{Im}\, G_{n,n}(E - i\eta)\,\mathrm{d}E\,; \quad (\eta \to 0)\,, \tag{14}$$

where μ is the Fermi energy. We shall not investigate the solution of (14), as this question has been investigated in [3, 13 14].

As far as the electron spectrum is defined by the poles of the Green's function [8], for a chain with an even number of atoms ($l = 2N$) according to (9), (10) it can be found from the solution of the equation $U_N(x) + U_{N-1}(x) = 0$,

$$E_s^{(2N)} = \frac{1}{2}\gamma \pm \sqrt{\gamma^2\delta^2 + 4\beta^2\cos^2\frac{2s\pi}{2N+1}}\,; \quad s = 1, 2, 3, \ldots, N \tag{15}$$

and for a chain with an odd number of atoms ($l = 2N + 1$) according to (11), (12) from the equation $U_N(x) = 0$, and $A_1 = 0$,

$$\left.\begin{aligned} E_s^{(2N+1)} &= \frac{1}{2}\gamma \pm \sqrt{\gamma^2\delta^2 + 4\beta^2\cos^2\frac{s\pi}{2N+1}}\,; \quad s = 1, 2, 3, \ldots, N\,,\\ E_0 &= \tfrac{1}{2}\gamma - \gamma\delta\,. \end{aligned}\right\} \tag{16}$$

The last level (16) is localized only on odd sites as it is clear from (11), (12).

3. Local Levels in the Finite Chain

The above-mentioned expressions for the Green's function (9) to (12) can be used for the solution of the influence of the local perturbation by using the method suggested in [15]. For this aim let us consider the Hamiltonian

$$H = H_0 + V\,; \qquad V = U_0 a_m^+ a_m\,, \tag{17}$$

where H_0 is given by (1). Solving Dyson's equation [15] for the perturbed Green's function

$$\hat{G}(E) = \hat{G}^{(0)}(E) + \hat{G}^{(0)}(E)\,\hat{V}\hat{G}(E) \tag{18}$$

we shall have [15]:

$$G_{n,n'}(E) = G^{(0)}_{n,n'}(E) + \frac{U_0 G^{(0)}_{n,m}(E)\,G^{(0)}_{m,n'}(E)}{1 - U_0 G^{(0)}_{m,m}(E)}. \tag{19}$$

From this solution it is clear that in addition to the poles of the ideal chain in

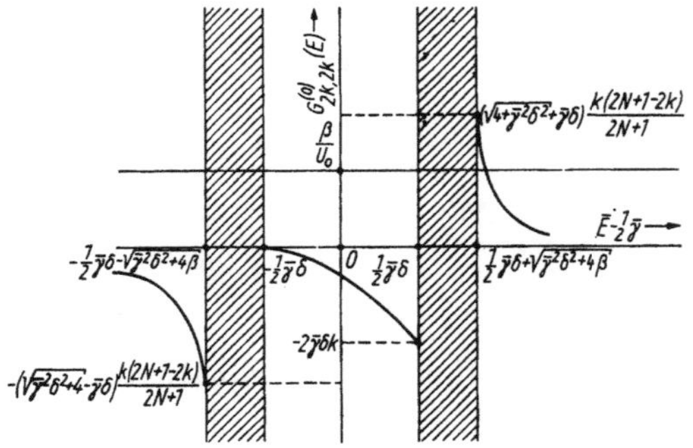

Fig. 1. Graphic solution of equation (20) in the case of a chain containing even numbers ($l = 2N$) of atoms. The local perturbation takes place on the even site $m = 2k$

case of $G^{(0)}_{n,n}(E)$ the perturbed Green's function can have new poles which can be found from the following equation:

$$1 - U_0 \operatorname{Re} G^{(0)}_{m,m}(E) = 0 \,. \tag{20}$$

The solution of (20) with the Green's functions (9) to (12) can be obtained graphically as it is shown on Fig. 1 for the case of a chain with an even number of atoms ($l = 2N$) when the perturbation is localized on the even site $m = 2k$ ($k = 1, 2, ..., N$). The analysis of the solution shows that in case of positive values of the perturbation

$$\left.\begin{aligned} &U_0 > U_1 \,, \\ &U_1 = \frac{\beta}{\sqrt{4 + \bar{\gamma}^2\delta^2} + \bar{\gamma}\delta} \, \frac{2N+1}{k(2N+1-2k)} \,, \quad \text{where} \quad \bar{\gamma} = \frac{\gamma}{\beta} \end{aligned}\right\} \tag{21}$$

the local level appears below the upper band which moves up from the latter if the perturbation increases. When $U_0 = U_1$ this level mixes with the upper band and when $0 < U_0 < U_1$ the local level does not appear. For negative values of the perturbation when their absolute value is great,

$$U_0 < U_2 \,; \qquad U_2 = -\frac{\beta}{2\bar{\gamma}\delta k} \,, \tag{22}$$

two localized levels appears: the first one below the bands when $U_0 \to \infty$ goes down and the second level which appears in the gap in case $U_0 = U_2$ splits off from the lower bound of the upper band and when $U_0 \to \infty$ approaches the upper bound of the lower band.

When

$$\left.\begin{aligned} &U_3 < U_0 < U_2 \,; \\ &U_3 = -\left(\sqrt{\bar{\gamma}^2\delta^2 + 4} - \bar{\gamma}\delta\right) \frac{k(2N+1-2k)}{2N+1} \end{aligned}\right\} \tag{23}$$

no level appears in the gap and only the level below the lower band appears which splits from its lower boundary in case $U_0 = U_3$.

In case $N \to \infty$ the results of (21) to (23) turn into the results obtained earlier in [5 to 7].

Using the expressions for the Green's function (10) to (12) one can obtained similar results in the remaining cases.

4. The Green's Function for the Ideal Semi-Infinite Chain and the Electron State Density at the Site

Let us consider the limiting case of a semi-infinite chain for the analysis of the density inside the band. For this aim let us have $N \to \infty$ in (9) to (12) and use the identity

$$\frac{U_{N-k}(x - i\eta)}{U_N(x - i\eta)} \xrightarrow[\eta \to 0]{N \to \infty} (x - \sqrt{x^2 - 1})^k . \tag{24}$$

In this limit the expressions for the Green's function will be equal for chains with an even and an odd number of atoms:

$$G^{(0)}_{2k,2k}(E) = \frac{A_1}{\beta^2} U_{k-1}(x) (x - \sqrt{x^2 - 1})^k ; \qquad N \to \infty , \tag{25}$$

$$G^{(0)}_{2k+1,2k+1}(E) = \frac{1}{A_2} [U_k(x) + U_{k-1}(x)] (x - \sqrt{x^2 - 1})^k (1 + x - \sqrt{x^2 - 1}) ; \quad N \to \infty \tag{26}$$

Let us use the connection between the electronic state density for the site with number n, ϱ_n, and the imaginary part of the Green's function

$$\varrho_n = \frac{2}{\pi l} \int\limits_{-\infty}^{\mu} \operatorname{Im} G_{n,n}(E - i\eta) \, dE ; \qquad (\eta \to 0; \quad n = 1, 2, 3, \dots, l) , \tag{27}$$

where μ is the Fermi energy and l the length of the chain. Substituting (25) and (26) into (27) we have

$$\varrho_{2k} = \frac{1}{2\pi N} \frac{A_1}{\beta^2} U^2_{k-1}(x) \sqrt{1 - x^2} ; \quad (N \to \infty; \quad k = 1, 2, 3, \dots) , \tag{28}$$

$$\varrho_{2k+1} = \frac{1}{2\pi N} \frac{1}{A_2} [U_k(x) + U_{k-1}(x)]^2 \sqrt{1 - x^2} ; \quad (N \to \infty; \; k = 0, 1, 2, \dots). \tag{29}$$

The graphs of these curves for the end site $n = 1$ and the one next to it, $n = 3, 5$, and the density of states at an odd site for the internal atom for the infinite Hubbard model which coincides with the density of states at the sites of the infinite chain are also shown in Fig. 2.

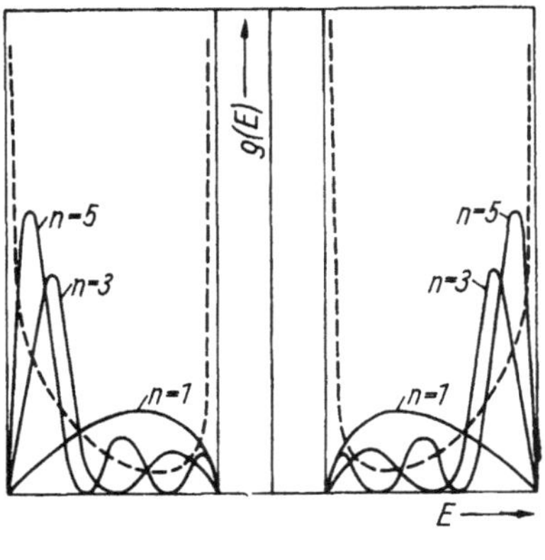

Fig. 2. The electron density of a semi-infinite chain on the end site ($n = 1$) and on the neighbouring one ($n = 3,5$)

From this figure it is clear that the density of state at the end atoms is greater in the middle of the band and for the internal atoms on the contrary at the end of the band. The further the curve of the density of states moves away from the end of the chain the more oscillating function it becomes; the number of oscillation is k $(k = 1, 2, 3, \ldots)$.

5. The Quasi-Local Impurity Levels in the Semi-Infinite Chain

Let us investigate the influence of the impurity atom on the distribution of the electronic density of states inside the permitted band of the semi-infinite Hubbard model. For this aim let us substitute in turn (23) and (26) into (19) and consider $n = n' = m$:

$$G_{mm} = G_{mm}^{(0)} + \frac{U_0 G_{mm}^{(0)2}}{1 - U_0 G_{mm}^{(0)}} = \frac{G_{mm}^{(0)}}{1 - U_0 G_{mm}^{(0)}}\,. \tag{30}$$

The density of states at the impurity site m will have two expressions, in case of perturbation of the site with an even number ($m = 2k$) and an odd number ($m = 2k + 1$), respectively,

$$\varrho_{2k} = \operatorname{Im} \frac{1}{2\pi N} G_{2k,2k} = \frac{\dfrac{A_1}{2\pi N\beta^2}\sqrt{1 - x^2}}{\sqrt{1 - x^2} + \dfrac{1}{U_{k-1}^2(x)}\left[T_k(x) + U_0 \dfrac{A_1}{\beta^2} U_{k-1}(x)\right]^2}, \tag{31}$$

$$\varrho_{2k+1} = \operatorname{Im} \frac{1}{2\pi N} G_{2k+1,2k+1} = \frac{1}{2\pi N} \frac{U_k(x) + U_{k-1}(x)}{A_2} \times$$

$$\times \frac{\sqrt{1 - x^2}\, U_{k-1}(x) + \sqrt{\dfrac{1 - x}{1 + x}}\, T_k(x)}{\left[T_k(x) - 2U_0 \dfrac{U_k(x) + U_{k-1}(x)}{A_2} U_{k-1}(x)\right]^2 + \left[\sqrt{\dfrac{1 - x}{1 + x}}\, T_k(x) - \sqrt{1 - x^2}\, U_{k-1}(x)\right]^2}\,. \tag{32}$$

The graphic dependence of the electronic density of states inside the band on the value of the perturbation U_0 at the end atom ($n = 1$) and the one next to it ($n = 2$) is shown in Fig. 3 and 4. As soon as U_0 increases the maximum of the density at the end atom moves up and in the lower band it moves quicker. When the maximum approaches the boundary of the band it begins to increase quickly. It may be shown that the lifetime of such a state also sharply increases [16]. These states were called quasi-localized in [8]. At some critical value of U_0 the quasi-localized level in the lower band reaches the upper boundary and after this turns into the local level and the density inside the bands falls sharply. Further growing of U_0 brings about the appearance of quasi-localized levels at the upper boundary of the upper band which approaches the boundary and gradually turns into the local one.

Fig. 3. The state density change for the end atom ($n = 1$) as a function of increasing perturbation

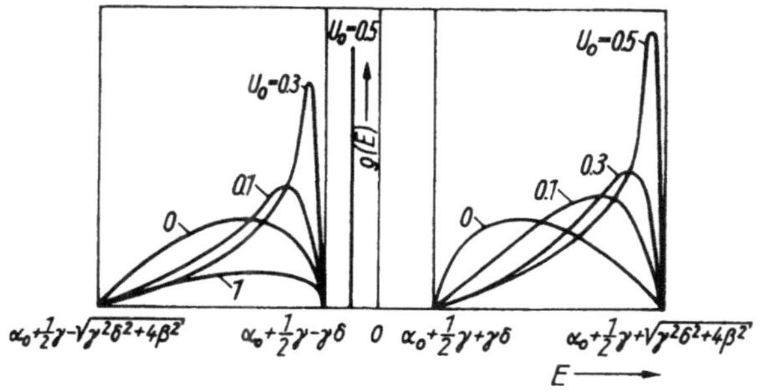

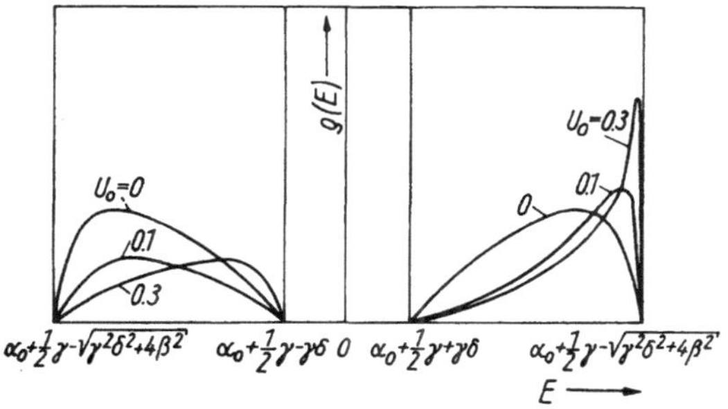

Fig. 4. The state density change on site $n = 2$ as a function of increasing perturbation U_0

Thus two localized levels can appear at the end atom when $U_0 > 0$ and their formation is much easier in the gap and it is caused by the asymmetry of the electronic density of states.

A quite contrary picture exists in the case of the site next to the end one because the density of states on this site has a maximum on the other boundaries when $U_0 = 0$.

In this case when U_0 increases the quasi-localized level appears immediately at the upper boundary of the upper band and the quasi-localized level does not appear in the lower band in spite of the fact that the maximum moves up as the absolute value of this maximum decreases.

References

[1] A. A. Ovchinnikov, I. I. Ukrainskii, and G. F. Kventsel, Uspekhi fiz. Nauk **108**, 81 (1972).
[2] I. A. Misurkin and A. A. Ovchinnikov, Teor. i. eksper. Khim. **4**, 3 (1968); Theor. Chim. Acta **13**, 115 (1969).
[3] I. A. Misurkin and A. A. Ovchinnikov, Theor. i. eksper. Khim. **3**, 431 (1967).
[4] A. I. Epstein, S. Estmad, A. F. Garito, and J. Huger, Phys. Rev. B **5**, 952 (1972).
[5] G. F. Kventsel, Teor. i eksper. Khim. **4**, 291 (1968).
[6] G. F. Kventsel, Ukr. fiz. Zh. **16**, 617 (1971).
[7] I. I. Ukrainskii and G. F. Kventsel, Theor. Chim. Acta **25**, 360 (1972).
[8] Yu. Kagan and Ya. A. Iosilevskii, Zh. eksper. i theor. Fiz. **42**, 259 (1962); **44**, 284 (1963).

[9] D. N. ZUBAREV, Uspekhi fiz. Nauk 71, 71 (1961).
[10] J. HUBBARD, Proc. Roy. Soc. **A276**, 238 (1963).
[11] I. M. RYZHIK and I. S. GRADSHTEIN, Tabilitsy integralov, summ, ryadov i proizvedenii, Moscow 1971 (p. 1047).
[12] I. D. MIKHAILOV, Fiz. Metallov i Metallovedenie **32**, 1141 (1971).
[13] I. I. UKRAINSKII and YU. A. KRUGLYAK, Ukr. fiz. Zh. **15**, 1068 (1970).
[14] YU. A. KRUGLYAK and I. I. UKRAINSKII, Internat. J. Quantum. Chem. **4**, 57 (1970).
[15] I. M. LIFSHITS, Uspekhi fiz. Nauk **83**, 617 (1964).
[16] R. S. GORDENKO and I. D. MIKHAILOV, Fiz. tverd. Tela **15**, 1181 (1973).

(Received April 22, 1974)

phys. stat. sol. (b) **65**, 381 (1974)

Subject classification: 13.5.1; 22.2; 22.4

Lehrstuhl II für Theoretische Physik der Technischen Universität Berlin (West)

Electronic Contribution to the Dynamical Screening in the Theory of Excitons in III–V and II–VI Semiconductors with Zincblende Structure

By

W. EKARDT

Starting from the Bethe-Salpeter equation formulation of Sham and Rice the influence of the dynamical screening in the theory of direct excitons in III–V and II–VI zincblende structure semiconductors is computed taking into account the wave vector dependence of the dielectric constant and higher reciprocal lattice vectors. With the aid of published results of the dielectric constant calculations in the R.P.A. for zincblende semiconductors the influence of the dynamical screening is calculated exactly within the ladder approximation up to the order of magnitude E_B/E_{gap}. The results agree with the general statement by Zimmermann that the binding energy should increase due to dynamical screening. Comparison with experiments becomes possible only through the inclusion of some further corrections.

Ausgehend von der Formulierung der Bethe-Salpeter-Gleichung nach Sham und Rice wird der Einfluß der dynamischen Abschirmung bei direkten Exzitonen in III–V- und II–VI-Zinkblendestruktur-Halbleitern unter Berücksichtigung der Wellenvektorabhängigkeit der Dielektrizitätskonstanten unter Einschluß der höheren Vektoren im reziproken Gitter berechnet. Mit Hilfe veröffentlichter Berechnungen der Dielektrizitätskonstanten in der R.P.A. für Halbleiter mit Zinkblende-Struktur gelingt es, den dynamischen Abschirm-Einfluß bis zur Größenordnung E_B/E_{gap} einschließlich im Rahmen der Leiternäherung exakt zu berechnen. Die Ergebnisse sind in Übereinstimmung mit der allgemeinen Aussage von Zimmermann, daß sich die Bindungsenergie vergrößern sollte. Der Vergleich mit dem Experiment ist nur möglich durch Berücksichtigung weiterer Korrekturen.

1. Introduction

In order to explain the deviations of Wannier exciton spectra from the hydrogen spectrum one has to start at the general many-particle derivations of the effective mass equation [1 to 3]. In this way one obtains a consistent procedure which takes into account higher corrections. The question is which corrections should be included. Noting that the basic structure of the spectrum is hydrogen-like a development of the irreducible particle–hole interaction in terms of the independent wave vectors of the scattering process, e.g. $\boldsymbol{K} - \boldsymbol{K}'$, $\boldsymbol{K}'$, and $\boldsymbol{q}$ (cf. below), seems reasonable. Since for the Wannier exciton the term E_B/E_{gap} is a suitable parameter we develop up to E_B/E_{gap}.

It becomes obvious then that one has to develop the irreducible particle–hole interaction up to terms proportional to (wave vector)^0.

In this way one obtains an exciton equation which includes three groups of corrections:

1. Corrections resulting from an improved irreducible particle–hole interaction:

a) analytical exchange interaction [2, 4 to 8],

b) nonanalytical exchange interaction [5 to 7],

c) wave vector dependence of the dielectric constant (spatial dispersion),

d) corrections resulting from a $\boldsymbol{K} \cdot \boldsymbol{p}$ development of conduction band and valence band Bloch factors,

e) local-field corrections (contributions from higher Fourier coefficients of the different terms involved),

f) dynamical electron–electron interaction: a consequence of the frequency dependence of the full screening function,

g) dynamical electron–phonon interaction.

2. Corrections of usually ignored contributions of the two-particle amplitude:

a) the influence of the hole–particle component leads to the so-called "non-Hermitean splitting term" [10] or "interband exchange" [11]; it has been treated in connection with the exciton for the first time by Horie [1] in order to study the connection between excitons and plasmons in semiconductors;

b) influence of the particle–particle and hole–hole terms: typical superconductivity terms.

3. Corrections resulting from an improved spectrum of the corresponding one-particle excitations:

Introduction of Shockley-type matrices [12, 13].

From all these corrections part 1.f) together with 1.c) will be treated in this paper. The wave vector dependence of the dielectric constant in the static part of the interaction will be studied in a paper to be published in connection with further corrections to the static part of the interaction.

For 1.f) Zimmermann [9] has shown a way to take into account correction beyond E_B/E_{gap}. On the other hand he has completely removed the wave vector dependence of the screening function and has not used concrete R.P.A. calculations of the dielectric constant. By contrast we are going to use the results of the R.P.A. from Vinsome and Richardson [14, 15] to calculate the dynamical correction.

The general result obtained by Zimmermann [9] about dynamical screening is corroborated: The binding energy is increased for the 1s states of $\Gamma^{3,4,5}$ by 1.5% for InSb up to 9.6% in the case of GaP (direct exciton in GaP). With increasing main quantum number the corrections decrease rapidly.

2. Derivation of the Basic Equation by the Method of Sham and Rice [3]

The starting point for the derivation is the Bethe-Salpeter equation obtained in [3]:

$$G_{0;\lambda\boldsymbol{q}}(2, 3) = G_{\boldsymbol{K}+\boldsymbol{q}}(2, 2')\, G_{\boldsymbol{K}}(3', 3)\, I(1', 2'; 3', 4')\, G_{0;\lambda\boldsymbol{q}}(4', 1')\,. \tag{1}$$

In (1) all variables appearing two times will be summed or integrated over, respectively. The zero-temperature formalism is used:

$$G_{0;\lambda\boldsymbol{q}}(2, 3) \equiv G_{0;\lambda\boldsymbol{q}}(\boldsymbol{r}_2, \omega_2; \boldsymbol{r}_3, \omega_3) =$$

$$= \sum_\alpha \left\{ \frac{\langle N0|\, \psi(\boldsymbol{r}_2)\, |N+1, \alpha\rangle \langle N+1, \alpha|\, \psi^+(\boldsymbol{r}_3)\, |N, \lambda, \boldsymbol{q}\rangle}{\omega_2 + i\delta - (E^\alpha_{N+1} - E^0_N)} - \right.$$

$$\left. - \frac{\langle N0|\, \psi^+(\boldsymbol{r}_3)\, |N-1, \alpha\rangle \langle N-1, \alpha|\, \psi(\boldsymbol{r}_2)\, |N, \lambda, \boldsymbol{q}\rangle}{-\omega_2 + i\delta + (E^0_N - E^\alpha_{N-1})} \right\}, \tag{2}$$

$$G_{\boldsymbol{K}}(2, 2') \equiv \sum_{\nu} \left\{ \frac{\varphi^{\mathrm{e}}_{\nu,\boldsymbol{K}}(\boldsymbol{r}_2)\, \varphi^{\mathrm{e}+}_{\nu,\boldsymbol{K}}(\boldsymbol{r}_{2'})}{\omega_{2'} + i\delta - \mathscr{E}^{\mathrm{e}}_{\nu,\boldsymbol{K}}} + \frac{\varphi^{\mathrm{h}}_{\nu,\boldsymbol{K}}(\boldsymbol{r}_{2'})\, \varphi^{\mathrm{h}+}_{\nu,\boldsymbol{K}}(\boldsymbol{r}_2)}{\omega_{2'} - i\delta - \mathscr{E}^{\mathrm{h}}_{\nu,\boldsymbol{K}}} \right\} 2\pi i \delta(\omega_2 - \omega_{2'}) \,, \tag{3}$$

$$\varphi^{\mathrm{e}}_{\nu,\boldsymbol{K}}(\boldsymbol{r}) = \langle N0|\, \psi(\boldsymbol{r})\, |N+1, \nu, \boldsymbol{K}\rangle, \quad \varphi^{\mathrm{h}}_{\nu,\boldsymbol{K}}(\boldsymbol{r}) = \langle N0|\, \psi^{+}(\boldsymbol{r})\, |N-1, \nu, -\boldsymbol{K}\rangle \,, \tag{4}$$

$$\mathscr{E}^{\mathrm{e}}_{\nu,\boldsymbol{K}} = E^{\nu,\boldsymbol{K}}_{N+1} - E^{0}_{N} \,, \qquad \mathscr{E}^{\mathrm{h}}_{\nu,\boldsymbol{K}} = E^{0}_{N} - E^{\nu,-\boldsymbol{K}}_{N-1} \tag{5}$$

($\psi(\boldsymbol{r})$ is the spinor field operator).

Using the ladder approximation including the exchange interaction but excluding the (irreducible) vertex corrections of 1. kind, 2. kind etc. we have

$$\begin{aligned} I(1', 2'; 3', 4') = &- 2\pi i \delta(\omega_{1'} + \omega_{2'} - \omega_{3'} - \omega_{4'})\, \delta(\boldsymbol{r}_{1'} - \boldsymbol{r}_{3'})\, \delta(\boldsymbol{r}_{2'} - \boldsymbol{r}_{4'}) \times \\ &\times V(\boldsymbol{r}_{1'}, \boldsymbol{r}_{2'}; \omega_{1'} - \omega_{3'}) + 2\pi i \delta(\omega_{1'} + \omega_{2'} - \omega_{3'} - \omega_{4'}) \times \\ &\times \delta(\boldsymbol{r}_{1'} - \boldsymbol{r}_{4'})\, \delta(\boldsymbol{r}_{2'} - \boldsymbol{r}_{3'}) \frac{e^2}{|\boldsymbol{r}_{1'} - \boldsymbol{r}_{2'}|} \,. \end{aligned} \tag{6}$$

Here $V(\boldsymbol{r}_{1'}, \boldsymbol{r}_{2'}; \omega_{1'} - \omega_{3'})$ is a two-point potential with the following Fourier transform:

$$V(\boldsymbol{k} + \boldsymbol{g}, \boldsymbol{k} + \boldsymbol{g}'; \omega) = \frac{|\boldsymbol{k} + \boldsymbol{g}|}{|\boldsymbol{k} + \boldsymbol{g}'|}\, \varepsilon^{-1}(\boldsymbol{k} + \boldsymbol{g}, \boldsymbol{k} + \boldsymbol{g}'; \omega)\, v(\boldsymbol{k} + \boldsymbol{g}') \,; \tag{7}$$

$$v(\boldsymbol{k} + \boldsymbol{g}') = \frac{4\pi e^2}{\Omega |\boldsymbol{k} + \boldsymbol{g}'|^2} \,, \tag{8}$$

$$\begin{aligned} \varepsilon^{-1}(\boldsymbol{k} + \boldsymbol{g}, \boldsymbol{k} + \boldsymbol{g}'; \omega) = \delta_{\boldsymbol{g},\boldsymbol{g}'} &+ \frac{|\boldsymbol{k} + \boldsymbol{g}|}{|\boldsymbol{k} + \boldsymbol{g}'|}\, v(\boldsymbol{k} + \boldsymbol{g}) \times \\ &\times \sum_{\lambda} \left\{ \frac{\langle N0|\, \varrho_{\boldsymbol{k}+\boldsymbol{g}'}\, |N, \lambda, \boldsymbol{k}\rangle \langle N, \lambda, \boldsymbol{k}|\, \varrho_{-\boldsymbol{k}-\boldsymbol{g}}\, |N0\rangle}{\omega + i\eta - \omega_{\lambda,\boldsymbol{k}}} - \right. \\ &\left. - \frac{\langle N0|\, \varrho_{-\boldsymbol{k}-\boldsymbol{g}}\, |N, \lambda, -\boldsymbol{k}\rangle \langle N, \lambda, -\boldsymbol{k}|\, \varrho_{\boldsymbol{k}+\boldsymbol{g}'}\, |N0\rangle}{\omega - i\eta + \omega_{\lambda,\boldsymbol{k}}} \right\} . \end{aligned} \tag{9}$$

The symbols in (7) to (9) have the usual meaning.

The equation for the exciton amplitude

$$\begin{aligned} f_{\lambda,\boldsymbol{q}}(\boldsymbol{r}_2, \boldsymbol{r}_3) &= \langle N0|\, \psi^{+}(\boldsymbol{r}_3)\, \psi(\boldsymbol{r}_2)\, |N.\, \lambda, \boldsymbol{q}\rangle = \\ &= \lim_{\eta \to +0} \frac{1}{2\pi i} \int\limits_{\frown} \mathrm{d}\omega_2\, \mathrm{e}^{i\eta\omega_2}\, G_{0;\lambda\boldsymbol{q}}(\boldsymbol{r}_2, \omega_2; \boldsymbol{r}_3, \omega_2 - \omega) \end{aligned} \tag{10}$$

with this irreducible particle–hole interaction is obtained by putting (1) into (10), substituting there (6), and finally taking into account (3) and (1).

The integration of the Bethe-Salpeter equation with (10) is readily performed using the residual theorem:

$$
\begin{aligned}
f_{\lambda\boldsymbol{q}}(\boldsymbol{r}_2, \boldsymbol{r}_3) = {} & \int \mathrm{d}\boldsymbol{r}_{2'}\, \mathrm{d}\boldsymbol{r}_{3'} \sum_{\substack{\tilde{\nu}\tilde{\boldsymbol{K}} \\ \nu\nu'\boldsymbol{K}}} V_+(\boldsymbol{r}_{2'}, \boldsymbol{r}_{3'}; \omega + \mathscr{E}_{\mathrm{h}}^{\tilde{\nu}\tilde{\boldsymbol{K}}} - \mathscr{E}_{\mathrm{e}}^{\nu, \boldsymbol{K}+\boldsymbol{q}})\, \langle N0|\, \psi^+(\boldsymbol{r}_{3'})\, |N-1, \tilde{\nu}, \tilde{\boldsymbol{K}}\rangle\, \langle N-1, \tilde{\nu}, \tilde{\boldsymbol{K}}|\, \psi(\boldsymbol{r}_{2'})\, |N, \lambda, \boldsymbol{q}\rangle \times \\
& \times \frac{\varphi_{\mathrm{e}}^{\nu, \boldsymbol{K}+\boldsymbol{q}}(\boldsymbol{r}_2)\, \varphi_{\mathrm{e}}^{+\nu, \boldsymbol{K}+\boldsymbol{q}}(\boldsymbol{r}_{2'})\, \varphi_{\mathrm{h}}^{\nu', \boldsymbol{K}}(\boldsymbol{r}_3)\, \varphi_{\mathrm{h}}^{+\nu', \boldsymbol{K}}(\boldsymbol{r}_{3'})}{-\omega + \mathscr{E}_{\mathrm{e}}^{\nu, \boldsymbol{K}+\boldsymbol{q}} - \mathscr{E}_{\mathrm{h}}^{\nu', \boldsymbol{K}}} + \int \mathrm{d}\boldsymbol{r}_{2'}\, \mathrm{d}\boldsymbol{r}_{3'} \sum_{\substack{\tilde{\nu}\tilde{\boldsymbol{K}} \\ \nu\nu'\boldsymbol{K}}} V_+(\boldsymbol{r}_{2'}, \boldsymbol{r}_{3'}; \omega + \mathscr{E}_{\mathrm{h}}^{\tilde{\nu}, \tilde{\boldsymbol{K}}} - \omega - \mathscr{E}_{\mathrm{e}}^{\nu', \boldsymbol{K}}) \times \\
& \times \langle N0|\, \psi^+(\boldsymbol{r}_{3'})\, |N-1, \tilde{\nu}, \tilde{\boldsymbol{K}}\rangle\, \langle N-1, \tilde{\nu}, \tilde{\boldsymbol{K}}|\, \psi(\boldsymbol{r}_{2'})\, |N, \lambda, \boldsymbol{q}\rangle\, \frac{\varphi_{\mathrm{h}}^{+\nu, \boldsymbol{K}+\boldsymbol{q}}(\boldsymbol{r}_2)\, \varphi_{\mathrm{h}}^{\nu, \boldsymbol{K}+\boldsymbol{q}}(\boldsymbol{r}_{2'})\, \varphi_{\mathrm{e}}^{+\nu', \boldsymbol{K}}(\boldsymbol{r}_3)\, \varphi_{\mathrm{e}}^{\nu', \boldsymbol{K}}(\boldsymbol{r}_{3'})}{\omega + \mathscr{E}_{\mathrm{e}}^{\nu', \boldsymbol{K}} - \mathscr{E}_{\mathrm{h}}^{\nu, \boldsymbol{K}+\boldsymbol{q}}} + \\
& + \int \mathrm{d}\boldsymbol{r}_{2'}\, \mathrm{d}\boldsymbol{r}_{3'} \sum_{\substack{\tilde{\nu}\tilde{\boldsymbol{K}} \\ \nu\nu'\boldsymbol{K}}} V_-(\boldsymbol{r}_{2'}, \boldsymbol{r}_{3'}; \mathscr{E}_{\mathrm{e}}^{\tilde{\nu}\tilde{\boldsymbol{K}}} - \mathscr{E}_{\mathrm{h}}^{\nu', \boldsymbol{K}} - \omega)\, \langle N0|\, \psi(\boldsymbol{r}_{2'})\, |N+1, \tilde{\nu}, \tilde{\boldsymbol{K}}\rangle\, \langle N+1, \tilde{\nu}, \tilde{\boldsymbol{K}}|\, \psi^+(\boldsymbol{r}_{3'})\, |N, \lambda, \boldsymbol{q}\rangle \times \\
& \times \frac{\varphi_{\mathrm{e}}^{\nu, \boldsymbol{K}+\boldsymbol{q}}(\boldsymbol{r}_2)\, \varphi_{\mathrm{e}}^{+\nu, \boldsymbol{K}+\boldsymbol{q}}(\boldsymbol{r}_{2'})\, \varphi_{\mathrm{h}}^{\nu', \boldsymbol{K}}(\boldsymbol{r}_3)\, \varphi_{\mathrm{h}}^{+\nu', \boldsymbol{K}}(\boldsymbol{r}_{3'})}{\omega - \mathscr{E}_{\mathrm{e}}^{\nu, \boldsymbol{K}+\boldsymbol{q}} + \mathscr{E}_{\mathrm{h}}^{\nu', \boldsymbol{K}}} + \int \mathrm{d}\boldsymbol{r}_{2'}\, \mathrm{d}\boldsymbol{r}_{3'} \sum_{\substack{\tilde{\nu}\tilde{\boldsymbol{K}} \\ \nu\nu'\boldsymbol{K}}} V_-(\boldsymbol{r}_{2'}, \boldsymbol{r}_{3'}; \mathscr{E}_{\mathrm{e}}^{\tilde{\nu}\tilde{\boldsymbol{K}}} - \mathscr{E}_{\mathrm{h}}^{\nu'\boldsymbol{K}}) \times \\
& \times \langle N0|\, \psi(\boldsymbol{r}_{2'})\, |N+1, \tilde{\nu}, \tilde{\boldsymbol{K}}\rangle\, \langle N+1, \tilde{\nu}, \tilde{\boldsymbol{K}}|\, \psi^+(\boldsymbol{r}_{3'})\, |N, \lambda, \boldsymbol{q}\rangle\, \frac{\varphi_{\mathrm{h}}^{+\nu, \boldsymbol{K}+\boldsymbol{q}}(\boldsymbol{r}_2)\, \varphi_{\mathrm{h}}^{\nu, \boldsymbol{K}+\boldsymbol{q}}(\boldsymbol{r}_{2'})\, \varphi_{\mathrm{e}}^{+\nu', \boldsymbol{K}}(\boldsymbol{r}_3)\, \varphi_{\mathrm{e}}^{\nu', \boldsymbol{K}}(\boldsymbol{r}_{3'})}{-\omega - \mathscr{E}_{\mathrm{e}}^{\nu', \boldsymbol{K}} + \mathscr{E}_{\mathrm{h}}^{\nu, \boldsymbol{K}+\boldsymbol{q}}} - \\
& - \int \mathrm{d}\boldsymbol{r}_{2'}\, \mathrm{d}\boldsymbol{r}_{3'} \sum_{\substack{\tilde{\nu}\tilde{\boldsymbol{K}} \\ \nu, \nu', \boldsymbol{K}, \lambda, \boldsymbol{k}}} V_+^{\lambda, \boldsymbol{k}}(\boldsymbol{r}_{2'}, \boldsymbol{r}_{3'};)\, \frac{\langle N0|\, \psi^+(\boldsymbol{r}_{3'})\, |N-1, \tilde{\nu}, \tilde{\boldsymbol{K}}\rangle\, \langle N-1, \tilde{\nu}, \tilde{\boldsymbol{K}}|\, \psi(\boldsymbol{r}_{2'})\, |N, \lambda, \boldsymbol{q}\rangle}{-\mathscr{E}_{\mathrm{e}}^{\nu', \boldsymbol{K}} + \mathscr{E}_{\mathrm{h}}^{\tilde{\nu}\tilde{\boldsymbol{K}}} - \omega_{\lambda, \boldsymbol{k}}} \times \\
& \times \frac{\varphi_{\mathrm{e}}^{\nu, \boldsymbol{K}+\boldsymbol{q}}(\boldsymbol{r}_2)\, \varphi_{\mathrm{e}}^{+\nu, \boldsymbol{K}+\boldsymbol{q}}(\boldsymbol{r}_{2'})\, \varphi_{\mathrm{e}}^{+\nu', \boldsymbol{K}}(\boldsymbol{r}_3)\, \varphi_{\mathrm{e}}^{\nu', \boldsymbol{K}}(\boldsymbol{r}_{3'})}{-\mathscr{E}_{\mathrm{e}}^{\nu, \boldsymbol{K}+\boldsymbol{q}} + \mathscr{E}_{\mathrm{h}}^{\tilde{\nu}, \tilde{\boldsymbol{K}}} + \omega - \omega_{\lambda, \boldsymbol{k}}} - \int \mathrm{d}\boldsymbol{r}_{2'}\, \mathrm{d}\boldsymbol{r}_{3'} \sum_{\substack{\tilde{\nu}\tilde{\boldsymbol{K}} \\ \nu, \nu'\, \boldsymbol{K}, \lambda, \boldsymbol{k}}} V_-^{\lambda, \boldsymbol{k}}(\boldsymbol{r}_{2'}, \boldsymbol{r}_{3'};) \times \\
& \times \frac{\langle N0|\, \psi(\boldsymbol{r}_{2'})\, |N+1, \tilde{\nu}, \boldsymbol{K}\rangle\, \langle N+1, \tilde{\nu}, \boldsymbol{K}|\, \psi^+(\boldsymbol{r}_{3'})\, |N, \lambda, \boldsymbol{q}\rangle\, \varphi_{\mathrm{h}}^{+\nu, \boldsymbol{K}+\boldsymbol{q}}(\boldsymbol{r}_2)\, \varphi_{\mathrm{h}}^{\nu, \boldsymbol{K}+\boldsymbol{q}}(\boldsymbol{r}_{2'})\, \varphi_{\mathrm{h}}^{\nu', \boldsymbol{K}}(\boldsymbol{r}_3)\, \varphi_{\mathrm{h}}^{+\nu', \boldsymbol{K}}(\boldsymbol{r}_{3'})}{(\mathscr{E}_{\mathrm{e}}^{\tilde{\nu}\tilde{\boldsymbol{K}}} - \mathscr{E}_{\mathrm{h}}^{\nu, \boldsymbol{K}+\boldsymbol{q}} + \omega_{\lambda, \boldsymbol{k}})\, (-\omega + \mathscr{E}_{\mathrm{e}}^{\tilde{\nu}, \tilde{\boldsymbol{K}}} - \mathscr{E}_{\mathrm{h}}^{\nu', \boldsymbol{K}} + \omega_{\lambda, \boldsymbol{k}})} + \\
& + \int \mathrm{d}\boldsymbol{r}_{2'}\, \mathrm{d}\boldsymbol{r}_{3'} \sum_{\nu\nu'\boldsymbol{K}} \frac{e^2}{|\boldsymbol{r}_{2'} - \boldsymbol{r}_{3'}|}\, \frac{\varphi_{\mathrm{e}}^{\nu, \boldsymbol{K}+\boldsymbol{q}}(\boldsymbol{r}_2)\, \varphi_{\mathrm{e}}^{+\nu, \boldsymbol{K}+\boldsymbol{q}}(\boldsymbol{r}_{2'})\, \varphi_{\mathrm{h}}^{\nu'\boldsymbol{K}}(\boldsymbol{r}_3)\, \varphi_{\mathrm{h}}^{+\nu', \boldsymbol{K}}(\boldsymbol{r}_{2'})}{\omega - \mathscr{E}_{\nu', \boldsymbol{K}+\boldsymbol{q}}^{\mathrm{e}} + \mathscr{E}_{\nu', \boldsymbol{K}}^{\mathrm{h}}}\, f_{\lambda, \boldsymbol{q}}(\boldsymbol{r}_{3'}, \boldsymbol{r}_{3'}) + \int \mathrm{d}\boldsymbol{r}_{2'}\, \mathrm{d}\boldsymbol{r}_{3'} \sum_{\nu\nu'\boldsymbol{K}} \frac{e^2}{|\boldsymbol{r}_{2'} - \boldsymbol{r}_{3'}|} \times \\
& \times \frac{\varphi_{\mathrm{h}}^{+\nu, \boldsymbol{K}+\boldsymbol{q}}(\boldsymbol{r}_2)\, \varphi_{\mathrm{h}}^{\nu, \boldsymbol{K}+\boldsymbol{q}}(\boldsymbol{r}_{2'})\, \varphi_{\mathrm{e}}^{+\nu', \boldsymbol{K}}(\boldsymbol{r}_3)\, \varphi_{\mathrm{e}}^{\nu', \boldsymbol{K}}(\boldsymbol{r}_{2'})}{-\omega - \mathscr{E}_{\nu', \boldsymbol{K}}^{\mathrm{e}} + \mathscr{E}_{\nu, \boldsymbol{K}+\boldsymbol{q}}^{\mathrm{h}}}\, f_{\lambda, \boldsymbol{q}}(\boldsymbol{r}_{3'}, \boldsymbol{r}_{3'})\,. \qquad (11)
\end{aligned}
$$

In equation (11) the following abbreviations are used:

$$V_+(\boldsymbol{r}_{2'}, \boldsymbol{r}_{3'}; \tilde{\omega}) = \sum_{\boldsymbol{k},\boldsymbol{g},\boldsymbol{g}'} \mathrm{e}^{i(\boldsymbol{k}+\boldsymbol{q})\boldsymbol{r}_{2'}} \frac{|\boldsymbol{k}+\boldsymbol{g}'|}{|\boldsymbol{k}+\boldsymbol{g}|} \varepsilon_+^{-1}(\boldsymbol{k}+\boldsymbol{g}, \boldsymbol{k}+\boldsymbol{g}'; \tilde{\omega}) \, \mathrm{e}^{i(\boldsymbol{k}+\boldsymbol{g}')\boldsymbol{r}_{3'}}, \quad (12)$$

$$\varepsilon_+^{-1}(\boldsymbol{k}+\boldsymbol{g}, \boldsymbol{k}+\boldsymbol{g}'; \tilde{\omega}) =$$

$$= \frac{1}{2}\delta_{\boldsymbol{g},\boldsymbol{g}'} + \frac{|\boldsymbol{k}+\boldsymbol{g}|}{|\boldsymbol{k}+\boldsymbol{g}'|} v(\boldsymbol{k}+\boldsymbol{g}) \sum_{\lambda} \frac{\langle N0|\, \varrho_{\boldsymbol{k}+\boldsymbol{g}'} \,|N, \lambda, \boldsymbol{k}\rangle \langle N, \lambda, \boldsymbol{k}|\, \varrho_{-\boldsymbol{k}-\boldsymbol{g}} \,|N0\rangle}{\tilde{\omega} + i\eta - \omega_{\lambda,\boldsymbol{k}}}, \quad (13)$$

(V_- corresponds with ε_-^{-1}) and

$$V_+^{\lambda,\boldsymbol{k}}(\boldsymbol{r}_{2'}, \boldsymbol{r}_{3'};) = \sum_{\boldsymbol{g},\boldsymbol{g}'} \mathrm{e}^{i(\boldsymbol{k}+\boldsymbol{g})\boldsymbol{r}_{2'}} v(\boldsymbol{k}+\boldsymbol{g}) \langle N0|\, \varrho_{\boldsymbol{k}+\boldsymbol{g}} \,|N, \lambda, \boldsymbol{k}\rangle \times$$
$$\times \langle N, \lambda, \boldsymbol{k}|\, \varrho_{-\boldsymbol{k}-\boldsymbol{g}'} \,|N0\rangle \, v(\boldsymbol{k}+\boldsymbol{g}') \, \mathrm{e}^{-i(\boldsymbol{k}+\boldsymbol{g}')\boldsymbol{r}_{3'}} \quad (14)$$

($V_-^{\lambda,\boldsymbol{k}}$ correspondingly).

Equation (11) is a convenient starting point for the calculation of all corrections up to $E_\mathrm{B}/E_\mathrm{gap}$ within the ladder approximation. In (11) all the positive infinitesimals have been disregarded after the frequency integration since they become significant only for $|\omega| > E_\mathrm{gap}$, E_gap being the principal gap. E.g.

$$(\omega - \omega_{\lambda,\boldsymbol{k}} - \mathscr{E}_\mathrm{e}^{\nu,\boldsymbol{K}+\boldsymbol{q}} + \mathscr{E}_\mathrm{h}^{\tilde{\nu},\tilde{\boldsymbol{K}}} + i\delta)^{-1} (-\omega_{\lambda,\boldsymbol{k}} - \mathscr{E}_\mathrm{e}^{\nu',\boldsymbol{K}} + \mathscr{E}_\mathrm{h}^{\tilde{\nu},\tilde{\boldsymbol{K}}} + i\delta)^{-1} =$$

$$= \left(\mathrm{P}\frac{1}{\omega - \omega_{\lambda,\boldsymbol{k}} - \mathscr{E}_\mathrm{e}^{\nu,\boldsymbol{K}+\boldsymbol{q}} + \mathscr{E}_\mathrm{h}^{\tilde{\nu},\tilde{\boldsymbol{K}}}} - i\pi\delta(\omega - \omega_{\lambda,\boldsymbol{k}} - \mathscr{E}_\mathrm{e}^{\nu,\boldsymbol{K}+\boldsymbol{q}} + \mathscr{E}_\mathrm{h}^{\tilde{\nu},\tilde{\boldsymbol{K}}})\right) \times$$
$$\times \left(\mathrm{P}\frac{1}{-\omega_{\lambda,\boldsymbol{k}} - \mathscr{E}_\mathrm{e}^{\nu',\boldsymbol{K}} + \mathscr{E}_\mathrm{h}^{\tilde{\nu},\tilde{\boldsymbol{K}}}} - i\pi\delta(-\omega_{\lambda,\boldsymbol{k}} - \mathscr{E}_\mathrm{e}^{\nu',\boldsymbol{K}} + \mathscr{E}_\mathrm{h}^{\tilde{\nu},\tilde{\boldsymbol{K}}})\right).$$

Since an equation for $|\omega| \leqq E_\mathrm{gap}$ is being looked for and all $\omega_{\lambda,\boldsymbol{k}} \geqq 0$ one finds that the imaginary parts cancel identically.

In this paper only the dynamical screening in the main part of the two-particle amplitude, namely the particle–hole component shall be calculated.

In order to be able to compare the results with experiments we have to note that the dynamical electron–phonon interaction, too, plays an important role [16 to 18]. This means one would have within the ladder approximation an irreducible particle–polaron–hole–polaron interaction, which including the exchange interaction within the ladder approximation has the following appearance:

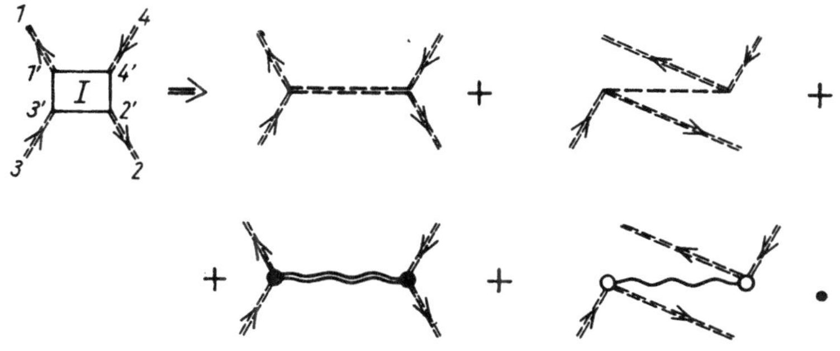

In the figure the graphical elements have the following meaning: dashed line: Coulomb interaction, double dashed line: screened Coulomb interaction (phonon propagation excluded), wavy line: unscreened phonon propagator, double wavy line: screened phonon propagator, circle: unscreened electron–phonon interaction, full circle: screened electron–phonon interaction.

Within the frame of these graphs ε^{-1} is approximately the screening function for the "Coulomb graph". For comparison with the experiment the corrections mentioned in the introduction are to be related to the group of these four graphs.

This will be evaluated quantitatively in a further paper.

Before calculating the dynamical correction quantitatively the following remark should be of some interest: The representation of the exciton amplitude resulting from (11)

$$\begin{aligned} f_{\lambda,\boldsymbol{q}}(\boldsymbol{r}_2, \boldsymbol{r}_3) = {} & \sum A_{\nu\nu'}(\boldsymbol{K}+\boldsymbol{q}, \boldsymbol{K})\, \varphi_{\mathrm{e}}^{\nu,\boldsymbol{K}+\boldsymbol{q}}(\boldsymbol{r}_2)\, \varphi_{\mathrm{h}}^{\nu',\boldsymbol{K}}(\boldsymbol{r}_3) + \\ & + \sum B_{\nu\nu'}(\boldsymbol{K}+\boldsymbol{q}, \boldsymbol{K})\, \varphi_{\mathrm{h}}^{+\nu,\boldsymbol{K}+\boldsymbol{q}}(\boldsymbol{r}_2)\, \varphi_{\mathrm{e}}^{+\nu',\boldsymbol{K}}(\boldsymbol{r}_3) + \\ & + \sum C_{\nu\nu'}(\boldsymbol{K}+\boldsymbol{q}, \boldsymbol{K})\, \varphi_{\mathrm{e}}^{\nu,\boldsymbol{K}+\boldsymbol{q}}(\boldsymbol{r}_2)\, \varphi_{\mathrm{e}}^{+\nu',\boldsymbol{K}}(\boldsymbol{r}_3) + \\ & + \sum D_{\nu\nu'}(\boldsymbol{K}+\boldsymbol{q}, \boldsymbol{K})\, \varphi_{\mathrm{h}}^{+\nu,\boldsymbol{K}+\boldsymbol{q}}(\boldsymbol{r}_2)\, \varphi_{\mathrm{h}}^{\nu',\boldsymbol{K}}(\boldsymbol{r}_3) \end{aligned} \tag{15}$$

has contributions with non-vanishing $C_{\nu\nu'}$ and $D_{\nu\nu'}$ only if one works from the beginning with a frequency-dependent two-point interaction. Otherwise, i.e. with a static two-point interaction, these terms cannot appear, since they correspond to poles in the same half-plane (upper or lower). Whereas the influence of the $B_{\nu\nu'}(\boldsymbol{K}+\boldsymbol{q}, \boldsymbol{K})$ component has already been studied extensively by Horie [1], Kübler [11], and Zimmermann [10], the influence of the $C_{\nu\nu'}(\boldsymbol{K}+\boldsymbol{q}, \boldsymbol{K})$ and $D_{\nu\nu'}(\boldsymbol{K}+\boldsymbol{q}, \boldsymbol{K})$ terms has been so far analysed only for spin systems by Teleman et al. [19]. But especially these terms may be of great importance for the excitonic phase, a problem being studied presently.

3. Dynamical Screening for the Main Part of the Exciton Amplitude

The equation for the main part of the exciton amplitude without exchange interaction reads[1])

$$\begin{aligned} f_{\lambda,\boldsymbol{q}}(\boldsymbol{r}_2, \boldsymbol{r}_3) = {} & \int \mathrm{d}\boldsymbol{r}_{2'}\, \mathrm{d}\boldsymbol{r}_{3'} \sum_{\nu\nu'\boldsymbol{K}\tilde{\nu}\tilde{\boldsymbol{K}}} V_{+}(\boldsymbol{r}_{2'}, \boldsymbol{r}_{3'}; \omega - \mathscr{E}_{\mathrm{e}}^{\nu,\boldsymbol{K}+\boldsymbol{q}} + \mathscr{E}_{\mathrm{h}}^{\tilde{\nu}\tilde{\boldsymbol{K}}}) \times \\ & \times \langle N0|\, \psi^{+}(\boldsymbol{r}_{3'})\, |N-1, \tilde{\nu}, \tilde{\boldsymbol{K}}\rangle\, \langle N-1, \tilde{\nu}, \tilde{\boldsymbol{K}}|\, \psi(\boldsymbol{r}_{2'})\, |N, \lambda, \boldsymbol{q}\rangle \times \\ & \times \frac{\varphi_{\mathrm{e}}^{\nu,\boldsymbol{K}+\boldsymbol{q}}(\boldsymbol{r}_2)\, \varphi_{\mathrm{h}}^{\nu',\boldsymbol{K}}(\boldsymbol{r}_3)\, \varphi_{\mathrm{e}}^{+\nu,\boldsymbol{K}+\boldsymbol{q}}(\boldsymbol{r}_{2'})\, \varphi_{\mathrm{h}}^{+\nu',\boldsymbol{K}}(\boldsymbol{r}_{3'})}{-\omega + \mathscr{E}_{\mathrm{e}}^{\nu,\boldsymbol{K}+\boldsymbol{q}} - \mathscr{E}_{\mathrm{h}}^{\nu',\boldsymbol{K}}} - \\ & - \int \mathrm{d}\boldsymbol{r}_{2'}\, \mathrm{d}\boldsymbol{r}_{3'} \sum_{\nu\nu'\boldsymbol{K}\tilde{\nu}\tilde{\boldsymbol{K}}} V_{-}(\boldsymbol{r}_{2'}, \boldsymbol{r}_{3'}; -\omega + \mathscr{E}_{\mathrm{e}}^{\tilde{\nu},\tilde{\boldsymbol{K}}} - \mathscr{E}_{\mathrm{h}}^{\nu',\boldsymbol{K}}) \times \\ & \times \langle N0|\, \psi(\boldsymbol{r}_{2'})\, |N+1, \tilde{\nu}, \tilde{\boldsymbol{K}}\rangle\, \langle N+1, \tilde{\nu}, \tilde{\boldsymbol{K}}|\, \psi^{+}(\boldsymbol{r}_{3'})\, |N, \lambda, \boldsymbol{q}\rangle \times \\ & \times \frac{\varphi_{\mathrm{e}}^{\nu,\boldsymbol{K}+\boldsymbol{q}}(\boldsymbol{r}_2)\, \varphi_{\mathrm{h}}^{\nu',\boldsymbol{K}}(\boldsymbol{r}_3)\, \varphi_{\mathrm{e}}^{+\nu,\boldsymbol{K}+\boldsymbol{q}}(\boldsymbol{r}_{2'})\, \varphi_{\mathrm{h}}^{+\nu',\boldsymbol{K}}(\boldsymbol{r}_{3'})}{-\omega + \mathscr{E}_{\mathrm{e}}^{\nu,\boldsymbol{K}+\boldsymbol{q}} - \mathscr{E}_{\mathrm{h}}^{\nu',\boldsymbol{K}}}. \end{aligned} \tag{16}$$

[1]) The exchange interaction is omitted because it does not contain any dynamical aspects; for comparison with experiment, however, it is absolutely necessary to be included.

The energy denominator in $V_{+,-}$ is developed as a geometrical series:

$$(-\omega_{\lambda,k} + \omega - \mathscr{E}_e^{\nu,K+q} + \mathscr{E}_h^{\tilde{\nu},\tilde{K}})^{-1} =$$
$$= \frac{-1}{\omega_{\lambda,k}}\left(1 + \frac{\omega - \mathscr{E}_e^{\nu,K+q} + \mathscr{E}_h^{\tilde{\nu},\tilde{K}}}{\omega_{\lambda,k}} + O\left(\left|\frac{E_B}{E_{gap}}\right|^2\right)\right) \tag{17}$$

with a similar expression for V_-.

For III–V compounds the first omitted term is of the order of magnitude of 0.8×10^{-6} in AlSb to 12.6×10^{-6} in GaP. The terms we studied, however, contribute remarkably (as will be shown). Putting (17) and the corresponding series for V_- into equation (16) one obtains

$$f_{\lambda,q}(r_2, r_3) = \sum_{\nu\nu' K} \frac{\varphi_e^{\nu,K+q}(r_2)\,\varphi_h^{\nu',K}(r_3)}{\omega - \mathscr{E}_e^{\nu,K+q} + \mathscr{E}_h^{\nu',K}} \int dr_{2'}\, dr_{3'}\, \varphi_e^{+\nu,K+q}(r_{2'})\, \varphi_h^{+\nu',K}(r_{3'}) \times$$
$$\times \sum_{\tilde{\nu}\tilde{K}} \{-V_+(r_{2'}, r_{3'}; 0)\, \langle N0|\, \psi^+(r_{3'})\, |N-1, \tilde{\nu}, \tilde{K}\rangle \times$$
$$\times \langle N-1, \tilde{\nu}, \tilde{K}|\psi(r_{2'})|N, \lambda, q\rangle + V_-(r_{2'}, r_{3'}; 0) \langle N0|\, \psi(r_{2'})|N+1, \tilde{\nu}, \tilde{K}\rangle \times$$
$$\times \langle N+1, \tilde{\nu}, K|\, \psi^+(r_{3'})\, |N, \lambda, q\rangle\} + \sum_{\nu\nu' K} \frac{\varphi_e^{\nu,K+q}(r_2)\,\varphi_h^{\nu',K}(r_3)}{\omega - \mathscr{E}_e^{\nu,K+q} + \mathscr{E}_h^{\nu',K}} \times$$
$$\times \int dr_{2'}\, dr_{3'}\, \varphi_e^{+\nu,K+q}(r_{2'})\, \varphi_h^{+\nu',K}(r_{3'}) \sum_{\tilde{\nu}\tilde{K}} \sum_{kgg'} (-)\, e^{i(k+g)\, r_{2'}}\, v(k+g) \times$$
$$\times \sum_{\lambda} \frac{\langle N0|\, \varrho_{k+g'}\, |N, \lambda, k\rangle\, \langle N, \lambda, k|\varrho_{-k-g}\, |N0\rangle}{-\omega_{\lambda,k}} \times$$
$$\times v(k+g)\, e^{-i(k+g')\, r_{3'}} \frac{\omega - \mathscr{E}_e^{\nu,K+q} + \mathscr{E}_h^{\tilde{\nu},\tilde{K}}}{\omega_{\lambda,k}} \times$$
$$\times \langle N0|\, \psi^+(r_{3'})\, |N-1, \tilde{\nu}, \tilde{K}\rangle\, \langle N-1, \tilde{\nu}, \tilde{K}|\, \psi(r_{2'})\, |N, \lambda, g\rangle +$$
$$+ \sum_{\nu\nu' K} \frac{\varphi_e^{\nu,K+q}(r_2)\,\varphi_h^{\nu',K}(r_3)}{\omega - \mathscr{E}_e^{\nu,K+q} + \mathscr{E}_h^{\nu',K}} \int dr_{2'}\, dr_{3'}\, \varphi_e^{+\nu',K+q}(r_{2'})\, \varphi_h^{+\nu',K}(r_{3'}) \times$$
$$\times \sum_{\tilde{\nu}\tilde{K}} \sum_{kgg'} e^{i(k+g)\, r_{2'}}\, v(k+g) \times$$
$$\times \sum_{\lambda} \frac{\langle N0|\, \varrho_{-k-g}\, |N, \lambda, -k\rangle\, \langle N, \lambda, -k|\, \varrho_{k+g'}\, |N0\rangle}{-\omega_{\lambda,k}} \times$$
$$\times v(k+g')\, e^{-i(k+g')\, r_{3'}} \frac{\omega - \mathscr{E}_e^{\tilde{\nu},\tilde{K}} + \mathscr{E}_h^{\nu',K}}{\omega_{\lambda,k}} \times$$
$$\times \langle N0|\, \psi(r_{2'})\, |N+1, \tilde{\nu}, \tilde{K}\rangle\, \langle N+1, \tilde{\nu}, \tilde{K}|\, \psi^+(r_{3'})\, |N, \lambda, q\rangle\,. \tag{18}$$

Taking into account

$$\sum_{\tilde{\nu}\tilde{K}} \langle N0|\, \psi^+(r_{3'})\, |N-1, \tilde{\nu}, \tilde{K}\rangle\, \langle N-1, \tilde{\nu}, \tilde{K}|\, \psi(r_{2'})\, |N, \lambda, q\rangle = f_{\lambda,q}(r_{2'}, r_{3'}) \text{ etc.},$$

one obtains the main part of the interaction, namely the static $V(r_{2'}, r_{3'}; 0)$ and there remain only additional dynamical terms.

Because of the time inversion symmetry in the field-free crystal we have

$$\sum_{kgg'} e^{i(k+g)r_{2'}} v(k+g) \sum_{\lambda} \frac{\langle N0| \varrho_{k+g'} |N, \lambda, k\rangle \langle N, \lambda, k| \varrho_{-k-g} |N0\rangle}{(\omega_{\lambda,k})^2} \times$$
$$\times v(k+g')\, e^{-i(k+g')r_{3'}} =$$
$$= \sum_{kgg'} e^{i(k+g)r_{2'}} v(k+g) \sum_{\lambda} \frac{\langle N0| \varrho_{-k-g} |N, \lambda, -k\rangle \langle N, \lambda, -k| \varrho_{k+g'} |N0\rangle}{(\omega_{\lambda,k})^2} \times$$
$$\times v(k+g')\, e^{-i(k+g')r_{3'}} . \tag{19}$$

Thus one can extract this sum as a common factor. On the other hand,

$$\mathscr{E}_{\mathrm{h}}^{\tilde{\nu},\tilde{K}} \langle N0| \psi^+(r_{3'}) |N-1, \tilde{\nu}, \tilde{K}\rangle \equiv \mathscr{E}_{\mathrm{h}}^{\tilde{\nu}\tilde{K}} \varphi_{\mathrm{h}}^{\tilde{\nu}\tilde{K}}(r_{3'}) = \varphi_{\mathrm{h}}^{\tilde{\nu}\tilde{K}}(r_{3'})\, \mathscr{H}(r_{3'}) . \tag{20}$$

Here $\mathscr{H}(r_{3'})$ is the (nonlocal) operator, the left-hand and right-hand eigenfunctions of which define the elementary excitations of the one-particle type. Similarly

$$\mathscr{E}_{\mathrm{e}}^{\tilde{\nu}\tilde{K}} \langle N0| \psi(r_{2'}) |N+1, \tilde{\nu}, \tilde{K}\rangle \equiv \mathscr{E}_{\mathrm{e}}^{\tilde{\nu}\tilde{K}} \varphi_{\mathrm{e}}^{\tilde{\nu}\tilde{K}}(r_{2'}) = \mathscr{H}(r_{2'}) \varphi_{\mathrm{e}}^{\tilde{\nu}\tilde{K}}(r_{2'}) . \tag{21}$$

After this substitution the sum $\sum_{\tilde{\nu}\tilde{K}}$ can be performed exactly and one obtains for the dynamical part of the interaction

$$- \sum_{\nu\nu' K} \frac{\varphi_{\mathrm{e}}^{\nu, K+q}(r_2)\, \varphi_{\mathrm{h}}^{\nu', K}(r_3)}{\omega - \mathscr{E}_{\mathrm{e}}^{\nu, K+q} + \mathscr{E}_{\mathrm{h}}^{\nu', K}} \int \mathrm{d}r_{2'}\, \mathrm{d}r_{3'} \times$$
$$\times \Bigg(\sum_{kgg'} e^{i(k+g)r_{2'}} \frac{|k+g'|}{|k+g|} \frac{|k+g|}{|k+g'|} \cdot v(k+g) \times$$
$$\times \sum_{\lambda} \frac{\langle N0| \varrho_{k+g'} |N, \lambda, k\rangle \langle N, \lambda, k| \varrho_{-k-g} |N0\rangle}{-(\omega_{\lambda,k})^2} v(k+g')\, e^{-i(k+g')r_{3'}} \Bigg) \times$$
$$\times \varphi_{\mathrm{e}}^{+\nu, K+q}(r_{2'})\, \varphi_{\mathrm{h}}^{+\nu', K}(r_{3'}) \left(\omega - \mathscr{E}_{\mathrm{e}}^{\nu, K+q} + \mathscr{E}_{\mathrm{h}}^{\nu', K} + \omega - \mathscr{H}(r_{2'}) + \mathscr{H}(r_{3'})\right) \times$$
$$\times f_{\lambda, q}(r_{2'}, r_{3'}) . \tag{22}$$

($\mathscr{H}(r_{3'}) \,.\, f(r_{2'}, r_{3'})$ is meant "symbolically"; it should be read $f(r_{2'}, r_{3'})\, \mathscr{H}(r_{3'})$).
Defining a matrix of the effective reciprocal energy gap as

$$\overline{\left(\frac{1}{\mathscr{E}_{\mathrm{gap}}^{k}(g, g')}\right)} =$$
$$= \frac{\dfrac{|k+g|}{|k+g'|} v(k+g) \sum_{\lambda} \dfrac{1}{\omega_{\lambda,k}}\, 2\, \dfrac{\langle N0| \varrho_{k+g} |N, \lambda, k\rangle \langle N, \lambda, k| \varrho_{-k-g'} |N0\rangle}{-\omega_{\lambda,k}}}{\dfrac{|k+g|}{|k+g'|} v(k+g) \sum_{\lambda} 2\, \dfrac{\langle N0| \varrho_{k+g} |N, \lambda, k\rangle \langle N, \lambda, k| \varrho_{-k-g'} |N0\rangle}{-\omega_{\lambda,k}}} \tag{23}$$

one can express finally the dynamical correction as

$$+ \sum_{\nu\nu'K} \frac{\varphi_e^{\nu, K+q}(r_2)\, \varphi_h^{\nu', K}(r_3)}{\omega - \mathscr{E}_e^{\nu, K+q} + \mathscr{E}_h^{\nu', K}} \int dr_{2'}\, dr_{3'}\, \varphi_e^{+\nu, K+q}(r_{2'})\, \varphi_h^{+\nu', K}(r_{3'}) \times$$
$$\times \left\{ \sum_{kgg'} e^{i(k+g)\, r_{2'}} \frac{|k+g'|}{|k+g|} \left(\delta_{g,g'} - \varepsilon^{-1}(k+g, k+g'; 0)\right) v(k+g')\, e^{-i(k+g')\, r_{3'}} \right\} \times$$
$$\times \overline{\left(\frac{1}{\mathscr{E}_{\text{gap}}^{k}(g, g')} \right)} \left(\omega - \mathscr{E}_e^{\nu, K+q} + \mathscr{E}_h^{\nu', K} + \omega - \mathscr{H}(r_{2'}) + \mathscr{H}(r_{3'})\right) f_{\lambda, q}(r_{2'}, r_{3'})\,. \tag{24}$$

4. Potential in the Two-Particle Equation

In (24) we restrict the sum over ν and ν' to the lowest S conduction bands and highest S′ valence bands, respectively, being degenerate in K-space at $K_c = K_v = 0$. Then we get in the usual way

$$\sum_{gg'K'je'jv'} U_{je, je'}^{K+q, K'+q}(g)\, U_{jv', jv}^{K', K}(-g')\, \frac{|K - K' + g'|}{|K - K' + g|} \times$$
$$\times \left(\delta_{g,g'} - \varepsilon^{-1}(K - K' + g, K - K' + g'; 0)\right) \frac{4\pi e^2}{\Omega\, |K - K' + g'|^2} \times$$
$$\times \overline{\left(\frac{1}{\mathscr{E}_{\text{gap}}^{K-K'}(g, g')} \right)} \left(\omega - \mathscr{E}_e^{je, K+q} + \mathscr{E}_h^{jv, K} + \omega - \mathscr{E}_e^{je', K'+q} + \mathscr{E}_h^{jv', K'}\right) \times$$
$$\times A_{je'jv'}(K' + q, K')\,. \tag{25}$$

In (25)

$$U_{je, je'}^{K+q, K'+q}(g) = \frac{1}{\Omega_0} \int_{\Omega_0} dr\, \left(u_{je, K+q}^{+}(r), u_{je', K+q}(r)\right) e^{+igr} \tag{26}$$

are the familiar Bloch integrals of the Bloch spinor products. For (25) we are going to use the following approximations:

1. Since for the off-diagonal elements of the screening function ε^{-1} no numerical values are available up to day for III–V and II–VI zincblende semiconductors (cf. [20]), only diagonal elements are taken into account.
2. In order to be able to develop in terms of $K - K'$ and K' one imagines pseudo-Bloch spinors being introduced [5]. Then the one-particle energy expression between brackets becomes a Shockley-type matrix, which is being approximated through the s-like diagonal part [12, 13].[2])
3. The development is performed up to the order (wave vector)^0. Thus we obtain

$$\sum_{K'je'jv'} \left\{ \delta_{je, je'} \delta_{jv, jv'} \left(1 - \varepsilon^{-1}(0; 0; 0)\right) \frac{4\pi e^2}{\Omega\, |K - K'|^2} \left(\frac{\omega - \mathscr{E}_G}{\tilde{\mathscr{E}}_G} - \frac{\hbar^2}{2\mu} \frac{(K - K')^2}{2\tilde{\mathscr{E}}_G} - \right.\right.$$
$$\left. - \frac{\hbar^2}{2\mu} \frac{(K - K')\, K'}{\tilde{\mathscr{E}}_G} - \frac{\hbar^2}{2\mu} \frac{K'^2}{\tilde{\mathscr{E}}_G} \right) + \sum_{g \neq 0} U_{jeje'}^{0,0}(g)\, U_{jv'jv}^{0,0}(-g) \times$$
$$\left. \times \left(1 - \varepsilon^{-1}(g, g; 0)\right) \frac{4\pi e^2}{\Omega\, |g|^2} \frac{\omega - \mathscr{E}_G}{\tilde{\tilde{\mathscr{E}}}_G(g)} \right\} A_{je'jv'}(K' + q, K')\,. \tag{27}$$

---

[2]) That means we have neglected only a higher correction term.

In (27) the following abbreviations have been used (we put $\boldsymbol{q} = 0$):

$$\overline{\left(\frac{1}{\mathscr{E}^0_{\text{gap}}(0,0)}\right)} \equiv \frac{1}{\tilde{\mathscr{E}}_{\text{G}}}, \qquad \overline{\left(\frac{1}{\mathscr{E}^0_{\text{gap}}(\boldsymbol{g},\boldsymbol{g})}\right)} \equiv \frac{1}{\tilde{\tilde{\mathscr{E}}}_{\text{G}}(\boldsymbol{g})}. \tag{28}$$

The calculation of these quantities is performed in the Appendix. After transforming as usual (27) back into the space of relative movement one obtains the desired expression containing the dynamical corrections:

$$\sum_{je'jv'} \Bigg\{ -\left(1 - \varepsilon^{-1}(0,0;0)\right) \left(\frac{\mathscr{E}_{\text{G}} - \omega}{\tilde{\mathscr{E}}_{\text{G}}} \frac{e^2}{r} + \frac{\hbar^2 e^2}{2\mu\tilde{\mathscr{E}}_{\text{G}}} \left[2\pi\delta(\boldsymbol{r}) + \frac{1}{r}\frac{1}{\hbar^2}\boldsymbol{p}^2 + \right.\right.$$

$$\left.\left. + \left(\frac{1}{\hbar}\boldsymbol{p}\frac{1}{r}\right)\frac{1}{\hbar}\boldsymbol{p}\right]\right) \delta_{jeje'}\delta_{jvjv'} - \sum_{\boldsymbol{g}\neq 0} \left(1 - \varepsilon^{-1}(\boldsymbol{g},\boldsymbol{g};0)\right) \times$$

$$\times U^{0,0}_{jeje'}(\boldsymbol{g})\, U^{0,0}_{jv'jv}(-\boldsymbol{g}) \frac{4\pi e^2}{|\boldsymbol{g}|^2} \frac{\mathscr{E}_{\text{G}} - \omega}{\tilde{\tilde{\mathscr{E}}}_{\text{G}}(\boldsymbol{g})} \delta(\boldsymbol{r}) \Bigg\} F^{\boldsymbol{q}}_{je'jv'}(\boldsymbol{r}) \tag{29}$$

(where $F^{\boldsymbol{q}}_{je'jv'}(\boldsymbol{r})$ is defined as usual [5]).

5. Evaluation for Direct Excitons in Zincblende Semiconductors

The local-field term in (29) is actually superfluous, since for Wannier-like excitons $\mathscr{E}_{\text{G}} - \omega = 0$ ($|\text{wave vector}|^2$). It will be shown below, too, that this term is by some orders of magnitude smaller than the $\boldsymbol{g} = 0$ term. Indeed, if one notes that the $\boldsymbol{K}$-linear terms could not be observed until now for nearly all the III–V compounds and therefore regards approximately the purely orbital functions at the point $\boldsymbol{K} = 0$ as functions of well, defined parity, then the local-field term transforms into the expression

$$-\sum_{\boldsymbol{g}\neq 0} \left(1 - \varepsilon^{-1}(\boldsymbol{g},\boldsymbol{g};0)\right) \frac{4\pi e^2}{|\boldsymbol{g}|^2} \frac{\mathscr{E}_{\text{G}} - \omega}{\tilde{\tilde{\mathscr{E}}}_{\text{G}}(\boldsymbol{g})} \langle S|\, e^{i\boldsymbol{g}\boldsymbol{r}}\, |S\rangle \langle Z|\, e^{-i\boldsymbol{g}\boldsymbol{r}}\, |Z\rangle\, \delta(\boldsymbol{r})\, \delta_{jeje'}\delta_{jvjv'}\,. \tag{30}$$

In (30) S is a function of the representation Γ_1 for T_d and Z the z-row of Γ_5 for T_d.

With the approximation (30) equation (29) is readily evaluated. Neglecting the local-field term one obtains within first order perturbation theory for $n = 1, 2, 3$ (state: ns-level of symmetry Γ_3, Γ_4, Γ_5):

$$\Delta E_{ns} = -\, Ry_0\left(1 - \varepsilon^{-1}_{\text{R.P.A.}}(0,0;0)\right) 2\varepsilon_0 \frac{Ry_0}{\tilde{\mathscr{E}}_{\text{G}}} \left(\frac{6}{n^3} - \frac{2}{n^4} + \frac{1}{4}\delta_{n,2} + \frac{32}{243}\delta_{n,3}\right). \tag{31}$$

In (31) we used the symbols: $Ry_0 = \mu e^4/2\hbar^2\varepsilon_0^2$; $\delta_{n,2}$, $\delta_{n,3}$ Kronecker symbol, ε_0 static dielectric constant, $\varepsilon_{\text{R.P.A.}}(0, 0; 0)$ static optical dielectric constant: high frequency dielectric constant in the R.P.A., $1/\mu = 1/m_e + \gamma_1/m_0$ is the mean reduced effective mass introduced already formerly where γ_1 is a Kohn-Luttinger parameter (cf. [12]). Evaluating (31) with different parameters one obtains the results listed in Table 1.

Table 1

Enhanced binding energy due to dynamical electron–electron interaction in zincblende semiconductors. We have due to various approximations no relevant splitting between the irreducible representations $\Gamma_{3,4,5}$

crystal	ε_0 [1])	ε_∞ [2])	Ry_0 [4]) (meV)	E_{gap} [3]) (meV)	$\tilde{E}_{gap}$ (meV)	ΔE_B (% Ry_0)		
						1s	2s	3s
AlSb	9.9	9.3	1.37	2220	5700	1.7	0.37	0.12
	11.22		1.07			1.33	0.28	0.12
GaP	11.1	8.4	8.3	2740	7330	8.85	1.93	0.73
	10.28		9.8			9.67	2.11	0.80
GaAs	12.5	11.1	4.18	1519	5980	6.35	1.39	0.53
	12.5		3.92		6280	5.68	1.24	0.47
	12.83		3.98		6370	5.76	1.26	0.48
	12.83		3.72		5980	5.80	1.26	0.48
GaSb	15.5	14.5	2.06	810	4950	4.78	1.04	0.39
	15.5		1.82			5.32	1.16	0.44
	15.69		1.94			4.30	1.00	0.38
	15.69		1.71			4.04	0.88	0.33
InP	12.1	9.8	4.94	1424	5680	7.55	1.65	0.62
	12.29		4.78			7.44	1.62	0.61
InAs	14.61	13.2	1.15	410	5320	2.33	0.51	0.19
			1.02			2.07	0.45	0.17
InSb	17.88	15.7	0.51	240	4380	1.56	0.33	0.13
ZnS	8.1	4.8	37.0	3800	8550	22.1	4.84	1.83
			49.6			29.8	6.50	2.46
ZnSe	8.7	5.5	22.5	2800	8600	14.55	3.24	1.25
	8.33		25.88			16.32	3.57	1.40
ZnTe	10.1	7.4	9.18	2390	6410	10.00	2.19	0.80
	9.86		11.33			12.08	2.62	0.97
CdTe	9.7	6.7	11.42	1610	6650	11.52	2.48	0.94
	10.23		9.06			9.52	2.08	0.76

[1]) First value from [12], second one from [21].
[2]) In [14] theoretically calculated R.P.A. values.
[3]) From [12] with the exception of InP (from [22]).
[4]) The various Ry_0 follow from various reduced masses and due to various dielectric constants.

Finally the local-field contribution has to be estimated. To do this one has to have an idea of the magnitude of $\tilde{\tilde{\mathscr{E}}}_G(\boldsymbol{g})$: In the Appendix it is shown that for studying orders of magnitude $\tilde{\tilde{\mathscr{E}}}_G(\boldsymbol{g})$ may be substituted by $\tilde{\tilde{\mathscr{E}}}_B$. But then one can find with the aid of calculated $\varepsilon_{R.P.A.}(\boldsymbol{g}, \boldsymbol{g}; 0)$ values [14] and the approximation $\varepsilon^{-1}_{R.P.A.}(\boldsymbol{g}, \boldsymbol{g}; 0) = 1/\varepsilon_{R.P.A.}(\boldsymbol{g}, \boldsymbol{g}; 0)$ an upper limit for the local-field contribution.

Table 2

$\varepsilon_{\text{R.P.A.}}(\boldsymbol{g}, \boldsymbol{g}; 0)$ [14]	$\lvert\boldsymbol{g}\rvert$
1.5	$\sqrt{3}$
1.4	2
1.1	$\sqrt{8}$
1	$>\sqrt{8}$

One then obtains as local-field correction (a lattice constant, a_0 Bohr radius)

$$|\Delta E'_{ns}| =$$

$$= \left| \sum_{\boldsymbol{g} \neq 0} (1 - \varepsilon^{-1}_{\text{R.P.A.}}(\boldsymbol{g}, \boldsymbol{g}; 0)) \frac{4\pi e^2}{|\boldsymbol{g}|^2} \frac{\mathscr{E}_{\text{G}} - \omega}{\tilde{\tilde{\mathscr{E}}}_{\text{G}}(\boldsymbol{g})} \langle S| \,\mathrm{e}^{i\boldsymbol{g}\boldsymbol{r}}\, |S\rangle \langle Z| \,\mathrm{e}^{-i\boldsymbol{g}\boldsymbol{r}}|Z\rangle \frac{1}{\pi n^3 a_0^3} \right| \leqq$$

$$\leqq \left\{ 8 \left(1 - \frac{1}{\varepsilon_{\text{R.P.A.}}(1, 1, 1)}\right) \frac{1}{3} + 6 \left(1 - \frac{1}{\varepsilon_{\text{R.P.A.}}(2, 0, 0)}\right) \frac{1}{4} + \right.$$

$$\left. + \, 12 \left(1 - \frac{1}{\varepsilon_{\text{R.P.A.}}(2, 2, 0)}\right) \frac{1}{8} + 0 \right\} \varepsilon_0 \left(\frac{a}{a_0}\right)^2 \frac{2}{\pi^2} \frac{\mathscr{E}_{\text{G}} - \omega}{\tilde{\mathscr{E}}_{\text{G}}} Ry_0 \leqq 10^{-6}\, Ry_0\,. \qquad (32)$$

Here we used for the estimate in III–V compounds:

$$\varepsilon_0 \approx 10^{+1}\,, \qquad E_{\text{B}}/\tilde{\mathscr{E}}_{\text{G}} \approx 10^{-3}\,, \qquad (a/a_0)^2 \approx 5 \times 10^{-4}\,.$$

Clearly the local-field contribution to the dynamical correction can be completely neglected. (The static component of the local-field correction and the local-field component of the exchange interaction, namely the analytical exchange interaction for Wannier excitons will be given in another paper; they are very significant.)

6. Summary

With the aid of integration method of the Bethe-Salpeter equation for the irreducible particle–hole interaction by Sham and Rice [3] — which, by the way, has already been used in [2] — an effective equation for the particle–hole pair has been derived, which allows to study the dynamical aspects of the Coulomb interaction up to the order of magnitude of $E_{\text{B}}/\mathscr{E}_{\text{gap}}$ in a simple way.

The result is as expected and allows a quantitative specification of the general statement by Zimmermann about the increasing binding energy due to dynamical screening. The published calculations by Vinsome and Richardson [14, 15] of the 15-band R.P.A. dielectric constant have been extremely useful for our calculation, since with their aid the "effective dielectric band gap" $\tilde{\mathscr{E}}_{\text{G}}$ could be computed (cf. Appendix). For comparison with experiments, all the above given corrections — within the ladder approximation — have of course to be taken into account. The results of such calculations shall be published in a further paper.

Appendix

For the sake of simplicity the effective reciprocal energy gap is calculated as the reciprocal effective energy gap; this corresponds to the same degree of accuracy as putting so far — with the exception [20] in diamond — ε^{-1} equal to $1/\varepsilon$.

Table 3

	GaP	GaAs	AlSb	InP	GaSb	InAs	InSb	CdTe	ZnS	ZnSe	ZnTe
Γ_{15c}	14.48	12.06	10.4	11.9	9.95	11.2	9.3				
Γ_{12c}	11.29	10.55	8.2	10.3	8.87	10.1	8.5	9.47	13.60	11.85	9.52
Γ_{1c}	10.67	9.40	7.1	9.6	5.73	8.5	6.4	8.00	13.04	11.02	7.70
Γ_{15c}	5.29	4.70	3.4	5.3	3.42	4.9	3.7	5.12	8.21	6.81	4.62
Γ_{1c}	2.65	1.54	2.5	1.4	0.99	0.5	0.45	1.62	3.95	2.59	1.93
Γ_{15v}	0.0	0.0	0.0	0.0	0.0	0.0	0.0	− 0.29	0.0	− 0.14	− 0.29
Γ_{1v}	−11.73	−12.46	−10.1	−11.1	−11.18	−11.5	−10.2	−10.81	−12.20	−12.59	−11.49
reference	[24a]	[24b]	[24c]	[24d]	[24e]	[24d]	[24d]	[23]	[23]	[23]	[23]

[24a]: p. 251, [24b]: p. 275, [24c]: p. 391, [24d]: p. 392, [24e]: p. 253.
The entries for CdTe, ZnS, ZnSe, and ZnTe are based on arithmetic mean values of the corresponding double group representations after [23].

Hence one has to calculate first

$$\overline{\mathscr{E}^{\boldsymbol{k}}_{\text{gap}}(0,0)} = \frac{v(\boldsymbol{k}) \sum\limits_{\lambda} \omega_{\lambda,\boldsymbol{k}} 2 \dfrac{\langle N0|\, \varrho_{\boldsymbol{k}}\, |N,\lambda,\boldsymbol{k}\rangle \langle N,\lambda,\boldsymbol{k}|\, \varrho_{-\boldsymbol{k}}\, |N0\rangle}{\omega_{\lambda,\boldsymbol{k}}}}{v(\boldsymbol{k}) \sum\limits_{\lambda} 2 \dfrac{\langle N0|\, \varrho_{\boldsymbol{k}}\, |N,\lambda,\boldsymbol{k}\rangle \langle N,\lambda,\boldsymbol{k}|\, \varrho_{-\boldsymbol{k}}\, |N0\rangle}{\omega_{\lambda,\boldsymbol{k}}}}. \tag{A1}$$

Substituting in (A1) the exact many-particle quantities by their R.P.A. values one obtains

$$\overline{\mathscr{E}^{\boldsymbol{k}}_{\text{gap}}(0,0)} = \frac{2v(\boldsymbol{k}) \sum\limits_{\text{c,v},\boldsymbol{q}} \left(\mathscr{E}^{\text{c}}(\boldsymbol{k}+\boldsymbol{q}) - \mathscr{E}^{\text{v}}(\boldsymbol{q})\right) \dfrac{|\langle \text{v},\boldsymbol{q}|\, \text{e}^{-i\boldsymbol{k}\boldsymbol{r}}\, |\text{c},\boldsymbol{q}+\boldsymbol{k}\rangle|^2}{\mathscr{E}^{\text{c}}(\boldsymbol{k}+\boldsymbol{q}) - \mathscr{E}^{\text{v}}(\boldsymbol{q})}}{2v(\boldsymbol{k}) \sum\limits_{\text{c,v},\boldsymbol{q}} \dfrac{|\langle \text{v},\boldsymbol{q}|\, \text{e}^{-i\boldsymbol{k}\boldsymbol{r}}\, |\text{c},\boldsymbol{q}+\boldsymbol{k}\rangle|^2}{\mathscr{E}^{\text{c}}(\boldsymbol{k}+\boldsymbol{q}) - \mathscr{E}^{\text{v}}(\boldsymbol{q})}}. \tag{A2}$$

Substituting still the numerator of the numerator $\mathscr{E}^{\text{c}}(\boldsymbol{k}+\boldsymbol{q}) - \mathscr{E}^{\text{v}}(\boldsymbol{g})$ by $\mathscr{E}^{\text{c}}(0) - \mathscr{E}^{\text{v}}(0)$ opens a way for the numerical evaluation of (A2). In fact we have

$$\varepsilon_{\text{R.P.A.}}(\boldsymbol{k}) - 1 = \sum_{\text{c}} (\varepsilon_{\text{R.P.A.}} - 1)_{\text{c}}\,, \tag{A3}$$

$$(\varepsilon_{\text{R.P.A.}} - 1)_{\text{c}} = \sum_{\text{v},\boldsymbol{q}} 20v(\boldsymbol{k}) \frac{|\langle \text{v},\boldsymbol{q}|\, \text{e}^{-i\boldsymbol{k}\boldsymbol{r}} |\text{c},\boldsymbol{k}+\boldsymbol{q}\rangle|^2}{\mathscr{E}^{\text{c}}(\boldsymbol{k}+\boldsymbol{q}) - \mathscr{E}^{\text{v}}(\boldsymbol{q})}. \tag{A4}$$

The quantities in (A4) are tabulated by Vinsome and Richardson so that they can be supposed to be known (within the 15-band model as shown in Table 2).

Since these band contributions — as stated in [14] — are given as k-independent, they are to be regarded as mean contributions for $k = 0$ and for different $k \neq 0$ as well, e.g. for $\boldsymbol{k} = \boldsymbol{g}$, if one notes that in Vinsome and Richardson's paper the relative band contributions are given. But only these are required in (A3) since in this formula, too, only the relative weights of a specific band of a specific semiconductor enter.

Since in [14] only the quantities (A4) are tabulated but no distinction is made for valence bands, one has to substitute in (A2)

$$\mathscr{E}^{\text{c}}(0) - \mathscr{E}^{\text{v}}(0) \Rightarrow \mathscr{E}^{\text{c}}(0) - \mathscr{E}^{\text{v}\Gamma_1,\Gamma_{15}}(0) \quad \text{with} \quad \mathscr{E}^{\text{v}\Gamma_1,\Gamma_{15}} = \tfrac{1}{4}\left(3\mathscr{E}^{\text{v}\Gamma_{15}}(0) + \mathscr{E}^{\text{v}\Gamma_1}(0)\right). \tag{A5}$$

Since in [14] the used band distances are not given, one has to look for these elsewhere. Table 3 gives the used band distances with corresponding references.

Acknowledgement

The author would like to thank Prof. Dr. A. Haug from Lehrstuhl II für Theoretische Physik for many helpful and stimulating discussions.

References

[1] C. Horie, Progr. theor. Phys. **21**, 113 (1959).
[2] Y. Abe, Y. Osaka, and A. Morita, J. Phys. Soc. Japan **17**, 1576 (1962).
[3] L. J. Sham and T. M. Rice, Phys. Rev. **144**, 708 (1966).
[4] Y. Abe, J. Phys. Soc. Japan **19**, 818 (1964).
[5] M. M. Denisov and V. P. Makarov, Preprint, Moscow 1971.

[6] M. M. Denisov and V. P. Makarov, Preprint No. 78, Moscow 1971 (in Russian).
[7] M. M. Denisov and V. P. Makarov, phys. stat. sol. (b) **56**, 9 (1973).
[8] Y. Onodera and Y. Toyozawa, J. Phys. Soc. Japan **22**, 833 (1967).
[9] R. Zimmermann, phys. stat. sol. (b) **48**, 603 (1971).
[10] R. Zimmermann, phys. stat. sol. **41**, 23 (1970).
[11] J. Kübler, phys. stat. sol. (b) **46**, 311 (1971).
[12] A. Baldereschi and N. O. Lipari, Phys. Rev. B **3**, 439 (1971).
[13] N. O. Lipari, Phys. Rev. B **4**, 4535 (1971).
[14] P. K. W. Vinsome and D. Richardson, J. Phys. C **4**, 2650 (1971).
[15] D. Richardson and P. K. W. Vinsome, Phys. Letters (Netherlands) **36 A**, 3 (1971).
[16] K. Shindo, J. Phys. Soc. Japan **29**, 287 (1970).
[17] S. D. Mahanti and C. M. Varma, Phys. Rev. B **6**, 2209 (1972).
[18] J. Sak, Phys. Rev. B **6**, 2226 (1972).
[19] E. Teleman, G. Ciobanu, and A. B. Fazakas, phys. stat. sol. (b) **52**, 375 (1972).
[20] J. A. van Vechten and R. M. Martin, Phys. Rev. Letters **28**, 446 (1972).
[21] E. Kartheuser, in: Polarons in Ionic Crystals and Polar Semiconductors, Ed. J. T. Devreese, North-Holland Publ. Co., 1972 (p. 717)
[22] J. U. Fischbach, G. Benz, N. Stath, M. H. Pilkuhn, and K. W. Benz, Solid State Commun. **11**, 721 (1972).
[23] I. B. Ortenburger and W. E. Rudge, IBM Res. RJ 1041, May 1972 (unpublished).
[24] F. Herman, R. L. Kortum, I. B. Ortenburger, and J. P. van Dyke, Electronic Structure and Optical Spectrum of Semiconductors, ARL 69-0080, Palo Alto (California), May 1969 (Technical Documentary Report, unpublished).

(Received May 31, 1974)

phys. stat. sol. (b) **65**, 397 (1974)

Subject classification: 14.4 and 14.4.1; 22.8.2

Institute „J. Stefan", University of Ljubljana

Proton Spin–Lattice Relaxation Study of NH_4^+ Motion in Ammonium-Doped Ferroelectric RbH_2PO_4

By

R. BLINC, G. LAHAJNAR, A. LEVSTIK, R. OSREDKAR, and L. A. SHUVALOV[1])

The temperature dependences of the proton Zeeman and dipolar spin–lattice relaxation rates have been measured in a $Rb_{1-x}(NH_4)_xH_2PO_4$ mixed single crystal both in the paraelectric and the ferroelectric phase. The correlation time for ammonium ion reorientation has been determined and it is found that ultraslow reorientational motion of the H_2PO_4 groups exists above room temperature.

Es werden die Temperaturabhängigkeiten der Protonen-Zeeman- und der Dipol-Spin-Gitterrelaxationsraten in einem einkristallinen $Rb_{1-x}(NH_4)_xH_2PO_4$-Mischkristall sowohl in der paraelektrischen als auch in der ferroelektrischen Phase untersucht. Die Korrelationszeit für die Reorientierung des Ammoniumions wird bestimmt und gefunden, daß ultralangsame Reorientierungsbewegung der H_2PO_4-Gruppen oberhalb Zimmertemperatur auftritt.

1. Introduction

It is well known that RbH_2PO_4 (RDP) is ferroelectric, whereas $NH_4H_2PO_4$ (ADP) is antiferroelectric [1]. Our knowledge of the role of the NH_4 groups in the phase transition mechanism in KH_2PO_4 (KDP) type crystals is however still rather limited. Deuteron quadrupole coupling data [2] have shown that the ND_4^+ ions are distorted both above and below T_c and that the deviation from tetrahedral symmetry increases on going into the antiferroelectric phase. While the ND_4 groups undergo hindered rotational reorientations both above and below the antiferroelectric transition temperature T_c, the quadrupole splitting of the ND_4 deuteron lines [3, 4] abruptly changes on going through T_c. Proton spin–lattice (T_1) relaxation data in ADP similarly [4] showed a discontinuous change in T_1 at T_c. In order to throw some additional light on the motion of the NH_4 groups in KDP type ferroelectrics and antiferroelectrics we decided to investigate the Zeeman and dipolar proton spin–lattice relaxation rates in an ammonium doped RbH_2PO_4 crystal, where — in contrast to ADP, ADA, and KDP — the phase transition seems to be of second order [5].

2. Experimental

The $(Rb)_{1-x}(NH_4)_xH_2PO_4$ single crystal investigated in this study had an ammonium concentration $x = 0.0035$ mole fraction as determined by chemical analysis. Dielectric measurements showed that the crystal undergoes an apparently continuous ferroelectric phase transition at $T_c = -131$ °C (Fig. 1). The Curie-Weiss constant for $T > T_c$ equals $C = 3270$ K.

---

[1]) Permanent address: Institute of Crystallography, Academy of Sciences, Moscow, SSSR.

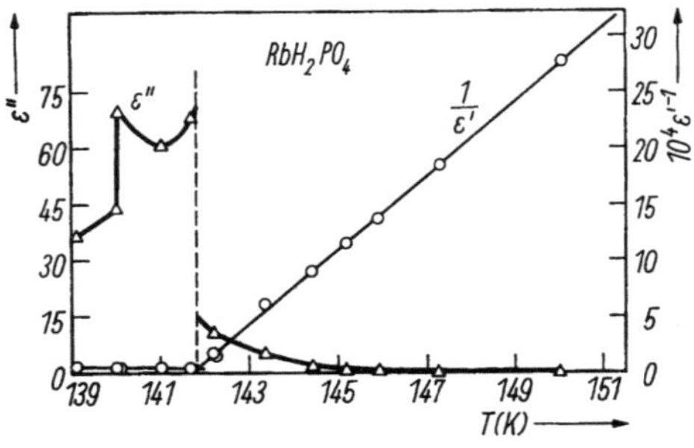

Fig. 1. Temperature dependence of the real and imaginary part of the dielectric constant in ammonium doped RbH_2PO_4

3. Theory

The spin Hamiltonian of our problem is a sum of a Zeeman term $\mathcal{H}_Z$ and the secular part of the magnetic dipolar interaction term $\mathcal{H}'_D$

$$\mathcal{H} = \mathcal{H}_Z + \mathcal{H}'_D \,. \tag{1}$$

It is well known that the density matrix of the proton spin system reaches in a time T_2 a quasi-equilibrium form

$$\varrho = 1 - \alpha \mathcal{H}_Z - \beta \mathcal{H}'_D \,, \tag{2}$$

where the constants α and β represent inverse Zeeman and dipolar spin temperatures which tend toward the lattice temperature β_L with time constants T_{1Z} and T_{1D}. The Zeeman and dipolar spin temperatures also tend to equalize under the effect of transitions induced by the non-secular part of $\mathcal{H}'_D$, but in a high external magnetic field the mixing time is much longer than T_{1Z} or T_{1D}, so that these two quantities can be studied independently.

In a mixed $Rb_{1-x}(NH_4)_xH_2PO_4$ crystal we have two different proton spin reservoirs: the "acid" protons and the "ammonium" protons. In view of the fast heat exchange between these two reservoirs, the observed spin–lattice relaxation rate is a weighted average of the relaxation rates of the "acid" and "ammonium" protons:

$$\frac{1}{T_1} = \frac{C(NH_4)}{T_1(NH_4)} + \frac{C(H_2)}{T_1(H_2)} \,. \tag{3a}$$

Here $C(NH_4)$ and $C(H_2)$ stand for the percentage of the "ammonium" and "acid" protons. In view of the fact that the dipolar interactions are much stronger within an NH_4 group than between the "acid" protons and that NH_4 motion is generally slower than "acid" proton motion, $T_1(NH_4)$ is – except at high temperatures – much shorter than $T_1(H_2)$, so that the ammonium reorientations will dominate the spin–lattice relaxation process even for a rather small concentration of NH_4 groups:

$$\frac{1}{T_1} \approx \frac{C(NH_4)}{T_1(NH_4)} . \tag{3b}$$

The Zeeman spin–lattice relaxation rate for a rotating NH_4 group has been evaluated by O'Reilly and Tsang [6]. For a powder sample they find

$$T_{1Z}^{-1} = C \frac{\tau}{1 + (\omega_0 \tau)^2} + \frac{4\tau}{1 + (2\omega_0 \tau)^2} , \tag{4a}$$

where

$$C = \frac{9\gamma^4\hbar^2}{10r^6} = 1.9 \times 10^{10}\ s^{-2} \tag{4b}$$

if an NH_4 inter-proton distance $r = 1.73$ Å is used. T_1 reaches a minimum at $\omega_0\tau = 0.62$. For a frequency of about 40 MHz the minimum value of T_{1Z} is about 10^{-2} s. This minimum occurs when the correlation time for NH_4 reorientations is of the order of 10^{-9} s.

Whereas the rates of NH_4 rotation can be obtained from the temperature and frequency dependence of T_{1Z}, the correlations between the motion of various NH_4 groups can be studied by measuring the dipolar spin–lattice relaxation rates T_{1D}.

For the case that the dipolar interactions can be approximated by a rapidly fluctuating local field, one obtains [7] for the case of *no* spatial correlations between the local field fluctuations the result

$$\frac{T_{1Z}}{T_{1D}} = 2 + \frac{1}{3}(\omega_0\tau)^2 . \tag{5}$$

For the case of *complete* spatial correlation, on the other hand, the result is

$$\frac{T_{1Z}}{T_{1D}} = 3 . \tag{6}$$

If, on the other hand, the local field fluctuations are so *slow* that the spin–spin relaxation time T_2 is shorter than the molecular correlation time τ, one finds [8] for the relaxation of the dipolar spin-temperature the result

$$\frac{1}{T_{1D}} = \frac{1-p}{\tau}, \tag{7}$$

where p measures the fractional change in the dipolar energy connected with a reorientational jump of the molecular group. For NH_4 reorientation, p has been evaluated by Goren and Pintar [9] as $p = -0.2$. It can be seen that in this last case, T_{1D} will generally be of the order of τ.

4. Results and Discussion

The temperature dependence of the proton T_{1Z} at $\nu_H = 60$ MHz, $\nu_H = 37$ MHz, and $\nu_H = 18$ MHz is shown on Fig. 2 together with the temperature dependence of T_{1D} at $\nu_0 = 18$ MHz. T_{1Z} first decreases and then increases with decreasing temperatures as predicted by equation (4a). As expected, the Zeeman spin–lattice relaxation rate is completely dominated by the NH_4 reorientations. This can be easily understood as measurements on a "pure" RbH_2PO_4 sample showed that $T_1(H_2) \gg 100$ s, whereas the minimum value of $T_1(NH_4)$ is — as mentioned before — of the order of 10^{-2} s. From the observed T_{1Z} minima the correlation time τ for NH_4 reorientation was obtained as a function of temperature. As it can be seen from Fig. 3, the NH_4 reorientations are thermally activated

$$\tau = \tau_0 \exp\frac{E}{kT} \tag{8}$$

and the activation energy is $E = 3.9$ kcal/mol. The value of τ decreases from 10^{-10} s at 300 K to 10^{-6} s at 120 K.

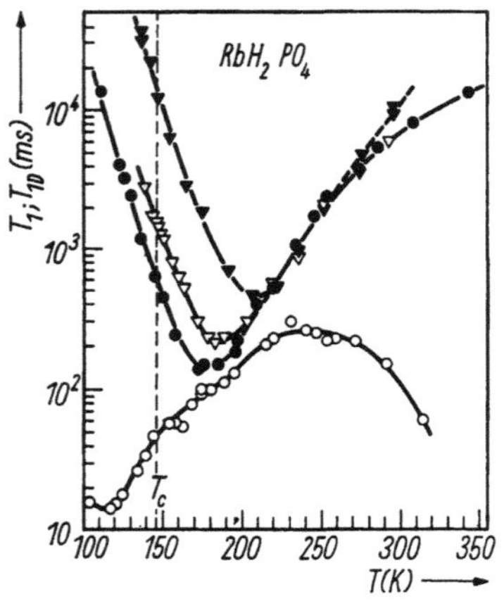

Fig. 2

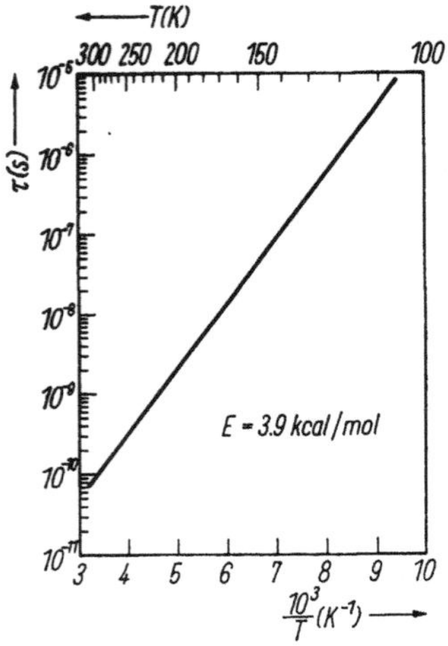

Fig. 3

Fig. 2. Temperature dependence of the proton Zeeman (T_{1Z}) and dipolar (T_{1D}) spin-lattice relaxation times in NH_4 doped RbH_2PO_4. T_{1Z}: ▼ $\nu_H = 60.1$ MHz, ▽ $\nu_H = 37$ MHz, ● $\nu_H = 18$ MHz; T_{1D}: ○ $\nu_0 = 18$ MHz

Fig. 3. Correlation time for NH^+ reorientations in NH_4 doped RbH_2PO_4 as a function of temperature

In contrast to ADP or ADA there is no discontinuity or change of slope in T_{1Z} or τ at T_c. The dipolar spin–lattice relaxation time T_{1D} is similarly continuous through T_c, but the slope of the T_{1D} versus $1/T$ plot changes on going through the phase transition.

The experimentally observed ratio of the Zeeman and dipolar spin–lattice relaxation times varies from $T_{1Z}/T_{1D} \approx 1.7$ at 200 K to $T_{1Z}/T_{1D} \approx 3 \times 10^2$ at 120 K. This means that the reorientations of the NH_4 groups are not — or only weakly — spatially correlated both above and below T_c. Together with the deuteron quadrupole coupling data [2] — which show a definite NH_4^+ contribution to the sublattice polarization in ADA^+ and ADP^+ — this result demonstrates that:

(i) the slow rotational reorientations of the NH_4 groups which determine the Zeeman proton-spin–lattice relaxation rate are *not* a part of the ferroelectric or antiferroelectric soft mode motion;

(ii) the modulation of the dipole–dipole intra-NH_4 interactions by the soft mode motion must be much smaller than the modulation due to the non-soft rotational reorientations.

In the ordered, low temperature phase, certain orientations of the distorted NH_4^+ group must be somewhat more probable than others. This results in a finite change in the ND_4^+ deuteron and nitrogen quadrupole coupling tensor at T_c but does not significantly affect the spectral density of the local magnetic field fluctuations at the proton Larmor frequency which determine T_{1Z}.

Above 250 K, the proton dipolar spin–lattice relaxation time starts to decrease with increasing temperature indicating that a new, rather slow motion ($\omega_0\tau \gg 1$) becomes rate determining. The same effect is observed in pure

RbH_2PO_4. We are thus forced to conclude that this high temperature decrease of T_{1D} is connected with the "acid" rather than with the NH_4 protons.

In view of the fact that we know [10] that the correlation times for motion of protons between the two potential minima in the O–H–O bonds is of the order of 10^{-12} s, this slow motion must be due to proton inter-bond motion or hindered rotation of the H_2PO_4 groups. The dipolar spin–lattice relaxation rate can be now analyzed with the help of equation (7). The correlation time for this motion is thus of the order of 0.1 s at room temperature.

References

[1] See for instance W. Känzig, Solid State Phys. **4**, 1 (1957).
[2] T. Chiba, J. Phys. Soc. Japan **38**, 491 (1965).
[3] V. Dimic, M. Osredkar, J. Slak, and A. Kandušar, phys. stat. sol. (b) **59**, 471 (1973).
[4] D. J. Genin, D. E. O'Reilly, and Tung Tsang, Phys. Rev. **167**, 445 (1968).
[5] B. A. Strukov, M. A. Korzhuev, and V. A. Kopcik, Izv. Akad. Nauk SSSR, Ser. fiz. **35**, 1846 (1971).
[6] D. E. O'Reilly and Tung Tsang, J. chem. Phys. **46**, 1291 (1967).
[7] M. Goldman, Spin Temperature and Nuclear Magnetic Resonance in Solids, Clarendon Press, Oxford 1970 (Chap. 3).
[8] C. P. Slichter and D. C. Ailion, Phys. Rev. **135**, A1099 (1964).
[9] S. D. Goren and M. M. Pintar, private communication.
[10] R. Blinc, J. Stepišnik, M. Jamšek-Vilfan, and S. Žumer, J. chem. Phys. **54**, 187 (1971).

(Received April 24, 1974)

phys. stat. sol. (b) 65, 403 (1974)

Subject classification: 13 and 18

Institut für Theoretische Physik der Universität Kiel

Thermodynamics of a Two-Band Hubbard Model

By

W.-H. STEEB and E. MARSCH

Using a variational principle the approximate free energy of a simple two-band Hubbard model is calculated. Numerical results are presented for the simple cubic lattice and these are given for the free energy, chemical potential, and energy gap as function of temperature. The values are compared with those of Dichtel et al. By introduction of a second band a lower transition temperature T_N is obtained. The system shows a second-order phase transition.

Mit Hilfe eines Variationsprinzips wird die approximative freie Energie von einem einfachen Zwei-Band-Hubbard-Modell berechnet. Die numerischen Rechnungen werden für das kubisch-einfache Gitter ausgeführt. Resultate werden angegeben für die freie Energie, das chemische Potential und die Energielücke als Funktion der Temperatur. Die Werte werden mit Auswertungen von Dichtel et al. verglichen. Durch die Einführung des zweiten Bandes wird eine niedrigere Übergangstemperatur erreicht. Der Phasenübergang ist von zweiter Ordnung.

1. Introduction

Dichtel et al. [1] have calculated by use of a variational principle thermodynamic quantities of the Hubbard model. They started with the well-known Hamiltonian

$$H_{\mathrm{H}} = \sum_{nm\sigma} T_{nm} d^+_{n\sigma} d_{m\sigma} + U \sum_n d^+_{n\uparrow} d_{n\uparrow} d^+_{n\downarrow} d_{n\downarrow} \tag{1}$$

and an inequality for the free energy. The calculations are performed for nearest neighbour hopping (hopping strength J), neutral model, and AB lattice. They found for all values of the coupling constant U two phases corresponding to two different values for the free energy of which the smaller one has to be taken as the physical free energy. It is shown that the condensed phase, as long as it exists, always leads to the lower value of the free energy. The condensed phase is purely antiferromagnetic and the Hartree phase is paramagnetic. The phase transition is of second order. In the half-filled case (neutral model) the transition temperature kT_N (Néel temperature) tends to $U/4$ for $U \gg J$. This is in consistence with results in the strong correlated limit [2]. Numerical calculation must be restricted to the range $U \leqq zJ$ (z number of nearest neighbours). Further, Dichtel et al. found that the chemical potential is independent of temperature.

The aim of this paper is to investigate the influence of a free band on the thermodynamic quantities where we use the same approximation (for the Hubbard term) like Dichtel et al. So we start with the Hamiltonian

$$H = H_{\mathrm{H}} + \sum_{K\sigma} (\varepsilon(K) + B)\, c^+_{K\sigma} c_{K\sigma} \,. \tag{2}$$

The first term of (2) is the Hubbard Hamiltonian. In Section 2 a short derivation of the variational principle (inequality for the grand thermodynamic potential) is given. In Section 3 the ansatz for the trial density matrix is made. A remark is made about the choice of the trial density matrix in the strongly correlated limit. In Section 4 the grand thermodynamic potential is calculated and the self-consistent equations are determined. Further we consider the chemical potential and the other quantities which we get from the self-consistent equations. Numerical calculations are performed in Section 5 and a discussion of the results is given. Our calculations are restricted to the simple cubic lattice and nearest-neighbour hopping for the Hubbard term.

2. The Variational Principle

The grand canonical density matrix, W_0, for a system with Hamiltonian H and number operator N_e at temperature $1/\beta$ and chemical potential μ can be derived by specifying it to be that which minimizes the grand thermodynamic potential

$$\Omega = \mathrm{Tr}\,\{W(H - \mu N_e)\} + \frac{1}{\beta}\,\mathrm{Tr}\,\{W \ln W\} \tag{3}$$

over all W. W is a linear, hermitian Hilbert-Schmidt operator with unit trace.

The condition that the first functional derivative vanishes leads uniquely to

$$W_0 = \frac{\exp\,[-\beta(H - \mu N_e)]}{\mathrm{Tr}\,\{\exp\,[-\beta(H - \mu N_e)]\}}. \tag{4}$$

The fact that W_0 gives a minimum follows from the form of the second functional derivative at $W = W_0$. This operator is positive definite, since W_0 is positive definite. So we get an inequality for the grand potential [3]:

$$\Omega \leqq \mathrm{Tr}\,\{W_t(H - \mu N_e)\} + \frac{1}{\beta}\,\mathrm{Tr}\,\{W_t \ln W_t\}\,, \tag{5}$$

where the trial density matrix W_t is a linear, hermitian positive definite operator with unit trace.

3. Choice of the Trial Density Matrix

In the following we consider only the Hubbard Hamiltonian because the second term in (2) can be calculated exactly. One chooses a trial density matrix

$$W_t = \frac{\exp\,(...)}{\mathrm{Tr}\,\{\exp\,(...)\}} \tag{6}$$

for which the trace

$$Z_t = \mathrm{Tr}\,\{\exp\,(\ldots)\} \tag{7}$$

can be evaluated exactly. One possibility is (in Bloch representation)

$$W_t = \frac{\exp\Big(-\beta \sum\limits_{K} [E_1(K)\,\bar{d}^+_{K\uparrow}\bar{d}_{K\uparrow} + E_2(K)\,\bar{d}^+_{K\downarrow}\bar{d}_{K\downarrow}]\Big)}{\mathrm{Tr}\Big\{\exp\Big(-\beta \sum\limits_{K} [E_1(K)\bar{d}^+_{K\uparrow}\bar{d}_{K\uparrow} + E_2(K)\,\bar{d}^+_{K\downarrow}\bar{d}_{K\downarrow}]\Big)\Big\}}. \tag{8}$$

The quasi-particle operators $\bar{d}^+$ and $\bar{d}$ obey also the ACR. The summation over K extends throughout the first Brillouin zone. $E_1(K)$ and $E_2(K)$ represent the energy spectrum of the quasi-particle (chemical potential included). They play the role of real variation parameters. $E_1(\boldsymbol{K})$ and $E_2(\boldsymbol{K})$ are determined by

$$\frac{\partial \Omega_{\text{app}}}{\partial E_1(\boldsymbol{K})} = \frac{\partial \Omega_{\text{app}}}{\partial E_2(\boldsymbol{K})} = 0 \,. \tag{9}$$

From (8) we obtain

$$Z_\text{t} = \text{Tr}\,\{\exp\,(\ldots)\} = \prod_{\boldsymbol{K}} (1 + \text{e}^{-\beta E_1(\boldsymbol{K})})\,(1 + \text{e}^{-\beta E_2(\boldsymbol{K})}) \tag{10}$$

and by use of an operator identity

$$\langle \bar{d}^+_{\boldsymbol{K}\uparrow} \bar{d}_{\boldsymbol{K}\uparrow} \rangle = \frac{1}{1 + \text{e}^{\beta E_1(\boldsymbol{K})}} \equiv f_1(\boldsymbol{K}) \,, \tag{11}$$

$$\langle \bar{d}^+_{\boldsymbol{K}\downarrow} \bar{d}_{\boldsymbol{K}\downarrow} \rangle = \frac{1}{1 + \text{e}^{\beta E_2(\boldsymbol{K})}} \equiv f_2(\boldsymbol{K}) \,, \tag{12}$$

where $f_1(\boldsymbol{K})$ and $f_2(\boldsymbol{K})$ are the Fermi distributions. The operators $\bar{d}^+$, $\bar{d}$ and d^+, d are connected by a Bogolyubov-Valentin transformation [4]. Since the Hubbard operator commutes with the number operator N_e, we require that also N_e is invariant under this transformation (in contrast to the BCS theory).

We make the ansatz

$$A = \text{e}^{iS} \bar{A}\, \text{e}^{-iS} \,; \qquad S = S^+ \,, \tag{13}$$

where

$$S = i \sum_{\boldsymbol{K}} \left(\frac{\alpha_{\boldsymbol{K}}}{2}\, \bar{d}^+_{\boldsymbol{K}\uparrow} \bar{d}_{\boldsymbol{K}+\boldsymbol{Q}\downarrow} - \frac{\alpha_{\boldsymbol{K}}}{2}\, \bar{d}^+_{\boldsymbol{K}+\boldsymbol{Q}\downarrow} \bar{d}_{\boldsymbol{K}\uparrow} \right) \tag{14}$$

with

$$\boldsymbol{Q} = \frac{1}{a}\,(\pi, \pi, \pi) \,. \tag{15}$$

We obtain from (13) and (14)

$$\begin{pmatrix} d_{\boldsymbol{K}\uparrow} \\ d_{\boldsymbol{K}+\boldsymbol{Q}\downarrow} \end{pmatrix} = \begin{pmatrix} \cos\dfrac{\alpha_{\boldsymbol{K}}}{2} & \sin\dfrac{\alpha_{\boldsymbol{K}}}{2} \\ -\sin\dfrac{\alpha_{\boldsymbol{K}}}{2} & \cos\dfrac{\alpha_{\boldsymbol{K}}}{2} \end{pmatrix} \begin{pmatrix} \bar{d}_{\boldsymbol{K}\uparrow} \\ \bar{d}_{\boldsymbol{K}+\boldsymbol{Q}\downarrow} \end{pmatrix} . \tag{16}$$

a is the lattice constant for the simple cubic lattice. $\alpha_{\boldsymbol{K}}$ plays also the role of a real variational parameter and is determined by

$$\frac{\partial \Omega_{\text{app}}}{\partial \alpha_{\boldsymbol{K}}} = 0 \,. \tag{17}$$

So we have three equations for the unknown quantities $E_1(\boldsymbol{K})$, $E_2(\boldsymbol{K})$, and $\alpha_{\boldsymbol{K}}$. A motivation of this ansatz is given in the next section.

A remark should be made about the strongly correlated limit. In this range holds $U \gg J$ and our approximation is unrealistic. Here we start with

$$W_\text{t} = \frac{\exp\left(-\lambda_1 \sum\limits_n \bar{d}^+_{n\uparrow} \bar{d}_{n\uparrow} \bar{d}^+_{n\downarrow} \bar{d}_{n\downarrow} - \lambda_2 \sum\limits_n \bar{d}^+_{n\uparrow} \bar{d}_{n\uparrow} - \lambda_3 \sum\limits_n \bar{d}^+_{n\downarrow} \bar{d}_{n\downarrow}\right)}{\text{Tr}\left\{\exp\left(-\lambda_1 \sum\limits_n \bar{d}^+_{n\uparrow} \bar{d}_{n\uparrow} \bar{d}^+_{n\downarrow} \bar{d}_{n\downarrow} - \lambda_2 \sum\limits_n \bar{d}^+_{n\uparrow} \bar{d}_{n\uparrow} - \lambda_3 \sum\limits_n \bar{d}^+_{n\downarrow} \bar{d}_{n\downarrow}\right)\right\}} \,, \tag{18}$$

where n runs over all lattice sites. d^+, d and $\bar{d}^+$, $\bar{d}$ are connected by

$$A = e^{iS}\,\bar{A}\,e^{-iS}\,, \tag{19}$$

where S is the Fourier transform of (14),

$$S = i\sum_{mn}\left(\frac{\varphi_{m-n}}{2}\,\bar{d}_{n\uparrow}^{+}\bar{d}_{m\downarrow}\,e^{i\boldsymbol{Q}\boldsymbol{R}_m} - \frac{\varphi_{m-n}}{2}\,\bar{d}_{m\downarrow}^{+}\bar{d}_{n\uparrow}\,e^{-i\boldsymbol{Q}\boldsymbol{R}_m}\right). \tag{20}$$

The variational parameters are λ_1, λ_2, λ_3, and φ_{m-n}.

4. Calculation of the Free Energy

Now we want to calculate the grand thermodynamic potential and the self-consistent equations of our extend model. We start with the Hamiltonian in Bloch representation

$$H = \sum_{\boldsymbol{K}\sigma} T(\boldsymbol{K})\,d_{\boldsymbol{K}\sigma}^{+}d_{\boldsymbol{K}\sigma} + \frac{U}{N}\sum_{\substack{\boldsymbol{K}_1\boldsymbol{K}_2\\ \boldsymbol{K}_3\boldsymbol{K}_4}} \delta(\boldsymbol{K}_1 - \boldsymbol{K}_2 + \boldsymbol{K}_3 - \boldsymbol{K}_4)\,d_{\boldsymbol{K}_1\uparrow}^{+}d_{\boldsymbol{K}_2\uparrow}d_{\boldsymbol{K}_3\downarrow}^{+}d_{\boldsymbol{K}_4\downarrow} +$$
$$+ \sum_{\boldsymbol{K}\sigma}(\varepsilon_{\boldsymbol{K}} + B)\,c_{\boldsymbol{K}\sigma}^{+}c_{\boldsymbol{K}\sigma} + 2\zeta S_z\,, \tag{21}$$

the trial density matrix

$$W_t = \frac{\left[\exp\left(-\beta\sum_{\boldsymbol{K}}[E_1(\boldsymbol{K})\,\bar{d}_{\boldsymbol{K}\uparrow}^{+}\bar{d}_{\boldsymbol{K}\uparrow} + E_2(\boldsymbol{K})\,\bar{d}_{\boldsymbol{K}\downarrow}^{+}\bar{d}_{\boldsymbol{K}\downarrow}] - \beta\sum_{\boldsymbol{K}}[\gamma_1(\boldsymbol{K})\,c_{\boldsymbol{K}\uparrow}^{+}c_{\boldsymbol{K}\uparrow} + \gamma_2(\boldsymbol{K})\,c_{\boldsymbol{K}\downarrow}^{+}c_{\boldsymbol{K}\downarrow}]\right)\right]}{\left[\mathrm{Tr}\left\{\exp\left(-\beta\sum_{\boldsymbol{K}}[E_1(\boldsymbol{K})\,\bar{d}_{\boldsymbol{K}\uparrow}^{+}\bar{d}_{\boldsymbol{K}\uparrow} + E_2(\boldsymbol{K})\,\bar{d}_{\boldsymbol{K}\downarrow}^{+}\bar{d}_{\boldsymbol{K}\downarrow}] - \beta\sum_{\boldsymbol{K}}[\gamma_1(\boldsymbol{K})\,c_{\boldsymbol{K}\uparrow}^{+}c_{\boldsymbol{K}\uparrow} + \gamma_2(\boldsymbol{K})\,c_{\boldsymbol{K}\downarrow}^{+}c_{\boldsymbol{K}\downarrow}]\right)\right\}\right]}, \tag{22}$$

and the Bogolyubov-Valentin transformation

$$\begin{pmatrix} d_{\boldsymbol{K}\uparrow} \\ d_{\boldsymbol{K}+\boldsymbol{Q}\downarrow}\end{pmatrix} = \begin{pmatrix} \cos\frac{\alpha_{\boldsymbol{K}}}{2} & \sin\frac{\alpha_{\boldsymbol{K}}}{2} \\ -\sin\frac{\alpha_{\boldsymbol{K}}}{2} & \cos\frac{\alpha_{\boldsymbol{K}}}{2}\end{pmatrix}\begin{pmatrix} \bar{d}_{\boldsymbol{K}\uparrow} \\ \bar{d}_{\boldsymbol{K}+\boldsymbol{Q}\downarrow}\end{pmatrix}. \tag{23}$$

With

$$\frac{\partial\Omega_{\text{app}}}{\partial E_1(\boldsymbol{K})} = \frac{\partial\Omega_{\text{app}}}{\partial E_2(\boldsymbol{K})} = \frac{\partial\Omega_{\text{app}}}{\partial\alpha_{\boldsymbol{K}}} = \frac{\partial\Omega_{\text{app}}}{\partial\gamma_1(\boldsymbol{K})} = \frac{\partial\Omega_{\text{app}}}{\partial\gamma_2(\boldsymbol{K})} = 0 \tag{24}$$

and

$$\frac{\partial\Omega_{\text{app}}}{\partial\mu} = -N_e\,; \qquad \frac{\partial\Omega_{\text{app}}}{\partial\zeta} = 2S_z \tag{25}$$

we obtain with

$$\Delta = -\frac{U}{2N}\sum_{\boldsymbol{K}}\sin\alpha_{\boldsymbol{K}}\left(f_1(\boldsymbol{K}) - f_2(\boldsymbol{K})\right) = +\frac{U}{2N}\sum_{\boldsymbol{K}}\langle d_{\boldsymbol{K}\uparrow}^{+}d_{\boldsymbol{K}+\boldsymbol{Q}\downarrow}\rangle \tag{26}$$

for the quasi-particle energy

$$E_{1,2}(\boldsymbol{K}) = -\mu + \frac{U}{2}(N_e - N_{fr}) \pm \sqrt{\left[T(\boldsymbol{K}) - \zeta + \frac{U}{2}(S_z - S_{zfr})\right]^2 + \Delta^2}\,, \quad (27)$$

$$\gamma_{1,2}(\boldsymbol{K}) = \varepsilon(\boldsymbol{K}) + B - \mu \pm \zeta \quad (28)$$

and for the grand thermodynamic potential

$$\Omega_{app} = -\frac{1}{\beta}\sum_{\boldsymbol{K}} \ln\left\{2\cosh(\beta E_{\boldsymbol{K}}) + 2\cosh\left[\beta\left(\mu - \frac{U}{2}\frac{N_e}{N} + \frac{U}{2}\frac{N_{fr}}{N}\right)\right]\right\} -$$

$$-\mu N + \frac{U}{2}(N_e - N_{fr}) + \frac{\Delta^2}{U}N - \frac{U}{4N}\left[(N_e - N_{fr})^2 - 4(S_z - S_{zfr})^2\right] + \Omega_{fr}\,, \quad (29)$$

where

$$\Omega_{fr} = -\frac{1}{\beta}\sum_{\boldsymbol{K}} \ln\{[1 + e^{-\beta(\varepsilon(\boldsymbol{K}) - \mu + \zeta)}][1 + e^{-\beta(\varepsilon(\boldsymbol{K}) - \mu - \zeta)}]\}.$$

Further we obtain the self-consistent equations

$$\Delta\left\{\frac{2}{U} - \frac{1}{N}\sum_{\boldsymbol{K}}\frac{1}{E_{\boldsymbol{K}}}\frac{\sinh(\beta E_{\boldsymbol{K}})}{\cosh(\beta E_{\boldsymbol{K}}) + \cosh\left[\beta\left(\mu - \frac{U}{2}\frac{N_e}{N} + \frac{U}{2}\frac{N_{fr}}{N}\right)\right]}\right\} = 0\,, \quad (30)$$

$$\left\{\frac{2(S_z - S_{zfr})}{N} + \frac{1}{N}\sum_{\boldsymbol{K}}\frac{T(\boldsymbol{K}) - U\dfrac{S_z - S_{zfr}}{N}}{E_{\boldsymbol{K}}}\times\right.$$

$$\left.\times\frac{\sinh(\beta E_{\boldsymbol{K}})}{\cosh(\beta E_{\boldsymbol{K}}) + \cosh\left[\beta\left(\mu - \frac{U}{2}\frac{N_e}{N} + \frac{U}{2}\frac{N_{fr}}{N}\right)\right]}\right\} = 0\,, \quad (31)$$

$$\left\{1 - \frac{N_e - N_{fr}}{N} + \frac{1}{N}\sum_{\boldsymbol{K}}\frac{\sinh\left[\beta\left(\mu - \frac{U}{2}\frac{N_e - N_{fr}}{N}\right)\right]}{\cosh(\beta E_{\boldsymbol{K}}) + \cosh\left[\beta\left(\mu - \frac{U}{2}\frac{N_e - N_{fr}}{N}\right)\right]}\right\} = 0\,, \quad (32)$$

where we use the abbreviations

$$E_{\boldsymbol{K}} = \sqrt{\left[T(\boldsymbol{K}) - \zeta - U\left(\frac{S_z - S_{zfr}}{N}\right)\right]^2 + \Delta^2}\,, \quad (33)$$

$$S_{zfr} = \tfrac{1}{2}\sum_{\boldsymbol{K}}\langle c^+_{\boldsymbol{K}\uparrow}c_{\boldsymbol{K}\uparrow} - c^+_{\boldsymbol{K}\downarrow}c_{\boldsymbol{K}\downarrow}\rangle\,, \quad (34)$$

$$N_{fr} = \sum_{\boldsymbol{K}\sigma}\langle c^+_{\boldsymbol{K}\sigma}c_{\boldsymbol{K}\eta}\rangle\,. \quad (35)$$

$\langle\cdots\rangle_{W_t}$ is the expectation value.

Analogous to Dichtel et al. [1] the gap equation (equation (30)) has two solutions where the trivial one $\Delta = 0$ is for the paramagnetic state how it can be found by calculation of the transverse susceptibility. A lower energy one, if it exists for $\Delta > 0$, corresponds to an antiferromagnetic state.

Now we want to give a motivation about the test Hamiltonian. To approximate the Coulomb interaction by a one-particle term one can introduce magnetic fields fixed at each lattice site $\boldsymbol{R}_n$. The most general form for such an operator is

$$H_{\text{field}} = \sum_n \underset{\sim}{d}_n^+ (\boldsymbol{H}_n \cdot \boldsymbol{\sigma})\, \underset{\sim}{d}_n = \sum_n \underset{\sim}{d}_n^+ \begin{pmatrix} H_n^z & H_n^x - iH_n^y \\ H_n^x + iH_n^y & -H_n^z \end{pmatrix} \underset{\sim}{d}_n\,, \tag{36}$$

where $\boldsymbol{\sigma}$ means the vector built up of the Pauli-spin matrices and $\underset{\sim}{d}_n^+$ the row vector $(d_{n\uparrow}^+, d_{n\downarrow}^+)$.

Calculating the free energy of the Hubbard model using the test operator for the Hubbard Hamiltonian

$$H_t = -J \sum_{n,j} (d_{n+j\uparrow}^+ d_{n\uparrow} + d_{n+j\downarrow} d_{n\downarrow}) - \sum_n \underset{\sim}{d}_n^+ (\boldsymbol{H} \cdot \boldsymbol{\sigma})\, \underset{\sim}{d}_n - \mu \sum_{n,\sigma} d_{n\sigma}^+ d_{n\sigma}\,, \tag{37}$$

the H_n should be variational parameters to minimize the free-energy functional. But this program is too difficult to perform, therefore we choose

$$H_{\text{field}} = \sum_n \underset{\sim}{d}_n^+ \begin{pmatrix} \zeta & \Delta e^{-i\boldsymbol{Q}\boldsymbol{R}_n} \\ \Delta e^{i\boldsymbol{Q}\boldsymbol{R}_n} & -\zeta \end{pmatrix} \underset{\sim}{d}_n\,. \tag{38}$$

After Fourier transformation we obtain

$$H_t = \sum_{\boldsymbol{K}} (d_{\boldsymbol{K}\uparrow}^+ d_{\boldsymbol{K}+\boldsymbol{Q}\downarrow}^+) \begin{pmatrix} -\mu - \zeta + \varepsilon(\boldsymbol{K}) & -\Delta \\ -\Delta & -\mu + \zeta + \varepsilon(\boldsymbol{K}+\boldsymbol{Q}) \end{pmatrix} \begin{pmatrix} d_{\boldsymbol{K}\uparrow} \\ d_{\boldsymbol{K}+\boldsymbol{Q}\downarrow} \end{pmatrix}. \tag{39}$$

A $\boldsymbol{K}$-dependent rotation (see equation (23)) makes H_t diagonal:

$$\begin{pmatrix} d_{\boldsymbol{K}\uparrow} \\ d_{\boldsymbol{K}+\boldsymbol{Q}\downarrow} \end{pmatrix} = \begin{pmatrix} \cos\dfrac{\alpha_{\boldsymbol{K}}}{2} & \sin\dfrac{\alpha_{\boldsymbol{K}}}{2} \\ -\sin\dfrac{\alpha_{\boldsymbol{K}}}{2} & \cos\dfrac{\alpha_{\boldsymbol{K}}}{2} \end{pmatrix} \begin{pmatrix} \bar{d}_{\boldsymbol{K}\uparrow} \\ \bar{d}_{\boldsymbol{K}+\boldsymbol{Q}\downarrow} \end{pmatrix}. \tag{40}$$

We obtain the form

$$H_t = \sum_{\boldsymbol{K}} [E_1(\boldsymbol{K})\, \bar{d}_{\boldsymbol{K}\uparrow}^+ \bar{d}_{\boldsymbol{K}\uparrow} + E_2(\boldsymbol{K})\, \bar{d}_{\boldsymbol{K}\downarrow}^+ \bar{d}_{\boldsymbol{K}\downarrow}]\,. \tag{41}$$

5. Numerical Results and Discussion

For numerical calculation we transform sums over the first Brillouin zone as

$$\frac{1}{N} \sum_{\boldsymbol{K}} \to \frac{a^3}{(2\pi)^3} \int_{1.\text{BZ}} d\boldsymbol{K} \tag{42}$$

and use the energy density of states established by Jelitto [5].

For our parameters U and B we choose

$$\frac{U}{J} = 6\,; \quad \frac{B}{J} = 0\,, \quad \text{and} \quad \frac{B}{J} = -1\,, \tag{43}$$

and $N = N_e$. Further the energy spectrum of the s-band is

$$\varepsilon(\boldsymbol{K}) = \frac{K^2 \hbar^2}{2m^*}\,, \tag{44}$$

where m^* is the band mass.

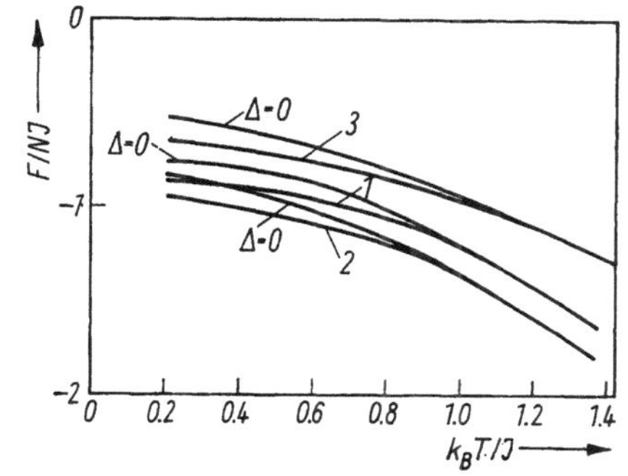

Fig. 1

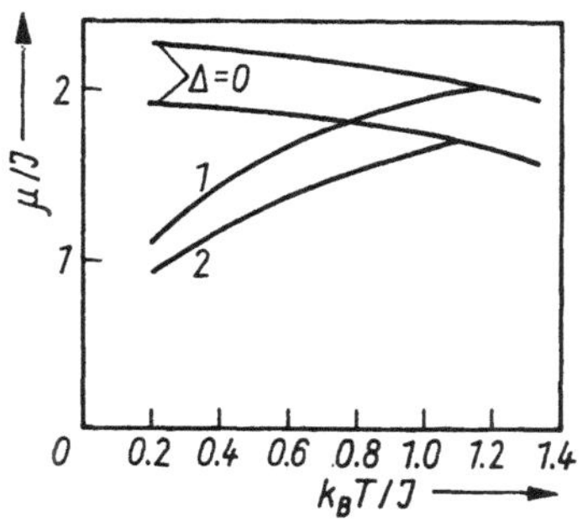

Fig. 2

Fig. 1. Free energy as a function of temperature. (1) $B/J = 0$; (2) $B/J = -1$; (3) without free band

Fig. 2. Chemical potential as a function of temperature. (1) $B/J = 0$; (2) $B/J = -1$

With the relation $J = \hbar^2/2ma^2$ we obtain

$$\varepsilon(\boldsymbol{K}) = (Ka)^2 \, J \, \frac{m}{m^*}. \tag{45}$$

As a first approximation we use $m/m^* = 1$. Further we set $\zeta = 0$. In case of the neutral model without free band the free energy of the condensed phase (also in the Hartree phase) has a single minimum at $S_{z_b} = 0$. In our model we take also $S_z = 0$. We calculate the free energy, the chemical potential, and the energy gap Δ as function of temperature.

From Fig. 1 we observe that the free energy is a continuous and concave function of the temperature T. The system for $\Delta > 0$ (antiferromagnetic case) has a lower free energy up to the Néel temperature T_N where the energy gap vanishes. In contrast to the result of Dichtel et al. the free energy is lower. The chemical potential (Fig. 2) is (for $\Delta > 0$ and $\Delta = 0$) a continuous and concave function of T.

In the neutral model without free band the chemical potential is independent of temperature. The energy gap (Fig. 3) as a function of T looks like the curve of Dichtel et al. By introduction of the second band we obtain a lower Néel temperature. The phase transition is of second order. For increasing ratios B/J the curves approach to the curve of Dichtel et al., although the transition temperature is lower. The results are only in the range $U \leqq zJ$ good for comparison with practical values.

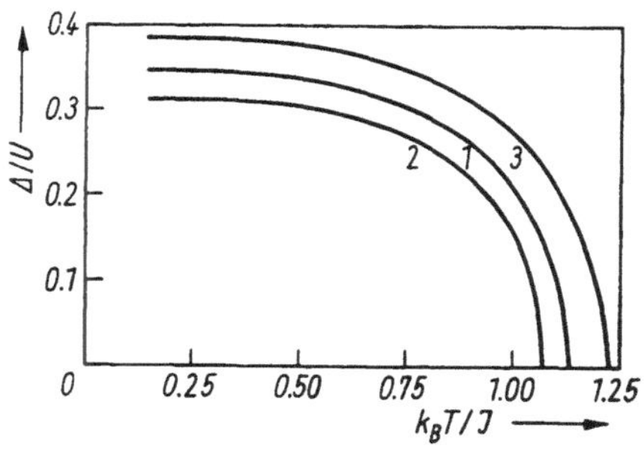

Fig. 3. Energy gap as a function of temperature. (1) $B/J = 0$; (2) $B/J = -1$; (3) without free band

For calculations in the range $U \gg J$ we must take the trial density matrix from (17). This leads to a transition temperature of J^2/U. In our approximation we can extend the results. Are there solutions of the free energy for $S_z \neq 0$ which are lower than for $S_z = 0$? This can lead to ferromagnetism. Further we have the requirement that Q is also a variation parameter. This yields new self-consistent equations. In this case we cannot use the density of states of Jelitto [5]. We also take no notice of s–d hybridization. These questions will be discussed in a following paper.

References

[1] K. Dichtel, R. J. Jelitto, and H. Koppe, Z. Phys. **251**, 173 (1972).
[2] T. A. Kaplan and R. A. Bari, J. appl. Phys. **41**, 875 (1970).
[3] N. D. Mermin, Ann. Phys. (U.S.A.) **21**, 99 (1963).
[4] H. G. Becker, Z. Naturf. **28a**, 332 (1973).
[5] R. J. Jelitto, J. Phys. Chem. Solids **30**, 609 (1969).

(Received May 30, 1974)

phys. stat. sol. (b) **65**, 411 (1974)

Subject classification: 1.5 and 2; 22.5.2

The Research Institute for Iron, Steel, and Other Metals, Tôhoku University, Sendai

Electron Diffraction Study of the Local Atomic Arrangement in Amorphous TlCl and CuCl Films

By

T. ICHIKAWA

Thin films of amorphous TlCl and CuCl have been prepared by low temperature condensation in high vacuum, and a transmission electron diffraction study of them has been undertaken. A radial distribution analysis from the diffuse halo patterns obtained has been made and the local atomic arrangements in amorphous TlCl and CuCl are discussed by comparison of the experimental $W(r)$ $(=\varrho(r)/\varrho_0)$ with $W(r)$'s calculated from disordered crystalline models.

Dünne Schichten von amorphem TlCl und CuCl werden durch Verdampfung im Hochvakuum auf kaltem Substrat hergestellt und Elektronenbeugungsuntersuchungen an ihnen durchgeführt. Eine Analyse der radialen Verteilung wird durchgeführt, und die Nahordnungen eines Atoms in amorphem TlCl und CuCl werden durch Vergleich mit den experimentellen Kurven und den mit ungeordneten Kristallmodellen berechneten Kurven diskutiert.

1. Introduction

It is well known that films of bismuth, gallium, iron, and nickel condensed onto a cold substrate in high vacuum show diffuse halo electron diffraction patterns characteristic of a highly disordered structure [1 to 3]. The radial distribution analyses were undertaken and the local atomic arrangements in the films were clarified [4, 5].

Some of the metal halides condensed onto cold substrates show diffuse halo patterns, too. Blackman and Khan [6] condensed thallium, cesium and rubidium halides at a slow rate of evaporation onto an amorphous substrate cooled by liquid oxygen and obtained films showing diffuse halo electron diffraction patterns. They pointed out that the centres of the diffuse halos coincided with either intense rings or centres of groups of intense rings of the respective crystalline structures. However, they described nothing about the local atomic arrangements in the films. In the present study, films of TlCl and CuCl showing diffuse halo patterns have been prepared by condensation onto a substrate at liquid helium temperature (LHeT) in high vacuum and the structures of the films have been investigated by comparison of $W(r)$ $(= \varrho(r)/\varrho_0)$ experimentally determined from the diffuse halo patterns with $W(r)$'s calculated from disordered crystalline models.

2. Experimental

TlCl of 99.9% purity and CuCl chemically purified were evaporated at the rate of ≈ 5 Å/s from a tungsten filament and were condensed on a substrate of Polyvinyl Formvar about 100 Å thick at LHeT in a vacuum of $\approx 1 \times 10^{-6}$ Torr in an electron diffraction camera equipped with a rotating sector. The thick-

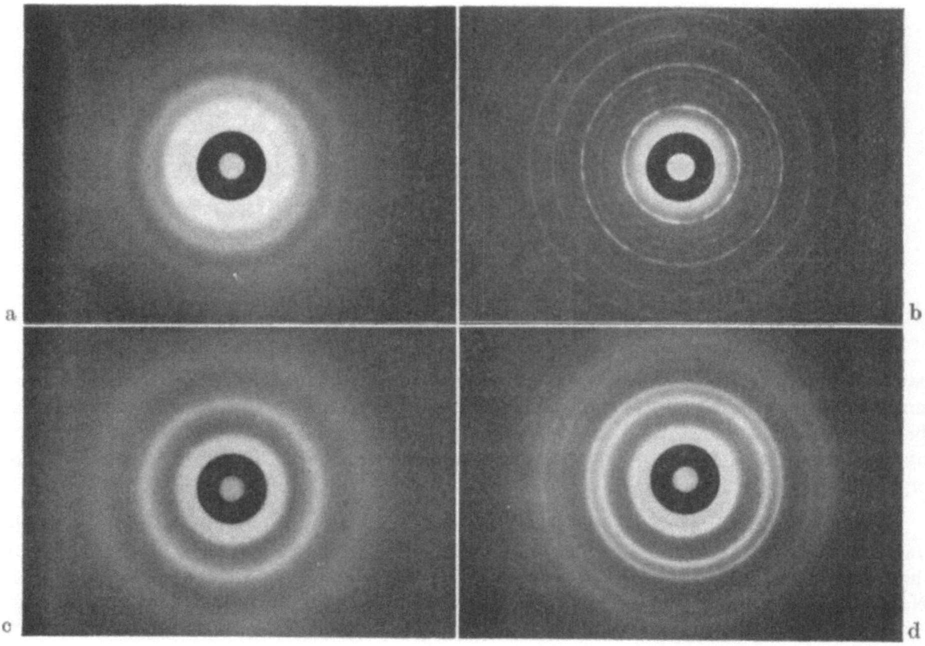

Fig. 1. a), b), c), and d) represent electron diffraction patterns of a TlCl film immediately after condensation at LHeT and after crystallization and annealing at RT, and those of a CuCl film immediately after condensation at LHeT and after crystallization and annealing at RT, respectively

nesses of TlCl and CuCl films were roughly estimated from the evaporated quantities to be about 150 and 200 Å, respectively. The procedures used for photographing of the electron diffraction patterns, for determination of s-values and for photometry were the same as those described in a previous paper [4].

Fig. 1a is a diffuse halo pattern of a TlCl film condensed at LHeT. On heating, the pattern rapidly changed to a ring pattern of the CsCl type structure, as shown in Fig. 1b. However, the transition temperature scattered from 130 to 150 °K probably due to slight differences of film thickness and the evaporation condition, such as the evaporation rate and the degree of vacuum during evaporation. Fig. 1c is a diffuse halo pattern of a CuCl film condensed at LHeT. On heating, the pattern gradually changed to a ring pattern of the zincblende (ZnS) type structure, as shown in Fig. 1d. Hence, the transition temperature could not be definitely determined, but the pattern was already that of the ZnS-type structure at 50 °K.

3. Radial Distribution Analysis

The theory of the radial distribution analysis of amorphous solids is given in detail in a standard text [7]. For the sake of introduction of notations used in the following discussion, it is briefly given here. In amorphous binary compounds, atomic densities $\varrho_{AA}(r)$, $\varrho_{AB}(r)$, $\varrho_{BA}(r)$ and $\varrho_{BB}(r)$ relating to A–A, A–B, B–A and B–B atomic pairs, respectively, are necessary to describe the structures. $\varrho_{ij}(r)$ is the atomic density of j-kind atoms at a distance r from an origin atom of i-kind. Determination of the four $\varrho_{ij}(r)$ requires three sets of diffraction data measured by three different waves such as X-rays, neutrons, and electrons which have different scattering mechanisms [8]. The present radial distribution functions (RDF's) for the TlCl and CuCl films, experimentally determined only from the diffuse electron diffraction patterns, are weighted combinations of the respective atomic densities as follows [9]:

$$4\pi r^2 \varrho(r) = 4\pi r^2 \varrho_0 + \frac{2r}{\pi} \int\limits_0^{s_{\max}} s I_{\mathrm{m}}(s)\, \mathrm{e}^{-Bs^2} \sin(sr)\, \mathrm{d}s \tag{1}$$

$$\begin{aligned} &= \tfrac{1}{4} \int\limits_{-\infty}^{\infty} 4\pi r' \{\varrho_{\mathrm{AA}}(r') - \varrho^0_{\mathrm{AA}}\}\, r T_{\mathrm{AA}}(r - r')\, \mathrm{d}r' + \\ &+ \tfrac{1}{4} \int\limits_{-\infty}^{\infty} 4\pi r' \{\varrho_{\mathrm{AB}}(r') - \varrho^0_{\mathrm{AB}}\}\, r T_{\mathrm{AB}}(r - r')\, \mathrm{d}r' + \\ &+ \tfrac{1}{4} \int\limits_{-\infty}^{\infty} 4\pi r' \{\varrho_{\mathrm{BA}}(r') - \varrho^0_{\mathrm{BA}}\}\, r T_{\mathrm{BA}}(r - r')\, \mathrm{d}r' + \\ &+ \tfrac{1}{4} \int\limits_{-\infty}^{\infty} 4\pi r' \{\varrho_{\mathrm{BB}}(r') - \varrho^0_{\mathrm{BB}}\}\, r T_{\mathrm{BB}}(r - r')\, \mathrm{d}r' , \end{aligned} \tag{2}$$

$$\left.\begin{aligned} T_{\mathrm{AA}}(r) &= \frac{4}{\pi} \int\limits_0^{s_{\max}} \frac{h^2(s)}{h^2(s)+1}\, \mathrm{e}^{-Bs^2} \cos(sr)\, \mathrm{d}s , \\ T_{\mathrm{AB}}(r) = T_{\mathrm{BA}}(r) &= \frac{4}{\pi} \int\limits_0^{s_{\max}} \frac{h(s)}{h^2(s)+1}\, \mathrm{e}^{-Bs^2} \cos(sr)\, \mathrm{d}s , \\ T_{\mathrm{BB}}(r) &= \frac{4}{\pi} \int\limits_0^{s_{\max}} \frac{1}{h^2(s)+1}\, \mathrm{e}^{-Bs^2} \cos(sr)\, \mathrm{d}s , \end{aligned}\right\} \tag{3}$$

$$\varrho_{ij}(r) = \varrho_{ij}(-r) ,$$

$$s = \frac{4\pi \sin\theta}{\lambda} ,$$

$$h(s) = \frac{f_{\mathrm{A}}(s)}{f_{\mathrm{B}}(s)} ,$$

where $\varrho(r)$ is the weighted atomic density at a distance r from an atom, ϱ_0 the mean atomic density, ϱ^0_{ij} the mean density of i–j atomic pair, $I_{\mathrm{m}}(s)$ the structure-sensitive intensity, $s_{\max}$ the maximum value of s below which $I_{\mathrm{m}}(s)$ is experimentally derived, B a damping factor which reduces ghost ripples due to the termination of $I_{\mathrm{m}}(s)$ and is usually chosen so that $\exp(-Bs^2_{\max}) = 0.1$, $T_{ij}(r)$ the weighting function of i–j pairs, 2θ the scattering angle, λ the electron wavelength and f_i the atomic scattering factor of i-kind atom. In the present study, TlCl and CuCl were assumed to be fully ionized compounds of Tl^+Cl^- and Cu^+Cl^-, and the atomic scattering factors of Tl^+, Cu^+ and Cl^- ions were used.

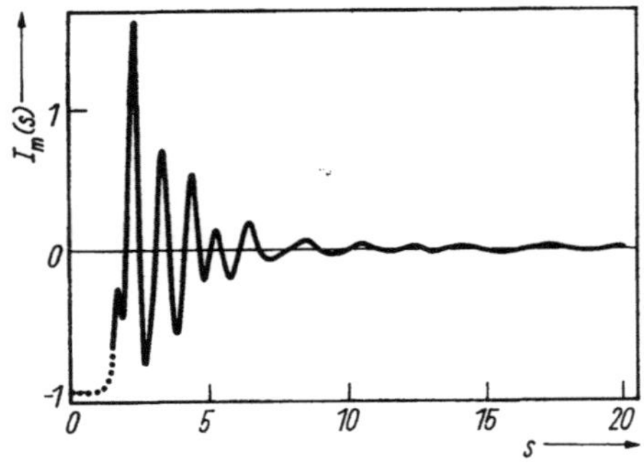

Fig. 2. $I_m(s)$ of TlCl film prepared by low temperature condensation at LHeT

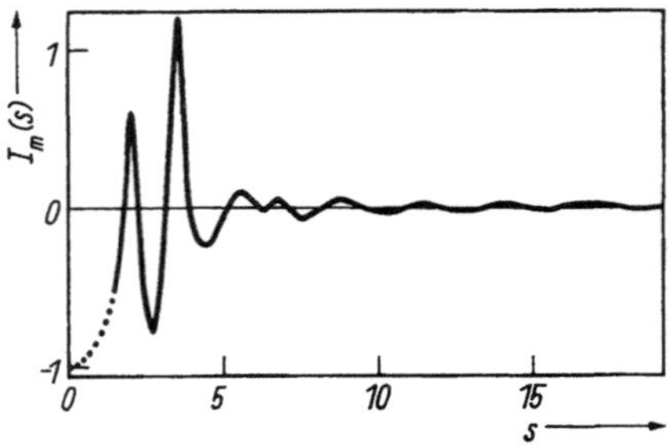

Fig. 3. $I_m(s)$ of CuCl film prepared by low temperature condensation at LHeT

The methods of derivation of $I_m(s)$ from diffuse halo patterns, those of removal of the contribution from the substrate and those of refinement of the derived $I_m(s)$ were the same as described in previous papers [4, 10]. Curves in Fig. 2 and 3 represent $I_m(s)$'s for the TlCl and CuCl films at LHeT, respectively. Values of $I_m(s)$ were derived from the diffuse halo patterns in the range from $s = 1.5$ to 20 Å^{-1} for the TlCl film and from $s = 1.5$ to 19.0 Å^{-1} for the CuCl film. Those in the range from $s = 0$ to 1.5 Å^{-1} were extrapolated so as to smoothly reach -1.0 at $s = 0$. The positions of the maxima of $I_m(s)$ of the TlCl and CuCl films are 2.2_3, 3.2_7, 4.3_5, 5.1_7, 6.4_0, 8.4_3, 10.5, 12.4, 14.1 and 17.2 Å^{-1} and 2.0_3, 3.5_2, 5.5_8, 6.7_4, 8.7_2, 11.5, 14.1 and 16.8 Å^{-1} in units of s, respectively.

Curves in Fig. 4 and 5 represent RDF's for the TlCl and CuCl films at LHeT, respectively, which were determined by numerically performing the Fourier integral of equation (1) with $B = 0.0058$ Å^2 and $\Delta s = 0.1$ Å^{-1}. As the mean densities of the present films were not measured, that of TlCl with the CsCl type structure and that of CuCl with the ZnS-type structure at room temperature (RT) were used as ϱ_0.

The experimental RDF often includes spurious details due to errors in diffraction data and due to the termination of the integral at s_{max}. Spurious details due to the former errors were reduced by averaging RDF's independently determined, and termination effects were sufficiently reduced by use of the damping factor.

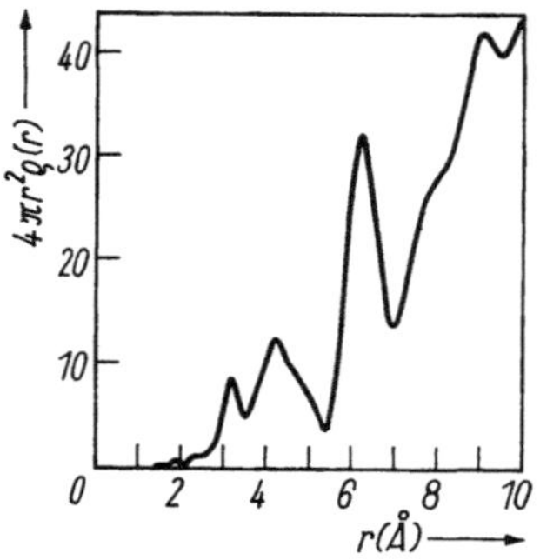

Fig. 4. $4\pi r^2\varrho(r)$ of TlCl film at LHeT

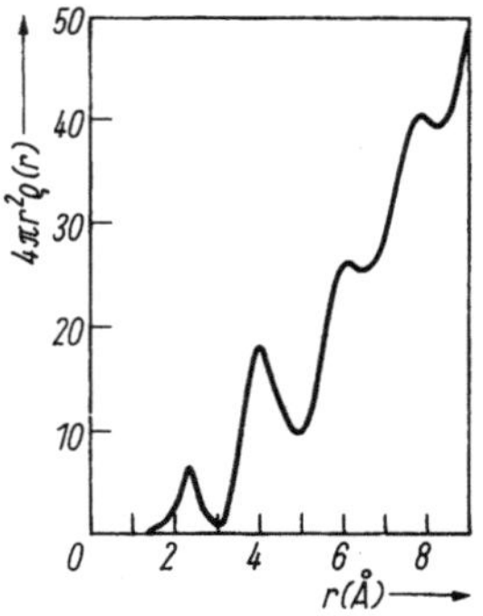

Fig. 5. $4\pi r^2\varrho(r)$ of CuCl film at LHeT

4. Discussion on the Local Atomic Arrangement

As described in Section 3, the determined RDF's in Fig. 4 and 5 are weighted combinations of the respective densities. Since the atomic scattering factors of Tl and Cu for electrons are larger than that of Cl, the contribution of $\varrho_{\text{TlTl}}(r)$ or $\varrho_{\text{CuCu}}(r)$ to the respective RDF's is larger, roughly, by a factor of

$$T_{\text{TlTl}}(r = 0)/T_{\text{TlCl}}(r = 0) = 4.1 \quad \text{or} \quad T_{\text{CuCu}}(0)/T_{\text{CuCl}}(0) = 1.8$$

than that of $\varrho_{\text{TlCl}}(r)$ or $\varrho_{\text{CuCl}}(r)$, and the contribution of $\varrho_{\text{ClCl}}(r)$ to the RDF's is smaller, roughly, by a factor of $T_{\text{ClCl}}(0)/T_{\text{TlCl}}(0) = 0.4$ or $T_{\text{ClCl}}(0)/T_{\text{CuCl}}(0) = 0.8$. Therefore, it is very important to take account of the enhancement or reduction of the contribution of $\varrho_{ij}(r)$ resulting from differences in the atomic scattering factor between the constituent atoms when the experimental RDF is interpreted. Since the features of the experimental $W(r)$ $\left(= \varrho(r)/\varrho_0\right)$ are insensitive to errors in ϱ_0 and to neglecting the intensity of small angle-scattering, the local atomic arrangement is discussed from features in the $W(r)$ in the present study.

4.1 TlCl

The experimental $W(r)$, shown by curve a in Fig. 6, has peaks at 3.1_7, 4.1_7, 6.1_9, 8.9_4 and 11.8_5 Å. The first three peaks are especially well-defined, which suggests the existence of local order in atomic arrangement in the TlCl film. It is very interesting to clarify what type of local order exists. Local order has so far been studied in many amorphous substances and it was clarified that the local order considerably resembles those found in crystalline states of the substances, e.g. tetrahedral units of atoms almost remain in amorphous Si and Ge [11]. In this stage, the local order in the TlCl film was investigated by comparison of the experimental $W(r)$ with $W(r)$'s calculated from structural models of heavily disordered CsCl and NaCl type structures, because TlCl has only the CsCl type structure in bulk at ordinary pressure, but the NaCl type structure appears in very thin deposited films [6]. $\varrho_{ij}(r)$'s of the models were computed in the same manner as described in the previous paper [4], and $W(r)$ was calculated from the $\varrho_{ij}(r)$ and $T_{ij}(r)$ by use of equation (2).

The first peak in the experimental RDF lies at 3.19 ± 0.02 Å, a value between the first nearest neighbor distance in the CsCl type structure at room temperature (RT), [3.33 Å] and that in the NaCl type structure at RT, [3.15 Å], which suggests that ions are surrounded with ions of different kind in the film state as well as in the crystalline state. Curves (b) and (c) in Fig. 6 represent $W(r)$'s calculated from disordered crystalline models of the CsCl and NaCl type structures, assuming a large Gaussian broadening of each shell of neighbor ions corresponding to heavy structural disorder.

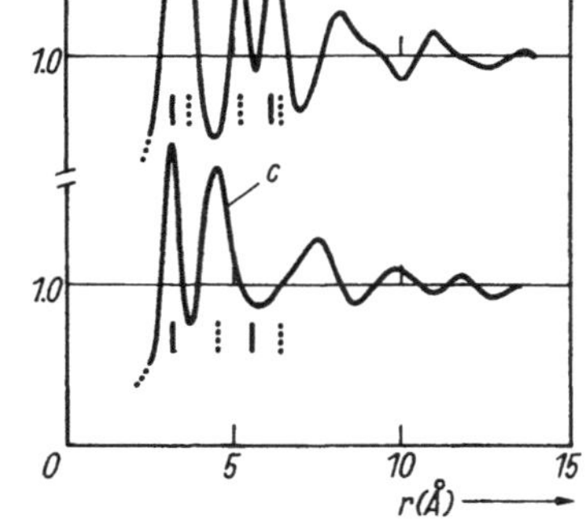

Fig. 6. Curves (a), (b) and (c) represent the experimental $W(r)$ of TlCl film at LHeT and $W(r)$'s calculated from models of the CsCl and NaCl type structures, respectively. The bars indicate interionic distances of the respective structures

Values of broadening were chosen so that the calculated $W(r)$ might resemble the experimental $W(r)$ as well as possible. The lattice parameters of the crystals were also adjusted so that the first nearest neighbor distances might agree with the first peak position of the experimental RDF. Interionic distances in the CsCl and NaCl type structure are marked by bars below the respective $W(r)$'s in Fig. 6. The solid and dotted bars represent distances between ions of different kind and ones between ions of the same kind, respectively.

The similarity of features of the experimental $W(r)$ curve with those of curve (b) is not good. The first peak of curve (b) is located around $r = 3.4$ Å and is very broad in comparison with that of the experimental $W(r)$. The broadness resulted from coalescence of peaks due to the first and second neighbors. The second peak is positioned at $r = 5.1_8$ Å and deep minimum exists at $r = 4.4$ Å in curve (b), but contrary to this a deep minimum and the second peak are observed at the respective distances in the experimental curve. The curve (c) seems to have a rough resemblance to the experimental $W(r)$ in the region below $r = 5$ Å, though the former second peak is located at a distance farther than the latter. However, no large peak is observed at all in curve (c) at $r = 6.1_8$ Å, where a definite peak appears in the experimental $W(r)$.

When the size of the crystal becomes very minute, a ring pattern changes to a diffuse halo pattern. Therefore, the film giving a diffuse halo pattern cannot be said to be necessarily in an amorphous state. However, the present TlCl film is surely in an amorphous state, because the features of the experimental $W(r)$ are different from those of $W(r)$ of the structural models of the CsCl and NaCl type structures. At present, we cannot help saying that the local order in the amorphous TlCl film is neither explainable by that of the disordered CsCl type structure nor by that of the disordered NaCl type.

4.2 CuCl

CuCl has the ZnS-type structure below 407 °C and the wurtzite type in the temperature range from 407 °C to the melting point [12]. Cu and Cl ions in the two structures are tetrahedrally surrounded with four Cl and Cu ions, respectively. Adjacent tetrahedra are linked with each other in a staggered or an eclipsed configuration and build up crystals. As shown in Fig. 7, adjacent two tetrahedra are not rotated about their common bond in the eclipsed configuration but the rotation of 60° occurs in the staggered one [13]. All the bonds in the ZnS-type structure form the staggered configuration. 3/4 of the bonds in the wurtzite type structure do so but 1/4 of the bonds form the eclipsed configuration.

The experimental $W(r)$ is shown by curve (a) in Fig. 8 and has two definite peaks at $r = 2.3_0$ and 3.9_1 Å and small peaks at $r = 5.8_5$, 7.5_2 and 9.4_0 Å. Curves (b) and (c) in Fig. 8 represent $W(r)$'s calculated from structural models of heavily disordered CuCl crystals with the ZnS- and wurtzite-type structures, respectively. The solid and dotted bars below the curves indicate distances between ions of the different kind and ones between ions of the same kind, respectively, in the respective structures. The first and second peaks in the two $W(r)$'s are positioned at $r = 2.3_0$ and 3.8_7 Å and coincide with those in the experimental $W(r)$. The two peaks are due to the first and second neighbors whose distances are that between a centre atom and a corner one in a tetrahedron and that between corner atoms. Hence, if tetrahedral units of ions exist in the present

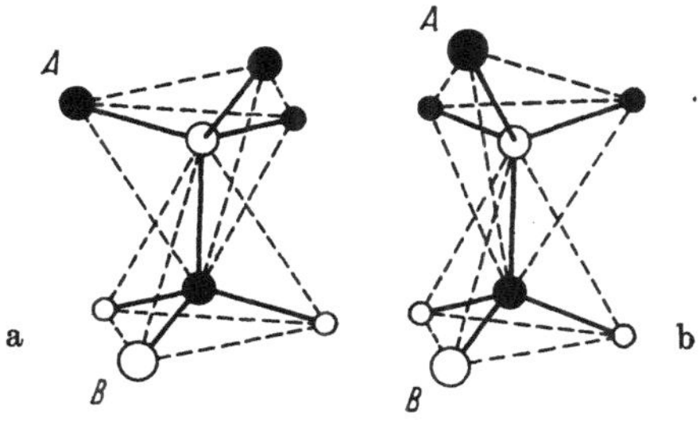

Fig. 7. Staggered (a) and eclipsed (b) configuration

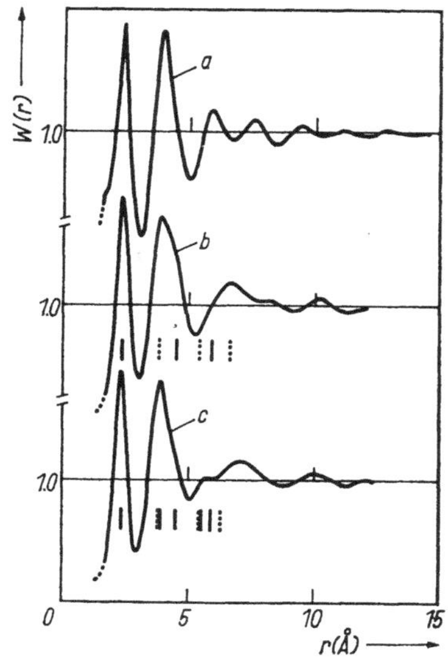

Fig. 8. Curves (a), (b) and (c) represent the experimental $W(r)$ of CuCl film at LHeT and $W(r)$'s calculated from models of the ZnS- and wurtzite-type structures, respectively. The bars indicate interionic distances of the respective structures

CuCl film, two peaks should be observed at $r = 2.3_0$ and 3.8_7 Å in the experimental $W(r)$. The coincidence of the first and second peak positions in the experimental $W(r)$ with those in the curves (b) and (c) means that many ions are tetrahedrally surrounded with four ions of different kind in the CuCl film as well as in the crystalline state.

It is an interesting problem to consider how tetrahedra link with adjacent ones. If all the adjacent tetrahedra are in the staggered configuration, the right side of the second peak swells as shown by curve (b) in Fig. 8, because the distance between atoms A and B is 4.49 Å. As the eclipsed configuration increases, the swell becomes inconspicuous, because the distance between atoms A and B is 3.92 Å in the configuration. Surely, the swell is smaller in the wurtzite-type structure than in the ZnS-type one. In addition, when the linkage with adjacent tetrahedra is heavily disturbed, the intermediate configuration between the staggered and eclipsed configurations may also arise. This makes the swell around $r = 4.5$ Å in the $W(r)$ of the ZnS-type structure smaller. The right side of the second peak has almost no swell in the experimental $W(r)$. Therefore, a most probable structural model of the amorphous CuCl film is considered as follows: tetrahedral units existing in the crystalline state remain even in the amorphous CuCl but the linkage with adjacent tetrahedra is much disturbed by rotation about their common bonds making various angles, which leads to the disappearance of long range order in the atomic arrangement. The similarity between the experimental $W(r)$ and $W(r)$ of the models is not good in the region beyond $r = 5$ Å. If the various rotations are considered and an assemblage of ions is constructed from tetrahedra, the disagreement may be reduced.

Acknowledgement

The author expresses his sincere thanks to Prof. Shiro Ogawa for his kind guidance and continuous encouragement throughout the present work.

References

[1] W. BUCKEL, Z. Phys. **138**, 136 (1954).
[2] S. FUJIME, Japan. J. appl. Phys. **5**, 1029 (1966).
[3] S. FUJIME, Japan. J. appl. Phys. **6**, 305 (1967).
[4] T. ICHIKAWA, phys. stat. sol. (a) **19**, 347 (1973).
[5] T. ICHIKAWA, phys. stat. sol. (a) **19**, 707 (1973).
[6] M. BLACKMAN and I. H. KHAN, Proc. Phys. Soc. 77, 471 (1961).
[7] R. W. JAMES, The Optical Principles of the Diffraction of X-Rays, G Bell & Sons; Ltd., London 1950.
[8] D. T. KEATING, J. appl. Phys. **34**, 923 (1963).
[9] T. ICHIKAWA, Japan. J. appl. Phys. **9**, 748 (1970).
[10] T. ICHIKAWA, phys. stat. sol. (b) **56**, 707 (1973).
[11] H. RICHTER and G. BREITLING, Z. Naturf. **13a**, 988 (1958).
[12] M. R. LORENZ and J. S. PRENER, Acta cryst. **9**, 538 (1956).
[13] R. GRIGOROVICI and R. MANAILA, Thin Solid Films **1**, 343 (1967/68).

(Received May 6, 1974)

phys. stat. sol. (b) 65, 419 (1974)

Subject classification: 14.2

Istituto di Fisica, Universitá di Salerno,
Gruppo Nazionale di Struttura della Materia del C.N.R.

Type II/1 and Type II/2 Superconductors

By

L. DE CESARE and F. MANCINI

The problem of a first-order transition at the lower critical field in type II superconductors is investigated by means of the boson formulation of superconductivity. The critical value $\varkappa_c$ of the Ginzburg-Landau parameter which separates type II/1 from type II/2 superconductors is theoretically determined and is in good agreement with the experimental value. The distribution of the magnetic field of an isolated flux line and the interaction potential between two flux lines are also computed in the entire region including the core.

Le problem d'une transition du premier ordre d'un supraconducteur du type II à $H_{c,1}$ est étudié au moyen de la formulation Bosonique de la supraconductivité. La valeur critique $\varkappa_c$ du paramètre de Ginzburg-Landau qui sépare les supraconducteurs du type II/1 de ceux du type II/2 est fixé par la théorie et est proche de la valeur experimentale. La distribution du champs magnétique d'une ligne de flux isolée et le potentiel d'interaction de deux lignes de flux sont calculées dans toute la region, y compris le coeur de la ligne de flux.

1. Introduction

There is now experimental evidence [1 to 5] that type II superconductors with a value of $\varkappa$, the Ginzburg-Landau parameter, less than a critical value $\varkappa_c$ exhibit a first-order transition between the Meissner state and the mixed state. By considering Ta and Nb systems, Auer and Ullmaier [5] have made magnetization measurements for a wide range of values of $\varkappa$ and they have been able to compute experimentally the temperature dependence of $\varkappa_c$.

From a theoretical point of view the situation, however, is not very well established. In the BCS theory it is known that in order to study problems related to the presence of a non-homogeneous ground state one must solve the Gorkov equations. Unfortunately, it is not easy to solve these equations and up to now most of the theoretical attempts to describe the mixed state of type II superconductors are based on the use of the Ginzburg-Landau equations. On the other hand the GL theory, which is strictly valid only at $T = T_c$, predicts a second-order phase transition at the lower critical field $H_{c,1}$, so that in order to perform a theoretical study on the possibility of a first-order phase transition of $H_{c,1}$ there are, in the framework of the BCS theory, only two choices: (i) Try to solve the Gorkov equations for large distances in order to have information about the asymptotic behaviour of the order parameter and vector potential; (ii) improve the approximation around $T = T_c$ by taking into account higher-order terms in the power series expansion of $\theta = 1 - T/T_c$.

The first approach has the merit that it can be studied without restrictions on values of temperature, but is analytically very complicated. Some approximate results have been obtained in [6, 7], where the authors have computed the asymptotic form of the magnetic field of an isolated flux line and the distance

r_0, where the magnetic field, $\boldsymbol{h}$, changes direction. Most of the studies have been made by using the simplified version of the Gorkov equations due to Eilenberger [8]. As a matter of fact, the first theoretical indication[1]) of an attractive interaction among the flux lines, which is now believed to be the cause of a first-order transition, is found in the paper of Eilenberger and Büttner [9]. Unfortunately, there has been some controversy [10 to 12] about the results of this paper just in the region $\varkappa < \varkappa_c$, where oscillations of the magnetic field were predicted. More accurate studies of the asymptotic behaviour of the vector potential and order parameter can be found in [12]. Though this approach is of theoretical interest, it has not furnished so far any result comparable with experimental data. Recently, the Gorkov equations for the case of an isolated flux line have been numerically solved by computer [13]: however, the accuracy reached so far does not permit the study of regions far from the centre of the flux line.

The second approach is based on the use of the generalized Ginzburg-Landau equations derived by Neumann and Tewordt [14], which give a first-order correction in θ to the GL equations. Several computations [15 to 17] have shown the existence of an attractive interaction among the flux lines for sufficiently small values of $\varkappa$, but they cannot predict any values for r_0; also, this approach is valid in a very narrow region close to T_c.

Using the boson formulation of superconductivity [18] we predicted in 1970 the existence of an attractive interaction among the flux lines [27, 22]; recently we have computed the magnetization curve for samples with different values of $\varkappa$ [19]; for $\varkappa$ sufficiently small we have found that around $H_{c,1}$ the magnetization curves have an inverted S-form, implying that the phase transition is first order. In order to make a more accurate analysis of the transition at $H_{c,1}$, the magnetization curves and the Gibbs free energy have been computed for many values of $\varkappa$ [21]. These studies have confirmed that for small values of $\varkappa$ a first-order transition occurs; furthermore, by computing the value of $\varkappa$ for which the S-shape form disappears, it has been computed the critical value $\varkappa_c$; the theoretical result is $\varkappa_c = 1.22$. In this article we compute the interaction energy between two flux lines and we show that there is a critical value $\varkappa_c$ such

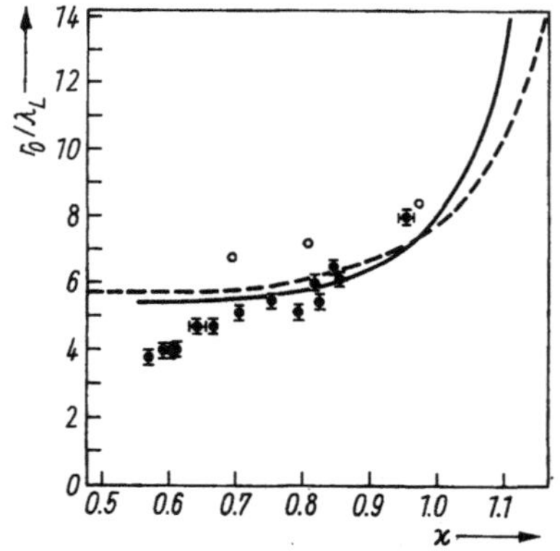

Fig. 1. The distance r_0 where the interaction energy between two flux lines becomes attractive is given as a function of the Ginzburg-Landau parameter $\varkappa$. The dashed curve denotes the theoretical results with $\delta = 0$; the solid curve denotes the theoretical results with $\delta \neq 0$. The circles represent the theoretical values for r_0 predicted by [7]. All these theoretical results have been computed at $t = 0.3$. The experimental data of [5] for the lattice parameter are also reported

[1]) Independently by the work of [9] and almost contemporaneously, the possibility of an attractive interaction between flux lines was predicted by means of the boson formulation (see [27]).

that for $\varkappa > \varkappa_c$ the interaction is always repulsive, while for $\varkappa < \varkappa_c$ the energy has a minimum for a certain distance between the flux lines. The theoretical value of $\varkappa_c$ is close to the experimental value of [5]. We also compute the characteristic value r_0, where the magnetic field changes direction, for many values of $\varkappa$; a comparison with the theoretical results of [7] is given in Fig. 1, where the experimental data for the lattice parameter d_0 of [5] are also reported. Finally, the distribution of the magnetic field of a single flux line is computed in the whole electromagnetic region, including the core.

2. Summary of Previous Results

In the last years we have been attempting to formulate a new approach to the theory of superconductivity by using some modern techniques of quantum field theory. This approach, which has been called the boson formulation of superconductivity, has been exposed in details in a series of articles [20, 22 to 25]. Recently, a review article has been published [18].

Here, we shall give a brief outline of the method. Starting from the BCS gauge-invariant Hamiltonian we have formulated the theory in operator form by making use of the Bethe-Salpeter equations. We have not made use of the Green's function method, since in this formalism we cannot apply those invariant operator transformations, boson transformations, which are an essential element of the boson method. These transformations, which change the ground state expectation value of the observables but leave invariant the equations of motion, allow us to study non-homogeneous situations, once that the dynamical problem has been solved in the homogeneous case.

In the case of time-independent situations, after the boson transformation, the ground state current, the order parameter, and the ground state energy take the following expressions [20, 24]:

$$\boldsymbol{J}(\boldsymbol{x}) = \frac{1}{4\pi e\lambda_L^2} \int d^3y\, c(\boldsymbol{x}-\boldsymbol{y})\,[\nabla f(\boldsymbol{y}) - e\boldsymbol{A}(\boldsymbol{y})]\,, \tag{2.1}$$

$$\Delta(\boldsymbol{x}) = e^{2if(\boldsymbol{x})}\,|\Delta(\boldsymbol{x})|\,, \tag{2.2}$$

$$W = \frac{1}{2e} \int d^3x\, [\nabla f(\boldsymbol{x}) - e\boldsymbol{A}(\boldsymbol{x})] \cdot \boldsymbol{J}(\boldsymbol{x})\,. \tag{2.3}$$

$\boldsymbol{A}$ is the transverse vector potential of the total electromagnetic field which contains the external field and the field due to the supercurrents. λ_L is the London penetration depth; $c(\boldsymbol{x})$ is the boson characteristic function defined in [20, 25, 18] and which is entirely fixed by dynamics. The function $f(\boldsymbol{x})$, half of the phase of the order parameter, has to be [22] a solution of the Laplace equation

$$\nabla^2 f(\boldsymbol{x}) = 0\,. \tag{2.4}$$

Combining (2.1) with the Maxwell equation $4\pi\boldsymbol{J} = \nabla\times\boldsymbol{H}$ we get the following equation for the vector potential:

$$\nabla^2\boldsymbol{A}(\boldsymbol{x}) - \frac{1}{\lambda_L^2}\int d^3y\, c(\boldsymbol{x}-\boldsymbol{y})\,\boldsymbol{A}(\boldsymbol{y}) + \frac{1}{e\lambda_L^2}\int d^3y\, c(\boldsymbol{x}-\boldsymbol{y})\,\nabla f(\boldsymbol{y}) = 0\,. \tag{2.5}$$

By solving (2.4) according to the appropriate boundary conditions of the problem under study we can describe different physical situations. For instance, the solution $f = \text{const}$ will correspond to the Meissner state. As shown in [22], the solution of (2.4) which corresponds to a single flux line is given by

$$f(\boldsymbol{x}) = \tfrac{1}{2}\,\varphi(\boldsymbol{x})\,, \tag{2.6}$$

where $\varphi(\boldsymbol{x})$ is the polar angle of the two-dimensional vector $\boldsymbol{x}$ lying in a plane normal to the direction of the external magnetic field (we use cylindrical coordinates with the centre at $\boldsymbol{x} = 0$).

By inserting (2.6) in (2.5) we can compute the vector potential and therefore the magnetic field due to a single flux line. Since the computation can be found in [24] we give here only the final result. Let us denote by $h(\boldsymbol{x})$ the magnitude of the magnetic field due to an isolated flux line with centre at the origin of our system of coordinates; then

$$h(\boldsymbol{x}) = \frac{\Phi}{2\pi} \int\limits_0^\infty \frac{k c(k)}{\lambda_L^2 k^2 + c(k)}\, J_0(kr)\, dk\,, \tag{2.7}$$

where r is the modulus of the vector $\boldsymbol{x}$; $\Phi = \pi/e$ (here $\hbar = c = 1$) is the unit quantum flux and J_0 is the Bessel function. $c(\boldsymbol{k})$ is the Fourier transform in the momentum space of $c(\boldsymbol{x})$. It should be noted that both the quantities λ_L and $c(k)$ depend on temperature.

3. Type II/1 and Type II/2 Superconductors

In the literature there is some controversy about the cause which should give rise to a first-order transition at $H_{c,1}$. Essentially two causes are considered: field reversal of the magnetic field of an isolated flux line, and an extra attractive term in the interaction energy which should exceed the electromagnetic repulsion. It is stated by many authors that there is no apparent reason why the two effects should be related to each other. According to the results of our calculation [18] the contrary is true. The interaction energy between two flux lines is given by [22, 23]

$$E(d) = \frac{\Phi}{4\pi}\, h(d)\,, \tag{3.1}$$

where d is the distance between the centres of the flux lines and h is the magnitude of the magnetic field of an isolated flux line given by (2.7). Equation (3.1) immediately relates the presence of an attractive interaction to the phenomenon of field reversal. A study of the distribution of the magnetic field of a single flux line will give the potential energy between two flux lines.

The boson characteristic function has been computed from the microscopic formulation for small values of k,[2]); the result is [25]

$$c(k) = 1 - \alpha(T)\, \xi_0^2(T)\, k^2 + O(k^4)\,, \tag{3.2}$$

where ξ_0 is the BCS temperature-dependent coherence length, $\xi_0(T) = v_F/\pi\Delta(T)$, and α is a function of temperature. At $T = 0\ °K$: $\alpha(0) = 2\pi^2/45$. Recently, we have computed $\alpha(T)$ in the entire domain $0 \leqq T \leqq T_c$ [26].

---

[2]) See Note added in proof.

On the other hand, we have shown [22, 24] that in order to have a stable vortex lattice, $c(k)$ must decrease at least as $1/k^2$ when $k \to \infty$. We shall thus write

$$c(k) = \frac{1}{1 + \alpha\xi_0^2 k^2 + \dots}, \tag{3.3}$$

where dots stand for higher-order terms in $(\xi_0 k)^2$. As first approximation we have considered the following expression:

$$c(k) = \frac{1}{1 + \alpha\xi_0^2 k^2}. \tag{3.4}$$

Already this simple form gives results which are in good quantitative agreement with the experiments. In [19] we have shown that (3.4) gives a form factor which is in very good agreement with the form factor of scattering of neutrons by $Nb_{0.27}Ta_{0.73}$ measured in [4]. In [28] we have also shown that (3.4) agrees with the experimental result of [5]; type II superconductors can be divided in two classes, type II/2 which exhibit a second-order phase transition at $H_{c,1}$, and type II/1 which exhibit a first-order transition. By defining $\varkappa_B = \lambda_L(T)/\xi_0(T)$ we found that the two classes are separated by a critical value of $\varkappa_B$ given by the expression

$$\varkappa_{B,c}(T) = 2\sqrt{\alpha(T)}. \tag{3.5}$$

At $T = 0$ °K (3.5) gives $\varkappa_{B,c} = 1.32$. If we use $\varkappa = 0.957\,\varkappa_B$, we find $\varkappa_c = 1.27$ which is not too far from the experimental value $\varkappa_c = 1.14$ of [5]. Equation (3.5) shows that a knowledge of the temperature dependence of α gives the temperature dependence of $\varkappa_{B,c}$ [26]. We would like to notice that the agreement with the experimental data shown by the results of [19, 28] is not a trivial one, since α is purely derived from the microscopic formulation [18].

In order to improve the agreement we have considered [19] the following expression

$$c(k) = \frac{1}{1 + \alpha\xi_0^2 k^2 + \delta\xi_0^4 k^4}, \tag{3.6}$$

where δ is a parameter which should be phenomenologically fixed. Using (3.6) we have computed [19] the magnetization curves for pure vanadium and niobium in the entire domain of H. By taking $\delta = 0.025$ we have found a very good agreement with the experimental magnetization curves of [29].

We shall now study the problem of phase transition at $H_{c,1}$ by using (3.6) for the boson characteristic function.[3] By putting (3.6) in (2.7) we can easily find that the distribution of the magnetic field of a single flux line has the following form:

$$h(r) = \frac{\Phi\varkappa_B^4}{2\pi\lambda_L^2\delta\varrho(\varkappa_B)}\left\{(\alpha_2 - \alpha_3)\,K_0\left(\frac{r}{\lambda_L}\sqrt{-\alpha_1}\right) + (\alpha_3 - \alpha_1)\,K_0\left(\frac{r}{\lambda_L}\sqrt{-\alpha_2}\right) + \right.$$
$$\left. + (\alpha_1 - \alpha_2)\,K_0\left(\frac{r}{\lambda_L}\sqrt{-\alpha_3}\right)\right\} \tag{3.7}$$

with

$$\varrho(\varkappa_B) = \alpha_1\alpha_2(\alpha_1 - \alpha_2) + \alpha_2\alpha_3(\alpha_2 - \alpha_3) + \alpha_3\alpha_1(\alpha_3 - \alpha_1). \tag{3.8}$$

[3]) A preliminary report of the results obtained has been presented in [30].

K_0 is the modified Bessel function of zero order; α_1, α_2, and α_3 are the roots of the cubic equation

$$\gamma x^3 + \beta x^2 + x + 1 = 0 , \tag{3.9}$$

where $\gamma = \delta/\varkappa_B^4$, $\beta = \alpha/\varkappa_B^2$. Expression (3.7) shows that if the roots α_i are all real, the magnetic field is always positive and will go exponentially to zero for $r \to \infty$; the interaction potential is always repulsive. Vice versa, if (3.9) admits complex solutions, then the magnetic field will be an oscillating damped function, and the interaction energy can become attractive for some range of values for r. Equations (3.7) and (3.9) immediately put an extreme condition on the values which δ can take. It is theoretically and experimentally known that for $\varkappa \to \infty$ the phase transition at $H_{c,1}$ is second order; if we impose this condition it can be easily seen that the following relation must hold:

$$\delta \leqq \tfrac{1}{4}\alpha^2 . \tag{3.10}$$

A study of (3.9) with the condition (3.10) will then give the theoretical value of $\varkappa_B$ which separates type II/2 from type II/1 [30]:

$$\varkappa_{B,c}^2 = \frac{\alpha(2\alpha^2 - 9\delta) + 2(\alpha^2 - 3\delta)^{3/2}}{\alpha^2 - 4\delta} . \tag{3.11}$$

Equation (3.11), which is very sensitive to the value of δ, gives a better way to fix the phenomenological parameter δ; from (3.11) $\varkappa_{B,c}$ decreases when δ increases, on the other hand the values of δ are limited by the condition (3.10). The value of δ which satisfies this condition and which gives a value for $\varkappa_c$ closer to the experimental value is then given by

$$\delta = \tfrac{1}{4}\alpha^2 . \tag{3.12}$$

It is worth noting that this choice corresponds to the following expression for the boson characteristic function:

$$c(k) = \frac{1}{(1 + \frac{1}{2}\alpha\xi_0^2 k^2)^2} . \tag{3.13}$$

With this choice the critical value $\varkappa_{B,c}$ is then given by

$$\varkappa_{B,c} = \tfrac{3}{2}\sqrt{\tfrac{3}{2}\alpha} . \tag{3.14}$$

At $T = 0$ °K: $\varkappa_c = 1.16$, which is very close to the experimental value $\varkappa_c = 1.14$ of [5].

For $\varkappa_B < \varkappa_{B,c}$ (3.9) admits one real solution and two complex conjugate solutions. Substituting these solutions into the (3.7) we find

$$h(r) = \frac{2\Phi\varkappa_B^4}{3\pi\lambda_L^2\alpha^2\tilde{\varrho}(\varkappa_B)} \left\{(B - A)\, K_0\left(\frac{r}{\lambda_L}a\right) + \sqrt{3}\,(A + B)\,\mathrm{Im}\, K_0\left(\frac{r}{\lambda_L}b\right) - \right.$$
$$\left. - (B - A)\,\mathrm{Re}\, K_0\left(\frac{r}{\lambda_L}b\right)\right\}, \tag{3.15}$$

where

$$a = d + A + B\,, \qquad b = b_1 + ib_2\,,$$

$$b_1 = b\cos\tfrac{1}{2}\theta\,, \qquad b_2 = b\sin\tfrac{1}{2}\theta\,,$$

$$d = \frac{4\varkappa_B^2}{3\alpha}\,, \qquad b = [A^2 + B^2 - AB + d(d - A - B)]^{1/4}\,,$$

$$\theta = \operatorname{arctg}\left(\frac{\sqrt{3}(A - B)}{2d - A - B}\right), \qquad \tilde{\varrho}(\varkappa_B) = (B - A)\,(A^2 + B^2 - AB)\,,$$

$$A = \left[\frac{\varkappa_B^4(2\varkappa_{B,c}^2 - \varkappa_B^2)}{\alpha^2\varkappa_{B,c}^2} - \frac{2\varkappa_B^4}{\varkappa_{B,c}\alpha^2}\sqrt{\varkappa_{B,c}^2 - \varkappa_B^2}\right]^{1/3},$$

$$B = \left[\frac{\varkappa_B^4(2\varkappa_{B,c}^2 - \varkappa_B^2)}{\alpha^2\varkappa_{B,c}^2} + \frac{2\varkappa_B^4}{\varkappa_{B,c}\alpha^2}\sqrt{\varkappa_{B,c}^2 - \varkappa_B^2}\right]^{1/3}.$$

At the centre of the flux line, (3.15) gives

$$h(0) = \frac{\Phi\varkappa_B^4}{3\pi\lambda_L^2\alpha^2(A^2 + B^2 - AB)}\left[2\ln\left(\frac{b}{a}\right) + \frac{\sqrt{3}(A + B)}{A - B}\,\theta\right]. \tag{3.16}$$

For large distances from the axis we can use the asymptotic representation of the Bessel function:

$$\mathrm{K}_0(z) = \sqrt{\frac{\pi}{2z}}\,\mathrm{e}^{-z}\left[1 - \frac{1}{8z} + \frac{9}{128}\frac{1}{z^2} + \cdots\right], \tag{3.17}$$

where dots stand for higher-order powers of $1/z$.

For $r > \lambda_L$ we then have the following distribution of the magnetic field:

$$h(r) = \frac{2\Phi\varkappa_B^4}{3\pi\lambda_L^2\alpha^2\tilde{\varrho}(\varkappa_B)}\sqrt{\frac{\pi\lambda_L}{2ra}}\,\mathrm{e}^{-ra/\lambda_L}\,f(r)\,[1 - g(r)]\,, \tag{3.18}$$

where

$$f(r) = (B - A)\left[1 - \frac{1}{8}\frac{\lambda_L}{ra} + \frac{9}{128}\frac{\lambda_L^2}{r^2a^2} + \cdots\right],$$

$$g(r) = F(r)\,\mathrm{e}^{rt/\lambda_L}\,[(B - A)\,G_1(r) + \sqrt{3}(A + B)\,G_2(r)]\,,$$

$$F(r) = \sqrt{\frac{a}{b}}\,\frac{1}{f(r)}\,, \qquad t = a - b_1\,,$$

$$G_1(r) = \cos\varphi_1 - \frac{1}{8}\frac{\lambda_L}{rb}\cos\varphi_2 + \frac{9}{128}\frac{\lambda_L^2}{r^2b^2}\cos\varphi_3 + \cdots\,,$$

$$G_2(r) = \sin\varphi_1 - \frac{1}{8}\frac{\lambda_L}{rb}\sin\varphi_2 + \frac{9}{128}\frac{\lambda_L^2}{r^2b^2}\sin\varphi_3 + \cdots\,,$$

$$\varphi_n = \frac{r}{\lambda_L}\,b_2 + \frac{2n - 1}{4}\,\theta\,.$$

As we see from (3.18), for $\varkappa_B < \varkappa_{B,c}$ the magnetic field of a single flux line is an oscillating damped function; the nodes of this function are given by the equation

$$g(r) = 1\,. \tag{3.19}$$

This equation has been numerically solved and the values of r_0/λ_L, where the magnetic field has its first zero, are given in Fig. 1. Due to the damping factor $\exp(-ra/\lambda_L)$ the amplitude of the magnetic field goes rapidly to zero after the first minimum. In Fig. 1 we also report the theoretical values of r_0/λ_L predicted by [7], and the experimental values for the lattice parameter d_0/λ_L, given in [5]. The theoretical values given in Fig. 1 have been computed at $t = T/T_c = 0.3$; we have used the value of $\alpha(T)$ predicted by the microscopic formulation [26]. From (3.19) we find that at $t = 0.3$, $r_0/\lambda_L \to \infty$ as $\varkappa_B$ approaches the value $\varkappa_{B,c} = 1.16$.

For $\varkappa_B = \varkappa_{B,c}$, (3.9) admits three real solutions, of which two are equal; in this case the magnetic field has the following value:

$$h(r) = \frac{\Phi}{2\pi\lambda_L^2}\left[K_0\left(\frac{3r}{\lambda_L}\right) - K_0\left(\frac{3r}{2\lambda_L}\right) + \frac{3}{2}\frac{r}{\lambda_L}K_1\left(\frac{3r}{2\lambda_L}\right)\right], \tag{3.20}$$

where K_1 is the modified Bessel function of first order. At $r = 0$,

$$h(0) = \frac{3\Phi}{4\pi\lambda_L^2}\left[1 + \frac{2}{3}\ln\left(\frac{1}{2}\right)\right]. \tag{3.21}$$

For $\varkappa_B > \varkappa_{B,c'}$, the roots of equation (3.9) are all real and are given by the following expressions:

$$\alpha_1 = -\frac{9}{2}\frac{\varkappa_B^2}{\varkappa_{B,c}^2}\left[1 - \cos\frac{\varphi}{3}\right],$$

$$\alpha_2 = -\frac{9}{2}\frac{\varkappa_B^2}{\varkappa_{B,c}^2}\left[1 + \frac{1}{2}\cos\frac{\varphi}{3} + \frac{\sqrt{3}}{2}\sin\frac{\varphi}{3}\right],$$

$$\alpha_3 = -\frac{9}{2}\frac{\varkappa_B^2}{\varkappa_{B,c}^2}\left[1 + \frac{1}{2}\cos\frac{\varphi}{3} - \frac{\sqrt{3}}{2}\sin\frac{\varphi}{3}\right],$$

where the angle φ is to be determined from the expression

$$\cos\varphi = 1 - \frac{2\varkappa_{B,c}^2}{\varkappa_B^2}.$$

Substituting these solutions in (3.7) we find the distribution of the magnetic field for type II/2 superconductors.

4. Conclusions

The problem of phase transition at the lower critical field has been studied by means of the boson formulation of superconduticvity. In this problem, as in all the other problems considered up to now [18] (for application of the boson method to the study of the Josephson effect see [31]), this approach offers the advantage with respect to the BCS formulation (see for example [9 to 17]) of a greater simplicity. Though we are using a very simple form for the boson characteristic function, the agreement with many different experiments [19, 21, 26, 28] is quite good. We feel that the real shape of the boson characteristic function is much more complicated and we think that the problem of the mixed state in type II superconductors can be solved by means of the complete knowledge of the boson characteristic function. Computations in this direction are in progress.

Acknowledgements

One of the authors (F. M.) wishes to thank Prof. L. Leplae and Prof. H. Umezawa for valuable discussions. He gratefully acknowledges the hospitality of the Physics Department of the University of Wisconsin-Milwaukee where this work was completed. He also is grateful to GNSM for travel grant.

References

[1] U. Krägeloh, phys. stat. sol. **42**, 559 (1970).
[2] U. Essmann and H. Träuble, Phys. Letters A **24**, 526 (1967).
[3] A. Seeger, Comments Solid State Phys. **3**, 97 (1970).
[4] J. Schelten, H. Ullmaier, and W. Schmatz, phys. stat. sol. (b) **48**, 619 (1971).
[5] J. Auer and H. Ullmaier, Phys. Rev. B **7**, 136 (1973).
[6] K. Dichtel, Phys. Letters A **35**, 285 (1971).
[7] E. H. Brandt, Phys. Letters A **39**, 193 (1972).
[8] G. Eilenberger, Z. Phys. **214**, 195 (1968).
[9] G. Eilenberger and H. Büttner, Z. Phys. **224**, 335 (1969).
[10] R. M. Cleary, Phys. Rev. Letters **24**, 940 (1970).
[11] M. C. Leung and A. E. Jacobs, Z. Phys. **253**, 89 (1972).
[12] M. C. Leung and A. E. Jacobs, J. low Temp. Phys. **11**, 395 (1973).
[13] L. Kramer and W. Pesch, Solid State Commun. **12**, 549 (1973).
[14] L. Tewordt, Phys. Rev. **137**, A1745 (1965).
L. Neumann and L. Tewordt, Z. Phys. **189**, 55 (1966).
[15] A. E. Jacobs, Phys. Rev. B **4**, 3029 (1971).
[16] A. Hubert, phys. stat. sol. (b) **53**, 147 (1972).
[17] A. E. Jacobs, J. low Temp. Phys. **10**, 137 (1973).
[18] F. Mancini, Ph. D. Thesis, University of Wisconsin-Milwaukee, 1972.
L. Leplae, F. Mancini, and H. Umezawa, Physics Reports **10C**, 152 (1974).
[19] F. Mancini, G. Scarpetta, V. Srinivasan, and H. Umezawa, Phys. Rev. B **9**, 130 (1974).
[20] L. Leplae, F. Mancini, and H. Umezawa, Phys. Rev. B **5**, 884 (1972).
[21] L. Leplae, V. Srinivasan, and H. Umezawa, Phys. Letters A **45**, 177 (1973).
[22] L. Leplae, F. Mancini, and H. Umezawa, Phys. Rev. B **2**, 3594 (1970).
[23] F. Mancini and H. Umezawa, Phys. Letters A **42**, 287 (1972).
[24] L. Leplae, F. Mancini, and H. Umezawa, Phys. Letters A **40**, 177 (1972); Phys. Rev B **6**, 4178 (1972).
[25] F. Mancini and H. Umezawa, Lett. Nuovo Cimento **7**, 125 (1973).
[26] F. Mancini, H. Umezawa, and G. Vitiello, Solid State Commun., **14**, 1123 (1974).
L. de Cesare and F. Mancini, Solid State Commun., in the press.
[27] L. Leplae, F. Mancini, and H. Umezawa, Bull. Amer. Phys. Soc. **15**, 39 (1970); Lett. Nuovo Cimento **3**, 153 (1970).
[28] F. Mancini, Phys. Letters A **45**, 179 (1973).
[29] S. T. Sekula and R. H. Kernohan, Phys. Rev. B **5**, 904 (1972).
[30] F. Mancini, in: Renormalization and Invariance in Quantum Field Theory, Plenum Press, New York 1974 (p. 159).
[31] L. Leplae, F. Mancini, and H. Umezawa, Nuovo Cimento **9B**, 233; **10B**, 267 (1972).

(Received October 29, 1973)

Note added in proof:

Recently, we have derived a set of exact equations which allow us to compute the boson characteristic function in the entire region of momenta (see E. Mancini, Some Exact Results in the Boson Formulation of Superconductivity, Physica, in the press; and G. F. De Angelis and F. Mancini, A Complete Computation of the Boson Characteristic Function at $T = 0\,°\mathrm{K}$, Physica, in the press).

phys. stat. sol. (b) 65, K1 (1974)

Subject classification: 10.1. and 20.1; 22.7

Physico-Technical Institute of Low Temperatures,
Academy of Sciences of the Ukrainian SSR, Kharkov

On the Effect of Single Twinning Interlayers on the Electric Conductivity and Magnetoresistance of Antimony Single Crystals

By

YU.A. BOGOD and V.B. KRASOVITSKII

Papers (1 to 3) reveal that the introduction of twins, whose sizes are small as compared to the crystal matrix, noticeably affects the galvanomagnetic properties of antimony and bismuth. The above papers studied the electric conductivity and Shubnikov-de Haas oscillations. The same papers investigated the diagonal and non-diagonal components of the magnetoresistance tensor (ϱ_{ik}) as a function of the direction of the magnetic field vector $\vec{H}$.

The present paper chiefly concerns the twin effect on the exponent n of the function $\varrho_{xx} = \alpha H^n$. To study this problem it is necessary to carry out successive measurements on absolutely identical samples[1)] having different numbers of twinning interlayers. Most appropriate for the experiment is a combined crystal which consists of several parts: one is "perfect" with no twins, another part contains one twin, then a part with two twins, etc. Such sample was prepared and studied in this investigation (Fig. 1). The original single crystal having a rather perfect structure (the dislocation density on the (111) plane found from etch pits did not exceed 10^2 cm^{-2}) was obtained by pulling from the melt[2)], and twins by slight sample bending in liquid nitrogen.

---

1) Identical should be the size, shape, surface state, purity, and structural perfection. The effect of these factors on the galvanomagnetic properties of Sb was discussed in detail in (4).

2) The authors are indebted to S.V. Tsivinskii for providing a sample of high perfection.

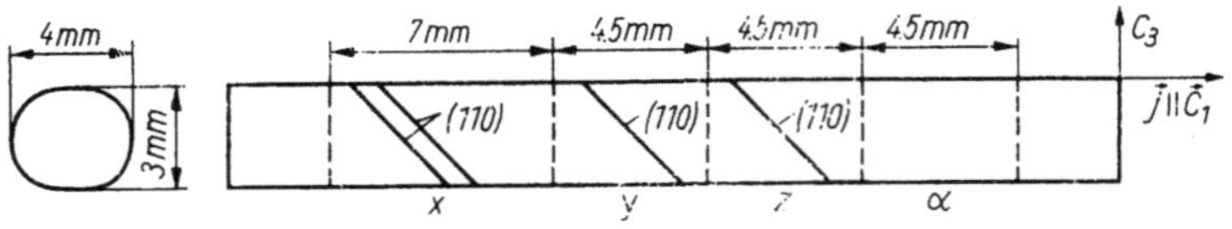

Fig. 1. Experimental geometry. (110) is the family of twinning planes whose traces on the (111) plane are perpendicular to the measuring current direction $\vec{j}$. General sizes for a twin and accommodation band: sample z ≈ 10 μm, sample y ≈ 100 μm, sample x ≈ 100 μm

The results are shown in Table 1 and in Fig. 2. The experiment evidence that the twin effect on the magnetoresistance (as well as on the exponent n) is most essential when the magnetic field vector lies in the twinning plane. This conclusion is valid for all possible orientations of twinning planes in both Sb and Bi (3). Besides, the extent of the twin effect on galvanomagnetic properties depends on the accommodation bandwidth (Fig. 2, curves for samples z and y).

To interpret the experimental results, one might assume that the charge carrier mobility μ_H^a in the accommodation band is much less than that in the crystal matrix, therefore the current in the magnetic field is concentrated in the interlayer region mainly. The comparison between the twin and sample thicknesses shows that in this case the ratio of magnetoconductivities should be $\sigma_H^a/\sigma_H \approx 50$. In addition, the conductivity σ_H^a must have a maximum (for the geometry in Fig. 1 this is near $\vec{H} \parallel \vec{C}_2$), i.e. an anisotropy which is generally determined by the mobility ratio.

Since additional minima on the sample rotation diagrams are observed only with $\vec{H}$ near the twinning planes, it is clear

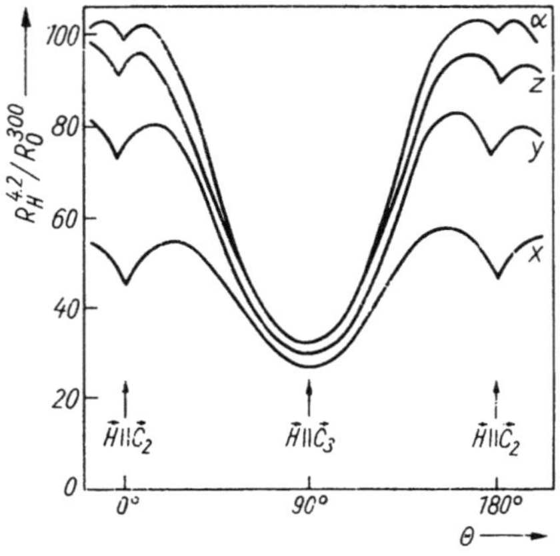

Fig. 2. Sample rotation diagrams. H = 20 kOe; $R_H^{4.2}$ is the magnetic field resistance at T = 4.2 ^{o}K; R_o^{300} is the zero magnetic field resistance at T = 300 ^{o}K

Table 1

sample	R_H/R_{300}	n	
		$\vec{H} \parallel \vec{C}_2$	$\vec{H} \parallel \vec{C}_3$
α	2100	1.97	2
z	2100	1.96	2
y	1970	1.93	2
x	1720	1.91	2

that these minima are determined rather by the magnetic field direction $\vec{H}$ (i.e. experimental geometry) than by the mobility ratio. Thus, we believe (1) that the effect of structural defects introduced into samples on magnetoresistance is chiefly due to the interaction of charge carriers which are drifting in crossed fields perpendicularly to $\vec{E}$ and $\vec{H}$ ($\vec{E}$ is the electric field vector) with the twinning planes; otherwise we observe the size effect at the twin boundaries.

A non-quadratic increase in magnetoresistance (Table 1) and a small (as compared to the transversal sample sizes) mean free path of charge carriers in Sb at helium temperatures (≈ 0.1 mm (4)) give evidence that the size effect is likely to be of different nature (5, 6).

References

(1) YU.A. BOGOD, B.I. VERKIN, V.B. KRASOVITSKII, and S.V. TSIVINSKII, Zh. eksper. teor. Fiz., Pisma 13, 491 (1971).

(2) YU.A. BOGOD and V.B. KRASOVITSKII, phys. stat. sol. (b) 49, K189 (1972).

(3) YU.A. BOGOD, V.B. KRASOVITSKII, and V.G. GERASIMECHKO, Preprint, FTINT AN UkrSSR, 1973.

(4) YU.A. BOGOD and V.B. KRASOVITSKII, Zh. eksper. teor. Fiz. 63, 1036 (1972).

(5) E.I. RASHBA, Zh. eksper. teor. Fiz. 48, 1427 (1965).

(6) G.I. BABKIN and V.YA. KRAVCHENKO, Zh. eksper. teor. Fiz. 60, 695 (1971).

(Received May 24, 1974)

phys. stat. sol. (b) 65, K5 (1974)
Subject classification: 14.4.1; 22.8.1

Fachbereich Physik (WE1A), Freie Universität Berlin

Temperature Dependence of the Electric Quadrupole Interaction in $PbZrO_3$ [1)]

By

H.v. EINSIEDEL and S.S. ROSENBLUM[2)]

We report here measurements of the temperature dependence of the electric quadrupole interaction at the Zr lattice site in polycrystalline $PbZrO_3$. The measurements were carried out over the range from room temperature to 533 K in which $PbZrO_3$ is known to undergo a series of transitions antiferroelectric → ferroelectric → paraelectric. The principal results of the investigation are the existence of:

1) a sharp transition at 508 K to the paraelectric phase,
2) a region showing marked temperature hysteresis between 453 and 508 K. Outside this range, hysteresis is no longer present.

Our measurements on $PbZrO_3$ are the direct analog of the measurements on $PbHfO_3$ which we reported recently (1). They were carried out by means of time-differential perturbed angular correlation (PAC) of γ-rays following the decay of Hf^{181} nuclei present as impurities in $PbZrO_3$. The samples were prepared by sintering pressed pellets of ZrO_2 containing 1% neutron irradiated HfO_2 with an equimolar quantity of PbO at 850 $^{\circ}$C. Since hafnium and zirconium show great similarity in their chemical behavior, this doping should not affect the properties of the $PbZrO_3$ and the Hf should reside at regular Zr lattice sites (2).

The perturbation of the intermediate 480 keV state ($I = 5/2$, $t_{1/2} = 10.95$ ns) in Ta^{181} by the $PbZrO_3$ lattice, was detected via the 133 to 482 keV cascade. This allowed the determination of the quadrupole interaction of the 480 keV state and thereby the electric field gradient (EFG) at the Zr site. The general theory of PAC is well known (3). For our case, the time dependent angular distribution is given by

$$W(\Theta, t) = 1 + A_2 G_{22}(t) P_2(\cos\Theta) \ , \qquad (1)$$

---

1) Work supported in part by the Bundesministerium für Wissenschaft und Technologie.

2) Present address: University of California, Los Alamos Scientific Laboratory, Los Alamos, New Mexico, U.S.A.

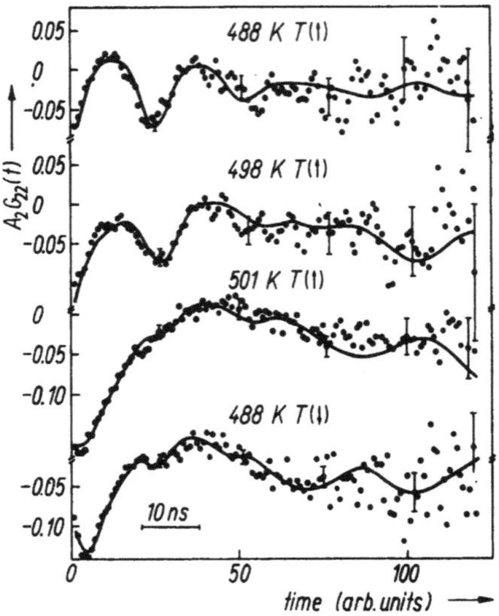

Fig. 1. Time-differential spectra least--squares fitted with equation (3). The upper three curves are taken with temperature increasing, whereas the bottom curve is taken with temperature falling

$$G_{22}(t) = \sigma_{20} + \sum_{n=1}^{3} \sigma_{2n} \cos \omega_n t \exp(-\frac{1}{2}\delta^2\omega_n^2 t^2) . \qquad (2)$$

In this equation, it is assumed that there is only one Ta^{181} field gradient. From least-squares fits of this formula to our data, we can determine $\nu_Q = eQV_{zz}/h$, $\eta = (V_{xx} - V_{yy})/V_{zz}$, and δ, a measure of the line broadening.

At all measured temperatures except in the region 453 to 508 K it was possible to fit the data with equation (2). In this region it was necessary to postulate the existence of two EFG's. The perturbation factor, $G_{22}(t)$, can then be written

$$G_{22}(t) = g^{(1)} G_{22}^{(1)}(t) + g^{(2)} G_{22}^{(2)}(t) \quad , \qquad (3)$$

where $g^{(1)}$ ($g^{(2)}$) gives the fraction of nuclei experiencing EFG 1 (2). Up to 453 K our results are in good agreement with previous work by Forker and Hammesfahr (4). Because they only assumed a single EFG with large δ our results differ significantly from their above this temperature. Typical data above 453 K are shown in Fig. 1. The data at 488 K show very clearly the temperature hysteresis.

In Fig. 2, we have plotted $g^{(1)}$ and $g^{(2)}$ of equation (3) as a function of temperature, for increasing and decreasing temperature. The very large hysteresis is probably a result of the varying activation energy of transition for different crystallites in the ceramic sample as has previously been reported for $PbHfO_3$ (1) as well as the appearance of the ferroelectric phase recently confirmed by Scott and Burns

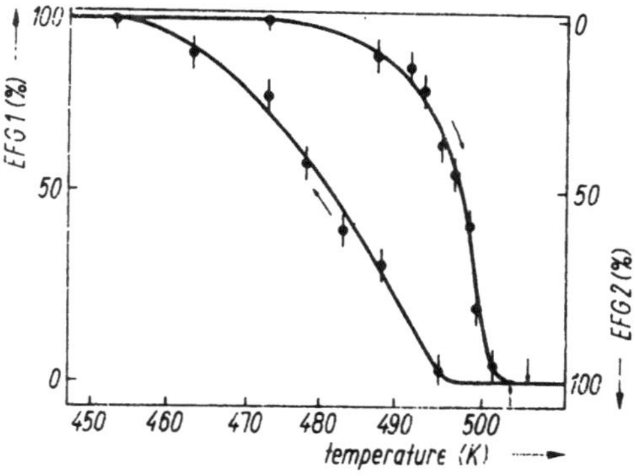

Fig. 2. Percent of EFG1 and EFG2 as a function of temperature showing hysteresis

(5). These workers found evidence for a ferroelectric phase between 479 and 501 K. In their measurements on single crystals of $PbZrO_3$ they also saw a large temperature hysteresis in the dielectric constant. In addition they found that the stability of this ferroelectric phase depended sensitively on the Pb content of the sample. This would explain the extremely broad temperature range of the hysteresis. $\nu_Q^{(1)}$ varies from 317 to 241 MHz in the region from 453 to 508 K with a constant asymmetry parameter, $\eta = 0.79 \pm 0.03$. $\nu_Q^{(2)}$ is 1/3 of $\nu_Q^{(1)}$ at the corresponding temperatures and varies from 101 to 77 MHz with $\eta = 0.45 \pm 0.07$. Above 508 K (Curie point) $\nu_Q^{(2)}$ changes (within 1 $^{\circ}$K) to the ν_Q of the cubic phase, which is approximately $\frac{1}{3}\ \nu_Q^{(2)}$. This nonvanishing EFG is probably due to lattice imperfections. There is no significant change in its EFG upon heating to 533 K and cooling afterwards to 508 K.

References

(1) H.v. EINSIEDEL and S.S. ROSENBLUM, J. Physique C 6, L292 (1973).

(2) L. GOULPEAU, S. LEMONTAGNER, and P. LIMOU, C.R. Acad. Sci. (France) 259, 1095 (1964).

(3) H. FRAUENFELDER and R.M. STEFFEN, Alpha-Beta- and Gamma-Ray Spectroscopy, Ed. K. SIEGBAHN, North-Holland Publ. Co., Amsterdam 1965 (p. 997).

(4) M. FORKER and A. HAMMESFAHR, Z. Phys. 255, 196 (1972).

(5) B.A. SCOTT and G. BURNS, J. Amer. Ceram. Soc. 55, 331 (1972).

(Received April 8, 1974)

phys. stat. sol. (b) 65, K9 (1974)

Subject classification: 18.2 and 18.4; 21.4

Institut für Strahlen- und Kernphysik der Universität Bonn

The Temperature Dependence of the Magnetic Hyperfine Field at ^{111}Cd in Magnetically Ordered Holmium

By

M. FORKER and J.B. FECHNER

In this letter we report about an investigation of the temperature dependence of the magnetic hyperfine (hf) field acting on ^{111}Cd impurities in metallic Ho. This study was performed using the time differential perturbed angular correlation (TDPAC) technique (1). Ho has a h.c.p. lattice structure ($c/a < 1.633$). At low temperatures two phases of spontaneous magnetic order have been found: Up to the Curie temperature $T_C = 20\ ^oK$ one observes a ferromagnetic screw-type structure, whereas between T_C and $T_N = 132\ ^oK$ the order of the 4f-moments is of the antiferromagnetic spiral type (2).

An impurity on a substitutional lattice position of magnetically ordered Ho will experience both a magnetic hf field and an axially symmetric field gradient (EFG), caused by the surrounding lattice ions and the conduction electrons. In the case of Ho the magnetic field and the EFG form a finite angle β, which should be close to the angle between the axis of easy magnetization and the c-axis ($\approx 80^o$). The perturbation of an angular correlation by such a combined, non-collinear hf interaction in a polycrystalline sample has been treated theoretically by Boström et al. (3). For an axially symmetric EFG the perturbation factor $G_{kk}(t)$ depends on the angle β and on the magnetic and electric interaction frequencies ($\omega_B = g\mu_n H_{hf}/\hbar$; $\omega_Q = eQV_{zz}/4I(2I-1)\hbar$).

In our experiment solid solutions of metallic Ho and radioactive, carrier-free ^{111}In were prepared by electron gun melting of the two components. The radioactive impurity concentration was less than 10^{-3} at%. A four detector TDPAC apparatus (4) was used to measure the time dependence of the angular correlation of the 173 to 247 keV cascade of ^{111}Cd, which is populated by the EC decay of ^{111}In, as a function of the source temperature. Fig. 1 shows as an example the time spectrum of $A_2G_{22}(t)$, measured at 79 ^oK. The interaction parameters ω_B, ω_Q, and β were

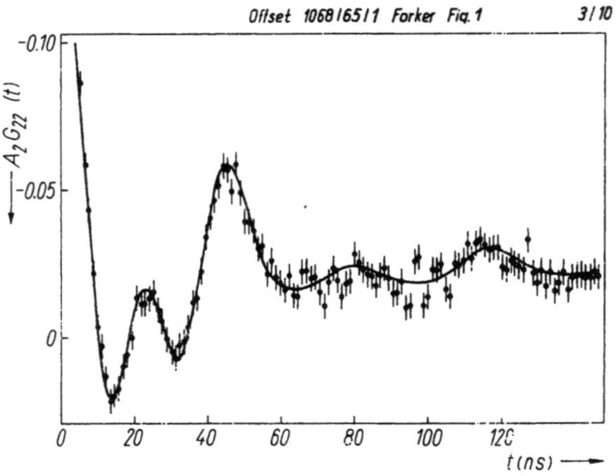

Fig. 1. TDPAC spectrum of ^{111}Cd in Ho at 79 ^{o}K

determined by use of a least squares fit program, which was based on the theoretical expression for $G_{kk}(t)$ derived in (3). Details on the data treatment can be found in (5, 6). A satisfactory fit to the time spectra $A_2G_{22}(t)$ could only be obtained by assuming that the Cd impurities occupy two different types of lattice site in Ho, where they are subject to different magnetic hf fields. At 4.2 ^{o}K the fields of the two sites, which are about equally populated, are 159(3) kG and 139(3) kG, respectively. The observation of two sites for Cd in Ho is unexpected, as in all experiments with impurities in Ho (7) and other comparable heavy RE metals (5, 6, 8) one has found only one impurity position. Most probably one of the sites in Ho is a substitutional lattice site, the other might be either an interstitial position or a regular site with a dislocation or impurity nearby.

Fig. 2 shows the relative temperature dependence of the magnetic hf field at the high and the low field site. For ω_Q and β the analysis gave $\omega_Q \approx 23$ MHz and $\beta \approx 85^{o}$ for the whole temperature range, with errors of about 15%. At both impurity sites the magnetic hf field has a nearly linear temperature dependence for temperatures $T/T_N > 0.3$ (straight solid line in Fig. 2, connecting the experimental points). At $T/T_N \approx 0.18$ there is a weak indication of the transition from the ferro- to the antiferromagnetic phase of Ho. The temperature dependence of the hf field in the antiferromagnetic phase deviates considerably from that of the spontaneous magnetization $\sigma(T/T_N)$ of a system with total angular momentum J = 8 (ground state of Ho^{3+}), which we have calculated in the molecular field approximation for a perfect, undoped lattice (solid curve in Fig. 2). The magnetization $\sigma_1(T/T_N)$ of the 12 nearest neighbours of a non-magnetic impurity in a hexagonal lattice with J = 8 is given by the dot-dashed curve in Fig. 2. $\sigma_1(T/T_N)$ has been calculated by extending

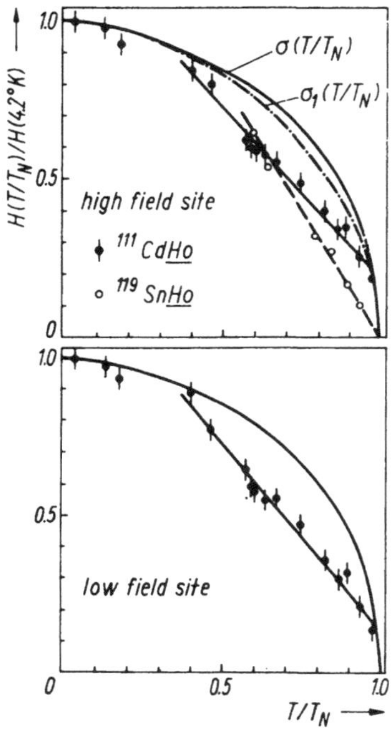

Fig. 2. Temperature dependence of the magnetic hf field of ^{111}Cd in holmium

the theory of van de Braak and Caspers (9) for a cubic structure to a hexagonal lattice. The helical spin order of Ho and the temperature dependence of the interlayer turn angle (2) was taken into account.

These calculations indicate that the spin polarization of the conduction electrons at the site of Cd in Ho, which is the main source of the magnetic hf field, decreases stronger with temperature than the host magnetization, even if the reduction of the nearest neighbour magnetization by the diamagnetic impurity is taken into account. Here it is interesting to note that the magnetic hf field of ^{119}Sn in Ho (7) (open circles and dashed line in Fig. 2) deviates even stronger from the magnetization curve, which indicates an influence of the impurity properties on the temperature dependence of the hf field.

Since Cd is a closed shell impurity, it appears improbable that the observed anomalous temperature dependence of the hf field is due to the existence of a local moment at the impurity (10). One possible explanation for the anomaly would be a decrease of the spin susceptibility of the conduction electrons with increasing temperature, caused either by changes in the density of states near the Fermi surface or by an increase of electron-electron scattering, which can lead to a disordering of the spins. The temperature dependence of the interlayer turn angle in Ho suggests the existence of such an effect, its size, however, is at present difficult to estimate.

References

(1) R.M. STEFFEN and H. FRAUENFELDER, in: Perturbed Angular Correlations, Ed. E. KARLSSON, E. MATTHIAS, and K. SIEGBAHN, North-Holland Publ. Co., Amsterdam 1964.

(2) W.C. KOEHLER, J. appl. Phys. 36, 1078 (1965).

(3) L. BOSTRÖM, E. KARLSSON, and S. ZETTERLUND, Phys. Scripta 2, 65 (1970).

(4) H.F. WAGNER and M. FORKER, Nuclear Instrum. and Methods 69, 197 (1969).

(5) M. FORKER and A. HAMMESFAHR, Z. Phys. 260, 131 (1973).

(6) M. FORKER and A. HAMMESFAHR, Z. Phys. 263, 33 (1973).

(7) A.S. KUCHMA and V.S. SHNIPEL, Soviet Phys. -J. exper. theor. Phys. 35, 556 (1972).

(8) L. BOSTRÖM, G. LILJEREN, B. JONSSON, and E. KARLSSON, Phys. Scripta 3, 175 (1971).

(9) H.P. VAN DE BRAAK and W.J. CASPERS, Z. Phys. 200, 270 (1967).

(10) V. JACCARINO, L.R. WALKER, and C.K. WERTHEIM, Phys. Rev. Letters 13, 752 (1964).

(Received June 7, 1974)

phys. stat. sol. (b) 65, K13 (1974)
Subject classification: 14.3; 10.2; 22.2.3

I. Physikalisches Institut der Rheinisch-Westfälischen Technischen Hochschule, Aachen

The Dynamical Conductivity for Ionized Impurity Scattering

By
E. GERLACH and M. RAUTENBERG

The frequency dependent conductivity of semiconductors has been studied by a large number of authors (1 to 15). It is well established that at low temperatures the basic scattering mechanism is due to ionized impurities. This leads to a frequency independent conductivity at low frequencies Ω and to an Ω^x law at high frequencies, where x is of the order of 3. In the intermediate region plasma effects are to be expected, which have been studied separately from ionized impurities, e.g. in reference (6).

In the following we apply a new type of transport theory which is based on the equivalence of energy loss and Joule heat (16 to 18). This method permits to cover the whole frequency range and includes plasma effects in the interaction between ionized impurities and electrons.

Instead of considering the conductivity it is more convenient to start from the resistivity ϱ_{total} which we write

$$\varrho_{total} = \varrho - \frac{i\,\Omega}{\varepsilon_o \Omega_p^2} , \tag{1}$$

where ϱ is due to scattering by ionized impurities and the second term on the right-hand side is the well-known free electron contribution which is due to the inertia, ($\Omega_p^2 = n e^2/(\varepsilon_o m)$, m effective mass, n electron concentration). The scattering of free electrons by plasmons will not be taken into account.

Now, in order to calculate ϱ we follow the concept of (16) and consider an oscillating point charge density ϱ_c, which represents our ionized impurity

$$\varrho_c(\vec{r}, t') = Z\, e\, \delta(x - A \sin \Omega t')\, \delta(y)\, \delta(z) \tag{2}$$

of amplitude A and frequency Ω in a sea of free electrons. The Fourier transform of ϱ_c is

$$\varrho_c(\vec{k}, \omega) = Z\, e \int dt'\; e^{-i(k_x A \sin \Omega t' - \omega t')} \tag{3}$$

As described in (1) ϱ_c gives rise to an induced electric field E_i at the position of the point charge, which is given by

$$E_i(A\sin\Omega t, t) = \frac{-i\,Z\,e}{\varepsilon_o(2\pi)^4}\int dt' \iint d^3k\,d\omega\,\frac{\vec{k}}{k^2}\left[\frac{1}{\varepsilon(k,\omega)} - 1\right] \times$$
$$\times\, e^{-i(k_x A\sin\Omega t' - \omega t')}\, e^{i(k_x A\sin\Omega t - \omega t)} , \tag{4}$$

where $\varepsilon(k,\omega)$ is the dielectric constant of the material excluding the impurities.

The energy loss of the oscillating point charge follows from the relation

$$-\frac{dW}{dt} = -E_i(A\sin\Omega t, t)(Z\,e\,A\Omega\cos\Omega t) , \tag{5}$$

where the last bracket on the right-hand side represents the velocity of the particle.

Identifying the energy loss with the Joule heat, the product of the current density and the macroscopic external field E, we obtain

$$-N\frac{dW}{dt} = n(-e)\,v\,E , \tag{6}$$

where N is the concentration of the impurities.

The connection of the current density n(-e)v with the field E will lead us to the resistivity. The Fourier transformation of E yields

$$E(\eta) = \int \frac{-N\,dW/dt}{n(-e)v}\, e^{i\eta t}\, dt =$$
$$= \frac{i(Z\,e)^2 N}{\varepsilon_o\, n(-e)(2\pi)^4\Omega^2}\iint d^3k\,d\omega\,\frac{k_x}{k^2}\left[\frac{1}{\varepsilon(k,\omega)} - 1\right] \times \tag{7}$$
$$\times\, G\left(k_x A, \frac{\omega-\eta}{\Omega}\right) G\left(-k_x A, \frac{-\omega}{\Omega}\right)$$

with the abbreviation

$$G(z,\gamma) = \int_{-\infty}^{\infty} e^{i(z\sin\Theta - \gamma\Theta)}\, d\Theta = 2\pi \sum_{-\infty<\varkappa<\infty} J_\varkappa(z)\,\delta(\varkappa - \gamma) . \tag{8}$$

Inserting G into (7) and carrying out the ω-integration we obtain

$$E(\eta) = \frac{i(Ze)^2 N}{\varepsilon_o n(-e)(2\pi)^2} \sum_{\varkappa,\lambda} (-1)^\lambda \int d^3k \, \frac{k_x}{k^2} J_\varkappa(k_x A) J_\lambda(k_x A) \, \times$$
$$\times \, \delta((\varkappa+\lambda)\Omega+\eta) \left[\frac{1}{\varepsilon(k, -\lambda\Omega)} - 1 \right] . \tag{9}$$

Now we compare the external field E with the product of the resistivity and the current density

$$E(\eta) = \varrho(\eta) n(-e) v = \varrho(\eta) n(-e) A\Omega\pi(\delta(\Omega-\eta) + \delta(\Omega+\eta)) . \tag{10}$$

In order to obtain a linear resistivity we select those terms of (9) which are linear in $A \sim v$, so that ϱ becomes independent of A. For this reason we restrict ourselves to products of J_o and J_1, and J_o and J_{-1}, and, further, put $k_x A \ll 1$. Integration up to infinity does not contradict this condition, since the integral is rapidly converging. This leads to

$$\varrho(\Omega) = \frac{i Z^2 N}{6\pi^2 \varepsilon_o n^2 \Omega} \int_0^\infty k^2 dk \left[\frac{1}{\varepsilon(k,\Omega)} - \frac{1}{\varepsilon(k,0)} \right] . \tag{11}$$

The real and the imaginary parts of the resistivity for the example of InSb are displayed in Fig. 1 (ε Lindhard's dielectric constant (16), $\varepsilon_L = 17.9$, $m/m_o = 0.013$, $n = 1.35 \times 10^{13}$ cm^{-3}, $N = 8.7 \times 10^{13}$ cm^{-3}, $Z = 1$).

In conclusion we find Re ϱ to be independent of Ω for $\Omega \ll \Omega_p/\sqrt{\varepsilon_L}$ and a Rutherford frequency behaviour (Re $\varrho \sim \Omega^{-1.5}$) for $\Omega \ll \Omega_p/\sqrt{\varepsilon_L}$. In the intermediate region, near the plasma frequency $\Omega_p/\sqrt{\varepsilon_L}$, the quantity Re ϱ has a peak which is due to plasmon excitation. This can immediately be seen from the fact that Im $1/\varepsilon(k,\Omega)$ in the integrand of (11) has a peak at the plasma frequency. The phase, Im ϱ, is peaked in the intermediate frequency range, too.

Going back to $\sigma_{total} = 1/\varrho_{total}$ we find from (1)

$$\mathrm{Re}\,\sigma_{total} = \frac{1}{\mathrm{Re}\,\varrho} \, \frac{1}{1 + \left(\frac{\mathrm{Im}\,\varrho - \Omega/(\varepsilon_o \Omega_p^2)}{\mathrm{Re}\,\varrho} \right)^2} \tag{12}$$

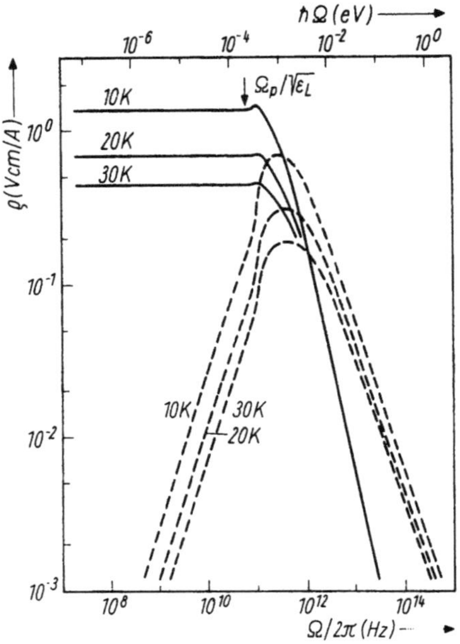

Fig. 1

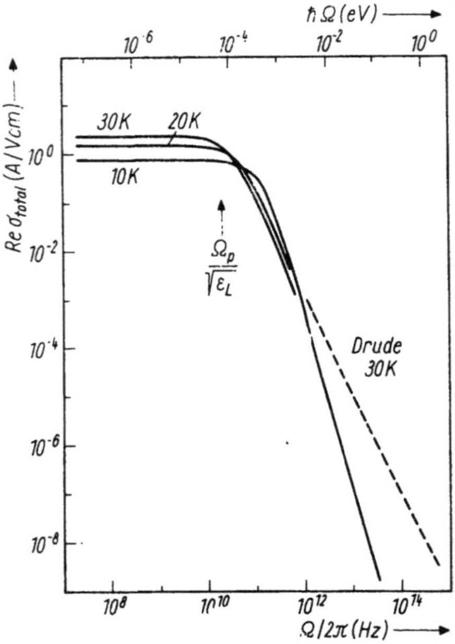

Fig. 2

Fig. 1. Real part (——) and imaginary part (– – –) of the frequency dependent resistivity ϱ of InSb due to ionized impurity scattering

Fig. 2. Real part of the total conductivity σ_{total} including ionized impurity scattering and the free electron contribution

which is proportional to $\Omega^{-3.5}$ for high frequencies, in agreement with earlier authors (2, 7).

The quantity Re σ_{total} is displayed in Fig. 2. In this representation the plasma peak is no longer pronounced, since it is situated on the descent of the curve. For $\Omega \approx 10^{11}\ s^{-1}$ the curves Re $\sigma_{total}(T)$ intersect, further they become independent of T for $\Omega \gg \Omega_p$. For comparison the Drude approximation is plotted for T = 30 K.

The authors thank Prof. P. Grosse and Dr. G. Bauer for helpful discussions. The numerical computations have been performed at the Rechenzentrum der RWTH Aachen.

References

(1) K.J. PLANKER and E. KNAUER, Z. angew. Phys. 12, 425 (1960).

(2) H. FRÖHLICH, Proc. Reading Conference, Ed. H.Y. FAN and M. BECKER, Butterworths, London 1951 (p. 132).

(3) H.Y. FAN, W. SPITZER, and R.J. COLLINS, Phys. Rev. 101, 566 (1956).

(4) R. WOLFE, Proc. Phys. Soc. A67, 74 (1954).

(5) H.J.G. MEYER, J. Phys. Chem. Solids 8, 264 (1959).

(6) R. VON BALTZ, phys. stat. sol. (b) 43, K133 (1971).

(7) R. VON BALTZ and W. ESCHER, phys. stat. sol. (b) 51, 499 (1972).

(8) A.A. KUKHARSKII and V.K. SUBASHIEV, Soviet Phys. - Semicond. 4, 234 (1970).

(9) H. LÖSCHNER and D. KRANZER, J. appl. Phys. 44, 3663 (1973).

(10) J.D. HOLM and K.S. CHAMPLIN, J. appl. Phys. 39, 275 (1968).

(11) K.S. CHAMPLIN, Phys. Rev. 130, 1374 (1963).

(12) D. KRANZER, phys. stat. sol. (b) 46, 591 (1971).

(13) E. CONWELL, Solid State Phys., Suppl. 9 (1967).

(14) E. HAGA and H. KIMURA, J. Phys. Soc. Japan 18, 777 (1963).

(15) B. JENSEN, J. Phys. Chem. Solids 34, 2235 (1973).

(16) E. GERLACH, phys. stat. sol. (b) 61, K97 (1974).

(17) E. GERLACH, phys. stat. sol. (b) 62, K43 (1974).

(18) J.C. DOUKHAN, R. DROPE, J.L. FARVACQUE, E. GERLACH, and P. GROSSE, phys. stat. sol. (b) 64, 237 (1974).

(Received July 18, 1974)

phys. stat. sol. (b) 65, K19 (1974)

Subject classification: 13.5.1 and 20.1; 22.3.1

State University of Chernovtsy

Exciton Absorption in GaS_xSe_{1-x} Solid Solutions

By

N.P. GAVALESHKO, M.V. KURIK, A.S. SAVTCHUK, and G.B. DELEVSKII

Complex semiconductor compounds which can be used for producing tunable lasers and other devices, have been of increasing interest recently. Among these are crystals of GaS_xSe_{1-x} solid solutions. These crystals have been found to show stimulated radiation when their composition is varied over a wide range (1). The information on the optical properties, particularly on the exciton absorption in GaS_xSe_{1-x} crystals, now available in literature (1 to 3), is not sufficient to make any conclusion with regard to excitons in solid solutions of such layer compounds.

The purpose of the present work is to investigate the absorption spectra of GaS_xSe_{1-x} single crystals and thus to study the nature of exciton-phonon interaction in connection with lattice disorder in such solid solutions.

The absorption spectra were measured using the standard photoelectric technique. The resolution of the experimental setup was better than 0.5 meV over the spectral range investigated. As a radiation source we used a xenon lamp, supplied with stabilized and rectified voltage.

GaS_xSe_{1-x} (x = 0.1, 0.3, 0.4, 0.5, 0.6, 0.7, 0.8, 0.9, 1.0) single crystals were grown by the Bridgman method. Compositions of the solid solutions were determined by chemical and X-ray structure analyses. The samples were plane parallel plates obtained by cleavage. The layer structure of the compounds made it possible to prepare samples with thicknesses as small as 0.5 μm, so that measurements far in the fundamental absorption region were possible. For the low temperature measurements the crystals were mounted "loosely".

Fig. 1 shows the spectral dependence of the absorption coefficient of GaS_xSe_{1-x} at 90 °K. The peculiar feature of the curves is the presence of a distinct band at the absorption edge, which is due to the formation of parabolic excitons in these crystals. With the S content the exciton band maximum shifts to shorter wavelengths. The detailed investigation of the composition dependence of the position of the ex-

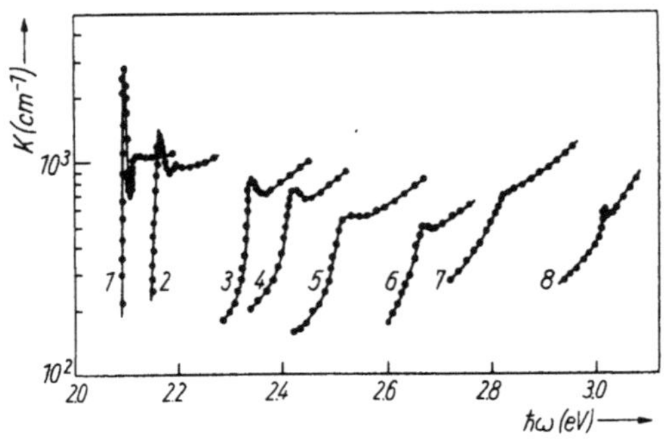

Fig. 1

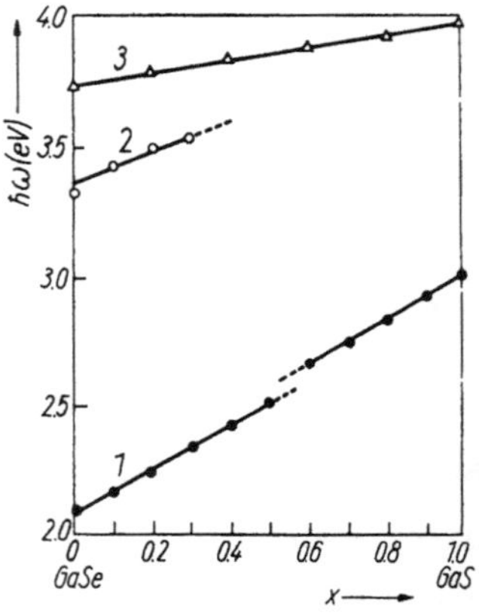

Fig. 2

Fig. 1. Absorption spectra of GaS_xSe_{1-x} single crystals at 90 ^oK for x (1) 0, (2) 0.1, (3) 0.3, (4) 0.4, (5) 0.5, (6) 0.6, (7) 0.8, (8) 1.0

Fig. 2. Change in the position of the absorption band maximum for parabolic excitons (1), for excitons far in the fundamental absorption region (2), and of the saddle point M_1 (3) with composition for GaS_xSe_{1-x} at 90 ^oK

citon absorption band maximum revealed a kink in the range $0.5 \leqq x \leqq 0.6$ (see Fig. 2), which is probably due to a phase transition in structure for this composition. Indeed, the X-ray structure analysis showed that GaS_xSe_{1-x} single crystals had a structure corresponding to a mixed γ, ε-modification for $x \leqq 0.5$, while for $x \geqq 0.6$ they had hexagonal β-type structure (4). These crystallographic modifications possess layers whith identical structure but differ in the manner of layer packing. It is worth noting that the change in exciton band maximum position with composition is the same for the two modifications.

In order to obtain data on exciton-phonon interaction in solid solutions we measured the temperature dependence of the exciton absorption band edge and its maximum position for temperatures 80 to 300 ^oK. It appeared that for the compositions $x \leqq 0.5$ the long wavelength exciton absorption band edge follows Urbach's rule. The analytical expression for the absorption has the form (reference (5))

$$K(\hbar\omega, T) = K_o \exp\left[-\sigma(\hbar\omega_o - \hbar\omega)/kT\right], \qquad (1)$$

where K_o and $\hbar\omega_o$ are the parameters for the given crystals, σ is dependent on temperature as follows:

$$\sigma = \sigma_o \frac{2kT}{\hbar\omega_p} \, \mathrm{th}\left(\frac{\hbar\omega_p}{2kT}\right), \tag{2}$$

where σ_o is connected with the exciton-phonon interaction constant and $\hbar\omega_p$ is the average energy of phonons involved in absorption band formation.

Fig. 3 shows the dependences of σ_o and $\hbar\omega_p$ on composition as well as the halfwidth and the magnitude of the absorption coefficient at the exciton band maximum. It is seen that σ_o and K_m decrease considerably with increasing x, while $\hbar\omega_p$ shows no change within the experimental error, and the halfwidth increases linearly with composition.

One can write empirical expressions for $\hbar\omega_m$ and σ_o dependence on composition for $x \leqq 0.5$

$$\hbar\omega_m(x) = \left[\hbar\omega_m\right]_{GaSe} + 0.86\,x , \tag{3}$$

$$\sigma_o(x) = 9.2\,x\,(x-1) + 2.7 . \tag{4}$$

Thus, we conclude that for GaS_xSe_{1-x} solid solutions ($x \leqq 0.5$) the position of the maximum and the halfwidth of the exciton absorption band vary linearly with composition. At the same time, σ_o and K_m show a considerable change with composition. This can probably be explained by the peculiarities in solid solution formation, namely that in such compounds the statistical disorder of composition within a crystal is important. Such disorder leads to additional exciton-phonon interaction. The observed change of σ_o with composition is mainly due to this phenomenon. Since GaSe and GaS has layer structure, it would manifest itself during solid solution formation (for example, the statistical disorder in atomic positions within the layer and between them can

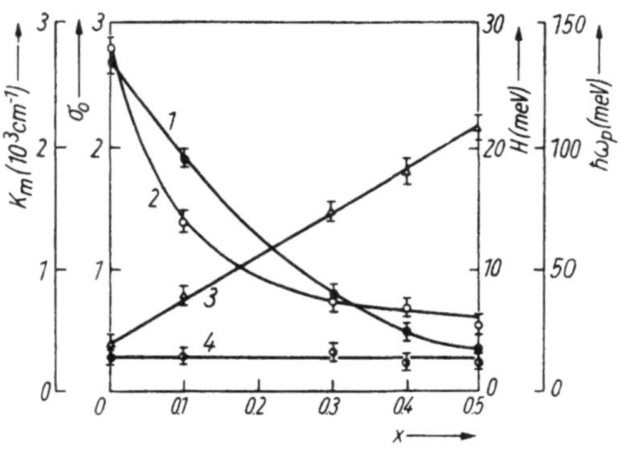

Fig. 3. Composition dependence of the basic parameters of GaS_xSe_{1-x}; (1) $\sigma_o(x)$, (2) $K_m(x)$, (3) H(x), (4) $\hbar\omega_p(x)$

be different), and can to a certain extent explain the difference in the dependence of various parameters on composition (see Fig. 3).

For compositions $x \geqq 0.6$ the absorption edge does not follow Urbach's rule, which indicates that optical transitions in the compounds of this modification are of different type, at least they are not direct transitions involving phonons, as is the case for compounds with $x \leqq 0.5$.

The dependence of the position of the exciton absorption band maximum on temperature is of the same character for all compositions, and is connected mainly with exciton-phonon interaction in the single oscillator crystal model approximation. The correspondence between the temperature dependence of the halfwidth H(T) and that of the maximum position $\hbar\omega_m(T)$, indicates that the same phonons are involved in exciton absorption band broadening and its shift and that their energy is close to that of the optical phonon in GaSe (6).

We should mention another important feature of GaS_xSe_{1-x} mixed crystal spectra. For very thin samples (about 0.5 μm) we observed a structure far in the fundamental absorption, which had been first observed in the absorption spectrum of GaSe crystal (7, 8) and was explained by the formation of hyperbolic excitons. The corresponding structure is found for compositions $0 \leqq x \leqq 0.3$ only. The structure corresponding to interband transitions at the saddle point M_1 is present for all compositions studied. Their position vs. x is shown in Fig. 2 (curves 2 and 3). It is seen that the position of exciton absorption bands at the edge and far in the fundamental absorption region changes with composition in the same way, except that the high energy bands are spread more quickly with increasing x. This fact as well as the difference in the dependences of the position of M_1 and of the exciton peak on composition far in the fundamental absorption region support the above-made conclusion that the structure of the high-energy spectrum for GaSe is connected with hyperbolic excitons (9). For solid solutions with $x \geqq 0.4$ and for pure GaS structure in the high energy absorption spectrum was not observed.

References

(1) G.B. ABDULLAEV, G.A. AKHUNDOV, A.A. AGAEVA, V.M. SALMANOV, and I.D. YAROSHETSKII, Fiz. Tekh. Poluprov. 7, 1108 (1973).

(2) N.A. GASANOVA, G.A. AKHUNDOV, and M.A. NIZAMETDINOVA, phys. stat.

sol. 17, K131 (1966).

(3) E. AULICH, J.L. BREBNER, and E. MOOSER, phys. stat. sol. 31, 129 (1969).

(4) F. JELLINEK and H. HAHN, Z. Naturf. 16a, 713 (1961).

(5) M.V. KURIK, phys. stat. sol. (a) 8, 9 (1971).

(6) M.I. KARAMAN, V.P. MUSHINSKII, B.S. RASBIRIN, A.I. STARUCHIN, and S.S. CHESOI, Fiz. Tekh. Poluprov. 6, 412 (1972).

(7) V.K. SUBASHIEV and LE-KHAC-BINH, Zh. eksper. teor. Fiz., Pisma 12, 139 (1970).

(8) V.K. SUBASHIEV, LE-KHAC-BINH, and L.S. CHERTKOVA, Solid State Commun. 9, 369 (1971).

(9) N.P. GAVALESHKO, M.V. KURIK, A.I. SAVTSHUK, G.B. DELEVSKII, and Z.D. KOVALYUK, Optika i Spektroskopiya, to be published.

(Received June 27, 1974)

phys. stat. sol. (b) 65, K25 (1974)

Subject classification: 6; 21.6

Physics Department, University of Delhi

Lattice Dynamics of Gold

By

S.N. KHANNA, J.C. UPADHYAY, and ASHOK JAIN

The phonon dispersion relations for gold in the high symmetry directions have recently been measured by Lynn et al. (1). To our knowledge, no pseudo- or model potential calculations of the dispersion relations for gold have been reported so far. In this paper we present a calculation of phonon dispersion relations for gold in certain symmetry directions, based on a new transition metal model potential (TMMP) proposed by Animalu (2) and compare the results with measurements of Lynn et al.

Starting from an expression for the Hamiltonian of the crystal and using the adiabatic and harmonic approximations, the phonon frequencies ω^2 are given by the solution of the determinant (reference (3))

$$\left| D_{\alpha\beta}(\vec{q}) - \omega^2 M I \right| = 0 , \tag{1}$$

where $D_{\alpha\beta}(\vec{q})$ represents a general element of the dynamical matrix, M is the ionic mass, I is the unit matrix, and $\vec{q}$ is the wave vector of the disturbance. Using Toya's (4) self-conistent field method, D can be expressed as a sum of three contributions:

(i) A Coulomb part $D^C_{\alpha\beta}(\vec{q})$ due to the electrostatic interaction between the ions.

(ii) An electronic part $D^E_{\alpha\beta}(\vec{q})$ due to the electron-ion-electron interaction.

(iii) A repulsive contribution $D^R_{\alpha\beta}(\vec{q})$ due to the exchange overlap between ion cores.

Along the principal symmetry directions (100), (110), and (111) equation (1) factors out and the solutions can be expressed as (reference (4))

$$\omega^2 = \omega_C^2 - \omega_E^2 + \omega_R^2 , \tag{2}$$

where $\omega_E^2 = \frac{1}{M} D^E$ etc. Since the ion cores in metals are small, we have neglected the contribution from ω_R^2 in our calculations on gold. The contribution ω_C^2 has been

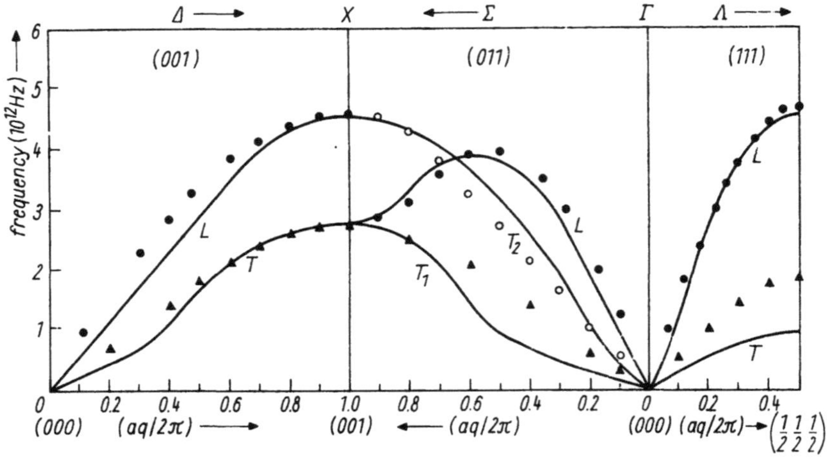

Fig. 1. Phonon dispersion curves for Au, calculated (continuous curves) and measured (points)

taken from the table for face-centered cubic metals given by Vosko et al. (5) (the plasma frequency $\nu_p = \omega_p/2\pi$ for gold is 3.609×10^{12} Hz). The problem of calculating the phonon frequencies ω^2 is thus reduced to the calculation of the electronic contribution given by reference (4)

$$\left[\omega_E^2\right]_{\alpha\beta} = \omega_p^2 \sum_{\vec{H}} \frac{(\vec{q}+\vec{H})_\alpha(\vec{q}+\vec{H})_\beta}{|\vec{q}+\vec{H}|^2} G(\vec{q}+\vec{H}) - \omega_p^2 \sum_{\vec{H}\neq 0} \frac{\vec{H}_\alpha \vec{H}_\beta}{|\vec{H}|^2} G(\vec{H}) , \tag{3}$$

where $\vec{H}$ is a reciprocal lattice vector and the function G can be expressed as (reference (6))

$$G(q) = \left(\frac{4\pi z e^2(1+\alpha_{eff})}{\Omega_q^2}\right)^{-2} \left|V^b(q)\right|^2 \frac{\varepsilon(q)-1}{\varepsilon(q)(1-f(q))} , \tag{4}$$

where z is the valence of the ion equal to 1 for gold, and Ω is the atomic volume. To evaluate G(q) we have computed the bare ion potential $V^b(q)$ using the expressions given in the Appendix A of reference (2). For $\varepsilon(q)$ we have use the Hartree dielectric function with a Hubbard-Sham modification for exchange and correlation introduced via f(q) (7). Knowing G(q), the phonon frequencies are calculated using equations (2) to (4). The calculated results together with the experimental data are displayed in Fig. 1.

Fig. 1 shows that our results are in good overall agreement with experimental results. The transverse branch in the (111) direction lies below the experimental data. It will be interesting to recalculate the phonon spectra using other dielectric functions available in the literature. Such calculations are now in progress.

Acknowledgements

We are thankful to Prof. L.S. Kothari for useful discussions and invaluable suggestions. One of us (S.N.K.) is thankful to C.S.I.R., New Delhi, for financial assistance.

References

(1) J.W. LYNN, H.G. SMITH, and R.M. NICKLOW, Phys. Rev. B 8, 3493 (1973).

(2) A.O.E. ANIMALU, Phys. Rev. B 8, 3542 (1973).

(3) M. BORN and K. HUANG, Dynamical Theory of Crystal Lattices, Oxford University Press, London 1954.

(4) T. TOYA, Catal. Rev. 7, 60 (1959).

(5) S.H. VOSKO, R. ROGER, and G.H. KEECH, Canad. J. Phys. 43, 1187 (1965).

(6) A.O.E. ANIMALU, Phys. Rev. B 8, 3555 (1973).

(7) J. HUBBARD, Proc. Roy. Soc. A 243, 336 (1967).

(Received June 17, 1974)

phys. stat. sol. (b) 65, K29 (1974)

Subject classification: 14.2

Saha Institute of Nuclear Physics, Calcutta

On the Elastic Constants of Flux Line Lattice in Type-II Superconductors

By

B.C. KHANRA and A.K. RAYCHAUDHURI

In the approximation of well separated flux lines with negligible or no overlap of the flux line cores, the elastic constants of the flux line lattice have been calculated by Matricon (1) and by Fetter et al. (2) from a two-body interaction potential between the flux lines. Brandt (3) also calculated the elastic constants of the flux line lattice in the limit of high Ginzburg-Landau (GL) parameter $\varkappa$ and magnetic field in the vicinity of h_{c_2}. Labusch (4, 5) reformulated the theory of Fetter et al. (2) in such a way that the elastic constants can be expressed in terms of B versus h curves for all values of $\varkappa$ and for all values of induction B. This method does not require the direct knowledge of the interaction energy. It may be of some interest to know the nature of the dependence of the elastic constants on the interaction energy of a flux line for the entire field ranging from h_{c_1} to h_{c_2}. In this work, we have calculated the elastic constants by Labusch's method for all values of $\varkappa$ and induction B. The interaction energy of a vortex with all the remaining vortices in the system is calculated by the circular cell approximation of Marcus (6). An approximate formula has been presented for the interaction energy for $\varkappa \geqq 5$. The nature of the dependence of the elastic constants on the interaction energy has been studied from h_{c_1} to h_{c_2}.

In order to calculate the elastic constants and the interaction energy one requires a very accurate formula for the free energy density of the superconductor. Recently Brandt (7) calculated the GL free energy density $I(B, \varkappa)$ for the stable hexagonal lattice symmetry and found for $B/\varkappa \geqq 0.15$ a simple approximate formula for $I(B,\varkappa)$ (in reduced units) as

$$I(B, \varkappa) = \frac{1}{2} + B^2 - \frac{(\varkappa - B)^2}{(2\varkappa^2 - 1)\beta_A + 1} - a(\varkappa)\, e^{-b(\varkappa)B/\varkappa} ,$$

where β_A is the Abrikosov geometrical factor (8), $a(\varkappa)$ and $b(\varkappa)$ are numerical constants (7). Very recently, Khanra and Raychaudhuri (9) used Marcus' circular

Table 1

Free energy F(B, $\varkappa$) per unit length of a unit cell (all in reduced units)

$\varkappa = 10$			$\varkappa = 1$		
B	F(B, $\varkappa$)[a)]	F(B, $\varkappa$)[b)]	B	F(B, $\varkappa$)[a)]	F(B, $\varkappa$)[b)]
10.8	6.315	6.315	1.000	9.425	9.425
8.888	5.620	5.620	0.888	9.111	9.078
6.920	4.391	4.390	0.781	8.820	8.750
5.000	3.192	3.191	0.692	8.581	8.486
2.958	1.920	1.920	0.500	8.074	7.955
2.222	1.462	1.462	0.413	7.849	7.741
1.250	0.864	*)	0.320	7.623	7.541
0.800	0.595	*)	0.222	7.416	7.377
0.200	0.265	*)	0.080	7.315	*)
0.008	0.182	*)	0.020	7.300	*)

a) Reference (9) and present calculation. b) Reference (7).
*) Brandt's approximate formula does not give accurate results for $B/\varkappa < 0.15$.

cell approximation (6) to calculate the free energy density. For comparing the results of Brandt with those of reference (9), we obtain the free energy F(B, $\varkappa$) per unit length of a Wigner-Seitz cell of radius R from I(B, $\varkappa$) by the circular cell approximation as $F(B,\varkappa) = \frac{2\pi}{\varkappa B} I(B,\varkappa)$ since $B = 2/\varkappa R^2$). The results are compared in Table 1 for two values of $\varkappa$. The agreement is excellent. For further calculations we shall use Brandt's formula.

The approximate interaction energy F_{int} per unit length of a vortex can be obtained by subtracting the free energy of an isolated vortex, ε, from the value of F(B, $\varkappa$). Since Brandt's formula is valid for $B/\varkappa \geqq 0.15$, this does not give the correct value of ε in the limit $B \to 0$. For ε we use the results reported in an earlier work (10). For $\varkappa \gg 1$ an approximate formula for ε is given by Hu (11) as $\varepsilon = \frac{2\lambda}{\varkappa^2} (\log \varkappa + 0.497)$. This is nothing but Abrikosov's formula, the constant within the bracket in the latter work being 0.08 instead of 0.497. Hu's formula is more accurate as can be found on comparing with the results presented in reference (10). Hence for $\varkappa \geqq 5$ we can express F_{int} (in reduced units) as

$$F_{int} = \frac{2\pi}{\varkappa B} \left[\frac{1}{2} + B^2 - \frac{(\varkappa - B)^2}{(2\varkappa^2 - 1)\beta_A + 1} - a(\varkappa) e^{-b(\varkappa)B/\varkappa} \right] - \frac{2\pi}{\varkappa^2} (\log \varkappa + 0.497) .$$

Because of the hexagonal symmetry of the flux lines and their properties in the case of a superconductor, there are three independent elastic constants. These are C_{11}, C_{44}, and C_{66}. For magnetic field in the neighbourhood of h_{c_1}, the shear modulus C_{66} in the xy-plane is given by (reference (4))

$$C_{66L} = \frac{1}{2}\int_0^B x^2 \frac{d^2 h(x)}{dx^2}\, dx,$$

where $h(x) = \partial I/\partial x$; and in the neighbourhood of h_{c_2} where the linearised GL equations are valid, C_{66} is given by

$$C_{66U} = \frac{0.48\varkappa^2(2\varkappa^2 - 1)}{\left[(2\varkappa^2 - 1)\beta_A + 1\right]^2}\,(1 - B/\varkappa)^2 ,$$

C_{11} and C_{44} are given by the relations $C_{11} = B^2 \frac{\partial^2 I}{\partial B^2} + C_{66}$ and $C_{44} = B\frac{\partial I}{\partial B}$, respectively.

We have calculated C_{66}, C_{11}, and C_{44} from the expression of the free energy density I given by Brandt. The interaction energy is calculated in the way already prescribed. In Fig. 1 we have plotted C_{11}, C_{44}, and C_{66} (C_{66L} and C_{66U}) for $\varkappa = 1$ and 10 as a function of the interaction energy of a vortex for the entire field from h_{c_1} to h_{c_2}. For higher $\varkappa$, $C_{11}/2\varkappa^2$ and $C_{44}/2\varkappa^2$ are found to merge to a single curve. In the figure the curves for C_{66L} and C_{66U} are extrapolated to beyond their range of applicability. Brandt's formula for the free energy density cannot be used to calculate the elastic constant for $B/\varkappa < 0.15$. Since the elastic constants must go

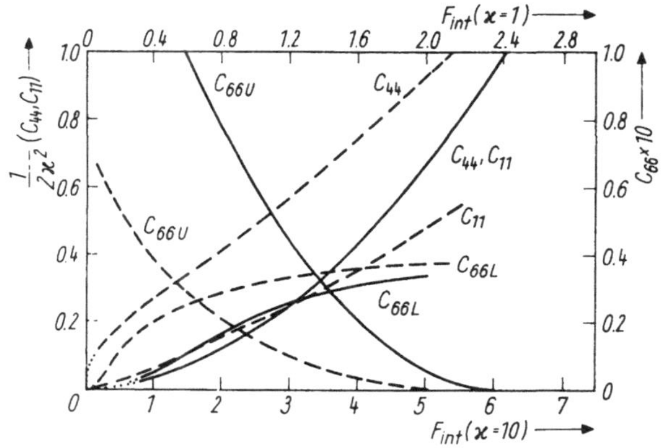

Fig. 1. Dependence of the elastic constants on the interaction energy of a vortex. The solid lines are for $\varkappa = 10$, and the dashed lines are for $\varkappa = 1$. The dotted portions are extrapolations from the values of F_{int} corresponding to $B/\varkappa = 0.15$. The lower x-axis reads the values of F_{int} for $\varkappa = 10$ and the upper one reads F_{int} for $\varkappa = 1$

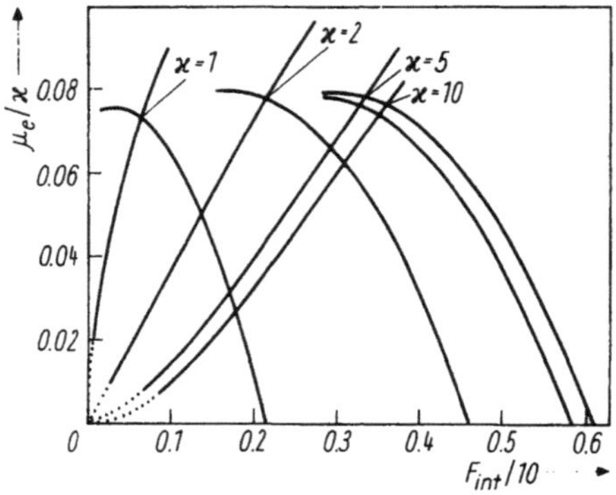

Fig. 2. Effective modulus μ_e for point pinning force as a function of the interaction energy for $\varkappa = 1$, 2, 5, and 10

to zero as $B/\varkappa \rightarrow 0$, the curves are extrapolated to $F_{int} = 0$, $C_{ij} = 0$ (shown by dots).

The elastic constant can next be used to calculate the effective modulus (5, 12), μ_e, required to determine the pinning force on a vortex by the crystal lattice defects. For point pinning forces μ_e is given by the relation (reference (12))

$$\mu_e = \frac{1}{1.07\pi^{1/2}} \left[\frac{1}{(C_{44} C_{11})^{1/2}} + \frac{1}{(C_{44} C_{66})^{1/2}} \right]^{-1}$$

while for line forces parallel to $\vec{B}$, $\mu_e = C_{66}$ and perpendicular to $\vec{B}$, $\mu_e \approx \frac{5}{4} C_{11}$. In Fig. 2 we have shown the nature of dependence of μ_e for point forces on the interaction energy of a vortex. Because of two different formulas for C_{66} applicable at two extreme regions of magnetic field, we have calculated μ_e with both C_{66L} and C_{66U}. In the figure we have extrapolated the curves slightly beyond the point where they meet. μ_e is the quantity which links the flux lattice system with the crystal lattice defects (12, 13).

In order to calculate the critical current density in type-II superconductors in the mixed state, one has to calculate the pinning force on a vortex and hence the interaction energy of the vortex with the crystalline lattice defects. But this vortex interacts with the other existing vortices in the sample. This interaction is taken into account by treating the flux line lattice as an elastic medium. The results presented in this work qualitatively show how the elastic constants depend on this interaction.

The computation has been performed on the CDC 3600 computer at the Tata Institute of Fundamental Research, Bombay.

References

(1) J. MATRICON, Phys. Letters (Netherlands) 9, 289 (1964).

(2) A.L. FETTER, P.C. HOHENBERG, and P. PINCUS, Phys. Rev. 147, 140 (1966).

(3) E.H. BRANDT, phys. stat. sol. 35, 1027 (1969); 36, 371 (1969).

(4) P. LABUSCH, phys. stat. sol. 19, 715 (1967).

(5) R. LABUSCH, phys. stat. sol. 32, 439 (1969).

(6) P.M. MARCUS, Low Temperature Physics, LT 9, Part A, Ed. J.G. DAUNT et al. Plenum Press, New York 1965 (p. 550).

(7) E.H. BRANDT, phys. stat. sol. (b) 51, 345 (1972).

(8) A.A. ABRIKOSOV, Soviet Phys. - J. exper. theor. Phys. 5, 1174 (1957).

(9) B.C. KHANRA and A.K. RAYCHAUDHURI, Phys. Letters A 46, 311 (1974).

(10) B.C. KHANRA and A.K. RAYCHAUDHURI, Phys. Letters A 42, 305 (1972).

(11) C.R. HU, Phys. Rev. B 6, 1756 (1972).

(12) A.M. CAMPBELL and J.E. EVETTS, Adv. Phys. 21, 199 (1972).

(13) R. LABUSCH, Crystal Lattice Defects 1, 1 (1969).

(Received July 4, 1974)

phys. stat. sol. (b) 65, K35 (1974)

Subject classification: 10.2 and 19; 22.5.3

Département de Chimie Physique, Université de Genève

A New O^- Center in $SrCl_2$ Crystals[1)]

By

J.M. DE SIEBENTHAL and H. BILL

This letter reports some results on an oxygen centre in $SrCl_2$ crystals (CaF_2 structure). The underlying investigation was undertaken in relation with Jahn-Teller studies of p-type electronic systems. The results further contribute to the knowledge on the color centers (1) in $SrCl_2$ crystals.

All of the samples involved in the experiments were prepared in our laboratory along the following lines of procedure. Analar grade $SrCl_2$ powder from British Drug House (Dorset, England) was dehydrated in vacuum and subsequently purified by heating it together with $PbCl_2$ to 700 ^oC in vacuum. Growth of the crystals took place in a Bridgeman type furnace under vacuum. Typical as-grown crystals were of good quality in volumes of 1.5 to 2 cm^3. They were perfectly transparent. Best conservation of the samples was obtained by stocking them in paraffin oil.

The specimen were cut to shape by cleavage technique and subsequently doped with oxygen by heating the crystals in an oxygen atmosphere. ^{17}O was introduced into some samples by using oxygen enriched to 20% by this isotope. After X-raying of the crystals several centers related to the presence of oxygen were observed.

The presently reported center consists of an O^- ion substituting for a Cl^- anion in the probably otherwise locally unperturbed crystal. This model is supported by the following results.

The center is paramagnetic (S = 1/2). Its EPR spectrum which shows up below approxi-

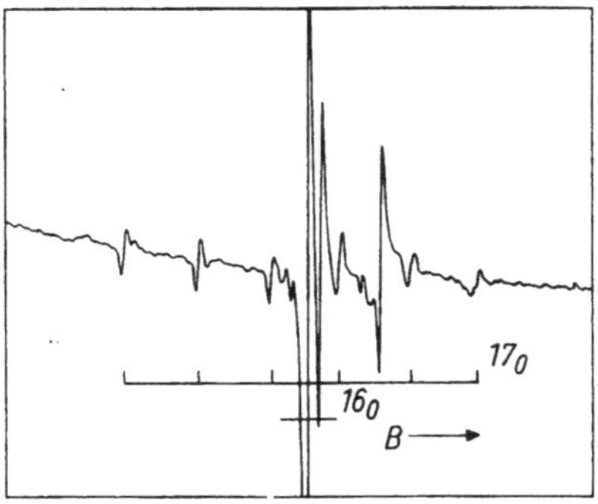

Fig. 1. EPR spectrum of the O^- center recorded with $\vec{B} \parallel \vec{C}_4$. The two lines situated above the one of ^{16}O are not due to this center

1) Work supported by the Swiss National Foundation for Scientific Research.

Table 1

Constants of the axially symmetric spin Hamiltonian of O^- in $SrCl_2$
(The principal axis is along a 1, 1, 1) crystal direction.)

$$g_{\parallel} = 2.0032 \pm 0.0003,$$
$$g_{\perp} = 2.0650 \pm 0.0003;$$

$$A_{\parallel} = 104.6 \pm 0.5\ G,$$
$$A_{\perp} = 18.5 \pm 1.5\ G.$$

mately 120 ^{o}K features an axial g-tensor. It does not exhibit any resolved hyperfine structure. The center involving ^{17}O, however, exhibits clearly the ^{17}O hf structure, as reproduced in the figure, demonstrating unambigously the presence of one O^- ion per center.

The experimentally determined constants of the appropriate spin Hamiltonian are given in the table.

The center was produced in $SrCl_2$ crystals purposely doped with Na, K, Rb to investigate the possibility of an alkali impurity cation being involved. The spectrum proved to be much weaker in these samples and the center is independent of the cationic impurity. At the same time these samples exhibited EPR spectra of other oxygen centers which depend on the alkali cations. Hence, the presently reported one is likely to be described by the given model. However, a missing Sr^{++} neighbor ion cannot be excluded totally at the present status of investigation. But it seems unlikely because the local charge state of the crystal would be very strongly disturbed. The O^- ion possesses a ^2P free ion ground state becoming $^2T_2$ in T_d symmetry of the locally unperturbed anion site. Thus the Jahn-Teller effect is expected to occur. This is observed in the case of SrF_2:O^- (2). There, the ion couples to the tetragonal distortion of E_g symmetry of the surrounding F^- octahedron. In the present case, the system couples to trigonal distortions. A second, equally likely model might describe the structure of the presently reported center. The ionic radius of the O^- ion is rather smaller than the radius of the Cl^- vacancy cavity it occupies. Small mutual repulsion and strong anharmonicity results. Hence, a rather large shift of the O^- ion equilibrium position away from the anion site is expected. Clearly, this model goes beyond the Jahn-Teller ansatz. Further work is in progress.

References

(1) See e.g. H. BILL, H. SUTER, and R. LACROIX, Phys. Letters (Netherlands) 22, 241 (1966).
S. LEFRANT, D. JUMEAU, and L. TAUREL, phys. stat. sol. (b) 50, K101 (1972).
E. RZEPKA, I. BALTOG, S. LEFRANT, M. YUSTE, and L. TAUREL, phys. stat. sol. (b) 57, 383 (1973).

(2) H. BILL and R.H. SILSBEE, to be published.

(Received June 11, 1974)

phys. stat. sol. (b) 65, K39 (1974)

Subject classification: 13 and 20; 21.7

Department of Physics, University of Warwick, Coventry

Enhanced Helicon Propagation in Zero-Gap Semiconductors at Low Temperatures

By

S.K. SHARMA

As a result of the anomalous screening in symmetry induced zero-gap semiconductors, their dielectric properties show markedly different behaviour than of normal metals and conventional semiconductors. It has been recently shown by Liu and Tan (1) that in the presence of a magnetic field the static dielectric constant of a zero-gap semiconductor at infinite wavelength ($\vec{q} = 0$) and zero temperature contains a magnetic field dependent term of the form $H^{-1/2}$. The purpose of the present communication is to show that this dependence of the static dielectric constant on the magnetic field gives rise to enhanced helicon propagation in zero-gap semiconductors at low temperatures where the dominant scattering mechanism is ionized impurity scattering.

The magnetic field dependent static dielectric constant of symmetry induced zero-gap semiconductors is given by (reference (1))

$$\varepsilon_s = \varepsilon_o + \lambda/H^{1/2} , \qquad (1)$$

where ε_o is the static dielectric constant in the absence of the magnetic field H. λ is a constant for a given material.

The relaxation time τ for ionized impurity scattering can be written as (reference (2))

$$\tau_i = \tau_o \varepsilon_s^2 . \qquad (2)$$

Using equation (1) in equation (2) one obtains

$$\tau_i = \tau_o \varepsilon_o^2 (1 + \lambda/\varepsilon_o H^{1/2})^2 . \qquad (3)$$

For the propagating helicon mode (i.e. the extraordinary wave) the imaginary part of the wave vector (which gives damping) is given by (reference (3))

$$K_{+Im} = - \left| \frac{\omega_p \omega^{1/2}}{C\omega_c^{1/2}} \right| \left| \frac{1}{2\omega_c \tau} \right| , \tag{4}$$

where ω_p is the plasma frequency, ω is the frequency of the wave, ω_c is the cyclotron frequency, and C is the velocity of light in vacuum. Substituting for τ from equation (3) in equation (4) one gets

$$K_{+Im} = - \left| \frac{\omega_p \omega^{1/2}}{C\omega_c^{1/2}} \right| \left| \frac{1}{2\omega_c \tau_o \varepsilon_o^2 (1 + \lambda/\varepsilon_o H^{1/2})^2} \right| . \tag{5}$$

To compare the damping of the helicon wave for zero-gap semiconductors and conventional semiconductors, we fix ω, ω_p, ω_c, τ_o, ε_o to be the same in both cases. Choosing α-Sn as a zero-gap semiconductor with $\lambda = 6.0 \times 10^3\ G^{1/2}$ (reference (4)), one gets from equation (5)

$$\frac{(K_{+Im})_{\alpha\text{-Sn}}}{(K_{+Im})_{\text{conv. semicond.}}} \approx 10^{-2} \qquad \text{for } H = 1\ \text{kG} .$$

Thus the magnetic field dependence of the dielectric constant in zero-gap semiconductors has reduced the damping of helicon waves by two orders of magnitude as compared to conventional semiconductors for usual parameters of interest. We therefore suggest that helicon wave propagation experiments should be carried out in zero-gap semiconductors at low temperatures to yield useful information about their yet unclear transport properties.

Acknowledgement

The work was supported by the Commonwealth Scholarship Commission in the United Kingdom.

References

(1) L. LIU and M. TAN, Phys. Rev. B 9, 632 (1974) and references cited therein.

(2) E.M. CONWELL, High Field Transport in Semiconductors, Academic Press, New York 1967.

(3) M.C. STEELE and B. VURAL, Wave Interactions in Solid State Plasmas, McGraw-Hill Book Co., New York 1969.

(4) L. LIU and M. TAN, Solid State Commun. 13, 1747 (1973).

(Received July 8, 1974)

phys. stat. sol. (a) 65, K43 (1974)

Subject classification: 13.4 and 14.3; 22.4.2

State University, Lvov

Temperature Dependence of the Intrinsic Concentration in Mercury Selenide

By

V.E. KREVS, R.V. LUTSIV, M.V. PASHKOVSKII, and P.P. PETROV

Many investigations deal with the band structure of mercury selenide (1 to 7). Mercury selenide like mercury telluride is a semimetal. The main band parameters of mercury selenide such as effective electron mass m_e^*, energy gap $\varepsilon_g = E_{\Gamma_6} - E_{\Gamma_8}$, matrix element P , are rather accurately known, while the values of the band overlap energy $\Delta\varepsilon$ significantly differ in the papers cited. The value of $\Delta\varepsilon$ determined from investigations of the galvanomagnetic effect and optical absorption is 0.07 eV (1, 4), while an analysis of the electrical properties of HgSe under hydrostatic pressure gives 0.007 eV (5). This strong divergence of $\Delta\varepsilon$ does not enable one to determine the temperature dependence of the intrinsic carrier concentration n_i.

This note is aimed at evaluating the overlap energy and at calculating the temperature dependence of the intrinsic carrier concentration in HgSe.

The $\Delta\varepsilon$ value was determined from the best fit of experimental and theoretical temperature dependences of the carrier concentration. The calculation was done on the basis of the electroneutrality equation

$$n - p = N_D \; , \tag{1}$$

where

$$n = \frac{3}{4\pi^2}\left(\frac{3}{2}\right)^{1/2}\left(\frac{k_o T}{P}\right)^3 \int_0^\infty \frac{t^{1/2}(t+\Phi_1)^{1/2}(2t+\Phi_1)dt}{1+\exp(t-\eta)} \; ,$$

$$\Phi_1 = \frac{|\varepsilon_g|}{k_o T} \; ,$$

k_o is the Boltzman constant, and $\Delta \gg |\varepsilon_g|$.

$$n = \frac{1}{2\pi^2}\left(\frac{3}{2}\right)^{1/2}\left(\frac{k_o T}{P}\right)^3 \left[1 + \frac{1}{2}\left(1 + \frac{\Delta}{|\varepsilon_g|}\right)^{-1}\right]^{-3/2} \int_0^\infty \frac{t^{1/2}(t+\Phi_1)^{1/2}(2t+\Phi_1)dt}{1+\exp(t-\eta)}$$

if one takes into account the finite value of the spin-orbit splitting Δ .

$$p = 4\pi \left(\frac{2k_o T m_p^*}{h^2} \right)^{3/2} \int_0^\infty \frac{t^{1/2}\, dt}{1 + \exp(t + \eta + \Phi_2)} ,$$

$$\eta = \frac{E_F}{k_o T} ,$$

$$\Phi_2 = \frac{\Delta \varepsilon}{k_o T} .$$

As HgSe is always n-type with large carrier mobility ratio $b = \mu_e/\mu_h$, the electron concentration n was determined from the experimental Hall coefficient in the impurity range according to the formula $n = 1/eR$. The ionized donor concentration N_D in the temperature range considered was supposed to be constant (8) and was determined from the Hall coefficient in the saturation range. Calculations were done on the "Minsk-22" computer and the following band parameters were used: $\varepsilon_g(4.2\ ^oK) = 0.24$ eV (5), $P = 8.4\times10^{-8}$ eV cm (4), $\Delta = 0.38$ eV (9), $d\varepsilon_g/dT = 6\times10^{-4}$ eV/deg (7). A heavy hole effective mass ratio of 0.55 was taken in the calculations (13). The results are shown on Fig. 1. The theoretical n(T) dependences were compared with the experimental ones for two samples containing different donor concentrations ($N_D = 3.48\times10^{17}$ cm^{-3} and $N_D = 1.1\times10^{17}$ cm^{-3} for curves 1 and 2, respectively).

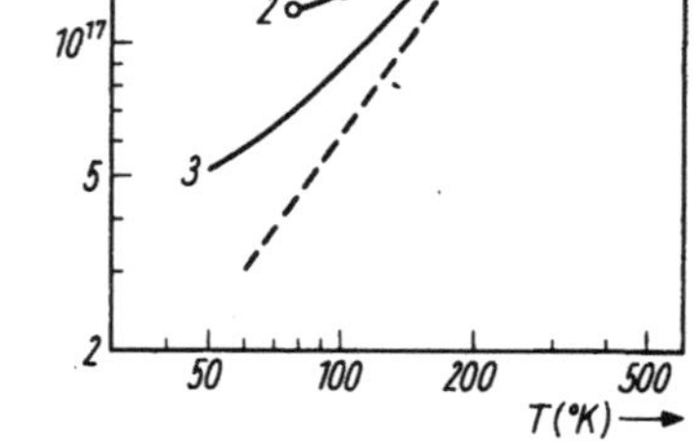

Fig. 1

The best fit was obtained with $\Delta\varepsilon = 0.018$ eV. This value is in good agreement with that given by Broerman (10). $d\varepsilon_g/dT = 5.6\times10^{-4}$ eV/deg leads to better agreement of theoretical and experimental data in the total temperature range considered. Taking into account the finite value of the spin-orbit splitting practically has no influence on the curves. The inaccuracy in the matrix element P leads to changes of the electron concentration, not exceeding the experimental errors.

The obtained values of $\Delta\varepsilon$ and $d\varepsilon_g/dT$ were used for calculating the temperature dependence of the intrinsic carrier concentration in mercury selenide (Fig. 1, curve 3). At 77 °K n is equal to 7.2×10^{16} cm^{-3} which is in good agreement with the electron concentration determined in (11). For T ≧ 100 °K the dependence is near to linear with slope 3/2.

The experimental curves at high temperatures form a system of almost parallel lines. In (12) this was taken as evidence that the intrinsic carrier concentration in semimetals depends upon the doping level and crystal lattice imperfections. However, we think that the concentration increase of ionized donors in the lattice shifts the intrinsic range in the direction of higher temperatures. This is evident from the character of the curves.

References

(1) T.C. HARMAN and A.J. STRAUSS, J. appl. Phys., Suppl. 32, 2265 (1961).
(2) C.R. WHITSETT, Phys. Rev. 138, A829 (1965).
(3) S.S. SHALYT and S.A. ALIEV, Fiz. tverd. Tela 46, 1879 (1964).
(4) V.P. DMITRIEV, P.S. KIREEV, and V.V. VOLKOV, Elektronnaya Tekh., Ser. 14, 3, 32 (1970).
(5) W. GIRIAT, Postepy fizyki 21, 591 (1970).
(6) D.G. SEILER, R.R. GALAZKA, and W.M. BECKER, Phys. Rev. B 3, 4274 (1971).
(7) L.G. VOLZENSKAYA, V.E. KREVS, and M.V. PASHKOVSKII, Ukr. fiz. Zh. 18, 835 (1973).
(8) B.L. GELMONT and M.I. DYAKONOV, Zh. eksper. teor. Fiz. 62, 713 (1972).
(9) L.M. BLIEK and G. LANDWEHR, phys. stat. sol. 33, K67 (1969).
(10) Y.G. BROERMAN, Phys. Rev. 183, 754 (1969).
(11) M. BAJ and S. POROWSKI, Materialy III. ogonokrajowego Seminarium zwazkow polprzewodnikowych A_2B_6, prace IF PAN 38, 107 (1972).
(12) P.S. KIREEV, V.P. DMITRIEV, I.M. ZUBAREVA, and V.V. PTASHINSKII, Izv. Akad. Nauk SSSR, Ser. neorg. Mater. 8, 1025 (1972).
(13) M.V. KURICH, Ukr. fiz. Zh. 14, 344 (1969).

(Received June 17, 1974)

phys. stat. sol. (b) 65, K47 (1974)

Subject classification: 13.5.1; 22.3

A. F. Ioffe Physico-Technical Institute, Academy of Sciences of the USSR, Leningrad

Lineshape of the Hyperbolic Exciton in GaSe

By

V.I. SOKOLOV and V.K. SUBASHIEV

Lately much attention has been paid to the optical properties of layer semiconductors, such as GaS, GaSe, and GaTe. These investigations yield information about the band structure of this new type of semiconductors. In (1, 2, 9) a discrete structure in the intrinsic absorption region of GaSe was revealed and identified as hyperbolic exciton absorption.[1)] Later in (3) the existence of a discrete structure (absorption peak) has been confirmed, but doubt was cast on the presence of saturation following this peak. In accordance with this fact and with the possibility of fitting experimental electroreflection data with a three-dimensional exciton structure, the authors of (3, 4) have contended that the absorption peak near 3.26 eV (295 $^{\circ}$K) could not be due to a saddle point exciton absorption. In (5) they came to the same conclusions on the basis of thermoreflection data.

As at present such contraversial points of view exist, we wanted to obtain additional experimental data.

In this note optical spectra of GaSe in the intrinsic absorption region are studied by direct methods. In (6) it was suggested that the hyperbolic exciton absorption line should have an asymmetrical shape with more abrupt slope on the high energy side. To test this statement it was interesting to carry out experiments at lower temperatures (liquid helium).

---

1) Excitons seen at the intrinsic absorption edge (M_0 excitons) are usually called parabolic. One means the parabolic dependence of energy on quasi-momentum in general direction with ellipsoidal surface of rotation. From this point of view the saddle point excitons ought to be named as parabolic because the dependence of energy on quasi-momentum is the same as for M_0 excitons. But there is a difference between the shapes of isoenergetic surface for M_0 and M_1 singularities (ellipsoidal and hyperboloidal). For this reason excitons of M_0 singularity could be named "elliptic", and excitons of M_1 singularity - "hyperbolic".

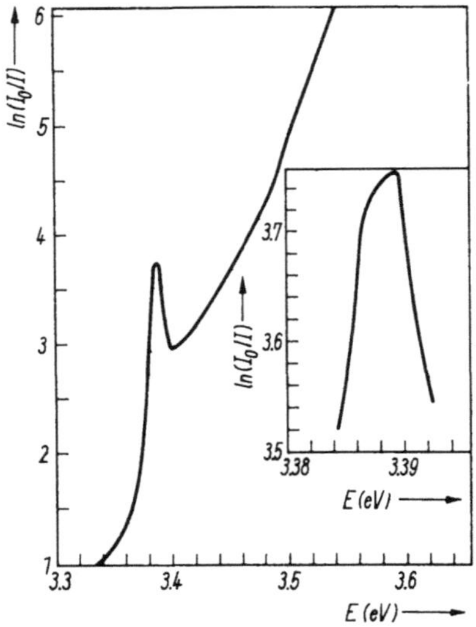

Fig. 1. The absorption spectrum of GaSe in the region of the hyperbolic exciton 4.2 °K

Fig. 1 shows the absorption spectrum in the region of the hyperbolic exciton at 4.2 °K. The spectrum was obtained with a resolving power of about 1 meV. The width of the line at T = 4.2 °K is 0.02 eV coinciding with the width at T = 77 °K. But the shape of the absorption line changes as compard with that at 77 °K, becoming steeper on the high energy side, especially in the region of the line maximum. However, this asymmetry is not as pronounced it would follow from Kane' s theory. This weakly pronounced asymmetry could be attributed to the fact that the line has been observed against a very strong and rapidly changing continuous absorption background. Therefore the hyperbolic exciton lineshape changes especially on the short-wavelength side.

The broad and temperature independent (in the region 4.2 to 77 °K) linewidth confirms that the hyperbolic exciton lifetime is limited by other than thermal mechanisms. According to (6), the linewidth depends on the effective mass ratio in the following way: for $m_3/m_1 = -320$, the linewidth is $\Delta E/Ry = 5.7$; for $m_3/m_1 = -40$, $\Delta E/Ry = 9.2$; and for $m_3/m_1 = -5$, $\Delta E/Ry = 15.4$. The linewidth is measured in units of the two-dimensional exciton Rydberg. From (2) it follows that Ry = 80 meV. If the linewidth is 20 meV, $\Delta E/Ry = 0,25$. Extrapolating the m_3/m_1 versus $\Delta E/Ry$ relationship, we obtain for our value of $\Delta E/Ry$ the ratio $m_3/m_1 \approx -1000$. Using for the transverse effective mass the value $\mu = 0.32$ (2), we obtain $\mu_3 \approx -300$ in free electron mass units. Such a large value of reduced longitudinal mass can be explained if one considers this mass to correspond to the motion of an electron-hole pair normal to layers bound by weak Van der Waals forces.

A suggestion is made on the basis of measurements of the absorption lineshape carried out at 77 °K (3) that the observed peak should be attributed to on elliptic M_0 exciton (the line with n = 1) with a binding energy of 10 meV. If this is so, the lines with n = 2, 3... should be resolved at low temperatures. However, there is no evidence of such lines at $T = 4.2\ ^{o}K$. In the case of the hyperbolic exciton the oscillator strength of these lines should be not more than 5% of that of the main line (5), therefore it is not surprising that these lines are not observed. Recently the authors of (8) have drawn the conclusion that the observed spectra of dR/dE and $d\varepsilon_2/dE$ are due to the hyperbolic exciton.

The authors thank Ya. W. Morosenko and I. N. Uraltsev for the help in measurements at $T = 4.2\ ^{o}K$.

References

(1) V. K. SUBASHIEV, LE-KHAK-BIN, and L. S. CHERTKOVA, Solid State Commun. 9, 369 (1971).

(2) V. I. SOKOLOV, D. B. KUSHEV, and V. K. SUBASHIEV, phys. stat. sol. (b) 50, K125 (1972).

(3) A. BALZAROTTI and M. PIACENTINI, Solid State Commun. 10, 421 (1972).

(4) A. BALZAROTTI, M. PIACENTINI, E. BURRATINI, and P. PICOZZI, J. Phys. C 4, L272 (1971).

(5) M. GRANDOLFO, F. SOMMA, and P. VECCIA, Phys. Rev. B 5, 428 (1972).

(6) E. KANE, Phys. Rev. 180, 852 (1969).

(7) D. F. BASSANI, D. L. GREENAWAY, and G. FISHER, Proc. Internat. Conf. Phys. Semicond., Paris 1964 (p. 51).

(8) S. KOHN, Y. PETROFF, and V. SHEN, First Internat. Conf. Modulation Spectroscopy, Tucson (Arizona) 1972.

(9) G. I. ABUTALYKOV and M. L. BELLE, Fiz. Tekh. Poluprov. 8, 368 (1974).

(Received Juli 31, 1974)

phys. stat. sol. (b) 65, K51 (1974)

Subject classification: 14

Department of Physics, Technion,Israel Institute of Technology, Haifa

Electrical Conductivity in the Cullen and Callen Chain

By

D. CABIB

In this letter we are concerned with an exact calculation of the linear response to an external electric field in a model that has been introduced by Cullen and Callen (1, 2) to explain the Verwey transition in magnetite (Fe_3O_4). The proposed Hamiltonian (CC Hamiltonian) is said to yield an adequate description of the motion of the extra d-electrons of the Fe^{2+} ions among the octahedral sites (B sites) which are responsible for the behaviour of Fe_3O_4, Hartree-Fock calculations of the one-particle energies have been performed recently (3 to 5) at finite temperature and the model has been shown to yield "multiple ordering" (3).

Very recent Mössbauer measurements (6) in magnetic indicate the existence of anomalies in the absorption intensitites at three different temperatures near the transition temperature T_V($\approx$119 $^{\circ}$K). This makes it very appealing to check the model and the experiment more quantitatively. For this purpose calculations of observable quantities as electrical conductivity, specific heat, etc. are crucial.

As a starting point we propose a calculation of the electrical conductivity in the one-dimensional CC model with very small bandwidth compared with Coulomb repulsion as it will be defined below. The theoretical interest of the calculation is in that the method employed (the transfer matrix method) is widespread in the study of the statistical mechanics of one-dimensional systems (Ising model (7, 8), extended Hubbard model (9, 10)), but it has never been used, as far as we know, to calculate the response to an electric field. (For the response to a magnetic field, see references (7, 8, 10).) A similar but more lengthy calculation could be performed in the extended Hubbard model (9), in a quite straightforward way, by using the same approach.

As far as comparison with experiment is concerned the calculation in one dimension cannot account for any sharp behaviour as is observed in the real system at T_V, but at low temperature it may give the right qualitative behaviour as we will mention below.

From Kubo's formula (11) the frequency-dependent conductivity is given by

$$\sigma(\omega) = \frac{1}{2N_a} \int_{-\infty}^{\infty} d\tau e^{i\omega\tau} \int_0^\beta d\lambda \operatorname{tr}(\varrho J e^{iH(\tau+i\lambda)} J e^{-iH(\tau+i\lambda)}), \tag{1}$$

H is the CC Hamiltonian (3) in one dimension:

$$H = b\sum_i (c_i^+ c_{i+1} + c_{i+1}^+ c_i) + U\sum_i n_i n_{i+1} , \tag{2}$$

$$\varrho = \frac{e^{-\beta(H-\mu N)}}{\operatorname{tr} e^{-\beta(H-\mu N)}} = \frac{e^{-\beta(H-\mu N)}}{Z} \tag{3}$$

with the boundary condition $c^+_{N_a+1} = c^+_1$. $n_i = c_i^+ c_i$ is the number operator of Wannier function at site i. μ is the chemical potential, $N = \sum_{i=1}^{N_a} n_i$, $\beta = 1/kT$; μ is to be determined by the condition

$$\operatorname{tr} \varrho N = \frac{N_a}{2} = \frac{\text{number of sites}}{2} . \tag{4}$$

J is the current operator ($\hbar = 1$) :

$$J = -ieba \sum_{i=1}^{N_a} (c_i^+ c_{i+1} - c_{i+1}^+ c_i) \tag{5}$$

and a is the lattice parameter. (For a rigorous definition of J see reference (12).)

The first term in (2) is the non-interacting energy (we use tight binding and the parameter b is equal to the bandwidth/4) and the second term is the Coulomb repulsion between electrons occupying nearest neighbor sites. Two electrons can never be found on the same site (reference (3)). We can now calculate (1) in the region $b/U \ll 1$. This condition allows us to disregard the band term in H in the lowest order of b/U, and makes it possible to use the transfer matrix method. (This same approximation has been used in the calculation of $\sigma(\omega)$ in the half-filled-band Hubbard model (reference (13)) and it gives the exact σ/b^2 in the limit $b \to 0$.) The region $b/U \ll 1$ is maybe not too realistic for magnetite (reference (3)) but it is exactly soluble and it could give the right insight because of the nature of its ground state.

The evaluation of Z is straightforward and in the thermodynamic limit it requires only the knowledge of λ_M, the maximum eigenvalue of the 2 x 2 matrix

$$T = \begin{pmatrix} 1 & e^{\beta\mu/2} \\ e^{\beta\mu/2} & e^{\beta(\mu-U)} \end{pmatrix} \tag{6}$$

which is the same as equation (12) of reference (10) if we allow only for spin up. μ is found to be equal to U by condition (4) and we obtain

$$Z = \lambda_M^{N_a}, = \left(1 + e^{\beta U/2}\right)^{N_a} . \tag{7}$$

The evaluation of the trace in (1) is somewhat more complicated. By use of (5) in (1) we obtain

$$\sigma(\omega) = -\frac{e^2a^2b^2}{2ZN_a}\int_{-\infty}^{\infty} d\tau e^{i\omega\tau}\int_0^{\beta} d\lambda \sum_{ij} \mathrm{tr}\Big\{e^{-\beta(H-\mu N)}(c_i^+c_{i+1} - c_{i+1}^+c_i)e^{iH(\tau+i\lambda)} \times \\ \times (c_j^+c_{j+1} - c_{j+1}^+c_j)e^{-iH(\tau+i\lambda)}\Big\}. \tag{8}$$

Now we note that the terms $c_i^+c_{i+1}c_j^+c_{j+1}$ and $c_{i+1}^+c_ic_{j+1}^+c_j$ do not contribute and that the terms $c_{i+1}^+c_ic_j^+c_{j+1}$ and $c_i^+c_{i+1}c_{j+1}^+c_j$ contribute to (8) only if $i = j$ because the eigenfunctions of H are simple Slater determinants with Wannier functions occupied. Equation (8) becomes

$$\sigma(\omega) = \frac{e^2a^2b^2}{2ZN_a}\int_{-\infty}^{\infty} d\tau e^{i\omega\tau}\int_0^{\beta} d\lambda \sum_{i} \mathrm{tr}\Big\{e^{-\beta(H-\mu N)}e^{-iHx}(c_{i+1}^+c_ie^{iHx}c_i^+c_{i+1} + \\ + c_i^+c_{i+1}e^{iHx}c_{i+1}^+c_i)\Big\} \tag{9}$$

with $x = \tau + i\lambda$.

Since

$$c_j^+(x) = e^{-iHx}c_j^+e^{iHx} = e^{-i\,U(n_{j-1}+n_{j+1})x}c_j^+ , \tag{10}$$

$$c_j(x) = e^{-iHx}c_je^{iHx} = e^{i\,U(n_{j-1}+n_{j+1})x}c_j , \tag{11}$$

$$c_j^+e^{iUn_jx} = c_j^+ \tag{12}$$

equation (9) becomes

$$\sigma(\omega) = \frac{e^2a^2b^2}{2ZN_a}\int_{-\infty}^{\infty}d\tau e^{i\omega\tau}\int_0^{\beta}d\lambda\sum_i \mathrm{tr}\Big\{e^{-\beta(H-\mu N)}\big(e^{-iU(n_{i+2}-n_{i-1})x}\,n_{i+1}(1-n_i) + e^{-iU(n_{i-1}-n_{i+2})x}\,n_i(1-n_{i+1})\big)\Big\}. \quad (13)$$

We evaluate now the first term of (13) and the reader can check easily that the second term is exactly equal to the first. We need the following five transfer matrices in addition to (6):

$$T_1 = \begin{pmatrix} 1 & e^{\beta\mu/2+iUx/2} \\ e^{\beta\mu/2} & e^{-\beta(U-\mu)+iUx/2}\end{pmatrix},\quad T_2 = \begin{pmatrix} 1 & e^{\beta\mu/2} \\ e^{\beta\mu/2+iUx/2} & e^{iUx/2}\end{pmatrix},\quad T_3 = \begin{pmatrix} 0 & e^{\beta\mu/2} \\ 0 & 0\end{pmatrix},$$
$$T_4 = T_1(-x),\quad T_5 = T_2(-x). \quad (14)$$

The first trace of (13) is therefore equal to

$$\mathrm{tr}\left(T^{N_a-5}\,T_1T_2T_3T_4T_5\right). \quad (15)$$

A straightforward evaluation of (15) in the limit $N_a \to \infty$ yields

$$\frac{1}{Z}\mathrm{tr}\left\{e^{-\beta(H-\mu N)}e^{-iU(n_{j+2}-n_{j-1})x}\,n_{j+1}(1-n_j)\right\} = \frac{\alpha(\alpha+\gamma^{-2})(1+\alpha\gamma^2)}{2(1+\alpha)^3}, \quad (16)$$

where $\alpha = e^{\beta\mu/2} = e^{\beta U/2}$, $\gamma = e^{iUx/2}$; $\mu = U$ to fulfill (4).

The final result is obtained by using (16) in (13) and performing the integrations:

$$\sigma(\omega) = \frac{e^2a^2b^2}{(1+e^{\beta U/2})^3}\left[\beta\delta(\omega) + \frac{\mathrm{singh}\,\beta U/2}{U}\left(\delta(\omega+U)+\delta(\omega-U)\right)\right]. \quad (17)$$

Note the striking similarity of equation (17) with the result in the halffilled-band Hubbard model, reference (13), without being identical to it. The same discussion of the physical meaning of their (reference (13)) result (6) is valid here. Let us point out the finiteness of the optical conductivity at T = 0, and the relation: (thermal gap in the dc conductivity) = 1/2 (optical gap). Here the optical gap is equal to U and it is the difference in energy between the ground state in which the electrons occupy alternating sites, and the first excited state, which has one electrons transferred to a nearest neighbor site.

In magnetite the ground state of (2) with $b/U \ll 1$ has all the electrons ordered on alternating sites of perpendicular chains of Fe ions and at low temperature the conduction is by charge transfer along these chains. Therefore a calculation of $\sigma(\omega)$ should give the same thermal activation energy and optical gap as in equation (17). An accurate estimate of the transition temperature T_V as a function of U in the three-dimensional model when $b/U \rightarrow 0$ would make it possible to test model itself against experiment, because then once U is fixed to fit the low temperature activation energy (see for example reference (14)), T_V can be determined.

The author would like to thank Dr. R. A. Buckwald for bringing the problem to his attention and Dr. Tsur Bernstein for encouragement.

References

(1) J.R. CULLEN and E. CALLEN, J. appl. Phys. 41, 879 (1970).

(2) J.R. CULLEN and E. CALLEN, Phys. Rev. Letters 26, 236 (1971).

(3) J.R. CULLEN and E. CALLEN, Phys. Rev. B 7, 397 (1973).

(4) B. LORENZ and D. IHLE, phys. stat. sol. (b) 54, 463 (1972).

(5) D. IHLE and B. LORENZ, phys. stat. sol. (b) 58, 79 (1973).

(6) R.A. BUCKWALD, private communication.

(7) A. ISIHARA, Statistical Physics, Academie Press, Inc., New York 1971.

(8) D. CABIB and S.D. MAHANTI, Progr. theor. Phys. 51, to be published,

(9) R.A. BARI, Phys. Rev. B 3, 2662 (1971).

(10) G. BENI and P. PINCUS, Phys. Rev. B 9, 2963 (1974).

(11) R. KUBO, J. Phys. Soc. Japan 12, 570 (1957).

(12) D. CABIB and T.A. KAPLAN, phys. stat. sol. (b) 58, 85 (1973).

(13) R. A. BARI and T.A. KAPLAN, Phys. Rev. B 6, 4623 (1972).

(14) D. ADLER, Rev. mod. Phys. 40, 714 (1968).

(Received July 15, 1974)

phys. stat. sol. (b) 65, K57 (1974)

Subject classification: 18.4; 22

Institute of Nuclear Research, Świerk Research Establishment, Otwock (a), Institute for Low Temperature and Structure Research, Polish Academy of Sciences, Wrocław (b), and Technical University, Wrocław (c)

On the Magnetic Transitions in the UP-USe System

By

J. LECIEJEWICZ (a), R. TROĆ (b), and T. PALEWSKI (c)

Early magnetic susceptibility measurements above 78 K have shown that the UP-USe (1) solid solutions exhibit magnetic behaviour not different from that one displayed by the UP-US (2), UAs-US (1), and UAs-USe (3) systems. However, neutron-diffraction studies indicated different magnetic phase relations existing in each of the above mentioned solid solutions (4 to 8). In the light of these findings a neutron-diffraction study has been extended to the UP-USe system. Preliminary results are now reported.

The samples were prepared by sintering of UP and USe mixtures in desired proportions at 1700 °C during one hour in an induction furnace under high vacuum. Compositions with x as listed in Table 1 were synthetized and used for measurements. The lattice constants were determined from powder X-ray patterns obtained in a standard Debye-Scherrer camera, 114 mm in diameter with tungsten as a reference sample. They follow Vegard's law in the composition range investigated.

Neutron diffraction experiments were performed on a DN-500 diffractometer at reactor EWA. The neutron wevelength was (1.32 ± 0.01) Å. Data were obtained at 4.2 K and at room temperature. The temperature dependence of the peak height for magnetic reflections characteristic of the particular magnetic structure gave the Néel or Curie points as well as the transition temperatures. The magnitude of magnetic moment at 4.2 K was determined from the intensities of magnetic reflections such as M100 (type I magnetic structure), M101/2 (type IA magnetic ordering) and M101 (ferromagnetic phase). The indices of magnetic reflections are based on a tetragonal body-centered unit-cell. In all cases the intensity of the very strong 110 nuclear peak was taken as reference to put magnetic intensities on absolute scale. For those compositions having the Néel points close to 90 K the values of magnetic moment could not be determined with reasonable accuracy since the M101/2 magnetic peak height versus temperature dependence exhibits at 82 K a steep decrease (Fig. 2).

The magnetic phase diagram for the UP-USe systems consists essentially of three phase ranges (Table 1):

1. Composition with x = 0.05 show type I antiferromagnetic ordering with a characteristic for UP "moment jump" transition at $T_e = (24 \pm 2)$K and Néel point at 121 K (Fig. 1).

2. $UP_{0.90}Se_{0.10}$ is already a type IA antiferromagnet exhibiting a peculiar temperature dependence of the M101/2 peak height (Fig. 2) connected with a change in the magnitude of magnetic moment at $T_e = (62 \pm 3)$K. We have found that the type IA ordering persists up to the composition $UP_{0.67}Se_{0.33}$, while the Néel points occur in a fairly narrow range of 90 to 100 K. The magnitudes of magnetic moment at 4 K is appreciably lower for the UP-USe compositions exhibiting IA type ordering than that one observed in $UP_{0.75}S_{0.25}$. For the former μ_s changes from 1.5 to 1.3 μ_B with an increase of Se content while for the latter composition which has also IA type magnetic structure μ_s amounts to 1.7μ_B (7).

3. The sample with x = 0.37 is ferromagnetic with a smooth temperature dependence of M101 peak height. The Curie point increases from 113 K for x = 0.37 to T_C = 120 K for $UP_{0.60}Se_{0.40}$ and 160 K for USe (9).

The striking feature in the magnetic behaviour of $UP_{1-x}Se_x$ solid solutions is the presence of a marked change in the magnitude of magnetic moment in all the compositions with IA type magnetic ordering. As to our knowledge such an effect in IA type of magnetic structure is being reported for the first time. We propose to call it a "step-like" transition. It seems to be similar to that one observed in UP (10) and $UP_{0.95}Se_{0.05}$ (Fig. 1), but it occurs at higher temperatures (at about 60 K) and is connected with a fairly large decrease in the magnitude of magnetic moment (Fig. 2).

Neutron diffraction patterns of IA type compositions obtained at 82 K show no traces of peaks belonging to other magnetic structures so that one can conclude that the transition at T_e is not accompanied by a change in the type of magnetic ordering. This resembles the case of UP as well as other type I antiferromagnets

Table 1

The crystal and magnetic data of the $UP_{1-x}Se_x$ system

composition $x \pm 0.02$	lattice constant ± 0.001	magnetic structure	ordered magnetic moment at 4.2 K (μ_B)	T_N, T_C (K)	moment-change temperature T_e (K)
0.05	5.591	SF type I	1.93 ± 0.05	121 ± 3	24 ± 2
			1.71 ± 0.05 at 82 K		
0.10	5.602	AF type IA	1.51 ± 0.05	97 ± 2	62 ± 3
0.20	5.620	AF type IA	1.56 ± 0.05	92 ± 3	55 ± 5
0.25	5.622	AF type IA	1.56 ± 0.05	93 ± 3	60 ± 3
0.30	5.633	AF type IA	1.46 ± 0.05	90 ± 3	60 ± 3
0.33	5.640	AF type IA	1.34 ± 0.05	87 ± 3	60 ± 3
0.37	5.645	F	1.4 ± 0.1	113 ± 3	-
0.40	5.650	F	1.2 ± 0.1	125 ± 3	-

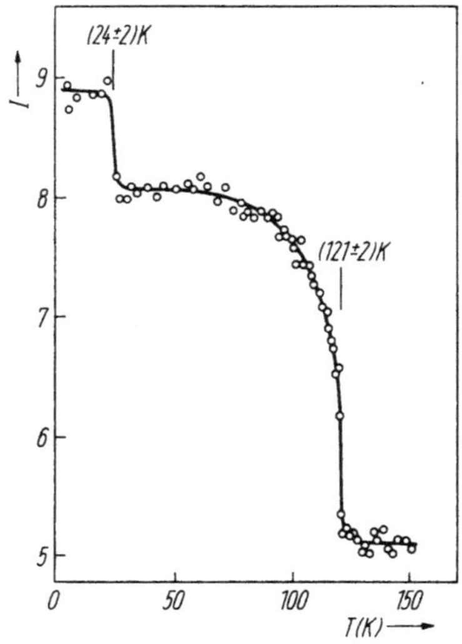

Fig. 1

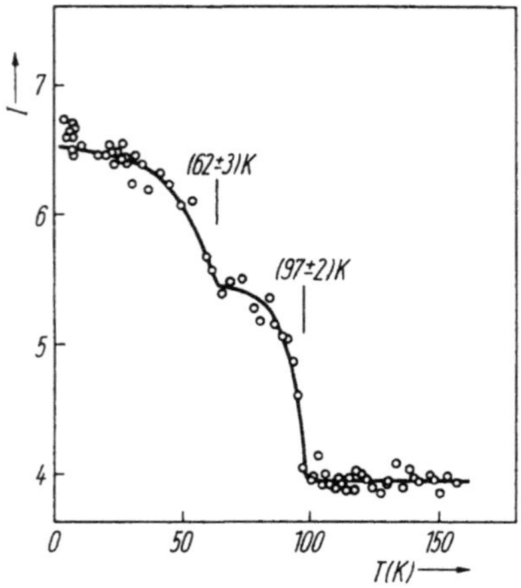

Fig. 2

Fig. 1. The variation of the peak height of the M100 magnetic reflection with temperature for $UP_{0.95}Se_{0.05}$. The magnetic structure is of the AF I type

Fig. 2. The M101/2 peak height versus temperature curve for $UP_{0.90}Se_{0.10}$. The antiferromagnetic ordering is of the IA type

such as $UP_{0.98}S_{0.02}$ (7) and $UP_{1-x}As_x$ solid solutions with $x \leqq 0.25$ (11).

It should be mentioned that the magnetic susceptibility measurements carried out for molten samples of $UP_{1-x}Se_x$ at temperatures above 78 K in magnetic field less than 15 kOe indicate that antiferromagnetic ordering is only present for compositions with $x \leqq 0.25$ (1). Recent magnetization measurements (12) performed on the same samples as used in the present study with $x \geqq 0.15$ from 4 K upwards in magnetic fields from 1 to 80 kOe gave two maxima on the σ versus T curves. The first one at about 100 K apparently connected with the Néel point, the second one at about 50 K most probably bound with the "step-like" transition as found in the course of neutron-diffraction investigation.

Acknowledgements

The authors would like to thank Prof. W. Trzebiatowski for his kind interest in this study. The technical assistance of Mr. A. Zawadzki, Mrs. B. Halkiewicz,

Mr. R. Chojnacki and Mr. S. Cichecki is gratefully acknowledged.

References

(1) W. TRZEBIATOWSKI and T. PALEWSKI, Bull. Acad. Polon. Sci., Ser. sci. Chim. 19, 83 (1971).

(2) W. TRZEBIATOWSKI and T. PALEWSKI, phys. stat. sol. 34, K51 (1969).

(3) W. TRZEBIATOWSKI, A. MISIUK, and T. PALEWSKI, Bull. Acad. Polon. Sci., Ser. sci. Chim. 15, 543 (1967).

(4) J. LECIEJEWICZ, A. MURASIK, R. TROĆ, and T. PALEWSKI, phys. stat. sol. (b) 46, 391 (1971).

(5) J. LECIEJEWICZ, R. TROĆ, A. MURASIK, and T. PALEWSKI, phys. stat. sol. (b) 48, 445 (1971).

(6) G.H. LANDER, M.H. MULLER, and J.F. REDDY, Phys. Rev. B 6, 1880 (1972).

(7) M. KUZNIETZ, G.H. LANDER, and Y. BASKIN, J. appl. Phys. 40, 1130 (1969).

(8) G.H. LANDER, M. KUZNIETZ, and D.E. COX, Phys. Rev. 188, 963 (1969).

(9) F.A. WEDGWOOD and M. KUZNIETZ, J. Phys. C. 5, 3012 (1972).

(10) N.A. CURRY, Proc. Phys. Soc. 89, 427 (1966).

(11) J. LECIEJEWICZ, A. MURASIK, and T. PALEWSKI, phys. stat. sol. (b) 46, K67 (1971).

(12) W. SUSKI, T. PALEWSKI, T. MYDLARZ, and H. REIZER-NETTER, Paper presented at the Internat. Conf. Magnetism, Moscow 1973.

(Received June 25, 1974)

phys. stat. sol. (b) 65, K63 (1974)

Subject classification: 13 and 18; 22

Instituto de Fisica, Universidade Federal de Pernambuco, Recife

Landau Level Splitting Due to the Spin-Orbit Interaction in Magnetic Semiconductors[1)]

By

L.C.M. MIRANDA

The effect of intense magnetic fields in semiconductors has been widely discussed for some time whereas to the knowledge of the present author there is no paper up to date devoted to this subject in the case of magnetic semiconductors. In dealing with magnetic semiconductors, however, the emphasis so far has been on the study of their optical and transport properties in the presence of magnetic fields sufficiently weak so that its effect on the carrier motion is negligible. Extensive reviews covering both the theoretical and experimental studies of magnetic semiconductors are now available (1, 2). Also most of the works on the subject has been focussed on the s-d exchange model (3) for the interaction between the carriers and the localized magnetic moments.

In this letter, a preliminary report on the effect of the spin-orbit interaction (4) between the localized spins and the particle orbits on the electronic properties is given for the case of an intense magnetic field. The field itself is supposed strong enough such that the particle states are described by the Landau levels.

The Hamiltonian for the free carrier system is

$$H_0 = \sum_{\alpha=n,\nu,k_z} \varepsilon_\alpha c_\alpha^+ c_\alpha , \tag{1}$$

where $\varepsilon_\alpha = \hbar\omega_c(n + \frac{|\nu| + \nu + 1}{2}) + \hbar^2 k_z^2/2m$ are the carrier energy levels in a uniform magnetic field (5), n and ν are the radial and orbital quantum numbers, respectively, k_z is the z-component of the carrier wave vector, and $\omega_c = eB/mc$ is the cyclotron frequency. We assume the external magnetic field applied along the z-axis. c_α and c_α^+ are the usual annihilation and creation operators for a particle in state α, respectively.

The spin-orbit interaction between an electron at a position $\vec{r}$ and the magnetic moments can be seen as follows: the localized moments produce a vector potential

---

1) Work partially supported by the CNP_q and BNDE (Brazilian Government).

$\vec{A}(\vec{r})$ such that (4, 6, 7)

$$\vec{A}(\vec{r}) = g\mu_B \frac{\vec{S} \times \vec{r}}{r^3}, \tag{2}$$

where $\vec{S}$ is the spin, μ_B is the Bohr magneton, and g is the g-factor. The interaction of a particle with momentum $\vec{p}$ with this field is given by

$$H_{so} = \frac{eg\mu_B}{mc} \frac{\vec{S} \cdot (\vec{r} \times \vec{p})}{r^3}. \tag{3}$$

Equation (3) describes the magnetic moment-carrier orbit interaction. It has recently been used in studying among other things, the indirect exchange between localized spins via conduction electrons and carrier effective mass in magnetic semiconductors (7). In particular, it has been proved to be important in the problems of spin wave amplification (6) and anomalous Hall effect (8).

In the case where the applied magnetic field is strong enough only to align the magnetic moments but sufficiently weak so that the particle motion perpendicular to the field is not quantized, one would expect equation (3) not to influence the energy bands radically. In fact, by using straight-forward perturbation theory, one can show that in this case the first nonvanishing correction to the particle energy comes from second-order perturbation theory. This can be physically understood by noticing that when the effect of the magnetic field on the particle motion is negligible, the angular momentum with respect to the axis is not a constant of motion. On the other hand, if the applied field is sufficiently strong such that the quantization of the particle motion perpendicular to the field axis has to be taken into account, the angular momentum with respect to this axis is a constant of motion and first-order perturbation theory on equation (3) gives a finite contribution.

Assuming that (3) is sufficiently weak to be treated as a perturbation, the first-order correction to ε_α is given by

$$\Delta\varepsilon_\alpha = \mathcal{N} \int d^3r\, \psi_\alpha^*(\vec{r}) H_{so} \psi_\alpha(\vec{r}), \tag{4}$$

where $\psi_\alpha(\vec{r})$ is the unperturbed particle wave function in the state α which in cylindrical coordinates is (reference (5))

$$\psi_\alpha(\vec{r}) = \frac{1}{\sqrt{2\pi L}} R_n^\nu(\rho) e^{i\nu\varphi} e^{ik_z z} \tag{5}$$

with

$$R_n^\nu(\rho) = N_n^\nu e^{-x/2} x^{|\nu|/2} L_n^{|\nu|}(x) \quad , \tag{6}$$

$$N_n^\nu = \frac{1}{r_c}\left[\frac{(n+|\nu|)!}{n!}\right] , \qquad x = \frac{\rho^2}{r_c^2} , \qquad r_c^2 = \frac{\hbar}{m\omega_c} . \tag{7}$$

$L_n^{|\nu|}(x)$ are the usual Laguerre polynomials. $\mathcal{N}$ is the number of magnetic moments seen by the electron, which in a first approximation, assuming that the electron motion perpendicular to the field is confined to a region of r_c (see equation (6)) is roughly

$$\mathcal{N} = \frac{N}{V} \pi r_c^2 L . \tag{8}$$

Retaining in equation (4) only the matrix elements which induce no transitions in the magnetic ions (the magnetic moments are aligned along the z-axis) it then follows from equations (4) to (8)

$$\Delta\varepsilon_\alpha = \eta\,\omega_c \frac{\nu}{n+|\nu|+1} , \tag{9}$$

where $\eta = \frac{g e^2 r_c^2}{m c^2} \frac{N}{V}$ S, with N/V being the density of localized moments.

Hence to first-order in perturbation theory the particle energy levels are given by

$$\varepsilon_\alpha^{(1)} = \hbar\omega_c\left[n + \frac{|\nu|+1}{2} + \nu\left(\frac{1}{2} + \frac{\eta}{2n+|\nu|+1}\right)\right] + \frac{\hbar^2 k_z^2}{2m} . \tag{10}$$

As seen from equation (10) the energy shift depends on the applied field. For large values of B, η vanishes. This is consistent with the fact that for B large the carrier motion is essentially one-dimensional along the z-direction, and therefore to first order it does not couple with the localized moments (zero angular momentum along the z-axis). In the case of small B, η apparently diverges. This divergence can be removed by adequately introducing the screening of the interaction (4, 7) with the result that η goes to zero as mentioned earlier. For typical values of B and N/V of the order of 10^5 G and 10^{23} cm^{-3}, respectively, one gets $\eta \approx 0.01$ to 0.1.

Finally, in the case of finite temperatures, $S^z = S$ in equation (9) should be replaced by $\langle S^z \rangle$ introducing therefore a temperature-dependent shift through

$$\langle S^z \rangle = S\, B_s\left(\frac{g\,\mu_B\,B}{k_B\,T}\right), \tag{11}$$

where

$$B_s(x) = \frac{2S+1}{2S}\coth\left(\frac{2S+1}{2S}x\right) - \frac{1}{2S}\coth\left(\frac{x}{2S}\right). \tag{12}$$

The shift of the Landau levels reported in this note could possibly be observed in cyclotron resonance experiments (9) in magnetic semiconductors like, for instance, doped EuO and $CdCr_2Se_4$.

The author acknowledges useful discussions with Dr. S.J. Allen, Jr. of Bell Laboratories.

References

(1) S. METHFESSEL and D.G. MATTIS, in: Handbuch der Physik, Vol. XVIII/I, Springer-Verlag, Berlin 1968.

(2) C. HAAS, Critical Revies in Solid State Phys. 1, 47 (1970).

(3) T. KASUYA, Critical Reviews in Solid State Phys. 3, 131 (1972).

(4) E.A. TUROV, in: Ferromagnetic Resonance, Ed. S.V. VONSOVSKII, Pergamon Press, Oxford 1966.

(5) L.D. LANDAU and E.M. LIFSHITS, Quantum Mechanics, Pergamon Press, Oxford 1958.

(6) M.D. COUTINHO FILHO, L.C.M. MIRANDA, and S.M. REZENDE, phys. stat. sol. (b) 57, 85 (1973).

(7) M.D. COUTINHO FILHO, Ph. D. Thesis, University of São Paulo, 1973.

(8) G.L. LAZAREV, Soviet Phys. - Solid State 14, 22 (1972).

(9) B. LAX and J.G. MAVROIDES, Solid State Phys. 11, (1968).

(Received June 5, 1974)

phys. stat. sol. (b) 65, K67 (1974)
Subject classification: 14.2; 21

Joint Institute for Nuclear Research, Dubna

Measurement of the Viscous Friction Coefficient of the Vortices of Superconducting Nb-80%Zr Alloy Near T_c

By

I.N. GONCHAROV and I.S. KHUKHAREVA

Based on the microscopic theory, different authors have recently obtained equations determining the electromagnetic dissipation due to a motion of vortices under the influence of transport current in type-II superconductors (SII) with $\varkappa \gg 1$ (1 to 6). In some cases the solutions of these equations made it possible to obtain accurately the dependence of the effective conductivity $\sigma_{eff} = \varrho^{-1} = (dE/dj)^{-1}$ on the temperature T and average magnetic field in the sample H. In particular, for $H/H_{c2}(t) \ll 1$ in the immediate vicinity of T_c $\sigma_{eff} = \beta(t)\,\varrho_n^{-1} H_{c2}^{GLAG}(t)/H$, where $t = T/T_c$; ϱ_n is a specific electroresistance in the normal state; the factor $\beta(t)$ is not the same in different papers (2, 3, 5).

According to (7), the viscosity coefficient $\eta(t) = \varrho_f^{-1} \varphi_o H c^{-2} = \beta(t)\,\varrho_n \times \varphi_o c^{-2} H_{c2}^{GLAG}(t)$, where φ_o is the flux quantum. As a rule, a linear dependence of $\varrho_f(H)$ is observed in weak magnetic fields. This dependence is usually expressed as $\varrho_f/\varrho_n = b(t) H/H_{c2}^{GLAG}(t)$ or $\varrho_f/\varrho_n = b_o(t) H/H_{c2}^{GLAG}(0)$. Assuming $H_{c2}^{GLAG}(t) = H_{c2}^{GLAG}(0) f(t)$ we obtain that $\beta(t) = b^{-1}(t)\left[b_o(t) f(t)\right]^{-1}$. Considering the temperature dependence of η it is more expedient to take into account $b_o^{-1}(t)$ than $b^{-1}(t)$ as is often done (see, for example, (8)). This approach may be explained thus: first, when calculating the experimental curve $\varrho_f(H)$ as a function of $H/H_{c2}(t)$ an additional error due to the spread $H_{c2}^{exper}(t)$ is introduced; second, for SII with paramagnetically limited critical field $H_{c2}^{exper}(t)$ is essentially lower than $H_{c2}^{GLAG}(t)$ (9, 10) (with the exception of the region in the immediate vicinity of T_c); third, taking into account only the temperature dependence $b^{-1}(t) = \beta(t)$, consideration is not given to an important distinguishing feature: at $t \to 1$ in spite of growing $\beta(t)$ the viscosity coefficient decreases, as a result of which a vortex velocity increases at a constant driving force.

The aim of this paper is to determine $\eta^{exper}(t)$ at temperatures close to

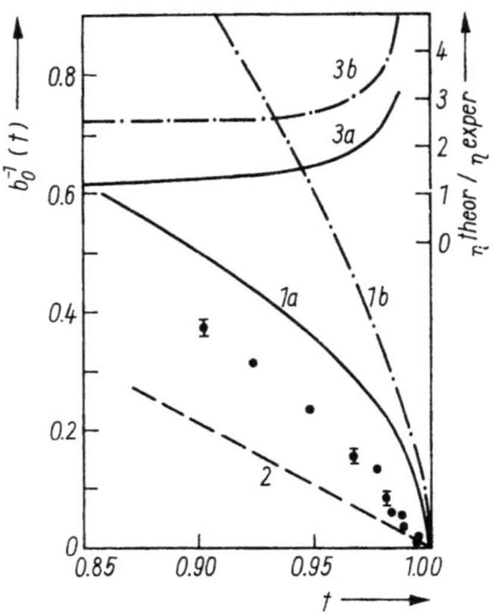

Fig. 1. Viscosity coefficient vs. temperature; • experimental values, (1, 2) theoretical curves (explanation in the text), (3a, 3b) ratio between the theoretical and experimental values of η (curves 1a, 1b)

T_c (up to t = 0.995) and to compare its behaviour with the results of the microscopic theory. Recrystallized Nb- 80 % Zr ribbon samples described in detail in (10) were used. The value of $\varrho_f(H)$ at fixed T and H was determined graphically by the slope of a linear part of the current-voltage characteristic. At all temperatures on the curves $\varrho_f(H)$ in the fields $H/H_{c2}(t) = 0.002$ to 0.25 the linear parts could clearly be seen from the slope of which $b_o(t)$ and b(t) were determined.

The plot of b_o^{-1} against the normalized temperatures is presented in Fig. 1. The same figure shows the theoretical curves. Curve 1a corresponds to $\beta(t) = 1.1(1-t)^{-1/2}$ from (3) in which consideration is given to the main part of the so-called anomalous term in σ_{eff}; this term arises only from a relaxation mechanism of energy dissipation for moving vortices in SII. Curve 1b corresponds to $\beta(t) = 1.1(1-t)^{-1/2} + 0.81\left[1 + \alpha_3(1-t)^{-1/10} + \alpha_4(1-t)^{-1/5}\right]$ from (5) in which account is taken of additional corrections for the main part.[1)] Finally, curve 2 corresonds to $\beta(t) = 1.47$ (2) and determines the lower bound of $\eta(t)$ since it has been obtained under the assumption that anomalous terms in σ_{eff} are absent.

For theoretical values of $b_o(t)$, $f(t) = (1-t)/0.693$, and for experimental ones $H_{c2}^{GLAG}(0) = 0.693\, T_c \left.\frac{dH_{c2}^{exper}}{dt}\right|_{T=T_c}$ (9, 10). For our samples $T_c = T(0.5\, R_n) =$

1) $\alpha_3 \approx \alpha_4 \approx 1$.

$= 7.85^{o}K$, $\Delta T = T(0.9\ R_n) - T(0.1\ R_n) = 0.03\ ^{o}K = 0.00038\ T_c$ and $H_{c2}^{exper}(t)$ is linear function in the whole investigated interval of temperatures.

As is seen from Fig. 1, the theoretical curves 1a and 1b lie above the experimental points by 20 to 40 % and 150 %, respectively, at t = 0.80 to 0.93. As far as the temperature dependence of η is concerned, at t < 0.93 the theory (5) gives the functional dependence (see curve 3b) which coincides with our results, but at $t \gtrsim 0.95$ it sharply disagrees with the experiment. On the curve $b^{-1}(t) \sim \eta(t)/H_{c2}(t)$ this discrepancy shows itself in the following way: the theoretical curves, corresponding to curves 1a and 1b in Fig. 1, steeply rise, and the experimental ones, having reached a maximum near t = 0.98, drop practically up to the values of curve 2.

In conclusion the authors would like to express their gratitude to V.V. Danilov, N.B. Kopnin, M.Yu. Kupriyanov, K.K. Likharev, and Yu.N. Ovchinnikov for fruitful discussions and to G.L. Dorofeev, A. Nikitiu, L.V. Petrova, and D. Frichovskii for their help in measurements.

References

(1) L.P. GORKOV and N.B. KOPNIN, Zh. eksper. teor. Fiz. 60, 2331 (1971).

(2) M.Yu. KUPRIYANOV and K.K. LIKHAREV, Zh. eksper. teor. Fiz., Pisma 15, 349 (1972).

(3) L.P. GORKOV and N.B. KOPNIN, Zh. eksper. teor. Fiz. 64, 360 (1973).

(4) A.I. LARKIN and Yu.N. OVCHINNIKOV, Zh. eksper. teor. Fiz. 64, 1096 (1973).

(5) Yu.N. OVCHINNIKOV, Zh. eksper. teor. Fiz. 65, 290 (1973).

(6) L.P. GORKOV and N.B. KOPNIN, Zh. eksper. teor. Fiz. 65, 396 (1973).

(7) Y.B. KIM, C.F. HEAMSTEAD and A.R. STRNAD, Phys. Rev. 139, A1163 (1965).

(8) N.Ya. FOGEL, Zh. eksper. teor. Fiz. 63, 1371 (1972).

(9) E. HELFAND and N.R. WERTHAMER, Phys. Rev. 147, 288 (1966).
R.R. HAKE, Phys. Rev. 158, 356 (1967).

(10) I.N. GONCHAROV and I.S. KHUKHAREVA, Zh. eksper. teor. Fiz. 62, 627 (1972).

(Received July 29, 1974)

phys. stat. sol. (b) 65, K71 (1974)
Subject classification: 13.1 and 20.1; 22.4.4

Institut für Festkörperphysik, Technische Universität Berlin

Vacuum-UV Reflectivity Measurements of $ZnSe_xTe_{1-x}$ Mixed Crystals at 80 K

By

G.H. HECKELMANN and T.S. WAGNER

Introduction First diffuse reflectivity measurements on polycrystalline powders of $ZnSe_xTe_{1-x}$ were reported by Larach et al. (1), recently Ebina et al. (2) published data from reflectivity measurements on $ZnSe_xTe_{1-x}$ single crystals at room temperature in the energy region 2 to 6.6 eV. No measurements at low temperature which will yield more exact information by the sharper structure of the spectra have been reported as yet.

Experiment In our experiment many crystals grown in our institute and crystals obtained from other sources[1)] have been studied. The crystals from our institute have been grown by a Piper-Polich technique (3) and a modified vapour phase method proposed by Aven et al. (4).

The first method was successful up to a ZnSe fraction of about 30% and gave big bulk crystals easy to cleave. The cleaved surfaces were optically flat and sufficiently homogeneous and had a minimum size of 6x6 mm^2. By the vapour phase technique crystals could be grown over the whole range of mole fraction x. These crystals had as-grown surfaces with a maximum size of 2x2 mm^2. The composition of the mixed crystals has been determined by a combined X-ray fluorescence and Debye-Scherrer method. The reflectivity measurements were made on as grown (111)- or on cleaved (110)-surfaces using a deuterium discharge lamp as light source combined with a 0.3 m Mc. Pherson vacuum monochromator. Using two photomultipliers coated with sodium salycilate for the detection of both the incident and the reflected beam continuous measurements employing lock-in technique were performed over the whole spectral range from 2 to 9 eV.

Results Some typical reflectivity spectra of $ZnSe_xTe_{1-x}$ at 80 K are shown in Fig. 1. The peaks in the spectrum of pure ZnTe are attributed to optical transitions

---

1) The crystals were obtained from Dr. Y.S. Park, Aerospace Research Laboratories Dayton, U.S.A. and Dr. S. Ibuki, Mitsubishi El. Corp., Osaka.

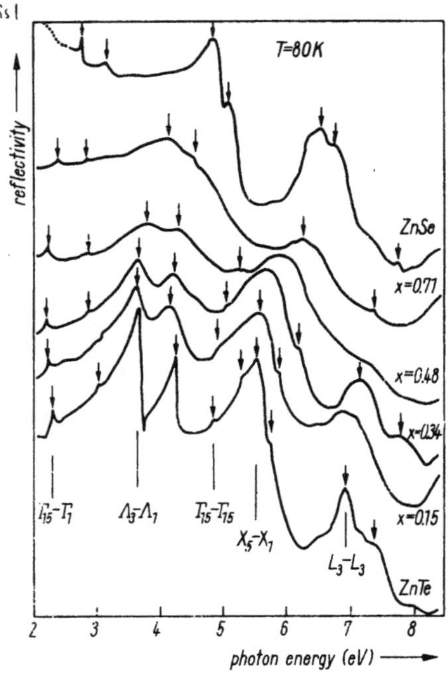

Fig. 1

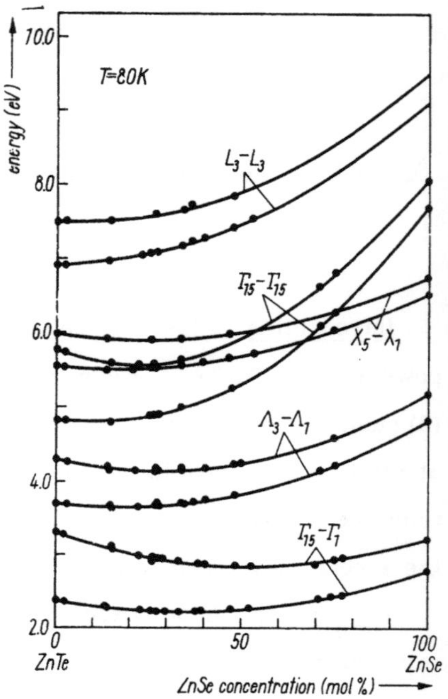

Fig. 2

Fig. 1. Reflectivity of $ZnSe_xTe_{1-x}$ mixed crystals for different mole fractions x at 80 K

Fig. 2. Variation of the energetic positions of reflectivity peaks as a function of the composition of $ZnSe_xTe_{1-x}$, assigned to different optical transitions at 80 K

as indicated in the lower part of Fig. 1 (5). All the assigned peaks can be followed over the whole composition range in the observed spectral region. The reflectivity data have been used to calculate ε_1 and ε_2 by means of a Kramers-Kronig analysis (6), but due to the small spectral range used in the measurements and the problematic extrapolation of the experimental data to higher energies, the calculated spectra of ε_1 and ε_2 are uncertain and were not used furthermore.

Fig. 2 shows the variation of the peak positions in the reflectivity spectra as a function of the composition. The curves for the transitions L_3-L_3 are extrapolated for energies above 8.0 eV to experimental data of pure ZnSe (7). All the curves show a continuous but nonlinear slope with different curvatures for the different transitions. This behaviour is predicted by Onodera and Toyozawa (8) and is the subject of theoretical work done by Richardson and Hill (9) and Maschke (10). A detailed quantitative interpretation of this aspect will be published later.

Acknowledgements

The authors wish to thank Dr. Y.S. Park and Dr. S. Ibuki for supplying some crystals and Dr. H.H. Landfermann (Technische Universität Berlin) for growing $ZnSe_xTe_{1-x}$ single crystals. We are greatful to Dr. U. Nielsen for his support in the numerical evaluation of our experimental curves at the DESY (Deutsches Elektronensynchrotron) in Hamburg.

References

(1) S. LARACH, R.E. SHRADER, and C.F. STOCKER, Phys. Rev. 108, 587 (1957).

(2) A. EBINA, H. YAMAMOTO and T. TAKAHASHI, Phys. Rev. B 6, 3786 (1972).

(3) W.W. PIPER and S.J. POLICH, J. appl. Phys. 32, 1278 (1961).

(4) M. AVEN, R.B. HALL, and J.S. PRENER, J. Electrochem. Soc. 115, 846 (1968).

(5) M. CARDONA, Phys. Rev. 131, 98 (1963).

(6) R. KLUCKER and U. NIELSEN, DESY-Report SR-73/5 (1973).

(7) M. AVEN, D.T.F. MARPLE, and B. SEGALL, J. appl. Phys. 32, 2261 (1961).

(8) Y. ONODERA and Y. TOYOZAWA, J. Phys. Soc. Japan 24, 341 (1968).

(9) D. RICHARDSON and R. HILL, J. Phys. C 5, 821 (1972).

(10) K. MASCHKE, private communication.

(Received May 14, 1974)

phys. stat. sol. (b) 65, K75 (1974)
Subject classification: 13.1 and 20.1; 22.4.4

Institut für Festkörperphysik, Technische Universität Berlin

Optical Reflectivity and Electronic Structure of ZnS_xSe_{1-x} Mixed Crystals

By

T.S. WAGNER, G.H. HECKELMANN, and H. NELKOWSKI

Introduction Two years ago optical investigations at room-temperature of ZnS_xSe_{1-x} mixed crystals were reported by our group (1). The results from the experiments and our interpretation at that time were in conflict with the theory of Onodera and Toyozowa (2) and were criticized by some theoretical physicists. For this reason the measurements on this system were repeated using a refined experimental technique at room temperature and 80 K.

Experiments In this experiment we have examined many crystals grown by vapour phase technique and by iodine transport reaction exhibiting cubic structure and optically flat as-grown (111) surfaces. The source materials were mixed ZnS and ZnSe powders. The mole fraction x varied in the region $0 \leqq x \leqq 1$. By X-ray structure analysis the compositions of the mixed crystals have been determined. The sample holder was a universal UHV-manipulator cooled by liquid nitrogen. The reflectivity measurements were performed with a deuterium discharge lamp and a 0.3 m Mc. Pherson vacuum monochromator. The angle of incidence was restricted to 15°. Two multipliers coated with sodium salycilate were employed to detect simultaneously both the incident beam and the reflected signal. These two signals were fed through separate amplifiers and divided continuously during the scan.

Results and discussion Fig. 1 shows typical reflection spectra of ZnS_xSe_{1-x} mixed crystals at 80 K for different compositions. All the spectra show three main structures: the low energetic, spin-orbit split exciton transitions at the band gap, the predominant maximum at approximately 5.5 eV - related to the L_3-L_1 transition - and a broad structure in the energy region at about 6.5 eV, attributed to transitions at or on the vicinity of the X-point (3).

The profiles of the spectra from the one pure compound to the other show a continuous variation. This is clearly seen regarding the spin-orbit splitting at the band gap at the L_3-L_1 transition, e.g. the spin-orbit split maximum in the spectrum of

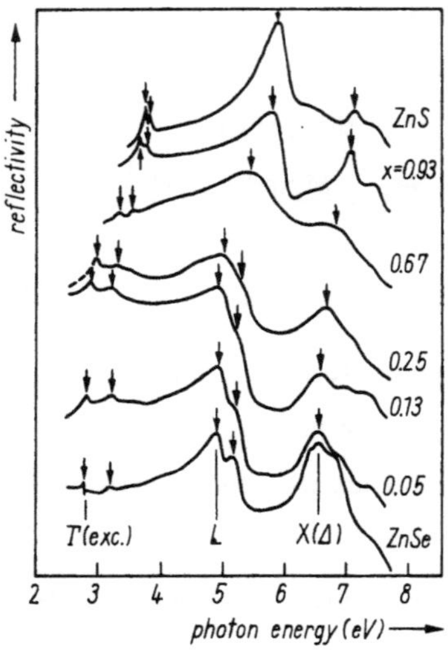

Fig. 1. Reflectivity of ZnS_xSe_{1-x} mixed crystals for different mole fractions x at 80 K

pure ZnSe shifts with growing S fraction and disappears at x = 0.25.

In contrast to the excitonic transitions at the band gap, the spin-orbit split maximum of the L_3-L_1 transition cannot be observed for x > 0.25 as the spin-orbit splitting of the L_3 valence band in ZnS is very small (0.05 eV) compared to the broad structure (>0.5 eV) of the L_3-L_1 transition.

Assigned to the different optical transitions the energetic positions of the peaks in the reflectivity spectra as a function of the composition are shown in Fig. 2 to 4. Considering the systematic shift of the different peaks with the composition, the system ZnS_xSe_{1-x} must be attributed to the amalgamation type. The bowing of the curves in Fig. 2 to 4 is in agreement with the theory of Onodera and Toyozawa (2) and will be discussed later together with experimental results of other mixed systems.

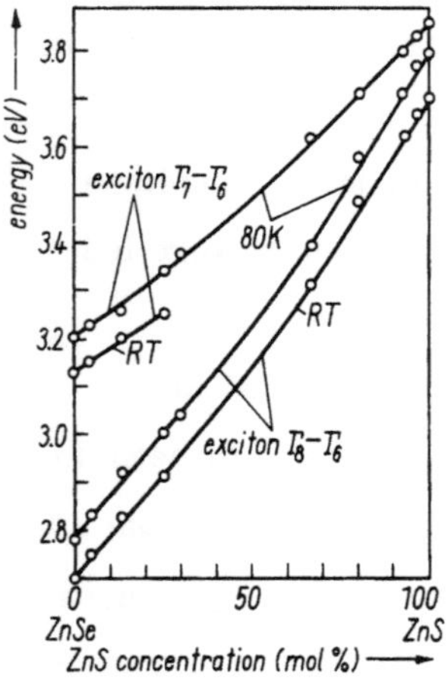

Fig. 2

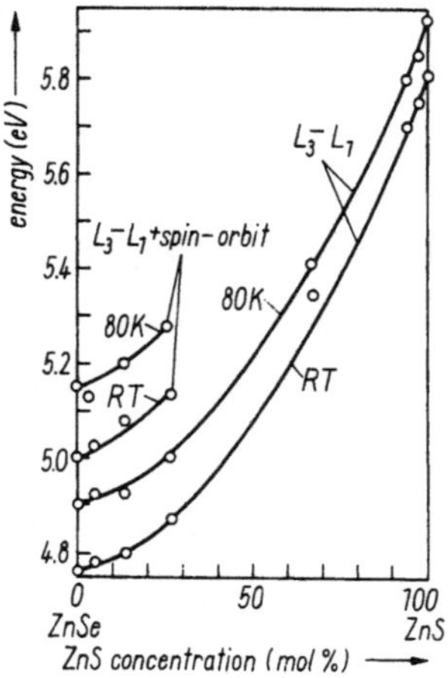

Fig. 3

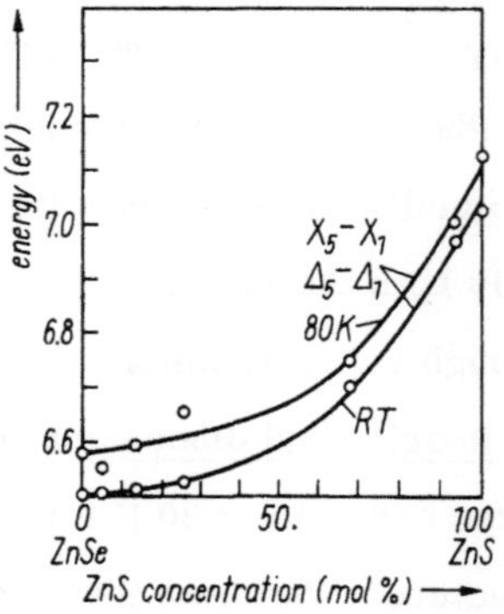

Fig. 4

Fig. 2. to 4. Variation of the energetic position of reflectivity peaks - assigned to the optical transitions Γ_8-Γ_6, Γ_7-Γ_6 (exc.), L_3-L_1 and X_5-X_1(Δ_5-Δ_1) as a function of the composition of ZnS_xSe_{1-x} mixed crystals at room temperature and 80 K

Acknowledgements

We are grateful for Dr. U. Nielsen, University Kiel, for his support during the computer calculations performed at the DESY, Deutsche Elektronen-Synchrotron, Hamburg. This experiment has been financially supported by the Deutsche Forschungsgemeinschaft.

References

(1) K.E. KIRSCHFELD, H. NELKOWSKI, and T.S. WAGNER, Phys. Rev. Letters 29, 66 (1972).

(2) Y. ONODERA and Y. TOYOZAWA, J. Phys. Soc. Japan 24, 341 (1968).

(3) J. TREUSCH, P. ECKELT, and O. MADELUNG, Proc. Internat. Conf. II-VI Semicond. Compounds, Providence (Rhode Island) 1967 (p. 588).

(Received July 2, 1974)

Pre-printed Titles

Papers to be published

in "physica status solidi (b) - basic research"

Vol. 65, No. 2

Original Papers

J. MONECKE, Freiberg

Single Particle Excitations in Mott Insulators

C. UHER and H.J. GOLDSMID, Kensington, New South Wales

Separation of the Electronic and Lattice Thermal Conductivities in Bismuth Crystals

J.F. HOULIHAN and L.N. MULAY, University Park, Pennsylvania

Correlation of Magnetic Susceptibility, Electrical Conductivity, and Structural Parameters of Ti_3O_5 via EPR Spectroscopy

E.G. SHAROYAN, O.S. TOROSYAN, E.A. MARKOSYAN, and V.T. GABRIELYAN, Erevan

EPR and Spin-Lattice Relaxation of Mo^{3+} Ions in Corundum

K. FALKOWSKI, J. SZUMOWSKI, B. KRUKOWSKA-FULDE, and R. JABŁOŃSKI, Warsaw

Spin-Lattice Coefficients of Gd^{3+} in CdF_2

S. ÇIRACI and W.A. TILLER, Stanford, California

Some Applications of the Bond Orbital Model

G.R. ZELLER, Aachen

Pseudopotential Calculation of the Piezoelectric Constant of GaAs

E.O. MANUCHARYANTS and I.P. ZVYAGIN, Moscow

Green's Functions Approach to the Theory of Hopping Conductivity in Disordered Semiconductors

I. BALSLEV and J. M. HVAM, Odense

Drift of Electron-Hole Drops in Exciton Density Gradients

P. I. KHADSHI and S. A. MOSKALENKO, Kishinev

Interserial Phototransitions of Excitons in Semiconductors

YU. A. BOGOD and V. B. KRASOVITSKII, Kharkov

Experimental Studies of Magnetoresistance Tensor Components in Bismuth at Low Temperatures

L. E. GUREVICH and R. G. TARKHANYAN, Leningrad and Erevan

Influence of Magnetic Field on Polaritons in Uniaxial Polar Semiconductors

H. R. ZELLER and P. BRÜESCH, Baden, Switzerland

Band Structure of $K_2[Pt(CN)_4]Br_{0.30} \cdot 3\,(H_2O)$

C. BENOIT and G. W. COHEN SOLAL, Montpellier

Induced Scattering of Neutrons by an Electromagnetic Field

K. I. GROZDEV and A. M. KOSEVICH, Kharkov

Effect of Impurities on the Phase Transition Temperature of a Two-Dimensional Ising Lattice with Interaction between Some of the Next-Nearest Neighbours

S. OLSZEWSKI and A. WIERZBICKI, Warsaw

Dynamical Observables in Cubic Crystals from the Standing LCAO Wave Functions

K. SAERMARK and J. LEBECH, Lyngby

Cyclotron Waves in Silver - Interference Effects

V. A. SOLOVEV, Moscow

The Stress Field Near the Dislocation Pile-Up Type Defects in Anisotropic Elasticity

V. T. BUBLICH, S. S. GORELICH, A. A. ZAITSEV, and A. Y. POLYAKOV, Moscow

Diffuse X-Ray Determination of the Energy of Mixing and Elastic Constants of Ga-Si Solid Solutions

C. A. POLDY and H. R. KIRCHMAYR, Wien

The Magnetic Properties and Indirect Exchange in $Gd(Ni,Cu)_2$

J. KALINOWSKI and J. GODLEWSKI, Gdańsk

Singlet Exciton-Charge Carrier Interaction in Anthracene Crystals

M. Z. ZGIERSKI, Cracow

Herzberg Teller Interaction and Vibronic Coupling in Molecular Crystals (III)

C.-U. TRAUTVETTER, Karlsruhe

Spin-Wave Spectra below Saturation Magnetization and at Various Orientations

E. THALMAYR, H. KAHLERT, and H. KUZMANY, Wien

Ohmic and Non-Ohmic Transport in Undeformed and Plastically Deformed Tellurium

R. PÄSSLER, Karl-Marx-Stadt

Quantum-Efficiency of Multiphonon Transitions according to the Static Coupling Scheme

A. M. DOBROTVORSKII and R. A. EVARESTOV, Leningrad

The Quasi-Molecular Large Unit Cell Model in the Theory of Deep Levels in Imperfect Crystals: Point Defects in Graphitic Boron Nitride

R. H. WILLIAMS and J. I. POLANCO, Coleraine, Ulster

The Electronic Structure of Chalcogenide Solids: A Photoemission Study of Sulphur

E. ASCHER, Carouge-Geneva

Relativistic Symmetries and Lower Bounds for the Magneto-Electric Susceptibility and the Ratio of Polarization to Magnetization in a Ferromagento-Electric

A. MYCIELSKI, A. AZIZA, J. MYCIELSKI, and M. BALKANSKI, Paris

Frequency-Dependent Relaxation Time of Electrons in PbSe from Infrared Measurements

J. ALIZON, G. BERTHET, J. P. BLANC, J. GALLICE, and H. ROBERT, Aubière

Spin-Spin and Spin-Lattice Equilibrium in $TCNQ^{-} \cdot NH_4^{+}$ Salt

F. BASSANI, J. J. FORNEY, and A. QUATTROPANI, Roma and Lausanne

Energy Level Structure of Biexcitons and Related Optical Transitions

F. BRIDGES, Santa Cruz, California

A Two-Multiplet Tunneling Model

R. KLUCKER, M. SKIBOWSKI, and W. STEINMANN, München

Anisotropy in the Optical Transitions from the π and σ Valence Bands of Graphite

A. J. NADOLNY, Clausthal

Photo-Induced Electron Spin Resonance (Photo ESR) in β-Rhombohedral Boron

M. D. COUTINHO FILHO, L. C. M. MIRANDA, and S. M. REZENDE, Recife, Brasil

Spin Wave Amplification in Ferromagnetic Semiconductors: Plasma Effects and s-d Interaction

B. P. PANDEY, Meerut, India

Phonon Dispersion in SiC and ZnS Polytypes

H. KÖHLER, Würzburg

Trigonal Warping of the Valence Band Fermi Surfaces of Tellurium

S. K. SRIVASTAVA, Allahabad

A Non-Parametric Model Pseudopotential and Electronic Transport Coefficients

D. J. H. COCKAYNE and V. VITEK, Oxford

Effect of Core Structure on the Determination of the Stacking-Fault Energy in Close-Packed Metals

M. SINGH and G. S. VERMA, Varanasi

Scattering of Phonons by Bound Holes and Phonon Conductivity of p-GaSb

Y. P. JOSHI, Varanasi

Resonant Scattering of Phonons by Two-Level Impurities

G. L. KRASKO and A. B. MAKHNOVETSKII, Moscow

The Pseudopotential Method and the Problem of Ordering in Alloys (I)

M. ŞCHWAB, W. BERKHAHN, H. G. KAHLE, A. SIMON, and W. WÜCHNER, Karlsruhe

Magnetic Ordering in $HoAsO_4$ Investigated by Means of Specific Heat and Spectroscopic Measurements

G. SPERLICH and W. D. LAZÉ, Darmstadt

Estimate of the d-d Overlap of V_2O_5 from ESR Measurements

H. J. WEBER and S. HAUSSÜHL, Köln

Electric-Field-Induced Optical Activity and Circular Dichroism of Cr-Doped $KAl(SO_4)_2 \cdot 12H_2O$

Papers to be published

in "physica status solidi (a) - applied research"

Vol. 25, No. 1

Review Article

H. KUZMANY, Wien

Acoustoelectric Interaction in Semiconductors

Original Papers

K. HIEBEL and M. WILKENS, Stuttgart

Point Defect Clusters in Neutron-Irradiated Molybdenum (II)

B. BARBARA, M. F. ROSSIGNOL, and J.X. BOUCHERLE, Grenoble

Influence of Magnetization Anisotropy and Cell Distortion on the Crystal Field Parameters Determined on a Single Crystal of $NdAl_2$

A.W. THOMPSON, Thousand Oaks, California

Yield Points during Cyclic Strain of a Cu-7.5% Al Alloy

N.L. KRAMARENKO, I.V. KURNOSOV, and YU.V. NABOIKIN, Kharkov

Structure Investigation of Liquid Crystals by Light Scattering

E.A. SILINSH, A.I. BELKIND, D.R. BALODE, A.J. BISENIECE, V.V. GRECHOV, L.F. TAURE, M.V. KURIK, J.I. VERTZYMACHA, and I. BOCK, Riga, Kiev and Prague

Photoelectrical Properties, Energy Level Spectra, and Photogeneration Mechanisms of Pentacence

R. VERGNE, J.L. PORTESEIL et Z. BLAŽEK, Grenoble et Prague

L'anisotropie de la désaimantation: élaboration d'un modèle; susceptibilités différentielles réversibles et irréversibles dans le domaine de Rayleigh.

T. TUOMI, K. NAUKHARINEN, and P. RABE, Otaniemi and Hamburg

Use of Synchrotron Radiation in X-Ray Diffraction Topography

C. LLINARES, J. P. DESFOURS, J. P. NADAI, C. GODART, A. PERCHERON, and J. C. ACHARD, Montpellier and Meudon-Bellevue
Electrical Transport Properties of EuO Single Crystal in Relation with Growth Parameters

YU. M. RAIROV, I. I. KHLEBNIKOV, and V. F. TSVETKOV, Leningrad
Investigation of Silicon Carbide Single Crystals Doped with Scandium

F. MILSTEIN, J. A. BALDWIN, Jr., and J. SPINGARN, Santa Barbara, California
Electrical and Thermal Behavior of Gd-C-Alloys in the Neighborhood of the Curie Temperature

D. SLOTFELDT-ELLINGSEN and B. FEDERSEN, Oslo
X-Ray-, NMR-, and DTA-Study of the TaD System and a Critical Evaluation of the TaH_x Phase Diagram

G. BAMBAKIDIS, R. J. SMITH, and D. A. OTTERSON, Cleveland, Ohio
Residual Resistivity of the Pd-Ag-H System

L. J. CHEN and G. THOMAS, Berkeley, California
Applications of Many Beam Systematic Diffraction Contrast in High Voltage Transmisson Electron Microscopy

J. P. DAY and A. L. RUOFF, Ithaca, New York
The Variation of the Elastic Constants of Lithium with Temperature and Pressure

W. HÄCKER, Stuttgart
Infrared Radiation from Breakdown Plasmas in Si, GaSb, and Ge: Evidence for Direct Free Hole Radiation

S. S. CHERN and F. A. KRÖGER, Los Angeles, California
The Relative Magnitude of the Diffusion Constants of Cadmium Vacancies and Indium-Cadmium Vacancy Pairs in CdS and CdTe

Y. N. LWIN, P. A. BROWN, J. J. JACKSON, and E. A. RYAN, Argonne, Illinois
The Initial Rate of Accumulation of Frenkel Defects in Platinum

E. P. WOHLFARTH, Grenoble
The Influence of Magnetoelasticity on the Debye Temperature of Ferromagnetic Alloys

R. KASSING and H. LENZ, Münster

Determination of the Capture Rate c_n of the Gold Acceptor Level from Single Injection n^+-i-n^+ Silicon SCLC Diodes

K. SOMOGYI, Budapest

Low-Temperature Behaviour of Solution-Grown $ZnGeP_2$ Crystals

J.E. PARROTT and L.PH. LEONIDOU, Cathays Park, Cardiff

A General Theory of p-n Junction Capacitance

R.O. SCATTERGOOD and D.J. BACON, Liverpool

Dislocation Shear Loops in Anisotropic Crystals

V.I. KHOKHLOV, YU.G. SIDOROV, and S.A. DVORETSKII, Novosibirsk

Electrophysical Properties of Non-Doped Epitaxial GaAs in the Range from 10 to 110 $^\circ$K

B.K. KARDASHEV, C.P. NIKANOROV, and O.A. VOINOVA, Leningrad

Internal Friction and Charged Dislocations in NaF Crystals

W. KELLER and A. MÜHLBAUER, München

Float-Zoned Silicon with Homogeneous Dopant Distribution

D. WELSCH, Dubna

Phase Fluctuations and Linewidths in a Solid State Anti-Stokes Raman Oscillator

T. BEDYŃSKA, Warsaw

X-Ray Diffraction in the Range of Two Diffracted Beams for Extremely Asymmetric Case

I. MUNTEANU and C. CONSTANTINESCU, Bucharest

Negative Magnetoresistance in Asymmetric p^+-i-n^+ Structures

E.W. KREUTZ and B. WAACK, Darmstadt

Thickness Dependence of the Galvanomagnetic Properties of InSb

Y. OBI, Sendai

On the Crystal Distortion in Several Spinel Systems

S. KÄSTNER and J. HESSE, Frankfurt/Main

The Influence of Dislocations on the Diffusion of Gold in Silicon

A.I. ULYANOV and A.V. DERYAGIN, Sverdlovsk

Temperature Dependence of Coercivity for Powdered Intermetallic Compounds of RCo_5 and R_2CO_{17}

A. BISCHOFF, Stuttgart

Untersuchung der Versetzungsverankerung an kaltverformtem Kupfer durch Modulmessung

A.G. KHACHATURYAN and V.M. AIRAPETYAN, Erevan

Spatially Periodic Distributions of New Phase Inclusions Caused by Elastic Distortions

K.D. GLINCHUK and A.V. PROKHOROVICH, Kiev

Temperature Quenching of the Copper-Induced 1.35 eV Emission Band in p-GaAs

H. KUBSCH, P. PAUFLER, and G.E.R. SCHULZE, Dresden

The Mobility of Grown-in Dislocations in the Intermetallic Compound $MgZn_2$ during Prismatic Slip

R. K. JAIN and R.J. VAN OVERSTRAETEN, Heverlee, Belgium

Calculation of the Diffusion InducedStressesin Silicon

19. Magnetic Resonance
20. Optical Properties
    20.1 Spectra. Optical Constants (X-Ray Spectra See 20)
    20.2 Lasers
    20.3 Luminescence (X-Ray Spectra See 20)
21. Metals. Alloys
    21.1 Transition Metals (Group Sc to Zn) and Their Alloys
        21.1.1 Fe and Fe Alloys
    21.2 Alkali Metals
    21.3 Alkali-Earth Metals
    21.4 Rare-Earth Metals
    21.5 Actinides
    21.6 Noble Metals
    21.7 Semi-Metals (Semiconducting Alloys See 22.7)
22. Semiconductors and Ionic Crystals
    22.1 Elements
        22.1.1 Germanium
        22.1.2 Silicon
        22.1.3 Group VI Elements
    22.2 III–V Compounds
        22.2.1 Arsenides
        22.2.2 Phosphides
        22.2.3 Antimonides
        22.2.4 Mixed Crystals
    22.3 IV–IV, II–V, and III–VI Compounds
        22.3.1 Mixed Crystals
    22.4 II–VI Compounds
        22.4.1 Sulphides
        22.4.2 Selenides
        22.4.3 Tellurides
        22.4.4 Mixed Crystals
    22.5 Halides
        22.5.1 Silver Halides
        22.5.2 Alkali Halides
        22.5.3 Alkali-Earth Halides
        22.5.4 Mixed Crystals
    22.6 Simple Oxides (CdO, ZnO See 22.4)
        22.6.1 Mixed Crystals
    22.7 Semiconducting Intermetallic Compounds
    22.8 Tertiary and Higher Compounds
        22.8.1 Salts of Oxy-Acids (Carbonates, Phosphates, Silicates Including Mica etc.)
        22.8.2 Spinels, Garnets, Ferrites
    22.9 Organic Semiconductors
23. Solidified Gases

Attention

The preceding classification scheme is closely akin to the scheme used by Physics Abstracts to classify the various fields of solid state physics. A paper will either be classified under one or more of the subsections or, if none of these are relevant, it will just appear under the general section heading (i.e. sections and subsections are equivalent in this sense).

The *subject classification* given at the head of each paper begins with the main subject matter of the paper, and this is followed, if necessary, by figures referring to subsidiary subject matter and the type of solids investigated.

The papers published in this issue are listed in the order of their subject classification at the end of the table of contents.

Manuscripts and letters for physica status solidi (b) — basic research and physica status solidi (a) — applied research

should be sent to

DDR, 102 Berlin, Neue Schönhauser Str. 20

or

Internationaler Buchversand GmbH, 1 Berlin 19, Eschenallee 21.

The object of "physica status solidi (b)" and "physica status solidi (a)" is to publish rapidly important and topical results concerning fundamental research in the field of solid state physics and aspects of application and experimental techniques of fundamental solid state research, respectively. The issues of "physica status solidi (b)" and "physica status solidi (a)" will be published on the 1st and the 16th day of each month, respectively. Either series will contain the following categories:

1. Review articles written by arrangement with the editors. They will be remunerated.
2. Original papers selected from those which have not been submitted for publication elsewhere, after being reviewed. (The latest date for acceptance of contributions is the 8th and the 23rd day of each month for "physica status solidi (b)" and "physica status solidi (a)", respectively, 50 days before the delivery of the journal.)
3. Short notes published in the off-set part of the journal. They are, if necessary after being reviewed, selected by the editorial staff. (The latest date for acceptance of manuscripts for this part is the 12th and the 27th day of each month for "physica status solidi (b)" and "physica status solidi (a)", respectively, 16 days before the delivery of the journal.)
4. Pre-printed titles of papers to be published in the next issue of both series.

Editorial Information

Manuscripts should be submitted preferentially in English. Besides, papers in German, French, and Russian are admitted. The final decision on the incorporation of the papers in the series (b) or (a) is taken by the editorial staff after consultation of the referee. Authors are asked to observe the following rules:

1. Manuscripts have to be written as concisely as possible. As most, 20 and 4 pages (28 lines per page; abstracts, tables, references, figures, and captions of figures included, where 3 figures are counted as about one page) are admitted for original papers and short notes, respectively. All manuscripts should be type-written on one side only. Two copies are required. Badly arranged manuscripts are returned.
2. The author's name, initials, and institution should be given.
3. One abstract of about ten lines in English and one in German, French, or Russian are required for original papers.
4. Because of lack of space conclusions (if any) must not have the character of a summary (i.e. substantially a repetition of the contents of the abstract).
5. Formulae should not be written too small and with pencil. Separate lines for formulae are desirable. Indistinct writing will lead to a return of the manuscript.
6. Drawings should be enclosed on separate sheets of paper, not exceeding the maximum and minimum sizes of 20×30 and $10 \times 15\ cm^2$, respectively; photographs should be submitted, on white paper, only for review articles and original papers. Duplicates are required.
7. Captions for the figures should be submitted on separate sheets of paper, and special instructions with regard to the size of illustrations should be given on these.
8. References should be numbered in order of appearance and should be given, on a separate sheet, in the form used in physica status solidi.
9. Units, symbols, etc. should be used in accordance with the recommendations given by the IUPAP.
10. Each paper should contain a proposal for the subject classification according to the classification scheme appearing on the second and third cover of each issue.

The authors will be sent galley proofs of review articles and original papers. Proof reading of short notes and pre-printed titles will be done by the editorial staff.

For each original paper and short note a total of 75 reprints is available free of charge. The number of reprints free of charge per review article is 25. Requests for additional reprints (against payment) should be given together with the submitted manuscript. **The deadline for orders of additional reprints is the 15th and the 30th day of each month** for "physica status solidi (b)" and "physica status solidi (a)", respectively, two weeks before the delivery of the journal.

Bei Fragen zur Produktsicherheit wenden Sie sich bitte an:
If you have any questions regarding product safety,
please contact:

Walter de Gruyter GmbH
Genthiner Straße 13
10785 Berlin
productsafety@degruyterbrill.com